Methods in Molecular Biology™

Series Editor
John M. Walker
School of Life Sciences
University of Hertfordshire
Hatfield, Hertfordshire, AL10 9AB, UK

For further volumes:
http://www.springer.com/series/7651

Wound Regeneration and Repair

Methods and Protocols

Edited by

Robert G. Gourdie

Virginia Tech Carilion Research Institute, Roanoke, VA, USA

Tereance A. Myers

School of Pharmacy, MCPHS University, Worcester, MA, USA

Humana Press

Editors
Robert G. Gourdie
Virginia Tech Carilion Research Institute
Roanoke, VA, USA

Tereance A. Myers
School of Pharmacy, MCPHS University
Worcester, MA, USA

ISSN 1064-3745 ISSN 1940-6029 (electronic)
ISBN 978-1-4939-5981-5 ISBN 978-1-62703-505-7 (eBook)
DOI 10.1007/978-1-62703-505-7
Springer New York Heidelberg Dordrecht London

Printed on acid-free paper

Humana Press is a brand of Springer
Springer is part of Springer Science+Business Media (www.springer.com)

Preface

In this book we survey classical and cutting-edge methods for studying wound healing and regeneration. Expert scientists, bioengineers, and physicians explain useful technologies, models, techniques, and critical new areas and approaches to clinical and commercial translation of research. We build on the excellent foundation provided by the 2003 publication on wound healing in the *Methods in Molecular Medicine* series edited by Drs. Luisa DePietro and Aime Burns. In our volume, cellular and molecular methods, genetic approaches, surgical techniques, clinical advances, drug discovery and delivery modalities, animal and humanized models, and new applications in the treatment of pathological wounds are covered in a variety of organs and tissues. The 33 chapters are divided among sections on tissue engineering and stem cell-based approaches to wound healing, skin, corneal, and cardiovascular models of injury repair, nonmammalian models, and pertinent aspects of commercialization and intellectual property. Most chapters contain step-by-step methodological descriptions, a detailed list of materials, and an exclusive collection of personal tips or "Notes" that the authors feel critical to success in their technique. Instructive diagrams and informative color photographs also accompany the chapters. Experienced professionals in the wound healing commercial space also provide fresh insight on biotechnology startups, intellectual property, and the patent process. This compendium should enable students, trainees, established basic and clinical researchers, biomedical engineers, and entrepreneurs to efficiently apply the trove of knowledge that has been compiled here by some of the foremost experts in wound healing today. In this way, we intend that the reader might discover and translate new knowledge in the field of wound repair and regeneration.

Tereance A. Myers received her doctorate from Tulane University in New Orleans, LA, and did postdoctoral studies in Wound Healing Biology at the Medical University of South Carolina. Dr. Myers currently works as an Assistant Professor of Pharmaceutical Sciences at Massachusetts College of Pharmacy and Health Sciences in Worcester/Boston, MA.

Robert G. Gourdie received his doctorate from the University of Canterbury in New Zealand and did postdoctoral studies in connexin biology at University College London, UK. Dr. Gourdie is currently Professor and Director of the Center for Heart and Regenerative Medicine at the Virginia Tech Carilion Research Institute and Director of Research for the Department of Emergency Medicine at the Carilion Clinic, Roanoke, Virginia. He is also Professor in the Wake Forest Virginia Tech School of Biomedical Engineering and Sciences at Virginia Tech in Blacksburg, Virginia.

Roanoke, VA, USA — *Robert G. Gourdie*
Worcester, MA, USA — *Tereance A. Myers*

Acknowledgements

Drs. Gourdie and Myers wish to thank Ms. Katherine E. Degen, a graduate student at Virginia Tech in biomedical engineering for her considerable assistance in the editing of this book.

Contents

Part V Non Mammalian and Mammalian Models of Regenerative Repair

Part VI Wound Healing Cell Biology

Part VII Commercialization of Wound Repair and Regenerative Discoveries

Contributors

CRISTIAN AGUILAR • *Department of Developmental and Cell Biology, University of California Irvine, Irvine, CA, USA*
ANTHONY ATALA • *Department of Urology, Wake Forest Institute for Regenerative Medicine, Winston-Salem, NC, USA*
FRANÇOIS A. AUGER • *Centre LOEX de l'Université Laval, Québec, QC, Canada; Médecine Régénératrice – Centre de Recherche FRSQ du Centre hospitalier universitaire (CHU) de Québec, Québec, QC, Canada; Département d'Ophtalmologie and Département de chirurgie, Faculté de médecine, Université Laval, Québec, QC, Canada*
STEPHEN F. BADYLAK • *Department of Surgery, University of Pittsburgh, Pittsburgh, PA, USA; McGowan Institute for Regenerative Medicine, University of Pittsburgh, Pittsburgh, PA, USA; Department of Bioengineering, University of Pittsburgh, Pittsburgh, PA, USA*
DAVID L. BECKER • *Department of Cell and Developmental Biology, University College London, London, UK*
W. MATTHIJS BLANKESTEIJN • *Department of Pharmacology, CARIM, Maastricht University, Maastricht, The Netherlands*
AMANDA R. BROCK • *Department of Biochemistry and Molecular Biology, The University of Texas Graduate School of Biomedical Sciences, The University of Texas MD Anderson Cancer Center, Houston, TX, USA*
SIRISHA BURRA • *Department of Biochemistry and Molecular Biology, The University of Texas Graduate School of Biomedical Sciences, The University of Texas MD Anderson Cancer Center, Houston, TX, USA*
MATTHEW CALEY • *Centre for Cutaneous Research, Blizard Institute of Cell and Molecular Science, Barts and The London School of Medicine and Dentistry, Queen Mary University of London, London, UK*
PATRICK CARRIER • *Centre LOEX de l'Université Laval, Québec, QC, Canada; CUO Recherche, and Médecine Régénératrice – Centre de Recherche FRSQ du Centre Hospitalier Universitaire (CHU) de Québec, Québec, QC, Canada*
ROBERT C. CASKEY • *Department of General Surgery, University of Pennsylvania School of Medicine, Philadelphia, PA, USA*
JERRY S. CHEN • *Department of Surgery, Stanford University, Stanford, CA, USA*
NEIL C. CHI • *Division of Cardiology, Department of Medicine, University of California, San Diego, La Jolla, CA, USA*
YING ANN CHIAO • *Division of Geriatrics, Gerontology and Palliative Medicine, Department of Medicine, Barshop Institute of Longevity and Aging Studies, The University of Texas Health Science Center, San Antonio, TX, USA; Department of Biochemistry, Barshop Institute of Longevity and Aging Studies, The University of Texas Health Science Center, San Antonio, TX, USA*
ALLISON J. COWIN • *Women's and Children's Health Research Institute (WCHRI), North Adelaide, SA, Australia; School of Pediatrics and Reproductive Health, University of South Australia, Adelaide, SA, Australia*

Jose S. Da Silva • *University of Miami, Miami, FL, USA*
Evangelos P. Daskalopoulos • *Department of Pharmacology, CARIM, Maastricht University, Maastricht, The Netherlands*
Joanne O. Davidson • *Department of Physiology, University of Auckland, Auckland, New Zealand*
Lisandra E. de Castro Brás • *Division of Geriatrics, Gerontology and Palliative Medicine, Department of Medicine, Barshop Institute of Longevity and Aging Studies, The University of Texas Health Science Center, San Antonio, TX, USA*
Michael S. Dickover • *Division of Cardiology, Department of Medicine, University of California, San Diego, La Jolla, CA, USA*
Jie Ding • *Wound Healing Research Group, Plastic Surgery Research Laboratory, Department of Surgery, University of Alberta, Edmonton, AB, Canada*
Máté Döbrössy • *University Medical Center Freiburg, Freiburg, Germany*
H. Paul Ehrlich • *Division of Plastic Surgery, Department of Surgery, M. S. Hershey Medical Center, Penn State University Medical College, Hershey, PA, USA*
Tuan A. Elstrom • *FirstString Research, Inc., Mount Pleasant, SC, USA; Regranion, LLC, Charleston, SC, USA*
Stephen A. Fann • *Department of General Surgery, Medical University of South Carolina, Charleston, SC, USA*
Michael J. Galko • *Department of Biochemistry and Molecular Biology, The University of Texas Graduate School of Biomedical Sciences, The University of Texas MD Anderson Cancer Center, Houston, TX, USA*
Erhe Gao • *Center for Translational Medicine, Temple University School of Medicine, Philadelphia, PA, USA*
David Gardiner • *Department of Developmental and Cell Biology, University of California Irvine, Irvine, CA, USA*
Lucie Germain • *Centre LOEX de l'Université Laval, Québec, QC, Canada; Médecine Régénératrice – Centre de Recherche FRSQ du Centre hospitalier universitaire (CHU) de Québec, Québec, QC, Canada; Département d'Ophtalmologie and Département de chirurgie, Faculté de médecine, Université Laval, Québec, QC, Canada*
Gautam S. Ghatnekar • *FirstString Research, Inc., Mount Pleasant, SC, USA; Halimed Pharmaceuticals, Charleston, SC, USA; Regranion, LLC, Charleston, SC, USA*
Daniel J. Gibson • *Department of Obstetrics and Gynecology, Institute for Wound Research, University of Florida, Gainesville, FL, USA*
Robert G. Gourdie • *Department of Regenerative Medicine and Cell Biology, Medical University of South Carolina, Charleston, SC, USA; Department of Bioengineering, Clemson University, Clemson, SC, USA; Virginia Tech Carilion Research Institute, Virginia Polytechnic Institute & State University, Roanoke, VA, USA; Virginia Tech-Wake Forest University School of Biomedical Engineering and Sciences, Virginia Polytechnic Institute & State University, Blacksburg, VA, USA*
Colin R. Green • *Department of Ophthalmology, Faculty of Medical and Health Sciences, University of Auckland, Auckland, New Zealand*
Sylvain L. Guérin • *Centre LOEX de l'Université Laval, Québec, QC, Canada; CUO-Recherche, and Médecine Régénératrice – Centre de Recherche FRSQ du Centre Hospitalier Universitaire (CHU) de Québec, Québec, QC, Canada;*

Département d'Ophtalmologie and Département de chirurgie, Faculté de médecine, Université Laval, Québec, QC, Canada
ALISTAIR J. GUNN • *Department of Physiology, University of Auckland, Auckland, New Zealand*
GEOFFREY C. GURTNER • *Hagey Laboratory for Pediatric and Regenerative Medicine, Division of Plastic and Reconstructive Surgery, Department of Surgery, Institute of Stem Cell Biology and Regenerative Medicine, Stanford University School of Medicine, Stanford, CA, USA*
SANDER O. HACKER • *Department of Laboratory Animal Resources, The University of Texas Health Science Center at San Antonio, San Antonio, TX, USA*
GANESH V. HALADE • *Division of Geriatrics, Gerontology and Palliative Medicine, Department of Medicine, Barshop Institute of Longevity and Aging Studies, The University of Texas Health Science Center, San Antonio, TX, USA*
MANJONG HAN • *Division of Cell and Molecular Biology, Tulane University, New Orleans, LA, USA*
PEIDONG HAN • *Division of Cardiology, Department of Medicine, University of California, San Diego, La Jolla, CA, USA*
JOSHUA M. HARE • *University of Miami, Miami, FL, USA*
DARIUSH HONARDOUST • *Wound Healing Research Group, Plastic Surgery Research Laboratory, Department of Surgery, University of Alberta, Edmonton, AB, Canada*
W. CHIP. HOOD • *Upstream Partners Inc., Duluth, GA, USA*
DAVID HUIZENGA • *Tao Life Sciences Inc., Anderson, SC, USA*
BEN J.A. JANSSEN • *Department of Pharmacology, CARIM, Maastricht University, Maastricht, The Netherlands*
L. JANE JOURDAN • *Department of Regenerative Medicine and Cell Biology, Medical University of South Carolina, Charleston, SC, USA; Department of Bioengineering, Clemson University, Clemson, SC, USA; Virginia Tech Carilion Research Institute, Virginia Polytechnic Institute & State University, Roanoke, VA, USA*
WALTER J. KOCH • *Center for Translational Medicine, Temple University School of Medicine, Philadelphia, PA, USA*
TIMOTHY J. KOH • *Department of Kinesiology and Nutrition, University of Illinois at Chicago, Chicago, IL, USA*
BRENDA R. KWAK • *Department of Pathology and Immunology, University of Geneva, Geneva, Switzerland; Division of Cardiology, Department of Internal Medicine, University of Geneva, Geneva, Switzerland*
PETER KWAN • *Firefighters' Burn Treatment Unit, Division of Plastic Surgery and Reconstructive Surgery, University of Alberta, Edmonton, AB, Canada; Division of Critical Care Medicine, University of Alberta, Edmonton, AB, Canada; Department of Surgery, University of Alberta, Edmonton, AB, Canada*
JANGWOO LEE • *Department of Developmental and Cell Biology, University of California Irvine, Irvine, CA, USA*
KENNETH W. LIECHTY • *Nemours Children's Hospital, Orlando, FL, USA*
MERRY L. LINDSEY • *Division of Geriatrics, Gerontology and Palliative Medicine, Department of Medicine, Barshop Institute of Longevity and Aging Studies, The University of Texas Health Science Center, San Antonio, TX, USA*
MICHAEL T. LONGAKER • *Hagey Laboratory for Pediatric and Regenerative Medicine, Division of Plastic and Reconstructive Surgery, Department of Surgery, Institute of Stem Cell Biology and Regenerative Medicine, Stanford University School of Medicine, Stanford, CA, USA; Department of Surgery, Stanford University, Stanford, CA, USA*

DANIEL J. LUTHER • *Department of Integrative Medical Sciences, Northeast Ohio Medical University, Rootstown, OH, USA*
YONGGANG MA • *Division of Geriatrics, Gerontology and Palliative Medicine, Department of Medicine, Barshop Institute of Longevity and Aging Studies, The University of Texas Health Science Center, San Antonio, TX, USA*
J. GARY MESZAROS • *Department of Integrative Medical Sciences, Northeast Ohio Medical University, Rootstown, OH, USA*
RITA E. MIRZA • *Department of Kinesiology and Nutrition, University of Illinois at Chicago, Chicago, IL, USA*
MOEIN MOMTAZI • *Firefighters' Burn Treatment Unit, Division of Plastic Surgery and Reconstructive Surgery, University of Alberta, Edmonton, AB, Canada; Division of Critical Care MedicineUniversity of Alberta, Edmonton, AB, Canada; Department of SurgeryUniversity of Alberta, Edmonton, AB, Canada*
MARY O. MORALES • *Department of General Surgery, Medical University of South Carolina, Charleston, SC, USA*
SANDRINE MOREL • *Department of Pathology and Immunology, University of Geneva, Geneva, Switzerland; Division of Cardiology, Department of Internal Medicine, University of Geneva, Geneva, Switzerland*
KURTIS E. MOYER • *Division of Plastic Surgery, Department of Surgery, M. S. Hershey Medical Center, Penn State University Medical College, Hershey, PA, USA*
KEN MUNEOKA • *Division of Cell and Molecular Biology, Tulane University, New Orleans, LA, USA*
RACHAEL Z. MURRAY • *Tissue Repair and Regeneration Program, Institute of Health and Biomedical Innovation, Queensland University of Technology, Brisbane, QLD, Australia*
ALLISON C. NAUTA • *Hagey Laboratory for Pediatric and Regenerative Medicine, Division of Plastic and Reconstructive Surgery, Department of Surgery, Institute of Stem Cell Biology and Regenerative Medicine, Stanford University School of Medicine, Stanford, CA, USA; Department of Surgery, Stanford University, Stanford, CA, USA*
LOUISE F.B. NICHOLSON • *Department of Anatomy with Radiology, Faculty of Medical and Health Sciences, University of Auckland, Auckland, New Zealand*
MARGARET L. NOVAK • *Department of Kinesiology and Nutrition, University of Illinois at Chicago, Chicago, IL, USA*
SIMON J. O'CARROLL • *Department of Anatomy with Radiology, Faculty of Medical and Health Sciences, University of Auckland, Auckland, New Zealand*
MICHAEL P. O'QUINN • *Department of Regenerative Medicine and Cell Biology, Medical University of South Carolina, Charleston, SC, USA*
STOJADINOVIC OLIVERA • *Wound Healing and Regenerative Medicine Research Program, Department of Dermatology and Cutaneous Surgery, University of Miami Miller Medical School, Miami, FL, USA*
EMILY L. ONGSTAD • *Department of Regenerative Medicine and Cell Biology, Medical University of South Carolina, Charleston, SC, USA; Department of Bioengineering, Clemson University, Clemson, SC, USA; Virginia Tech Carilion Research Institute, Virginia Polytechnic Institute & State University, Roanoke, VA, USA*
D.P. ORGILL • *Division of Plastic Surgery, Brigham and Women's Hospital and Harvard Medical School, Boston, MA, USA*
JOSEPH A. PALATINUS • *Department of Regenerative Medicine and Cell Biology, Medical University of South Carolina, Charleston, SC, USA*

MATTHEW PEAKE • *School of Biology, Newcastle University, Newcastle upon Tyne, UK*
G. PIETRAMAGGIORI • *Plastic and Reconstructive Surgery, University Hospital of Lausanne, Lausanne, Switzerland*
JAN PRUSZAK • *University of Freiburg, Freiburg, Germany*
VERONICA RODRIGUEZ-RIVERA • *Department of General Surgery, Medical University of South Carolina, Charleston, SC, USA*
SASHWATI ROY • *Department of Surgery, Comprehensive Wound Center, The Ohio State University, Columbus, OH, USA*
S. SCHERER • *Plastic and Reconstructive Surgery, University Hospital of Lausanne, Lausanne, Switzerland*
GREGORY S. SCHULTZ • *Department of Obstetrics and Gynecology, Institute for Wound Research, University of Florida, Gainesville, FL, USA*
CHANDAN K. SEN • *Department of Surgery, Comprehensive Wound Center, The Ohio State University, Columbus, OH, USA*
BRIAN SICARI • *Department of Cellular and Molecular Pathology, University of Pittsburgh School of Medicine, Pittsburgh, PA, USA*
JENNIFER SIMKIN • *Division of Cell and Molecular Biology, Tulane University, New Orleans, LA, USA*
PHIL STEPHENS • *Wound Biology Group, Cardiff Institute of Tissue Engineering and Repair Tissue Engineering and Reparative Dentistry, School of Dentistry, Cardiff University, Cardiff, Wales, UK*
ERIK G. STRUNGS • *Department of Regenerative Medicine and Cell Biology, Medical University of South Carolina, Charleston, SC, USA*
LOUIS TERRACIO • *Department of Basic Science and Craniofacial Biology, NYU College of Dentistry, New York, NY, USA*
CHARLES K. THODETI • *Department of Integrative Medical Sciences, Northeast Ohio Medical University, Rootstown, OH, USA*
MARJANA TOMIC-CANIC • *Wound Healing and Regenerative Medicine Research Program, Department of Dermatology and Cutaneous Surgery, University of Miami Miller Medical School, Miami, FL, USA*
EDWARD E. TREDGET • *Wound Healing Research Group, Plastic Surgery Research Laboratory, Department of Surgery, University of Alberta, Edmonton, AB, Canada; Firefighters' Burn Treatment Unit, Division of Plastic Surgery and Reconstructive Surgery, University of Alberta, Edmonton, AB, Canada; Division of Critical Care Medicine, University of Alberta, Edmonton, AB, Canada; Department of Surgery, University of Alberta, Edmonton, AB, Canada*
NEILL TURNER • *Department of Surgery, University of Pittsburgh, Pittsburgh, PA, USA*
YAN WANG • *Department of Biochemistry and Molecular Biology, The University of Texas Graduate School of Biomedical Sciences, The University of Texas MD Anderson Cancer Center, Houston, TX, USA*
JAMES M. WATERS • *Women's and Children's Health Research Institute (WCHRI), North Adelaide, SA, Australia*
TRACI A. WILGUS • *Department of Pathology, The Ohio State University, Columbus, OH, USA*
BRIAN C. WULFF • *Department of Pathology, The Ohio State University, Columbus, OH, USA*
MINGQUAN YAN • *Division of Cell and Molecular Biology, Tulane University, New Orleans, LA, USA*

ERIC M. YOST • *Department of General Surgery, Medical University of South Carolina, Charleston, SC, USA*
MICHAEL J. YOST • *Department of General Surgery, Medical University of South Carolina, Charleston, SC, USA*
LING YU • *Division of Cell and Molecular Biology, Tulane University, New Orleans, LA, USA*
ROGELIO ZAMILPA • *Division of Geriatrics, Gerontology and Palliative Medicine, Department of Medicine, Barshop Institute of Longevity and Aging Studies, The University of Texas Health Science Center, San Antonio, TX, USA*
KARINE ZANIOLO • *Centre LOEX de l'Université Laval, Québec, QC, Canada; CUO-Recherche, and Médecine Régénératrice – Centre de Recherche FRSQ du Centre Hospitalier Universitaire (CHU) de Québec, Québec, QC, Canada*
JIANHUA ZHANG • *Division of Geriatrics, Gerontology and Palliative Medicine, Department of Medicine, Barshop Institute of Longevity and Aging Studies, The University of Texas Health Science Center, San Antonio, TX, USA*
RUILIN ZHANG • *Division of Cardiology, Department of Medicine, University of California, San Diego, La Jolla, CA, USA*
YUANYUAN ZHANG • *Department of Urology, Wake Forest Institute for Regenerative Medicine, Wake Forest University School of Medicine, Winston-Salem, NC, USA*

Part I

Tissue Engineered Wound Repair Models

Chapter 1

An In Vivo Model System for Evaluation of the Host Response to Biomaterials

Brian Sicari, Neill Turner, and Stephen F. Badylak

Abstract

We describe an in vivo model system designed to evaluate the host response to implanted biomaterials: *The partial thickness rat abdominal wall defect model.* The model allows for determination of the temporal and spatial distribution of the cellular and vascular response, the remodeling of the implanted material and surrounding host soft tissue, and the function of the remodeled tissue over time.

Key words Soft tissue reconstruction, Biomaterials, Abdominal wall defect

1 Introduction

The fields of regenerative medicine, tissue engineering, polymer science, and biomaterials continue to develop novel scaffold materials, many of which attempt to facilitate a constructive host tissue remodeling response following implantation. Naturally occurring and synthetic biomaterial scaffolds with and without cellular components are currently being developed and evaluated in both preclinical and clinical settings [1–3]. A well-recognized sequela following implantation is the foreign body inflammatory response. Strategies to alter the acute and/or chronic inflammatory host response involve modifications to existing biomaterials or development of new materials. Such strategies and materials require a reliable in vivo test system. The partial thickness rat abdominal wall defect model discussed in this chapter allows for a systematic evaluation of the host response to an implanted test article.

The successful development and clinical translation of novel biomaterials or scaffolds requires a thorough understanding of their elicited in vivo biological responses. Biocompatibility reflects the interaction between biomaterials and host tissue. Its assessment is one of the critical concerns in biomaterials research and required

Robert G. Gourdie and Tereance A. Myers (eds.), *Wound Regeneration and Repair: Methods and Protocols*, Methods in Molecular Biology, vol. 1037, DOI 10.1007/978-1-62703-505-7_1, © Springer Science+Business Media New York 2013

by the FDA [4–6]. Biocompatibility refers to both: (a) biosafety, which refers to an appropriate host response lacking toxicity, mutagenesis, and/or carcinogenesis, and (b) biofunctionality, which is the ability of the biomaterial to perform its intended task [7, 8]. The partial thickness rat abdominal wall defect model discussed in this chapter allows for a systematic evaluation of a biomaterial's biocompatibility in terms of the nature of the elicited host response and the biomaterial's constructive remodeling potential.

1.1 Default Mammalian Injury Response Versus Constructive Tissue Remodeling

The default mammalian response to injury is highly conserved among tissue types and occurs in three overlapping yet distinct phases: the inflammatory phase, the proliferative phase, and the remodeling phase [9]. The inflammatory phase occurs immediately after tissue damage and is marked by activation of the coagulation cascade and the subsequent influx of innate inflammatory cells that facilitate the removal of cellular debris and infection prevention. Specifically, hemostasis is achieved through the deposition of a provisional fibrin matrix which acts as a scaffold for infiltrating neutrophils and macrophages [10]. The proliferative phase is characterized by angiogenesis as well as the migration, proliferation and differentiation of different cell types [11, 12]. During the remodeling phase macrophages and endothelial cells are subject to apoptosis while fibroblasts and myofibroblasts produce and deposit a collagenous extracellular matrix. This relatively acellular fibrous matrix is the precursor of scar tissue which typically replaces site appropriate functional tissue within the injury site.

In contrast, host constructive tissue remodeling is a process by which the default healing response to injury is modified from one that results in scar tissue formation to one that results in the formation of site appropriate functional tissue. Mechanisms behind a constructive remodeling response to injury include the participation of an activated host progenitor cell population and a predominant M2/Th2 restricted immune response among others [13–15].

The implantation of a biomaterial has the potential to significantly alter the default mammalian response to injury. For example, following injury and implantation, biomaterials are able to affect: hemostasis and provisional matrix formation; the phenotype of infiltrating inflammatory cells; host cell migration, proliferation, and differentiation; as well as fibrosis and fibrous capsule development [7, 8]. The manner in which the injury response is altered depends on the biomaterial's origin and composition, contact duration, degradation rate, surface structure, cellularity, size, and implantation site among others [5, 7, 8].

Implantation of nondegradable synthetic or chemically crosslinked biomaterials typically intensifies the host's inflammatory response to injury and leads to a foreign body reaction (FBR) and an increased deposition of scar tissue [16, 17]. During a FBR persistent inflammatory stimuli, such as the continual presence of a nondegradable biomaterial, lead to chronic inflammation and the formation of multinucleate foreign body giant cells [18]. Foreign body giant cells are formed through the fusion of macrophages located at the surface of the biomaterial and further exacerbate the host's inflammatory response through a process known as "frustrated phagocytosis" and the associated production of reactive oxygen intermediates and pro-inflammatory cytokines [19, 20]. A FBR typically results in the deposition of a 50–200 μm thick collagenous capsule surrounding the implant [21, 22]. This dense accumulation of fibrous tissue confines the implant and prevents integration with the surrounding tissue.

While the implantation of nondegradable synthetic scaffolds intensifies the host's default fibrotic response to injury, the implantation of degradable biologic scaffolds composed of xenogeneic extracellular matrix (ECM) results in constructive tissue remodeling [23, 24] (Fig. 1d). Stated differently, the implementation of a properly decellularized and non-cross-linked ECM scaffold modifies the default host response to tissue injury and results in site appropriate and functional tissue replacement.

1.2 Role of Macrophages in the Host Response

Macrophages play a pivotal role in the host response to biomaterial implantation. In fact, macrophage depletion prevents the degradation of naturally occurring biologic scaffolds in vivo [25]. Macrophages have been shown to respond to many different implanted biomaterials, including those composed of metals [26], ceramics [27], polymers [28], and biologic proteins such as collagens and xenogeneic ECM [25, 29]. As stated above, macrophage mediated foreign body giant cell formation and pro-inflammatory cytokine production are frequently associated with the implantation of nondegradable or synthetic biomaterials. However, several recent studies suggest that macrophages can facilitate constructive and site-appropriate tissue remodeling in response to biomaterial implantation [25, 30]. The phenotype of responding macrophages has been found to be an important determining factor in the success of an implanted biomaterial and its ultimate remodeling outcome [13, 14].

1.2.1 Macrophage Heterogeneity

Macrophages are a plastic cell population capable of acquiring diverse phenotypes. They can be classified as either classically activated and pro-inflammatory (M1) or alternatively activated and immunomodulatory (M2) [31]. Classically activated macrophages, designated M1 after the Th1/Th2 nomenclature, produce pro-inflammatory cytokines and reactive oxygen intermediates; and are responsible for pathogen killing, cell clearance, and the general propagation of a pro-inflammatory response [32]. Alternatively activated macrophages, designated M2, deposit collagen; produce anti-inflammatory cytokines; and facilitate tissue remodeling and repair [33]. Although histologically indistinguishable, M1 and M2

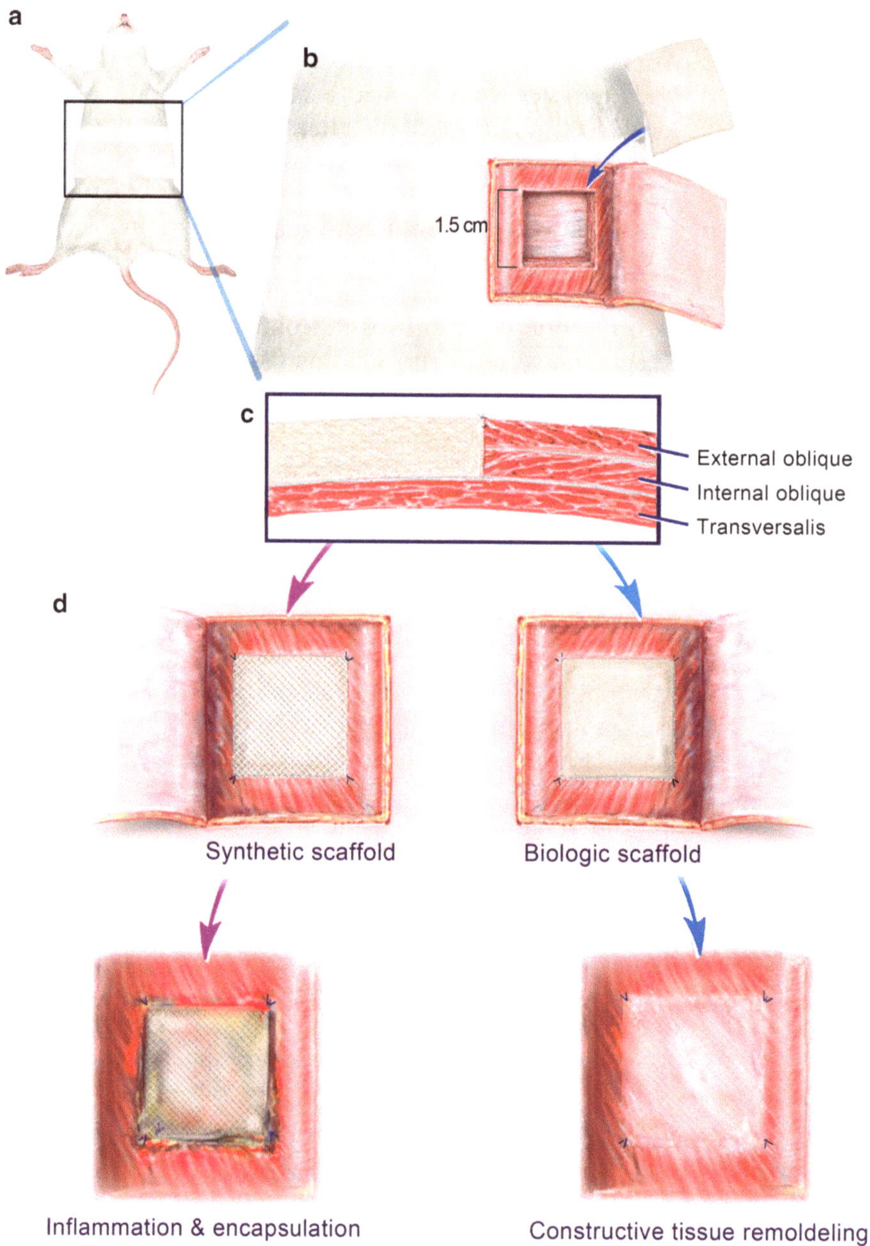

Fig. 1 Schematic of the partial thickness abdominal wall defect procedure showing preparation of the rat abdominal wall (**a**); the defect site (**b**); a cross-sectional diagram of the abdominal wall following test article implantation (**c**); and an example of gross morphology depicting an encapsulation versus constructive tissue remodeling host response (**d**)

macrophages can be identified and distinguished by their species specific and distinct cytokine, gene expression, and cell surface marker profiles [34, 35].

The role of a heterogeneous host macrophage response to injury has recently been reviewed [36]. Immediately following acute skeletal muscle injury, pro-inflammatory cytokines polarize responding host macrophages towards a predominant M1 phenotype. To facilitate appropriate skeletal muscle regeneration, a shift in macrophage polarization takes place and after approximately 3 days macrophages within the injury site display a predominant M2 phenotype [14]. Furthermore, when an M2 macrophage response is inhibited, the host injury response is characterized by a severe lack of muscle regeneration and an accentuation of inflammation and necrosis [37]. In the context of biomaterial implantation, recent studies have shown that a predominant M2 macrophage response to degradable biologic scaffolds is associated with a constructive remodeling outcome. Conversely, a predominant M1 response typically correlates with a FBR, the presence of foreign body giant cells, and a poor remodeling outcome in nondegradable biologic scaffolds [14, 25]. In this regard, it is possible that the remodeling outcome of an implanted biomaterial may be predicted based on the host's immune response and the phenotype of responding macrophages. The partial thickness rat abdominal wall defect model described in this chapter allows for the systematic evaluation of the host response to biomaterials. The model permits the examination of the temporal and spatial distribution of the cellular response; the presence of a FBR; the phenotype of responding macrophages; the remodeling of the implanted material and surrounding host tissue; and function of the remodeled tissue over time.

1.3 Experimental Design Considerations for In Vivo Models of Biomaterial Implantation

The term biomaterial encompasses a diverse range of products and materials. Devices composed of such biomaterials are designed for a similarly diverse range of clinical uses. As a result a number of in vivo models have been developed to evaluate the safety and biocompatibility of biomaterials. Some of these models are designed to evaluate specific aspects of a biomaterial or clinical use, such as mechanical strength, biodegradability, or chemical leaching of materials designed for use in the cornea [38] or dura mater [39]. However, general compatibility and/or safety studies, and those investigating the host response, typically utilize either a subcutaneous [40, 41] or intramuscular [42, 43] implantation model.

1.3.1 Selection of a Model

In vivo studies are a required element for FDA certification of a biomaterial [6]. In addition, agencies such as the FDA and its worldwide counterparts, have established a series of standard reference experiments with which to evaluate biomaterial safety. These preclinical tests, such as the USP implantation test, British standard implantation test, and ISO 10993 implantation test, typically use rabbits as the animal model [44]. However, it is increasingly becoming common for rodents, particularly rats, to be used as a pre-evaluation model prior to undertaking the required reference studies.

The rat subcutaneous and intramuscular implantation models present an ideal environment for evaluating the host response to biomaterials. Implantation of devices at these sites is relatively straight forward requiring only basic surgical skill. Both models also easily facilitate the use of multiple sites along the host allowing different materials to be evaluated in the same animal. In addition, the subcutaneous and muscle tissues are highly vascularized allowing a rapid host response to the implanted test article. Both models also allow for a range of study designs to be investigated, from short term degradation and biochemical studies to long term remodeling and functional outcome studies. The partial thickness rat abdominal wall defect model does, however, have a number of advantages over other intramuscular and subcutaneous models. The partial thickness model is particularly suited for long term studies up to 2 years in duration since the material implant can be more securely anchored to the surrounding host tissue. This ability to be securely anchored to surrounding muscle tissue also makes this model more suitable for mechanical testing and evaluation of function, particularly of degradable and ECM-based scaffolds designed to be replaced with functional host tissue over time. The subcutaneous model is more suitable for shorter term studies and for nondegradable materials due to the risk of implant migration within the tissue.

1.3.2 Size of Biomaterial

In addition to selecting the most appropriate implantation site another important consideration is the size of the implanted device. Tissue reaction to implanted devices can be elicited by the material directly or by substances that may be contained within the material as a consequence of the manufacturing process. For example a material that is chemically pure but does not resorb will not markedly affect the surrounding host tissue environment whereas a biodegradable material will release degradation products that will influence the surrounding host tissue. Similarly impurities such as plasticizers, preservatives, and/or stabilizers may leach from the biomaterial causing the surrounding host tissue to react. The dimensions of the material will in part determine the contribution of these components to the host response. If the implanted device is too small, particularly with biodegradable materials, the degradation may be so rapid that the surrounding tissue does not have sufficient time to respond to the implant in a reliable and reproducible manner. Similarly, if the implant device is too big, it may cause irritation of the surrounding tissues promoting a heightened inflammatory response that is not representative of the true host response to the biomaterial but a consequence of the mechanical stimulation. For biocompatibility studies it is therefore best to utilize implants that will induce minimal mechanical effects upon the surrounding host tissue, but which still allow an appropriate host response. For example it has been observed that while a particulate (i.e., comminuted) version of a material may prove biocompatible, a large film of the same material can promote carcinogenesis [45]. As a result, for the partial thickness rat abdominal wall defect model, the recommended size of the implant is between 1 × 1 cm and 2 × 2 cm.

1.3.3 Duration of Biomaterial Implantation

Depending on the properties of the biomaterial to be tested a range of sacrifice intervals may be required to fully understand and characterize the host response to the implant. Nondegradable materials, by their very nature, do not usually promote the constructive remodeling of the defect site into site appropriate tissue; indeed, a chemically pure, nondegradable material, will illicit little response from the tissue other than the standard foreign body response. As such, studies of these types of material longer than 28 days would provide little additional information about the host response since the foreign body reaction will largely be completed. In addition, the acute inflammatory response generated by implanted biomaterials generally peaks within the first 7 days following implantation before gradually decreasing [5, 8]. Similarly, the polarization of macrophages and the switch from a pro-inflammatory M1 phenotype to a remodeling M2 phenotype again occurs within the first 2–4 days, with the clearest differences in host response being detected between days 7 and 14 [14, 36]. However, longer term studies may be useful in measuring changes in the mechanical strength of the biomaterial implant in response to the tissue microenvironment.

For biodegradable materials or scaffolds derived from naturally occurring materials such as extracellular matrix, the replacement of these materials with site appropriate tissue is an essential component inherent in their design. As a consequence, long term implantation studies are essential in evaluating a scaffolds ability to promote constructive tissue remodeling. Typically, the constructive remodeling of the defect site does not fully become apparent until the scaffold material has completely degraded. The degradation products of these scaffold materials are the stimulus to attract tissue resident or circulating stem and progenitor cells, stimulate angiogenesis, and promote an M2 macrophage response [14, 46, 47]. While early sacrifice points, up to 14 days, will allow the study of the host immune response to the biomaterials, time points less than 28 days will provide little information about the potential of these materials to promote constructive tissue remodeling. Depending on the degradation properties of the biomaterials being studied sacrifice points of 6 months to 1 year may be required to allow sufficient time for the material to degrade and for the constructive remodeling response to reach equilibrium. Time points between 28 days and 6 months allow the constructive remodeling response to be monitored and the temporal and spatial changes associated with the remodeling process to be assessed, which is particularly important when comparing biomaterials with similar compositions but different structural properties.

1.3.4 Evaluating the Host Response to a Biomaterial

Evaluation of the Host Response

The partial thickness rat abdominal wall defect model allows for the systematic evaluation of the host response to biomaterial implantation. Hematoxylin and Eosin or Masson's trichrome trichrome stained histology sections from the biomaterial-host tissue interface can provide a semiquantitative examination of the host response. Histomorphologic analysis reveals (a) the rate of biomaterial degradation, (b) amount and spatial distribution of host cell infiltration, (c) the presence or absence of foreign body giant cells and fibrous encapsulation, (d) vascularity of the remodeled tissue, (e) muscle ingrowth, and (f) the degree of organization of the replacement tissue [17]. With appropriate immunolabeling, the phenotype of macrophages responding to a biomaterial scaffold may also be evaluated. An early favorable host response would show a predominant M2 macrophage cellular infiltrate along with the absence of foreign body giant cells. At later time points, biomaterial degradation and host mediated deposition of organized tissue along with muscle ingrowth and the lack of fibrous encapsulation would be indicative of a favorable host response.

Evaluation of Constructive Tissue Remodeling

Constructive tissue remodeling is a process by which the default host scarring response to injury is modified and instead, results in the formation of site appropriate and functional tissue. Following the implantation of a biomaterial scaffold, constructive tissue remodeling in the partial thickness rat abdominal wall defect model would result in functional abdominal wall musculature. At all time points, the model allows for the evaluation of site appropriate constructive remodeling through the immunolabeling of the remodeled tissue for markers of soft tissue differentiation. The presence of fast and slow muscle fibers along with blood vessels, and nerve fibers within the remodeled tissue are indicative of site appropriate skeletal muscle. Functionally innervated and vascularized skeletal muscle should also be able to contract. The abdominal wall defect model allows for the systematic evaluation of functional constructive remodeling through the examination of the in situ contractile force generated by the remodeled tissue [24].

2 Materials

2.1 Partial Thickness Rat Abdominal Wall Defect Model

2.1.1 Surgical Procedure

1. Sprague-Dawley rats weighing 300–500 g.
2. Isoflurane (2 % in O_2).
3. Betadine.
4. 4-0 prolene or similar, nonabsorbable suture.
5. 3-0 Vicryl, or similar, resorbable suture.
6. Heating pad.
7. Appropriate analgesia such as 0.01–0.05 mg/kg Buprenorphine Hydrochloride.

2.1.2 Euthanasia and Sample Harvest

1. Isoflurane (5 % in O_2).
2. Cork board for pinning the explanted tissue.
3. 10 % neutral buffered formalin.
4. Paraffin.

2.2 Histomorphologic Analysis

1. Hematoxylin and Eosin staining materials.
2. Masson Trichrome staining materials.

2.3 Immunolabeling Studies

2.3.1 Macrophage Phenotype

1. Primary Antibodies.
   (a) Goat anti-CD206.
   (b) Rabbit anti-CCR7.
   (c) Mouse anti-CD68.
2. Secondary Antibodies
   (a) Alexa Fluor 350 anti-mouse.
   (b) Alexa Fluor 488 anti-goat.
   (c) Alexa Fluor 568 anti-rabbit.
3. Xylene.
4. Ethanol.
5. Citric Acid (pH 6.0).
6. Tris buffered saline (pH 7.4).
7. Blocking Buffer: 2 % horse serum, 1 % bovine serum albumin, 0.1 % Triton-X, and 0.1 % Tween 20 in TBS.
8. Draq5 in TBS.

2.3.2 Vascularity and Innervation

1. Xylene.
2. Ethanol.
3. 1 mM EDTA (pH 8.0).
4. Phosphate buffered saline (pH 7.0).
5. Blocking solution: 2 % normal horse serum and 2 % normal bovine serum.
6. Primary Antibodies.
   (a) Rabbit anti-CD31.
   (b) Mouse anti-beta-III tubulin.
7. Secondary Antibodies
   (a) Biotinylated anti-rabbit.
   (b) Biotinylated anti-mouse.
8. Horseradish peroxidase detection solution.
9. Diaminobenzidine.
10. Hematoxylin counter staining materials.

2.3.3 Myofibers

1. Xylene.
2. Ethanol.
3. Citric Acid (pH 6.0).
4. Tris buffered saline (pH 7.4).
5. PBS.
6. 0.1 % trypsin and 0.1 % calcium chloride in TBS.
7. Blocking Buffer: 2 % normal horse serum and 2 % normal bovine serum in TBS.
8. Antibody solution: 2 % normal horse serum and 2 % normal bovine serum in TBS.
9. Primary Antibodies:
   (a) Mouse anti-slow myosin.
   (b) Alkaline phosphatase conjugated anti-fast myosin.
10. Secondary:
    (a) Biotinylated anti-mouse.
    (b) Alkaline phosphatase detection substrate.
11. Mouse horseradish peroxidase detection solution.
12. Diaminobenzidine.
13. Hematoxylin staining materials.

2.4 Functional Testing

2.4.1 In Situ Contractile Force Testing

1. 2 % Isoflurane in O_2.
2. 70 % Ethanol.
3. Silk sutures.
4. Platinum electrodes.
5. Force transducer.
6. Analog to digital data acquisition card.
7. LabVIEW software.
8. S88X Grass stimulator.

2.4.2 In Vitro Contractile Force Testing

1. Krebs' buffer (121.0 mM NaCl, 5.0 mM KCl, 0.5 mM $MgCl_2$, 1.8 mM $CaCl_2$, 24 mM $NaHCO_3$, 0.4 mM NaH_2PO_4, and 5.5 mM glucose, pH 7.4) oxygenated with 95 % oxygen and heated to 37 °C.
2. Organ bath (Radnoti glassware).
3. Platinum field stimulating electrodes.
4. S88X Grass stimulator.
5. Data recording system.
6. CO_2 for euthanasia.

7. 70 % Ethanol.
8. Non-stretching suture (e.g., silk or stainless steel).
9. Micropositioning mounting bracket.

3 Methods

3.1 Partial Thickness Rat Abdominal Wall Defect Model

The partial thickness rat abdominal wall defect is created as previously described by Badylak et al. [48]. Sprague-Dawley rats weighing 300–500 g can be purchased from Charles River Laboratories (Wilmington, Massachusetts) or similar vendors. Animals are then housed in standard cages and receive food and water ad libitum. The housing environment is maintained at a temperature of 68–76 °F. All procedures described are performed in accordance with the National Institutes of Health guidelines for care and use of laboratory animals.

3.1.1 Surgical Procedure

1. Animals are anesthetized with 2 % isoflurane in oxygen (*see* **Note 1**).
2. Prepare the ventral abdomen by clipping or shaving (Fig. 1a) and scrub with Betadine (povidone–iodine) before placing sterile drapes around the surgical site (*see* **Note 2**).
3. Make a ventral midline abdominal incision through the epidermis and bluntly dissect the dermis and subcutaneous tissue from the underlying muscle tissues on one side of the midline for a distance of approximately 4 cm to the anterior axillary line, thus exposing the oblique musculature while leaving the epidermis (*see* **Note 3**).
4. Retract the incision in the ventral midline of the abdomen to expose the ventral lateral abdominal wall adjacent to the linea alba, including the musculotendinous junction of the abdominal wall musculature.
5. Excise a 1.5 cm × 1.5 cm partial thickness defect consisting of the internal and external oblique layers of the abdominal wall musculature while leaving the underlying transversalis and peritoneum intact (Fig. 1b). Uniformity of the defect size and shape is ensured by implanting test articles with a fixed size and shape in each animal.
6. Replace the defect with a 1.5 cm × 1.5 cm piece of the biomaterial being tested (Fig. 1c).
7. Place a single interrupted 4-0 prolene or similar, nonabsorbable suture at each of the four corners of the test article to secure the biomaterial to the adjacent abdominal wall musculature. Suturing the test article to the native adjacent musculature subjects it to a physiologic mechanical load while allowing the demarcation of the implant upon necropsy (*see* **Note 4**).

8. Close the skin using 3-0 Vicryl, or similar, resorbable suture in an interrupted fashion and allow the animal to recover from anesthesia on a heating pad.
9. Each animal should receive appropriate analgesia such as 0.01–0.05 mg/kg Buprenorphine Hydrochloride immediately post-operatively and for at least 2 days afterward. In addition, a prophylactic broad spectrum antibiotic may also be given such as 2–3 mg/kg gentamicin (*see* **Note 5**).

3.1.2 Euthanasia and Sample Harvest

On the predetermined sacrifice date, each animal is anesthetized using isoflurane (5 % in oxygen) and then euthanized by intracardiac injection of saturated potassium chloride to induce cardiac arrest prior to implant harvest.

1. Following euthanasia, make a ventral midline incision through the epidermis and bluntly dissect the dermis and subcutaneous tissues from the underlying musculature.
2. Explant the full thickness abdominal wall containing the implant site (corners demarcated by the sutures) along with a small amount of adjacent native tissue.
3. In order to preserve its native tissue morphology, pin the explanted tissue to a corkboard and fix in 10 % neutral buffered formalin before standard paraffin embedding and histological processing.

3.2 Histomorphologic Analysis

For histomorphologic analysis, tissue sections are stained with either Hematoxylin and Eosin or Masson's trichrome trichrome before coverslipping. Semiquantitative Histomorphologic analysis reveals (a) the rate of biomaterial degradation, (b) amount of host cell infiltration, (c) the presence or absence of foreign body giant cells and fibrous encapsulation, (d) vascularity of the remodeled tissue, (e) muscle ingrowth, and (f) the degree of organization of the replacement tissue [17]. Depending on the time point selected for analysis, different criteria may be used to generate a histological score of the tissue. An example of a histomorphological scoring system can be seen in Tables 1 and 2. Histological scoring should be performed by at least three investigators who are blinded to the nature of the implant being tested. Statistical analysis should be performed using a suitable nonparametric test such as Fisher's exact test, Mann–Whitney *U*-test, or Kruskal–Wallis test.

3.3 Immunolabeling Studies

3.3.1 Macrophage Phenotype

To determine the phenotype of macrophages responding to the implantation of the test article, immunofluorescent labeling of markers known to be strong indicators of M1 or M2 macrophages can be performed. CCR7 and CD206 are often used as markers of M1 and M2 macrophages, respectively, with CD68 used as a pan-macrophage marker in this rat model.

Table 1
Scoring criteria for histomorphologic analysis at 14 days post-implantation

	0	1	2	3
Cellular infiltration score	0 Cells per HPF	1–75 Cells per HPF	75–150 Cells per HPF	More than 150 cells per HPF
Multinucleated giant cell score	More than 5 multinucleate giant cells per HPF	2–5 Multinucleate giant cells per HPF	1 Multinucleate giant cell per HPG	No multinucleate giant cells per HPF
Vascularity	0–1 Blood vessels per HPF	2–5 Blood vessels per HPF	6–10 Blood vessels per HPF	More than 10 blood vessels per HPF
Connective tissue organization	Original scaffold intact	Original scaffold disrupted, poorly organized new ECM present	Moderately organized connective tissue present	Dense, highly organized connective tissue present
Encapsulation	Dense tissue encapsulation	Moderate encapsulation	Slight encapsulation	No encapsulation
Scaffold degradation	No degradation	Scaffold mostly present	Some scaffold present	No scaffold present

Cell counts represent the number of cells per high power field (HPF). Higher scores are indicative of a more positive constructive host remodeling response

Table 2
Scoring criteria for histomorphologic analysis at 35 days post-implantation

	0	1	2	3
Multinucleated giant cell score	More than 5 multinucleate giant cells per HPF	2–5 Multinucleate giant cells per HPF	1 Multinucleate giant cell per HPF	No multinucleate giant cells per HPF
Connective tissue organization	Original scaffold intact	Original scaffold disrupted, poorly organized new ECM present	Moderately organized connective tissue present	Dense, highly organized connective tissue present
Muscle ingrowth	No muscle ingrowth	Any muscle cells present in scaffold	Muscle cells present in center of scaffold	Organized muscle present
Encapsulation	Dense tissue encapsulation	Moderate encapsulation	Slight encapsulation	No encapsulation
Scaffold degradation	No degradation	Scaffold mostly present	Some scaffold present	No scaffold present

Cell counts represent the number of cells per high power field (HPF). Higher scores are indicative of a more positive constructive host remodeling response

1. Paraffin embedded tissue sections (5 μm) should be deparaffinized with xylene and rehydrated through a series of graded ethanol washes.
2. Process slides for epitope retrieval by boiling at approximately 98 °C for 20 min in 10 mM citric acid (Spectrum C1285 or equivalent), pH 6.0.
3. Allow the slides to cool and then wash in Tris buffered saline (TBS) pH 7.4 for 5 min. Repeat three times.
4. Perform blocking of nonspecific antigens by incubating tissue sections in 2 % horse serum, 1 % bovine serum albumin, 0.1 % Triton-X, and 0.1 % Tween 20 in TBS for 1 h at room temperature.
5. Incubate tissue sections in primary antibodies: mouse anti-CD68 (1:50); goat anti-CD206 (1:50); and rabbit anti-CCR7 (1:250). Dilute in blocking buffer and incubate overnight at 4 °C (*see* **Note 6**).
6. Wash slides in TBS for 5 min. Repeat three times.
7. Incubate tissue sections in fluorescently labeled secondary antibodies: Alexa Fluor 350 anti-mouse (1:25); Alexa Fluor 488 anti-goat (1:200); and Alexa Fluor 568 anti-rabbit (1:200). Dilute antibodies in blocking buffer and incubate for 1 h at room temperature.
8. Wash slides in TBS for 5 min. Repeat three times.
9. To stain nuclei, incubate tissue sections in Draq5 (1:500) diluted in TBS for 5 min at room temperature (*see* **Note 7**).
10. Wash slides in TBS for 5 min. Repeat three times.
11. Apply mounting media and coverslip each slide prior to imaging (*see* **Note 8**).

3.3.2 Vascularity and Innervation

To determine if the remodeling tissue is vascularized and innervated, immunohistochemical labeling of CD31 for endothelial cells and beta-III tubulin for neurons is performed.

1. Paraffin embedded tissue sections (5 μm) should be deparaffinized with xylene and rehydrated through a series of graded ethanol washes.
2. Process slides for epitope retrieval by boiling at approximately 95 °C for 20 min in 1 mM EDTA, pH 8.0.
3. Allow the slides to cool and then wash slides in phosphate buffered saline (PBS) pH 7.0 for 5 min. Repeat three times.
4. Perform blocking of nonspecific antigens by incubating tissue sections in blocking buffer: 2 % normal horse serum and 2 % normal bovine serum in PBS for 30 min at room temperature.

5. Incubate tissue sections in the primary antibodies diluted in blocking solution. For endothelial cells: rabbit anti-CD31 (1:100); for neurons: mouse anti-beta-III tubulin (1:100). Incubate at 4 °C overnight.
6. Wash slides in PBS for 5 min. Repeat three times.
7. In order to quench endogenous peroxidase activity, incubate tissue sections in 3 % H_2O_2 in Methanol in for 30 min at room temperature.
8. Wash slides in PBS for 5 min. Repeat three times.
9. Incubate tissue sections with the secondary antibody diluted in blocking solution. For CD31: biotinylated anti-rabbit (1:200); for beta-III tubulin: biotinylated anti-mouse (1:200). Incubate at room temperature for 60 min.
10. Wash slides in PBS for 5 min. Repeat three times.
11. Incubate tissue sections in a horseradish peroxidase detection solution for 30 min at 37 °C.
12. Wash slides in PBS for 5 min. Repeat three times.
13. Apply Diaminobenzidine (DAB) to detect positive staining cells.
14. Wash slides in water for 5 min. Repeat three times.
15. Counterstain with hematoxylin and coverslip prior to imaging.

3.3.3 Myofibers

For studies investigating constructive remodeling of a biomaterial implant, identification of new skeletal muscle tissue may be of interest. To examine the composition of any skeletal muscle tissue that has formed staining for fast and slow myosin heavy chains may be performed to identify fast-twitch and slow-twitch muscle fibers.

1. Paraffin embedded tissue sections (5 μm) should be deparaffinized with xylene and rehydrated through a series of graded ethanol washes.
2. Process slides for antigen retrieval by boiling at approximately 98 °C for 20 min in 10 mM citric acid, pH 6.0.
3. Allow the slides to cool and the wash slides in Tris buffered saline (TBS) pH 7.4 for 5 min. Repeat three times (*see* **Note 9**).
4. To unmask antigens, incubate tissue sections in 0.1 % trypsin and 0.1 % calcium chloride in TBS for 10 min at 37 °C.
5. Wash the slides in TBS for 5 min. Repeat three times.
6. Perform blocking of nonspecific antigens by incubating tissue sections in blocking buffer: 2 % normal horse serum (Vector Labs; S-2000 or equivalent) and 2 % normal bovine serum in TBS for 10 min at room temperature.
7. Incubate tissue sections in first primary antibody diluted in 4 % normal goat serum in TBS: mouse anti-slow myosin (1:4,000). Incubate for 30 min at room temperature.

8. Wash slides in PBS for 5 min. Repeat three times.
9. Incubate tissue sections in secondary antibody diluted in 4 % normal goat serum in TBS: biotinylated anti-mouse (1:200), Incubate for 60 min at room temperature.
10. Wash slides in PBS for 5 min. Repeat three times.
11. Incubate tissue sections in mouse horseradish peroxidase detection solution (Vector; PK-6100 or equivalent) for 30 min at 37 °C
12. Wash slides in PBS for 5 min. Repeat three times.
13. Apply DAB to detect positive staining cells.
14. Wash slides in PBS for 5 min. Repeat three times.
15. Perform blocking of nonspecific antigens by incubating tissue sections in blocking buffer: 2 % normal horse serum and 2 % normal bovine serum in TBS for 10 min at room temperature.
16. Incubate tissue sections in the second primary antibody diluted in 4 % normal goat serum in TBS: alkaline phosphatase conjugated anti-fast myosin (1:200). Incubate for 60 min at room temperature.
17. Wash slides in PBS for 5 min. Repeat three times.
18. Apply alkaline phosphatase detection substrate to detect positive staining (*see* **Note 10**).
19. Counterstain with hematoxylin and coverslip prior to imaging.

3.4 Functional Testing

3.4.1 In Situ Testing

1. Anesthetize each animal with 2 % isoflurane in oxygen by inhalation or equivalent.
2. Prepare the ventral abdomen by clipping or shaving and cleanse with 70 % ethanol.
3. Create a midline epidermal incision and bluntly dissect the overlying fascia from the underlying body wall at the site of test article implantation (demarcated by the non-resorbable marker sutures).
4. Dissect a flap of body wall tissue containing the remodeled site plus 2 mm of adjacent native musculature. Carefully preserve the origin and rib attachment of the external oblique muscle to maintain blood flow and innervation to the remodeled test article site. Isolate the tissue flap by removing the muscle fibers from all remaining sides of the implantation site including the insertion site at the linea alba and the underlying surface adjacent to the transversalis fascia and peritoneum (Fig. 2).
5. Position the isolated tissue flap such that the direction of the muscle contraction is aligned along an axis from the costochondral arch toward the linea alba.

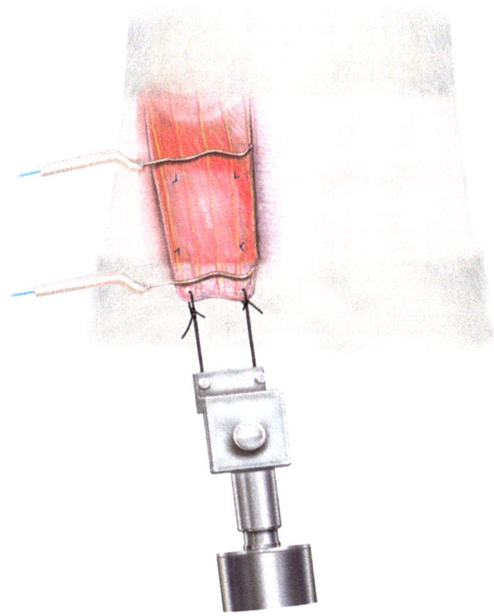

Fig. 2 Schematic of in situ contractile force testing. A flap of tissue is created which contains the site of test article implantation. The dense connective tissue of the insertion site at the linea alba is connected to the force transducer with silk suture. Stimulatory platinum electrodes are placed across the tissue flap

6. Place platinum electrodes across the flap of tissue perpendicular to the direction of muscle contraction and at both the proximal and distal ends of the scaffold placement site as identified by the Prolene sutures (Fig. 2).
7. Secure the distal (previous insertion end) tissue to a force transducer with silk suture. An analog to digital data acquisition receives signals from the force transducer. Custom-designed programs can be created in LabVIEW software to monitor and record the forces measured by the force transducer.

3.4.2 In Situ–Contractile Force Testing

1. Stimulate the isolated tissue flap with a S88X Grass stimulator. Optimum length is determined as the length at which the maximum twitch force is obtained using a 2 ms pulse at 50 pulses per second (pps). The maximum tetanic force (Pt-MAX) for the test site (Pt,S) and the contralateral native tissue (Pt,N) as a function of stimulation rate is determined by delivering 5, 10, 15, 20, 30, 40, 60, and 75 pps in 1 s trains with 2 min of rest between each test.
2. Normalize the Pt,S to Pt,N and to the native Pt-MAX (Pt,N-MAX). The contractile force for each stimulation frequency is then calculated for each tetanic force and specific force is calculated for Pt-MAX by normalizing with the cross-sectional area of the tissue.

3.4.3 In Situ–Fatigue Resistance Testing

1. Determine the fatigue resistance of the tissue at the test site by comparing the force generated at time $t=0$ to $t=2$ min. After a 5 min rest period, deliver pulses at the frequency at which the forces plateau during the tetanic force testing, in trains of 330 ms duration at a rate of 1 train per second, for a total of 2 min. The Fatigue Index is calculated as the ratio of the residual force at $t=2$ to the initial force at $t=0$.

3.4.4 In Vitro Testing

An alternative to in situ muscle testing is in vitro testing of isolated muscle strips in a tissue organ bath. In vitro testing allows a more extensive and precise quantification of muscle properties to be performed, although careful setup is required to ensure that accurate measurements are obtained.

1. Assemble the tissue organ bath following the manufacturer's instructions (Radnoti glassware). The system should be filled with Krebs' buffer (121.0 mM NaCl, 5.0 mM KCl, 0.5 mM $MgCl_2$, 1.8 mM $CaCl_2$, 24 mM $NaHCO_3$, 0.4 mM NaH_2PO_4, and 5.5 mM glucose, pH 7.4) oxygenated with 95 % oxygen and heated to 37 °C. The system should be assembled and allowed to equilibrate for at least 2 h prior to starting the experiment in order to ensure adequate oxygenation and temperature stabilization. A pair of platinum field stimulating electrodes should be secured within the organ bath and connected to an electrical stimulator (S88X). A calibrated force transducer should be connected to a data recording system.
2. Animals for in vitro muscle testing should be euthanized by inhaled CO_2 (*see* **Note 11**).
3. Prepare the ventral abdomen by clipping or shaving and cleanse with 70 % ethanol.
4. Create a midline epidermal incision and bluntly dissect the overlying fascia from the underlying body wall at the site of test article implantation (demarcated by the non-resorbable marker sutures).
5. The body wall containing the biomaterial implant should be carefully dissected to minimize damage to the muscle tissue within the implant site. Excessive pulling on the muscle or touching of the muscle directly should be avoided. In general, a 1 cm border of native tissue should be dissected with the implant to preserve the muscle tissue as much as possible.
6. The explanted body wall tissue should be gently stretched and pinned to a dissecting dish and placed immediately into oxygenated Krebs' buffer at 24–25 °C and allowed to equilibrate for 10 min.
7. Identify the orientation of the muscle fibers within the implant site or if unclear within the surrounding body wall tissue.

A muscle strip should be prepared with the muscle fibers running longitudinally that is approximately 1.5 cm in length and between 2 and 4 mm in width (*see* **Note 12**).

8. The muscle strip should be secured at both ends with tissue clips and placed between two field stimulating electrodes within a 50 ml organ bath containing oxygenated Krebs' solution. One end of the muscle strip should be secured to a force transducer, mounted in a micropositioner mounting bracket with non-stretching suture, e.g., silk or stainless steel.
9. Adjust the angle of the micropositioner so that the muscle strip is running parallel to the stimulating electrodes and so that the force transducer is directly over the muscle strip.
10. Raise the micropositioner so that the muscle strip is vertical but not under tension. The suture should have a small amount of slack. Make sure that there is enough room to increase the height of the micropositioner (at least 20 mm).
11. Determine the optimal voltage to produce a maximal twitch response by stimulating the muscle strips with a 0.2 ms duration pulse at 1 min intervals with increasing voltage (0–24 V in 1 V increments) until the twitch response reaches a maximum and plateaus (optimal voltage is within the plateau region).
12. Determine the optimal muscle length by adjusting the pretension on the muscle through micropositioner adjustment. The muscle should be subjected to an initial pre-tension of 0.5 g and allowed to equilibrate for 10 min. Following equilibration the muscle strip should be stimulated with a 0.2 ms stimulus at the optimal voltage. Increase the tension using the micropositioner in 0.05 g increments, at 1 min intervals until a maximal twitch response is obtained. Record the optimal length of the muscle strip using calipers. This optimal length should be maintained throughout the rest of the experiment.
13. Determine maximum isometric tetanic contractile force by stimulating the muscle strip at optimal voltage and optimal length with increasing stimulus frequency. The muscle strips should be stimulated with 0.2 ms pulses at frequencies between 20 and 200 Hz with train duration of 1.5 s. A two minute rest period should follow each stimulus.
14. Determine muscle fatigue resistance by measuring and comparing the force generated at time $t=0$ to $t=2$ min. After a 5 min rest period, administer a constant train of pulses at the frequency at which the tetanic force plateaued during the isometric force testing, for a total of 2 min. The fatigue index is calculated as the ratio of the residual force at $t=2$ to the initial force at $t=0$ (*see* **Note 13**).

15. Immediately following the experiment record the mass of the isolated muscle strip.
16. Calculate the specific force generated for each muscle strip by dividing the muscle mass by the product of optimal fiber length (Lf) and muscle density (1.06 mg/mm^3). Optimum fiber length for rat external and internal oblique muscle can be obtained from a previously published study [49] (*see* **Note 14**).

4 Notes

1. Isoflurane is known to cause postoperative immunosuppression in mice [50]. Although effects are not known in rats, injectable anesthetics such as Ketamine + xylazine may also be used if inhaled anesthesia is unavailable.
2. Aseptic surgical technique is very important in this model. Great care must be taken to prevent hair follicles or other contaminants from compromising the implantation site, thus altering the endogenous host inflammatory response.
3. This model allow for the opportunity for bilateral implantation studies to be performed allowing two different biomaterials to be evaluated for mechanical properties in the same animal. Care should be taken when performing bilateral studies investigating host response in case the biomaterial being used has systemic effects on the immune system which may affect the contralateral implant.
4. A minimal amount of suture material placed only at the corners of the implant avoids eliciting a host response to the suture that would obscure the innate host response to the test article.
5. It is common for rats to be given an analgesic such as acetaminophen or Ibuprofen in the drinking water. However, this type of analgesia usually involves a nonsteroidal anti-inflammatory drug (NSAID). Since these drugs possess anti-inflammatory properties they will affect the host immune response to the implanted biomaterials to a degree. As a result it is recommended that an opioid analgesic be given.
6. Although neither CCR7 nor CD206 is independently specific for macrophages, in combination with the pan macrophage marker CD68 they provide a high sensitivity for detection of M1 and M2 polarized macrophages in vivo. In addition, several alternative markers for polarized macrophages have been reported [51]. While these may have a higher specificity there may also be a contrasting decrease in sensitivity meaning that not all the macrophages may be detected.
7. Draq5 is a novel bisalkylaminoanthraquinone dye with a high affinity for DNA. The emission spectrum for Draq5 is in the

deep red range (655–800 nm) which may require specialized filters to view. Alternative nuclear stains may be used but may not provide suitable spectral separation to allow all four fluorophores to be imaged.

8. Certain antifade agents may alter the fluorescent signal from the fluorophores, particularly when using the range of fluorophores required in this protocol. We routinely use Dako Fluorescent mounting media without any problems. Similarly, if using a non-hardening mountant, the choice of sealant such as nail varnish, may also cause quenching of the fluorescent signal due to solvents in the sealant solution.
9. It is essential for this protocol that TBS be used for preparation of all buffers and blocking solutions, since the phosphate in PBS will quench the alkaline phosphatase conjugated to the fast myosin antibody.
10. Staining times should be monitored closely to ensure a balance in intensity between the red alkaline phosphatase substrate and the DAB substrate while minimizing background staining. DAB staining should be allowed to develop a darker brown color than would normally be expected since some substrate will be lost with the subsequent fast myosin staining. Alkaline phosphatase detection may take between 10 and 15 min to develop an intense staining.
11. Avoid injectable euthanasia agents such as potassium chloride or sodium pentobarbital which can diffuse into muscle tissues and affect contractile responses.
12. The mass of the muscle strip should be less than 25 mg. Larger muscle strips may develop an anoxic core and diminished functional output because perfusion by the organ bath is insufficient to support the metabolic demands of the muscle.
13. A disadvantage of in vitro fatigue testing is that it does not account for issues relating to innervation, the neuromuscular junction or circulation which may all influence muscle fatigue in vivo.
14. Calculation of specific force using this method approximates the true specific force of the muscle. Damage to the muscle fibers as a result of the dissection will reduce the overall contractile force generated by the muscle. Muscle strips can be stained with a 0.2–1.0 % solution of Procion Orange to identify and quantify the presence of damaged myofibers [52].

References

1. Badylak SF, Hoppo T, Nieponice A, Gilbert TW, Davison JM, Jobe BA (2011) Esophageal preservation in five male patients after endoscopic inner-layer circumferential resection in the setting of superficial cancer: a regenerative medicine approach with a biologic scaffold. Tissue Eng Part A 17: 1643–1650

2. Atala A, Bauer SB, Soker S, Yoo JJ, Retik AB (2006) Tissue-engineered autologous bladders for patients needing cystoplasty. Lancet 367: 1241–1246
3. Teng YD, Lavik EB, Qu X, Park KI, Ourednik J, Zurakowski D, Langer R, Snyder EY (2002) Functional recovery following traumatic spinal cord injury mediated by a unique polymer scaffold seeded with neural stem cells. Proc Natl Acad Sci USA99:3024–3029
4. Williams DF (2008) On the mechanisms of biocompatibility. Biomaterials 29:2941–2953
5. Mikos AG, McIntire LV, Anderson JM, Babensee JE (1998) Host response to tissue engineered devices. Adv Drug Deliv Rev 33:111–139
6. Chapekar MS (1996) Regulatory concerns in the development of biologic-biomaterial combinations, United States food and drug administration. J Biomed Mater Res 33:199–203
7. Ratner BD (1996) Biomaterials science: an introductions to materials in medicine. Elsevier Academic Press, Amsterdam, 484 pp
8. Anderson JM (2001) Biological responses to materials. Annu Rev Mater Res 31:81–110
9. Singer AJ, Clark RA (1999) Cutaneous wound healing. N Engl J Med 341:738–746
10. Clark RA, Lanigan JM, DellaPelle P, Manseau E, Dvorak HF, Colvin RB (1982) Fibronectin and fibrin provide a provisional matrix for epidermal cell migration during wound reepithelialization. J Invest Dermatol 79:264–269
11. Witte MB, Barbul A (1997) General principles of wound healing. Surg Clin North Am 77:509–528
12. Nissen NN, Polverini PJ, Koch AE, Volin MV, Gamelli RL, DiPietro LA (1998) Vascular endothelial growth factor mediates angiogenic activity during the proliferative phase of wound healing. Am J Pathol 152:1445–1452
13. Brown BN, Valentin JE, Stewart-Akers AM, McCabe GP, Badylak SF (2009) Macrophage phenotype and remodeling outcomes in response to biologic scaffolds with and without a cellular component. Biomaterials 30: 1482–1491
14. Badylak SF, Valentin JE, Ravindra AK, McCabe GP, Stewart-Akers AM (2008) Macrophage phenotype as a determinant of biologic scaffold remodeling. Tissue Eng Part A 14:1835–1842
15. Beattie AJ, Gilbert TW, Guyot JP, Yates AJ, Badylak SF (2009) Chemoattraction of progenitor cells by remodeling extracellular matrix scaffolds. Tissue Eng Part A 15:1119–1125
16. Anderson JM, Rodriguez A, Chang DT (2008) Foreign body reaction to biomaterials. Semin Immunol 20:86–100
17. Valentin JE, Badylak JS, McCabe GP, Badylak SF (2006) Extracellular matrix bioscaffolds for orthopaedic applications: a comparative histologic study. J Bone Joint Surg Am 88: 2673–2686
18. Dadsetan M, Jones JA, Hiltner A, Anderson JM (2004) Surface chemistry mediates adhesive structure, cytoskeletal organization, and fusion of macrophages. J Biomed Mater Res A 71:439–448
19. Henson PM (1971) The immunologic release of constituents from neutrophil leukocytes. II. Mechanisms of release during phagocytosis, and adherence to nonphagocytosable surfaces. J Immunol 107:1547–1557
20. Hernandez-Pando R, Bornstein QL, Aguila LD, Orozco EH, Madrigal VK, Martinez CE (2000) Inflammatory cytokine production by immunological and foreign body multinucleated giant cells. Immunology 100:352–358
21. Kovacs EJ (1991) Fibrogenic cytokines: the role of immune mediators in the development of scar tissue. Immunol Today 12:17–23
22. Williams GT, Williams WJ (1983) Granulomatous inflammation—a review. J Clin Pathol 36:723–733
23. Turner NJ, Yates AJ, Weber DJ, Qureshi IR, Stolz DB, Gilbert TW, Badylak SF (2010) Xenogeneic extracellular matrix as an inductive scaffold for regeneration of a functioning musculotendinous junction. Tissue Eng Part A 16:3309–3317
24. Valentin JE, Turner NJ, Gilbert TW, Badylak SF (2010) Functional skeletal muscle formation with a biologic scaffold. Biomaterials 31:7475–7484
25. Valentin JE, Stewart-Akers AM, Gilbert TW, Badylak SF (2009) Macrophage participation in the degradation and remodeling of extracellular matrix scaffolds. Tissue Eng Part A 15: 1687–1694
26. Takebe J, Champagne CM, Offenbacher S, Ishibashi K, Cooper LF (2003) Titanium surface topography alters cell shape and modulates bone morphogenetic protein 2 expression in the J774A.1 macrophage cell line. J Biomed Mater Res A 64:207–216
27. Lu J, Descamps M, Dejou J, Koubi G, Hardouin P, Lemaitre J, Proust JP (2002) The biodegradation mechanism of calcium phosphate biomaterials in bone. J Biomed Mater Res 63:408–412
28. Labow RS, Sa D, Matheson LA, Santerre JP (2005) Polycarbonate-urethane hard segment type influences esterase substrate specificity for human-macrophage-mediated biodegradation. J Biomater Sci Polym Ed 16:1167–1177
29. Khouw IM, van Wachem PB, de Leij LF, van Luyn MJ (1998) Inhibition of the tissue reaction to a biodegradable biomaterial by monoclonal antibodies to IFN-gamma. J Biomed Mater Res 41:202–210

30. Brodbeck WG, Patel J, Voskerician G, Christenson E, Shive MS, Nakayama Y, Matsuda T, Ziats NP, Anderson JM (2002) Biomaterial adherent macrophage apoptosis is increased by hydrophilic and anionic substrates in vivo. Proc Natl Acad Sci USA99:10287–10292
31. Gordon S, Taylor PR (2005) Monocyte and macrophage heterogeneity. Nat Rev Immunol 5:953–964
32. Mantovani A, Sica A, Locati M (2005) Macrophage polarization comes of age. Immunity 23:344–346
33. Mosser DM (2003) The many faces of macrophage activation. J Leukoc Biol 73:209–212
34. Mantovani A, Sica A, Sozzani S, Allavena P, Vecchi A, Locati M (2004) The chemokine system in diverse forms of macrophage activation and polarization. Trends Immunol 25:677–686
35. Stout RD, Jiang C, Matta B, Tietzel I, Watkins SK, Suttles J (2005) Macrophages sequentially change their functional phenotype in response to changes in microenvironmental influences. J Immunol 175:342–349
36. Tidball JG, Villalta SA (2010) Regulatory interactions between muscle and the immune system during muscle regeneration. Am J Physiol Regul Integr Comp Physiol 298: R1173–R1187
37. Ruffell D, Mourkioti F, Gambardella A, Kirstetter P, Lopez RG, Rosenthal N, Nerlov C (2009) A CREB-C/EBPbeta cascade induces M2 macrophage-specific gene expression and promotes muscle injury repair. Proc Natl Acad Sci USA106:17475–17480
38. Nguyen C, Boldea RC, Roy S, Shaarawy T, Uffer S, Mermoud A (2006) Outflow mechanisms after deep sclerectomy with two different designs of collagen implant in an animal model. Graefes Arch Clin Exp Ophthalmol 244: 1659–1667
39. Barbolt TA, Odin M, Leger M, Kangas L, Hoiste J, Liu SH (2001) Biocompatibility evaluation of dura mater substitutes in an animal model. Neurol Res 23:813–820
40. VandeVord PJ, Matthew HW, DeSilva SP, Mayton L, Wu B, Wooley PH (2002) Evaluation of the biocompatibility of a chitosan scaffold in mice. J Biomed Mater Res 59:585–590
41. Jeyanthi R, Rao KP (1990) In vivo biocompatibility of collagen-poly(hydroxyethyl methacrylate) hydrogels. Biomaterials 11:238–243
42. Buchen SY, Cunanan CM, Gwon A, Weinschenk JI, Gruber L, Knight PM (2001) Assessing intraocular lens calcification in an animal model. J Cataract Refract Surg 27:1473–1484
43. Gosain AK, Riordan PA, Song L, Amarante MT, Kalantarian B, Nagy PG, Wilson CR, Toth JM, McIntyre BL (2004) A 1-year study of osteoinduction in hydroxyapatite-derived biomaterials in an adult sheep model: part II. Bioengineering implants to optimize bone replacement in reconstruction of cranial defects. Plast Reconstr Surg 114:1155–1163, discussion 64–65
44. Gad SC (2002) Safety evaluation of medical devices. Marcel Dekker, New York, NY
45. Bischoff F, Bryson G (1964) Carcinogenesis through solid state surfaces. Prog Exp Tumor Res 5:85–133
46. Agrawal V, Johnson SA, Reing J, Zhang L, Tottey S, Wang G, Hirschi KK, Braunhut S, Gudas LJ, Badylak SF (2010) Epimorphic regeneration approach to tissue replacement in adult mammals. Proc Natl Acad Sci USA107:3351–3355
47. Li F, Li W, Johnson S, Ingram D, Yoder M, Badylak S (2004) Low-molecular-weight peptides derived from extracellular matrix as chemoattractants for primary endothelial cells. Endothelium 11:199–206
48. Badylak S, Kokini K, Tullius B, Simmons-Byrd A, Morff R (2002) Morphologic study of small intestinal submucosa as a body wall repair device. J Surg Res 103:190–202
49. Brown SH, Banuelos K, Ward SR, Lieber RL (2010) Architectural and morphological assessment of rat abdominal wall muscles: comparison for use as a human model. J Anat 217: 196–202
50. Markovic SN, Knight PR, Murasko DM (1993) Inhibition of interferon stimulation of natural killer cell activity in mice anesthetized with halothane or isoflurane. Anesthesiology 78:700–706
51. Martinez FO, Gordon S, Locati M, Mantovani A (2006) Transcriptional profiling of the human monocyte-to-macrophage differentiation and polarization: new molecules and patterns of gene expression. J Immunol 177: 7303–7311
52. Petrof BJ, Shrager JB, Stedman HH, Kelly AM, Sweeney HL (1993) Dystrophin protects the sarcolemma from stresses developed during muscle contraction. Proc Natl Acad Sci 90:3710

Chapter 2

Urothelial Cell Culture

Yuanyuan Zhang and Anthony Atala

Abstract

This chapter reviews the use of urothelial cells as a means to enhance tissue regeneration and wound healing in urinary tract system. It addresses the properties of urothelial cells, including their role as a permeability barrier to protect underlying muscle tissue from the caustic effects of urine and as one of the main cell types, along with smooth muscle cells, that are used in urethral or bladder tissue engineering today. This description includes a general overview of various isolation techniques and culture methods that have been developed to improve urinary tract reconstruction in vivo and aid the characterization of growth factor expression in vitro. The chapter then describes various applications using urothelial cells, including production of multilayer urothelial sheets, tissue engineered bladder mucosa, tissue engineered urethra, and tissue engineered bladder. It also outlines the advantages of sandwich and layered coculture of these cells and the effects of epithelial–stromal cell interactions during tissue regeneration or wound healing processes in the urinary tract.

Key words Urothelial cells, Bladder, Ureter, Tissue engineering, Scaffold, Urinary reconstruction, Epidermal growth factor, Serum-free medium

1 Introduction

Culture of urothelial cells (UC) provides an in vitro model system to help advance our understanding of the cellular mechanisms of urothelial development as well as epithelial–stromal interactions and cell–cell signaling in the pathogenesis of bladder cancer, interstitial cystitis, and urinary tract infection. Additionally, autologous UC provide a critical cell source for urinary tract reconstruction and regeneration using tissue engineering technology.

The urothelium forms a barrier between urine and the underlying muscle and connective tissue. It consists of multiple cell layers: a basal layer containing progenitor cells that attaches to the connective tissue substratum, an intermediate cell layer that is one to two layers thick, and a superficial cell layer composed of highly differentiated umbrella cells that line the luminal surface of the bladder.

Robert G. Gourdie and Tereance A. Myers (eds.), *Wound Regeneration and Repair: Methods and Protocols*, Methods in Molecular Biology, vol. 1037, DOI 10.1007/978-1-62703-505-7_2, © Springer Science+Business Media New York 2013

A few decades ago, UC could be harvested and maintained in culture for limited periods, but their growth and expansion was not possible. Their growth was further hampered by inadequate media that contained serum, which facilitated fibroblast overgrowth. With improvements in culture technology, clonal growth of bladder and ureter UC has been achieved with the development of serum-free media and techniques which target the basal cells, such as scraping of the mucosa, to facilitate harvest. Additionally, the optimization of nutrients in the media, including a reduction in calcium concentration and the omission of serum, favors proliferation and formation of a monolayer of UC rather than differentiation, which extends the life of the culture and permits serial propagation [1–4]. Recently, stratified urothelial sheets and three-dimensional (3D) growth of urothelial structures have been generated for urological tissue engineering purposes [5–7]. Stratified urothelial cell-sheets can be produced for potential use in demucosalized gastrointestinal flaps [8–14]. Three dimensional urothelial structures can be formed using a combination of culture techniques, i.e., coculture of UC and smooth muscle cells (SMC) seeded on porous scaffolds in a bioreactor under dynamic culture conditions [15–18], and these structures can potentially be used in the engineering of urinary tract organs [8, 17, 19–25].

2 Materials

2.1 Culture Medium and Supplements

1. Keratinocyte serum-free media (KSFM).
2. Recombinant epidermal growth factor (EGF).
3. Bovine pituitary extract (BPE) Cholera toxin.
4. Soy Bean Trypsin Inhibitor.
5. Dispase II for detaching cell sheets from culture dishes.
6. Collagenase IV.
7. 0.2 % Ethylenediaminetetraacetic acid (EDTA).
8. 0.05 % Trypsin.
9. Dulbecco's Modified Eagle's Medium (DMEM).
10. Fetal bovine serum (FBS).
11. Penicillin–streptomycin solution.
12. Peracetic acid (PAA).
13. Calcium ($CaCl_2$) Stock solution 0.5 M.
14. Hank's Balanced Salt Solution (HBSS).
15. A pair of microsurgery forceps and scalpel.
16. NIH 3T3 fibroblasts.
17. Triton X-100.

2.2 Coculture Model and Conditioned Medium

1. 0.20 μm sterile filter.
2. T25 Primaria cell culture flasks.
3. Dynamic culture system with an orbital shaker.

2.3 RT-PCR

1. RNA isolation from cell cultures is performed using the RNA isolation kit from 5 PRIME Inc.
2. Reverse transcription reactions are performed using a High-Capacity cDNA.
3. Transcription Kit.
4. Primers for uroplakin Ia and III, CK7, CK 13, CK 20, were designed and purchased from Eurofins MWG Operon.
5. Taq DNA Polymerase.
6. PCR buffer.
7. dNTPs.
8. ddH_2O.
9. Agarose.
10. PCR tubes.
11. Thermocycler.
12. System for electrophoresis of PCR products.

2.4 Western Blot

1. RIPA buffer, composed of 50 mM Tris–HCl pH 7.4, 150 mM NaCl, 2 mM EDTA, 1 % NP-40, 0.1 % SDS.
2. SDS-PAGE gels, 10–12.5 % polyacrylamide.
3. 10× TGS running buffer.
4. Transfer buffer (48 mM Trizma Base, 39 mM Glycine, 20 % methanol pH 9.2).
5. Transfer apparatus.
6. PVDF membranes.
7. Blocking solution (5 % nonfat dry milk powder (Carnation) in Tris Buffered Saline with Tween20 (TBST).
8. Washing solution (TBST).
9. Antibodies to uroplakin Ia and III, CK7, CK 13, CK 19, CK 20, α-smooth muscle actin, desmin, myosin, smoothelin.
10. Secondary antibodies: horseradish peroxidase (HRP)-conjugated secondary antibodies to the species of each primary antibody in use.
11. Chemiluminescent detection kit—Western Lightning Chemiluminescence Reagent.

2.5 Immuno-fluorescence

1. Chambered slides for cell culture.
2. Antibodies to uroplakin Ia and III, CK7, CK 13, CK 19, CK 20, AE1/AE3; α-smooth muscle actin, desmin, myosin, smoothelin.
3. Fluorescent labeled secondary antibodies.

2.6 Nuclear Staining

1. 4′-6-Diamidino-2-phenylindole.
2. Propidium iodide (PI).

2.7 Biomaterials for Urothelial Cell Growth

1. Natural collagen matrices derived from porcine bladder mucosa and lamina propria (i.e., bladder submucosa, BSM) and porcine small intestine submucosa (SIS).
2. Home-made BSM (or SIS) inserts: These 1 cm disks can be manufactured in a manner such that the BSM (or SIS) is suspended over a circular polypropylene frame (border 5 mm on top and 2 mm on bottom), with the mucosal surface upwards, to create a double well culture disk with the BSM (or SIS) acting as the separating membrane. The mucosal surface of BSM (or SIS) forms the base of the upward facing well while its serosal surface forms the top of the bottom well. The upper well holds 500 μl of media and the bottom well holds 200 μl. Following seeding of cells, BSM (or SIS) disks are placed in a 12- or 6-well cell culture dish filled with media to allow free contact of the media with both sides of the BSM (or SIS) [8, 18].

2.8 Animals for Isolation of Urothelial Cells

Nu/nu athymic nude mice.

2.9 Roszell's Procedure for Rat Urothelial Cell Culture

This procedure was developed for large-scale in vitro growth and serial cultivation of normal diploid rat bladder epithelial cells. Primary cultures are initiated by attachment of bladder mucosal explants to type I collagen gels. A rapid outgrowth of epithelial cells from the explants occurs when these are cultured in a hormone-supplemented medium with epidermal growth factor. These primary outgrowths are passaged by nonenzymatic dispersion with 0.1 % ethylenediaminetetraacetic acid followed by replating onto new gels. The capacity for routine serial passaging and maintenance of rat bladder epithelial cells requires the presence of epidermal growth factor, a requirement not observed with human urothelial cells. The characteristics of the cultured rat bladder epithelial cells are similar to those observed in human urothelial cells in terms of ultrastructural and phase-contrast morphologic properties, junctional complexes, desmosomes, stratification, and an apical glycocalyx; the absence of stromal cell contamination; and the ability to be serially passaged. Spontaneous cell-line formation is often observed with the rat bladder epithelial cells, but has not been found with the human urothelial cells. With the method that we have developed, the number of rat bladder epithelial cells generated from a single bladder of a 4–6 week old rat is increased 100-fold from about 7×10^5 cells to 7×10^7 viable cells within 3 weeks of culture. The capability to culture normal, primary rat bladder epithelial cells on this scale has not been reported previously and will facilitate comparative studies of the biological and molecular characteristics of the mammalian urothelium. Furthermore, this culture system will be useful for carcinogenesis studies, including metabolic activation of carcinogens and cellular transformation in vitro [26].

3 Methods

The culture of UC requires different approaches for cell isolation, expansion, and optimal culture media for UC growth in vitro depending on whether the cells are derived from humans or animals. The processes of both monolayer and multilayer culture of UC and methods for UC growth on matrices in vitro and in vivo for tissue engineering will be introduced.

3.1 Isolation and Culture Methods for Urothelial Cells from Different Species

Urothelial cells from different species require different methods for cell isolation and cultivation with distinctive culture media. For isolation from urothelium, UC can be retrieved by enzymes, such as dispase, trypsin, or collagenase IV. Such methods are most fitting for smaller animals (e.g., rats or mice) in which mechanical manipulation is complex. For human urothelium and that from large animals (i.e., pig and dog), UC can be obtained by a combination of microdissection and enzymatic retrieval methods or by explant culture. Generally, human UC are much easier to culture than cells from rodents, dogs and pigs. Human UC tend to grow well in KSFM for over 10 passages, while large animal UC often require the addition of 1–5 % serum and stop growing by passage 3–5. It has been shown that rat UC can proliferate for up to 18 passages when conditioned medium derived from an immortalized cell line is added [9].

3.1.1 Human Urothelial Cell Culture

All of the following culture protocols should be performed in a class II laminar flow biosafety cabinet (hood) using aseptic techniques. A bladder tissue specimen (size <2 × 2 cm^2) should be obtained and placed in a sterile 50 ml centrifuge tube containing either 20 ml chilled phosphate-buffered saline (PBS) or KSFM containing penicillin and streptomycin, and the tube should be kept on ice. The tissue should be processed as soon as possible once it has been obtained (*see* **Note 1**). To begin the process, the bladder tissue should be transferred to a sterile 100 mm petri dish and washed with PBS or KSFM containing penicillin and streptomycin to remove red blood cells. The bladder mucosa should then be dissected away from the underlying submucosa with sterile surgical scissors (*see* **Note 2**).

Primary Explant Method

1. Mince urothelial mucosa into 1 mm pieces. Place 6–9 pieces of tissue, mucosal side down and without medium, on a small culture dish (30 mm dish or 6-well plate). This positioning increases the chances of establishing a successful primary urothelial culture.
2. Place the tissue into an incubator with a humidified atmosphere containing 95 % air and 5 % carbon dioxide (CO_2) at 37 °C for 60 min to allow the tissue fragments to attach to the dish.
3. Prepare KSFM with a calcium concentration of 0.09 mmol/l, supplemented with epidermal growth factor (EGF) (5 ng/ml), bovine pituitary extract (BPE) (50 mg/ml), cholera toxin (CT) at a final concentration of 30 ng/ml, penicillin (100 U/ml), and streptomycin (1 mg/ml).
4. Add 3 ml KSFM to the culture dish slowly. The outgrowth of cells from each piece of bladder tissue can usually be observed after 48 h of incubation (*see* **Note 3**).
5. Examine the cultured cells daily by phase-contrast microscopy to assess growth and morphology and to make certain the medium is clear of any contamination.

Enzyme-Based Method

1. Mince urothelium into 2–3 mm fragments and digest in 1 % collagenase type IV (with no trypsin activity) at 37 °C for 1–2 h.
2. To remove collagenase, centrifuge the digested cells, wash them in Hank's Balanced Salt Solution (HBSS), and centrifuge again.
3. Resuspend the cells in KSFM and culture the primary UC in 5 ml (T25) Primaria tissue culture flasks at 37 °C.
4. Change the medium every 48 h. When cells are 95 % confluent, they should be passaged (*see* **Note 4**). To subculture the cells, use cold PBS (4 °C) to wash the culture after aspirating the culture medium and then add 0.2 % EDTA plus 0.05 % Trypsin for 2–3 min. Add soy bean trypsin inhibitor solution to stop the trypsin activity, rinse the cells off the dish and move them to a centrifuge tube, and spin down the cells at 1,500 K for 5 min (*see* **Note 4**).
5. For subculture, plate the cells at a ratio of 1:5 with a cell concentration of 200×10^4 cells in 1 ml medium per flask; after 24 h, when the cells have attached to the flask, add an additional 4 ml fresh KFSM.

Scraping Method

1. Transfer the bladder specimen to a 100 mm petri dish in KSFM. Usually, for a specimen sized 1×1–2×2 cm^2, 3.5 ml of medium is adequate. Detach UC from the specimen by scraping it gently with a sterile surgical blade.

2. Transfer the scraped cell suspension to either a 6-well or a 24-well cell culture vessel (0.5 ml in each well) by means of a 5 ml pipette or a 5 ml syringe with a 23G needle attached.
3. Add another 0.5 ml fresh KFSM to each well and incubate for 48 h, then change medium (*see* **Note 5**).

3.1.2 Isolation and Culture of Urothelial Cells Derived from Large Animals

The methods of isolation and culture of UC from large animals (such as pigs and dogs) are similar to the methods used in human UC. The procedure is a combination of the microdissection and enzymatic techniques as described below. UC isolated from large animals have been used as an autologous cell source in the production of tissue engineered bladder or urethra in the same animals [20, 27–30].

Isolation and Primary Culture of Urothelial Cells from Large Animals

1. Wash the bladder specimen as described previously for the microdissection technique.
2. Microdissection step: strip the urothelial mucosa and connective tissues from muscle tissue using a pair of microsurgery forceps and a scalpel.
3. Enzymatic step: Transfer the scraped mucosa to a 100 mm culture dish containing an 8×8 cm^2 sterile Styrofoam piece with 10 sharp metal pins along each edge. With the mucosa side up, the tissue should be stretched across the pins and then incubated overnight at 4 °C in KSFM containing 2.5 mg/ml dispase.
4. Remove the medium-dispase solution and scrape the uroepithelial cells from the underlying connective tissue with two flexible cell scrapers.
5. Transfer the scraped cells to a culture dish, resuspend them in 20 ml of 0.25 % trypsin 1 mM EDTA, and incubate at 37 °C for 30 min.
6. Bring the cell suspension up to 50 ml with KSFM containing 5 % fetal bovine serum in a sterile conical tube and spin down at $500 \times g$ for 5 min to pellet the cells and remove the trypsin.
7. Carefully remove the supernatant and wash the cells in KSFM with a calcium concentration of 0.09 mmol/l that has been supplemented with recombinant epidermal growth factor, bovine pituitary extract, cholera toxin, and 1 % penicillin–streptomycin solutions (P/S).
8. Resuspend the cells at a final concentration of $6–8 \times 10^5$ cells/ml and plate them in KSFM with 2 % FBS (*see* **Note 6**).

3.1.3 Rodent Urothelial Cells In Vitro

Rodents are a commonly used animal model for biomedical research. However, the use of cultured UC from rodents, such as rat urothelial cells (RUC), has been limited due to difficult isolation procedures and the inability to maintain them in a long-term culture. The main approach presently used for the culture of RUC is to plate the cells in serum-containing medium (5 % FBS) on a collagen-coated culture dish (Biocompare) or on a feeder layer of lethally irradiated NIH 3 T3 fibroblasts that can support growth of cultured cells in vitro. However, these techniques also present problems, as the RUC culture is often contaminated with other cell types such as fibroblasts, smooth muscle cells, or capillary endothelium. These stromal cells will eventually overgrow and finally replace UC in the serum-containing medium after only a few passages.

We have modified Roszell's procedure [26] and developed a simple technique to isolate rat urothelium by enzymatic release of RUC from an "everted bladder" and to maintain the primary culture in KSFM [8]. In this study, the conditions for the growth of RUC in long-term culture were investigated systematically. We show that a mixture of conditioned medium (CM) obtained from NIH 3T3 fibroblasts and KSFM (CM-KSFM) [31] yielded large quantities of normal RUC without stromal cell contamination. RUC could be subcultured up to 18 times in CM-KSFM during an observation period of up to 5 months. Additionally, rat urothelium stratification can be induced with or without a feeder layer in vitro, which provides potential application as an autograft for urothelial replacement in bladder augmentation studies in a rat model. This technique also serves as a tool for research on various bladder diseases including tumorigenesis and urinary tract infection. Below, we describe our method for RUC isolation.

Rat Urothelial Cell Isolation and Culture

1. Euthanize the rats with an intraperitoneal injection of sodium pentobarbital (40–50 mg/kg). After the whole bladder is excised, apply a modified Roszell's procedure [26] to evert the bladder to expose the urothelial surface.
2. Reinsert the bladder neck into the lumen, and surgically close to form an "everted bladder ball" so that only the urothelial surface is exposed.
3. Immerse the everted bladder either in 4 ml of 1 % collagenase IV at 37 °C on a shaker for 60 min, or in 10 ml of 0.1 % disodium EDTA at 48 °C for 4 h.
4. Gently scrape the bladder mucosa from the muscle tissue following digestion using a forceps with coarse tips.
5. Collect, wash, and plate UC on T25 Primaria cell culture flasks at a cell density of 5×10^6 cells/ml in 5 ml mixed media containing KSFM and conditioned medium derived from a 3T3 cell culture.
6. The conditioned medium (CM) is derived from NIH 3T3 cell culture. To produce it, culture 3T3 cells in DMEM supplemented with 10 % FBS and 1 % penicillin–streptomycin and

then collect the culture medium when the cells reach 60–80 % confluence (about 12 h).

7. Filter-sterilize the conditioned medium with a 0.20 μm filter to prevent fibroblast contamination.
8. Culture the primary RUC using the mixed media composed of fresh KFSM and conditioned media (CM-3T3) (1:1 ratio).
9. Examine the cultured urothelial cells daily by phase-contrast microscopy to assess cell growth and morphology, and to check for any contamination.

3.2 Stratified Urothelial Sheet Culture

Viable cultured urothelial cell sheets can be applied for urothelium regeneration on demucosalized gastrointestinal segments through grafting techniques. There are two main techniques to produce these urothelial cell sheets in culture: (1) cultured cell sheets can be gently detached from a culture dish with 2.5 % dispase and (2) temperature-responsive culture dishes can be used. When using the dispase method, in vitro cultured human stratified urothelium shows complete differentiation of its superficial cells. It retains the same ultrastructure of barrier characteristics (such as microvillus and tight junctions as observed via electron microscope) against principal urine components [10]. The cell-sheets appear to be stable in genotype and no chromosomal aberrations have been found. Cells in the sheets express the urothelial specific cell marker uroplakin and cytokeratins 7, 8, 17, and 18.

3.2.1 Production of Stratified Urothelium In Vitro

1. Plate UC into culture flasks at an average cell density of 1×10^5 cells per cm^2 in KSFM and incubate with 5 % CO_2 in 95 % air at 37 °C.
2. Change KSFM on alternate days.
3. Culture UC up to 100 % confluence, and then enrich the serum-free medium with calcium to a final concentration of 1.5 mM.
4. When stratification is induced, usually within 10 days, remove the culture medium, and detach the in vitro urothelial cell sheet construct from the culture flask with 2.5 % dispase II at 37 °C for 30 min. An entire urothelial cell sheet that is about half the size of the culture dish area can usually be easily detached (*see* **Note** 7).

3.3 In Vitro Coculture Methods

Coculture of urothelial and smooth muscle cells on bio-scaffolds is usually applied in the production of tissue-engineered urinary tract organs [8, 19, 20], stem cell differentiation protocols [17, 32, 33] or in in vitro bladder models for urinary tract infection. In order to create a layered construct that includes both a urothelial layer and a smooth muscle layer, various coculture techniques can be used, including the layered, sandwich, and mixed coculture techniques.

In vitro observations with these three different coculture techniques suggest that the layered and sandwich methods of seeding have distinct advantages over the mixed coculture technique in terms of cell stratification, cell-matrix penetration, and cell differentiation. In order to induce a stratified urothelium bladder SMC structure, cells are usually seeded on biomaterials such as BSM and SIS with coculture techniques [8, 19, 27, 28, 34, 35].

3.3.1 Layered Coculture Method

1. Set some SIS disks in the wells of a 24-well plate. Seed bladder SMC (1×10^5 cells per cm) onto the mucosal side of the SIS using DMEM plus 10 % FBS. After incubation for 1 h, UC can be seeded on top of the SMC at the same cell concentration (*see* **Note 8**).
2. On day 3, distinct cell sorting will be noted, in which the SMC grow on the surface of the SIS with early cell-matrix penetrance, and the UC will grow on top of the smooth muscle cells as a separate population of cells. This cell sorting can be confirmed by immunohistochemical analysis. UC on the top portion of the culture stain positive for AE1/AE3, while SMC in the bottom portion are positive for α-smooth muscle actin, desmin, myosin, and smoothelin.
3. On day 7, the culture will show further stratification and increased matrix penetrance.
4. For the remainder of the culture period, cell growth and matrix penetrance progresses while cell sorting is maintained.
5. By day 28, a well-developed, pseudostratified layer of UC is present that is 3–4 layers thick. The smooth muscle cell layer is 5–7 layers thick at this time.
6. The vast majority of SMC will no longer be on the surface of the SIS; rather, they penetrate the matrix of SIS membrane and proliferate under its surface within the membrane. In several areas, SMC can usually be seen traversing into deep portions of the SIS membrane [8, 27, 28].

This matrix penetrance by SMC is distinctly different from the pattern of membrane ingress that is observed when SMC are grown alone on SIS. The degree and pattern of cell-matrix penetrance by smooth muscle cells with the sandwich coculture technique were similar to those observed with the layered coculture technique, although some of SMC were still located on the serosal surface of SIS.

3.3.2 Sandwich Coculture Method

1. Set SIS disks into the wells of 6-well plates. Seed SMC on the serosal surface of the SIS and add 0.5 ml DMEM with 10 % FBS on the top portion of the disk.
2. Follow by seeding the mucosal surface of the SIS with UC in mixed culture media (KSFM:DMEM with 10 % FBS, 1:1) 24 h later (*see* **Note 6**).

3. Organized, layered growth of UC on one side of the SIS and growth of SMC on other side of the SIS will be readily evident on day 3.
4. Cell growth will progress during the 28-day period of observation (as with the layered coculture technique).
5. By day 28, there will be a well-defined pseudostratified layer of UC on the mucosal side of the SIS disk that is 3–4 cells thick, and there should be little to no evidence of matrix penetration by the UC [8, 19, 36].

On the serosal surface of SIS, the SMC layer will be about 5–7 cells thick, and there is usually significant penetrance of the SIS membrane by the SMC.

3.4 Three-Dimensional (3D) Culture of Urothelial Structures

Generation of 3D cultures of urothelial structures requires a combination of three techniques: coculture of UC and SMC [8, 19, 36], culture on a biomaterial with porous microstructure [17, 18, 33, 34], and the use of a bioreactor to provide dynamic culture conditions [16, 17, 33].

3.4.1 Biomaterials

A prior study, published in 2009 [18], demonstrated that recellularization of biological collagen acellular matrices with porous microstructure using cell seeding technology provides a promising option for promoting tissue regeneration. A strategy for performing tissue reconstruction using tissue engineering techniques is to repopulate a scaffold with cells isolated from the patient's own tissues to provide an autologous repair of the tissue defect. For example, in urethral reconstruction using tissue engineering techniques, a tissue-engineered tube composed of differentiated UC on the urethral lumen side and SMC on the submucosal side can be used. To produce such a scaffold, the acellular matrix has to be porous, contain infinitesimal heterogeneous cellular compounds, and retain most of the extracellular matrix components to permit cell seeding.

Bladder Submucosa Matrix (BSM) and Small Intestinal Submucosa (SIS)

An ideal biological collagen matrix for urethral tissue engineering would have high porosity for cell seeding, be degradable, histocompatible, and have the least xenogenic cellular compounds retained within the matrix to minimize the potential for inflammation. Most importantly, the matrix needs to have a three-dimensional (3-D) structure with high porosity, but at the same time, it must maintain a nearly normal tensile strength.

Two naturally derived matrix materials, i.e., BSM and SIS, meet these criteria and have been used in a number of urological applications both in vitro and in vivo. BSM is composed of a slender basement membrane with high density of collagen and a lamina propria with areolar connective tissue. BSM has been valuable in reconstruction of urethra in animal models and in humans. On the other hand, SIS is a xenogenic, acellular, collagen rich membrane with inherent growth factors that have previously been shown to promote in vivo bladder regeneration. Our previous study evaluated in vitro use of SIS to support the individual and combined growth of bladder UC and SMC for potential use in tissue engineering [8].

Although no in vitro cell culture substrate can fully mimic the in vivo state, SIS has significant advantages over conventional plastic and other coated surfaces because it provides a unique environment that promotes cell-cell and cell-matrix interactions. The presence of UC significantly impacts the pattern of SMC growth on SIS since active penetrance of the membrane only occurs when UC are grown in conjunction with SMC.

Biomaterial Porosity

A 3-D scaffold with higher porosity and large pore size (50–200 mM) promotes cell proliferation, migration and infiltration into the matrix, and appears to allow abundant cell loading onto the scaffold, thereby promoting in vivo tissue regeneration and wound healing in a nu/nu athymic mouse model [17, 18, 33]. Such a scaffold would also allow the host cells to participate in the tissue remodeling processes by infiltration or migration into the matrix from the wound edges. Hence, a 3-D porous matrix would be potentially beneficial in the reconstruction of bladder or urethra tissues.

3-D Porous Matrix Method: Obtaining Porcine BSM

1. Clean fresh porcine bladders upon receipt and manually remove muscle layers.
2. Retain and wash BSM in distilled water.
3. For decellularization, transfer BSM to a 500 ml bottle filled with distilled water and place at 4 °C on a rotary shaker at 200 rpm for 2 days.
4. Discard the distilled water and oxidize the BSM by soaking it in 5 % peracetic acid (PAA) for 4 h.
5. Next, treat the BSM with a solution containing 1 % Triton X-100 for 2 days, and then wash once more with distilled water for another 2 days.
6. Finally, disinfect the BSM using 0.1 % PAA in 20 % alcohol for 2 h, rinse three times with sterile distilled water for 10 min each, and store in sterile distilled water at 4 °C until further use.

The methods for producing SIS scaffolds are nearly identical to those shown above for BSM.

3.5 Dynamic Culture

Dynamic culture methods using a bioreactor that can provide media perfusion or rotation have frequently been used for seeded scaffolds. Dynamic culturing mimics the physiological environment and promotes cell adhesion, proliferation, infiltration and differentiation. The advantages associated with dynamic culture conditions include: (1) even distribution of nutrition and oxygen leads to uniform growth of cells on the matrix; (2) increased synthesis of endogenous ECM; (3) physiologically relevant mechanical forces on the cultured cells (i.e., shear stress, pressure and hydrodynamic compression).

Bioreactors have become an important tool to improve bladder tissue engineering under physiologic conditions. Recently, Farhat et al. developed a urinary bladder bioreactor with a hydrodynamic chamber to produce stretch and strain on cell-seeded scaffolds [15, 16]. This bioreactor helps engineered tissues to better adapt to the changing environment when implanted in vivo. It enhances epithelial-stromal and cell-ECM interactions, which are necessary for building bladder tissues but which cannot be achieved using static cultures in plates. The use of a bladder bioreactor system may accelerate tissue organization and maturation in vivo, and may shorten the time required to achieve a fully functioning organ.

3.6 In Vivo Urothelium Formation

Cell-based approaches to engineer human bladder tissue have been reported [23], and bioengineering has allowed creation of functional neo-bladder tissues in several animal models [19, 20, 28, 30, 32, 35]. Such tissue-engineered bladder is generated from autologous cells derived from a biopsy of tissue. After bladder tissues are obtained via biopsy, UC and SMC are isolated and expanded in culture, and then seeded onto a biodegradable scaffold.

A clinical experience involving engineered bladder tissue for cystoplasty was conducted starting in 1998. A small pilot study of seven patients was reported, using a collagen scaffold seeded with cells either with or without omentum coverage, or a combined PGA-collagen scaffold seeded with cells and omental coverage. The patients reconstructed with the engineered bladder tissue created with the PGA-collagen cell-seeded scaffolds with omental coverage showed increased compliance, decreased end-filling pressures, increased capacities and longer dry periods over time [23]. It is clear from this experience that the engineered bladders continued their improvement with time, mirroring their continued development. Although the experience is promising in terms of showing that engineered tissues can be implanted safely, it is just a start in terms of accomplishing the goal of engineering fully functional bladders. This was a limited clinical experience, and the technology is not yet ready for wide dissemination, as further experimental and clinical studies are required. FDA Phase 2 studies have now been completed.

To identify whether urothelial cells maintain their specific phenotypes including gene and protein expression after isolation, culture and coculture with SMC, Reverse Transcriptase-Polymerase Chain Reaction (RT-PCR), Western blot, and immunefluorescence analysis are performed. Urothelial cell transcripts including uroplakin (Up)-Ia, Up-III, cytokeratin (CK)-7 and CK-13 are regularly used. For RNA extraction experiments, UC are cultured in KSFM in 10 cm dishes for 14 days. Five micrograms of RNA, extracted using Trizol (Invitrogen, Carlsbad, CA) reagent, is used for cDNA synthesis using the Superscript II RT enzyme to the manufacturer's instructions. Briefly, RNA is incubated with random hexamers, nucleotides, reverse transcriptase enzyme, and reaction buffer in a 20 μl volume for synthesis of cDNA. One tenth of the reaction volume is taken for PCR using specific primer pairs.

For Western blot analysis, cells are harvested from culture dishes and lysed. Proteins extracted from whole cells are run using sodium dodecyl sulfate-polyacrylamide gel electrophoresis to separate the proteins and then transferred to nitrocellulose membranes overnight at 4 °C. Mouse anti-human CK7, CK13, Ck19, CK20, AE1/AE3, and uroplakin Ia/III, are used as the primary antibodies to probe the membrane and peroxidase labeled goat anti-mouse IgG is used as the secondary antibody for detection. Protein bands are detected with an enhanced chemiluminescence assay.

For immunofluorescence, urothelial cells are assessed for the expression of 7 types of cell markers. Urothelium specific markers include uroplakin Ia and III, a tissue specific and differentiation dependent transmembrane protein of the urothelial luminal surface. Epithelial cell markers include CKs 7, 13, 17, 19, and 20. All markers should be assessed for all 3 urothelial layers. All surface marker assays should be performed at least 3 times to ensure consistent results.

4 Notes

1. To enhance the success rate of primary cultures of UC, avoid the use of electrical knives (electrocautery) when bladder tissues are harvested from patients [9]. In addition, any foreign objects such as bladder stones and catheters should be removed 1 week before the tissue biopsy is procured, as these foreign bodies can interfere with the outcome of urothelial cell culture.
2. Microdissection to strip off the urothelium is one of the methods reported to be successful in large mammals such as humans, pigs, and dogs. This technique, however, is difficult in small animals such as rats or mice.
3. Do not touch the culture dish or flask during the first 48 h to allow the explants attach to the plastic.
4. Use Soy Bean Trypsin Inhibitor instead of FBS to stop trypsin activity when UC are subcultured. This can prevent any fibroblast contamination in the culture.
5. Culture of UC is not always successful, and urothelial cells from various areas of the urinary tract respond differently to culture

protocols. For example, urothelial cells from the ureter are more easily cultured than urothelial cells obtained from bladder [9].

6. Generally speaking, UC derived from pig and dog can be cultured up to only passage 3–4 and then they will senesce. However, human UC can be cultured for over 10 passages. The culture medium needs to be optimized for culture of UC from larger animals.
7. Detached urothelial cell sheets can be used for histology or tissue repair. Do not keep these in the incubator for more than 3 h; otherwise, the cell sheet will shrink.
8. Air bubbles must be aspirated with a 1 ml tubercle syringe with a fine needle.

Acknowledgments

The authors would like to thank Dr. Jennifer Olson for editorial assistance with this manuscript.

References

1. Cilento BG, Freeman MR, Schneck FX, Retik AB, Atala A (1994) Phenotypic and cytogenetic characterization of human bladder urothelia expanded in vitro. J Urol 152:665–670
2. Freeman MR, Yoo JJ, Raab G, Soker S, Adam RM, Schneck FX, Renshaw AA, Klagsbrun M, Atala A (1997) Heparin-binding EGF-like growth factor is an autocrine growth factor for human urothelial cells and is synthesized by epithelial and smooth muscle cells in the human bladder. J Clin Invest 99:1028–1036
3. Southgate J, Harnden P, Selby PJ, Thomas DF, Trejdosiewicz LK (1999) Urothelial tissue regulation. Unraveling the role of the stroma. Adv Exp Med Biol 462:19–30
4. Southgate J, Hutton KA, Thomas DF, Trejdosiewicz LK (1994) Normal human urothelial cells in vitro: proliferation and induction of stratification. Lab Invest 71:583–594
5. Atala A, Aminutes M, Harty JI (1992) Diethylstilbestrol in treatment of postorchiectomy vasomotor symptoms and its relationship with serum follicle-stimulating hormone, luteinizing hormone, and testosterone. Urology 39:108–110
6. Atala A, Bauer SB, Hendren WH, Retik AB (1993) The effect of gastric augmentation on bladder function. J Urol 149:1099–1102
7. Atala A (1998) Tissue engineering in urologic surgery. Urol Clin North Am 25:39–50
8. Brandt WD, Matsui W, Rosenberg JE, He X, Ling S, Schaeffer EM, Berman DM (2009) Urothelial carcinoma: stem cells on the edge. Cancer Metastasis Rev 28:291–304
9. Li HL, Chen XG, Zhang FC, Ma J, Xu CS (2008) Expression profiles of the extracellular matrix-associated genes during rat liver regeneration. Yi Chuan 30:333–340
10. Sugasi S, Lesbros Y, Bisson I, Zhang YY, Kucera P, Frey P (2000) In vitro engineering of human stratified urothelium: analysis of its morphology and function. J Urol 164: 951–957
11. Ludwikowski B, Zhang YY, Frey P (1999) The long-term culture of porcine urothelial cells and induction of urothelial stratification. BJU Int 84:507–514
12. Staack A, Hayward SW, Baskin LS, Cunha GR (2005) Molecular, cellular and developmental biology of urothelium as a basis of bladder regeneration. Differentiation 73:121–133
13. Scriven SD, Trejdosiewicz LK, Thomas DF, Southgate J (2001) Urothelial cell transplantation using biodegradable synthetic scaffolds. J Mater Sci Mater Med 12:991–996
14. Feil G, Maurer S, Nagele U, Krug J, Bock C, Sievert KD, Stenzl A (2008) Immunoreactivity of p63 in monolayered and in vitro stratified human urothelial cell cultures compared with native urothelial tissue. Eur Urol 53: 1066–1072
15. Gabouev AI, Schultheiss D, Mertsching H, Koppe M, Schlote N, Wefer J, Jonas U, Stief CG (2003) In vitro construction of urinary

bladder wall using porcine primary cells reseeded on acellularized bladder matrix and small intestinal submucosa. Int J Artif Organs 26:935–942
16. Farhat WA, Yeger H (2008) Does mechanical stimulation have any role in urinary bladder tissue engineering? World J Urol 26:301–305
17. Eeles RA, Kote-Jarai Z, Al Olama AA, Giles GG, Guy M, Severi G, Muir K, Hopper JL, Henderson BE, Haiman CA, Schleutker J, Hamdy FC, Neal DE, Donovan JL, Stanford JL, Ostrander EA, Ingles SA, John EM, Thibodeau SN, Schaid D, Park JY, Spurdle A, Clements J, Dickinson JL, Maier C, Vogel W, Dork T, Rebbeck TR, Cooney KA, Cannon-Albright L, Chappuis PO, Hutter P, Zeegers M, Kaneva R, Zhang HW, Lu YJ, Foulkes WD, English DR, Leongamornlert DA, Tymrakiewicz M, Morrison J, Ardern-Jones AT, Hall AL, O'Brien LT, Wilkinson RA, Saunders EJ, Page EC, Sawyer EJ, Edwards SM, Dearnaley DP, Horwich A, Huddart RA, Khoo VS, Parker CC, Van As N, Woodhouse CJ, Thompson A, Christmas T, Ogden C, Cooper CS, Southey MC, Lophatananon A, Liu JF, Kolonel LN, Le Marchand L, Wahlfors T, Tammela TL, Auvinen A, Lewis SJ, Cox A, FitzGerald LM, Koopmeiners JS, Karyadi DM, Kwon EM, Stern MC, Corral R, Joshi AD, Shahabi A, McDonnell SK, Sellers TA, Pow-Sang J, Chambers S, Aitken J, Gardiner RA, Batra J, Kedda MA, Lose F, Polanowski A, Patterson B, Serth J, Meyer A, Luedeke M, Stefflova K, Ray AM, Lange EM, Farnham J, Khan H, Slavov C, Mitkova A, Cao G, Easton DF (2009) Identification of seven new prostate cancer susceptibility loci through a genome-wide association study. Nat Genet 41:1116–1121
18. Bendall SC, Hughes C, Campbell JL, Stewart MH, Pittock P, Liu S, Bonneil E, Thibault P, Bhatia M, Lajoie GA (2009) An enhanced mass spectrometry approach reveals human embryonic stem cell growth factors in culture. Mol Cell Proteomics 8:421–432
19. Song WS, Yoo JJ, Koo KH, Yoon KS, Kim YM, Kim HJ (2004) Subchondral fatigue fracture of the femoral head in military recruits. J Bone Joint Surg Am 86-A:1917–1924
20. Oberpenning F, Meng J, Yoo JJ, Atala A (1999) De novo reconstitution of a functional mammalian urinary bladder by tissue engineering. Nat Biotechnol 17:149–155
21. Bahrenberg G, Brauers A, Joost HG, Jakse G (2000) Reduced expression of PSCA, a member of the LY-6 family of cell surface antigens, in bladder, esophagus, and stomach tumors. Biochem Biophys Res Commun 275:783–788
22. Pariente JL, Kim BS, Atala A (2001) In vitro biocompatibility assessment of naturally derived and synthetic biomaterials using normal human urothelial cells. J Biomed Mater Res 55:33–39
23. Amiel GE, Komura M, Shapira O, Yoo JJ, Yazdani S, Berry J, Kaushal S, Bischoff J, Atala A, Soker S (2006) Engineering of blood vessels from acellular collagen matrices coated with human endothelial cells. Tissue Eng 12: 2355–2365
24. Hipp J, Andersson KE, Kwon TG, Kwak EK, Yoo J, Atala A (2008) Microarray analysis of exstrophic human bladder smooth muscle. BJU Int 101:100–105
25. Amiel GE, Yoo JJ, Atala A (2000) Renal therapy using tissue-engineered constructs and gene delivery. World J Urol 18:71–79
26. Johnson MD, Bryan GT, Reznikoff CA (1985) Serial cultivation of normal rat bladder epithelial cells in vitro. J Urol 133:1076–1081
27. Chen LL, Trent JC, Wu EF, Fuller GN, Ramdas L, Zhang W, Raymond AK, Prieto VG, Oyedeji CO, Hunt KK, Pollock RE, Feig BW, Hayes KJ, Choi H, Macapinlac HA, Hittelman W, Velasco MA, Patel S, Burgess MA, Benjaminutes RS, Frazier ML (2004) A missense mutation in KIT kinase domain 1 correlates with imatinib resistance in gastrointestinal stromal tumors. Cancer Res 64:5913–5919
28. Andl T, Murchison EP, Liu F, Zhang Y, Yunta-Gonzalez M, Tobias JW, Andl CD, Seykora JT, Hannon GJ, Millar SE (2006) The miRNA-processing enzyme dicer is essential for the morphogenesis and maintenance of hair follicles. Curr Biol 16:1041–1049
29. Fraser M, Thomas DF, Pitt E, Harnden P, Trejdosiewicz LK, Southgate J (2004) A surgical model of composite cystoplasty with cultured urothelial cells: a controlled study of gross outcome and urothelial phenotype. BJU Int 93:609–616
30. Cartwright L, Farhat WA, Sherman C, Chen J, Babyn P, Yeger H, Cheng HL (2006) Dynamic contrast-enhanced MRI to quantify VEGF-enhanced tissue-engineered bladder graft neovascularization: pilot study. J Biomed Mater Res A 77:390–395
31. Zhang YY, Ludwikowski B, Hurst R, Frey P (2001) Expansion and long-term culture of differentiated normal rat urothelial cells in vitro. In Vitro Cell Dev Biol Anim 37: 419–429
32. Frimberger D, Morales N, Shamblott M, Gearhart JD, Gearhart JP, Lakshmanan Y (2005) Human embryoid body-derived stem cells in bladder regeneration using rodent model. Urology 65:827–832
33. Amante FH, Haque A, Stanley AC, Rivera Fde L, Randall LM, Wilson YA, Yeo G, Pieper C, Crabb BS, de Koning-Ward TF, Lundie RJ,

Good MF, Pinzon-Charry A, Pearson MS, Duke MG, McManus DP, Loukas A, Hill GR, Engwerda CR (2010) Immune-mediated mechanisms of parasite tissue sequestration during experimental cerebral malaria. J Immunol 185:3632–3642

34. Cen L, Liu W, Cui L, Zhang W, Cao Y (2008) Collagen tissue engineering: development of novel biomaterials and applications. Pediatr Res 63:492–496

35. Jack GS, Almeida FG, Zhang R, Alfonso ZC, Zuk PA, Rodriguez LV (2005) Processed lipoaspirate cells for tissue engineering of the lower urinary tract: implications for the treatment of stress urinary incontinence and bladder reconstruction. J Urol 174:2041–2045

36. Abdel-Hamid NM (2009) Premalignant variations in extracellular matrix composition in chemically induced hepatocellular carcinoma in rats. J Membr Biol 230:155–162

Chapter 3

Cell-Populated Collagen Lattice Contraction Model for the Investigation of Fibroblast Collagen Interactions

H. Paul Ehrlich and Kurtis E. Moyer

Abstract

The fibroblast-populated collagen lattice (FPCL) was intended to act as the dermal component for "skin-equivalent" or artificial skin developed for skin grafting burn patients. The "skin-equivalent" was clinically unsuccessful as a skin graft, but today it is successfully used as a dressing for the management of chronic wounds. The FPCL has, however, become an instrument for investigating cell–connective tissue interactions within a three-dimensional matrix. Through the capacity of cell compaction of collagen fibrils, the FPCL undergoes a reduction in volume referred to as lattice contraction. Lattice contraction proceeds by cell-generated forces that reduce the water mass between collagen fibers, generating a closer relationship between collagen fibers. The compaction of collagen fibers is responsible for the reduction in the FPCL volume. Cell-generated forces through the linkage of collagen fibers with fibroblast's cytoskeletal actin-rich microfilament structures are responsible for the completion of the collagen matrix compaction. The type of culture dish used to cast FPCL as well as the cell number will dictate the mechanism for compacting collagen matrices. Fibroblasts, at moderate density, cast as an FPCL within a petri dish and released from the surface of the dish soon after casting compact collagen fibers through cell tractional forces. Fibroblasts at moderate density cast as an FPCL within a tissue culture dish and not released for 4 days upon release show rapid lattice contraction through a mechanism of cell contraction forces. Fibroblasts at high density cast in an FPCL within a petri dish, released from the surface of the dish soon after casting, compact a collagen lattice very rapidly through forces related to cell elongation. The advantage of the FPCL contraction model is the study of cells in the three-dimensional environment, which is similar to the environment from which these cells were isolated. In this chapter methods are described for manufacturing collagen lattices, which assess the three forces involved in compacting and/or organizing collagen fibrils into thicker collagen fibers. The clinical relevance of the FPCL contraction model is related to advancing our understanding of wound contraction and scar contracture.

Key words Fibroblasts, Collagen lattice, Wound contraction, Scar contraction

1 Introduction

The organization of connective tissue by resident fibroblasts plays a critical role in embryonic tissue development as well as in tissue repair through the generation of a scar or the closure of an open wound by wound contraction. The cultured fibroblast-populated

Robert G. Gourdie and Tereance A. Myers (eds.), *Wound Regeneration and Repair: Methods and Protocols*,
Methods in Molecular Biology, vol. 1037, DOI 10.1007/978-1-62703-505-7_3, © Springer Science+Business Media New York 2013

collagen lattice (FPCL) contraction model has long been a tool to examine the interactions between cells and collagen in an effort to increase our understanding of complex connective tissue interactions within a three-dimensional matrix. Examining cells within a two-dimensional format, monolayer culture, has greatly advanced our knowledge and understanding of the biochemistry and physiology of a variety of cell types including dermal fibroblasts; however, applying the third-dimensional aspect seen with FPCL studies, that knowledge has been expanded even further. The FPCL contraction model investigates fibroblast interactions with a newly polymerized collagen matrix in a three-dimensional lattice. The contraction of the collagen matrix by resident cells is due to the capacity of fibroblasts to compact collagen fibrils through specific cellular mechanisms as well as by reducing extracellular fluid volume. Creating this three-dimensional system in a consistent manner requires the adherence to specific principles through the control of multiple variables. Altering fibroblast density, the concentration and/or the type of collagen present (type I versus III), the type of culture medium, as well as the type of culture dish employed have all resulted in differences in the mechanism employed by fibroblasts to compact their surrounding collagen matrix.

Here, the one focus will be on how altering these variables in casting an FPCL can change the mechanism responsible for the compaction of the collagen lattice. This is important because the rearrangement of collagen fibrils through compaction by resident cells such as fibroblasts and smooth muscle cells can produce a matrix environment with characteristics similar to dermal skin as well as blood vessel walls. As already mentioned, variations in the components in the casting of an FPCL dictate the mechanism utilized by resident cells to contract that matrix. The mechanisms typically utilized by these resident cells to rearrange collagen fibrils include cell contractile forces, cell locomotion forces, and cell elongation forces. These forces rely on actin–myosin filament sliding through myosin ATPase activity.

Eugene Bell and coworkers wrote the classic paper that introduced the free-floating FPCL contraction model [1]. His objective was to develop a “skin-equivalent” for use as a synthetic skin graft in the treatment of severely burned patients. The free-floating FPCL was the dermal component of the “skin-equivalent” and therefore referred to as the “dermal-equivalent.” In Bell’s original paper, cultured dermal fibroblasts were developed from a biopsy taken from a male Sprague Dawley rat. Each milliliter of the “dermal-equivalent” or FPCL was cast with 100,000 rat dermal fibroblasts, 1.25 mg of acid, soluble native rat tail tendon (RTT) collagen, and serum-enriched culture medium. Over a 2-day period the FPCL contracted, forming a dense collagen matrix with elongated fibroblasts. A second skin biopsy was performed on the same rat, but a culture of keratinocytes was generated from this biopsy to create the "skin-equivalent." A suspension of keratinocytes was placed on the surface of a 2-day free-floating FPCL composed of rat dermal fibroblasts as described above. This casted FPCL, with seeded keratinocytes, floated on the surface of the culture medium. The developing epidermal layer was situated at the air interface of the culture medium surface. The newly developed "skin-equivalent" contained collagen from pooled RTT as well as fibroblasts and keratinocytes obtained from the same rat that was to receive the "skin-equivalent" graft. Once matured, the skin-equivalent was grafted as an autograft onto a full-thickness fresh excisional wound made on the back of the cell-donor rat. This "skin-equivalent" graft was accepted as a skin graft in the rat model [2]. Unfortunately, the success of the "skin-equivalent" graft in rats was not transferable to establishing an accepted "skin-equivalent" graft for patients. Initially, the "skin-equivalent" or free-floating FPCL with a keratinocyte layer on top was marketed as a testing device for evaluating the safety of skin products [3]. Though it failed as a synthetic skin graft for the treatment of burn wounds, the "skin-equivalent" eventually found a clinical application. The "skin-equivalent," now called Apligraf®, is used clinically as an engineered wound care dressing, promoting the closure of chronic wounds [4].

The original dermal-equivalent as described by Bell, the free-floating FPCL, has four components: primary human cultured dermal fibroblasts; McCoy's 5a medium, supplemented with 10 % fetal bovine serum (FBS); native acid solubilized RTT collagen; and a petri dish, not a tissue culture dish, which minimizes the attachment of the lattice to the surface of the dish [1]. Fibroblasts in monolayer culture released by limited trypsin digestion were counted, suspended in serum-supplemented culture medium, combined with soluble native collagen, and then poured into the petri dish. The dish was transferred to an incubator set at 37 °C with a water-saturated atmosphere and 5 % CO_2. The mixture rapidly polymerized forming a disc shape within the petri dish. Initially, fibroblasts entrapped within the collagen matrix had a spherical shape, Fig. 1 left. At 24 h the spherical shaped fibroblasts had elongated and spread out taking on a spindle-type shape, Fig. 1 right. To accomplish this change in morphology, the fibroblasts pull the surrounding fine collagen fibrils towards their cell body as their lamellipodia project outward from the cell body, making numerous attachments to the polymerized collagen lattice, see Fig. 1. During this period of cell elongation, the thickness of the lattice in the *Z*-axis decreases in size, resulting in the FPCL disc becoming thinner.

Modifications in the casting of the FPCL have revealed distinctive cellular morphologies and mechanisms for rearranging collagen fibrils within their surrounding collagen matrix. The three variations

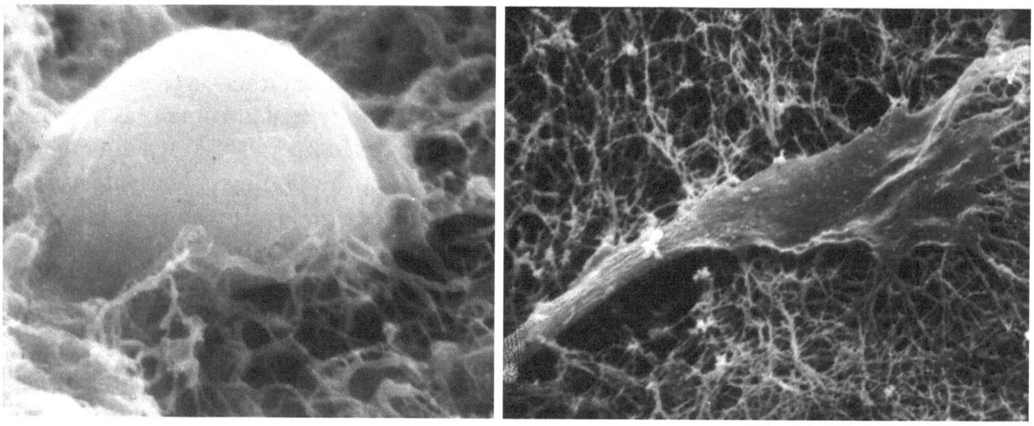

Fig. 1 Scanning electron micrograph of a fibroblast incorporated within an FPCL. The fibroblast on the *left* is shown 30 min after casting in a collagen lattice and the fibroblast on the *right* is 24 h after casting. Note how the cell was initially spherical in shape and 24 h later the cell is elongated. There is attachment of collagen fibers to the fibroblast surface within the first 30 min

most commonly utilized in the casting of the FPCL are (1) moderate-cell-density, free-floating FPCL contraction model as introduced by Bell and coworkers [1]; (2) high-cell-density, free-floating FPCL contraction model [5, 6]; and (3) the moderate-cell-density attached-delayed-released (ADR)-FPCL that is released 4 days after casting [7]. The measurement of lattice contraction is determined by changes in their area in the *XY* plane. In the ADR-FPCL the lattices are cast in tissue culture dishes, rather than a petri dish. While these lattices remain attached to their underlying surface in the tissue culture dish, lattice contraction is initially limited to the *Z*-axis as they become thinner discs. Rapid contraction occurs in the *XY* plane upon physically releasing the lattices from the surface of the dish with a spatula on day 4. The majority of rapid contraction occurs within the first 10 min as the resident elongated cells physically contract their thick cytoplasmic stress fibers [7]. In the high-cell-density free-floating FPCL there is rapid contraction within the first 4 h after casting unlike the moderate-cell-density free-floating FPCL [5].

In the initial paper by Bell, increasing the collagen concentration reduced the rate and degree of lattice contraction, while increasing the number of fibroblasts increased the rate and degree of lattice contraction. Serum was required to optimize lattice contraction. The initial contraction of a free-floating FPCL occurs in the *Z*-axis, which commences within the first hour after casting, while contraction in the *X*–*Y*-axis begins 4–6 h after casting and continues at a constant rate over the next 48 h, where the area of the lattice can be reduced as much as 90 % [1]. After 48 h, the lattice contraction continues, but at a reduced rate.

The casting of an ADR-FPCL, in the absence of serum, upon its release at 4 days, fails to contract [7]. The introduction of serum restores lattice contraction. When an ADR-FPCL adheres to its underlying substratum, the resident fibroblasts develop tension and by day 4 the fibroblast population has transformed into a myofibroblast population, expressing α smooth muscle actin (SMC) [8]. Microfilaments within a fibroblast under tension undergo a condensation into stress fibers, the defining morphology of myofibroblasts. Upon the release of an adherent lattice, the resident myofibroblasts undergo observable cell contraction, resulting in the rapid compaction of the collagen lattice. This mechanism is in contrast to the free-floating FPCL, which rely on resident fibroblasts to contract the collagen through cell locomotion forces and no change in cell shape.

Modulating the casting of the FPCL enables the differences between fibroblast interactions with their surrounding connective tissue matrix to be examined. Each milliliter of the original Bell free-floating FPCL contraction model contained 50,000–100,000 fibroblasts in petri dishes released soon after casting. FPCL initially contracts in the *Z*-axis, becoming a thinner disc in the first 4–6 h and contract at a steady rate for the next 2 days without a change in the elongated fibroblast morphology (see Fig. 2a). As shown in Fig. 2b, polarized light microscopy reveals birefringence of collagen fibers aligned parallel with the elongated lamellipodia of the fibroblast. The mechanism for free-floating FPCL contraction is through the passing or the translocation of collagen fibrils over the fibroblast plasma membrane surface, known as cell locomotion, resulting in a parallel packing of collagen fibrils generating a uniform collagen matrix [9]. With the free-floating FPCL, contraction forces related to cell locomotion or cell tractional forces are responsible for the compaction of collagen fibrils into more uniform organized collagen fibers.

As already described, altering the technique of FPCL casting spawns a change in the mechanism for lattice contraction. An ADR-FPCL utilizes cell contractile forces rather than cell locomotion forces for the compaction of collagen fibrils. The moderate-density ADR-FPCL contraction model introduced by Tomasek and coworkers [7] utilizes 100,000 cultured human dermal fibroblasts with of 1.25 mg of collagen in 1 mL of culture medium containing 10 % serum. Lattices cast in tissue culture dishes remained attached to the dish surface for 4 days. Over time, fibroblasts residing in an ADR-FPCL are subjected to tension, which results in the condensation of their fine cytoplasmic microfilaments, characteristic of fibroblasts, into thick cytoplasmic stress fibers, characteristic of myofibroblasts. At 4 days, the fibroblast population has transformed into a myofibroblast population. The release of the attached myofibroblast-populated collagen lattice leads to its rapid contraction, mostly completed

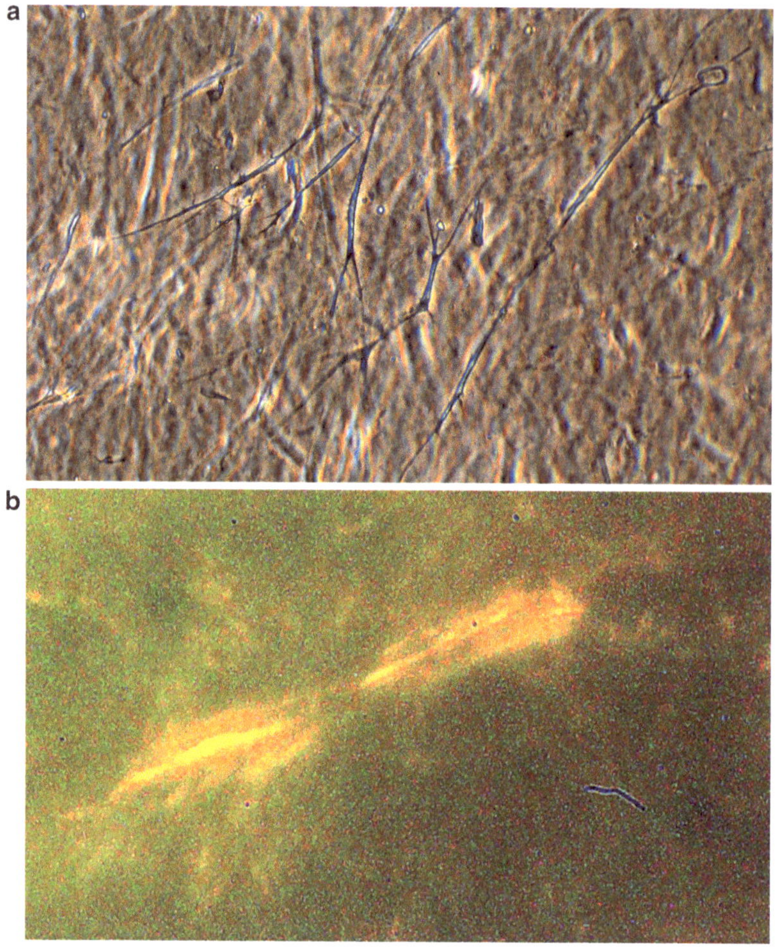

Fig. 2 Free-floating FPCL at 24 h. Panel **a** is a phase contrast micrograph, showing elongated fibroblasts within the collagen matrix. Panel **b** is an unstained FPCL, viewed with polarized light, showing birefringence of collagen fibers parallel to the cell's lamellipodia or extended cell arms

within 10 min after release. The contraction of the resident myofibroblasts triggers the compaction of the collagen fibrils. Unlike free-floating FPCLs there is minimal organization of collagen fibrils into uniform fibers as demonstrated by polarized light birefringence. This model facilitates the study of cell contractile forces compacting the connective tissues matrix as opposed to cell locomotion forces.

Casting high-cell-density free-floating FPCLs containing more than 500,000 cells per mL of lattice enables the study of the interactions of fibroblast cell surface lamellipodia protrusions within a three-dimensional collagen matrix. High-density free-floating FPCLs contract rapidly, where lattice contraction begins in less than 2 h and is completed in 4–6 h [6]. The contraction of the collagen matrix in high-density free-floating FPCLs is through cell elongation and retraction, with the sending out of lamellipodia followed by the retraction of lamellipodia, which results in the pulling of collagen fibrils back towards the cell body. There is no reorganization of collagen fibrils, which remain arranged in random arrays in contrast to the moderate-density free-floating FPCL. This three-dimensional model of lamellipodia protrusion interactions with a collagen matrix is more biological-like compared to the study of fibroblast lamellipodia protrusions along the surface of a plastic tissue culture dish.

2 Materials

Culture medium: The most commonly used culture medium employed for making FPCLs is Dulbecco's modification of Eagle's medium (DMEM) supplemented with 10 % FBS. To avoid premature polymerization of the native soluble collagen, the culture medium should be kept cold. Lattice contraction requires the inclusion of added FBS [1]. In the absence of FBS little, if any, lattice contraction occurs.

Collagen: Collagen can be purchased from a variety of sources. RTT collagen solution purchased from Aldrich-Sigma, St Louis, MO, or Advanced BioMatrix, Inc., San Diego, CA, is supplied in a 20 mM acetic acid solution at 4 mg/mL and is maintained under cold conditions. Collagen under cold acidic conditions is retained in solution in its native state. In order for collagen to polymerize, it has to be at a neutral pH and at 37 °C. The addition of 17 μL of 1N NaOH per mL of native collagen solution in 20 mM acetic acid neutralizes the collagen solution. To avoid subjecting cells to acidic conditions the collagen solution is neutralized just before introducing cells.

It is not difficult to generate your own RTT collagen solution and avoid the NaOH neutralization step. Tendons are isolated from rat tails. A common pair of pliers is used to break off and free individual tail vertebrae, teasing out arrays of tail tendon. Scissors free the tendons from the vertebrae and then they are transferred to a flask with a stir bar, 0.1 M acetic acid at 4 °C, and stirred with a stir motor. A ratio of 1 mg of tendon per 1 mL 0.1 M acetic acid is stirred for 2–7 days until the solution develops uniform viscosity. The viscous solution is centrifuged at 10,000 × *g* for 30 min, the supernatant saved, and the pellet discarded. NaCl is added to the measured volume of the supernatant at 10 % w/v, stirred at 4 °C for 2 h, after which it is allowed to stand for 2 h without stirring at 4 °C. The precipitated collagen is collected by again centrifuging at 10,000 × *g* for 10 min, the supernatant discarded, and the pellet is now saved. The pellet is dissolved in the same initial volume of 0.1 M acetic acid as used to free the collagen from the tendons. The collagen solution is transferred to dialysis tubing and exhaustively dialyzed against running cold tap water overnight.

The pelleted collagen is again collected by centrifugation at 10,000 × *g* for 10 min. The pellet is taken up in 0.1 M acetic acid as stated above and stirred at 4 °C until the collagen pellet is completely solubilized. The viscous solution is exhaustively dialyzed against at least four changes of 10 mM HCl at a ratio of 1 volume of dialysate in the tubing versus 10 volumes of 1 mM HCl at 4 °C with stirring. The solution is cleared by centrifugation at 10,000 × *g* for 30 min to remove any insoluble collagen debris. The supernatant is saved, frozen, and lyophilized. The dried collagen is weighed and for every 5 mg of collagen, 1 mL of ice-cold 10 mM filter-sterilized HCl is added, and the solution vigorously stirred at 4 °C overnight in a sealed container to avoid contamination. The viscous solution is kept at 4 °C and it should never be frozen. It is stable when kept in the refrigerator for more than a year and it will readily polymerize into a gel. Having the collagen in 10 mM HCl eliminates the need to neutralize the collagen solution with NaOH, since HCl has no buffering capacity. When the collagen is mixed with culture medium and cells, there is no need to neutralize the culture medium collagen mixture.

Culture dish: The type of culture dish used in casting FPCLs determines a free-floating FPCL from an ADR-FPCL. For free-floating moderate-density and high-density FPCL, the petri dish is preferred because of the ease in releasing the lattice from the surface of the dish. To cast ADR-FPCLs, a tissue culture dish is used because the lattice remains attached to the dish surface until the FPCLs are released with a spatula at 4 days.

3 Methods

Components to make a 4.0 mL fibroblast–collagen mixture to cast moderate-density FPCLs:

(a) Using a purchased collagen solution in 20 mM acetic acid:

1.25 mL of RTT collagen solution in 4 mg/mL 20 mM acetic acid at 4 °C.

1.35 mL of 2× DMEM at 4 °C.

0.017 mL 1N NaOH.

0.4 mL of FBS.

1.0 mL of DMEM with 10 % FBS and 400,000 fibroblasts at room temperature.

Mix completely with agitation (vortex for 2–5 s or until the solution is uniform in color and viscosity).

Every milliliter of FPCL contains 100,000 cells, 10 % FBS, and 1.25 mg of collagen.

(b) Components to cast a 4.0 mL FPCL, using RTT collagen at 5 mg/mL in 10 mM HCl:

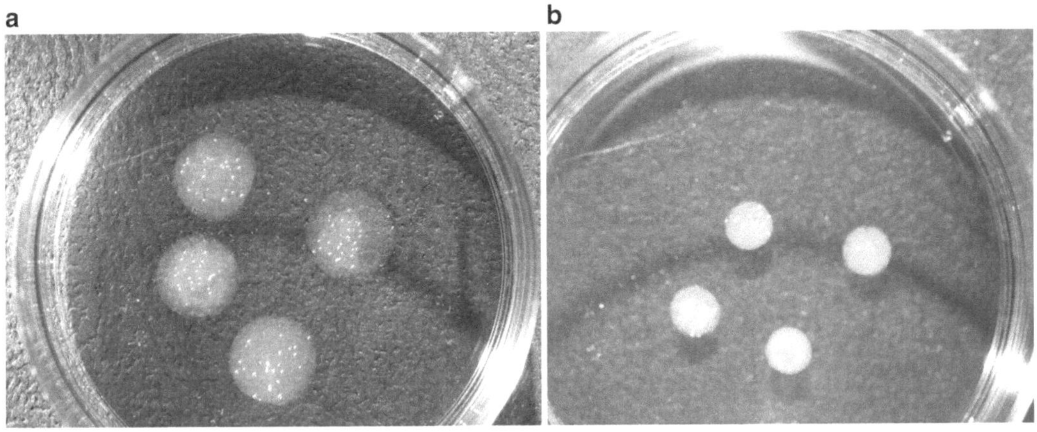

Fig. 3 Four FPCLs are cast in a 35 mm petri dish. Panel **a** shows four 200 μL FPCL beads in culture medium 10 min after casting and immediately after their release from the petri dish surface. Panel **b** shows those same FPCLs 24 h later, demonstrating a dramatic decrease in lattice size from FPCL contraction

1.0 mL of RTT collagen solution of 5 mg/mL in 10 mM HCl at 4 °C.

1.6 mL of 2× DMEM at 4 °C.

0.4 mL of FBS.

1.0 mL of DMEM with 10 % FBS and 400,000 fibroblasts at room temperature.

Mix completely with agitation until the solution is uniform in color and viscosity.

Every milliliter of FPCL contains 100,000 cells, 10 % FBS, and 1.25 mg of collagen.

3.1 To Cast a Free-Floating FPCL

Fibroblasts in monolayer culture are released from tissue culture by limited trypsin digestion and the freed cells suspended in DMEM with 10 % FBS serum. The fibroblasts are counted and 50,000–100,000 cells are utilized for every milliliter of FPCL. The final homogenous solution will contain 1.25 mg of collagen with 50,000–100,000 fibroblasts in 1.0 mL of DMEM supplemented with 10 % FBS. To reduce the number of dishes, while maintaining the identical conditions with sufficient number of collagen lattices for determining significance, three or four 200 μL aliquots of the fibroblast–collagen suspension, as individual beads, are pipetted into a 35 mm or four 200 μL aliquots into a 60 mm petri dish. Within 10 min at room temperature, the fibroblast–collagen suspension will have polymerized enough so that the petri dish can be transferred to an incubator set at 37 °C for an additional 10 min to further stabilize the collagen matrix. The dishes are removed from the incubator and 1 mL, to the 35 mm dish, or 3 mL, to the 60 mm dish, of complete serum containing culture medium is added. In some situations, the FPCL beads may need to be gently pushed off the surface or edges of the dish to induce free floating. Released FPCL will float in culture medium.

The petri dish with the floating collagen lattices as shown in Fig. 3a is returned to the incubator. The advantage of this technique is the uniformity of the cast FPCL and culture medium in each dish, where three or four individual lattices are cast from the same collagen–cell–medium suspension and maintained in the same culture medium. Measurements of lattice size can start as early as 4 h after casting the lattices. Digital camera photographs are taken with a ruler in place and the areas calculated for each lattice at each time point using the NIH-downloaded ImageJ program. The mean contraction plus the standard deviation of the three or the four lattices per dish are determined and used to plot lattice size changes over time from cell locomotion forces rearranging collagen fibrils [9].

3.2 To Cast an Attached-Delayed-Released FPCL

The culture medium, cell number, and collagen solution are the same as described above for the free-floating FPCL. However, instead of the 35 or the 60 mm petri dish a tissue culture dish is employed. The cell–collagen–culture medium aliquots contain a decreased volume of 180 μL, because the tissue culture dish surface is hydrophilic compared to the petri dish and therefore the FPCL beads are more spread out, taking up a greater area. Reducing the bead volume to 180 μL will maintain the same-size FPCL beads in the petri dish reducing the chances of the FPCL beads coalescing together. Tissue culture dishes with attached FPCL beads are incubated for 20 min in a 37 °C incubator. As to minimally disturb the attached FPCL, 1 or 3 mL of culture medium is carefully added to each dish. At day 4, attached FPCL beads are photographed with a ruler in place. The FPCL beads are carefully freed from the dish surface with a spatula and the dish with the three or the four floating lattice beads is returned to the incubator for 10 min. The incubated released FPCL beads are rephotographed with the same ruler in place. All images are subjected to ImageJ evaluations, where the change in area assesses lattice contraction by a mechanism of cell contraction forces [10].

3.3 To Cast a High-Cell-Density FPCL

The culture medium, native collagen solution, and petri dish components are all the same as described above for the moderate-density free-floating FPCL. However, instead of a moderate concentration of cells (50,000–100,000 cells) per mL of FPCL, the high-density FPCLs are cast with 500,000–1,000,000 cells per mL of FPCL. To a 35 or a 60 mm petri dish, respectively, three or four high-cell-density collagen–culture medium mixtures with a volume of 200 μL are placed as separate beads on the dish surface. The dishes with FPCL beads are transferred to an incubator set at 37 °C for 10 min to stabilize the collagen lattices. The addition of 1 or 3 mL of culture medium is made to each dish, which is then returned to the incubator. Measurements of lattice contraction can

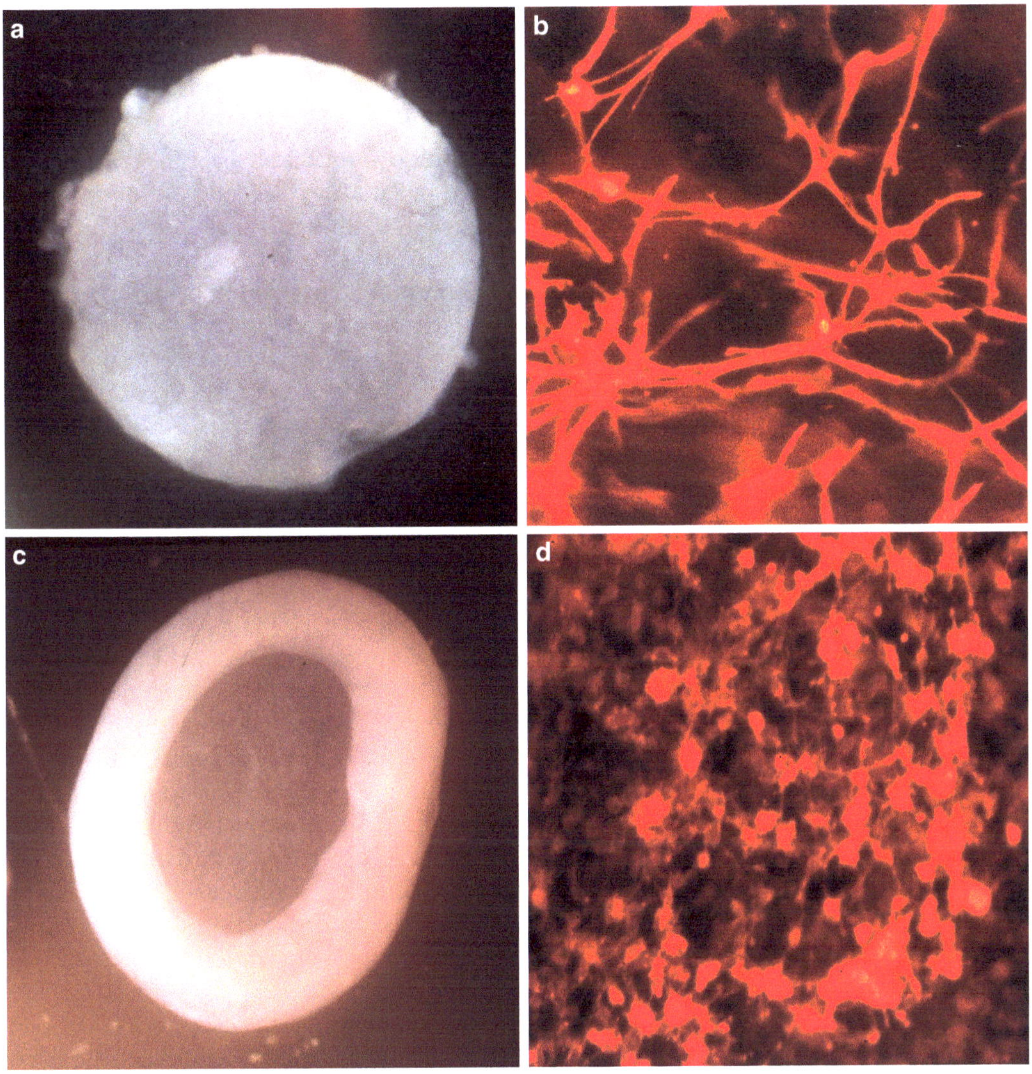

Fig. 4 Moderate-density versus high-density FPCL contraction and cell morphology. Panel **a** is an FPCL at 24 h cast with 100,000 fibroblasts. Panel **b** depicts the fibroblasts from the FPCL in panel **a** stained with phalloidin, demonstrating filamentous actin filaments. Panel **c** is an FPCL at 4 h cast with 1,000,000 fibroblasts. Panel **d** depicts the fibroblasts from the FPCL in panel **c** showing round phalloidin-stained fibroblasts

start as early as 1 h after casting a high-cell-density FPCL (Fig. 4). Digital camera photographs are taken with a ruler in place every hour up to 4–6 h and the area of images evaluated. Lattice contraction proceeds by a mechanism of forces related to cell elongation and retraction within the collagen lattice [6].

The fibroblasts in the high-density cast FPCLs are attempting to spread out and elongate within the collagen matrix, but the limited space for cells to spread causes the retention of a rounded shape with very short lamellipodia protrusions. Due to the high density of cells the enormous number of short lamellipodia protrusions is enough to pull collagen fibrils towards the cell body, resulting in the compaction of the collagen matrix.

3.4 Fibroblasts or Other Cell Types

Other cell types incorporated in collagen lattices that generate lattice contraction include smooth muscle cells [11], keratinocytes [12], Sertoli cells [13], and melanoma cells [14]. All of these cells produce lattice contraction. Inflammatory cells like neutrophils and mast cells do not cause lattice contraction [15, 16]. Inflammatory cells that leave the circulation pass through the blood vessel endothelial cell-lining, coming in contact with a collagen-rich matrix. Neutrophils and mast cells rapidly pass through that connective tissue matrix because their interaction with the collagen connective tissue matrix does not engage with their cytoskeletal contractile apparatus. They interact with a connective tissue matrix without distorting it.

4 Troubleshooting

One of the major problems with casting FPCLs is the premature polymerization of collagen that produces multiple nonuniform FPCLs that appear as fragmented collagen satellites. The most common reason for this failure to cast uniform collagen lattices is the premature polymerization of the collagen solution due to warming of the collagen–culture medium mixture before thorough blending and transfer to a dish have occurred. Maintaining the collagen solution in a cold state while quickly blending the components together and immediately pipetting the mixture into dishes will prevent the premature polymerization of a native collagen solution.

When contained in a collagen matrix, the capacity of fibroblast proliferation is dependent upon what species of animal the fibroblasts were derived from. When incorporated into the collagen matrix, human fibroblasts show a 4-day delay in cell division. In contrast, mouse and rat fibroblasts divide within the first 24 h. The human fibroblasts, like rodent fibroblasts, synthesize DNA and are in a 4N state by 24 h, but human fibroblasts are trapped in the G2/M phase of the cell cycle and require 3 days to undergo cell division [17]. Differences in collagen types influence the rate and degree of FPCL contraction as well. FPCLs composed exclusively of type III collagen contract at a faster rate and to a greater degree than lattices composed of type I collagen. In addition, FPCLs composed of type II collagen contract slower than FPCLs composed of type I collagen [18]. FPCLs composed of collagen isolated from hypertrophic scars undergoing contracture contract faster and to a greater degree than FPCLs composed of collagen isolated from normal dermis [19].

Free-floating FPCLs treated with the growth factor TGF-β show enhanced lattice contraction as compared to untreated cells [20]. Likewise, in ADR-FPCLs, the inclusion of TGF-β enhanced lattice contraction [21]. These findings suggest that TGF-β promotes cell contraction activities, associated with the myofibroblasts, as well as with cell translocation or locomotion forces associated with free-floating FPCL contraction. It has also been reported that platelet-derived growth factor (PDGF) enhances high-cell-density FPCL contraction [5]. There are no reports of PDGF promoting moderate or ADR-FPCL contraction. These findings support the possibility that PDGF enhances fibroblast lamellipodia protrusions. Lamellipodia protrusion activity is associated with cell migration and it is reported that PDGF enhances cell migration [22]. Gap junctional intercellular communication (GJIC) has been shown to promote FPCL contraction and disrupting GJIC will retard FPCL contraction [23]. The mechanism of promotion of lattice contraction through GJIC is unclear. Fibroblast plasma membrane fluidity influences that rate of FPCL contraction. Saturated fatty acids, fatty acids containing no double bonds, can be incorporated within the plasma membranes of fibroblasts in monolayer culture through adding fatty acid-saturated albumen. As the saturated fatty acid chain length increases, the plasma membrane fluidity decreases and FPCL contraction is retarded. Likewise, decreasing the length of the saturated fatty acid in the fibroblast cell membrane increases its fluidity and increases the rate of lattice contraction [24].

References

1. Bell E, Ivarsson B, Merrill C (1979) Production of a tissue-like structure by contraction of collagen lattices by human fibroblasts of different proliferative potential in vitro. Proc Natl Acad Sci USA 76:1274–1278
2. Bell E, Parenteau N, Gay R, Nolte C, Kemp P, Bilbo P, Ekstein B, Johnson E (1991) The living skin equivalent: its manufacture, its organotypic properties and its responses to irritants. Toxicol In Vitro 5:591–596
3. Clark RA, Folkvord JM, Hart CE, Murray MJ, McPherson JM (1989) Platelet isoforms of platelet-derived growth factor stimulate fibroblasts to contract collagen matrices. J Clin Invest 84:1036–1040
4. Brem H, Balledux J, Bloom T, Kerstein MD, Hollier L (2000) Healing of diabetic foot ulcers and pressure ulcers with human skin equivalent: a new paradigm in wound healing. Arch Surg 135:627–634
5. Bell E, Ehrlich HP, Buttle DJ, Nakatsuji T (1981) Living tissue formed in vitro and accepted as skin-equivalent tissue of full thickness. Science 211:1052–1054
6. Ehrlich HP, Rittenberg T (2000) Differences in the mechanism for high versus moderate density fibroblast populated collagen lattice contraction. J Cell Physiol 185:432–439
7. Tomasek JJ, Haaksma CJ, Eddy RJ, Vaughan MB (1992) Fibroblast contraction occurs on release of tension in attached collagen lattices: dependency on an organized actin cytoskeleton and serum. Anat Rec 232:359–368
8. Gabbiani G, Ryan GB, Majno G (1971) Presence of modified fibroblasts in granulation tissue and their possible role in wound contraction. Experientia 27:549–555
9. Ehrlich HP, Rajaratnam JBM (1990) Cell locomotion forces versus cell contraction forces for collagen lattice contraction: an in vitro model of wound contraction. Tissue Cell 22:407–417
10. Vaughan MB, Howard EW, Tomasek JJ (2000) Transforming growth factor-beta1 promotes the morphological and functional differentiation of the myofibroblast. Exp Cell Res 257:180–189

11. Ehrlich HP, Griswold TR, Rajaratnam JBM (1986) Studies of vascular smooth muscle cells and dermal fibroblasts in collagen matrices: effects of heparin. Exp Cell Res 164:154–163
12. Souren JM, Ponec M, van Wijk R (1989) Contraction of collagen by human fibroblasts and keratinocytes. In Vitro Cell Dev Biol 25:1039–1045
13. Borland K, Ehrlich HP, Muffly K, Dills WL Jr, Halls PF (1986) The interaction of rat Sertoli cells with a collagen lattice in vitro. In Vitro Cell Dev Biol 22:661–669
14. Friedl P, Maaser K, Klein CE, Niggemann B, Krohne G, Zänker KS (1997) Migration of highly aggressive MV3 melanoma cells in 3-dimensional collagen lattices results in local matrix reorganization and shedding of alpha2 and beta1 integrins and CD44. Cancer Res 57:2061–2070
15. Lackie JM, Brown AF (1983) Adhesion and the locomotion of neutrophils on surfaces and in matrices. Agents Actions Suppl 12:73–90
16. Moyer KE, Saggers GC, Ehrlich HP (2004) Mast cells promote fibroblast populated collagen lattice contraction through gap junction intercellular communication. Wound Repair Regen 12:269–275
17. Greco RM, Ehrlich HP (1992) Differences in cell division and thymidine incorporation with rat and primate fibroblasts in collagen lattices. Tissue Cell 24:843–851
18. Ehrlich HP (1988) The modulation of contraction of fibroblast populated collagen lattices by types I, II, and III collagen. Tissue Cell 20:47–50
19. Ehrlich HP (1988) Wound closure: evidence of cooperation between fibroblasts and collagen matrix in scar contracture. Eye 2:149–157
20. Montesano R, Orci L (1988) Transforming growth factor beta stimulates collagen-matrix contraction by fibroblasts: implications for wound healing. Proc Natl Acad Sci USA 85:4894–4897
21. Hinz B, Celetta G, Tomasek JJ, Gabbiani G, Chaponnier C (2001) Alpha-smooth muscle actin expression up regulates fibroblast contractile activity. Mol Biol Cell 12:2730–2741
22. Seppä H, Grotendorst G, Seppä S, Schiffmann E, Martin GR (1982) Platelet-derived growth factor in chemotactic for fibroblasts. J Cell Biol 92:584–588
23. Bowman N, Donahue H, Ehrlich HP (1998) Gap junctional intercellular communication contribution to the contraction of rat osteoblast populated collagen lattices. J Bone Miner Res 13:1700–1706
24. Rittenberg T, Ehrlich HP (1992) Free fatty acids and dialyzed serum alterations of fibroblast populated collagen lattice contraction. Tissue Cell 24:243–251

Chapter 4

A Tissue-Engineered Corneal Wound Healing Model for the Characterization of Reepithelialization

Karine Zaniolo, Patrick Carrier, Sylvain L. Guérin, François A. Auger, and Lucie Germain

Abstract

Progress in tissue engineering has led to the discovery of technologies allowing reconstruction of autologous tissues from the patient's own cells and the development of new in vitro models to study cellular and molecular mechanisms implicated in wound healing. The outer surface of the eye, the cornea, is involved in the sense of sight, thus an adequate reepithelialization process after wounding is essential in order to maintain corneal function. In this chapter, protocols to generate a new in vitro three-dimensional human corneal wound healing model suitable for studying the different components that play important roles in corneal reepithelialization are described in details. The methods include extraction and culture of human corneal epithelial cells (HCECs), human corneal fibroblasts, a complete description of the cornea reconstructed by tissue-engineering as well as the corneal wound healing model.

Key words Cornea, Corneal epithelial cells, Fibroblasts, Tissue-engineering, Reepithelialization

Abbreviations

alhcDME-Ham	Air–liquid corneal epithelial cell culture medium
hcDME-Ham	Complete human corneal epithelial cell culture medium
HCEC	Human corneal epithelial cells
DME	Dulbecco's modified Eagle's medium
DMSO	Dimethyl sulfoxide
ECM	Extracellular matrix
fDME	Human corneal fibroblast culture medium
Ham	Ham's F12 medium
i3T3	Irradiated Swiss 3T3
PBS-P/G/F	Phosphate buffered saline–Penicillin G/Gentamicin/Fungizone
T3	3,3′,5′-Triiodo-L-thyronine

Karine Zaniolo and Patrick Carrier have contributed equally to this work.

Robert G. Gourdie and Tereance A. Myers (eds.), *Wound Regeneration and Repair: Methods and Protocols*, Methods in Molecular Biology, vol. 1037, DOI 10.1007/978-1-62703-505-7_4, © Springer Science+Business Media New York 2013

1 Introduction

The cornea is localized at the outer surface of the eye. It is highly specialized and unique as a translucent organ. The cornea is continually subjected to abrasive forces and occasional mechanical or chemical trauma because of its anatomical localization. Damages to the cornea can result in scarring or opacification that may lead to visual defects and even complete loss of vision. A complete reepithelialization and the reorganization of a mature smooth epithelium, as well as a structured stroma, is essential in restoring the imaging properties of the cornea [1, 2]. Therefore, a better understanding of the mechanisms of corneal wound healing could help develop new therapeutic approaches to improve corneal wound management.

Several models of wound healing have been developed in order to investigate the corneal mechanisms of reepithelialization and to screen for growth factors susceptible to stimulate an adequate healing [3–11]. The limitations of cell monolayers in vitro models are the lack of epithelial–mesenchymal interactions and the limited epithelium thickness. Studies of corneal wound healing in animal models are expensive and interindividual variability among animals is inherent to in vivo experiments. Progress in tissue engineering resulted in the development of tissue-engineered corneas. These tissues were designed to mimic their in vivo counterpart in terms of cell phenotype and tissue architecture [12]. The three-dimensional tissue-engineered corneas were intented for basic research or clinical purposes [13–15].

The concept of the self-assembly approach consists in reconstructing an organ in a fashion resembling its formation in vivo. Appropriate culture conditions induce cells to secrete a significant amount of ECM as it occurs during organogenesis [16, 17]. Corneas reconstructed by the self-assembly approach exhibit a well-developed epithelium (*see* Fig. 2) that expresses differentiation markers (such as keratin 3 (K3)), a stroma, and a well-organized basement membrane [18, 19]. The tissue-engineered cornea also enables the development of a new in vitro model to study cellular and molecular mechanisms involved in corneal wound healing.

This chapter first details methods for the extraction and culture on plastic of the cell types (human corneal epithelial cells (HCECs), human corneal fibroblasts and human dermal fibroblasts) that are required in order to produce a tissue-engineered cornea in vitro. In the second part, a detailed description of the methods for the generation of tissue-engineered corneas by the self-assembly approach and the tissue-engineered corneal wound healing model are presented.

2 Materials

2.1 Culture Media

2.1.1 Tissue Transport Medium

1. Preparation of tissue transport medium components (for final concentration *see* Table 1):
   (a) DME-Ham. Dulbecco's modified Eagle's medium (DME): Ham's F12 medium 3:1, 3.07 g/L $NaHCO_3$ (36.54 mM), 24.3 mg/L adenine (0.18 mM), 312.5 μL/L 2N HCl. Dissolve in apyrogenic ultrapure water. Adjust pH to 7.1. Sterilize by filtration through a 0.22 μm low-binding disposable filter. Aliquot and store at 4 °C.
   (b) Fetal calf serum. Thaw in cold water. Inactivate in hot water (56 °C) for 30 min. Distribute in single-use aliquots and store at −20 °C.
   (c) Penicillin G and Gentamicin. Dissolve 50,000 IU/mL of Penicillin G and 12.5 mg/mL of Gentamicin sulfate in apyrogenic ultrapure water to make a 500× stock solution. Sterilize by filtration through a 0.22 μm low-binding disposable filter, distribute in single-use aliquots and store at −80 °C.
   (d) Fungizone. Dissolve 0.25 mg/mL of amphotericin B (0.27 mM), in apyrogenic ultrapure water to make a 500× stock solution. Sterilize by filtration through a 0.22 μm low-binding disposable filter, distribute in single-use aliquots and store at −80 °C.
2. Thaw all components at 4 °C. To make 1 L (Table 1).
3. Store the transport medium at 4 °C.

2.1.2 Human Fibroblast Culture Medium (fDME)

1. Preparation of fDME components (for final concentration *see* Table 2):
   (a) DME. Dulbecco's modified Eagle's medium (DME), 3.7 g/L (44 mM) $NaHCO_3$. Dissolve in apyrogenic ultrapure water. Adjust pH to 7.1. Sterilize by filtration through a 0.22 μm low binding disposable filter, aliquot, and store at 4 °C.

Table 1
Tissue transport medium

Component	Quantity (mL)	Final concentration
DME-Ham	900	90 % v/v
Fetal calf serum	100	10 % v/v
Penicillin G–Gentamicin 500×	2	Penicillin G 100 IU/mL Gentamicin 25 μg/mL
Fungizone 500×	2	0.5 μg/mL

Table 2
fDME

Component	Quantity (mL)	Final concentration
DME	900	90 % v/v
Fetal calf serum	100	10 % v/v
Penicillin G–Gentamicin 500×	2	Penicillin G 100 IU/mL Gentamicin 25 μg/mL

Table 3
Complete hcDME-Ham

Component	Quantity (mL)	Final concentration
DME-Ham	900	90 % v/v
Fetal clone II	100	10 % v/v
Insulin 1,000×	1	5 μg/mL
Hydrocortisone 500×	2	0.4 μg/mL
Cholera toxin 1,000×	1	10^{-10} M
Epidermal growth factor 1,000×	1	10 ng/mL
Transferrin/T3 1,000×	1	Transferrin: 5 μg/mL T3: 2×10^{-9} M
Penicillin G–Gentamicin 500×	2	Penicillin G 100 IU/mL Gentamicin 25 μg/mL

(b) Fetal calf serum (*see* Subheading 2.1.1, **item 1b**).

(c) Penicillin G–Gentamicin (*see* Subheading 2.1.1, **item 1c**).

2. Thaw all components at 4 °C. To make 1 L (Table 2).
3. The fDME can be stored at 4 °C for 10 days.

2.1.3 Complete Human Corneal Epithelial Cell Culture Medium (Complete hcDME-Ham)

1. Preparation of hcDME-Ham components; thaw all components described (for final concentration, *see* Table 3):

(a) DME-Ham (*see* Subheading 2.1.1, **item 1a**).

(b) Fetal clone II serum. Thaw in cold water. Inactivate in hot water (56 °C) for 30 min. Distribute in single-use aliquots and store at −20 °C.

(c) Insulin. Dissolve 250 mg in 50 mL 0.005 N HCl (125 μL 2N HCl/50 mL apyrogenic ultrapure water) to make a 1,000× stock solution (0.87 mM). Sterilize by filtration through a 0.22 μm low binding disposable filter, distribute in single-use aliquots and store at −80 °C.

(d) Hydrocortisone. Dissolve 25 mg in 5 mL of 95 % ethanol (4.8 mL 99 % ethanol /0.2 mL apyrogenic ultrapure water). Complete to 125 mL with DME-Ham (*see* Subheading 2.1.1; **item 1a**) to make a 500× stock solution (0.53 mM). Sterilize by filtration through a 0.22 μm low binding disposable filter, distribute in single-use aliquots and store at −80 °C.

(e) Cholera toxin. Dissolve 1 mg in 1 mL of apyrogenic ultrapure water. Complete to 118.18 mL with DME-Ham (*see* Subheading 2.1.1, **item 1a**) supplemented with 10 % (v/v) Fetal clone II (*see* Subheading 2.1.3, **item 1b**) to make a 1,000× stock solution (10^{-7} M). Sterilize by filtration through a 0.22 μm low binding disposable filter, distribute in single-use aliquots and store at −80 °C.

(f) Epidermal growth factor. Dissolve 500 μg in 2.5 mL of 10 mM HCl. Complete to 50 mL with DME-Ham (*see* Subheading 2.1.1, **item 1a**) supplemented with 10 % (v/v) Fetal clone II (*see* Subheading 2.1.3, **item 1b**) to make a 1,000× stock solution. Sterilize by filtration through a 0.22 μm low binding disposable filter, distribute in single-use aliquots and store at −80 °C.

(g) Penicillin G–Gentamicin (*see* Subheading 2.1.1, **item 1c**).

(h) Transferrin: vial of 600 mg/20 mL. 3,3',5' triiodo-L-thyronine (T3): Solution #1. Dissolve 6.8 mg of T3 in a small volume (about 3 mL) of 0.02N NaOH (10 μL 10N NaOH/4.99 mL apyrogenic ultrapure water). Complete to 50 mL with apyrogenic ultrapure water. Sterilize by filtration through a 0.22 μm low binding disposable filter, distribute in single-use aliquots and store at 4 °C. Mix the transferrin vial with 1.2 mL of solution #1 and complete to 120 mL with apyrogenic ultrapure water to make a 1,000× stock solution. Sterilize by filtration through a 0.22 μm low binding disposable filter, distribute in single-use aliquots and store at −80 °C.

2. Complete hcDME-Ham can be stored at 4 °C for 10 days.

2.1.4 Air–Liquid Human Corneal Epithelial Cell Culture Medium (alhcDME-Ham)

1. Thaw all components described in Subheading 2.1.3, **item 1**, at 4 °C, with the exception of epidermal growth factor. To make 1 L, refer to Table 4 (*see* **Note 1**).
2. alhcDME-Ham can be stored at 4 °C for 10 days.

2.1.5 Freezing Medium

1. Preparation of freezing medium components.

(a) Fetal calf serum (*see* Subheading 2.1.1, **item 1b**).

Table 4
Alhc-DME

Component	Quantity (mL)	Final concentration
DME-Ham	900	90 % v/v
Fetal clone II	100	10 % v/v
Insulin 1,000×	1	5 μg/mL
Hydrocortisone 500×	2	0.4 μg/mL
Cholera toxin 1,000×	1	10^{-10} M
Transferrin/T3 1,000×	1	Transferrin: 5 μg/mL T3: 2×10^{-9} M
Penicillin G–Gentamicin 500×	2	Penicillin G 100 IU/mL Gentamicin 25 μg/mL

Table 5
Freezing medium

Component	Quantity (mL)	Final concentration
Fetal calf serum	9	90 % v/v
DMSO	1	10 % v/v

(b) Dimethyl sulfoxide (DMSO). Distribute the stock solution (99.7 %) in single-use aliquots and store at −20 °C (*see* **Note 2**).

2. Thaw all components at 4 °C. To make 10 mL (Table 5).
3. Keep on ice or at 4 °C (*see* **Note 3**).

2.2 Monolayer Culture of Human Dermal Fibroblasts

2.2.1 Extraction and Culture of Human Fibroblasts

1. Source of human fibroblasts: normal adult breast skin removed by surgery (*see* **Note 4**).
2. Sterile container.
3. Transport medium (*see* Subheading 2.1.1).
4. 50 mL tubes.
5. Phosphate buffered saline–Penicillin G/Gentamicin/Fungizone (PBS-P/G/F): 137 mM NaCl, 2.7 mM KCl, 6.5 mM Na_2HPO_4, 1.5 mM KH_2PO_4. Dissolve in apyrogenic ultrapure water to make a 10× stock solution. Store at room temperature. To dilute 10× PBS to 1×, add apyrogenic ultrapure water. Verify pH is 7.4. Store at room temperature. Before use, add Penicillin G–Gentamicin 500× stock solution

(*see* Subheading 2.1.1, **item 1c**) and Fungizone 500× stock solution (*see* Subheading 2.1.1, **item 1d**) by diluting these additives to 1×.
6. Petri dish (size: 100 × 15 mm).
7. Thermolysin. Dissolve 500 μg/mL in Hepes 1×. Sterilize by filtration through a 0.22 μm low binding disposable filter and store at 4 °C (*see* **Note 3**).
8. Collagenase H. Dissolve 0.125 U/mL in fDME (*see* Subheading 2.1.2). Sterilize by filtration through a 0.22 μm low binding disposable filter (*see* **Note 3**).
9. Dissecting curved forceps.
10. Scalpel and blade (size: 22).
11. Trypsinization unit (Celstir suspension culture flask).
12. Parafilm.
13. Tissue culture flask, 75 cm^2.
14. fDME (*see* Subheading 2.1.2).

2.2.2 Subculture of Human Dermal Fibroblasts (Passage)

1. Complete fDME-Ham (*see* Subheading 2.1.2).
2. Trypsin/EDTA. 2.8 mM D-glucose, 0.05 % (w/v) trypsin 1–500, 0.00075 % (v/v) phenol red (Phenol red solution 0.5 %, sterile-filtered, , 100,000 IU/L penicillin G, 25 mg/L gentamicin, 0.01 % (w/v) EDTA (EDTA, Disodium salt). Dissolve in 1× PBS. Adjust pH to 7.45. Sterilize by filtration through a 0.22 μm low binding disposable filter, distribute in single-use aliquots and store at −20 °C.
3. 50 mL tube.
4. Tissue culture flask, 75 cm^2.

2.2.3 Cryopreservation of Human Dermal Fibroblasts

1. Freezing medium (*see* Subheading 2.1.5)
2. Sterile cryogenic vials.
3. Freezing container filled with 99 % ethanol and precooled at −20 °C.

2.2.4 Thawing of Human Dermal Fibroblastes

1. Complete fDME-Ham (*see* Subheading 2.1.2).
2. Tissue culture flasks, 75 cm^2.

2.3 Monolayer Culture of Human Corneal Fibroblasts

2.3.1 Extraction and Culture of Human Corneal Fibroblasts

1. Source of human corneal fibroblasts: normal cornea from post-mortem donors unsuitable for transplantation (Banque d'yeux du Centre Universitaire d'Ophtalmologie (CUO), Quebec, Canada) (*see* **Note 4**).
2. Sterile container.
3. Transport medium (*see* Subheading 2.1.1).
4. 50 mL tubes.

5. PBS-P/G/F (*see* Subheading 2.2.1, **item 5**).
6. Petri dish (size: 100 × 15 mm).
7. Dispase II. Dissolve 2 mg/mL dispase in Hepes. Make a 10× stock solution in apyrogenic ultrapure water: 0.1 M Hepes, 67 mM KCl, 1.42 M NaCl. Adjust pH to 7.3. Protect from light and store at 4 °C. Hepes 1×: dilute the 10× stock solution to 1× with apyrogenic ultrapure water and add 1 mM $CaCl_2$. Adjust pH to 7.45. Protect from light (*see* **Note 5**) and store at 4 °C.
8. Collagenase H. Dissolve 0.125 U/mL in fDME (*see* Subheading 2.1.2). Sterilize by filtration through a 0.22 μm low binding disposable filter (*see* **Note 3**).
9. Dissecting curved forceps.
10. Scalpel (size: 4) and blade (size: 22).
11. Parafilm.
12. Tissue culture flask, 25 cm².
13. fDME (*see* Subheading 2.1.2).

2.3.2 Subculture of Human Corneal Fibroblasts (Passage)

1. Complete fDME-Ham (*see* Subheading 2.1.2).
2. Trypsin/EDTA (*see* Subheading 2.2.2, **item 2**).
3. 50 mL tube.
4. Tissue culture flasks, 75 cm².

2.3.3 Cryopreservation of Human Corneal Fibroblasts

1. Freezing medium (*see* Subheading 2.1.5).
2. Sterile cryogenic vials.
3. Freezing container, filled with 99 % ethanol and precooled at −20 °C.

2.3.4 Thawing of Human Dermal Fibroblasts

1. Complete fDME-Ham (*see* Subheading 2.1.2).
2. Tissue culture flask, 75 cm².

2.4 Monolayer Culture of Human Corneal Epithelial Cells

2.4.1 Extraction and Culture of Human Corneal Epithelial Cells

1. Source of human corneal epithelial cells: human post-mortem donor corneas unsuitable for transplantation (Banque d'yeux du Centre Universitaire d'Ophtalmologie (CUO), Quebec, Canada) (*see* **Note 4**).
2. 35 and 100 mm petri dishes (size: 35 × 10 mm, and size: 100 × 15 mm).
3. Sterile 3 in. × 3 in. gauze.
4. Scalpel (#4) and blades (size: 11, and size: 22).
5. Dissecting curved forceps.
6. Dissecting curved scissor.
7. 15 and 50 mL tubes.

8. PBS-P/G/F (*see* Subheading 2.2.1, **item 5**).
9. 7.5 mm diameter trephine.
10. Dispase II (*see* Subheading 2.3.1, **item 7**).
11. Complete hcDME-Ham (*see* Subheading 2.1.3).
12. Tissue culture flask, 75 cm^2.
13. Irradiated Swiss 3T3 (i3T3) (ATCC, #CCL-92). To obtain about $8–10 \times 10^6$ cells, seed 1×10^6 cells in a 75 cm^2 culture flask (Tissue culture flask) with 20 mL of DME. Incubate for 4 days in 8 % CO_2, 100 % humidity atmosphere at 37 °C. Irradiate at 6,000 rads with a Gammacell irradiator (60Co source) (*see* **Note 6**).
14. Dissecting microscope.

2.4.2 Subculture of Human Corneal Epithelial Cells (Passage)

1. Complete hcDME-Ham (*see* Subheading 2.1.3).
2. Irradiated Swiss 3T3 (i3T3) (*see* Subheading 2.4.1, **item 13**).
3. Trypsin/EDTA (*see* Subheading 2.2.2, **item 2**).
4. 50 mL tubes.
5. Tissue culture flasks, 75 cm^2.

2.4.3 Cryopreservation of Human Corneal Epithelial Cells

1. Freezing medium (*see* Subheading 2.1.5).
2. Sterile cryogenic vials.
3. Freezing container filled with 99 % ethanol and precooled at −20 °C.

2.4.4 Thawing of Human Corneal Epithelial Cells

1. Complete hcDME-Ham (*see* Subheading 2.1.3).
2. Tissue culture flasks, 75 cm^2.
3. i3T3 (*see* Subheading 2.4.1, **item 13**).

2.5 Human Cornea Reconstruction by the Self-Assembly Approach

1. Confluent corneal fibroblasts between their 5th and 7th passages.
2. Confluent human fibroblasts between their 5th and 7th passages.
3. Human corneal epithelial cells at 80 % confluence between their 1st and 3rd passages (cocultured with i3T3, *see* Subheading 2.4.1, **item 13**).
4. Ascorbic acid. Dissolve 10 mg/mL of ascorbic acid in DME (*see* Subheading 2.1.1, 1a) to make a 200× stock solution. Sterilize by filtration through a 0.22 μm low binding disposable filter (*see* **Note 7**).
5. Culture media:
   (a) fDME (*see* Subheading 2.1.2) containing 50 μg/mL ascorbic acid.

(b) Complete hcDME-Ham (*see* Subheading 2.1.3) containing 50 μg/mL ascorbic acid.

(c) alhcDME-Ham (*see* Subheading 2.1.4) containing 50 μg/mL ascorbic acid.

6. Ingots (stainless steel grade # 316).
7. Merocel®, Medtronic.
8. Anchoring rings (stainless steel grade # 316). Dimension: 3 cm diameter, 7/8 in. wide, 1/8 in. height.
9. Dissecting curved forceps.
10. Tissue culture flasks, 25 cm^2.
11. Cell culture dish (size: 100 × 20 mm).
12. Petri dish (size: 100 × 15 mm).
13. Soldering iron.
14. Anchoring paper. Cut a circle with a 60 mm diameter in a Wathman sheet. Remove the concentric inside disk of 25 mm diameter after cutting.
15. Air–liquid stand (homemade acrylic stand).

2.6 Human Tissue-Engineered Corneal Wound Healing Model

1. Human tissue-engineered cornea reconstructed by self-assembly approach with epithelial cells after a week of epithelial maturation at the air–liquid interface (*see* Subheading 2.5).
2. Culture media: alhcDME-Ham (*see* Subheading 2.1.4) containing 50 μg/mL ascorbic acid.
3. 6 mm diameter biopsy punch.
4. Dissecting curved forceps.
5. Petri dishes (size: 100 × 15 mm).
6. One human tissue-engineered stroma comprising one dermal and one corneal fibroblast sheet that were superimposed and cultured for an additional week in fDME (*see* Subheading 2.1.2) containing 50 μg/mL ascorbic acid.
7. Anchoring paper. Cut a circle with a 60 mm diameter in a Wathman sheet. Remove the concentric inside disk of 25 mm diameter after cutting.

3 Methods

3.1 Monolayer Culture of Human Dermal Fibroblasts

3.1.1 Extraction and Culture of Human Dermal Fibroblasts

1. Human fibroblasts were obtained from normal adult breast skin (*see* **Note 4**).
2. Transport and conservation: In the surgery room, put the skin specimen into a sterile container filled with cold (4 °C) transport medium.

3. All further manipulations are performed under a sterile laminar flow hood cabinet.
4. Wash the skin specimen in a 50 mL tube containing 30 mL PBS-P/G/F. Agitate vigorously. With sterile forceps, transfer the skin specimen in another PBS-P/G/F tube. Repeat this step three times.
5. Spread out the skin specimen, epidermis on the top, into a 100 mm petri dish.
6. Cut the skin in 3 mm × 100 mm pieces with scalpel (blade 22).
7. Add 10 mL of warm (37 °C) thermolysin. Seal the petri dish with parafilm.
8. Incubate 2–3 h at 37 °C.
9. With two curved forceps, separate the epidermis from the dermis. Put dermal pieces within a trypsinization unit containing 20 mL of warm (37 °C) collagenase H. Incubate under agitation, overnight at 37 °C.
10. Collect the cellular suspension. Put into a 50 mL tube and add 20 mL of fDME. Wash the trypsinization unit with 10 mL of fDME and add it to the tube (total 40 mL).
11. Count the cells and measure the viability by trypan blue staining.
12. Centrifuge cell suspension at 300 × *g* for 10 min at room temperature.
13. Seed 5×10^5 fibroblasts in a 75 cm^2 culture flask with 15 mL of fDME. Incubate in 8 % CO_2, 100 % humidity atmosphere at 37 °C. Change culture medium three times a week.
14. When the fibroblasts reach confluence, subculture (*see* Subheading 3.1.2) or freeze (*see* Subheading 1) cells.

3.1.2 Subculture of Human Dermal Fibroblasts (Passage)

For a 75 cm^2 culture flask of fibroblasts:

All further manipulations are performed under a sterile laminar flow hood cabinet.

1. Remove medium.
2. Wash the culture flask with 2 mL of Trypsin/EDTA and remove it.
3. Add 3 mL of Trypsin/EDTA. Incubate at 37 °C until the cells are detached from the flask.
4. Add 3 mL of fDME. Collect the cellular suspension. Put into a 15 mL tube. Wash the flask with 2 mL of fDME, collect the cellular suspension and add it to the 50 mL tube (total 8 mL).
5. Count the cells and measure the viability by trypan blue staining. The cell viability should be higher than 95 %. Usually, a confluent culture flask contains four millions cells.

6. Centrifuge cell suspension at 300 × *g* for 10 min at room temperature.
7. Resuspend the cell pellet in a given volume of fDME.
8. Seed 5×10^5 fibroblasts in a 75 cm^2 culture flask with 15 mL of fDME. Incubate in 8 % CO_2, 100 % humidity atmosphere at 37 °C. Change culture medium three times a week.

3.1.3 Cryopreservation of Human Dermal Fibroblasts

All further manipulations are performed under a sterile laminar flow hood cabinet.

1. Resuspend fibroblasts in a given volume of cold (4 °C) freezing medium in order to obtain 2×10^6 fibroblasts/mL (*see* **Note 2**).
2. On ice, aliquot in cryogenic vials. Put them in a Nalgene freezing container filled with 99 % ethanol that has previously been cooled at −20 °C.
3. Freeze overnight at −80 °C, in the freezing container.
4. Within 24 h, transfer and store in liquid nitrogen.

3.1.4 Thawing of Human Fibroblasts

1. Put the cryogenic vial in 37 °C water until a small cluster of ice remains.
2. All further manipulations are performed under a sterile laminar flow hood cabinet.
3. Using a pipette, put 1 mL of cold (4 °C) fDME in the cryogenic vial. When the ice is thawed, put cryogenic vial content in a tube containing 9 mL of cold fDME.
4. Count the cells and measure the viability by trypan blue staining.
5. Centrifuge cell suspension at 300 × *g* for 10 min at room temperature.
6. Resuspend cell pellet in a given volume of warm (37 °C) fDME.
7. Seed 5×10^5 fibroblasts in a 75 cm^2 culture flask with 15 mL of fDME Ham's. Incubate in 8 % CO_2, 100 % humidity atmosphere at 37 °C. Change culture medium three times a week.

3.2 Monolayer Culture of Human Corneal Fibroblasts

3.2.1 Extraction and Culture of Human Corneal Fibroblasts

1. Human corneal fibroblasts were obtained from post-mortem donor corneas unsuitable for transplantation (Banque d'yeux du Centre Universitaire d'Ophtalmologie (CUO), Québec, Canada) (*see* **Note 4**).

   All further manipulations are performed under a sterile laminar flow hood cabinet.
2. Human corneal fibroblasts were isolated from the stromal portion of the cornea (*see* Subheading 3.3.1, **steps 1–12**).
3. Put the stroma in a 35 mm petri dish and cut into small pieces with a #22 scalpel blade (*see* **Note 8**).

4. With a curved forcep, collect the pieces and put stromal pieces within a trypsinization unit containing 20 mL of warm (37 °C) collagenase H (125 U/mL). Incubate under agitation, 3 h at 37 °C.
5. Collect the cellular suspension. Put into a 50 mL tube and add 10 mL of fDME. Wash the trypsinization unit with 10 mL of fDME and add it to the tube (total 40 mL).
6. Count the cells and measure the viability by trypan blue staining.
7. Centrifuge the cell suspension at 300 × *g* for 10 min at room temperature.
8. Seed 5×10^5 fibroblasts in a 75 cm^2 culture flask with 15 mL of fDME.
9. Incubate in 8 % CO_2, 100 % humidity atmosphere at 37 °C. Change culture medium three times a week.
10. When the fibroblasts reach confluence, subculture (*see* Subheading 3.2.2) or freeze (*see* Subheading 3.2.3) cells.

3.2.2 Subculture of Human Corneal Fibroblasts (Passage)

1. Follow **steps 1–8** of Subheading 3.1.2.

3.2.3 Cryopreservation of Human Fibroblasts

1. Follow **steps 1–4** of Subheading 3.1.3.

3.2.4 Thawing of Human Fibroblasts

1. Follow **steps 1–6** of Subheading 3.1.4.

3.3 Monolayer Culture of Human Corneal Epithelial Cells

3.3.1 Extraction and Culture of Human Corneal Epithelial Cells

1. Human corneal epithelial cells are obtained from human post-mortem donors corneas unsuitable for transplantation (Banque d'yeux du Centre Universitaire d'Ophtalmologie (CUO), Québec, Canada) (*see* **Note 4**)

   All further manipulations are performed under a sterile laminar flow hood cabinet.
2. Wash the eye in a 50 mL tube containing 30 mL PBS-P/G/F. Agitate gently 1–2 min. With sterile forceps, transfer the eye in another tube filled with PBS-P/G/F. Repeat this step three times.
3. Place the eye into a 100 mm petri dish.
4. Surround the eye with folded sterile gauze. This helps in holding the eye without having to touch it.
5. With the #11 scalpel blade, make a small opening of 2–3 mm.
6. With curved scissors, cut-out the cornea in order to obtain only the limbus and central cornea. Avoid leaving sclera to eliminate conjunctival epithelial cell contamination.

7. With two curved forceps, peel-off the iris. Do this step while holding the cornea in the air to avoid any damage to the epithelium during the procedure.
8. Place the limbus into a 35 mm petri dish. The limbal ring is obtained by separating the limbus from the central cornea with a 7.5 mm diameter trephine.
9. Add 5 mL of cold (4 °C) dispase II. Seal the petri dish with parafilm.
10. Incubate overnight at 4 °C.
11. With 2 curved forceps, mechanically separate the epithelium and endothelium (with Descemet's membrane) from the stroma under a dissecting microscope.
12. Keep the corneal stroma for corneal fibroblast isolation (*see* Subheading 3.2.1).
13. Put the epithelium in a 35 mm petri dish and cut into small pieces with a #22 scalpel blade (*see* **Note 8**).
14. With a plastic pipette, collect the pieces with 2 × 5 mL of warm (37 °C) complete hcDME-Ham. Put into a 15 mL tube.
15. Centrifuge at 300 × *g* for 10 min at room temperature.
16. Seed 1×10^6 epithelial cells and 1.5×10^6 i3T3 by 75 cm^2 culture flask with 20 mL of complete hcDME-Ham. Incubate in 8 % CO_2, 100 % humidity atmosphere at 37 °C. Change culture medium three times a week.
17. When the human corneal epithelial cells reach 80 % confluence, subculture (*see* Subheading 3.2.2) or freeze (*see* Subheading 3.2.3) cells.

3.3.2 Subculture of Human Corneal Epithelial Cells (Passage)

For a 75 cm^2 culture flask of human corneal epithelial cells:

All further manipulations are performed under a sterile laminar flow hood cabinet.

1. Remove medium.
2. Wash the culture flask with 2 mL of warm (37 °C) Trypsin/EDTA and remove it.
3. Add 8 mL of Trypsin/EDTA. Incubate at 37 °C until the cells are detached from the flask.
4. Add 8 mL of complete hcDME-Ham (37 °C). Collect the cell suspension. Put into a 50 mL tube. Wash the flask with 2 mL of complete hcDME-Ham, collect the cell suspension and add it to the 50 mL tube (total 18 mL).
5. Count the cells and measure the viability by trypan blue staining. The cell viability should be superior to 80 %.
6. Centrifuge cell suspension at 300 × *g* for 10 min at room temperature.

7. Resuspend the cell pellet in a given volume of complete hcDME-Ham.
8. Seed 5×10^5 to 1×10^6 keratinocytes and 1.5×10^6 i3T3 by 75 cm^2 culture flask with 20 mL of complete hcDME-Ham. Incubate in 8 % CO_2, 100 % humidity atmosphere at 37 °C. Change culture medium three times a week.

3.3.3 Cryopreservation of Human Corneal Epithelial Cells

1. Follow **steps 1–4** of Subheading 3.1.3.

3.3.4 Thawing of Human Corneal Epithelial Cells

1. Follow **steps 1–6** of Subheading 3.1.4 but seed 5×10^5 to 1×10^6 keratinocytes and 1.5×10^6 i3T3 by 75 cm^2 culture flask with 20 mL of complete hcDME-Ham.

3.4 Human Tissue-Engineered Cornea Reconstruction by the Self-Assembly Approach

All further manipulations are performed under a sterile laminar flow hood cabinet.

3.4.1 Assembly of Fibroblast Sheets for Stromal Reconstruction

1. Seed 2×10^5 corneal fibroblasts by 25 cm^2 culture flask in 5 mL of fDME containing 50 μg/mL ascorbic acid (*see* **Note 7**). Incubate in 8 % CO_2, 100 % humidity atmosphere at 37 °C. Change culture medium three times a week.
2. Seed 2×10^5 dermal fibroblasts by 25 cm^2 culture flask in 5 mL of fDME containing 50 μg/mL ascorbic acid (*see* **Note 7**). Incubate in 8 % CO_2, 100 % humidity atmosphere at 37 °C. Change culture medium three times a week.
3. After 40 days, open the top of the flask with a soldering iron.
4. Using curved forceps, detach carefully one corneal fibroblast sheet. Transfer and inverse it into a 100 mm cell culture dish (upper surface of the fibroblast sheet must be in contact with the bottom of the cell culture dish).
5. Anchor peripherally the fibroblast sheet with ingots. Move one-by-one ingots towards the tissue periphery to flatten the fibroblast sheet.
6. Detach a dermal fibroblast sheet, transfer and inverse it on the first corneal fibroblast sheet. Repeat **step 5** of this section, resulting in human tissue-engineered stroma.
7. Place a Merocel® sponge (that has been cut to fit within the ingots) on top of the superimposed fibroblast sheets. Keep Merocel® sponge in place with ingots.
8. Add 25 mL of fDME containing 50 μg/mL ascorbic acid (*see* **Note 7**). Incubate in 8 % CO_2, 100 % humidity atmosphere at 37 °C.
9. After 2 days, remove Merocel® sponge and ingots. Incubate in 8 % CO_2, 100 % humidity atmosphere at 37 °C. Change culture medium three times a week.

3.4.2 Assembly of Tissue-Engineered corneas

All further manipulations are performed under a sterile laminar flow hood cabinet.

1. One week after the stacking of fibroblast sheets, remove culture medium and seed 1.5×10^6 human corneal epithelial cells within the anchoring ring (from a suspension of 3.75×10^6 cells/mL of complete hcDME-Ham). Incubate in 8 % CO_2, 100 % humidity atmosphere at 37 °C.
2. Three hours later, add 25 mL of complete hcDME-Ham containing 50 μg/mL ascorbic acid (*see* **Note** 7). Incubate in 8 % CO_2, 100 % humidity atmosphere at 37 °C. Change culture medium three times a week.
3. Five hours later, anchoring ring can be removed.

3.4.3 Maturation of the Tissue-Engineered Cornea: Culture at the Air–Liquid Interface

All further manipulations are performed under a sterile laminar flow hood cabinet.

1. One week after the seeding of human corneal epithelial cells, remove culture medium and ingots.
2. Using curved forceps, detach carefully the reconstructed cornea from the bottom of the cell culture dish.
4. Place the reconstructed cornea on the anchoring paper centering it in order that the area with corneal epithelial cells is on top of the aperture.
5. Put air–liquid stand in a petri dish.
6. Lift tissue-engineered cornea together with the anchoring paper and transfer it on the air–liquid stand.
7. Add 20 mL of alhcDME-Ham containing 50 μg/mL ascorbic acid (*see* **Notes** 7 and **9**). Incubate in 8 % CO_2, 100 % humidity atmosphere at 37 °C. Change culture medium three times a week.

3.5 Human Tissue-Engineered Corneal Wound Healing

3.5.1 Reconstructed Cornea Wounding

All further manipulations are performed under a sterile laminar flow hood cabinet.

1. After a week of epithelial maturation at the air–liquid interface, using curved forceps, put in a petri dish the tissue-engineered cornea with its anchoring paper.
2. Wound the tissue-engineered cornea using a 6 mm diameter punch (*see* Fig. 1).
3. Carefully detach the wounded tissue-engineered cornea from the bottom of the petri dish and place it over a second human tissue-engineered stroma also placed on an anchoring paper (one dermal and one corneal fibroblast sheet that were superimposed and cultured for an additional week in fDME containing 50 μg/mL ascorbic acid (Fig. 2) (*see* **Notes** 7).
4. Put a air–liquid stand in a petri dish.

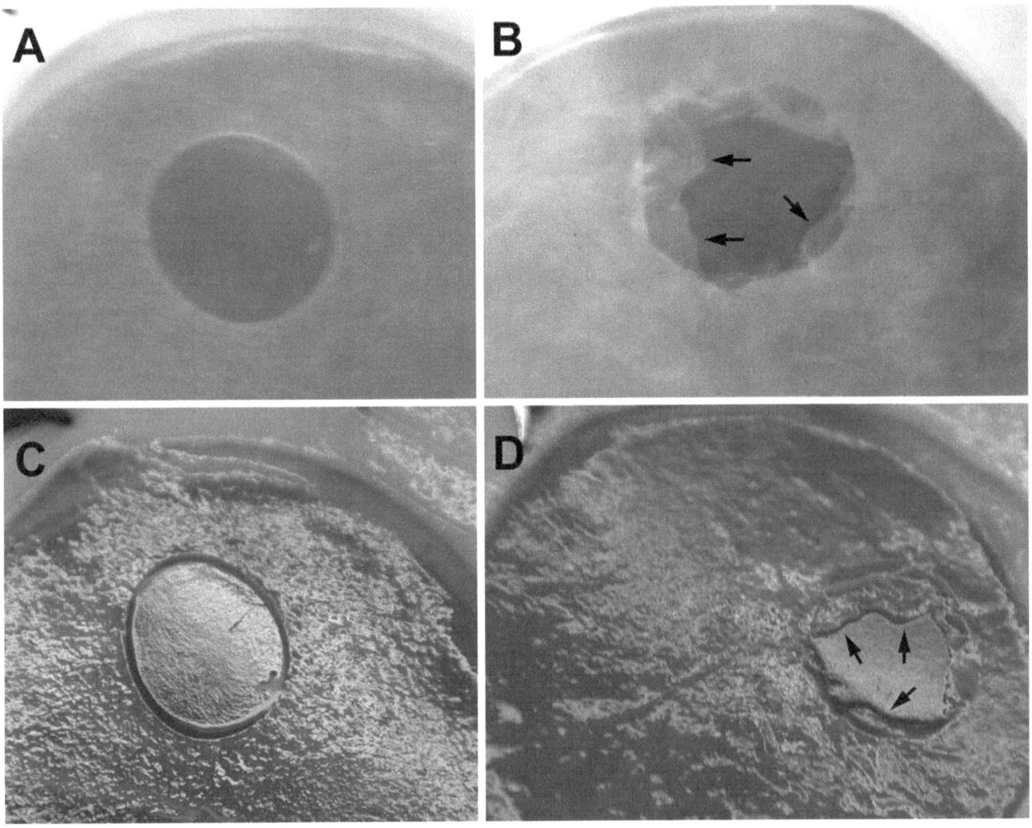

Fig. 1 Macroscopic aspect of the wound in the human tissue-engineered corneal wound healing model (hTECWH). hTECWH immediately (**a**, **c**) and 2 days (**b**, **d**) after wounding with a 6 mm punch biopsy. Two days after wounding, the reepithelialization progressing from the wound margin toward the center can be observed macroscopically (*arrows*). Note that in (**c**) and (**d**), the angle of the camera and the omission of the flash allowed us to properly visualize the extent of reepithelialization from the surrounding epithelium. Taken from Carrier et al. [18] with permission

5. Lift the wounded reconstructed cornea and the second tissue-engineered stroma together with the two anchoring papers and transfer them on an air–liquid stand.
6. Add 20 mL of alhcDME-Ham containing 50 μg/mL ascorbic acid (*see* **Notes** 7 and **9**). Incubate in 8 % CO_2, 100 % humidity atmosphere at 37 °C. Change culture medium three times a week.

Note: This human tissue-engineered corneal wound healing model can be cultivated at the air–liquid interface for 3–5 days at the end of which the reepithelialization is complete and the surface can be photographed for reepithelialization rate measurement before processing for biopsies.

The three-dimensional in vitro wound healing model described in the present study comprises a corneal epithlium on a stroma.

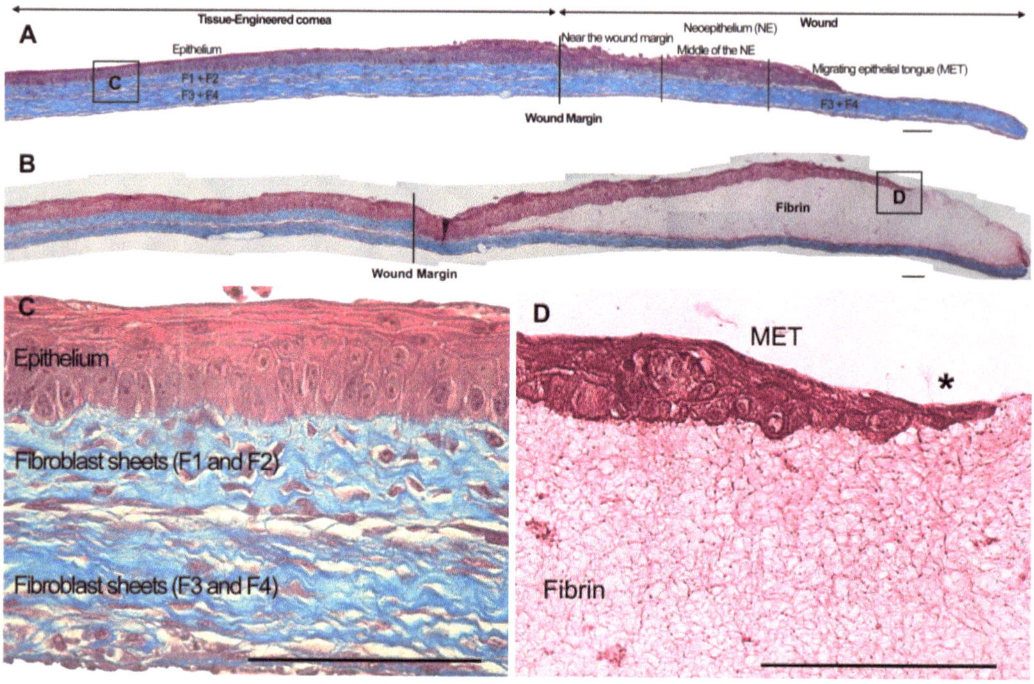

Fig. 2 Histology of the in vitro human tissue-engineered corneal wound healing model (hTECWH) 3 days after wounding and treated (**b**, **d**) or not (**a**, **c**) with fibrin. Sections were stained with Masson's trichrome (cells are *pink* and collagen *bluish*). (**a**) Composite picture showing a complete view of the hTECWH. (**b**) When a fibrin clot was added onto the wounds, reepithelialization was accelerated. (**c**) Higher magnification shows the histological organization of the unwounded side of the hTECWH. (**d**) Suprabasal epithelial cells at the tip of the migrating epithelial tongue (MET) (*) elongated over the basal cells to make contact with the fibrin matrix. Scale bar: 100 μm. Taken from Carrier et al. [18] with permission

It is devoid of any synthetic material. This, fully human, tissue-engineered corneal wound healing model allows reepithelialization by corneal epithelial cell migration over a natural matrix after wounding. This model, produced from living fibroblasts and untransformed human corneal epithelial cells, mimics many aspects of the reepithelialization process including cell migration, proliferation and the restoration of a stratified epithelium (Figs. 1 and 2). It is interesting to note the special pattern of migrating cells that produce convex fronts (Fig. 1). This particular characteristic was observed in corneal healing in vivo [20], ex vivo in human [21] and in animal models [22] as well as in vitro after wounding confluent epithelial cell cultures on plastic [23].

This completely biological wound healing model is very promising to dissect the mechanisms involved in the corneal reepithelialization process. Moreover, the positive or negative effect of an exogenous factor such as growth factors or drugs on corneal reepithelialization can be quantified. This model is also suitable for toxicological and pharmacological studies.

4 Notes

1. Serum must be added first followed by insulin. Insulin must be added with a sterile plastic pipette.
2. DMSO is an oxidative agent toxic for cells, especially at temperature above 10 °C. Thus, it must be used at 4 °C.
3. This solution must be prepared the day of its use.
4. Protocols were approved by the institution's committee for the protection of human subjects.
5. When exposed to light, Hepes buffer may undergo degradation and become toxic.
6. i3T3 should be kept 1 week in 8 % CO_2, 100 % humidity atmosphere at 37 °C. However, the cell yield may fall by approximately 10–15 % per day.
7. Ascorbic acid must be prepared immediately before use and protected from light.
8. Change scalpel blade and clean other instruments (forceps) between eye specimens.
9. The lower surface of the reconstructed cornea must be in direct contact with culture medium.

Acknowledgments

The authors would like to thank current and former members of the LOEX laboratory who contributed to develop the foregoing protocols. This work was supported by the Canadian Institutes for Health Research (CIHR) and the Vision Research Network of the Fonds de la Recherche en Santé du Québec (FRSQ). L.G. is the holder of the Canadian Research Chair from the CIHR on Stem Cells and Tissue Engineering.

References

1. Taliana L, Evans MD, Dimitrijevich SD, Steele JG (2000) Vitronectin or fibronectin is required for corneal fibroblast-seeded collagen gel contraction. Invest Ophthalmol Vis Sci 41:103–109
2. Agrawal VB, Tsai RJ (2003) Corneal epithelial wound healing. Indian J Ophthalmol 51:5–15
3. Nelson JD, Silverman V, Lima PH, Beckman G (1990) Corneal epithelial wound healing: a tissue culture assay on the effect of antibiotics. Curr Eye Res 9:277–285
4. Simmons SJ, Jumblatt MM, Neufeld AH (1987) Corneal epithelial wound closure in tissue culture: an in vitro model of ocular irritancy. Toxicol Appl Pharmacol 88:13–23
5. Boisjoly HM, Laplante C, Bernatchez SF et al (1993) Effects of EGF, IL-1 and their combination on in vitro corneal epithelial wound closure and cell chemotaxis. Exp Eye Res 57:293–300
6. Grant MB, Khaw PT, Schultz GS, Adams JL, Shimizu RW (1992) Effects of epidermal growth factor, fibroblast growth factor, and transforming growth factor-beta on corneal cell chemotaxis. Invest Ophthalmol Vis Sci 33:3292–3301

7. Maldonado BA, Furcht LT (1995) Epidermal growth factor stimulates integrin-mediated cell migration of cultured human corneal epithelial cells on fibronectin and arginine-glycine-aspartic acid peptide. Invest Ophthalmol Vis Sci 36:2120–2126
8. Kim MJ, Jun RM, Kim WK et al (2001) Optimal concentration of human epidermal growth factor (hEGF) for epithelial healing in experimental corneal alkali wounds. Curr Eye Res 22:272–279
9. Zieske JD, Hutcheon AE, Guo X, Chung EH, Joyce NC (2001) TGF-β receptor types I and II are differencially expressed during corneal epithelial wound repair. Invest Ophthalmol Vis Sci 42:1465–1471
10. Brazzell RK, Stern ME, Aquavella JV, Beuerman RW, Baird L (1991) Human recombinant epidermal growth factor in experimental corneal wound healing. Invest Ophthalmol Vis Sci 32:336–340
11. Burling K, Seguin MA, Marsh P et al (2000) Effect of topical administration of epidermal growth factor on healing of corneal epithelial defects in horses. Am J Vet Res 61:1150–1155
12. Rama P, Matuska S, Paganoni G et al (2010) Limbal stem-cell therapy and long-term corneal regeneration. N Engl J Med 363: 147–155
13. Proulx S, Uwamaliya J, Carrier P et al (2010) Reconstruction of a human cornea by the self-assembly approach of tissue engineering using the three native cell types. Mol Vis 16: 2192–2201
14. Shortt AJ, Secker GA, Notara MD et al (2007) Transplantation of ex vivo cultured limbal epithelial stem cells: a review of techniques and clinical results. Surv Ophthalmol 52:483–502
15. Gaudreault M, Carrier P, Larouche K, Leclerc S, Giasson M, Germain L et al (2003) Influence of sp1/sp3 expression on corneal epithelial cells proliferation and differentiation properties in reconstructed tissues. Invest Ophthalmol Vis Sci 44:1447–1457
16. Auger FA, Rémy-Zolghadri M, Grenier G, Germain L (2000) The self-assembly approach for organ reconstruction by tissue engineering. E-Biomed J Regen Med 1:75–86
17. Germain L, Berthod F, Moulin V, Goulet F, Auger FA (2004) Principles of living organs reconstruction by tissue engineering. In: Yaszemski MJ, Trantolo DJ, Lewandrowski K-W, Hasirci V, Altobelli DE, Wise DL (eds) Tissue engineering and novel delivery systems. Marcel Dekker, New York, pp 197–228, Chapter 10
18. Carrier P, Deschambeault A, Talbot M, Giasson CJ, Auger FA, Guérin SL, Germain L (2008) Characterization of wound reepithelialization using a new human tissue-engineered corneal wound healing model. Invest Ophthalmol Vis Sci 49:1376–1385
19. Carrier P, Deschambeault A, Audet C, Talbot M, Gauvin R, Giasson CJ, Auger FA, Guérin SL, Germain L (2009) Impact of cell source on human cornea reconstructed by tissue engineering. Invest Ophthalmol Vis Sci 50:2645–2652
20. Dua HS, Forrester JV (1987) Clinical patterns of corneal epithelial wound healing. Am J Ophthalmol 104:481–489
21. Zagon IS, Sassani JW, McLaughlin PJ (2000) Reepithelialization of the human cornea is regulated by endogenous opioids. Invest Ophthalmol Vis Sci 41:73–81
22. Petroutsos G, Guimaraes R, Giraud J, Pouliquen Y (1983) Antibiotics and corneal epithelial wound healing. Arch Ophthalmol 101:1775–1778
23. Jumblatt MM, Neufeld AH (1986) A tissue culture assay of corneal epithelial wound closure. Invest Ophthalmol Vis Sci 27:8–13

Part II

Stem Cells and Wound Repair

Chapter 5

Adult Stem Cells in Small Animal Wound Healing Models

Allison C. Nauta, Geoffrey C. Gurtner, and Michael T. Longaker

Abstract

This chapter broadly reviews the use of stem cells as a means to accelerate wound healing, focusing first on the properties of stem cells that make them attractive agents to influence repair, both alone and as vehicles for growth factor delivery. Major stem cell reservoirs are described, including adult, embryonic, and induced pluripotent cell sources, outlining the advantages and limitations of each source as wound healing agents, as well as the possible mechanisms responsible for wound healing acceleration. Finally, the chapter includes a materials and methods section that provides an in-depth description of adult tissue harvest techniques.

Key words Adult stem cells, Adipose derived stromal cells, Bone marrow derived mesenchymal cells

1 Introduction

Although many species (such as salamanders, the Australian skink, and the flatworm) are capable of complete regeneration in response to injury, most adult human organs lack this ability and generally replace lost or damaged tissue with more disorganized, weaker tissue. Adult skin is not capable of replacing itself with architecturally organized collagen. However, the process of wound repair is generally efficient, involving the interplay of complex and highly evolved biological pathways.

Evolution has favored accelerated wound closure over perfecting the architecture of regenerated tissue. Therefore, in healthy individuals, wounds tend to close quickly and with the appearance of scar. However, in unhealthy patients, the process of wound healing can be delayed, causing the development of chronic ulcers that can take months to years to heal. This delay in wound healing can result in prolonged hospital stays and periods of disability, as well as exacerbation of chronic disease. Additionally, when wounds finally heal, scar formation can be severe, resulting in functional impairment and cosmetically unfavorable outcomes. As a result of these clinical obstacles, researchers have devoted a tremendous

Robert G. Gourdie and Tereance A. Myers (eds.), *Wound Regeneration and Repair: Methods and Protocols*, Methods in Molecular Biology, vol. 1037, DOI 10.1007/978-1-62703-505-7_5, © Springer Science+Business Media New York 2013

amount of resources to developing gene therapies that either [1] accelerate healing or [2] diminish the appearance of scar.

Gene therapy, the insertion of a target gene into a recipient cell, initially emerged for the purpose of treating congenital defects [1]. However, as gene therapy has evolved, the technology has gained popularity for broader applications, such as the development of wound healing therapies. In wound repair, the goal is to transiently increase the expression of specific growth factors to improve angiogenesis, increase cell proliferation and improve cell migration until complete wound closure. The skin is a convenient model for studying the effects of gene therapy because the process of wound healing occurs over a relatively short time period, making transient over- or under-expression ideal. Additionally, changes in wound physiology are easily monitored due to the location of the skin on the surface of the body.

Stem cells are attractive candidates for gene therapy and tissue regeneration due to their ability to self-renew and differentiate into various specialized cell types. Stem cells also have the tendency to home to areas of tissue injury [2]. As a result, these cells can serve as vehicles for growth factor and cytokine delivery to the wound bed.

1.1 Gene Therapy: Stem Cell Sources

1.1.1 Embryonic Stem Cells

The fertilized oocyte produces two generations of totipotent cells, cells that are capable of forming all tissues of the embryo and trophoblasts of the placenta. Beyond these two generations, the progeny become specialized to form the blastocyst and inner cells mass (ICM), which is comprised of pluripotent cells capable of producing the three primary germ cell layers—endoderm, mesoderm, and ectoderm—but not the placenta or supporting tissues [3]. Embryonic stem cells (ESCs) are derived from the ICM and have been recognized as a promising source of cells for tissue regeneration.

Embryonic stem cells derived from mice, monkeys, and humans have been studied in culture and are theoretically capable of maintaining pluripotency indefinitely if maintained under proper culture conditions [3–5]. If culture conditions are altered, ESCs can directionally differentiate to cells of any of the three germ layers, traditionally through the induction of embryoid body formation. With the use of strictly maintained culture conditions that mimic in vivo environmental conditions, research groups have been able to show commitment to all three germ layers [6–8].

To specifically study ESC differentiation into epidermis (ectodermal lineage), research groups have found that exposure to in vivo-like conditions is essential, using bone morphogenic protein-4 signaling [9] and exposure to extracellular matrix (ECM) [10]. ESC differentiation to epidermal lineage can be directed by several cytokines and growth factors alone and in combination: Fibroblast Growth Factor 2 (FGF-2), Transforming Growth Factor Beta 1 (TGF-β1), Epidermal Growth Factor, Platelet Derived Growth

Factor (PDGF), vitamin C, and retinoic acid [11]. Keratinocytes derived from ESCs have the ability to terminally differentiate and form epidermis-like tissue, and early ectodermal progenitors expressing K17 capable of contributing to appendages or the periderm have been identified [12, 13]. Using an organotypic culture model, ESCs have been stimulated to produce a stratified epidermal-like structure that not only histologically resembles mouse dermis and epidermis but also approximates the dynamic developmental program observed in vivo [9].

Despite these advances, there are still many barriers to the use of embryonic stem cells in skin tissue regeneration. For example, researchers have been unable to define culture conditions that can reliably maintain a homogeneous population of ESCs, progenitor cells capable of becoming more than one cell type, or terminally differentiated cells in vitro. This information is essential to the development of ESC-derived cell therapy because of the risk of teratoma formation following cell injection [14]. Some researchers suggest that sequential treatment with growth factor combinations can enrich for epidermal stem cell lineages and those specific populations can be isolated using cell marker/separation techniques. Similar protocols have been used to purify ESC-derived cardiomyocytes and oligodendrocytes [15, 16].

Although advances have been made in expanding and purifying cultures of mouse ESC-derived epithelial cells and their progenitors, ethical and legal barriers to the use of human ESC-derived cells will likely delay or prevent the use of these cells in clinical trials. The use of human ESCs in clinical therapeutics may require immunosuppression to prevent rejection. Furthermore, traditional in vitro methods of maintaining undifferentiated human ESCs involve the use of mouse fibroblast feeders. This culture method has limited translational application, and more recent methods that substitute human fetal fibroblasts, adult human epithelial cells, foreskin cells, or Matrigel/laminin matrix conditioned by mouse embryonic fibroblasts (MEFs) still raise concerns regarding good manufacturing practice (GMP) conditions [17–19].

1.1.2 Induced Pluripotent Stem Cells

In response to the ethical and legal limitations of embryonic stem cell based therapy, many stem cell researchers have shifted focus to reprogramming differentiated cells to pluripotency. In 2006, Takahashi and Yamanaka published a landmark paper that defined a set of four transcription factors—octamer 3/4 (Oct4), SRY box-containing gene 2 (Sox2), Kruppel-like factor 4 (Klf4), and c-Myc—that, when used in combination, are sufficient to reprogram adult mouse fibroblasts into ES-like induced pluripotent stem cells (iPS cells) [20]. Subsequently, the same defined factors were shown to be sufficient for reprogramming human cells [21]. Further research aimed at the refinement of reprogramming mechanisms has shown that other combinations of transcription factors are possible and that the only gene that is essential to reprogramming in both adult and human cells (as in, cannot be substituted for) is Oct4 [22].

Induced pluripotent stem cell technology has been validated with other cell types and holds great promise for translational therapy. Theoretically, adult cells could be harvested, reprogrammed, and directionally differentiated for use in patient-specific therapies. Patient-specific iPS technology would provide the opportunity for cell-based therapy without the need for immunosuppressants. In cases of a known genetic defect, iPS gene therapies could restore cellular function and be used to overcome illness. Moreover, iPS cells are unique pluripotent stem cells that avoid the ethical and legal arguments that prevent ESCs from becoming frontrunner gene therapy candidates. However, most protocols developed for directed differentiation of iPS cells are based on ESC differentiation protocols, suggesting that these technologies should be investigated in parallel.

Induced pluripotent stem cells are attractive to the field of regenerative medicine, but there are still major obstacles preventing the use of these cells in humans. First generation methods of reprogramming somatic cells involve lentiviral vector transfer, which, due to the risk of integration and insertional mutagenesis, would likely prevent iPS research from reaching the level of FDA approval for clinical trials. Nonviral methods of exogenously expressing transcription factors, such as nanoparticle, protein, and minicircle vectors [23–25], have been developed to address these concerns; however, these technologies are in their infancy and are not yet efficient. Small molecules can be used to both modulate epigenetic changes and inhibit cellular differentiation pathways. Valproic acid, DNA methyltransferase inhibitors, and the histone methyltransferase inhibitor BIX-01294 have been shown to both increase the efficiency of reprogramming and serve as surrogates for one or more transcription factors [26–30].

Induced pluripotent stem cells are defined by the ability to produce teratoma with in vivo injection [31]. This characteristic is as much a flaw as it is an asset, as uncontrolled differentiation could lead to tumor development rather than the generation of specific tissue types. Researchers have attempted to address this concern by bypassing the pluripotent state through conversion of one adult cell type to another. Vierbuchen et al. demonstrated that direct conversion from one adult cell type to another was possible through exposure to high expression levels of lineage-specific transcription factors [32]. However, it is possible that a small number of rare multipotent cells are responsible for the group's findings, and lentiviral vectors were used to induce conversion from one cell type to another [32].

Finally, the generation of pluripotent cells from adult cells is inefficient, as only a small number of cells become germ-line-competent iPS cells [31]. There is also a subset of cells that become only partially reprogrammed, making these cells incapable of producing cells of all three germ layers without continuous exposure to reprogramming factors [20]. Some researchers believe that all cells are capable of being reprogrammed and that only a few complete the process, while others believe that only an "elite" group of cells possess reprogramming ability [33].

1.1.3 Adult Tissue-Derived Stem Cells

While ESCs and iPS cells are considered pluripotent, possessing the ability to produce tissue of all three germ layers, there are subsets of cells in the adult human that retain some degree of multipotency, the ability to differentiate into more than one tissue type. These cells have been identified in various niches throughout the human body and contribute to the regeneration of their respective organ tissues. These cells are referred to as adult tissue derived stem cells, and are thought to be promising sources for the development of patient-specific treatments. The utility of these sources is determined by the accessibility, yield, and ease of harvest, the ability to expand and maintain cells in culture, and the capacity of the cells to contribute to tissue regeneration after implantation.

Two promising sources of adult stem cells have been found in the bone marrow and in the stromal vascular portion of adipose tissue. Bone marrow derived mesenchymal cells (BM-MSCs) are fibroblast-like cells found in the bone marrow that are thought to provide microenvironmental support for hematopoietic cells. These cells have the ability to adhere to plastic in culture and differentiate into fat, cartilage, and bone [34, 35].

It is now believed that fibroblast-like mesenchymal cells exist throughout the body and can be found in most connective tissue and organs. Adipose derived stromal cells (ASCs) in particular have been isolated from the subcutaneous fat of both mice and humans and have been shown to possess similar characteristics to BM-MSCs. These cells reach all the aforementioned criteria and are arguably even more attractive candidates for cell-based therapy, as a large number of cells could be isolated through the minimally invasive technique of lipoaspiration [36].

1.2 Tissue Regeneration Theory: Mechanisms of Repair

1.2.1 Differentiation

Differentiation is a physiological process by which an unspecialized cell becomes more specialized through the upregulation and/or downregulation of specific genes. Differentiation capacity is defined by a cell's ability to become various tissue types; unipotent cells are only capable of maintaining one cell type, whereas multipotent cells are capable of becoming multiple cell types, and pluripotent cells can differentiate into cells of all three primary germ layers. Differentiation in vitro can either occur spontaneously or in a directed and controlled manner. Embryonic stem cells (ESCs) have the tendency to spontaneously differentiate if cultured under suboptimal and crowded conditions. After differentiation, cell sorting can select for specific cells for use in targeted cell replacement therapies. If directed and controlled differentiation of ESCs is desired, the cells are traditionally converted to embryonic bodies (EBs), then exposed to growth factors and cytokines that direct the cells to the desired cell lineage.

1.2.2 Cell Fusion

Cell fusion is an unlikely method by which stem cells contribute to tissue regeneration. There are some reports in the literature showing that implanted autologous adult stem cells are capable of fusing with cells in the native tissue, producing a tetraploid cell. However, it is unclear how these cells replicate, and the incidence of cell fusion is thought to be extremely low [37].

1.2.3 Transdifferentiation

Transdifferentiation is the conversion of one specialized cell type to another. Although many authors have attempted to demonstrate this process in vivo, few publications are able to definitively show that adult stem cells—or any partially or fully differentiated cells—are capable of contributing to the architecture of regenerative tissue. It is more likely that implanted cells survive in the healing environment for a set period of time; during this time, these cells secrete growth factors and cytokines that both augment the healing process and activate local and distant endogenous stem cell reservoirs that contribute to tissue regeneration.

1.2.4 Paracrine Effects

Many reports suggest that cell-based therapies may contribute to tissue regeneration mainly through paracrine mechanisms. In cardiac studies, transplanted cells were shown to release growth factors and cytokines that stimulate native endothelial cell proliferation, an observation that is supported by studies in other tissue models [38, 39].

1.3 Small Animal Wound Healing Models

Although in vitro studies can provide valuable information, traditional culture conditions lack the three dimensional quality and dynamic paracrine interactions that exist in the wound healing environment. Newer "organotypic" in vitro systems simulate the stratified squamous epithelial environment of native skin and have provided a more biologically relevant model for studying wound repair [45, 46].

Despite the advances in the development of "living skin substitutes," small animal wound healing models remain necessary to study the complexity of mammalian tissue repair. Specifically, the most widely used species used as a model for studying human disease is the laboratory mouse. Although there are many documented differences between the structure and physiology of mouse and human skin [47–50], wound healing models designed

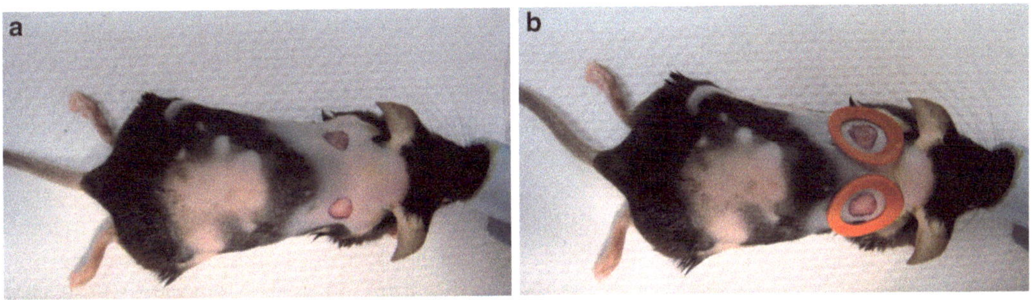

Fig. 1 Excisional Wound Healing Model. (**a**) Full thickness 6 mm skin wounds are placed at the same level on the dorsum of mice. (**b**) Rubber washers are placed around the perimeter of the wounds and are sutured in place to splint the skin and isolate reepithelialization

with these differences in mind can provide valuable translational information.

Many small animal wound healing models have been designed to study skin pathology, and Wong et al. has provided a comprehensive overview of these protocols [51]. One of the most commonly used wound healing models is the excisional wound healing model. This model can be used to evaluate cell-based therapies for wound healing and skin regeneration. Although excisional wound healing models have been employed for many years, the protocols have evolved to better approximate human physiology. While human skin heals primarily by reepithelialization, mouse skin has greater laxity; therefore, 70–80 % of mouse wound healing occurs through contraction [50]. To control for contraction and isolate epithelialization, a rubber washer can be sutured around the perimeter of the wound to splint the skin [50] (Fig. 1). Once wounds are created, cell based therapies can be applied, and wounds are then followed over the course of healing to track phenotypic wound healing changes.

1.4 Delivery of Cell-Based Gene Therapy

The skin's major function is to act as a barrier; therefore, foreign gene transfer to native cells is challenging. Rather than employing ex vivo and in vivo methods of gene transfer to endogenous fibroblasts and keratinocytes, stem cell-based gene therapy focuses on the transfer of genes to cells that are known to release growth factors and cytokines and act through paracrine mechanisms to stimulate chemotaxis and accelerate healing.

1.5 Delivery of DNA to Cells

Traditional methods of gene transfer involve the use of modified viruses. However, as previously described, new technologies have allowed for gene transfer and circumvent the risk of insertional mutagenesis. Continued advances in gene transfer technology will hopefully eliminate the need for modified virus transfection and allow for the use of gene therapies in clinical trials.

1.6 In Vivo Cell Survival

Even if safe, effective, and efficient gene delivery is achieved, environmental barriers prevent stem cell therapies from achieving maximum efficacy in vivo. The wound bed is an extremely hostile environment for implanted cells. Hypoxia, mechanical forces, inflammation, ischemia, infection, and low pH are all environmental conditions that cause apoptosis in the wound bed.

It is generally thought that strategies developed to increase stem cell survival following in vivo implantation would augment the efficacy of cell-based therapies. However, a recent publication suggests that an appropriate balance between cell survival and apoptosis may be a more appropriate goal. In a recent publication, Fang et al. describe the role of caspases 3 and 7 in the recruitment of progenitor or stem cells in wound healing and tissue regeneration. Caspases 3 and 7 are proteases that are activated in apoptosis and stimulate cell death. This group describes a "Phoenix Rising" pathway that stimulates these caspases to activate downstream Prostaglandin E2, which has been shown to promote progenitor and stem cell recruitment [52]. This finding is particularly interesting given the belief that one of the major limitations of gene therapy is in vivo cell viability. If both paracrine signals from live implanted cells and "Phoenix Rising" activation of caspases contribute to wound healing, it would be interesting to investigate an appropriate balance between cell survival and cell death.

Despite the known antioxidant properties of stem cells [53, 54], implanted cells show low survival rates in vivo. Cell based therapies have limited delivery efficiency, and failure to protect cells from the acute inflammatory environment and oxidative stress causes cell death [55]. Technologies aimed at improving in vivo stem cell delivery and survival could maximize the influence of cell-based therapies on wound healing. Increased survival could be achieved through over-expression of anti-apoptotic genes (and conversely, knockdown of apoptosis-related genes), treatment of stem cells with novel small molecules that act through anti-apoptotic mechanisms, or through improved delivery agents that mechanically protect the cells from injury.

Current delivery systems rely largely on cell injection, topical administration, and spray-based preparations. However, novel biogels, microparticles, and foams may contour better to the surface of complex wounds, obviating the need for foreign materials that can act as fomites and increase infection rates. Ideal products for stem cell delivery would be available off-the-shelf and take little preparation time. These preparations would be of particular utility in acute traumatic wounds that may require immediate inflammatory control.

The wound bed is a diverse and dynamic environment, with cells at the center of the wound bed behaving quite differently from cells at the wound margin; likewise, cells deep in the wound bed behave quite differently from cells at the surface [56].

Increasingly complex micro-electro-mechanical (MEMS) technologies applied to wounds could allow for precise detection of changes in oxygenation, pH, temperature, and ion levels with improving spatial resolution. Through microfluidic channels, this technology could allow for precise delivery of cell and growth factor combinations to specific regions of the wound bed.

Smart biomaterials may allow the recruitment of local and systemic sources of stem cells, exploiting the body's innate regenerative capacity. These scaffolds may become increasingly biomimetic, incorporating into the wound bed or dissolving when necessary [57].

Finally, composite dressings with layer-specific stem cells may allow the delivery of dermal mesenchymal-based or epidermal ectodermal-based therapies to deep and superficial wounds.

1.7 Replacement Skin Engineering

Although the gold standard for skin coverage to treat extensive skin injury is the skin graft, this procedure is imperfect, as it requires either harvest from the patient (creating an issue of supply) or the use of cadaver skin. A more ideal method of therapy would not only provide coverage from an abundantly available source, but also encourage the reconstitution of normal skin architecture. Therefore, many scientists have focused on engineering replacement skin seeded with cells that release paracrine factors. Cells can be derived locally (such as fibroblasts, keratinocytes, melanocytes, adipocytes, hair follicle cells, and skin progenitor cells) or systemically (such as cells in the bone marrow system). Studies indicate that progenitor cells produce a more favorable outcome than differentiated keratinocytes in bioengineered skin [58].

Although these advances are encouraging, currently used skin substitutes are not fully functional. These tissues lack key components of normal skin such as hair follicles, sweat and sebaceous glands. Adult hair follicle dermal cells could be another potential source of cells for creation of skin substitutes. When grafted in combination with hair follicle buds, fresh prepared dermal cells reconstitute haired skin in immunodeficient mice. This result depends on the concentration of dermal papilla cells contained in the dermal cell preparation [59]. The inherent inductive properties of these cells result in new follicle creation, while stem cell capabilities allow these cells to produce dermal replacement [60].

1.8 Methods of (Adult) Stem Cell Isolation

Although this review broadly discusses all stem cell sources that are investigated in the development of wound healing technology, a detailed discussion of the methods involved in isolating all cell types would be exhaustive and outside the scope of this chapter. Therefore, the methods described will focus mainly on adult tissue derived stem cells that are used routinely in small animal wound healing models. For a comprehensive review of iPS cell isolation, please reference the recent review by Maherali et al. [40].

2 Materials

2.1 Mouse Bone Marrow Derived Mesenchymal Stem Cells (BM-MSCs)

1. 5 Mice at 5–6 weeks of age.
2. Ice.
3. 50 mL polypropylene conical tubes.
4. 100 % alcohol.
5. Spray bottle containing 70 % alcohol.
6. Culture media containing Dulbecco's Modified Eagle Medium with GlutaMAX™ with 10 % Fetal Bovine Serum and 1 % Penicillin–Streptomycin.
7. BD Luer-Lok™ 3 mL syringes.
8. 16.5 Gauge needles.
9. 6 Well culture plate for serial Betadine washes.
10. Ice cold Phosphate Buffered Saline pH 7.4.
11. Betadine Topical Microbicide Solution.
12. 10-Blade disposable scalpels.
13. Surgical instruments, including forceps and surgical scissors.
14. Corning® 9 in. glass disposable Pasteur pipettes.
15. Fisherbrand* sterile polystyrene disposable 10 mL surgical pipettes.
16. 10 cm sterile plastic tissue culture plates.

2.2 Mouse Adipose Derived Stromal Cells (mASCs)

1. 3–5 Mice at 5–6 weeks of age.
2. Ice.
3. Cold sterile PBS 7.4.
4. Betadine topical microbicide solution.
5. 6-well Sterile culture plate.
6. 50 mL sterile polypropylene conical tubes.
7. 100 μm cell strainer.
8. Tissue culture media (DMEM + GlutaMAX™, 10 % FBS, 1 % Penicillin Streptomycin).
9. Surgical instruments including forceps and scissors.
10. Styrofoam shipping container lid.
11. Tacks.
12. 100 % alcohol.
13. Spray bottle containing 70 % alcohol.
14. Collagenase type II.
15. 1× Hanks Balanced Salt Solution (HBSS).
16. Corning® 9 in. glass disposable Pasteur pipettes.
17. Sterile Polystyrene 10 mL sterile surgical pipettes.

3 Methods

3.1 Mouse Bone Marrow Derived Mesenchymal Stem Cells (BM-MSCs)

There are many protocols in the literature describing techniques of adult tissue-derived stem cell isolation. Mouse bone marrow derived MSCs are particularly challenging to isolate and maintain in culture due to low yield and the expansion of non-MSC lineages with increasing passage number. Xu et al. recently published an updated protocol that controls for these variables, involving mechanical crushing of bones, collagenase digestion, and immunodepletion using three antibodies [41]. Although this protocol may be an improvement over traditional methods, the protocol is fairly labor intensive. It is generally accepted that BM-MSCs may be defined by their adherence to plastic [42]. Therefore, the following protocol describes the materials and methods of the traditional protocol used in our laboratory to isolate bone marrow derived MSCs.

1. Prepare 6 well plate with serial betadine washes (100 % betadine in first well, 50 % betadine/50 % PBS in second well, and 100 % PBS in remaining 4 wells). Place on ice.
2. Fill one 50 mL conical tube with cold PBS and place on ice for bone harvest.
3. Fill one 50 mL conical tube with 100 % alcohol to cleanse instruments between animals.
4. Euthanize one mouse with CO_2 asphyxiation and neck dislocation (or as indicated in APLAC-approved protocol).
5. Place mouse supine on a clean benchtop and spray with 70 % ethanol. Elevate the skin off the peritoneum using forceps and create a small midline abdominal incision with surgical scissors. Using scissors and continuing to separate the skin from the abdominal viscera, extend the excision caudally to one hindlimb, as far as the paw. Deglove the left hindlimb by separating the skin and subcutaneous tissue from the bone, muscle, and tendons. Perform this procedure with care, avoiding contamination with fur (Fig. 2a). After degloving the limb, use scissors to dislocate the limb at the hip and ankle joint. Harvest the limb, place on ice (Fig. 2b), and repeat this procedure on the contralateral hindlimb.
6. Once both limbs are isolated, for each, dislocate the tibia and fibula from the femur at the patellar joint using a 10 blade scalpel. Stabilize the tibia with forceps and use the scalpel to scrape muscle and connective tissue from the bone (Fig. 2c). Once isolated, wash the bone with serial betadine washes in the 6-well plate. Place the tibia in the 50 mL conical tube with cold PBS. Repeat this procedure with the femur.

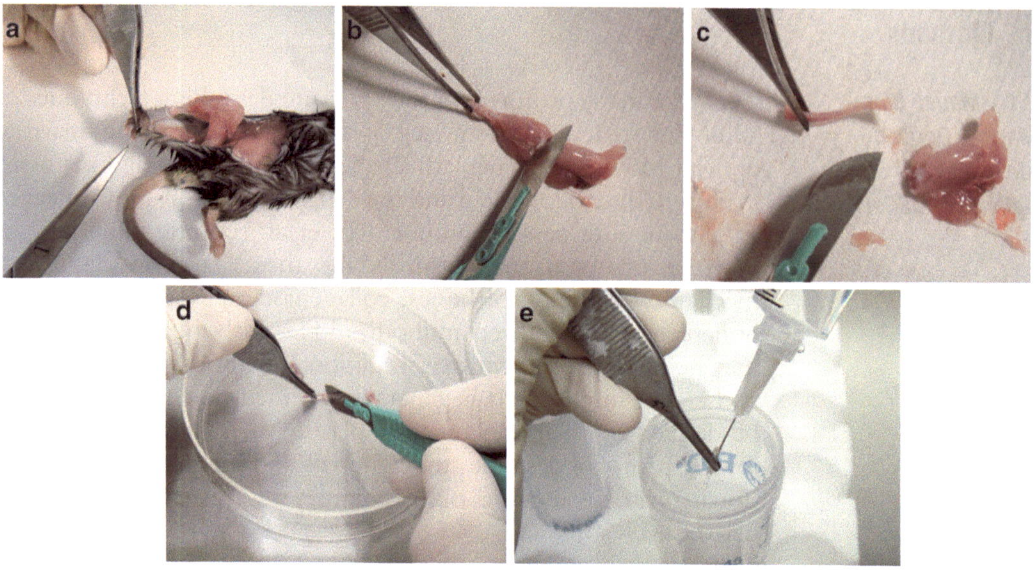

Fig. 2 Mouse Bone Marrow Derived Mesenchymal Stem Cell (BM-MSC) Harvest. (**a**) Exposed intact hindlimb with skin dissection. Once the entire limb is exposed, dislocate at the hip with a single cut of the scissors (not shown). (**b**) Harvested hindlimb. (**c**) Harvested tibia, separated from femur at patellar joint and cleaned by scraping muscle and soft tissue down to bone. (**d**) Tibia in cell culture dish. Expose bone marrow cavity by cutting either end of the bone with a 10-blade scalpel. (**e**) The bone marrow contents of the tibia are flushed into a 50 mL conical for collection. After harvest is complete, the bone marrow contents are centrifuged, and the cell pellet is collected for plating in cell culture medium

7. Repeat all aspects of this procedure with the remaining mice. Take great care to wash instruments in alcohol between animals and prevent contamination.
8. Once all bones are harvested, place 50 mL conical tube with harvested bones in the sterile tissue culture hood.
9. Empty bones and PBS into one 10 cm plastic culture plate. With clean forceps, remove one bone from the culture plate and place in another sterile 10 cm culture plate. Using a sterile 10-blade disposable scalpel, cut off the ends of the bone to expose the bone marrow (Fig. 2d).
10. Fill a sterile Laur-Lok syringe with ice cold PBS and load a 16.5 gauge needle on the syringe. Place the tip of the needle in the hollow bone marrow cavity and, holding the bone over an empty, sterile 50 mL conical tube, inject ice cold PBS into the cavity, flushing the bone marrow contents and collecting the cells in the conical tube (Fig. 2e).
11. Repeat this procedure with all harvested bones, refilling the syringe as needed.

12. Once all harvested bones are flushed, centrifuge the bone marrow contents for 5 min at 1,000 rpm and 4 °C.
13. After centrifugation, aspirate PBS from the conical tube in the tissue culture hood, taking great care to leave an intact pellet at the base.
14. Resuspend the tissue pellet in 10 mL of culture media and plate in 10 cm sterile plastic cell culture plate. Incubate for 24 h.
15. At 24 h, aspirate the media and wash the cells with warm PBS twice. Replace culture media. Repeat this procedure every 24 h until the cells reach confluence.

3.2 Mouse Adipose Derived Stromal Cells (mASCs)

There are many protocols that describe isolation techniques of mouse adipose derived stromal cells. However, most techniques share the same steps of isolation, digestion, neutralization, and centrifugation [43]. Our laboratory has found the below protocol to be both reliable and reproducible.

1. Prepare 6 well plate with serial betadine washes (100 % betadine in first well, 50 % betadine/50 % PBS in second well, and 100 % PBS in remaining 4 wells). Place on ice.
2. Fill one 50 mL conical tube with cold PBS and place on ice for fat harvest.
3. Fill one 50 mL conical tube with 100 % alcohol to cleanse instruments between animals.
4. Euthanize one mouse with CO_2 asphyxiation and neck dislocation (or as indicated in APLAC-approved protocol).
5. Place mouse supine on a clean paper towel wrapped around Styrofoam shipping container lid and spray animal with 70 % ethanol. Elevate the skin off the peritoneum using forceps and create a small midline abdominal incision with surgical scissors. Using scissors and continuing to separate the skin from the abdominal viscera, extend the excision caudally to one hindlimb, as far as the paw (Fig. 3a).
6. Stabilize lower extremity in place by placing a tack through the foot and securing into the Styrofoam surface. Extend the midline skin incision to the axilla using scissors and dissect skin away from the peritoneum. Secure the skin away from the peritoneum with a tack through the tissue and Styrofoam (Fig. 3b). Dissect and harvest groin fat from the surface of the peritoneum and the skin that wraps around the dorsum of the animal (Fig. 3c).
7. Once all fat is isolated from the groin, perform serial betadine washes and place fat in 50 mL conical tube with ice cold PBS.

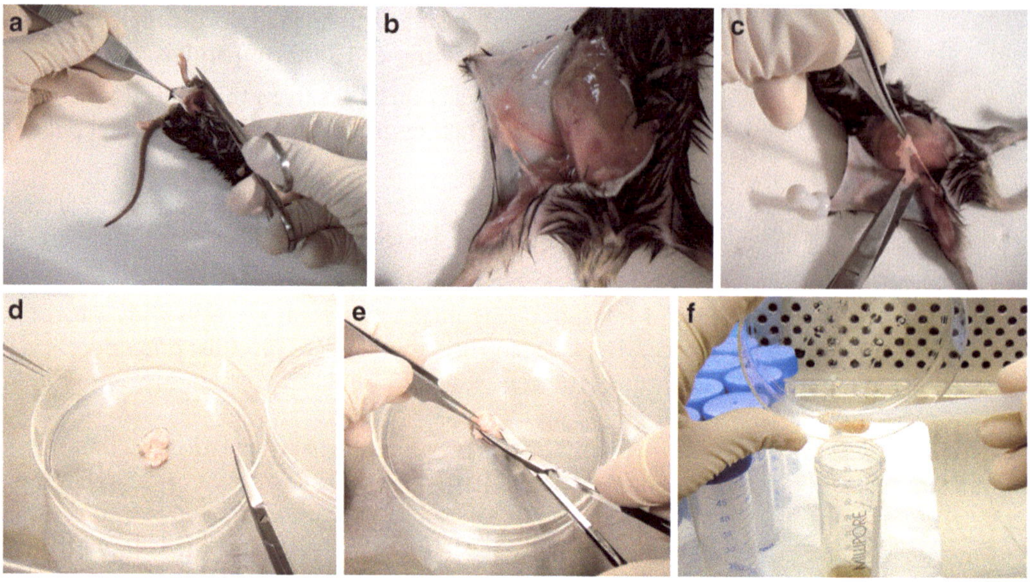

Fig. 3 Mouse Adipose Derived Stromal Cell (ASC) Harvest. (**a**) After a midline incision is made with scissors, incision is extended caudally to distal tibia. (**b**) Skin dissected away from hindlimb, exposing groin adipose tissue. (**c**) Groin adipose tissue is harvested using scissors and forceps. (**d**) Adipose tissue is collected in a tissue culture dish in the cell culture hood. (**e**) Adipose tissue is minced in the tissue culture dish using surgical scissors. (**f**) Minced adipose tissue is collected in a 50 mL conical tube containing collagenase type II solution

8. Repeat this procedure on the contralateral side of the animal and for the remaining 2–4 mice.
9. Once all adipose tissue is isolated, place the 50 mL conical containing harvested tissue in the sterile tissue culture hood (still on ice).
10. Prepare solution of collagenase type II in HBSS. Weigh 22.5 mg of collagenase and dissolve in 30 mL of HBSS in a sterile 50 mL conical tube. Filter contents through a Sterigrip cell strainer for use in cell culture. Place sterile conical containing this solution in a water bath at 37 °C to activate the collagenase.
11. Empty the contents of the 50 mL conical tube containing harvested tissue into one 10 cm sterile plastic culture dish (Fig. 3d).
12. Aspirate the PBS from the culture plate, taking great care not to aspirate the fat itself. Wash the tissue twice with ice cold PBS, aspirating the PBS between washes.
13. Mince the washed fat pads with clean surgical scissors until tissue is completely minced (Fig. 3e). Remove the 50 mL conical tube containing collagenase solution from the water bath and empty the minced adipose tissue into the collagenase solution. Use a cell scraper to collect remaining tissue from the culture plate (Fig. 3f).

14. Secure the top of the 50 mL conical containing the tissue in collagenase solution. Shake vigorously, then place in a shaking water bath set at 37 °C.
15. Allow the tissue to digest at 37 °C in the shaking water bath. Every 10 min for a total of 30 min, remove the 50 mL conical from the water bath and spray with 70 % alcohol and wipe clean with a paper towel. Remove the top of the conical to release air pressure and replace, securing the top tightly.
16. After 30 min, spray the 50 mL conical with 70 % alcohol and place in culture hood. Neutralize the digestion mixture with 30 mL of media, which will require transfer of half of the contents of the tube to another 50 mL conical.
17. Centrifuge the two 50 mL conical tubes, each containing approximately 30 mL, for 5 min at 1,000 rpm and 4 °C.
18. Aspirate the supernatant (floating lipid-like faction first, followed by underlying liquid), leaving approximately 1 mL of supernatant and the tissue pellet at the bottom of each tube.
19. Resuspend each pellet with 10 mL of culture media. Strain the contents of each conical into one sterile 50 mL conical with tissue strainer. Centrifuge the strained tissue media for 5 min at 1,000 rpm and 4 °C.
20. Aspirate culture media, leaving tissue pellet at the bottom of the 50 mL conical. Resuspend pellet in 10 mL of culture media and plate on a 10 cm sterile plastic culture plate. Incubate for 24 h at 37 °C.
21. Twenty-four hours later, aspirate media and wash cells with warm PBS twice. Replace media. Repeat this process every 24 h until cells reach confluence.

3.3 Human Adipose Derived Stromal Cells (hASCs)

Isolation protocols for human adipose derived stromal cells (hASCs) are similar to the procedure involved in mASC harvest. Briefly, adipose tissue is isolated from lipoaspirate obtained from white adipose tissue. The tissue must be washed in betadine and PBS before digested in collagenase type II to break down the extracellular matrix (ECM). After the tissue is digested, the tissue is neutralized and centrifuged, resuspended in culture media, and strained before it is resuspended and plated for cell culture. These cells are then routinely washed until reaching confluence. Although the cells plated represent many cell types at first, it is generally accepted that ASCs are adherent to plastic, and cells from hematopoietic lineages are eliminated as cells are passaged [44]. Once the cells are grown out in culture, these cells can be used in immunodeficient mouse wound healing models.

4 Conclusion

Diverse cell types such as ESCs, adult tissue-derived stem cells, resident tissue stem cells, and iPS cells have enormous potential for tissue generation, as the cells both contribute to regeneration and release growth factors to affect wound healing through paracrine mechanisms. Further investigation into the experimental and clinical applications of cell-based therapies in wound healing is necessary to identify the ideal source of stem cells and the most efficacious mode of delivery. Autologous adult tissue derived stem cells hold great promise for the development of patient-specific therapies, while other technologies will likely require many more years of dedicated research to reach the bedside.

References

1. Hernandez A, Evers BM (1999) Functional genomics: clinical effect and the evolving role of the surgeon. Arch Surg 134:1209–1215
2. Yoder MC, Mead LE, Prater D, Krier TR, Mroueh KN, Li F, Krasich R, Temm CJ, Prchal JT, Ingram DA (2007) Redefining endothelial progenitor cells via clonal analysis and hematopoietic stem/progenitor cell principals. Blood 109:1801–1809
3. Thomson JA, Itskovitz-Eldor J, Shapiro SS, Waknitz MA, Swiergiel JJ, Marshall VS, Jones JM (1998) Embryonic stem cell lines derived from human blastocysts. Science 282: 1145–1147
4. Evans MJ, Kaufman MH (1981) Establishment in culture of pluripotential cells from mouse embryos. Nature 292:154–156
5. Reubinoff BE, Pera MF, Fong CY, Trounson A, Bongso A (2000) Embryonic stem cell lines from human blastocysts: somatic differentiation in vitro. Nat Biotechnol 18:399–404
6. Fuegemann CJ, Samraj AK, Walsh S, Fleischmann BK, Jovinge S, Breitbach M (2010) Differentiation of mouse embryonic stem cells into cardiomyocytes via the hanging-drop and mass culture methods. Curr Protoc Stem Cell Biol. Chapter 1, Unit 1F.11
7. Troy TC, Turksen K (2002) ES cell differentiation into the hair follicle lineage in vitro. Methods Mol Biol 185:255–260
8. Sulzbacher S, Schroeder IS, Truong TT, Wobus AM (2009) Activin A-induced differentiation of embryonic stem cells into endoderm and pancreatic progenitors-the influence of differentiation factors and culture conditions. Stem Cell Rev 5:159–173
9. Coraux C, Hilmi C, Rouleau M, Spadafora A, Hinnrasky J, Ortonne JP, Dani C, Aberdam D (2003) Reconstituted skin from murine embryonic stem cells. Curr Biol 13:849–853
10. Turksen K, Troy TC (1998) Epidermal cell lineage. Biochem Cell Biol 76:889–898
11. Segre J (2003) Complex redundancy to build a simple epidermal permeability barrier. Curr Opin Cell Biol 15:776–782
12. Troy TC, Turksen K (2005) Commitment of embryonic stem cells to an epidermal cell fate and differentiation in vitro. Dev Dyn 232:293–300
13. McGowan KM, Coulombe PA (1998) Onset of keratin 17 expression coincides with the definition of major epithelial lineages during skin development. J Cell Biol 143:469–486
14. Yanai J, Doetchman T, Laufer N, Maslaton J, Mor-Yosef S, Safran A, Shani M, Sofer D (1995) Embryonic cultures but not embryos transplanted to the mouse's brain grow rapidly without immunosuppression. Int J Neurosci 81:21–26
15. Glaser T, Perez-Bouza A, Klein K, Brüstle O (2005) Generation of purified oligodendrocyte progenitors from embryonic stem cells. FASEB J 19:112–114
16. Klug MG, Soonpaa MH, Koh GY, Field LJ (1996) Genetically selected cardiomyocytes from differentiating embronic stem cells form stable intracardiac grafts. J Clin Invest 98: 216–224
17. Xu C, Inokuma MS, Denham J, Golds K, Kundu P, Gold JD, Carpenter MK (2001) Feeder-free growth of undifferentiated human embryonic stem cells. Nat Biotechnol 19: 971–974
18. Richards M, Fong CY, Chan WK, Wong PC, Bongso A (2002) Human feeders support prolonged undifferentiated growth of human

inner cell masses and embryonic stem cells. Nat Biotechnol 20:933–936
19. Amit M, Margulets V, Segev H, Shariki K, Laevsky I, Coleman R, Itskovitz-Eldor J (2003) Human feeder layers for human embryonic stem cells. Biol Reprod 68:2150–2156
20. Takahashi K, Yamanaka S (2006) Induction of pluripotent stem cells from mouse embryonic and adult fibroblast cultures by defined factors. Cell 126:663–676
21. Takahashi K, Tanabe K, Ohnuki M, Narita M, Ichisaka T, Tomoda K, Yamanaka S (2007) Induction of pluripotent stem cells from adult human fibroblasts by defined factors. Cell 131:861–872
22. Kiskinis E, Eggan K (2010) Progress toward the clinical application of patient-specific pluripotent stem cells. J Clin Invest 120:51–59
23. Jia F, Wilson KD, Sun N, Gupta DM, Huang M, Li Z, Panetta NJ, Chen ZY, Robbins RC, Kay MA, Longaker MT, Wu JC (2010) A nonviral minicircle vector for deriving human iPS cells. Nat Methods 7:197–199
24. Lee CH, Kim JH, Lee HJ, Jeon K, Lim H, Choi H, Lee ER, Park SH, Park JY, Hong S, Kim S, Cho SG (2011) The generation of iPS cells using non-viral magnetic nanoparticle based transfection. Biomaterials 32:6683–6691
25. Zhou H, Wu S, Joo JY, Zhu S, Han DW, Lin T, Trauger S, Bien G, Yao S, Zhu Y, Siuzdak G, Schöler HR, Duan L, Ding S (2009) Generation of induced pluripotent stem cells using recombinant proteins. Cell Stem Cell 4:381–384
26. Mikkelsen TS, Hanna J, Zhang X, Ku M, Wernig M, Schorderet P, Bernstein BE, Jaenisch R, Lander ES, Meissner A (2008) Dissecting direct reprogramming through integrative genomic analysis. Nature 454:49–55
27. Huangfu D, Maehr R, Guo W, Eijkelenboom A, Snitow M, Chen AE, Melton DA (2008) Induction of pluripotent stem cells by defined factors is greatly improved by small-molecule compounds. Nat Biotechnol 26:795–797
28. Huangfu D, Osafune K, Maehr R, Guo W, Eijkelenboom A, Chen S, Muhlestein W, Melton DA (2008) Induction of pluripotent stem cells from primary human fibroblasts with only Oct4 and Sox2. Nat Biotechnol 26:1269–1275
29. Shi Y, Do JT, Desponts C, Hahm HS, Schöler HR, Ding S (2008) A combined chemical and genetic approach for the generation of induced pluripotent stem cells. Cell Stem Cell 2:525–528
30. Lyssiotis CA, Foreman RK, Staerk J, Garcia M, Mathur D, Markoulaki S, Hanna J, Lairson LL, Charette BD, Bouchez LC, Bollong M, Kunick C, Brinker A, Cho CY, Schultz PG, Jaenisch R (2009) Reprogramming of murine fibroblasts to induced pluripotent stem cells with chemical complementation of Klf4. Proc Natl Acad Sci U S A 106:8912–8917
31. Okita K, Ichisaka T, Yamanaka S (2007) Generation of germline-competent induced pluripotent stem cells. Nature 448:313–317
32. Vierbuchen T, Ostermeier A, Pang ZP, Kokubu Y, Südhof TC, Wernig M (2010) Direct conversion of fibroblasts to functional neurons by defined factors. Nature 463:1035–1041
33. Yamanaka S (2009) Elite and stochastic models for induced pluripotent stem cell generation. Nature 460:49–52
34. Friedenstein AJ, Petrakova KV, Kurolesova AI, Frolova GP (1968) Heterotopic of bone marrow. Analysis of precursor cells for osteogenic and hematopoietic tissues. Transplantation 6:230–247
35. Dominici M, Le Blanc K, Mueller I, Slaper-Cortenbach I, Marini F, Krause D, Deans R, Keating A, Prockop D, Horwitz E (2006) Minimal criteria for defining multipotent mesenchymal stromal cells. The International Society for Cellular Therapy position statement. Cytotherapy 8:315–317
36. Zuk PA, Zhu M, Mizuno H, Huang J, Futrell JW, Katz AJ, Benhaim P, Lorenz HP, Hedrick MH (2001) Multilineage cells from human adipose tissue: implications for cell-based therapies. Tissue Eng 7:211–228
37. Dimmeler S, Zeiher AM (2009) Cell therapy of acute myocardial infarction: open questions. Cardiology 113:155–160
38. Wang Y, Hu X, Xie X, He A, Liu X, Wang JA (2011) Effects of mesenchymal stem cells on matrix metalloproteinase synthesis in cardiac fibroblasts. Exp Biol Med (Maywood) 236:1197–1204
39. Jung H, Kim HH, Lee DH, Hwang YS, Yang HC, Park JC (2011) Transforming growth factor-beta 1 in adipose derived stem cells conditioned medium is a dominant paracrine mediator determines hyaluronic acid and collagen expression profile. Cytotechnology 63:57–66
40. Maherali N, Hochedlinger K (2008) Guidelines and techniques for the generation of induced pluripotent stem cells. Cell Stem Cell 3:595–605
41. Xu S, De Becker A, Van Camp B, Vanderkerken K, Van Riet I (2010) An improved harvest and in vitro expansion protocol for murine bone marrow-derived mesenchymal stem cells. J Biomed Biotechnol 2010:105940
42. Lennon DP, Caplan AI (2006) Isolation of rat marrow-derived mesenchymal stem cells. Exp Hematol 34:1606–1607

43. Yu G, Wu X, Kilroy G, Halvorsen YD, Gimble JM, Floyd ZE (2011) Isolation of murine adipose-derived stem cells. Methods Mol Biol 702:29–36
44. Mitchell JB, McIntosh K, Zvonic S, Garrett S, Floyd ZE, Kloster A, Di Halvorsen Y, Storms RW, Goh B, Kilroy G, Wu X, Gimble JM (2006) Immunophenotype of human adipose-derived cells: temporal changes in stromal-associated and stem cell-associated markers. Stem Cells 24:376–385
45. Chioni AM, Grose R (2008) Organotypic modelling as a means of investigating epithelial-stromal interactions during tumourigenesis. Fibrogenesis Tissue Repair 1:8
46. Garlick JA (2007) Engineering skin to study human disease–tissue models for cancer biology and wound repair. Adv Biochem Eng Biotechnol 103:207–239
47. Paus R, Müller-Röver S, Van Der Veen C, Maurer M, Eichmüller S, Ling G, Hofmann U, Foitzik K, Mecklenburg L, Handjiski B (1999) A comprehensive guide for the recognition and classification of distinct stages of hair follicle morphogenesis. J Invest Dermatol 113:523–532
48. Porter RM (2003) Mouse models for human hair loss disorders. J Anat 202:125–131
49. Azzi L, El-Alfy M, Martel C, Labrie F (2005) Gender differences in mouse skin morphology and specific effects of sex steroids and dehydroepiandrosterone. J Invest Dermatol 124:22–27
50. Galiano RD, Michaels JT, Dobryansky M, Levine JP, Gurtner GC (2004) Quantitative and reproducible murine model of excisional wound healing. Wound Repair Regen 12:485–492
51. Wong VW, Sorkin M, Glotzbach JP, Longaker MT, Gurtner GC (2011) Surgical approaches to create murine models of human wound healing. J Biomed Biotechnol 2011:969618.
52. Li F, Huang Q, Chen J, Peng Y, Roop DR, Bedford JS, Li CY (2010) Apoptotic cells activate the "phoenix rising" pathway to promote wound healing and tissue regeneration. Sci Signal 3:ra13
53. Kim WS, Park BS, Kim HK, Park JS, Kim KJ, Choi JS, Chung SJ, Kim DD, Sung JH (2008) Evidence supporting antioxidant action of adipose-derived stem cells: protection of human dermal fibroblasts from oxidative stress. J Dermatol Sci 49:133–142
54. Valle-Prieto A, Conget PA (2010) Human mesenchymal stem cells efficiently manage oxidative stress. Stem Cells Dev 19:1885–1893
55. Schäfer M, Werner S (2008) Oxidative stress in normal and impaired wound repair. Pharmacol Res 58:165–171
56. Lund AW, Yener B, Stegemann JP, Plopper GE (2009) The natural and engineered 3D microenvironment as a regulatory cue during stem cell fate determination. Tissue Eng Part B Rev 15:371–380
57. Rustad KC, Wong VW, Sorkin M, Glotzbach JP, Major MR, Rajadas J, Longaker MT, Gurtner GC (2011) Enhancement of mesenchymal stem cell angiogenic capacity and stemness by a biomimetic hydrogel scaffold. Biomaterials
58. Dunnwald M, Tomanek-Chalkley A, Alexandrunas D, Fishbaugh J, Bickenbach JR (2001) Isolating a pure population of epidermal stem cells for use in tissue engineering. Exp Dermatol 10:45–54
59. Lichti U, Anders J, Yuspa SH (2008) Isolation and short-term culture of primary keratinocytes, hair follicle populations and dermal cells from newborn mice and keratinocytes from adult mice for in vitro analysis and for grafting to immunodeficient mice. Nat Protoc 3:799–810
60. Gharzi A, Reynolds AJ, Jahoda CA (2003) Plasticity of hair follicle dermal cells in wound healing and induction. Exp Dermatol 12:126–136

Chapter 6

Novel Animal Models for Tracking the Fate and Contributions of Bone Marrow Derived Cells in Diabetic Healing

Robert C. Caskey and Kenneth W. Liechty

Abstract

There is a vast wealth of information to be gained by tracking both the fate and contribution of individual cell types to the wound healing response. This is particularly important in research focused on impaired healing, such as diabetic wound healing, where the number or function of one or more specific cell types may be abnormal and contribute to the observed healing derangements. Specifically, diabetic wounds have been shown to have an overactive inflammatory response and decreased angiogenesis. The ability to track specific cell types participating in these responses would dramatically improve our understanding of the cellular derangements in diabetic healing. In this chapter, we review two novel chimeric models based on the leptin deficient Db/Db mouse. The use of these models allows for the tracking of bone marrow derived inflammatory and progenitor cell populations as well as the determination of the molecular contributions of these cell populations to the wound healing response.

Key words Diabetes, Wound healing, Chimeric, Angiogenesis, Inflammation, Progenitor cells

1 Introduction

1.1 The Diabetic Wound

Diabetes has reached pandemic proportions both in the USA and worldwide [1]. Complications of diabetes represent a significant clinical problem, with total healthcare expenditures for diabetes care in the USA alone reaching $116 billion in 2007 [2]. A significant proportion of this expense was for the treatment of chronic lower extremity wounds and their complications which are the leading cause of hospital admissions for diabetic patients [2]. This is even further highlighted by the fact that approximately 84 % of all lower extremity amputations performed in the USA are preceded by a diabetic foot ulcer [2, 3]. In 2007, diabetic wounds also accounted for an additional $58 billion in lost revenue secondary to disability or early mortality [1]. These statistics have not changed in recent years despite multiple advancements in the management

Robert G. Gourdie and Tereance A. Myers (eds.), *Wound Regeneration and Repair: Methods and Protocols*, Methods in Molecular Biology, vol. 1037, DOI 10.1007/978-1-62703-505-7_6, © Springer Science+Business Media New York 2013

of chronic wounds including biological dressings, hyperbaric oxygen, and recombinant growth factor therapy [3–5].

It has become increasingly apparent that the impairment in diabetic wound healing is a multifactorial process with derangements in multiple mechanisms necessary for healing. Two of the mechanisms that are critically affected are the processes of inflammation and neovascularization. Diabetic wounds from both humans and laboratory animals have been shown to display a dysregulated inflammatory response and decreased neovascularization when compared to the wounds of nondiabetics [6–8]. Investigations into these processes have demonstrated derangements in many of the cell types involved in both the inflammatory cascade and new blood vessel formation within the diabetic wound [9–11]. However, despite vast amounts of research into these cell populations, little is truly known regarding the fate and contribution(s) of these cells within the diabetic wound.

Chimeric animal models have been used extensively in cancer research to track cells involved in tumorigenesis and metastasis, as well as angiogenesis [12]. They have also been used in wound healing research [12–14], but their application to diabetic wound healing has been minimal. In this chapter, we will discuss two such chimeric models, the Db/Db-GFP chimera and the Db/Db-Tie-2/GFP chimera, and their potential to further our understanding of the cellular response in impaired diabetic wound healing.

1.2 Inflammation in the Chronic Diabetic Wound

Chronic wounds of every etiology, from ischemia to burns, have been found to be associated with a dysregulated inflammatory response [15]. In particular, a chronic and hyperactive inflammatory response has been described in the chronic wounds of diabetics. The etiology of this inflammation and the mechanisms by which it contributes to the diabetic wound phenotype are poorly understood [16, 18]. Two potential etiologies merit further discussion. One explanation is that diabetic skin may be primed at baseline to mount a hyperactive immune response when it is wounded. Proponents of this explanation have demonstrated that unwounded diabetic skin has a pro-inflammatory gene expression profile at baseline when compared to nondiabetic skin [16, 17]. The mechanisms responsible for this phenotype are unknown; however, one recent study has provided some fresh insight. Grice et al. demonstrated that the leptin deficient Db/Db mouse has 40 times the bacterial load on its skin at baseline when compared to the skin of heterozygote mice [16]. The majority of this bacterial load was composed of *Staphylococcus* spp. More importantly, microarray analysis also demonstrated that the increased bacterial load was significantly correlated with an increase in the baseline expression of genes related to the gene ontology (GO) categories "immune response" (IR), "defense response" (DR), and "response to wounding" (WR) in unwounded diabetic skin [16]. While these studies have not been repeated in human diabetic skin, unwounded human diabetic skin has repeatedly been shown to have a higher and more *Staphylococcus spp.* predominant bacterial load when compared to that of nondiabetics [19]. This may, in part, explain why diabetic skin displays more inflammation at baseline. However, the relationship between this elevated baseline inflammation and the dysregulated inflammation observed in the chronic diabetic wound has yet to be determined.

The second potential etiology of increased inflammation in the chronic diabetic wound is vastly different from the first. Pradhan et al. have hypothesized that chronic baseline inflammation in diabetic skin prevents the initiation of an appropriate inflammatory response during the acute phase of wound healing. This results in a wound with increased susceptibility to both infection and hypoxia [20]. Over time, as bacterial load and reactive oxygen species build up in the wound environment, a threshold is reached, resulting in a delayed but hyperactive inflammatory response [21]. Again, a causal relationship between this inflammatory response and the inflammation present in the chronic diabetic wound has yet to be established.

These two potential etiologies share a final common pathway or pathophysiology, with dysregulation of inflammation in the diabetic wound leading to even further derangements in wound healing. Often, these derangements are enough to derail the wound from progressing through the normal sequence of inflammation, proliferation, remodeling and maturation, resulting in a chronic wound. What occurs during the critical period between acute and chronic wound is poorly understood and likely involves many interrelated mechanisms. Relevant to our discussion here is the concept that high levels of inflammation have been shown to prevent the migration and function of multiple cell populations necessary for normal healing [10, 11, 18]. In order to better delineate how inflammation and cellular dysfunction interact with each other in the diabetic wound, new and better diabetic animal models are needed which allow investigators to both track the inflammatory cells and determine their molecular contributions. To highlight the potential uses of the Db/Db-GFP chimeric mouse, we will briefly discuss one inflammatory cell type important to the healing process: the macrophage.

Due to their central and regulatory role in the normal wound, macrophages have been the target of multiple studies investigating the pathogenesis of disordered inflammation within the diabetic wound [9, 10, 22]. From these studies we know that diabetes affects the cellular functions of macrophages in multiple ways. First and foremost, diabetes and/or hyperglycemia have repeatedly been shown to directly inhibit the macrophage's ability to perform phagocytosis [9, 23]. This most basic function of macrophages is of vital importance to the wound repair process. Increasing evidence demonstrates that the clearance of not only bacteria, but also of neutrophils and other apoptotic cells from the early wound is essential for progression out of the inflammatory stage of healing [24, 25]. However, the macrophage potentially serves many other complex functions within the healing wound. These include the production of growth factors and the regulation of proliferation.

In healing wounds, macrophages have been demonstrated to orchestrate both inflammation and repair [25, 26]. They may accomplish this through existing in three or more different phenotypic states throughout the wound healing process [27]. Recently, the potential functions of these phenotypes have been described in a study by Lucas et al. [26]. In this study, conditional depletion of macrophages was performed during the different phases of healing and the wounds followed. Depletion of macrophages during the inflammatory stage of healing resulted in reduced granulation tissue formation and impaired epithelialization within the wound. Depletion during the proliferative stage, however, resulted in a hemorrhagic wound which failed to close [26]. In contrast, macrophage depletion during the maturation phase of healing had little effect on wound closure [26]. These and other studies are beginning to demonstrate that the transition between the different macrophage phenotypes is necessary for normal healing to occur [25]. Due to its central role within the healing wound, alterations in macrophage function may play an important role in the diabetic wound healing impairment. However, many studies, including the one mentioned above, have yet to be performed in a diabetic animal model. In addition, the macrophage is only one example of the many inflammatory cells which migrate into the healing wound, with other cells, including neutrophils and lymphocytes, playing important roles as well.

1.3 Vasculogenesis in the Chronic Diabetic Wound

Inflammatory cells are only one set of bone marrow derived cells to arrive at the healing wound. Since their discovery over 20 years ago, there has been increasing interest in the bone marrow derived endothelial progenitor cell (EPC) and its function in multiple physiological processes, including wound healing. It has long been known that diabetes is associated with pathologic angiogenesis. This ranges from the hyper-proliferative response seen in diabetic retinopathy, to the diminished neovascularization seen in diabetic wounds [28]. Neovascularization encompasses the two separate but related processes of angiogenesis and vasculogenesis. In angiogenesis, new blood vessels branch from preexisting vessels via the migration and proliferation of mature resident endothelial cells. In vasculogenesis, the new blood vessels are instead formed by collections of bone marrow derived EPCs. The formation of granulation tissue is largely dependent on vasculogenesis, and has been shown to be critical for healing full thickness, excisional wounds [13, 29, 30]. Of the two neovasculogenesis processes, it is vasculogenesis which has been shown to be particularly deficient in diabetics. Many of these defects within vasculogenesis center on the diabetic endothelial progenitor cell population [11, 28, 31, 32].

Many studies have demonstrated the full spectrum of defects within the EPCs from diabetics. Diabetics have repeatedly been shown to have a decreased number of circulating EPCs, when compared to nondiabetics [11, 32]. Diabetics have also been shown to have significantly decreased levels of stromal cell derived factor-1α (SDF-1α), a key signaling molecule involved in the recruitment of EPCs into the wound [32–34]. Combined, these two deficits result in a significantly decreased number of EPCs being present in the healing diabetic wound [11]. The few EPCs that actually arrive to the diabetic wound have further been found to be deficient in multiple necessary enzymes [32, 35]. One of these enzymes, manganese superoxide dismutase, allows the EPC to resist hypoxia by reducing reactive oxygen species. This is a critical function for cells which are responsible for establishing the new vascular network which will ultimately bring oxygen to the wound [35]. These defects in diabetic EPC number and function demonstrate why diabetic wounds are particularly prone to hypoxia [30]. This hypoxia can lead to the rapid build-up of reactive oxygen species within a wound which is already suffering from high levels of dysregulated inflammation [21]. A further cascade of inflammation and inflammation-induced cellular dysfunction can follow and vastly increases the wound's chance of transforming into a chronic wound.

Functional EPCs, while unlikely to correct the entire spectrum of diabetic healing defects, may prevent the acute diabetic wound from transforming into a chronic wound. Multiple studies have reported improved healing in diabetic animal models by correcting one or more of the EPC defects discussed above [32, 35, 36]. In a study by Gallagher et al. both EPC mobilization from the bone marrow and homing to the wound were positively manipulated with a synergistic improvement in wound healing. Likewise, Marrotte et al. improved diabetic healing by treating diabetic wounds with diabetic EPCs that had been transfected with an adenovirus overexpressing manganese superoxide dismutase.

Many of the studies discussed in this section have made use of genetically modified nondiabetic mice which specifically allow for the tracking of EPCs. These mice have greatly expanded our knowledge of general endothelial progenitor cell biology over the past 10 years; however, our understanding of the defects and functions of these cells within diabetic skin and wounds is limited. In order to further this understanding, animal models need to be developed that allow for the tracking of EPCs within a diabetic environment.

1.4 Chimeric Models: Powerful Tools for Tracking Individual Cell Populations

The multiple varieties of transgenic mice available have significantly advanced our understanding of cell biology and provided valuable tools with which to study various diseases. The creation of chimeric animals allows the investigator to study specific cell populations

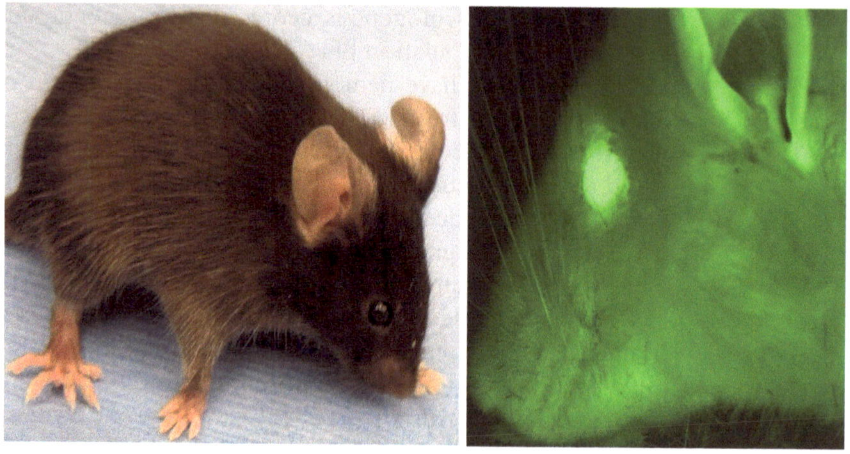

Fig. 1 The GFP mouse. Every cell from these animals with the exception of red blood cells fluoresces green

within an animal model of a specific disease process. This is no better exemplified than by the history of the GFP mouse (Fig. 1). The GFP mouse expresses green fluorescent protein in every cell with the exception of red blood cells. Hayakawa et al. utilized bone marrow from this mouse, transplanted into C57BL/6 mice, to make the first chimeric mouse made by bone marrow transplantation [14]. This model allowed investigators, for the first time, to track the fate of bone marrow derived cells in a wound healing study. Based on these initial studies, the investigators concluded that, as opposed to bone marrow cells injected at the time of wounding, the transplanted bone marrow cells preserve both their functionality and natural behavior [14]. This was an important distinction to make and, as such, has served as a guiding principle for the multitude of chimeric studies which have followed.

Since the Hayakawa et al. study, many subsequent studies have been performed using GFP chimeras to investigate multiple disease processes from wound healing to nephropathy [32, 37–39]. Of these, the report by Möller et al. most truly demonstrates the power of chimeric models in tracking individual cell populations. In this study, the investigators used GFP chimeric mice to track macrophage turnover within peripheral nerves and dorsal root ganglions. They did this over a 36 week period and discovered up to a 75 % turnover rate within these nerves and ganglions [38].

Despite the fact that GFP chimeras allow for tracking of all bone marrow derived cells, most of the wound healing studies which have utilized GFP chimeras to date have focused on the EPC population [32, 36, 37]. However, there are transgenic mice which allow for much more specific tracking of this population, the most common of which are the Tie-2/LacZ and Tie-2/GFP mice (Fig. 2). These mice express either ß-galactosidase (LacZ) or green fluorescent protein (GFP) under the direct

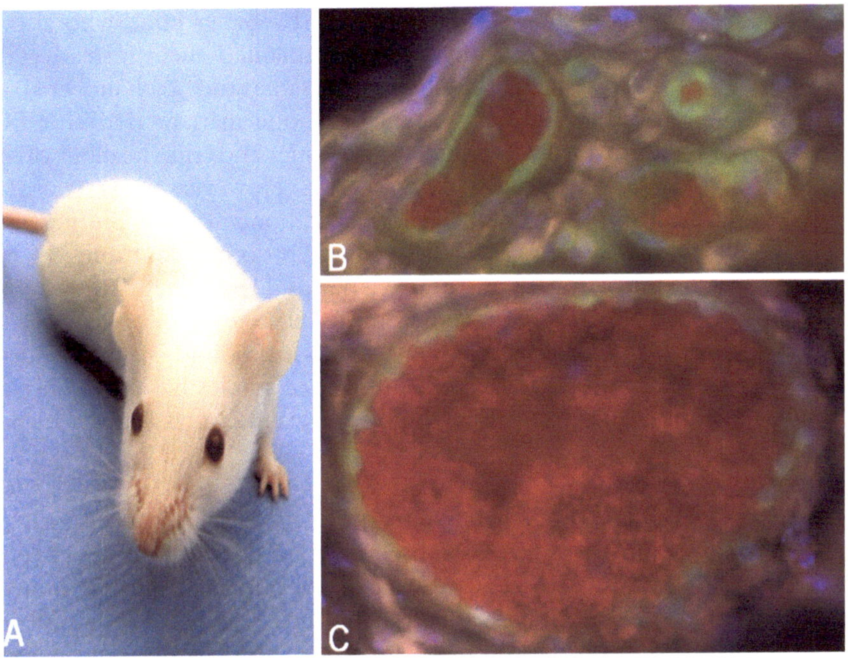

Fig. 2 (**a**) The Tie-2/GFP mouse. (**b** and **c**) Merged fluorescence microscopy images (merged DAPI, GFP, and red channel) taken from a frozen section of a 7-day-old wound on a Tie-2/GFP mouse. GFP positive endothelial cells are appreciated lining many blood vessels of varying caliber (40×)

control of the tyrosine kinase with immunoglobulin and epidermal growth factor homology-2 (Tie-2) promoter. The Tie-2 gene is only expressed in functionally active endothelial cells and is the only known receptor that can interact with all four of the identified angiopoietins (Ang-1-Ang-4) [30]. These mice were originally developed to investigate the embryogenesis of the vascular system and, due to this design, every endothelial cell of arterial, venous and even lymphatic origin will be positive for LacZ or GFP [40]. While this is a useful property for studying the embryogenesis of the vascular system, it prevents the specific tracking of bone marrow derived endothelial cells to sites of injury in these animals.

Chimeric mice made using bone marrow derived from either Tie-2/GFP or Tie-2/LacZ mice have been used extensively in tumor angiogenesis research and to a lesser extent in wound healing research [12, 13]. Using immunodeficient mice transplanted with Tie-2/LacZ bone marrow, Asahara et al. demonstrated that bone marrow derived EPCs contribute to both physiological neovascularization and the neovascularization related to tumor growth [12]. In another study which used FVB/NJ mice transplanted with Tie-2/GFP bone marrow, O'Niell et al. reported that bone marrow derived cells did not contribute to new endothelial cells within the capillary beds of muscle exposed to hypoxia [37].

This finding was later refuted by Bauer et al. who also used the same FVB/NJ mice but transplanted them instead with Tie-2/LacZ bone marrow [29]. This same study by Bauer et al. also demonstrated the importance of bone marrow derived EPCs in the formation of granulation tissue and the healing of excisional wounds in normal mice [29]. However, using C57BL/6 mice transplanted with Tie-2/Lac-Z bone marrow, Bluff et al. reported that bone marrow derived EPCs do not contribute to new vessel formation in incisional wounds [13]. These studies and others have greatly furthered our understanding of bone marrow derived EPCs and the role they play during wound healing in normal mice; however, only a small number of studies have used chimeric models to investigate pathological healing, such as that which occurs in diabetics.

1.5 Diabetic Chimeric Models Based on the Db/Db Background

Chimeric diabetic animal models have been previously described and a few have been used in wound healing studies [11, 32]. Most of these studies, because of their importance to the subject matter, have been previously mentioned in this chapter. Albiero et al. transplanted GFP bone marrow into C57BL/6 mice and induced diabetes with streptozotocin for their wounding studies [11]. The novel wound healing study by Gallagher et al. also used GFP chimeric mice but on both FVB and C57BL/6 backgrounds and then induced diabetes in these mice by giving them streptozotocin [32]. Also in the same study, Gallagher et al. pharmacologically induced diabetes in Tie-2/GFP mice, but only used these mice to study EPC populations within the peripheral blood [32]. This was appropriate given that one would be unable to differentiate bone marrow derived endothelial cells from resident endothelial cells in these mice. These studies have certainly contributed to our understanding of the role of bone marrow derived EPCs in healing in an acute model of Type I diabetes. Chimeric animal models that better approximate the pathophysiology of type II diabetes and its complications are lacking. The development of such animals will allow investigators to examine the tracking and fate of all bone marrow derived cells, from endothelial progenitor cells to those involved in the inflammatory response, in a more clinically relevant model of diabetes.

For our chimeric models we chose the leptin deficient Db/Db mice as our background animal. This differs greatly from the previous studies mentioned above which used streptozotocin to induce diabetes. Although it is an accurate animal model of hypoinsulinemia and insulin dependent (Type 1) diabetes mellitus, streptozotocin induction has some shortfalls which limit its application in studying diabetic wound healing [41, 42]. This method of pharmacologically inducing diabetes was originally described in rats nearly 50 years ago, and has historically been used to study the beta islet cells of the pancreas [43]. Yet, despite this, many investigators

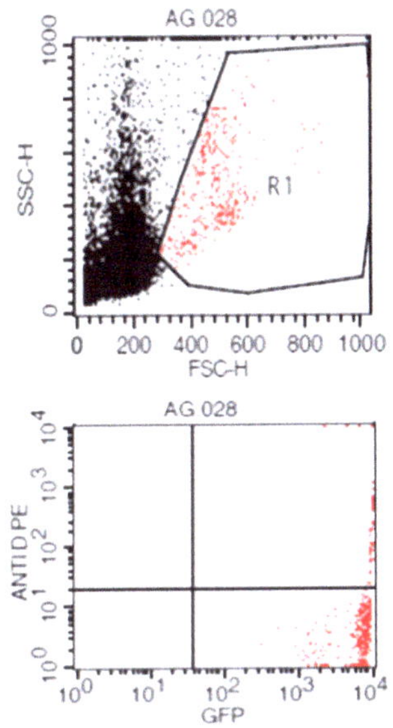

X Parameter GFP (Log) Y Parameter ANTI D PE (Log)
Quad Location 37, 20

Quad	Events	% Gated	X Geo Mean	Y Geo Mean
UL	0	0.00	***	***
UR	37	10.60	8892.17	232.46
LL	0	0.00	***	***
LR	312	89.40	6034.09	2.98

Fig. 3 FACS analysis of blood from a Db/Db-GFP chimera 4 weeks following bone marrow transplant. *X* axis: GFP positive. *Y* axis: Anti d antibody tagged with PE. Gating on the leukocyte population (R1) demonstrated greater than 80 % GFP positivity

have translated this methodology into mice for use in wounding studies. In contrast, the genetically leptin deficient Db/Db mouse is an accurate model of non-insulin dependent (Type 2) diabetes. As a result of their leptin deficiency, these mice are highly polyphagic and usually develop a diabetic phenotype represented by hyperglycemia, insulin resistance and obesity, by 6 weeks of life or earlier. When used in wound healing studies, our lab and others have consistently demonstrated the superiority and reliability of the Db/Db mouse. This is especially true in the studies evaluating excisional wound healing [34, 41, 42, 44].

In addition to using streptozotocin to induce diabetes, previous diabetic chimera studies examining wound healing have focused the EPC population only. While we agree that these cells are of significant importance to the diabetic healing phenotype, we feel there has been neglect of other cells populations, such as the monocytes/macrophages, which are also known to be important to the healing process. Experiments such as those performed by Lucas et al. furthered the case for the importance of macrophages in wound healing, yet reciprocal studies in a diabetic animal model have not been conducted to date [26]. Performing such studies using the Db/Db—GFP chimeric mouse described here (Fig. 3) would allow investigators to differentiate between bone marrow derived and resident tissue monocytes. The Db/Db-GFP chimera may even allow investigators to determine the cellular source of inflammation within the chronic diabetic wound. In their paper on the source of inflammation in the diabetic wound, Pradhan et al.

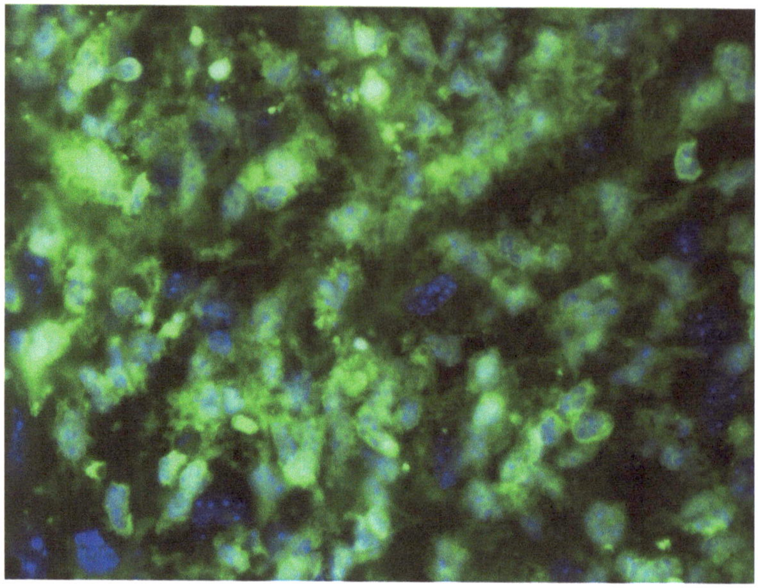

Fig. 4 Merged DAPI and GFP images from Fluorescence microscopy of the wound bed of a Db/Db-GFP chimera 7 days following creation of an 8 mm wound. Scattered mononuclear cells and plentiful polymorphonuclear (PMNs) cells are easily appreciated with the wound base (40×)

stated that immunotracking studies of inflammatory cells were lacking [20]. The use of the Db/Db-GFP chimeric model would allow investigators to accomplish this. The Db/Db-GFP chimera also allows for the identification of all the basic inflammatory cell types (macrophage, neutrophil, lymphocyte), which are GFP positive, as they migrate into the wound. Histology or immunohistochemical techniques can then be used to identify the inflammatory cell type (Fig. 4). The combination of these techniques with laser capture microscopy and subsequent real-time PCR, would then allow for the determination of the molecular contribution of these cells to the wound healing process.

Previous studies tracking bone marrow derived EPCs in diabetic chimeric mice have either utilized donor bone marrow from GFP mice followed by streptozotocin induction or have similarly induced diabetes in transgenic Tie-2/GFP mice. To our knowledge, such tracking studies have yet to be evaluated by transplantation of either GFP or Tie-2/GFP bone marrow into an established murine model of type II diabetes, such as the Db/Db mouse. The lack of development of such animals may, in part, be related to concerns of high mortality rates following irradiation of the Db/Db mice. In addition, the Tie-2/GFP mouse is on a FVB background whereas the Db/Db mouse is on a C57BL/6 background, and has raised concerns of rejection of the bone marrow transplant [37]. Despite these concerns, we have successfully created

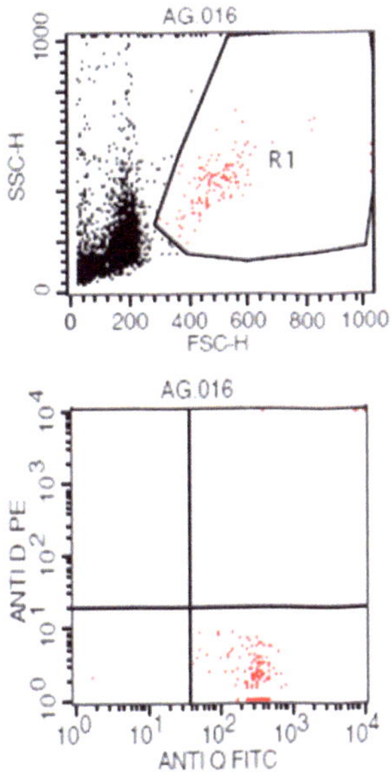

X Parameter: ANTI Q FITC (Log) Y Parameter: ANTI D PE (Log)

Quad Location: 37, 20

Quad	Events	% Gated	X Geo Mean	Y Geo Mean
UL	1	0.18	29.96	153.99
UR	44	7.76	4728.42	4278.62
LL	4	0.71	6.06	4.22
LR	518	91.36	268.41	2.61

Fig. 5 FACS analysis of blood sample from a Db/Db-Tie-2/GFP chimera 4 weeks after bone marrow transplant. Cells were double stained with Anti q (FITC) antibody (*x* axis) and Anti d (PE) antibody (*y* axis). Gating on the leukocyte population (R1) demonstrated greater than 90 % chimerism of H2K^q positive cells

Db/Db-Tie-2/GFP chimeras without any pharmacological immunosuppression (Fig. 5). When these mice are wounded, GFP positive endothelial cells which also express CD 31 can be found in the granulation tissue bed of the wounds (Fig. 6). With the superiority of Db/Db mice in the study of excisional wounding and the importance of EPCs in the healing of such wounds, this chimera allows investigators to better characterize the role these cells play in the diabetic healing phenotype. As with the GFP chimeras, immunohistochemistry combined with laser capture microscopy and real-time PCR allows for the determination of the molecular contribution and function of these cells within the diabetic wound.

To our knowledge this is the first description of chimeric mice being made by bone marrow transplantation into a Db/Db background. The development of chimeric models in the Db/Db mouse, combined with the large number of transgenic mice available today for transplantation, can allow for the examination of a number of different cells types and factors in the diabetic environment, despite concerns about rejection between strains. Ultimately these models will help to better define the individual and interactive roles these cells play in the diabetic wound healing impairment and present investigators with needed therapeutic targets for the improvement and prevention of the diabetic wound healing impairment.

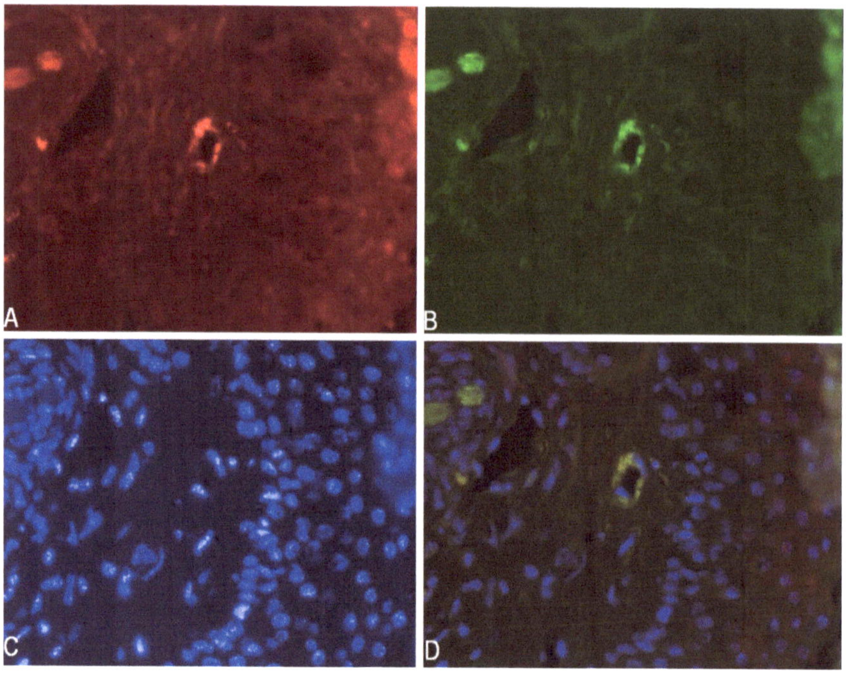

Fig. 6 Immunohistochemistry and fluorescence microscopy of the wound bed of a Db/Db-Tie-2/GFP chimera 7 days following the creation of an 8 mm wound (40×). Staining for CD 31 (**a**), GFP (**b**), and DAPI (**c**) are shown. The merged image (**d**) demonstrates specific endothelial cell GFP and CD31 staining

2 Materials

2.1 Db/Db-GFP or Db/Db-Tie-2/GFP Chimeric Mice

1. *Recipient Mice*:
   (a) Db/Db mice (Jackson Lab BKS.Cg-Dock7^{m} +/+ Lepr^{db}/J).
   (b) Db/+ Breeder mice (Jackson Lab Dock7^{m} +/+ Lepr$^{db)}$.
2. *Donor Mice*:
   (a) GFP mice (Jackson Lab CByJ.B6-Tg (UBC-GFP)30Scha/J) or
   (b) Tie-2(Tek)-GFP mice (Jackson Lab Tg(TIE2GFP)287Sato/J).
3. Phosphate Buffered Saline (PBS).
4. Sterile instruments (Scissors, forceps).
5. 26¾ Gauge needles.
6. 1 cc syringes.
7. 3 cc syringes.
8. 70 % Ethanol for surgical prep.

9. Sulfatrim (Bactrim) Solution (200 mg Sulfamethoxazole/40 mg Trimethoprim).
10. Ficoll Paque.
11. Irradiator.

3 Methods

3.1 Db/Db-GFP or Db/Db-Tie-2/GFP Chimeric Mice

1. On the day prior to irradiation move all recipient mice into autoclaved cages and change their water out for sulfatrim water at a dose of 5 mL of 200/40 solution per 200 mL of distilled water. Agitate the bottles frequently to ensure the antibiotic is solubilized in the water.
2. The next morning irradiate all recipient mice with half of total radiation dose (we use 750 cGy per dose for a total of 1,500 cGy for Db/Db mice and 800 cGy per dose for a total of 1,600 cGy for Db/+ mice-(*see* **Note 1**).
3. Repeat this same dose of radiation approximately 6–8 h after the first dose.
4. The following morning, euthanize donor mice (one at a time).
5. After euthanasia, thoroughly spray the hind limbs, back, and perineum of the donor mouse with ethanol (one donor mouse will give enough cells for two recipient mouse).
6. Using the sterile instruments, harvest the bilateral tibias and femurs of each mice being careful not to break the bones. Clean all muscle and other tissue from the bones as thoroughly as possible.
7. Place all harvested bones in cold, sterile PBS in a 50 mL conical tube.
8. After all bones have been harvested, wash them within the 50 mL conical tube using sterile PBS for a total of five, 50 mL washes. Using sterile technique suction away the PBS each time.
9. Place the bones on a large sterile cell culture plate and cut the ends of the tibias and femurs using sterile scissors (one at a time).
10. Partially fill another sterile cell culture plate with sterile PBS and then, using the 3 cc syringe and 26¾ gauge needle, flush the bones (one at a time) with this PBS into the cell culture plate. Flush all of bones until clear then discard.
11. After you have flushed all the bones, aspirate and flush the PBS -bone marrow cell mixture multiple times into the same cell culture plate using the same needle and syringe to remove any cell clumps.

12. Filter all cells through a 70 μm filter into a new sterile 50 mL conical tube.
13. Place 3 mL of Ficoll-Paque into a sterile 15 mL conical tube.
14. Gently layer the filtered cells from number 12 on top of the Ficoll.
15. Using a centrifuge with the brake removed, spin number 14 at $400 \times g$ for 8 min at 4 °C.
16. Using a 1 cc pipette, aspirate the buffy coat from number 15 and transfer to new sterile 50 mL conical tube.
17. Fill number 16 to the top with sterile PBS and re-spin at $400 \times g$ for 10 min (put brake back on centrifuge).
18. Using sterile technique, aspirate the supernatant from 17 and resuspend the cells in approximately 2 mL of sterile PBS (use more or less depending on intended number of recipient mice).
19. Count the cells using a hemocytometer and then dilute using sterile PBS so that the concentration is between 5 and 10 million cells per 200 μL (this is the volume given to each recipient mouse).
20. Once donor cell preparation is complete, warm each recipient mouse for 5 min under a heat lamp.
21. Immediately before injection, aspirate 200 μL of the cells into the 1 cc syringe ensuring there are no clumps of cells.
22. Cleanse the recipient mouse tail with an isopropyl wipe. Then give the 200 μL of bone marrow cells to the mouse via tail vein injection. Use new needle and syringe for each recipient mouse.
23. Following cell injection, immediately return each mouse to a clean, autoclaved cage.
24. Recipient mice should receive Bactrim solution for a total of 2 weeks. Agitate the solution multiple times per day to ensure they are getting an adequate amount. Replace with new solution as needed (*see* Subheading 4).
25. Four weeks following transplant each recipient mouse should undergo retro-orbital bleed for FACS analysis. Bleed approximately 200 μL into heparinized PBS at a concentration of 1 unit of heparin per mL of PBS. A 20–30 % mortality among recipient mice is to be expected (*see* **Note 2**).

4 Notes

1. Notes on the General Care of Db/Db and Chimeric Mice. Db/Db are immunosuppressed at baseline. They also stool frequently and some do so even more after being irradiated. Therefore frequent changing of their cages is important,

especially after they have been irradiated and transplanted. Their immunosuppression may also affect their necessary radiation dose. It has been our experience that this dose differs by irradiator type and geographical location. Therefore, investigators may find it useful to create a dose response curve in both the diabetic and heterozygote mice before attempting bone marrow transplantation. The dose given is meant to be a lethal dose and the correct dose will be one which kills a large majority of mice by day 7–10 post irradiation. Earlier deaths are due to too strong a dose while later deaths are likely the result of a sub-lethal dose which only temporally made the mouse more susceptible to infection.

Once any attempted transplantations have been performed we recommend changing the cage of the recipient mice at least every other day. Change their cages out for new autoclaved cages spraying your gloved hands with ethanol prior to changing. Also, frequent monitoring of their Sulfatrim water solution cannot be stressed enough. The solution should be agitated multiple times a day to make sure the mice are getting an adequate dose. We also recommend that the investigators use distilled water for mixing the antibiotic solution. This is because water bottles made with tap water by animal care facility staff may sit exposed to open air for an unknown amount of time. This allows for, among other things, fungus spores to contaminate the water. Normally when this happens it would be inconsequential. However, when antibiotic solution is then added to this water it allows for the rapid growth of fungus within the water. Even when distilled water is used this can be a problem. The solution should be changed out for fresh solution as frequently as needed and at least once a week. Despite all these measures, investigators should expect a 20–30 % mortality rate among recipient mice (mostly during the first 2 weeks post-transplant) and plan accordingly.

2. Notes for Fluorescent Cell Sorting and Determination of Chimerism.

   Four weeks after bone marrow transplant we recommend determination of chimerism using fluorescence-activated cell sorting (FACS). When doing this, chimeric Db/Db mice should be bled no more than 200–300 μL of blood via retro-orbital bleed. These mice are somewhat fragile so we also recommend resuscitation of each mouse with up to 1 mL of warmed normal saline or lactated ringers subcutaneously immediately following retro-orbital bleed. This resuscitation can also be performed if mice are determined to not be drinking their Sulfatrim solution in the immediate days following irradiation. A protocol for FACS analysis is beyond the scope of this chapter however the antibodies we use deserve mention. We use antibodies to MHC Class I (H2K). Db/Db mice are on a Bl6 background and are H2K^d (we use BD #553457, BD Biosciences, San Jose, CA). Cells from GFP mice fluoresce on their own so logically no antibodies are needed for them. Tie-2/GFP mice are on a FVB background and are H2K^q (we use BD# 553597, BD Biosciences, San Jose, CA). These can be combined with antibodies for specific cell types at the discretion of the investigator.

References

1. Wild S et al (2004) Global prevalence of diabetes: estimates for the year 2000 and projections for 2030. Diabetes Care 27:1047–1053
2. (2011) National diabetes information clearinghouse: National diabetes statistics fact sheet. http://diabetes.niddk.nih.gov/dm/pubs/statistics/index.aspx, Accessed 2 Aug 2011
3. Bermudez DM, Liechty KW (2010) Mesencymal stem cells in diabetic wound healing. In: Chandan S (ed) Advances in wound care, vol 1. Mary Ann Liebert, New York, pp 477–482
4. Kranke P et al (2004) Hyperbaric oxygen therapy for chronic wounds. Cochrane Database Syst Rev (1). Art No. CD004123
5. Steed DL (2006) Clinical evaluation of recombinant human platelet-derived growth factor for the treatment of lower extremity ulcers. Plast Reconstr Surg 117:143S–149S
6. Nilsson J et al (2008) Inflammation and immunity in diabetic vascular complications. Curr Opin Lipidol 19:519–524
7. Falanga V (2005) Wound healing and its impairment in the diabetic foot. Lancet 366: 1736–1743
8. Brem H, Tomic-Canic M (2007) Cellular and molecular basis of wound healing in diabetes (comment). J Clin Invest 117:1219–1222
9. Maruyama K et al (2007) Decreased macrophage number and activation lead to reduced lymphatic vessel formation and contribute to the impaired cutaneous healing response associated with diabetes mellitus. Am J Path 170: 1178–1191
10. Khanna S et al (2010) Macrophage dysfunction impairs resolution of inflammation in the wounds of diabetic mice. PLos One 5(3):e9539
11. Albiero M et al (2011) Defective recruitment, survival and proliferation of bone marrow-derived progenitor cells at sites of delayed diabetic wound healing in mice. Diabetologia 54:945–953
12. Asahara T et al (1999) Bone marrow origin of endothelial progenitor cells responsible for postnatal vasculogenesis in physiological and pathological neovascularization. Circ Res 85: 221–228
13. Bluff JE et al (2007) Bone marrow-derived endothelial progenitor cells do not contribute significantly to new vessels during incisional wound healing. Exp Hematol 35:500–506
14. Hayakawa J et al (2003) Generation of a chimeric mouse reconstituted with green fluorescent protein-positive bone marrowcells: a useful model for studying the behavior of bone marrow cells in regeneration in vivo. Int J Hematol 77:456–462
15. Loots MA et al (1998) Differences in cellular infiltrate and extracellular matrix of chronic diabetic and venous ulcers versus acute wounds. J Invest Dermatol 111:850–857
16. Grice E et al (2010) Longitudinal shift in diabetic wound microbiota correlates with prolonged skin defense response. PNAS 107(33): 14799–14804
17. Dandona P et al (2004) Endothelial dysfunction, inflammation and diabetes. Rev Endocr Metab Disord 5:189–197
18. Oren et al (2007) Cellular dysfunction in the diabetic fibrobalst: impairment in migration, vascular endothelial factor production, and response to hypoxia. Am J Pathol 162: 303–312
19. Roukis TS (2010) Bacterial skin contamination before and after surgical preparation of the foot, ankle and lower leg in patients with diabetes and intact skin versus patients with diabetes and ulceration: a prospective controlled therapeutic study. J Foot Ankle Surg 49(4): 348–356
20. Pradhan L et al (2011) Gene expression of pro-inflammatory cytokines and neuropeptides in diabetic wound healing. J Surg Res 167(2): 336–342
21. Mustoe TA et al (2006) Chronic wound pathogenesis and current treatment strategies: a unifying hypothesis. Plast Reconstr Surg 117(Suppl 7):64–72
22. Leibovich SJ, Ross R (1975) The role of the macrophage in wound repair. A study with hydrocortisone and antimacrophage serum. Am J Pathol 78:71–100
23. Abrass CK, Hori M (1984) Alterations in Fc receptor function of macrophages from

streptozotocin-induced diabetic rats. J Immunol 133:1307–1312
24. Meszaros AJ et al (1999) Macrophage phagocytosis of wound neutrophils. J Leukoc Biol 65:35–42
25. Koh TJ, DiPietro LA (2011) Inflammation and wound healing: the role of the macrophage. Expet Rev Mol Med 13:e23
26. Lucas T et al (2010) Differential roles of macrophages in diverse phases of skin repair. J Immunol 184:3964–3977
27. Riches DWH (1996) Macrophage involvement in wound repair, modeling and fibrosis. In: Clark RAF (ed) The molecular and cellular biology of wound repair, 2nd edn. Plenum Press, New York, NY, pp 95–142
28. Martin A et al (2003) Abnormal angiogenesis in diabetes mellitus. Med Res Rev 23: 117–145
29. Bauer SM et al (2006) The bone marrow-derived endothelial progenitor cell response is impaired in delayed wound healing from ischemia. J Vasc Surg 43:134–141
30. Kämpfer H et al (2001) Expressional regulation of angiopoetin-1 and -2 and tie-1 and -2 receptor tyrosine kinases during cutaneous wound healing: a comparative study of normal and impaired repair. Lab Invest 81:361–372
31. Velazquez OC (2007) Angiogenesis and vasculogenesis: inducing the growth of new blood vessels and wound healing by stimulation of bone marrow-derived progenitor cell mobilization and homing. J Vasc Surg 45(Suppl A):A39–A47
32. Gallagher KA et al (2007) Diabetic impairments in NO-mediated endothelial progenitor cell mobilization and homing are reversed by hyperoxia and SDF 1α. J Clin Invest 117:1249–1259
33. Yin Y et al (2010) SDF-1α involved in mobilization and recruitment of endothelial progenitor cells after arterial injury in mice. Cardiovasc Pathol 19:218–227
34. Bermudez DM et al (2011) Inhibition of stromal cell-derived factor-1α further impairs diabetic wound healing. J Vasc Surg 53:774–784
35. Marrotte EJ et al (2010) Manganese superoxide dismutase expression in endothelial progenitor cells accelerates wound healing in diabetic mice. J Clin Invest 120:4207–4219
36. Gallagher KA et al (2006) Hyperbaric oxygen and bone marrow-derived endothelial progenitor cells in diabetic wound healing. Vascular 14:328–337
37. O'Neill TJ et al (2005) Mobilization of bone marrow-derived cells enhances the angiogenic response to hypoxia without transdifferentiation into endothelial cells. Circ Res 97: 1027–1035
38. Möller M et al (2010) On the longevity of resident endoneurial macrophages in the peripheral nervous system: a study of physiological macrophage turnover in bone marrow chimeric mice. J Peripher Nerv Syst 15(4):357–365
39. Imasawa T (2003) Roles of bone marrow in glomerular diseases. Clin Exp Nephrol 7:179–185
40. Motoike T et al (2000) Universal GFP reporter for the study of vascular development. Genesis 28:75–81
41. v Michaels J et al (2007) Db/Db mice exhibit severe wound-healing impairments compared with other murine diabetic strains in a silicone-splinted excisional wound model. Wound Repair Regen 15(5):665–670
42. Fang RC et al (2010) Limitations of the Db/Db mouse in translational wound healing research: Is the NONcNZO10 polygenic mouse model superior? Wound Repair Regen 18(6):605–613
43. Rakieten N et al (1963) Studies on the diabetogenic action of streptozotocin. Cancer Chemother Rep 29:91
44. Badillo AT et al (2007) Lentiviral gene transfer of SDF-1 alpha to wounds improves diabetic wound healing. J Surg Res 143(1):35–42

Chapter 7

Neural Repair with Pluripotent Stem Cells

Máté Döbrössy and Jan Pruszak

Abstract

The nervous system is characterized by its complex network of highly specialized cells that enable us to perceive stimuli from the outside world and react accordingly. The computational integration enabled by these networks remains to be elucidated, but appropriate sensory input, processing, and motor control are certainly essential for survival. Consequently, loss of nervous tissue due to injury or disease represents a considerable biomedical challenge.

Stem cell research offers the promise to provide cells for nervous system repair to replace lost and damaged neural tissue and alleviate disease. We provide a protocol-based chapter on fundamental principles and procedures of pluripotent stem cell (PSC) differentiation and neural transplantation. Rather than detailed methodological step-by-step descriptions of these procedures, we provide an overview and highlight the most critical aspects and key steps of PSC neural induction, subtype specification in different in vitro systems, as well as neural cell transplantation to the central nervous system. We conclude with a summary of suitable readout methods including in vitro phenotypic analysis, histology, and functional analysis in vivo.

Key words Neural stem cells, Pluripotent stem cells, Differentiation, Induction, Specification, Transplantation, Histological and functional analysis

1 Introduction

1.1 Stem Cell Research in the Context of Neurological Disease

The human nervous system is considered to be the most complex organ of the human body. It is composed of more than one hundred billion (10^{11}) neuronal cells, made up of at least hundreds subclasses of neurons and functionally connected to one another by a myriad of synapses. Each single neuron bears 10^3–10^4 of these electrochemical cell–cell connections, receiving a range of inputs mainly via its dendrites, processing and integrating these inputs in a more or less elusive manner, and eventually transmitting its output via the singular axon or neurite.

The brain and spinal cord are delicate organs. While they are anatomically very well protected in their respective encasements, i.e., the skull and the vertebral canal, respectively, their injury or disease has devastating effects: The central nervous system (CNS)

Robert G. Gourdie and Tereance A. Myers (eds.), *Wound Regeneration and Repair: Methods and Protocols*, Methods in Molecular Biology, vol. 1037, DOI 10.1007/978-1-62703-505-7_7, © Springer Science+Business Media New York 2013

is the main center for the organism's perception of the environment as well as its ability to act on and within it. Moreover, our emotions and our cognitive abilities depend on well-balanced function and interaction of specialized areas of the brain. It thereby represents the underpinning of our basic vital functions as well as of the abilities and processes that define us as human beings and individuals.

In addition to the nerve cells themselves, a number of other cell types are present in the CNS, among them the insulating myelin ensheathing cells (oligodendrocytes) as well as a functionally diverse array of astroglia and immune modulatory cells (e.g., microglia). Further specialized phenotypes include ependymal cells, vascular endothelial cells, and somatic neural stem cells.

Importantly, despite the presence of neural stem cells in circumscribed regions of the CNS (subventricular zone of the lateral ventricles, subgranular zone of the hippocampal dentate gyrus, filum terminale) its capacity for spontaneous regeneration and self-repair is naturally limited. Damage to the CNS and loss of neural tissue can occur in a multitude of ways; among others, by means of injury, stroke, and slow but progressive degeneration. Nervous system disease translates itself to being a major cost factor in current health care [1]. Contributing to this is the increased overall life expectancy. Parkinson's and Alzheimer's diseases, correlated with age, are the most common neurological diseases, and despite considerable progress regarding the early intervention in stroke patients, the long-term sequelae of stroke represent debilitating factors that affect a growing number of people.

Here, we present approaches aimed at restoring neural function through administration of neuronal cells in vitro-derived from PSCs with the goal of repairing the neural circuitry damaged by degenerative processes or trauma. The goal of such research is not only to administer cells as vectors to release trophic or protective factors (chaperone effect [2]) but also, if possible, to achieve neural circuit reconstruction. To this end, traditionally we have experimentally exploited the plasticity of embryonic primary cells; however, the present and the future lies with the unique potential of truly pluripotent cells, i.e., their ability to self-renew in a potentially unlimited manner and their capacity for differentiation toward all cell types of the body. These properties are in contrast to somatic stem cells which exhibit more restricted expansion capacity and inherent differentiation spectra. Human PSCs are derived either from the inner cell mass of the blastocyst (embryonic stem cells, ES cells) [3] or by epigenetic reprogramming of somatic cells (induced pluripotent cells, iPS cells), the latter enabled by landmark work of Shinya Yamanaka and colleagues [4, 5]. The advent of reprogramming approaches has enabled the development of autologous approaches, circumventing issues of immune rejection as well as some of the ethical concerns associated with ES cells. Given all that, PSCs are expected to represent a plentiful source of cells for regenerative approaches. How can we best realize the potential of PSCs for neural repair?

(#1) We have to guide the differentiation of PSCs toward the phenotype of interest.

(#2) We have to apply these cells in a way that ensures functional integration.

Over the past couple of years, significant progress has been made in the fast-moving stem cell and neuroregeneration field. Related to #1, we present robustly established protocols for neural induction of PSCs, as well as differentiation toward specific neural subtypes. Related to #2, we present standardized protocols for nervous system transplantation and functional analysis of stem cell-derived neurons in vitro and in vivo. Our goal with this chapter is as follows:

1. To provide a comprehensive overview of the concepts and chief methodologies applied in neural regeneration research using stem cells.
2. To exemplify the main protocols, yet refer to the details of the primary literature as needed.
3. To provide a scaffold/framework for the better interpretation of papers published in the field and facilitate practical work in this area intended by the reader.

1.2 Neural Repair from Its Beginnings to Clinical Applications

While the earliest scientific work studying CNS repair using tissue transplantation dates back to the beginning of the twentieth century, the modern era of intracerebral transplantation started in the late 1970s by Stenevi, Björklund, and Svendgaard with their work on catecholamine grafts into rodent forebrain [6]. Correspondingly, animal models of neurodegenerative diseases became available: Ungerstedt developed the 6-hydroxydopamine-based striatal dopamine depleting lesion as a rodent model for Parkinson's disease (PD) [7], while Coyle and Schwarcz lesioned striatal projection neurons by injecting excitotoxins such as quinolinic acid into the striatum to simulate Huntington's disease (HD) pathology [8], for instance. In parallel, cell suspension protocols were developed using primary rodent embryonic tissue cell suspensions prepared from the ventral mesencephalon or the ganglionic eminence of embryonic day (E)12 to E15 embryos, depending on whether aiming for cell replacement of dopaminergic (DA) neurons in the PD or for that of GABAergic neurons in the HD animal model. Over the last decades, a substantial amount of indisputably robust scientific data has been generated using primary cells supporting the functional relevance of the neural grafts, as well as establishing feasible protocols for these neurodegenerative disease entities [9–12].

In Parkinson's, despite several outstanding questions, numerous groups entered the clinical phases of cell therapy from the 1980s onwards. Pioneering and promising clinical trials using primary mesencephalic tissue were initiated by Lindvall and colleagues [13, 14]. These first clinical trials were open-labeled with carefully chosen patients and small group sizes. Other parallel studies, utilized autologous adrenal grafts containing dopamine producing cells to eliminate the need for immunosuppressants after transplantation and to circumvent ethical concerns regarding use of fetal tissue. However, the long-term graft survival and functional outcome of adrenal grafts turned out to be poor in the several hundred patients transplanted in numerous trials [15], underlining the critical importance of being rooted in a strong scientific fundament of experimental research before proceeding to clinical translation. From the late 1980s on, a number of small open-label clinical trials with two to six patients each were carried out using primary human DA transplants obtained from fetal ventral midbrain. These early studies showed that fetal DA neurons can survive long term for more than 10 years in patients and are capable of improving the classical symptoms of Parkinson's disease. Following this, supposedly more sophisticated double-blinded clinical trials with PD patients were initiated which have raised several new safety questions [16, 17]. However, some of these did not necessarily use state-of-the-art methods including variation in terms of tissue preparation, surgical approaches, and immunosuppressive paradigms [18]. Although most clinical trials on fetal tissue transplantation for PD revealed encouraging results, the side effects and inconsistent efficacy, especially observed in the controlled trials [19–22], underscore the need for a new look at identifying crucial parameters that have an impact on clinical outcome. Indeed, a clinical trial run by a consortium of centers has recently been awarded European Union funding to address critical questions concerning patient selection, tissue preparation, delivery, immunosuppression, and off-medication dyskinesia (*see* http://www.transeuro.org.uk). A key objective of this study will be to generate a protocol that can serve as a template for future clinical trials in the cell therapy field including stem cell-based therapies for PD.

In Huntington's, the first reports from clinical trials using ganglionic eminence (GE) transplantation were published in 2000 and 2002, respectively. Bachoud-Levy and colleagues grafted GE tissue pieces into five patients exhibiting moderate to severe features of HD, with the first safety and efficacy reports published 2 years following the transplantation [23, 24]. Three patients had increasing striatal glucose uptake after grafting within the striatum, and these patients had improved or at least stabilized Unified Huntington's Disease Rating Scale (UHDRS) motor and neuropsychological tests scores. A follow-up publication from Bachoud-Levy's group recently reported on outcomes 6 years post-grafting. Two patients continued to significantly decline, while three patients, who had initially improved, could maintain or even improve their initial clinical level [25]. Similarly, another long-term report on two patients receiving enzymatically dissociated GE tissue 5 years previously described one patient as improving and the other as deteriorating [26]. The limited publications available suggest that cell replacement therapy in HD can, in some cases, alter the natural progression of the disease.

Thus, at least for these disease entities there is substantial experimental proof-of-principle evidence for successful cell therapeutic intervention. Other rather circumscribed disease candidates in which cell-based therapeutic interventions have a history of being explored include spinal cord injury [27] and retinal disease [28]. In contrast, the nature of stroke leads to loss of a much broader range of cell types in the affected tissue area, and the time window for therapeutic intervention is far narrower [29]. The array of cell types therapeutically applied in animal models of stroke has also been more varied, ranging from rodent hippocampal neurons over mesenchymal cell preparations to human bone marrow-derived stem cells [30, 31]. The emergence of the pluripotent stem cell (PSC) field has given a new lease of life to neural transplantation in general as it has the potential to overcome issues of tissue availability, standardization, and associated ethical questions [2]. Generating cells of the desired phenotype and numbers under highly controlled and standardized Good Manufacturing Practice (GMP) conditions could open up transplantation to a larger number of patients. However, new cell candidates must achieve the "gold standards" set by established cell sources such as primary fetal cells in safety and functional recovery in order to be considered for clinical application.

1.3 Pluripotent Stem Cells in Neural Differentiation and Transplantation

To some extent, PSC-based therapies have already become a clinical reality. Treatment of complete, subacute spinal cord injured patients with hESC-derived oligodendroglial precursor cells represents the first clinical trial of human ES cell-derived neural cells, being run at Phase-I level in the USA, executed by the California-based company Geron (http://clinicaltrials.gov/ct2/show/NCT01217008). Also, the US company Advanced Cell Technology (ACT) has gained Food and Drug Administration (FDA) approval for initiation of hESC-derived retinal pigment epithelial (RPE) cell transplantation in early onset macular dystrophy (Stargardt's type) and macular degeneration (http://clinicaltrials.gov/ct2/show/NCT01344993; http://clinicaltrials.gov/ct2/show/NCT01345006). Approval of these safety studies has also been given by the regulatory agencies in the UK.

We argue that a high degree of caution and scientific scrutiny are warranted when deriving novel experimental cell therapeutic approaches for neurological diseases, and one has to critically take into consideration a range of scientific, biomedical, and ethical issues. Even at the stage of basic scientific research with potential future therapeutic impact, one should critically consider whether the disease in question would be amenable to stem cell-based efforts or cell-based strategies in general. A focal disease pattern in which only a circumscribed brain region or specific neuronal subset is affected may lend itself better to cell replacement strategies. For example, the widespread and diffuse neurodegeneration present in Alzheimer's disease may make it difficult to generate the diversity of cell types lost as well as to deliver the cell therapeutic product appropriately to the affected brain tissue regions and to ultimately achieve functional and structural restoration.

Moreover, socioeconomic realities have to be considered, as some of the biotechnological processes may have significant cost and/or logistic demands that may impede any future translation to the benefit of a broader patient collective. The previously mentioned economically high-risk clinical trials with human ES cells would hardly have been possible without considerable support by federal or state grants or loans.

Also, one has to take ethical aspects into consideration [32]: ES cells are originally derived from in vitro-fertilized oocytes generated for implantation in the uterus. Excess non-implanted blastocysts are used to isolate the inner cell mass for the generation of long-term expandable human ES cell lines. Some stress the importance of attributing equal rights to this early stage embryo as to any later stage of the individual. Others do not consider the blastocyst-stage embryo as an individual at all, while others again acknowledge that while this is controversial and constitutes an ethical dilemma, the promise of saving and facilitating many lives in the future may justify even the possibly outrageous act of destroying said blastocyst. Pragmatists argue that, certainly, instead of routinely discarding the "leftover" blastocyst-stage embryos, one should use it for potential biomedical benefit.

iPS cells are considered to carry less of an ethical burden and represent a promising alternative. Their generation in 2006, initially in mouse, by Takahashi and Yamanaka spurred from the quest to identify factors that were able to reprogram somatic cell DNA toward pluripotency, as had previously been observed in the transfer of a somatic cell nucleus to an enucleated oocyte (nuclear transfer, nuclear reprogramming, cloning) [4]. They were able to translate this to human cell systems shortly after [5]. A second paper by Thomson and colleagues confirmed this revolutionary finding with a slightly modified combination of factors [33]. Surprisingly enough, in these studies a rather limited set of stemness-related transcription factors, after transient expression over a couple of weeks, was sufficient to convey epigenetic rearrangements similar to the oocyte cytoplasm and consequently render somatic cells pluripotent. By means of converting patient-derived skin fibroblasts to pluripotency, this enables the generation of individualized cell lines for autologous transplantation and in vitro disease-specific cellular models. Fundamental research with both ES and iPS cells, will still be required to elucidate the subtle differences between both cell types and to refine epigenetic (re)programming strategies [34, 35]. A remaining challenge with either type of PSC is that growth and differentiation of phenotype have to be well controlled and directed to the cell type of therapeutic interest.

2 Protocols of Neural Induction, Differentiation, and Transplantation

2.1 PSC Neural Induction and Patterning Protocols

How does one guide a PSC exclusively down the road toward the desired phenotype? The most promising hPSC induction and targeted differentiation protocols have exploited insights into normal embryological development and apply growth and patterning factors in a way that mimics ontogenesis. For instance, stemming from the classic knowledge of early neuralizing factors [36, 37] recombinant human Noggin was applied to drive hESC to neural fate [38]. In addition, there are a number of alternative protocols, including the formation of embryoid bodies, i.e., proliferative cell clusters containing cell types of all three germ layers which are then further induced toward (neuro) ectoderm. Other efficient yet rather poorly defined and xeno-product-containing protocols include stromal feeder-based neural induction protocols (using murine stromal feeder cell lines such as PA6 [39] or MS5 [40]). During neural differentiation of hPSCs in most protocols a varying degree of the formation of early columnar neuroepithelial cells can be observed, so-called neuroepithelial "rosettes" (Fig. 1), typically expressing the neuroectodermal transcription factor Pax6, anterior homeodomain protein Otx2, and the intermediate filament protein nestin, among others.

For subsequent patterning strategies, for example, toward motor neurons that are located ventrally in the spinal cord, sonic hedgehog (Shh) as a known ventralizing factor is applied, combined with higher concentrations of retinoic acid to caudalize toward spinal cord fates [41, 42]. Following such reasoning, concentrations and times of exposure are usually determined empirically. In that manner, different protocols have been devised to direct the differentiation of PSCs toward a range of neural phenotypes, among them DA neurons [38, 40, 43–45], oligodendrocytes [46], motoneurons [47], or GABAergic neurons [48]. However, in addition to the application of gradients or different concentrations of induction and patterning factors, the microenvironment within the dish has also to be taken into account to appropriately modulate shape phenotype establishment, growth control, and tissue morphogenesis [49].

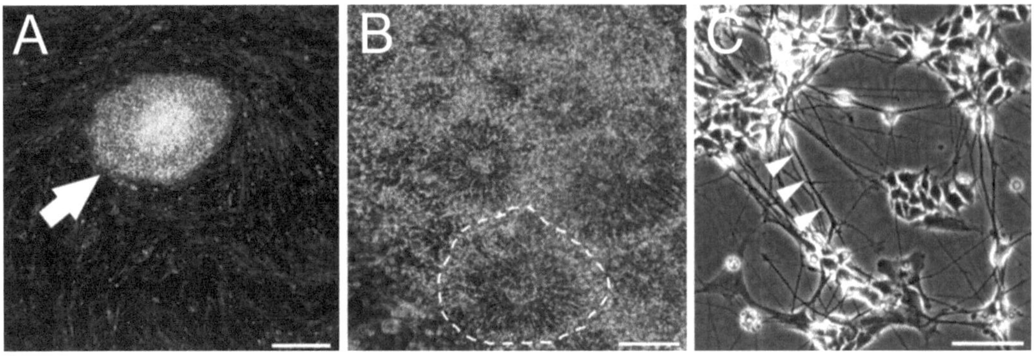

Fig. 1 Principle of in vitro differentiation from pluripotency to neurons. (**a**) Human PSC colony (*arrow*) on a bed of fibroblast feeder. (**b**) Columnar neuroepithelial rosettes characteristic of neural induction stage (*dashed line*). (**c**) Neuronal process extension (*arrowheads*) and maturation at neuronal differentiation stage [Differentiation of the human iPS cell line IMR90-4DL1; WiCell]

Cell–cell interactions by means of direct cell contact (surface molecules) and diffusible (autocrine, exocrine, paracrine) factors do greatly influence cell development in vivo, and similarly so in these artificial in vitro systems. In this section, we provide example protocols for the efficient neural induction and subsequent neuronal subtype differentiation of hPSC.

2.2 Example Protocol 1: Neural Induction of Pluripotent Stem Cells

2.2.1 Materials

PSC lines (e.g., H7 or H9 ES cell lines; WiCell, Madison, WI, USA). *Note*: Pay careful attention to governmental and institutional guidelines for work with human stem cell lines.

1. Sterile, cell culture grade DI water.
2. Ca^{2+}-, Mg^{2+}-free Dulbecco's Phosphate-buffered saline (D-PBS) (Life Technologies).
3. Ca^{2+}-, Mg^{2+}-free Hank's balanced salt solution (HBSS).
4. Knockout DMEM.
5. DMEM/F12.
6. Penicillin–streptomycin (Pen-Strep).
7. Knockout serum replacement (KOSR).
8. MEM-NEAA 100× L-glutamine.
9. N2 supplement.
10. Basic fibroblast growth factor (bFGF).
11. Epidermal growth factor (EGF) recombinant human Noggin.
12. Dorsomorphin.
13. SB431542.
14. Rock inhibitor Y-27632 dihydrochloride.
15. Dibutyryl-cAMP.

16. Glial cell line-derived neurotrophic factor (GDNF).
17. Ascorbic acid.
18. Sonic hedgehog.
19. FGF8.
20. Tumor growth factor 3beta.
21. Dickkopf-1 (Dkk1).
22. All-trans retinoic acid (RA).
23. TrypLE Express.
24. Laminin.
25. Poly-L-ornithine, 0.01 % solution.
26. 0.1 % Gelatin.
27. Matrigel matrix.
28. Culture vessels (six-well plates, 24-well plates, Petri dishes).
29. 0.22 μm filter units for aseptic preparation of solutions and media.
30. 15 and 50 mL conical plastic tubes.
31. Pipette tips (10–1,000 μL).
32. Serological pipettes (5–25 mL).
33. Microdissection tool for harvesting of neuroepithelial rosettes (e.g., a 1 mL syringe with a 27.5 Gauge needle).
34. Cell culture lab (incl. laminar flow hood, CO_2 incubator, table centrifuge, and bright field/phase contrast inverted microscope).
35. Access to fluorescence microscope for immunocytochemical analysis.

2.2.2 Stock Solutions

Note: growth factor stock solutions are prepared in D-PBS or dH_2O as specified by supplier. If recommended take up in 0.1 % bovine serum albumin. Store frozen.

Stocks for basic neural induction and differentiation		
Noggin	Stock 10 μg/mL	Final conc.: 300 ng/mL
SB-431542	Stock 10 mM	Final conc.: 10 μM
Dorsomorphin	Stock 0.2 mM	Final conc.: 0.2 μM
bFGF	Stock 10 μg/mL	Final conc.: 4–8 ng/mL; 20 ng/mL
BDNF	Stock 50 μg/mL	Final conc.: 20 ng/mL
Ascorbic acid	Stock 200 mM	Final conc.: 200 μM
GDNF	Stock 10 μg/mL	Final conc.: 10 ng/mL
cAMP	Stock 200 mM	Final conc.: 500 μM

Stocks of patterning factors (selection)		
Sonic hedgehog	Stock 50 μg/mL	Final conc.: 200 ng/mL
FGF8	Stock 50 μg/mL	Final conc.: 100 ng/mL
TGF3b	Stock 2 μg/mL	Final conc.: 1 ng/mL
Dickkopf-1 (Dkk1)	Stock 50 μg/mL	Final conc.: 100 ng/mL
RA	Stock 10 mM	Final conc.: 0.1–0.5 μM

2.2.3 Media Preparation

hPSC medium	
80 % DMEM/F12	386.5 mL
20 % KOSR	100 mL
MEM NEAA	5 mL
200 mM L-Glutamine	2.5 mL
Pen-Strep	5 mL
1 mM beta-mercaptoethanol	1 mL

Filter and store for up to 10 days at 4 °C
Add freshly before use: bFGF to a final concentration of 4–8 ng/mL

N2-based differentiation medium	
DMEM/F12	485 mL
N2 supplement	5 mL
MEM NEAA	5 mL
Pen-Strep	5 mL

Ascorbic acid at 200 μM final concentration is added to N2-medium right before use

2.2.4 Coating of Plates and Tissue Culture Dishes

Gelatin-coated plates

Add 0.1 % gelatin solution to the dish, ensuring that the entire surface is covered. Leave dish coating at room temperature (RT) for ca. 15 min, and aspirate the solution before plating of cells.

Poly-ornithine/laminin-coated plates

The solution of 0.01 % poly-L-ornithine is pipetted into the dish, covering the entire surface. Coat for at least 2 h or overnight. After three washes with D-PBS, coat with a 1 μg/mL solution of laminin for at least 2 h or overnight. Remove the solution; wash three times with D-PBS and once with water. Dry briefly and preferably use right away.

Matrigel-coated plates

Slowly thaw an aliquot of matrigel at 4 °C or on ice to avoid formation of a gel. Dilute to 10 % final concentration with cooled DMEM/F12 medium and immediately distribute into dishes, for example, 1.2 mL to each well of a six-well plate at RT. Remove matrigel solution after an hour, carefully wash once with medium and plate the cells (see below).

Procedure

1. Human PSCs are cultured under standard conditions (*see* refs. 3, 38, 50). This is traditionally conducted on a bed of mitotically inactivated (human or mouse) embryonic feeder cells. Alternatively, a number of feeder-free protocols exploiting feeder-conditioned medium instead, or high concentration supplementation with bFGF and cells grown on matrices such as Matrigel can be applied. Cells are best propagated by manual picking/selection. Critical attention must be paid to whether the conditions used in the lab allow for long term, karyotypically normal expansion. Principles of neural induction differentiation protocols are exemplified by the following steps applied in our laboratories (adapted from refs. 38, 51, 52).
2. A full six-well plate of hPSC (ES or iPS) grown on feeder fibroblasts is harvested by using TrypLE incubation for approximately 5–10 min (under close microscopic observation), resulting in the detachment of hPSC from the substrate and dissolution of colonies. The stem cells are carefully detached using gentle trituration with a P1000 pipet and collected into a 15 mL conical tube with excess of hPSC medium (without bFGF).
3. After a spinning step (120 × *g*), the pellet is taken up in 10 mL of hPSC medium (without bFGF) containing 10 μM ROCK inhibitor Y-27632 (preventing cell death/anoikis of single cells) and plated on 0.1 % gelatin-coated, tissue culture-treated dishes for 30 min, allowing the remaining fibroblast feeders to attach.
4. The single cell suspension of floating hPSCs purified in such manner is collected, spun down at 120 × *g* and plated at a density of ca. 20,000 cells/cm^2 into 10 % Matrigel-coated dishes in hPSC medium with 10 μM Y-27632 and 8 ng/mL bFGF.
5. After 2 days, medium is changed to N2 medium containing neuralizing agents such as 0.2–1 μM Dorsomorphin (a small molecule as a more cost effective alternative to 300 ng/mL of recombinant noggin) as well as 10 μM SB431542 (for dual Smad inhibition according to Chambers et al. [51]).

6. Media is changed every 2–3 days in vitro, and cells are differentiated in this manner for 10–12 days upon which neural markers and occasional neural rosettes arise.
7. Patterning toward phenotype of interest is usually initiated at this stage (*see* Subheading 2.2).
8. For terminal differentiation, cells are harvested en bloc (tissue pieces) or alternatively after careful harvesting and transfer by gentle enzymatic digestion with TrypLE to conditions on poly-ornithine/laminin with N2-medium containing 20 ng/mL bFGF (for optional expansion) and/or for neuronal differentiation with ascorbic acid, BDNF, GDNF, cAMP after removal of the mitogen.
9. An optional expansion step of the generated neural precursor cells with bFGF and EGF at 20 ng/mL each in N2-based medium on poly-ornithine-/laminin-coated dishes can be inserted between procedure **steps** 7 and **8**.

2.3 Notes on Subtype Specification and In Vitro Phenotypic Readout

As pointed out earlier, phenotypic patterning exploits insights into normal development. Consequently, patterning molecules of known relevance from in vivo or primary tissue studies are being applied depending on the phenotype of interest. *See* above (Subheading 2.2.1: Preparation—Stock Solutions) for working concentrations commonly used.

1. Joint application of fibroblast growth factor-8 and Shh (or the small molecule alternative of 1 μM purmorphamine [47]) leads to the ventro-dorsal and rostro-caudal establishment of ventral midbrain coordinates, essential for DA neuronal development. Originally this was applied by Lee et al. in murine ES cell systems but has been equally successfully applied in hPSC DA differentiation [38, 40, 43]. Appropriate markers for midbrain DA phenotypic readout include tyrosine hydroxylase, Pitx3, Lmx1a, and Nurr1, among other.
2. While inhibitory GABAergic neurons frequently comprise a considerable fraction of a number of neural differentiation protocols, the targeted differentiation of specific subtypes of therapeutic relevance such as striatal medium-spiny GABAergic neurons lost in HD has proven more difficult. Aubry et al. [48] utilized Shh as well as Wnt-pathway modulating agents to induce ventral telencephalic cell fates. As a readout, DARPP32-positivity is indicative, as well as expression of GAD67, calbindin, and calretinin of GABAergic neuronal differentiation.
3. Oligodendrocyte differentiation protocols exploit growth factors known to promote oligodendroglial phenotype such as Platelet-derived growth factor (PDGF) as part of the differentiation cocktail [46].

4. In analogy to the aforementioned principle, as Lower Motoneurons are located in the ventral spinal cord, ventralizing factors such as Shh are combined with the caudalizing effect of RA to derive MNs from hPSCs. In that manner, PSCs can be directed toward Olig2 expressing spinal neural precursors and ultimately HB9 (MNR2) expressing motoneurons [47, 53].
5. Significant progress has also been made with the derivation of RPE cells to be applied in ophthalmological disease [54].

2.4 Neural Cell Transplantation to the Central Nervous System

As described earlier (*see* Subheading 1.2), optimized protocols of administering cells to the CNS are particularly well established in models of neurodegenerative diseases such as PD and HD. In both cases the therapeutic strategy employed, experimentally as well as clinically [25, 26, 55], is that the cells (1) is introduced with the precision of stereotactical surgical methods (Fig. 2) either homotopically (i.e., at the site of neuronal degeneration) or ectopically (i.e., at a site other than where the neuronal loss occurs); and (2) is identical/similar in phenotype to that of the lost neurons, for example, DAergic in PD and GABAergic in HD. An alternative to the direct stereotactical cell delivery is the intravenous injection of cells into the blood stream followed by migration to the target area, and such methods are being explored to mitigate cell loss following stroke, among other. Although the mechanisms of action of such an approach are unknown, it is believed that the impact of the cells can occur through indirect, immunomodulatory, or trophic mechanisms, for example [56, 57]. For spinal cord injury, cell administration is frequently performed by direct injection of, for example, oligodendroglial cells in or immediately around the area of an acute or subacute lesion [46]. The experimental establishment of animal models of neurodegenerative disease is described in detail in several previous publications [58–61], below we summarize a typical protocol for the harvesting, preparation, and stereotactical transplantation of PSC-derived neural cells into an animal model of HD. The cell source might vary depending on the candidate disease, but the overall methodology would be very similar also when grafting into other animals models.

2.5 Example Protocol 2: Neural Transplantation Procedure

Note: Experiments involving animals must adhere to national and institutional regulations and require approval by the respective authorities. Critically consider the scientific and biomedical rationale for the number of animals used.

2.5.1 Materials and Preparation

Toxin preparation

1. For the PD model: 6-Hydroxydopamine
2. For the HD model: Quinolinic acid
3. Saline
4. Ascorbic acid
5. Phosphate-buffered saline

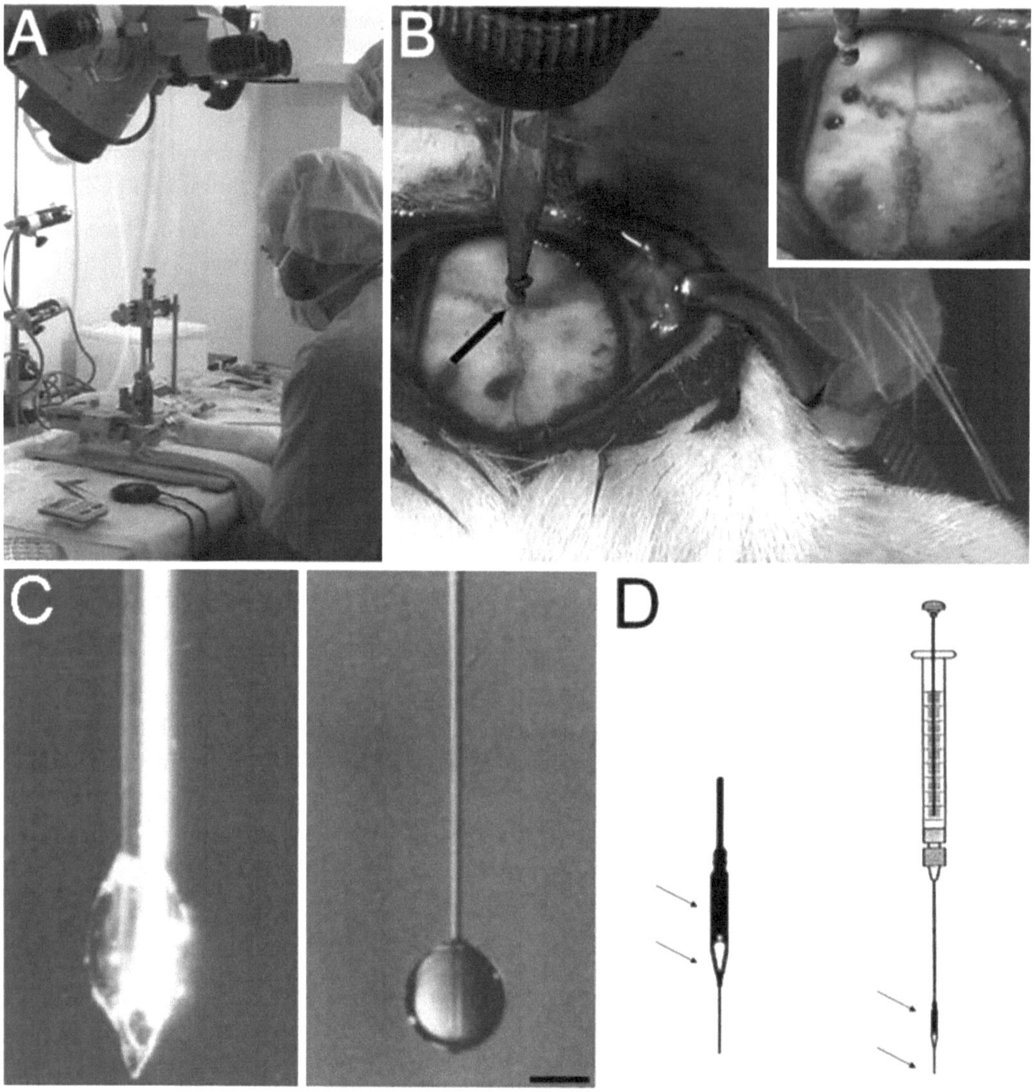

Fig. 2 Lesion and transplantation of cells into animals. A stereotactic approach (**a**) is used to establish the animal model into which cells are transplanted into as this ensures the required precision concerning the placement of the cells. The bore holes are prepared using a fixed or hand held drill where the coordinates are taken from the bregma (**b**; bregma indicated by the *arrow*). The cells are delivered into the brain using either a steel cannula (**c**, *left*) or a glass capillary (**c**, *right*) which causes less tissue damage. When using the latter approach, also known as microtransplantation, a glass capillary of 50 μm inner diameter is connected with a short polyethylene tube to a 2–5 μL Hamilton syringe (*arrows*, **d**). In either case, the steel cannula or the capillary is loaded with the cell suspension and used to penetrate the brain to the required depth as determined by the stereotactic atlas. Images (**c**) and (**d**) have been adapted from ref. 71

6. 10 M sodium hydroxide
7. Precision balance
8. 1.5 mL microcentrifuge tubes
9. Sonicator, vortex
10. pH meter, litmus paper

Surgery

1. Stereotactic frames
2. Micropump
3. Steel cannulae
4. 2 and 10 μL Hamilton syringe
5. Glass capillaries (I.D. 50–70 μm)
6. Polythene tubing (I.D. 0.28 mm)
7. Paracetamol tablets and diazapam (postoperative pain management)
8. Surgery tools: scalpel, suture, scissors, etc.
9. Trypan blue stain 0.4 %
10. Cell source: appropriately patterned and differentiated neural cells

OHDA preparation

Powder form of 6-OHDA is weighed out and stored in 1.5 mL microcentrifuge tubes in the fridge prior to use. When required for the lesion, appropriate amount of the toxin is freshly made up with 0.2 % ascorbic acid to a working dilution of 3.6 μg/μL. The solution is protected from light and kept on ice, and is discarded after 3 h maximum.

Stock (0.12 M) Quinolinic acid preparation

Quinolinic acid (QA) has a molecular weight of 167.12. The aim is to prepare 6.25 mL of stock solution using 125 mg of research grade quinolinic acid. Dissolve 125 mg of QA in 750 μL PBS (pH = 7.4), add 50 μL of 10 M sodium hydroxide. Sonicate the above solution for 15 min. Add 3,200 μL PBS. The total volume at this stage will be 4 mL, and this is will permit the use of a pH meter. Add 50 μL of 10 M sodium hydroxide to bring the solution to pH = 7.4; if pH needs to be adjusted use sodium hydroxide or concentrated hydrochloric acid. Add 2,200 μL of PBS to obtain the required concentration of 0.12 M QA. Check pH again, and if needed adjust to pH = 7.4. Aliquot 50 μL of QA into microcentrifuge tubes, label and store in freezer at −20 °C. The stock can be stored at −20 °C safely for 12 months; beyond this time point a new batch should be made up.

Note: The final pH is essential and the toxin going into solution depends on this. Instead of directly measuring the pH with litmus paper (which in case of small volumes could itself absorb a significant quantity of solution and change the concentration), one should dip a needle tip into the toxin and bring the tip into contact with the litmus paper.

2.5.2 Procedure

Surgery and lesioning

Anesthesia is induced by placing the animal into an induction box with 3–4 % isoflurane volatilized by O_2 used as the carrier gas. Once in deep anesthesia, the animal is transferred into the stereotactic frame ready for surgery. The lesion coordinates are set according to bregma as a reference point for the antero-posterior (AP) and medio-lateral (ML) coordinates and the dura as reference for the dorso-ventral (DV) coordinate, using the brain atlas Paxinos and Watson for rats [62], or Paxinos and Franklin for mice [63]. The bore holes are made with a drill either hand held or fixed to the stereotactic frame.

PD model

Typically a 6-OHDA dose of 3.6 mg/μL in saline containing 0.2 % (w/v) ascorbic acid is used to induce the loss of DA neurons from the substantia nigra. The toxin is administered into the brain at the level of the medium forebrain bundle (example for rats given below) that contains the DA projections from the substantia nigra to striatum. Alternative targets are the striatum itself leading to partial dopamine loss from the striatum, or the substantia nigra itself. 6-OHDA is taken up into a 10 μL Hamilton and the syringe is lowered into the brain tissue.

Track #	TB (mm)	AP (mm)	ML (mm)	DV (mm)	Volume (μL)
1	-2.3	-4.4	-1.2	-7.8	2.5
2	+3.4	-4.0	-0.8	-8.0	3

The injection rate is 1.0 μL/min and the cannula is kept in place for an additional 4 min before it is slowly retracted.

HD model

QA-mediated excitotoxic lesioning of the striatum requires precision delivery (assured by the stereotactic frame) of the toxin at the planned coordinates (see below for typical values for rats), of the required volume and at the required pace making it necessary to use:

1. A thin steel cannula (recommended size: 30 gauge) that penetrates the brain with minimal co-lateral tissue damage, and which is fixed to the stereotactic arm.
2. A micropump capable of delivering sub-microliter volumes of toxin over minutes.
3. Polythene tubing with an inner diameter of approximately 0.28 mm.

The polythene tube is used to connect up the steel cannula with a Hamilton syringe which is placed in the micropump. The Hamilton syringe-tubing-cannula system is loaded with saline to render it airtight. Prior to lesioning, an air bubble is introduced into the tube to mark the interface between the saline and the toxin. This is done to visually confirm the flow of the solution through the system.

Typical coordinates of QA striatal lesion:

Track #	TB (mm)	AP (mm)	ML (mm)	DV (mm)	Volume (μL)
1	0.0	+1.0	+2.9	−5.0/−4.0	0.2
2	0.0	−0.4	+3.3	−5.07/−4.0	0.2

Under this two track and two deposits/track protocol a total of four deposits of 0.20 μL (total of 0.8 μL) of QA is released into the striatum.

Post-surgery care

Surgery is a major intervention and the health status of the animals must be monitored afterwards. Postoperative anesthesia must be in accordance with local rules and regulations dealing with animal welfare. Typically, the administration of paracetamol in the format of soluble tablet in the animals' water bottle a day before and for 48 h after surgery is advised; however, the intermuscular injection of diazepam into a hind leg muscle, and the injection of saline/glucose bolus subcutaneously in the scruff immediately after the surgery might be also necessary.

Transplantation

1. At late stage of differentiation (e.g., day in vitro 35 for GABAergic neurons), cells are harvested from the multi-well dishes by gentle enzymatic dissociation with TrypLE.
2. Optional cell sorting/purification step: Massive tumor overgrowth observed in earlier studies [44, 48] can be circumvented by adding a sorting step for the $CD15^{-}/CD24^{high}/CD29^{low}$ subset [64].
3. Viability of the embryonic cells is assessed by Trypan Blue exclusion prior to grafting using a hemocytometer. The single-cell suspension is adjusted to a concentration as required by the particular study, but a typical working concentration is 100,000 cells/μL. The total volume prepared in transplantation medium depends on the number of animals grafted and the number of cells available.
4. The transplantation procedure takes place immediately following the preparation of the cell suspension. Typically, in our laboratories, transplantation is done using a glass capillary with a diameter of 50–70 μm connected via an adaptive polyethylene tube to the cannula of a 5 μL Hamilton microsyringe or a 10 μL

Hamilton microsyringe with 26-Gauge steel cannula. While the former method has been reported to result in better cell survival [65, 66], the second approach may allow for inclusion of smaller aggregates which has been interpreted as advantageous in a recent human-to-rodent transplantation study [67].

5. First, host animals are anesthetized using isoflurane and positioned in a rodent stereotactic frame. The animal's head is prepared appropriately with the bore-hole drilled at the coordinates required by the protocol (obtained from stereotactic atlases for mice [63] or rats [62]). The cell suspension is aspirated into the glass capillary/cannula and the instrument is lowered into the respective target area.
6. Cells are expelled at a rate of 1 μL/min from the instrument, and the capillary or cannula is kept in place for 2 min, before slowly retracting.
7. If the protocol requires several cell deposits at different coordinates, then the procedure is repeated until the final deposit has been made. Typically, between 200,000 and 500,000 cells are grafted, but this needs to be carefully assessed depending on the cell source and its proliferative properties.
8. Following grafting, the instruments are gently removed, the exposed area carefully rinsed with the sterile saline solution. While still in the ear bars, the scalp wound is closed with wound clips, and treated with antibiotic ointment. As analgesic, the animals can be given s.c. Temgesic (0.05 mg buprenorphine-hydrochloride/kg bodyweight) in the neck.

3 Overview of Suitable Readout Methods of Neural Repair

3.1 Notes on In Vivo Functional Analysis and Behavior

Functional analysis is indispensible if the investigator is interested in potential in vivo therapeutic relevance of the cells. Cells can survive, migrate, differentiate into various phenotypes, and even morphologically and anatomically integrate in the host CNS to some degree, but its therapeutic use is compromised if it is not capable of having an impact on the host's behavior as assessed by motor and/or cognitive tests.

Functional testing of animals can be divided into two large categories: Manual and Operant testing. These two labels refer to various aspects of the nature of the tests such the degree of involvement of the investigator with the experimental animal, the number of trials per session, methods of data collection, etc. However, they are not indicative of the general faculty of the animal that is being tested, because with both type of tests the investigator can study motor, sensorimotor, or cognitive behavior. Manual tasks are the most common and widespread behavioral procedures, and many nonbehavioral scientists may familiarize themselves with them as they do not require extensive specialist knowledge and limited equipment (drug-induced rotation, in particular). There are four categories of manual test to study graft-mediated recovery:

1. Simple motor tasks: These tests are predominantly observational tests where the animals are placed either in an open field, in a specially designed apparatus or in holding cages, and their behavior is recorded. Behavioral patterns such as locomotor activity (spontaneous or activated), explorative profiles, or drug-induced rotation can be carried out with the help of video recording or tracking equipment. Simpler analysis methods might include placing ink on the feet of the animals to analyze the walking pattern revealed by footprints on paper [68, 69].
2. Sensorimotor tasks: These tasks involve more interaction between the animals and their environment. Essentially, the objective of the tasks is to analyze how, and to what degree the animals are relying on their sensory inputs. The animal models of Parkinson's and Huntington's diseases we use rely on unilateral dopamine depletion and unilateral striatal lesion. As such, these tasks are very useful in revealing lateralized neglect and deficits, and have been shown to be sensitive to transplantation mediated recovery. Examples: Stepping test, Cylinder test, Corridor test, and Disengage test [70–72].
3. Motor skills tasks: These tasks combine both motor aspects and interaction with the apparatus, but in addition, require motor learning as well. Performance on these tasks depends on balance, equilibrium, or fine, skilled digit control. Examples: Paw-reaching, Beam balance, and Rotorod [73–76].
4. Cognitive, learning, and memory tasks: There are a variety of mazes as manual test that are capable of examining working and reference memory, hippocampal, and striatal learning. Examples are T-maze, Y-maze, or the 8-arm radial maze [77].

Operant testing is the reserve of behaviorally specialized laboratories and is generally not used to simple screen to see whether a graft has functional effect or not. For a review on operant testing *see* ref. 78.

3.2 Notes on Postmortem Histological Analysis

Overview

Postmortem histological (Fig. 3) assessment is paramount to every stem cell transplantation study as it offers the opportunity to confirm the survival of the cells, their distribution and phenotype, their migratory properties, as well as the number of cells and the volume of the graft, for example. The quantitative and qualitative description of the grafted cells can then feedback into the interpretation of the presence or absence of any graft-mediated functional effects observed during the behavioral assessment period. We briefly summarize typical methodological steps for the postmortem assessment of striatal grafts in animal models of HD or PD, for instance.

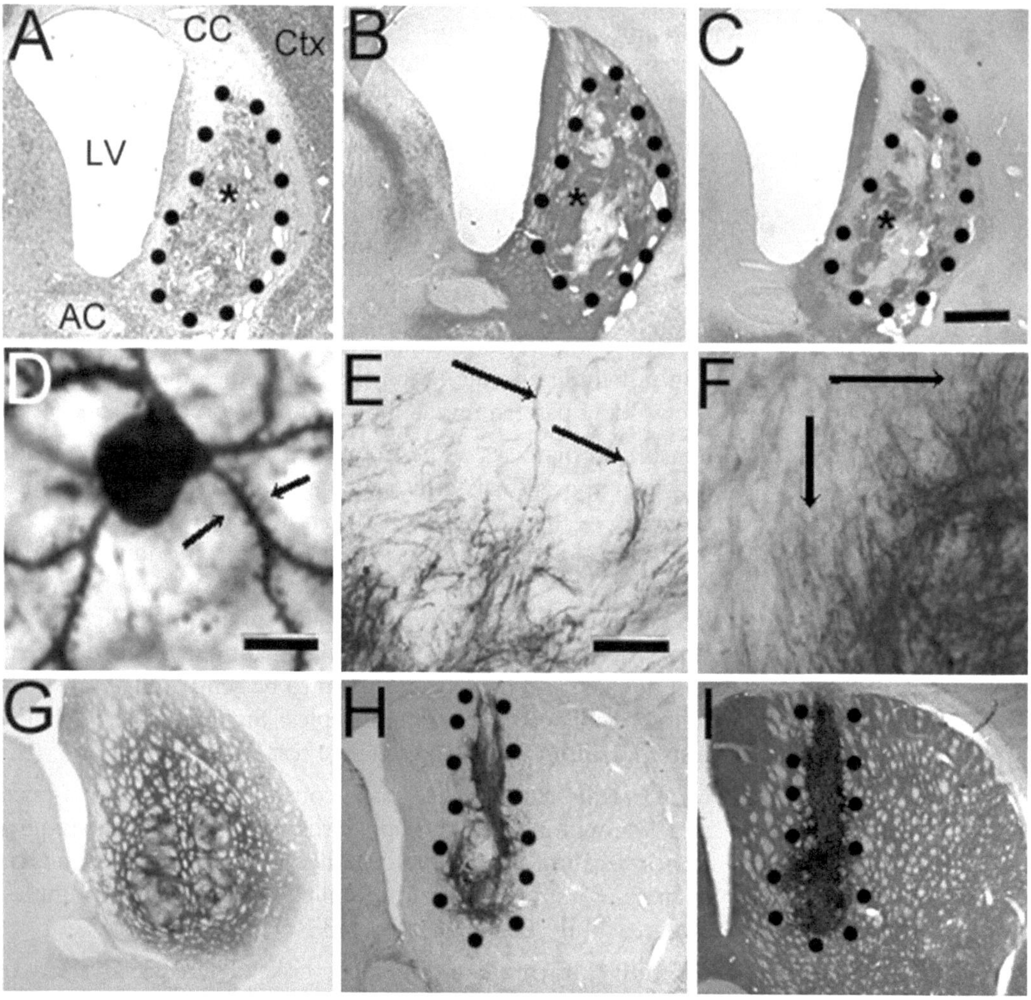

Fig. 3 Assessment of transplanted cells by histological methods. In the rodent model of HD primary tissue grafts (here: E14 whole ganglionic eminence) can be visualized using various markers such as the neuronal marker NeuN (**a**), the enzyme tyrosine hydroxylase (labeling DA afferents, **b**), or the striatal tissue marker dopamine-, cAMP-regulated phosphoprotein of 32,000 kDa, DARPP-32 (**c**). Morphology and ultrastructural properties of the grafted neurons can be studied if the cells express particular markers, e.g., β-Galactosidase allowing the identification of dendritic spines (indicated by *arrows*, **d**). Using tissue from donor animals that express GFP in specific neuronal subpopulations is also an option (*arrows* show fibers of graft origin projecting into the host, **e**). Fiber outgrowth from grafted human neural stem cells can also be studied using antibodies specific to human neurofilament (*arrows*, **f**). DA cells grafted into models of PD are typically visualized by tyrosine hydroxylase showing graft derived fiber outgrowth (**g**). When using xenografts, for example human tissue into the rat, markers for human nuclei (**h**), or human neurofilament (**i**) can be used. *LV* lateral ventricle, *CC* corpus callosum, *Ctx* cortex, *AC* anterior commisure. *Dots* delineate the grafted tissue in all cases, the *asterisk* identifies the graft. Scale bars: panels **a**–**c** and **g**–**i**, 1 μm; panel **d**, 10 μm; panels **e**–**f**, 50 μm. Images have been adapted from refs. 73, 75, 79, 83. Images (**a**–**f**) have been adapted from refs. 79, 83

Perfusion, fixation, sectioning

1. At the end of the study animals are terminally anesthetized by intraperitoneal injection of an overdose of ketamine and rompun and transcardially perfused with 300 mL 0.1 M PBS followed by 300 mL 4 % paraformaldehyde in 0.1 M PBS.
2. The brains are removed and postfixed in 4 % paraformaldehyde for 1.5–24 h, then immersed in 30 % sucrose in 0.1 M PBS for 24 h until they sink for cryoprotection.
3. Typically, tissue is serially sectioned on a freezing microtome at a thickness of 40 μm through both the striatum and substantia nigra and collected as six series. Cutting of the brain can occur also on a vibratome, or a cryostat, in which case the preparations will vary.

Staining and immunohistochemical analysis

1. Free floating sections are washed 3× with PBS, pre-incubated for 2 h in 3 % bovine serum albumine/0.3 % Triton X-100/ PBS and incubated overnight at room temperature with the selected antibodies (for examples of antibodies used *see* Table 1). Typically, we would use NeuN, a marker of mature neurons; DARPP32, a marker of striatal tissue [79, 80]; and tyrosine hydroxylase, a marker used to identify DA neurons and axonal input [81]. Human cells in rodent brain can further be identified by human nuclei/human nuclear antigen (HNA) as well as the surface marker human neural cell adhesion molecule (NCAM).
2. After three washes with PBS, sections are incubated for 2 h with biotinylated secondary antibody for diaminobenzidine-based (DAB) staining or with fluorescent-labeled secondary antibody. Visualization of the biotinylated antibody is carried out with the ABC-kit (Vector, Germany) and DAB (Merck, Germany). Sections are mounted on slides, dehydrated and cover slipped, and subjected to image analysis using a light/ fluorescent microscope and appropriate software.
3. Double-/triple-labeling studies use fluorescent secondary antibodies to describe the co-expression of markers, while other quantitative measurements are carried out with DAB-stained sections. The extent of the lesion or the volume of the graft is estimated by tracing the outlines of the cross section areas throughout the anterior–posterior axis of the region of interest, e.g., the striatum, on the appropriate stained sections: DARPP-32 to identify the striatal-like sections in the graft. The volumes are then calculated by taking into account the sum of the areas and correcting for section thickness (M) and sample frequency (f): Volume $(\text{mm}^3) = \sum$ of areas $(\text{mm}^2) \times (M) \times (f)$.

Table 1
Typical markers used in vitro and in histological assessment of cell transplantation

Target protein/molecule	Target cells	Reason
hNu/HNA	Cells of human origin, human nuclei	To distinguish more clearly between the rat and the human (grafted) neurons and to identify neurons that might have migrated away from grafted region
hNF	Neurofilament of human origin	To follow the projections originating from cells of human origin
human NCAM, CD56	Human neural cells and neurons including processes	
Ki67	Cell division/mitotic marker	To stain cells that are still mitotic/dividing
CD29, beta-1-integrin	Proliferative neuroepithelial cells; also neural crest cells (plus a range of neural cell types)	High levels are present on proliferative stem cells, coexpressing CD15
SSEA1/CD15	Proliferative neuroepithelial cells	Neural stem cells are positive for CD15, coexpressing high levels of CD29
FORSE1	Proliferative neuroepithelial cells	Proliferative neural cells. Telencephalic precursors
Nestin	Neural precursor and stem cells	Neural precursor and stem cells
CD24	Neuroblasts (plus other cell types)	Neuroblasts exhibit high levels of CD24 and Dcx
Dcx	Migrating precursors, neuroblasts	
TUJ1	Early neuronal marker with III β-tubulin epitope	To identify neurons at various differentiation stages
MAP2	Neurons	Mature neurons
NeuN	Mature neurons	To identify host and graft neurons

Synaptophysin	Neurons	Synaptic marker identifying mature neurons
GAD65 or GAD67	GABA neurons	To identify GABAergic neurons
ChAT	Cholinergic neurons	To identify cholinergic neurons
DARPP-32	Striatal-like neurons	To identify striatal neurons
TH	Dopaminergic (DA) neurons	To identify DA neurons
GFAP	Astrocytes	To identify astrocytes within the graft
APC or MBP	Oligodendrocytes	To identify oligodendrocytes within the graft
Mac-3, CD3, B-220	Microglia, T-, B-cells, respectively	To monitor the host's immune response to the graft

4. The cell counting method will depend on the number of cells there are to count. If only an estimate is possible because the numbers range in the thousands per graft per section (e.g., using a general neuronal marker), one needs to apply unbiased stereological counting methods [82]. However, if the cell population of interest (e.g., DARPP-32 or TH-positive cells) ranges in the hundreds then the cells can be counted manually using a light microscope. Counting should be done across the anterior–posterior axis of the region of interest on sections where the distance between consecutive sections and the frequency of sections is known. The Abercrombie formula of cell estimate is used as this corrects for the possibility of over counting: Cell number, P, is calculated by the formula: $P = 1/f \times A \times M/(D + M)$, where A is the cell count from the entire graft, f is the frequency of sections, M is the section thickness, and D is the average cell diameter.

Acknowledgements

Hereditary Disease Foundation, USA (MD).

Emmy Noether Program of the German Research Foundation (DFG); Neurex Welcome back support (JP).

References

1. Gustavsson A, Svensson M, Jacobi F, Allgulander C, Alonso J, Beghi E, Dodel R, Ekman M, Faravelli C, Fratiglioni L, Gannon B, Jones DH, Jennum P, Jordanova A, Jönsson L, Karampampa K, Knapp M, Kobelt G, Kurth T, Lieb R, Linde M, Ljungcrantz C, Maercker A, Melin B, Moscarelli M, Musayev A, Norwood F, Preisig M, Pugliatti M, Rehm J, Salvador-Carulla L, Schlehofer B, Simon R, Steinhausen H-C, Stovner LJ, Vallat J-M, den Bergh PV, van Os J, Vos P, Xu W, Wittchen H-U, Jönsson B, Olesen J (2011) Cost of disorders of the brain in Europe 2010. Eur Neuropsychopharmacol 21(10):718–779
2. Singec I, Jandial R, Crain A, Nikkhah G, Snyder EY (2007) The leading edge of stem cell therapeutics. Annu Rev Med 58:313–328
3. Thomson JA, Itskovitz-Eldor J, Shapiro SS, Waknitz MA, Swiergiel JJ, Marshall VS, Jones JM (1998) Embryonic stem cell lines derived from human blastocysts. Science 282(5391):1145–1147
4. Takahashi K, Yamanaka S (2006) Induction of pluripotent stem cells from mouse embryonic and adult fibroblast cultures by defined factors. Cell 126(4):663–676
5. Takahashi K, Tanabe K, Ohnuki M, Narita M, Ichisaka T, Tomoda K, Yamanaka S (2007) Induction of pluripotent stem cells from adult human fibroblasts by defined factors. Cell 131(5):861–872
6. Stenevi U, Björklund A, Svendgaard NA (1976) Transplantation of central and peripheral monoamine neurons to the adult rat brain: techniques and conditions for survival. Brain Res 114(1):1–20
7. Ungerstedt U, Arbuthnott GW (1970) Quantitative recording of rotational behavior in rats after 6-hydroxy-dopamine lesions of the nigrostriatal dopamine system. Brain Res 24(3):485–493
8. Coyle JT, Schwarcz R (1976) Lesion of striatal neurones with kainic acid provides a model for Huntington's chorea. Nature 263(5574):244–246
9. Nikkhah G, Cunningham MG, Cenci MA, McKay RD, Björklund A (1995) Dopaminergic microtransplants into the substantia nigra of neonatal rats with bilateral 6-OHDA lesions. I. Evidence for anatomical reconstruction of the nigrostriatal pathway. J Neurosci 15(5 Pt 1):3548–3561

10. Nikkhah G, Bentlage C, Cunningham MG, Björklund A (1994) Intranigral fetal dopamine grafts induce behavioral compensation in the rat Parkinson model. J Neurosci 14(6): 3449–3461
11. Björklund A, Dunnett SB, Stenevi U, Lewis ME, Iversen SD (1980) Reinnervation of the denervated striatum by substantia nigra transplants: functional consequences as revealed by pharmacological and sensorimotor testing. Brain Res 199(2):307–333
12. Björklund A, Stenevi U, Dunnett SB, Iversen SD (1981) Functional reactivation of the deafferented neostriatum by nigral transplants. Nature 289(5797):497–499
13. Lindvall O, Rehncrona S, Brundin P, Gustavii B, Astedt B, Widner H, Lindholm T, Björklund A, Leenders KL, Rothwell JC, Frackowiak R, Marsden D, Johnels B, Steg G, Freedman R, Hoffer BJ, Seiger A, Bygdeman M, Strömberg I, Olson L (1989) Human fetal dopamine neurons grafted into the striatum in two patients with severe Parkinson's disease. A detailed account of methodology and a 6-month follow-up. Arch Neurol 46(6):615–631
14. Lindvall O, Widner H, Rehncrona S, Brundin P, Odin P, Gustavii B, Frackowiak R, Leenders KL, Sawle G, Rothwell JC (1992) Transplantation of fetal dopamine neurons in Parkinson's disease: one-year clinical and neurophysiological observations in two patients with putaminal implants. Ann Neurol 31(2):155–165
15. Lindvall O (1991) Transplants in Parkinson's disease. Eur Neurol 31(Suppl 1):17–27
16. Freeman TB, Olanow CW, Hauser RA, Nauert GM, Smith DA, Borlongan CV, Sanberg PR, Holt DA, Kordower JH, Vingerhoets FJ (1995) Bilateral fetal nigral transplantation into the postcommissural putamen in Parkinson's disease. Ann Neurol 38(3):379–388
17. Olanow CW, Goetz CG, Kordower JH, Stoessl AJ, Sossi V, Brin MF, Shannon KM, Nauert GM, Perl DP, Godbold J, Freeman TB (2003) A double-blind controlled trial of bilateral fetal nigral transplantation in Parkinson's disease. Ann Neurol 54(3):403–414
18. Isacson O (2009) Cell therapy ahead for Parkinson's disease. Science 326(5956):1060
19. Mendez I, Sanchez-Pernaute R, Cooper O, Viñuela A, Ferrari D, Björklund L, Dagher A, Isacson O (2005) Cell type analysis of functional fetal dopamine cell suspension transplants in the striatum and substantia nigra of patients with Parkinson's disease. Brain 128(Pt 7):1498–1510
20. Mendez I, Viñuela A, Astradsson A, Mukhida K, Hallett P, Robertson H, Tierney T, Holness R, Dagher A, Trojanowski JQ, Isacson O (2008) Dopamine neurons implanted into people with Parkinson's disease survive without pathology for 14 years. Nat Med 14(5):507–509
21. Li J-Y, Englund E, Holton JL, Soulet D, Hagell P, Lees AJ, Lashley T, Quinn NP, Rehncrona S, Björklund A, Widner H, Revesz T, Lindvall O, Brundin P (2008) Lewy bodies in grafted neurons in subjects with Parkinson's disease suggest host-to-graft disease propagation. Nat Med 14(5):501–503
22. Kordower JH, Chu Y, Hauser RA, Freeman TB, Olanow CW (2008) Lewy body-like pathology in long-term embryonic nigral transplants in Parkinson's disease. Nat Med 14(5):504–506
23. Bachoud-Lévi AC, Rémy P, Nguyen JP, Brugières P, Lefaucheur JP, Bourdet C, Baudic S, Gaura V, Maison P, Haddad B, Boissé MF, Grandmougin T, Jény R, Bartolomeo P, Dalla Barba G, Degos JD, Lisovoski F, Ergis AM, Pailhous E, Cesaro P, Hantraye P, Peschanski M (2000) Motor and cognitive improvements in patients with Huntington's disease after neural transplantation. Lancet 356(9246):1975–1979
24. Bachoud-Lévi A-C, Hantraye P, Peschanski M (2002) Fetal neural grafts for Huntington's disease: a prospective view. Mov Disord 17(3):439–444
25. Bachoud-Lévi A-C, Gaura V, Brugières P, Lefaucheur J-P, Boissé M-F, Maison P, Baudic S, Ribeiro M-J, Bourdet C, Remy P, Cesaro P, Hantraye P, Peschanski M (2006) Effect of fetal neural transplants in patients with Huntington's disease 6 years after surgery: a long-term follow-up study. Lancet Neurol 5(4):303–309
26. Bachoud-Lévi A-C (2009) Neural grafts in Huntington's disease: viability after 10 years. Lancet Neurol 8(11):979–981
27. Sahni V, Kessler JA (2010) Stem cell therapies for spinal cord injury. Nat Rev Neurol 6(7):363–372
28. Stout JT, Francis PJ (2011) Surgical approaches to gene and stem cell therapy for retinal disease. Hum Gene Ther 22(5):531–535
29. Savitz SI, Rosenbaum DM, Dinsmore JH, Wechsler LR, Caplan LR (2002) Cell transplantation for stroke. Ann Neurol 52(3):266–275
30. Jeong S-W, Chu K, Jung K-H, Kim SU, Kim M, Roh J-K (2003) Human neural stem cell transplantation promotes functional recovery

in rats with experimental intracerebral hemorrhage. Stroke 34(9):2258–2263
31. Kelly S, Bliss TM, Shah AK, Sun GH, Ma M, Foo WC, Masel J, Yenari MA, Weissman IL, Uchida N, Palmer T, Steinberg GK (2004) Transplanted human fetal neural stem cells survive, migrate, and differentiate in ischemic rat cerebral cortex. Proc Natl Acad Sci USA 101(32):11839–11844
32. Hyun I (2010) The bioethics of stem cell research and therapy. J Clin Invest 120(1):71–75
33. Yu J, Vodyanik MA, Smuga-Otto K, Antosiewicz-Bourget J, Frane JL, Tian S, Nie J, Jonsdottir GA, Ruotti V, Stewart R, Slukvin II, Thomson JA (2007) Induced pluripotent stem cell lines derived from human somatic cells. Science 318(5858):1917–1920
34. Sendtner M (2011) Regenerative medicine: Bespoke cells for the human brain. Nature 476(7359):158–159
35. Kim K-S (2011) Converting human skin cells to neurons: a new tool to study and treat brain disorders? Cell Stem Cell 9(3):179–181
36. Spemann H, Mangold H (2001) Induction of embryonic primordia by implantation of organizers from a different species. 1923. Int J Dev Biol 45(1):13–38
37. Gaulden J, Reiter JF (2008) Neur-ons and neur-offs: regulators of neural induction in vertebrate embryos and embryonic stem cells. Hum Mol Genet 17(R1):R60–R66
38. Sonntag K-C, Pruszak J, Yoshizaki T, van Arensbergen J, Sanchez-Pernaute R, Isacson O (2007) Enhanced yield of neuroepithelial precursors and midbrain-like dopaminergic neurons from human embryonic stem cells using the bone morphogenic protein antagonist noggin. Stem Cells 25(2):411–418
39. Kawasaki H, Mizuseki K, Nishikawa S, Kaneko S, Kuwana Y, Nakanishi S, Nishikawa SI, Sasai Y (2000) Induction of midbrain dopaminergic neurons from ES cells by stromal cell-derived inducing activity. Neuron 28(1):31–40
40. Perrier AL, Tabar V, Barberi T, Rubio ME, Bruses J, Topf N, Harrison NL, Studer L (2004) Derivation of midbrain dopamine neurons from human embryonic stem cells. Proc Natl Acad Sci USA 101(34):12543–12548
41. Jessell TM (2000) Neuronal specification in the spinal cord: inductive signals and transcriptional codes. Nat Rev Genet 1(1):20–29
42. Okada Y, Shimazaki T, Sobue G, Okano H (2004) Retinoic-acid-concentration-dependent acquisition of neural cell identity during in vitro differentiation of mouse embryonic stem cells. Dev Biol 275(1):124–142
43. Lee SH, Lumelsky N, Studer L, Auerbach JM, McKay RD (2000) Efficient generation of midbrain and hindbrain neurons from mouse embryonic stem cells. Nat Biotechnol 18(6):675–679
44. Roy NS, Cleren C, Singh SK, Yang L, Beal MF, Goldman SA (2006) Functional engraftment of human ES cell-derived dopaminergic neurons enriched by coculture with telomerase-immortalized midbrain astrocytes. Nat Med 12(11):1259–1268
45. Brederlau A, Correia AS, Anisimov SV, Elmi M, Paul G, Roybon L, Morizane A, Bergquist F, Riebe I, Nannmark U, Carta M, Hanse E, Takahashi J, Sasai Y, Funa K, Brundin P, Eriksson PS, Li J-Y (2006) Transplantation of human embryonic stem cell-derived cells to a rat model of Parkinson's disease: effect of in vitro differentiation on graft survival and teratoma formation. Stem Cells 24(6):1433–1440
46. Sharp J, Keirstead HS (2007) Therapeutic applications of oligodendrocyte precursors derived from human embryonic stem cells. Curr Opin Biotechnol 18(5):434–440
47. Li X-J, Du Z-W, Zarnowska ED, Pankratz M, Hansen LO, Pearce RA, Zhang S-C (2005) Specification of motoneurons from human embryonic stem cells. Nat Biotechnol 23(2):215–221
48. Aubry L, Bugi A, Lefort N, Rousseau F, Peschanski M, Perrier AL (2008) Striatal progenitors derived from human ES cells mature into DARPP32 neurons in vitro and in quinolinic acid-lesioned rats. Proc Natl Acad Sci USA 105(43):16707–16712
49. Pruszak J, Isacson O (2009) Molecular and cellular determinants for generating ES-cell derived dopamine neurons for cell therapy. Adv Exp Med Biol 651:112–123
50. Karki S, Pruszak J, Isacson O, Sonntag KC (2006) ES cell-derived neuroepithelial cell cultures. J Vis Exp 1:118
51. Chambers SM, Fasano CA, Papapetrou EP, Tomishima M, Sadelain M, Studer L (2009) Highly efficient neural conversion of human ES and iPS cells by dual inhibition of SMAD signaling. Nat Biotechnol 27(3):275–280
52. Morizane A, Doi D, Kikuchi T, Nishimura K, Takahashi J (2011) Small-molecule inhibitors of bone morphogenic protein and activin/nodal signals promote highly efficient neural induction from human pluripotent stem cells. J Neurosci Res 89(2):117–126
53. Hedlund E, Hefferan MP, Marsala M, Isacson O (2007) Cell therapy and stem cells in animal models of motor neuron disorders. Eur J Neurosci 26(7):1721–1737
54. Vugler A, Carr A-J, Lawrence J, Chen LL, Burrell K, Wright A, Lundh P, Semo M, Ahmado A, Gias C, da Cruz L, Moore H, Andrews P, Walsh J, Coffey P (2008) Elucidating the phenomenon of HESC-

derived RPE: anatomy of cell genesis, expansion and retinal transplantation. Exp Neurol 214(2):347–361
55. Björklund A, Lindvall O (2000) Cell replacement therapies for central nervous system disorders. Nat Neurosci 3(6):537–544
56. Liu YP, Lang BT, Baskaya MK, Dempsey RJ, Vemuganti R (2009) The potential of neural stem cells to repair stroke-induced brain damage. Acta Neuropathol 117(5):469–480
57. Lindvall O, Kokaia Z (2011) Stem cell research in stroke: how far from the clinic? Stroke 42(8):2369–2375
58. Betarbet R, Sherer TB, Greenamyre JT (2002) Animal models of Parkinson's disease. Bioessays 24(4):308–318
59. Hirsch EC, Höglinger G, Rousselet E, Breidert T, Parain K, Feger J, Ruberg M, Prigent A, Cohen-Salmon C, Launay JM (2003) Animal models of Parkinson's disease in rodents induced by toxins: an update. J Neural Transm Suppl 65:89–100
60. Ramaswamy S, McBride JL, Kordower JH (2007) Animal models of Huntington's disease. ILAR J 48(4):356–373
61. Döbrössy M, Büchele F, Nikkhah G (2011) Excitotoxic lesions of the rodent striatum. In: Animal models of movement disorders, vol 2. Neuromethods, Humana Press, pp 21–35. http://link.springer.com/protocol/10.1007%2F978-1-61779-301-1_2
62. Paxinos G, Watson C (1998) The rat brain in stereotaxic coordinates. Academic, New York
63. Paxinos G, Franklin K (2001) The mouse brain in stereotaxic coordinates. Academic, New York
64. Pruszak J, Ludwig W, Blak A, Alavian K, Isacson O (2009) CD15, CD24, and CD29 define a surface biomarker code for neural lineage differentiation of stem cells. Stem Cells 27(12):2928–2940
65. Nikkhah G, Cunningham MG, Jödicke A, Knappe U, Björklund A (1994) Improved graft survival and striatal reinnervation by microtransplantation of fetal nigral cell suspensions in the rat Parkinson model. Brain Res 633(1–2):133–143
66. Nikkhah G, Olsson M, Eberhard J, Bentlage C, Cunningham MG, Björklund A (1994) A microtransplantation approach for cell suspension grafting in the rat Parkinson model: a detailed account of the methodology. Neuroscience 63(1):57–72
67. Rath A, Klein A, Papazoglou A, Pruszak J, Garcia J, Krause M, Maciaczyk J, Dunnett SB, Nikkhah G. Survival and functional restoration of human fetal ventral mesencephalon following transplantation in a rat model of Parkinson's Disease. Cell Transplant. 2012 Sep 7. [Epub ahead of print]
68. Nikkhah G, Rosenthal C, Falkenstein G, Roedter A, Papazoglou A, Brandis A (2009) Microtransplantation of dopaminergic cell suspensions: further characterization and optimization of grafting parameters. Cell Transplant 18(2):119–133
69. Klein A, Wessolleck J, Papazoglou A, Metz GA, Nikkhah G (2009) Walking pattern analysis after unilateral 6-OHDA lesion and transplantation of foetal dopaminergic progenitor cells in rats. Behav Brain Res 199(2):317–325
70. Döbrössy MD, Dunnett SB (2007) The corridor task: striatal lesion effects and graft-mediated recovery in a model of Huntington's disease. Behav Brain Res 179(2):326–330
71. Jiang W, Büchele F, Papazoglou A, Döbrössy M, Nikkhah G (2011) Multi-tract microtransplantation increases the yield of DARPP-32 positive embryonic striatal cells in a rodent model of Huntington's disease. Cell Transplant 20(10):1515–1527
72. Döbrössy MD, Dunnett SB (2005) Training specificity, graft development and graft-mediated functional recovery in a rodent model of Huntington's disease. Neuroscience 132(3):543–552
73. Cordeiro KK, Jiang W, Papazoglou A, Tenório SB, Döbrössy M, Nikkhah G (2010) Graft-mediated functional recovery on a skilled forelimb use paradigm in a rodent model of Parkinson's disease is dependent on reward contingency. Behav Brain Res 212(2):187–195
74. Döbrössy MD, Dunnett SB (2003) Motor training effects on recovery of function after striatal lesions and striatal grafts. Exp Neurol 184(1):274–284
75. Falkenstein G, Rosenthal C, Reum T, Morgenstern R, Döbrössy M, Nikkhah G (2009) Pattern of long-term sensorimotor recovery following intrastriatal and—accumbens DA micrografts in a rat model of Parkinson's disease. J Comp Neurol 515(1):41–55
76. Nikkhah G, Rosenthal C, Falkenstein G, Samii M (1998) Dopaminergic graft-induced long-term recovery of complex sensorimotor behaviors in a rat model of Parkinson's disease. Zentralbl Neurochir 59(2):97–103
77. Dunnett SB, Isacson O, Sirinathsinghji DJ, Clarke DJ, Björklund A (1988) Striatal grafts in the ibotenic acid-lesioned neostriatum: functional studies. Prog Brain Res 78:39–45
78. Döbrössy M, Brooks S, Truemann R, Brasted P, Dunnett S (2009) Operant analysis of fronto-striatal function in rodents. In: Methods of behaviour analysis in neuroscience, pp 215–246. http://www.ncbi.nlm.nih.gov/books/NBK5215
79. Döbrössy MD, Dunnett SB (2006) Morphological and cellular changes within

embryonic striatal grafts associated with enriched environment and involuntary exercise. Eur J Neurosci 24(11):3223–3233
80. Graybiel AM, Liu FC, Dunnett SB (1989) Intrastriatal grafts derived from fetal striatal primordia. I. Phenotypy and modular organization. J Neurosci 9(9):3250–3271
81. Liu FC, Graybiel AM, Dunnett SB, Baughman RW (1990) Intrastriatal grafts derived from fetal striatal primordia: II. Reconstitution of cholinergic and dopaminergic systems. J Comp Neurol 295(1):1–14
82. Peterson DA (1999) Quantitative histology using confocal microscopy: implementation of unbiased stereology procedures. Methods 18(4):493–507
83. Döbrössy M, Klein A, Janghra N, Nikkhah G, Dunnett SB (2011) Validating the use of M4-BAC-GFP mice as tissue donors in cell replacement therapies in a rodent model of Huntington's disease. J Neurosci Methods 197(1):6–13

Chapter 8

Cell-Based Therapies for Myocardial Repair: Emerging Role for Bone Marrow-Derived Mesenchymal Stem Cells (MSCs) in the Treatment of the Chronically Injured Heart

Jose S. Da Silva and Joshua M. Hare

Abstract

Accumulating data support the use of bone marrow (BM)-derived MSCs in animal models (e.g., swine) to restore cardiac function and tissue perfusion in chronic cardiac injury. Based on results obtained in swine, we are currently conducting phase I/II clinical trials to address the safety, cell type, cell dose, delivery technique, and efficacy of MSCs in patients with chronic heart failure. MSCs for these trials are isolated from harvested BM and then processed and expanded for intracardiac injection. The BM-MSCs in use for the clinical trials are of clinical grade having been processed successfully in an FDA-approved cGMP facility.

Key words BM MSC, FDA, MI, HF, cGMP, cGTP, IND, HCT/P, CMC

1 Introduction

Cell-based therapies and regenerative medicine represent an exciting and emerging field where cells (either autologous or allogeneic preparations) are used for the treatment of damaged cells, tissues, or organs due to diseases or injury [1]. Since the first bone marrow transplant was conducted in 1968 [2–4], BM stem cells (SC), marrow-derived stromal cells, and MSCs [5] have been the focus of extensive biologic and translational investigations [6]. MSCs offer therapeutic potential due to their ability to self-renew, give rise to specialized cell types, be immune-privileged, be immunosuppressive, and be expanded extensively in in vitro adherent cultures without loss of function and phenotype [7–10]. While the greatest limitation to their use is their low abundance in the bone marrow, this is offset by their proliferative capacity *ex vivo*. Today, the therapeutic method of choice to treat end stage heart failure (HF) is heart transplant despite its inherent hurdles [11, 12];

Robert G. Gourdie and Tereance A. Myers (eds.), *Wound Regeneration and Repair: Methods and Protocols*, Methods in Molecular Biology, vol. 1037, DOI 10.1007/978-1-62703-505-7_8, © Springer Science+Business Media New York 2013

nevertheless, cell-based therapies could offer a major new alternative therapeutic solution for patients with myocardial infarction (MI), heart failure, or congenital heart disease by rescuing or regenerating the myocardium [13–16]. Before these therapeutic agents can be used in humans, an investigational new drug (IND) application is required to be submitted to the Food Drug Administration (FDA) since their licensing is subjected to the Public Health Act.

Before submitting an IND application to the FDA, safety and effectiveness of the cell therapeutic should be addressed in preclinical studies. These studies are performed both in vitro and in vivo and the proof-of-concept requires appropriate selection of an animal model. The most important safety aspect that must be addressed in preclinical studies is the purity of the cell therapeutic followed by the cell type and its potency. Cell therapeutics represent a high safety risk due to potential contamination from other cell sources, microorganisms, or adventitious agents due to the manipulation required and the different reagents used during culture. Moreover, genetic aberration can arise during long-term cell culture. Other areas to investigate during the preclinical studies are the route of infusion, dose, engraftment, engraftment duration, bio-distribution, and tumorigenesis.

Several studies have demonstrated the use of MSCs effectively in vivo [17]. As a primary example, we have conducted a randomized blinded, placebo-controlled trial of BM-MSCs in adult mini swine to assess whether autologous MSCs safely reduce infarct size and improve cardiac function in post myocardial infarction (MI) heart failure (HF). This study demonstrated that (1) autologous MSCs can be safely isolated from BM obtained from pigs in a porcine model of ischemic HF; (2) MSCs can be safely cultured and expanded (4–7 passages) *ex vivo* in order to obtain adequate numbers of cells necessary for the infusion; (3) the phenotype and purity of the isolated cells were MSCs which were assessed by flow cytometry ($CD45^-/CD90^+$) and CFU-F assay; (4) the MSCs can be safely delivered epicardially by injections into and surrounded akinetic or hypokinetic areas of the heart; and (5) MSCs in two different cell doses (low $= 20 \times 10^6$ and high $= 200 \times 10^6$) effectively showed over time: (a) a decrease in scar size, (b) a decrease in the circumferential extent of the infarct scar, (c) an increase in wall thickness in the infarct region, (d) increased improvement in contractile function, (e) increased perfusion, (f) improved global left ventricular (LV) function, and (g) an increase in ejection fraction (EF) compared to the placebo group. The use of MSCs did not foster neoplastic processes for the duration of the study [17]. Even though, most of the questions about safety and efficacy were answered, there are concerns about the safety of using late passage MSCs since genetic abnormalities may occur as well as histological proof of MSC engraftment and heart tissue calcification at the site of infusion [17].

The results of these studies were used as preclinical data for the IND submission to the FDA of the approved clinical trial PROMETHEUS (A Phase I/II, Randomized, Double-Blinded, Placebo-Controlled Study of the Safety and Efficacy of Intramyocardial Injection of Autologous Human Mesenchymal Stem Cells (MSCs) in Patients With Chronic Ischemic Left Ventricular Dysfunction Secondary to Myocardial Infarction (MI) Undergoing Cardiac Surgery for Coronary Artery Bypass Grafting) (clinical trials.govNCT00587990) [18, 19]. The study enrolled patients who meet the following enrollment criteria:

1. Diagnosis of chronic ischemic heart failure caused by MI.
2. Required cardiac surgery for Coronary Artery Bypass Grafting CABG.
3. Ejection fraction between 15 and 50 %.
4. Presence of an akinetic or dyskinetic region by standard imaging [18].

A BM aspirate was obtained approximately 5 weeks before treatment in order to scale up the autologous MSCs. The BM aspirate was sent to our cGMP (current Good Manufacturing Practice) facility where the MSCs were scaled up and processed according to FDA cGMP/cGTP guidelines prior to infusion at the time of surgery. Patients were randomized to three different groups: placebo, low-dose (20×10^6), and high-dose (200×10^6) MSC groups.

As previously mentioned, the use of cellular therapies in preclinical and clinical trials is regulated by the FDA. Laboratory practices which are intended to collect data for preclinical research or marketing permits for products regulated by the FDA fall under the guidelines of current Good Laboratory Practices (cGLP) stated in the code of federal regulations 21 CFR Part 58, which assures the integrity and the quality of the safety data. Manufacturing of biologics falls under cGMP guidelines as stated in the code of federal regulations 21 CFR parts 600–680. GMPs are practices that are used during the manufacturing of drug products for administration to humans or animals or for the manufacturing of biological products for administration to humans. Under the cGMPs there is a subset of regulations which deals with the prevention of the introduction, transmission or spread of communicable diseases when manipulating Human Cells, Tissues, and Cellular and Tissue-Based Products (HCT/P) and it is defined as current Good Tissue Practices (cGTP) and are stated in 21 CFR 1271.

The manufacturing, processing, or scale up of human cell therapies present many challenges which include the natural characteristics of the different components used to generate the final product, such as the source of the actual cells and the reagents and supplies that will come in contact with them, the probability of adventitious agent contamination, the need for aseptic processing which includes the processing area, and the inability to sterilize the final product (cells) [21]. Another factor that should be taken into account is the product transport and distribution since due to the nature of the living cells, they will require the appropriate transport temperature, device of transport, expiration time, and stability of the product as well as the availability of release tests results [20]. The manufacturing issues described above are detailed and resolved in the Chemistry, Manufacturing and Controlled (CMC) section of the IND.

The detailed description on the manufacturing process in the CMC will provide a good assessment of the identity, quality, purity, and potency of your product. The CMC should have the following manufacturing key points [20]:

1. A list and a description of all the components and materials used during the manufacture of the cellular product [21], such as cells (autologous or allogeneic), cell source and cell type, if the cells are mobilized or activated, collection or cell recovery, transport method and conditions if shipped for processing, donor screening, and testing [22], cell bank systems (Master Cell Bank and Working Cell Bank) if applicable.
2. A list of any reagents used, most importantly, the CMC should detail the following information: (a) vendor information, (b) concentration information of the reagent at the step that is going to be used, (c) source (human, porcine, bovine, horse, etc.) with their respective Certificate of Analysis (COA), be of clinical grade, FDA approved or cleared, or be a licensed product; if it is of bovine source, have all the requirements for the ingredients of animal origin used for production of biologics [23] and if it is not FDA approved or cleared, should have a qualification program for additional testing in order to assure the safety and quality of the reagent (sterility, endotoxin, mycoplasma, and adventitious agents) and have functional analysis, purity testing, and assays (e.g., residual solvent testing) in order to show the lack of dangerous substances [20].
3. The final product should be tested to see if reagents used during the manufacturing, which are toxic, are present.
4. The lack of use of beta-lactam antibiotics, a rational if used and present in the final product, and the actions to be taken to remove traces of antibiotics in the final product.
5. A list with concentration and source information of all excipients used that will be present in the final product [24].
6. A description of all procedures used during the collection, production, and purification of the cell product with a schematic of the manufacturing process (Fig. 1), and the in-process and final product testing [20]: (a) description of the method of cell collection, cell processing, and culture conditions, (b) if irradiation

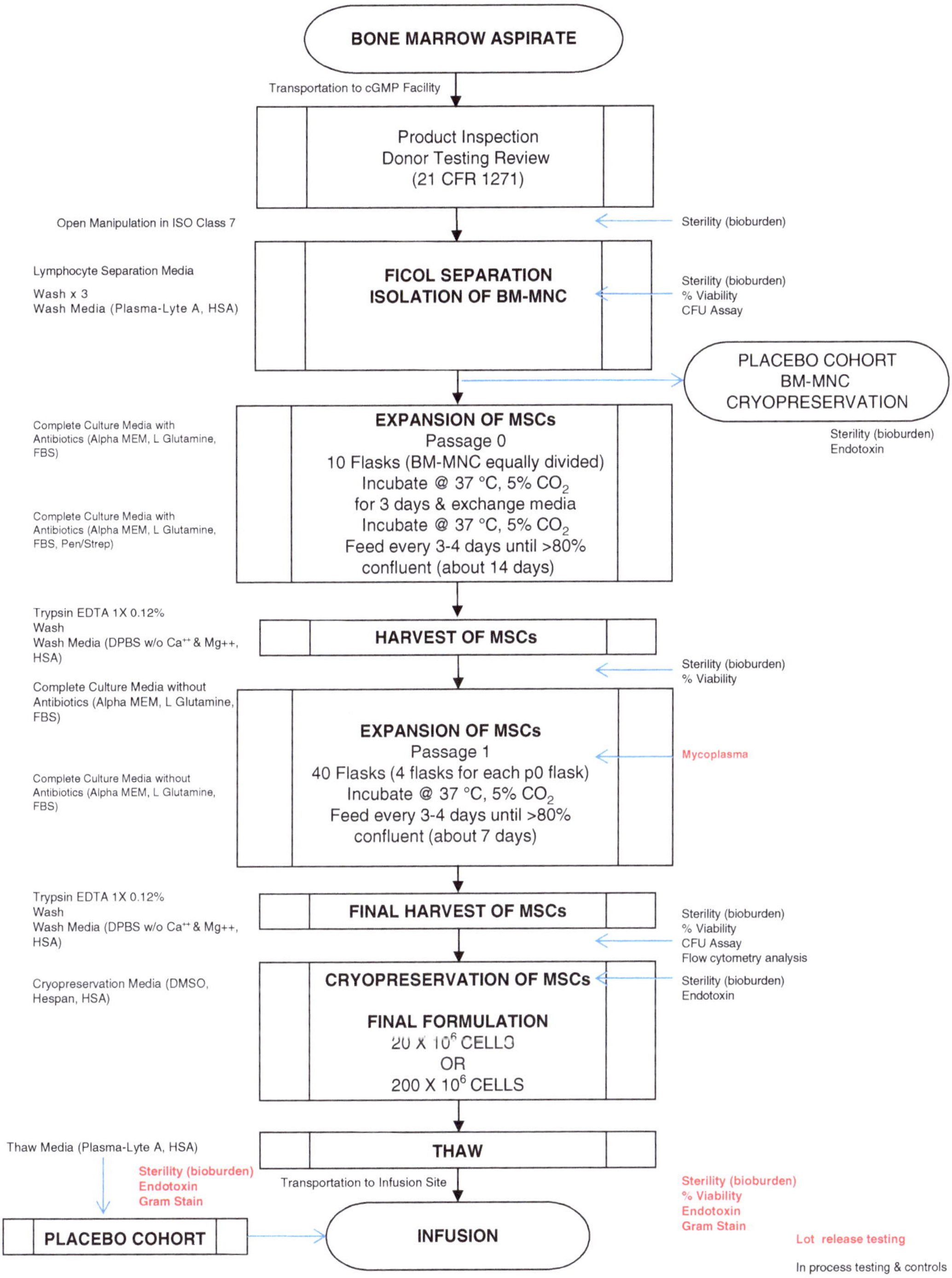

Fig. 1 cGMP manufacturing overview

is needed, (c) description of the final harvest, (d) duration of processing, (e) description storage conditions if they are needed, and (f) description of the final formulation.

7. An in-process and final product testing: (a) microbial testing: bacterial (anaerobic and aerobic) and fungal testing [25, 26], (b) duration of the test, (c) mycoplasma testing [27], and (d) adventitious agents testing.
8. A final product identity testing.
9. A purity testing [28] such as entoxing testing [29–31] and gram stain testing.
10. A potency assay.
11. A minimum release criteria for cell viability.
12. A minimum viable cell number (cell number/dose).
13. A final product release criteria testing (viability, sterility, mycoplasma, endotoxin, and cell dose) performed in each lot of the manufacture product and should be available prior to the infusion of the cell product.
14. A product stability testing (in-process and final product testing) [32].
15. A product tracking system [33].
16. An in-process and final product labeling [34–36].
17. A description of types of container and closures [37].

2 Materials

2.1 Isolation of Bone Marrow Mononuclear Cell by Density Gradient

Equipment

1. Microscope.
2. Biological Safety Cabinet (BSC).
3. Centrifuge.
4. Vacuum Pump.

Supplies

1. 4×4 gauze, sterile.
2. Conical tubes, 50 and 250 ml, sterile.
3. Pipettes, 1, 10, 25, 50 ml and aspirating pipettes, sterile.
4. Pipette-aid.
5. Surgical gowns, shoe covers, head cover, and mask, all sterile.
6. Surgical gloves, sterile.
7. Vacuum collection flask with two associated sterile tubing sets, sterile.
8. Pipette tips, sterile.
9. Microscope slides for cytospin.

Reagents

1. Lymphocyte Separation Media (LSM).
2. Plasma-Lyte A, 1×.
3. Trypan Blue.
4. Human Serum Albumin (HSA), 25 %.

2.2 Culture and Expansion of Human Marrow Mesenchymal Stem Cells

Equipment

1. Microscope.
2. CO_2 Incubators.
3. Biological Safety Cabinet (BSC).
4. Centrifuge.
5. Vacuum Pump.
6. Hemocytometer.
7. Water bath, set a 37 °C (to thaw reagents required to prepare the necessary media).
8. Pipette-aid.

Supplies

1. Gauze 4×4, sterile.
2. Conical centrifuge tubes, 50 and 250 ml, sterile.
3. Pipettes, 1, 10, 25, 50 ml and aspirating pipettes, all sterile.
4. Surgical gowns, shoe covers, head cover, and mask, all sterile.
5. Surgical gloves, sterile.
6. Vacuum collection flask with two associated sterile tubing sets, sterile.
7. Pipette tips, 200 μl, sterile.
8. Culture flasks, Nunc Δ T185 Cell Culture Treated, sterile.
9. Cell Scraper, sterile.

Reagents

1. Alpha Minimal Essential Media (MEM), 1×.
2. Penicillin–streptomycin–glutamine (100×), liquid.
3. Fetal Bovine Serum (FBS), Gamma Irradiated.
4. Trypsin–EDTA, 1×; 0.12 % Gamma Irradiated.
5. Phosphate-Buffered Saline (DPBS), Ca and Mg free.
6. L-Glutamine 200 mM (100×).
7. HSA, 25 %.
8. Trypan Blue.

2.3 Cryopreservation of Human Mesenchymal Stem Cells

Equipment

1. Microscope.
2. BSC.
3. Centrifuge.
4. Vacuum Pump.
5. Hemocytometer.
6. Control Rate Freezer.
7. Tube Heat Sealer.

Supplies

1. 4 × 4 gauze, sterile.
2. Conical tube, 50, sterile.
3. Pipettes, 2, 5, 10, 25 ml, and aspirating pipettes, sterile.
4. Pipette-aid.
5. Surgical gowns, shoe covers, head cover, and mask, all sterile.
6. Vacuum system or vacuum collection flask with two sterile tubing sets.
7. Cryovials, 2 ml.
8. Cryocyte Freezing Bags.

Reagents

1. Cryoserve, dimethyl sulfoxide (DMSO).
2. DPBS, Ca and Mg free.
3. 25 % HSA, USP grade.
4. Hespan, B Braun.
5. Trypan Blue, DI water, for irrigation.

2.4 Thawing of Mesenchymal Stem Cell Products

Equipment

1. BSC.
2. Centrifuge.
3. Water bath, set at 37 °C.
4. Pipette-aid.
5. Microscope.

Supplies

1. Sample Site Coupler (2).
2. Cobe couplers (3).
3. Needles, 16 g for products, 18–20 g for solutions.

4. Syringes, 5, 10, 30, 60 ml.
5. Microcentrifuge tube, orange cap.
6. Anaerobic BBL Culture System.
7. Alcohol wipes.
8. Steri-Drape.
9. Gauze.
10. Scissors.

Reagents

1. Plasma-Lyte A, 1×.
2. HSA, 25 %.
3. Trypan Blue.
4. Water, for irrigation.

3 Methods

3.1 Isolation of Bone Marrow Mononuclear Cell by Density Gradient

Wash Media Preparation

1. Plasma-Lyte A, 1×.
2. 25 % HSA, 1 % final concentration.
3. Label the media with the name of the media, storage temperature, date of preparation, date of expiration (24–48 h following preparation), and initials of the personnel preparing the media.

Product Processing

1. Record the total volume of the bone marrow aspirate. Remove samples for counting, 100 μl, and sterility.
2. Transfer bone marrow aspirate into an appropriate number of 250 ml conical tubes. Dilute the bone marrow aspirate 1:1 with Plasma-Lyte A in each of the 250 ml tubes and mix well. After dilution, aliquot the bone marrow into 50 ml conical tubes with 30 ml of marrow per tube.
3. Aliquot 15 ml of LSM into each of the 50 ml tubes.
4. Overlay 30 ml of the diluted bone marrow aspirate on top of LSM (30 ml of diluted bone marrow aspirate over 15 ml of LSM).
5. Carefully, transfer the cells to a centrifuge and spin at 800 × *g* for 30 min, at room temperature, with the centrifuge brake "OFF."
6. After centrifugation, using a sterile transfer pipette, collect the interface from each 50 ml conical tube and transfer it into clean 50 ml conical tubes.

7. Bring the volume in each tube up to 50 ml with Wash Media.
8. Centrifuge at 500 × *g* for 10 min, with brake set to "Low."
9. Using an aspirating pipette, remove almost all supernatant, leaving 1–2 ml in each tube.
10. Gently resuspend the cell pellet in each conical tube. Combine cells from all the conical tubes into a single clean 50 ml conical. Bring the total volume up to 50 ml with Wash Media.
11. Centrifuge at 500 × *g*, for 10 min, at room temperature, with the centrifuge brake set to "Low."
12. Using an aspirating pipette, remove almost all supernatant, leaving 1–2 ml of the supernatant in the conical. Gently resuspend the cell pellet and bring the volume up to 50 ml using fresh Wash Media.
13. Perform a cell count and viability.
14. Allow the cells to settle for 10 min and remove 3 ml of supernatant for sterility testing.
15. Remove another sample containing 1×10^6 cells for the CFU assay.
16. Continue with further product processing. The product might be used for culture and expansion of human bone marrow-derived mesenchymal stem cells (MSC) or cryopreserved.

3.2 Culture and Expansion of Human Bone Marrow Mesenchymal Stem Cells

Wash Media Preparation

1. DPBS (Ca and Mg free).
2. 25 % HSA, final concentration 1 %.
3. Label the media with name of the media, storage temperature (4 °C), date of preparation, date of expiration (14 days), initials of the personnel preparing the media.

Complete Culture Media with Antibiotics

1. Alpha MEM media.
2. Penicillin–streptomycin–glutamine (100×), liquid.
3. FBS, gamma irradiated and heat inactivated, final concentration 20 %.
4. Label the media with name of the media, storage temperature (4 °C), date of preparation, date of expiration (14 days), initials of the personnel preparing the media.

Complete Culture Media without Antibiotics

1. Alpha MEM media.
2. L-Glutamine 200 mM (200×), liquid.
3. FBS, gamma irradiated and heat inactivated, final concentration 20 %.

4. Label the media with name of the media, storage temperature (4 °C), date of preparation, date of expiration (14 days), initials of the personnel preparing the media.

Cell Culture and Expansion (P0)

1. Once the MNCs are prepared, label ten flasks using "in-process" label as follows:
   (a) Passage number: Initial plating is P0 then increase after every harvest (i.e., P1, P2, etc.).
   (b) Recipient name.
   (c) Recipient ID.
   (d) Date: Day of plating.
   (e) Product number (on a separate label specific to product #).
2. Seed each T-185 flask by dividing the mononuclear cells equitably among the ten flasks and bring to a final volume of 25 ml with complete culture media with Antibiotics.
3. Prior to placing the flasks in the incubator, check that the media covers the entire surface of the flask and the outside of the culture flask is wiped with 70 % alcohol. This should be done every time a flask is returned to the incubator.
4. Incubate in tissue culture incubator, with 5 % CO_2, for at least 72 h, to allow the cells to adhere to the flask.
5. After 72 h from the initial plating, remove the flask from the incubator and remove the unattached cells by tilting the flask while holding it upright and aspirating off all of the media from the bottom corner of the flask.
6. Add 25 ml of complete culture media with antibiotics.
7. Before replacing the cap on the flask, ensure the cap filter is clean and dry. This should be done every time the flask is recapped.
8. Place the flask in a tissue culture incubator, at 37 °C and 5 % CO_2.

Feeding the Cells

1. Cells should be fed every 3–4 days until harvest, when attached cells are confluent. For the initial culture period, complete culture media with antibiotics must be used. After the first harvest culture media without antibiotics will be used, see below.
2. Remove the flasks from the incubator. Observe under the microscope to determine the extent to which the cells are confluent. The cells will be split when they are >80 % confluent. In addition, examine each tissue culture flask for absence of contamination. If contamination is present, discard the flask or the whole preparation.

3. Using vacuum aspiration, remove the culture media from the T-180 flasks.
4. Add fresh 25 ml of complete culture media with antibiotics.
5. Wipe the outside of each flask with a sterile alcohol wipe, before placing it back into the 37 °C incubator with 5 % CO_2, for further culture.

Harvesting Cells using Trypsin–EDTA

1. When the cells are approximately >80 % confluent (≈day 14 of culture), observe each tissue culture flask under the microscope to examine the cells for absence of contamination. If contamination is present discard the flask or the whole preparation.
2. Remove the media from each flask using vacuum aspiration.
3. Add 25 ml of Wash Media and swirl it around the flask.
4. Aspirate off the Wash Media and add 10 ml of Trypsin–EDTA to each flask. Return the flask to the 37 °C CO_2 incubator for no more than 8 min. Make sure to monitor the cells at 4 min intervals under the microscope, as leaving the Trypsin in the flask for a prolonged period of time is damaging the cells.
5. If there are still attached cells, use a sterile cell scraper to gently lift the adhered cells by moving the scraper across the bottom of each flask.
6. Neutralize Trypsin activity by adding 15 ml of complete culture media without antibiotics to each flask. Swirl the media around the flask to make sure all the cells are in suspension.
7. Transfer the detached cells from each flask to a sterile 50 ml conical tube, using a 25 ml pipette.
8. Add 25 ml of Wash Media to the flask and swirl to remove any remaining cells. Using a 25 ml pipette, collect the Wash Media from the flask and add it to the 50 ml conical tube with the detached cells.
9. Centrifuge the 50 ml conical tubes at 500 × *g* for 10 min, at room temperature, with brake set to "Low."
10. Using a vacuum system, aspirate the supernatant to leave 1–2 ml of media and resuspend the pellet in the 50 ml conical tube. Bring the volume up to 40 ml with complete culture media without antibiotics.
11. Take a 100 μl sample and perform a cell count and viability (*see* **Note 1**).
12. After 10 min of settling, remove a 1 ml sample from each 50 ml conical for sterility. Place all samples in a separate 50 ml conical tube. A 3 ml sterility sample will be collected from the pooled media (1 ml each, for aerobic, anaerobic and fungal testing).

Cell Culture, Expansion (P1) and Final Harvest

1. Label four T-185 flasks per each harvested P0 flask, as follows (*see* **Note 2**):
   (a) Passage number: Initial plating is P0 then increase after every harvest (i.e., P1, P2, etc.).
   (b) Recipient name.
   (c) Recipient ID.
   (d) Date: Day of plating.
   (e) Product number (on a separate label specific to product).
2. Seed each T-185 flask by equitably dividing the cell suspension from each of the 50 ml conical tubes into four tissue culture flasks, i.e., 12.5 ml of cell suspension in each flask. Bring to a final volume of 25 ml with complete culture media without antibiotics.
3. Culture the cells for an additional 7 days, feeding the cells 3–4 days after seeding or until cells are >80 % confluent (*see* **Note 3**). Prior to final harvest, remove a 1 ml sample (from the supernatant, i.e., media without cells) from each tissue culture flask. Combine all samples in a single 50 ml conical tube. A 1 ml sample will be sent to an approved vendor for mycoplasma testing.
4. When cells are >80 % confluent, harvest the cells as described in before. After the first centrifugation, transfer each pellet into one 50 ml conical.
5. Resuspend cells in 50 ml of Wash Media and centrifuge at $500 \times g$ for 10 min.
6. Aspirate supernatant, pull all cell pellets together into a single 50 ml conical and resuspend in another 50 ml of Wash Media.
7. Take a 100 μl sample and perform a cell count and viability.
8. If the cell number is less than 250×10^6 total cells, culture and expansion must continue.
9. If cell number is sufficient, proceed to cryopreservation. Remove 2×10^6 cells for CFU-F analysis and 1×10^6 cells for flow cytometric analysis.
10. After 10 min of settling, remove a 3 ml sample (1 ml each for aerobic, anaerobic and fungal).
11. Proceed to the cryopreservation of MSCs.

3.3 Cryopreservation of Human Mesenchymal Stem Cells

Wash Media Preparation

1. DPBS buffer, Ca and Mg free.
2. 25 % HSA, 1 % final concentration.
3. Label the media with name of the media, storage temperature, date of preparation, date of expiration (2 weeks following preparation), and initials of the personnel preparing the media.

Cryopreservation Media without DMSO (Freeze Media 1)

1. Hespan (6 % Hetastarch in 0.9 % sodium chloride).
2. 25 % HSA, 2 % final concentration.
3. Label the media with name of the media, storage temperature, date of preparation, date of expiration (24–48 h following preparation), and initials of the personnel preparing the media.

Cryopreservation Media with 10 % DMSO (Freeze Media 2)

1. Hespan (6 % Hetastarch in 0.9 % sodium chloride).
2. 25 % HSA, 2 % final concentration.
3. DMSO, 10 % final concentration.
4. Label the media with name of the media, storage temperature, date of preparation, date of expiration (24–48 h following preparation), and initials of the personnel preparing the media (*see* **Note 4**).

3.4 Cryopreservation of BM MNC (Placebo)

1. Centrifuge MNCs obtained as a result of the process in Subheading 3.1 in a 50 ml conical tube at 500 × *g*, for 10 min, at room temperature, with break set at "Low."
2. Remove the supernatant with a sterile aspirating pipette. Resuspend the cells with 10 ml amount of Freeze Media 1. Add an equal volume (10 ml) of Freeze Media 2. This will result in a final concentration of 5 % DMSO.
3. After a 5 min settling time in an ice bath, remove 4 ml of the supernatant: 1 ml for each, aerobic, anaerobic and fungal and 1 ml for endotoxin (*see* **Note 5**).
4. Replace 4 ml removed in the previous step with 2 ml Freeze Media 1 and 2 ml of Freeze Media 2.
5. Cell suspension in a final volume of 20 ml will be cryopreserved in 50 ml Cryocyte bag.
6. Label the product.
7. Cryopreserve the cells with a controlled rate freezer.

3.5 Cryopreservation of MSCs

1. Centrifuge the MSCs obtained before in a 50 ml conical at 500 × *g* for 10 min, at room temperature, with break set at "Low."
2. Remove the supernatant with an aspirating pipette.
3. Calculate the total resuspension volume, in order to resuspend the cells to a final concentration of 15×10^6 cells/ml (stock cell suspension) (*see* **Note 6**).
4. Resuspend the cells in up to 50 % of the total resuspension volume with Freeze Media.
5. Add Freeze Media 2 up to the total resuspension volume, for a final concentration of 15×10^6 cells/ml (*see* **Note 7**).

6. After a 5 min settling time in an ice bath, remove 1 ml of the supernatant for each: aerobic, anaerobic and fungal cultures, and endotoxin (*see* **Note 8**).
7. Replace the 4 ml that have been removed with 2 ml of Freeze Media 1 and 2 ml Freeze Media 2.
8. The cells will be aliquoted into 50 ml Cryocyte bags, for cryopreservation as follows:
   (a) For the 20 × 106 Cells Study Cohort: The first bag to be frozen will contain 30 × 106 cells in 20 ml of cryopreservation media (1:1 Freeze Media 1 and Freeze Media 2). The second bag will be 250 × 106 cells in 20 ml of cryopreservation media (1:1 Freeze Media 1 and Freeze Media 2), and the remainder will be frozen in a third bag in 20 ml of cryopreservation media (1:1 Freeze Media 1 and Freeze Media 2).
   (b) For the 200 × 106 Cells Study Cohort: Freeze one bag with 300 × 106 in 20 ml of cryopreservation media (1:1 Freeze Media 1 and Freeze Media 2), and the remainder in a second bag in 20 ml of cryopreservation media (1:1 Freeze Media 1 and Freeze Media 2).
9. Label each bag.
10. Cryopreserve the cells using a controlled rate freezer.
11. Following cryopreservation, transfer the metal cassettes holding the product bags to liquid nitrogen storage. Document the location of the product in the liquid nitrogen storage tank.

3.6 Labeling the Product Designated for Distribution or Storage

1. The label must be attached to the container either prior to cryopreservation or before delivery for transplant, if the final product is designated for transplant immediately following processing.
2. After the product is labeled, a second individual must verify the information reflected on the product label.
3. Make a copy of the final product label and place it in the appropriate batch record.
4. After labeling and cryopreservation of the product, product bags are maintained in a designated liquid nitrogen storage tank (vapor phase) until such time when the product is requested and issued for transplant.

3.7 Thawing of Mesenchymal Stem Cell Products

Thawing Media

1. Plasma-Lyte A
2. 25 % HSA, 1 % final concentration
3. Label the media with name of the media, storage temperature date of preparation date of expiration (48 h following preparation), and initials of the personnel preparing the media.

Product Thawing

1. Identify the product number and name to be thawed, and locate it in the liquid nitrogen (LN2) storage tank. Remove the metal cassette(s) with the product bag(s) from the LN2 storage tank (*see* **Note 9**).
2. Place a metal cassette with product in a steri-drape isolation bag, and place in the water bath for a few seconds.
3. Gently but quickly, remove the Cryocyte bag with the product from the metal cassette.
4. Verify the product label attached to the product bag.
5. Place the product bag in the steri-drape isolation bag (use two bags).
6. Place the Cryocyte bag with the product (now encased in two isolation bags) in the water bath.
7. Thaw the product, gently massaging the bag in the water bath (*see* **Note 10**). After the product is completely thawed, transfer the product bag to the BSC, take it out of the steri-drape isolation bag, and wipe it using an alcohol wipe.
8. Using the product label, confirm the total product volume in the bag. This step is necessary to determine the volume of thawing media (must equal to product volume) required for washing (e.g., when the final product volume is 20 ml, 20 ml of the thawing media is required).

3.8 Washing the Product in a 50 ml Cryocyte Bag (See Note 11)

1. Using sterile alcohol wipe, clean the injection port of the Cryocyte bag holding the product. Aseptically, insert a COBE coupler into an already clean port.
2. Aseptically, connect the other side of the COBE coupler to a sterile 60 cc syringe containing 20 ml of Thawing Media.
3. Slowly introduce the Thawing media into the product bag, via the 60 cc syringe (*see* **Note 12**).
4. After all the Thawing media is added to the bag, remove the contents of the bag using the syringe and transfer to a 50 ml conical tube.
5. Rinse the Cryocyte bag using an additional 10 ml of thawing media. Add the rinse to the same 50 ml conical. At the point, the total volume in a 50 ml conical tube will be 50 ml.
6. At this time remove 100 μl of the cell suspension to perform a cell count and test for viability.
7. Centrifuge the conical at 500 × *g* for 10 min, at 22 °C.
8. Remove 5 ml of the supernatant for sterility (1 ml each for aerobic, anaerobic and fungal cultures, 1 ml for endotoxin, and 2 drops for gram stain). Discard the remaining supernatant.
9. For a 20×10^6 cell/dose cohort, adjust the cell concentration to 4×10^6cells/ml so as to deliver 20×10^6 cells in a total volume

of 5.5 ml and slowly, resuspend the pellet using a sterile 10 ml regular tip pipette (*see* **Note 13**).

10. For a 200×10^6 cell/dose cohort, adjust the cell concentration to 4×10^7 cells/ml so as to deliver 200×10^6 cells in a total volume of 5.5 ml and slowly, resuspend the pellet using a sterile 10 ml regular tip pipette (*see* **Note 14**).
11. Continue to disperse the pellet until it is completely resuspended, and there are no clumps. Avoid formation and/or introduction of air bubbles.

3.9 Product Lot Release

1. In accordance with established regulatory requirements, any given product designated for transplant, must meet predetermined lot release criteria.
2. The following lot release testing must be performed, unless specified otherwise:
    (a) Sterility testing, preliminary result: Anaerobic culture, 1 ml aerobic culture, 1 ml fungal culture, 1 ml, gram stain, 2 drops for 2 slides (1 drop/slide).
    (b) Endotoxin (LAL) content, 1.0 ml.
    (c) Mycoplasma, by PCR (*see* **Note 15**).

The list below contains the list of tests performed "for information only" and those in which the final results will not be available at the time of product release:

(a) MSC dose (thawed and washed product).
(b) Sterility (thawed and washed product), final result: Aerobic, Anaerobic, Fungal.
(c) CD 105/CD45 cell quantification (MSCs before the addition of Cryoprotectant).
(d) CFU-F quantification (MSCs before the addition of Cryoprotectant).

3.10 Labeling the Final Product Designated for Transplant

1. After the product is labeled, a second individual must verify the information reflected on the product label.
2. A copy of the final product label must be made and saved in the appropriate batch record.
3. The label must be affixed to the product container (50 ml conical) before the product leaves the cGMP facility.

4 Notes

1. Cells harvested from each of the ten tissue culture flasks must be (a) kept in separate 50 ml conical tubes and (b) counted separately.
2. 40 Tissue culture flasks must be prepared.

3. During P1, complete culture media without antibiotics must be used to culture the cells.
4. Once prepared, cryopreservation media must be maintained at 4–8 °C until and while being used.
5. Endotoxin samples will be analyzed in house using the FDA-approved PTS system.
6. The final resuspension volume is made up of equal volumes of Freeze Media 1 and Freeze Media 2 (1:1).
7. The final volume in the Cryocyte bag must not exceed 20 ml.
8. Endotoxin samples will be analyzed using the FDA-approved PTS system.
9. A single product (single lot) may be cryopreserved in several bags (sub-lots), and, consequently, stored in several metal cassettes.
10. The product is considered thawed when crystals can no longer be observed in a Cryocyte bag. This process should not take longer than 5 min.
11. Thawing media must be added slowly, while continuously mixing the product bag. Continuous mixing increases cell viability.
12. While introducing the Thawing media, gently mix the contents of the bag.
13. The cells must be resuspended in at least 5.5 ml of Thawing media. If there are more than 20×10^6 total cells, resuspend in the final volume to the final concentration of 4×10^6 cells/ml. If the target cell number is not available, the cells must be resuspended in 5.5 of Thawing Media.
14. The cells must be resuspended in at least 5.5 ml of Thawing media. If there are more than 200×10^6 total cells, resuspend in the final volume to the final concentration of 4×10^7 cells/ml. If the target cell number is not available, the cells must be resuspended in 5.5 of Thawing Media.
15. If the test result is positive, testing will be repeated with a standard 28 day culture.

Acknowledgements

This work is funded by National Institutes Health grants U54-HL081028 (Specialized Center for Cell-Based Therapy), P20-HL101443, and R01-grants HL084275, HL110737-01, HL107110, and HL094849 to Dr. Hare.

References

1. Ratcliffe E, Thomas RJ, Williams DJ (2011) Current understanding and challenges in bioprocessing of stem cell-based therapies for regenerative medicine. Br Med Bull 100(1):137–155
2. Astori G et al (2010) Bone marrow derived stem cells in regenerative medicine as advanced therapy medicinal products. Am J Transl Res 2(3):285–295
3. Bach FH et al (1968) Bone-marrow transplantation in a patient with the Wiskott-Aldrich syndrome. Lancet 2(7583):1364–1366
4. Gatti RA et al (1968) Immunological reconstitution of sex-linked lymphopenic immunological deficiency. Lancet 2(7583):1366–1369
5. Deans RJ, Moseley AB (2000) Mesenchymal stem cells: biology and potential clinical uses. Exp Hematol 28(8):875–884
6. Williams AR et al (2011) Intramyocardial stem cell injection in patients with ischemic cardiomyopathy: functional recovery and reverse remodeling. Circ Res 108(7):792–796
7. Kemp KC, Hows J, Donaldson C (2005) Bone marrow-derived mesenchymal stem cells. Leuk Lymphoma 11(46):1531–1544
8. Togel F, Westenfelder C (2007) Adult bone marrow-derived stem cells for organ regeneration and repair. Dev Dyn 236(12):3321–3331
9. Biehl JK, Russell B (2009) Introduction to stem cell therapy. J Cardiovasc Nurs 24(2):98–103
10. Boyle AJ, McNiece IK, Hare JM (2010) Mesenchymal stem cell therapy for cardiac repair. Methods Mol Biol 660:65–84
11. Husnain K, Haider MA (2004) Bone marrow cell transplantation in clinical perspective. J Mol Cell Cardiol 38:225–235
12. Muller YD et al (2011) Transplantation tolerance: clinical potential of regulatory T cells. Self Nonself 2(1):26–34
13. Christoforou N, Gearhart JD (2007) Stem cells and their potential in cell-based cardiac therapies. Prog Cardiovasc Dis 49(9):396–413
14. Kajstura J et al (2005) Bone marrow cells differentiate in cardiac cell lineages after infarction independently of cell fusion. Circ Res 96(1):127–137
15. Nunes SS et al (2011) Stem cell-based cardiac tissue engineering. J Cardiovasc Transl Res 4(5):592–602
16. Kanashiro-Takeuchi RM, Schulman IH, Hare JM (2011) Pharmacologic and genetic strategies to enhance cell therapy for cardiac regeneration. J Mol Cell Cardiol 51(4):619–625
17. Schuleri KH et al (2009) Autologous mesenchymal stem cells produce reverse remodelling in chronic ischaemic cardiomyopathy. Eur Heart J 30(22):2722–2732
18. http://clinicaltrials.gov/ct2/show/NCT00587990?term=NCT00587990&rank=1
19. Hare JM (2009) Translational development of mesenchymal stem cell therapy for cardiovascular diseases. Tex Heart Inst J 36(2):145–147
20. U.S. Department of Health and Human Services, Food and Drug Administration, Center for Biologics Evaluation and Research (2008) Guidance for FDA reviewers and sponsors. Content and review of chemistry, manufacturing, and control (CMC) information for human somatic cell therapy investigational new drug applications (INDs)
21. 21 CFR 312.23(a)(7)(iv)(b)
22. 21 CFR 1271
23. 9 CFR 113.53
24. 21 CFR 211.84(a)
25. 21 CFR 610.12
26. United States Pharmacopeia (USP) <71> Sterility tests
27. CBER (1993) Points to consider in the characterization of cell lines used to produce biologicals
28. 21 CFR 600.3(r)
29. 21 CFR 610.13
30. 21 CFR 610.9
31. FDA (1987) FDA guideline on validation of the Limulus Amebocyte Lysate (LAL) Test as end-product endotoxin test for human and animal parenteral drugs, biological products, and medical devices
32. 21 CFR 312.23(a)(7)(ii)
33. 21 CFR 1271.290(b)
34. 21 CFR 312.6(a)
35. 21 CFR 1271.250
36. 21 CFR 1271.90
37. FDA (1999) Guidance for industry: container closure systems for packaging human drugs and biologics; chemistry, manufacturing, and controls documentation, May 1999

Chapter 9

A Model System for Primary Abdominal Closures

Michael J. Yost, Mary O. Morales, Veronica Rodriguez-Rivera, Eric M. Yost, Louis Terracio, and Stephen A. Fann

Abstract

The foreign body response to medical devices and materials implanted in the human body, including scarring, fibrous encapsulation, and potential rejection, is a longstanding and serious clinical issue. There are no widely acceptable or safe therapies for ameliorating the foreign body response. Clinical complications resulting from the response include disfigurement of silicone prostheses and loss of function of devices such as implanted pacemakers, stents, and shunts. Cellularized implants and stem cells placed in the body are also subject to the foreign body response with the added issue that the regenerative repair intended to be prompted by the graft may be inhibited. Beneficial modification of the body's reaction to implanted materials, medical devices, engineered constructs, or stem cells would be a fundamentally important therapeutic advance.

As part of investigating the cellular response, we have developed a model which uses cells isolated from skeletal muscle biopsy, cultured, and proliferated in vitro. These satellite cells, which are mononucleated progenitor cells, reside between the plasma membrane of the muscle fiber and the basal membrane that encompasses the fiber. While usually quiescent, these cells become activated following muscle damage. Once activated, the satellite cells proliferate, migrate to injured muscle, and participate in repair by fusing with existing muscle fibers or by differentiating into new skeletal muscle fibers. Satellite cells have been shown to be heterogeneous populations of stem cells and progenitor cells. We have developed an explant method for isolating, sorting, enriching, and culturing these cells for use in skeletal muscle regenerative medicine to determine if the foreign body response can be inhibited by manipulating the cell–cell communication.

Key words Satellite cells, Muscle precursor cells, Skeletal muscle repair

1 Introduction

An enabling technology of fundamental significance to surgery would be one that improved the integration and tolerance of implanted materials, cells, and engineered devices in the human body. The long-term goal here is to prevent the "scarring-up" of implants and cellularized devices. When a foreign substance is inserted into soft tissues, the body reacts by up-regulating the acute inflammatory cascade. The end result is that a fibrous capsule

Robert G. Gourdie and Tereance A. Myers (eds.), *Wound Regeneration and Repair: Methods and Protocols*, Methods in Molecular Biology, vol. 1037, DOI 10.1007/978-1-62703-505-7_9, © Springer Science+Business Media New York 2013

of scar tissue is formed around this material and contracts. As this process negatively affects implants, it has been the subject of numerous investigations [1–11]. Histological evaluation of capsules in different settings has demonstrated the universality of the foreign body reaction, with its associated deposition of contractile and fibrotic tissue and in extreme cases, mineralization of the scar-like deposit [12]. Surgical removal of the fibrous capsule is often required to correct this problem. However, this repeat procedure has potential complications, including postoperative hemorrhage, skin damage, and scar contracture.

In this chapter we discuss primary abdominal closures using the rat model system. Rats are an excellent species to study muscle development and wound healing since their structure and processes are similar to humans. Muscle growth and wound repair consist of a complex biological process in which satellite muscle cells play an important role. There are many different skeletal muscle abnormalities that can occur due to either from genetic defects or traumatic injury of some kind. Repair of muscle damage is centered on satellite muscle cells [13–16]. Satellite muscle cells are usually quiescent in adult muscle and become activated following trauma to the muscle or surrounding area to help in wound repair [13–16]. The isolation of satellite cells and their expansion in culture will help in the discovery of a novel tissue engineering repair system that utilizes biological substitutes.

In our model, satellite muscle cells are isolated from adult rat muscle biopsy taken from the animals' hind limb. The cells are explanted onto tissue culture dishes and allowed to proliferate until they can be stained and sorted for cells that are positive for integrin alpha 7. Integrin alpha 7 subunit is a protein that is expressed in muscle cells that are involved in myogenic differentiation, therefore, when there is an increase in integrin alpha 7 there is also an increase in muscle cell proliferation and adhesion. Once satellite cells are activated, they migrate into the damaged area and form myofibers to help promote repair of the injured site.

2 Materials

2.1 Preparation Equipment for Surgery

1. Isoflurane.
2. Isoflurane vaporizer.
3. Supply gas (oxygen).
4. Supply gas regulator.
5. Flowmeter (0–1,000 ml/min).
6. Induction chamber.
7. Connection tubing and valves.
8. Facemask or nosecone.

9. Scavenging method.
10. Disposable surgical drape.
11. Sterile towels.
12. Sterile drape with 3″ fenestration.
13. Sterile 4×4 gauze sponges.
14. Betadine swabs.
15. Electric razor.
16. Sterile surgical gloves.
17. Sterile surgical instrument packet containing: scalpel handle, tissue forceps, needle holder, two hemostatic forceps, surgical scissors.
18. Surgical blade #15.
19. 4-0 Prolene suture and 4-0 vicryl suture.
20. Sterile Autoclip wound closing system with 9 mm stainless steel wound clips.
21. 8-Week-old Sprague Dawley rats.
22. Topical first aid antibiotic ointment.
23. T/pump and warming blanket.
24. Repair material.

2.2 Explant of Muscle Biopsy

1. Sterile 50 ml conical tubes.
2. Phosphate-buffered saline (PBS) containing 200 U/ml penicillin G and 200 μg/ml streptomycin.
3. Phosphate-buffered saline.
4. 100 mm tissue culture dishes.
5. Sterile glass coverslips.
6. Fine sterile forceps.
7. Sterile scalpel handle with #15 surgical blade.
8. Ice bucket.
9. Sterile surgical scissors.
10. Explant medium: Dulbecco's modified Eagle's medium (DMEM), 25 % fetal bovine serum (FBS), 200 U/ml penicillin G and 200 μg/ml streptomycin, 0.1 % gentamycin, and 5μg/ml amphotericin B.
11. SAT medium (muscle culture selective medium): Ham's F10, 20 % FBS, 2.5 ng/ml bFGF human recombinant, 100 U/ml penicillin G, 100 μg/ml, streptomycin, 0.1 % gentamycin, and 2.5 μg/ml amphotericin B.
12. 0.25 % Trypsin/EDTA.
13. 37 °C incubator with 5 % CO_2.

2.3 Cell Sorting and FACS Analysis

1. Accutase.
2. Flow Buffer: PBS without Ca^{2+} and Mg plus 1 % Bovine Serum Albumin.
3. Blocking solution (flow buffer + 10 % normal goat serum).
4. Alpha 7-PE antibody (MBL, Clone 3C12).
5. Mouse IgG1-PE antibody.
6. Ice bucket.
7. Aluminum foil.
8. Flow tubes.

3 Methods

3.1 Surgical Isolation of Muscle Biopsy

1. General anesthesia is achieved on 8-week-old Sprague Dawley Rats using the isoflurane inhalation method.
2. After general anesthesia is obtained, the hind limb of the rat is prepped by clipping fur down to the skin (*see* **Note 1**).
3. The skin is then prepped in standard surgical manner with betadine scrub solution in triplicate.
4. The animal is then placed on the surgical table on top of a warming blanket that is covered with a sterile drape.
5. Sterile towels are draped over the site of incision to define the surgical field consisting of the hind limb.
6. A small incision is created using a sterile scalpel with a no. 15 blade (*see* **Note 2**).
7. Using sterile surgical scissors, dissect away the skin layer from the muscle so you have a clear view of the muscle (Fig. 1a).

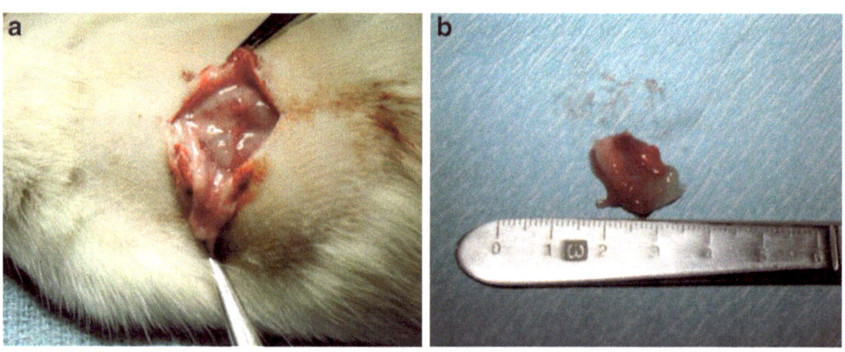

Fig. 1 Isolation of muscle biopsy procedure. (**a**) An incision is made in the left hind limb of the rat. (**b**) A 5 mm × 5 mm × 4 mm = 0.1 cm^3 section of the muscle and invested fascia lata is removed

8. Using sterile scissors, cut a 5 mm×5 mm×4 mm = 0.1 cm^3 section of the muscle and invested fascia lata and put directly into a 50 ml conical tube containing sterile PBS/antibiotic solution (Fig. 1b).
9. The muscle biopsy is immediately placed in a sterile 50 ml conical tube containing PBS with appropriate antibiotics and kept on ice until all samples are collected and tissue is explanted.
10. The skin is closed using 4-0 prolene interrupted stitch and then further secured with wound clips.
11. A topical antibiotic ointment is then applied to the wound and the animal is returned to a clean cage.

3.2 Isolation of Satellite Muscle Cells

1. Wash the muscle tissue that was obtained from the surgery two times in cold PBS containing antibiotics to remove any blood and debris from the sample (*see* **Note 3**).
2. The tissue is then transferred to a culture dish that has a small amount of media to prevent the tissue from drying out (*see* **Note 4**).
3. Cut the tissue into 1–2 mm pieces using a sterile sharp scalpel blade.
4. Once the tissue is cut, using sterile fine forceps, arrange the tissue on 100 mm tissue culture dish creating 5–6 clusters that contain 5–6 pieces of tissue.
5. To prevent tissue from drying out, place a small drop of media on each cluster and then using sterile fine forceps, cover the tissue clusters with sterile glass coverslip.
6. Carefully, add 8–10 ml of Explant media to the tissue culture dish making sure that the coverslips do not float.
7. Incubate dishes in 37 °C incubator with 5 % CO_2 for 1–2 days without disturbing them (*see* **Note 5**).
8. After 3 days refeed the dishes with Explant media. You should start to see cells crawling out of the tissue (*see* **Note 6**).
9. In about 7 days, remove coverslips and tissue pieces with sterile fine forceps and rinse the tissue culture plates in PBS containing 2× antibiotic/antimycotics and passage cells with trypsin into 150 mm tissue culture dishes in SAT media and allow them to grow to about 50 % confluence.

3.3 Flow Cytometry and Cell Sorting

1. Collect the satellite cells using Accutase (*see* **Notes 7, 8**).
2. Centrifuge cells at 800 ×*g* for 5 min.
3. The supernatant is aspirated off the cell pellet and the cell pellet is resuspended in 3–5 ml flow buffer.
4. The cells are then counted and centrifuged at 800 ×*g* for 5 min.

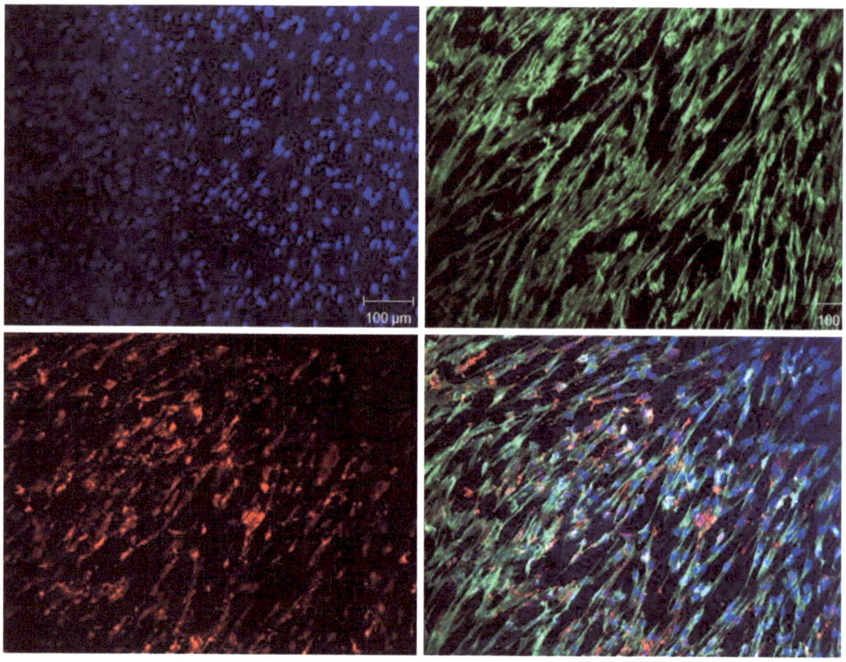

Fig. 2 Confocal Microscopy images using a 10× objective shows the sorted satellite cells after further expansion cultured for 3 days on collagen hydrogels at a concentration of 14 mg/ml. The *blue* corresponds to DAPI stain which indicates the cell's nucleus. The *green* color corresponds to phalloidin that binds to the f actin composing the cytoskeleton of the cells. The *red* color is immunohistochemical labeling of integrin alpha7

5. Discard the supernatant and resuspend the cell pellet in flow buffer at a concentration of 5×10^6 cells/ml.
6. The cell suspension is then split into flow tubes at a concentration of $2.5–5 \times 10^5$ cells/tube and 10 μl of blocking solution is added.
7. The tubes are mixed well and incubated for 10 min on ice. For staining rat cells, 20 μl of integrin alpha 7-PE (MBL, Clone 3C12) or 20 μl of a 1:10 dilution of Mouse IgG1-PE as a control are added to the respective tubes.
8. The tubes are then incubated for 30 min on ice mixing every 10–15 min.
9. After 30 min, add 1 ml of flow buffer and centrifuge as before.
10. The cells that are stained with integrin alpha 7-PE are resuspended at $10–12 \times 10^6$ cells/ml [11]. The control cells that are stained with mouse IgG1-PE are resuspended at 1×10^6 cells/0.5 ml to use as an isotype control to help set gates for sorting (*see* **Note 9**).
11. Collect both the positive and negative fractions into media containing serum to plate for expansion and further analysis (Fig. 2) (*see* **Note 10**).

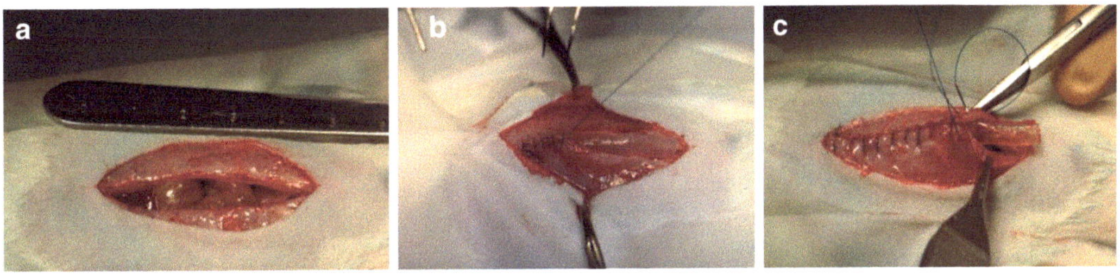

Fig. 3 Creation and repair of abdominal closure (**a**) The 3 cm midline abdominal wall detect. (**b**) Repair material has been placed in the detect. (**c**) The detect is closed with repair material in place using 4.0 prolene suture

3.4 Creation and Repair of Abdominal Closure

1. General anesthesia is achieved on 8-week-old Sprague Dawley Rats using the isoflurane inhalation method.
2. After general anesthesia is achieved, the abdomen is prepped by clipping abdominal fur down to the skin.
3. The skin is then prepped in standard surgical manner with betadine scrub solution in triplicate.
4. The animal is then placed on the surgical table on top of a warming blanket that is covered with a sterile drape.
5. Sterile towels are draped over the site of incision to define the surgical field consisting of the anterior abdominal wall from xiphoid to pubis longitudinally and laterally to table margins.
6. A midline incision is made through the skin and subcutaneous tissue.
7. Using sterile surgical scissors, the subcutaneous tissue is dissected free from the abdominal wall giving wide access to the anterior surface of the musculature.
8. A midline incision is the made through the linea alba and the peritoneal cavity is entered.
9. An omental flap is created by dissecting the greater omentum off the greater curve of the stomach based on the right gastroepiploic artery.
10. With the omentum adequately mobilized, a 3 cm midline abdominal wall defect is created (Fig. 3a).
11. The repair material is then placed in the wound and sewn into the defect using an interrupted suture of 4.0 prolene with suture bites taken 4 mm from the edge of the defect (Fig. 3b, c).
12. The mobilized skin flaps are returned to the midline and closed with 4-0 Vicryl subcuticular stitch to minimize dead space.
13. To insure wound remains closed, stainless steel wound clips are placed over the suture area if needed.
14. Topical first aid antibiotic ointment is placed on the incision and the animal is allowed to ascend from anesthesia and placed in a clean cage.

4 Notes

1. First make sure to check the animal is fully sedated before beginning the procedure. We generally check toe and tail reflex.
2. If there is any bleeding during the surgical procedure, make sure to use sterile gauze and firmly put pressure on the site. Bleeding should stop fairly quickly.
3. Make sure to keep everything as sterile as possible by doing everything in the tissue culture hood.
4. It is very important that the tissue does not dry while processing it.
5. Be patient. It may be several days before you see anything, but once they start growing, they grow pretty rapidly.
6. Make sure to change the media every 2–3 days and make sure they do not get 100 % confluent.
7. Make sure to work as quickly as possible. Have all the buffers made up prior to collecting cells.
8. Make sure to use Accutase to lift the cells off the tissue culture dishes. Accutase will not destroy cell surface receptors which are critical for the staining and cell sorting process. EDTA makes the cells too sticky and they tend to clump together making both counting and sorting difficult.
9. Make sure to keep the tubes covered with foil after adding the labeled antibody.
10. Count the cells after they come off the cell sorter so that you can plate accordingly.

Acknowledgements

This publication was made possible by Grant Number DE019355 from NIH.

References

1. Fann SA et al (2006) A model of tissue-engineered ventral hernia repair. J Invest Surg 19:193–205
2. Schnur PL et al (1996) Silicone analysis of breast and periprosthetic capsular tissue from patients with saline or silicone gel breast implants. Plast Reconstr Surg 98:798–803
3. Batra M, Bernard S, Picha G (1995) Histologic comparison of breast implant shells with smooth, foam, and pillar microstructuring in a rat model from 1 day to 6 months. Plast Reconstr Surg 95:354–363
4. Emery JA et al (1994) The synovial structure of breast-implant-associated bursae. Mod Pathol 7:728–733
5. McConnell JP et al (1997) Determination of silicon in breast and capsular tissue from patients with breast implants performed by

inductively coupled plasma emission spectroscopy: comparison with tissue histology. Am J Clin Pathol 107:236–246
6. Brohim RM et al (1992) Early tissue reaction to textured breast implant surfaces. Ann Plast Surg 28:354–362
7. Bucky LP et al (1994) The capsule quality of saline-filled smooth silicone, textured silicone, and polyurethane implants in rabbits: a long term study. Plast Reconstr Surg 93: 1123–1133
8. Smahel J, Hurwitz PJ, Hurwitz N (1993) Soft tissue response to textured silicone implants in an animal experiment. Plast Reconstr Surg 92: 474–479
9. Picha GJ, Goldstein JA, Stohr E (1990) Natural-Y meme polyurethane versus smooth silicone: analysis of the soft-tissue interaction from 3 days to 1 year in the rat animal model. PlastReconstr Surg 85:903–916
10. Bern S, Burd A, May JW Jr (1992) The biophysical and histologic properties of capsules formed by smooth and textured silicone implants in the rabbit. Plast Reconstr Surg 89: 1037–1044
11. Barone FE et al (1992) The biomechanical and histopathologic effects of surface texturing with silicone and polyurethane in tissue implantation and expansion. Plast Reconstr Surg 90:77–86
12. del Rosario AD et al (1994) True synovial metaplasia of breast implant capsules: a light electron microscopic study (Abstract). Mod Pathol 7:14A
13. Logan MS, Propst JT et al (2010) Human satellite progenitor cells for use in myofascial repair. Ann Plast Surg 64:794–799
14. Adams GR (2006) Satellite cell proliferation and skeletal muscle hypertrophy. Appl Physiol Nutr Metab 31:782–790
15. Propst JT, Fann SA, Franchini JL et al (2009) Focused in vivo genetic analysis of implanted engineered myofascial constructs. J Invest Surg 22:35–45
16. Yost MJ, Baicu CF, Stonerock CE et al (2004) A novel tubular scaffold for cardiovascular tissue engineering. Tissue Eng 10:273–284

Part III

Skin and Corneal Wound Repair Models

Chapter 10

Alternatives for Animal Wound Model Systems

Phil Stephens, Matthew Caley, and Matthew Peake

Abstract

In this chapter a review of animal model systems already being utilized to study normal and pathologic wound healing is provided. We also go into details on alternatives for animal wound model systems. The case is made for limitations in the various approaches. We also discuss the benefits/limitations of in vitro/ex vivo systems bringing everything up to date with our current work on developing a cell-based reporter system for diabetic wound healing.

Key words Animal models, Wound healing, Diabetic wound healing, In vitro models, Limitations, Benefits

1 Introduction

Wound healing is a complex process involving interactions between multiple cell types from the skin and infiltrating inflammatory cells. The sequence of events that take place during wound healing is reasonably well understood, as macroscopic changes in cell numbers and tissue structure are easily observed. However, on a molecular level the wound healing process is still being delineated. Further complications come with chronic, dysfunctional wounds associated with the aged, where dysregulation of the normal tissue repair processes is even less well understood. In order to better understand the intricacies of these processes models of part, or all, of the wound healing process have been developed over many decades. Such models of wound healing are generally split into two groups, namely in vivo and in vitro model systems each offering their specific advantages and also shortcomings (for an overview *see* Table 1).

Robert G. Gourdie and Tereance A. Myers (eds.), *Wound Regeneration and Repair: Methods and Protocols*, Methods in Molecular Biology, vol. 1037, DOI 10.1007/978-1-62703-505-7_10, © Springer Science+Business Media New York 2013

Table 1
A comparison of the advantages and disadvantages of in vivo vs. in vitro wound model systems

Model system	Advantages	Disadvantages
Animal models	Enables the investigation of the complex interplay between multiple cells populations/body systems involved with tissue repair Enables the investigation of multiple elements of the wound repair process Permits the study of a functional immune system Enables the placement of multiple wounds within one animal (i.e., control vs. test) Easy access to the wound bed to allow the testing of agents Enables the selective deletion of specific genes to determine their effect on wound healing Can model different wound healing causes (e.g., surgery, burns, crushing) Widely utilized models of chronic wound healing (existing data)	The complexity of the whole animal system sometimes prevents the clear analysis of distinct contribution of tissues/cells during the healing process Often need to sacrifice animals at discrete time points rather than allowing continual analysis of the same animal Hard to control the size/depth/dimensions of the wound Animal skin is not an exact replica of human skin Still no accurate, reproducible chronic wound model representation (chronic wound inducers often lead to other, unwanted side effects) Need for high numbers to reach statistical significance Expensive Painful for the animals Lack of reproducibility in different laboratories Immune responses may differ from humans Scaling up to large-scale studies is very expensive and time consuming Genetic manipulations are more complex and time consuming

Lab-based model systems	Simple system allowing the discrete analysis of the contribution that individual cell populations/components make to the wound healing process Possible to utilized human disease and normal cells alongside each other in experiments Can established reproducible human disease cell lines for extended investigation Cheap to establish and work with More complex 3D model systems can be established which mirror the in vivo situation Possible to undertake continual analysis of the same system over multiple time points without any need to "sacrifice" the experiment Can easily manipulate the model systems (e.g., chemical, genetic) to test reagents Can run 1,000 s of experiments to enhance statistical significance Can be utilized for high-throughput screening Reproducible No animal suffering	Only enables the limited investigation of the interplay between multiple cells populations involved with tissue repair (no whole body response) Even the most complex 3D systems do not truly replicate the structure/function of human skin and so may have limited clinical relevance Some model systems (e.g., ex vivo culture of skin) have a limited time span over which they can be effectively utilized Lack of a "complete" immune system limits the study of the immune component of any wound healing response Cell culture can induce cellular artifacts which can skew outcomes

2 In Vivo Models of Wound Healing

Many animal models for wound healing exist. At their most simple they involve wounding a laboratory animal and observing wound closure over time; however, they can be made more complicated through the manipulation (physical/chemical/biological) of the wound tissue environment. Various different species are used regularly for skin wounding experiments including rats, mice, rabbits, and pigs. As well as these "normal" laboratory animals, genetically modified animals are also used to model human diseases such as diabetes and the effect this would have on wound healing. Using these animal models different types of wound systems can be established.

2.1 Full Thickness Excision Wounds

These wounds are generated by the surgical removal of all the layers of the skin (epidermis, dermis, subcutaneous fat and in the case of the mouse the subcutaneous smooth muscle layer) from the animal [1] (for a recent review *see* ref. 2). Such models allow the investigation of hemorrhage, inflammation, granulation tissue formation, re-epithelialization, the formation of new blood vessels, and wound tissue remodeling. Multiple wounds may be generated on each animal allowing direct comparison between treatments. When mice are used with this wound healing model, wounds of up to 6 mm are typically generated with up to six wounds per mouse. Wounds are generated under anesthetic, hair is removed from the back of the mouse, and the loose skin on the mouse back is lifted and cut generating a wound. Wound area may be recorded over time giving a wound healing rate. For histological analysis animals are sacrificed at desired time points after wounding and the wounded skin is removed for analysis. The ability to generate multiple wounds on each animal increases the statistical value of this model and the ability to easily access the wound bed to apply topical agents is of real benefit. However, the wounds generated in this model of wound healing are often not uniform as they are generated by hand. The depth and size of wounds, though similar, are not identical introducing an additional variable into this model system. Also the complex nature of this type of model system is sometimes its downfall in that the investigation of distinct elements of the tissue repair process cannot be undertaken in isolation.

2.2 Punch Wounds

Punch wounds generated in the ears of rabbits have been used as a model for full thickness wound healing for many years. Wounds are generated using biopsy punches cutting through the full thickness of the ear. The wound includes damage to the epidermis, dermis, and cartilage that make up the ear. Rabbits have the ability to regenerate the tissue of the ear from the margin of the punch wound inwards [3–6]. The wounds heal over a period of weeks and can be taken at different time points for histological analysis. As the wounds are straight through the ear the variation in depth possible in excision wounds on the back of animals is not observed in this model, and as the wounds are generated using a biopsy punch, the size of the wounds is reproducible. Multiple punch wounds can be generated in each ear allowing for experimental controls. As well as its use in rabbits this model of wound healing has also been used in mice [7–9]. Modifications to this model of wound healing have been proposed as a chronic wound model (*see* Subheading 3).

2.3 Partial Thickness Wounds

Unlike the full thickness and ear punch wounds described above, this model of wound healing aims to look primarily at epidermal wound healing [10–12]. Wounds are generated using a dermatome, using the same technique as is used to harvest skin for split thickness skin grafts. The dermatome is used to remove a set thickness of skin generating a partial thickness wound with the epidermis and some of the dermis removed. The depth of wound is somewhat reproducible if the dermatome is used correctly; however, it is possible to generate an uneven wound with the dermatome leaving patches of epidermis within the wound. Removal of this tissue causes initial bleeding as the blood supply to the epidermis is disrupted. This model of wound healing generates a larger wound by area and is ideal for use on larger experimental animals, such as pigs. Multiple wounds may be generated on each animal allowing for experimental treatments and controls on the same animal. Partial thickness wounds can also be created as part of a burn model [13–15]. As with the full thickness model, differences in wound size and depth are problematic with these models as the application of a burn or use of a dermatome is not a precise instrument and different users may generate different wound depths. Any variation in wound depth and size would alter wound healing rates.

2.4 Suction Blisters

This model of wound healing uses dry suction to separate the dermis and epidermis at the level of the basement membrane. Vacuum suction is applied to the skin. The suction slowly separates the epidermis from the dermis, with fluid filling the generated inter-dermal space. The wounds generated by this technique are of a uniform size (the size of the suction device) and of a uniform depth (down to the basement membrane). This method of wound generation has been used on experimental animal species [16, 17] and also on human volunteers [18, 19]. The blister formed can be de-roofed leaving an exposed wound. As the dermis is not damaged this is a good model of re-epithelialization, with reproducible wounds and the potential to generate multiple wounds per animal.

2.5 Water Scald Burns

Wounds generated by scalding the skin of animals to generate blisters have been utilized in different species as a model of wound healing [20, 21]. Blisters are generated by exposing a fixed area of skin to hot water. The burning process is halted by the application of ice-cold water to the site of injury. A blister forms at the site of injury and, as in the suction blister model, the blister may be de-roofed to expose the wound. By keeping the temperature of the water and exposure time constant, repeatable wounds can be formed using this technique. With this model it is possible to control the depth of the partial thickness wound by altering exposure time. This model requires a water-tight seal to be generated around the desired wound area. In smaller animals it is harder to generate multiple wounds on a single animal with this technique. To determine the required exposure time to generate the desired depth of wound, multiple trials must be undertaken for each new species or strain of experimental animal.

2.6 Thermal Burns

Another method of generating wounds with thermal damage is through the direct application of heat to the skin [21–23]. Partial thickness wounds can be reproducibly generated by applying heat to the skin for timed periods using a metal template. As with the suction blister and scald wound models, a blister is formed and can be de-roofed to expose the dermis and leave an open wound. The depth of wound can be controlled by increase or decrease in either the time the heated template is applied to the skin or in the heat of the template. As in the water scald model, a series of trial experiments is required to determine the optimum conditions for generating a wound of a specified depth. Multiple wounds may be generated in each animal and the size and shape of the wounds can be easily altered and is reproducible. This model of wound healing has been utilized in multiple different species of laboratory animal and is often used in the porcine model [21, 24–26]. The two thermally generated wound models both lead to denatured proteins in the wound environment due to the applied heat, these denatured proteins are not found in wounds generated by physical trauma and their influence on wound healing should be considered.

3 Chronic Wound Animal Model Systems

The percentage of people classed as aged is increasing annually, such that by 2050 it is estimated that between 40 and 50 % of the population of the western world and the Pacific Rim will be over the age of 60 and on a world basis the percentage of the population above the age 60 is likely to increase to 20 % in 2050 and to 27 % by 2100 [27]. Chronic degenerative disorders are a common, costly problem associated with age-related decline in repair and regeneration processes [28]. Indeed, impaired wound healing,

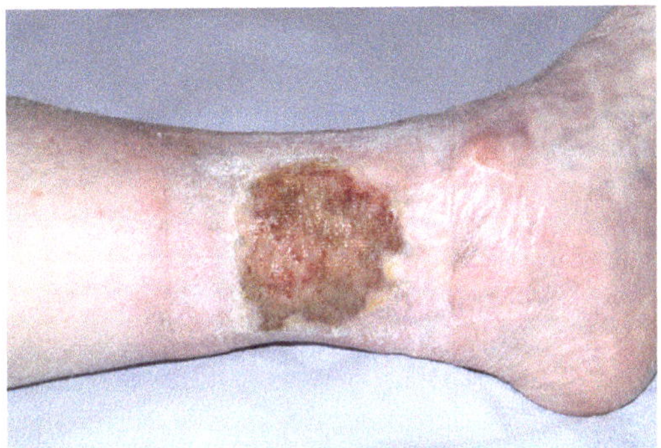

Fig. 1 A typical chronic, non-healing wound (venous leg ulcer). Image courtesy of Professor Keith Harding, School of Medicine, Cardiff University

which currently affects 3 % of the population over 60, is consequently a significant quality of life issue. These chronic wounds encompass a spectrum of diseases and exist in three principal forms (pressure sores, venous ulcers, and diabetic ulcers) (Fig. 1). They represent a significant cause of distress and disability among the aged and currently cost the National Health Service in the UK an estimated £2–3 billion annually and healthcare providers in the USA in excess of $25 billion annually (with them affecting in 6.5 million patients in the USA [29]). These wounds are characterized by impaired re-epithelialization, defective extracellular matrix (ECM) remodeling and critically, chronic inflammation within the dermis. In order to study these chronic wounds and in turn develop potential treatments for them, there is a desperate need to develop model systems that can reproducibly replicate the chronicity associated with this non-healing situation.

3.1 Diabetic Wound Animal Models

It is possible to artificially generate diabetes in animal models either through the use of chemical treatments to kill the β-cells of the islets of Langerhans [30, 31] or exposing the mice to a high-fat diet [152]. It has been reported that these models of induced diabetes demonstrate impaired wound healing compared to control animals [30, 32–36]. Furthermore, induced diabetic animals have been used to study different wound treatments that may be useful in treating diabetic wounds [32, 33]. However, there are disadvantages to the use of such model systems. The use of agents such as alloxan and streptozotocin to chemically induce diabetes may also have other side effects associated with the use of chemotherapy agents potentially limiting the value of this model. Hence, recent attention has turned to the use of genetically engineered murine model systems.

Genetically diabetic animals are available for use in wounding studies, with one of the most widely used diabetic models being the db/db diabetic mouse [37]. This mouse lacks the receptor for leptin normally present within the hypothalamus. Although recent work would suggest that there are limitations to using this particular murine model system [38]. Genetically diabetic mice have impaired wound healing [39–41] and have been suggested to be a useful model for the study of diabetic wounds. Indeed a number of investigations have demonstrated that this impaired healing can be reversed by the addition of growth factors [42–45], vitamins [46], erythropoietin [47], by blocking processes such as lipid peroxidation [48] and blocking the action of advanced glycation end products [49]. However, what is typical of all these wounds is that they only demonstrate a short-term impairment in the wound repair process and fail to replicate a true chronic wound which can persist for tens of years in human subjects. Hence these diabetic wound models are really models of impaired acute wound healing rather than true chronic wounds. Furthermore, the mouse (like many of the species used such as rat, rabbit, and hamster) has a subcutaneous *panniculus carnosus* muscle (not found in humans) which contributes to the repair process via contraction and collagen formation. Attempts to limit the involvement of the contractile component of murine wound healing have involved the use of wound splinting to minimize wound contraction in this murine wound healing model (in order to make it more similar to wound healing in humans [1]).

3.2 Ischemic Wound Models

Both venous leg and pressure ulcers have an underlying tissue ischemia linked with their formation. Chronic wound models that can generate an ischemic environment have therefore been developed to model the human disease state. Two methods for generating ischemic wounds are widely used: a back and an ear model. The ischemic back wound [50] is based on a skin flap generated by an H-shaped incision wound in the back of a rat. The generated wound shows reduced blood perfusion along the horizontal wound creating an ischemic environment. The vertical wounds do not have a reduced blood flow so can act as controls for wounding experiments. This is a relatively short-term wound model as the skin returns to normal levels of blood perfusion within 16 days. The model can be used to study the effect of potential treatments on chronic wound healing [51]. A similar model in pigs has also been reported [52]. A second model of ischemic wounds involves the generation of an ischemic environment in a mouse, rat, or rabbit ear and the subsequent wounding of the ear [53–58]. The surgical division of the rostral and central arteries of the ear generates an ischemic environment in one ear of the animal. This has been shown to generate an ischemic environment in the ear for up to 28 days. In this model one ear is ischemic and the other acts as a control and when performed in the rabbit multiple punch wounds may be generated in each ear. Whilst these models do generate impaired healing they do not create wounds that fail to heal. They provide useful tools for studying impaired wound healing but more accurately model delayed acute wound healing than true chronic wounds.

3.3 Other Animal Chronic Wound Models

Unlike chronic wounds in human subjects, a number of current animal "chronic wound" model systems rely on either crushing or inducing burns/chemical insults in the skin of the animal [59–62]. Importantly, the severity of the induction of such ulcers in the subjects would likely cause substantial pain and suffering for the animal. Some attempts have been made to induce impaired healing through deliberate infection by the addition of bacterial species or the creation of bacterial biofilms to these wounds but invariably they rarely show a delay in healing and are therefore more suited to test the anti-bacterial properties of reagents rather than their pro-healing properties [63–65]. The animal with the skin structure closest to humans is the pig. However, the pig has the disadvantage of very rapid healing and contraction formation that varies depending where on the body the wound is placed [66]. Furthermore, the pig model demonstrates the limitations of trying to establish chronic wound animal models in that infection with bacteria often at concentrations rarely, if ever, demonstrated in chronic wound patients are required to cause a significant effect on the healing process [67, 68]. Other pig models require the animal to be irradiated thereby creating major logistical problems in the utilization of this type of model [69]. Finally, a number of genetically altered mice animals have also been reported to have dysfunctional wound healing (e.g., Slug –/– [70], integrin beta 6 –/– [71], estrogen receptor –/– [72], telomerase –/– [73]) but again these are models of delayed healing rather than a good representation of a true chronic wound.

4 The Need for Alternatives to Animal Model Systems

Over the past decade of so it has become increasing obvious that there is an immediate need for research into alternative model systems as it is clear that no suitable chronic wound animal model currently exists. Whilst current animal models reproduce some of the characteristics of chronic wounds none truly reproduce the dysfunctional wound healing responses that occur in aged human subjects. Indeed as the aged population increases there is a drive to develop and test more and more products to keep up with health care and patient demands. The increasing size of this market ($billions worldwide and growing [29]) means that more and more animal testing will have to be undertaken. However, as a result of the limitations of the animal models testing is also undertaken on human subjects. Whilst this is undoubtedly the best possible model system ethical issues and cost must be taken into consideration. Therefore, there is clearly a need to develop simple, reproducible in vitro model systems to permit rapid, low cost testing of materials, reagents, and drugs in order to reduce unnecessary animal experimentation. It is now readily appreciated that there are concerns over using both animal and human experimental models and that alternative models are a necessity.

5 In Vitro Models of Wound Healing

Animal models of wound healing generate complex wounds involving multiple cell types. In some circumstances the complexity of the wounding response can make it impossible to accurately measure the effect of a treatment or environmental change and difficult to determine which cells are specifically affected. Simpler models of wound healing are therefore useful for looking at specific parts of the wound healing response. Cells cultured either alone or in multicellular complex cultures have been used to study wound healing. As it is possible to use human cells in these models possible differences between wound responses in different species can be reduced. It is also possible to use cells from different human disease states or to genetically modify the cells to study the effect these changes may have on the wounding response.

5.1 The Monolayer Scratch Wound

One of the simplest in vitro wound models is the monolayer scratch assay, allowing the quantification of cellular migration/wound repopulation. When a confluent monolayer of cells is wounded the cells at the edge of the wound reorientate, proliferate, and migrate into the space closing the generated wound. By imaging the wound at the beginning of cell migration, and at regular intervals during cell migration, it is possible to determine the rate of wound closure. Differences in cellular migration can be observed using this method; for example keratinocytes migrate as a contiguous sheet [74], whereas fibroblasts migrate individually [75]. Scratch wound models can be used by themselves to study specific responses to wounding, or in conjunction with animal models, with the scratch wounds allowing greater control of the wound environment [76–80]. By altering the environment, the cells or the media used to generate the scratch wound model it is possible to study the effect that different stimuli or disease states have on wound healing [79, 81]. Scratch wounds have been used to study chronic wounds either by the addition of deleterious compounds known to be found in chronic wounds [82] or by using cells isolated from chronic wounds in the scratch assay itself [83, 84]. Overall, scratch wounds are a low cost, versatile, albeit simple model of wound healing.

5.2 The Fibroblast Populated Collagen Lattice Model

Wound repair within the dermis requires the replacement and remodeling of ECM by the dermal fibroblasts. This process can be studied in vitro using a fibroblast-populated collagen lattice (FPCL; [85–93]). Fibroblasts are introduced into a solution of collagen (typically type I collected from rat tail tendons). The mixture of cells, media, and collagen is cast in a petri dish and allowed to polymerize to form a cell-encapsulated three-dimensional lattice. Depending on the geometric constraints applied to the system, two potential models can be obtained; an *attached* or a *floating* FPCL. Attached FPCLs remain fixed to the base and sides of the culture well, whereas floating FPCLs become detached from the culture well and float in the medium. These differences have profound effects on collagen distribution and on fibroblast morphology, number and DNA synthesis [94–97] with the two models thought to represent models for different stages of the wound healing process.

For the free-floating collagen lattice system, the area of the FPCL is reduced over time and this can be measured giving a value for the rate of fibroblast-mediated collagen reorganization of the surrounding ECM. By altering the cells used to form such FPCLs different disease states and wound healing situations can be modeled. For example, the rate of lattice reorganization by fetal and oral fibroblasts is greater than that compared to skin fibroblasts [98–102] reflecting the distinct differences in wound healing (scarring) outcome observed between these sites. Furthermore, the utilization of chronic wound and aged cells within these systems has demonstrated inherent dysfunctionality with respect to ECM reorganization in these cells obtained from non-healing wounds [103, 104]. It is also possible to wound these systems and manipulate the wound space (i.e., filling it with other ECM molecules, biomaterials, and inductive factors) to directly study wound repopulation in three dimensions [105]. However, despite the great advance these systems have brought in terms of wound healing models, they were still deemed insufficient because of their lack of true skin architecture and vital cell/cell interactions.

5.3 Organotypic Coculture Model

Single cell in vitro models do not allow any of the interactions between different cell types that are so important within the wound healing response within tissues. To study the interactions between dermal and epidermal cells without using animals, cocultures of keratinocytes and fibroblasts can be utilized. Growing epidermal keratinocytes on a collagen matrix containing fibroblasts generates such epidermal equivalents [106]. The keratinocytes are exposed to the air/liquid interface and form stratified layers as seen in the epidermis and are supplied with nutrients from the underlying matrix [107–112]. The use of de-epidermalized dermis (DED) as an alternative to fibroblast populated collagen gels has been demonstrated to generate a well-differentiated epidermis [113]. The orgonotypic cultures generated using DEDs are histologically closer to human epidermis and ideal for studying alterations in keratinocyte differentiation.

Commercially available epidermal equivalents are also available (www.skinethic.com, www.mattek.com): modeling full thickness skin, epidermis, and epithelium from different parts of the body.

Epidermal equivalents can be modified by using cells from different sources such as Langerhans cells, endothelial cells, melanocytes, or adipose cells [114–116] or by the addition of hair follicles [117] to the culture. Wound healing can be studied in the skin equivalents by wounding the skin model and recording the cellular responses [105, 118, 119]. Organotypic coculture models add a level of complexity to in vitro wound modeling, allowing the investigation of complex cell–cell interactions. Most of the organotypic models described to date do not involve a functional vasculature or immune response although reports of more advanced skin equivalent systems are emerging [114, 115, 120, 121]. Despite the utilization of the skin equivalent for over 30 years there is, however, little published work on the development of the system as a chronic wound healing model (either through the utilization of chronic wound cells or the addition of exogenous agents); rather most work has focused on the use of such tissue engineered constructs for clinical benefit. However, organotypic models do offer a simplified (yet slight more complex) system to study areas of wound healing such as cell/cell signaling and differential cellular migration.

5.4 Skin in Culture

In vitro studies of the epidermis began with explant and organ culture work [122]. In explant culture epithelial cells grow from the cut edges of the skin segment (epiboly) onto the substrate. Epithelial growth occurs faster than the outgrowth of fibroblasts, although ultimately the fibroblasts take over the culture [123]. Organ cultures are usually established to maintain full thickness skin in vitro. The survival of the original skin sample in explant or organ culture is characterized by a series of degenerative then regenerative events. There are normally signs of necrosis before the end of the first week in culture and the tissue rarely remains viable for more than 14 days [124, 125]. Despite this limited time over which investigations can be conducted, such model systems have been used to study wound healing [126–132] and limited investigations have studied the role of chronic wound cell populations [133].

5.5 Cell Lines as Models of Disease

Normal human somatic cells have a finite lifespan in culture before they enter replicative senescence [134]. Primary cells taken from a patient can therefore only be used for a limited period of time before fresh patient samples are required. To overcome this problem cell lines have been generated for numerous different cell types. Initially cell lines were derived from spontaneously immortalized cells such as the HeLa cell line [135]. An alternative approach for cell line generation has been the use of oncogenic viral proteins. The introduction of viral genes such as Simian virus 40 (SV40), T antigen, Epstein Barr virus (EBV), Adenovirus E1A and E1B, and human Papilomavirus (HPV) E6 and E7 can lead to the immortalization of primary cells. The expansion of tumor biopsies into cell lines or the expression of viral oncogenes can also generate useful immortalized cell lines; however, the cells are phenotypically different from the primary cells they were derived from. While this may be useful in the study of cancer biology, where the phenotypic changes in the cancer cell lines may relate to changes that occur in tumor formation, these cell lines are less useful for the study of wound healing.

Another method for generating cell lines from somatic cells is the over-expression of the enzyme telomerase. Telomerase is a DNA polymerase that specifically restores the ends of telomeres that are damaged at cell division. Telomerase adds repeats of TTAGGG onto the existing telomere ends [136–138]. Telomerase is made up of two components: a protein component human telomerase reverse transcriptase (hTERT) and a RNA component human telomerase RNA component (HTERC or hTR). Most somatic cells do not express sufficient hTERT to maintain their telomeres at a constant length resulting in the shortening of their telomeres at each cell division, growth arrest, and eventual senescence [139]. The gene for hTERT was isolated and cloned in late 1990s, allowing for the over-expression of hTERT in human cells [140, 141]. The forced over-expression of hTERT by retroviral infection of somatic cells has now allowed the generation of many stable cell lines that maintain their previous phenotype [138, 140–143]. Importantly hTERT immortalized cells do not have a tumor-like phenotype: they remain diploid, contact inhibited, anchorage dependent, and differentiated. Telomerization now allows the production of a range of apparently normal cell lines that have, at least for practical purposes, indefinite growth potential and a phenotype essentially identical to that observed with normal primary cell strains.

6 Development of the Cardiff In Vitro Human Chronic Wound Reporter Assay

In order to gain a better understanding of chronic wounds an immortalized cell line retaining a chronic wound-specific phenotype would provide a useful tool reducing the requirement for fresh patient samples. Such a cell line could be used in the in vitro models of wound healing described previously allowing for analysis of chronic wound healing without the need for animal models.

6.1 The Importance of Fibroblasts

Previous studies into the molecular and cellular responses of fibroblasts in human wound healing states have demonstrated that fibroblasts from tissues that exhibit altered wound healing in vivo (i.e., oral tissues and chronic, non-healing venous leg ulcers) exhibit distinct phenotypic responses in vitro [83, 98, 99, 101–103, 144, 145]. In addition to the fibroblasts' ability to repopulate and remodel wounds following injury (via matrix metalloproteinase (MMP) production, and activation), they influence repair via their ability to control re-epithelialization and angiogenesis [145]. In support of the importance of fibroblasts influencing wound healing outcome, alterations in MMP production in diabetic fibroblasts and within the diabetic skin have been demonstrated [146]. Furthermore, recent investigations have demonstrated that chronic wound fibroblasts demonstrate premature, telomere-independent senescence which impacts on their ability to drive repair of the wound due to a lack of production of several key chemokines [83]. Thus fibroblasts are considered an important therapeutic target in wound healing. Indeed, strategies have been developed to both stimulate healing (e.g., growth factors [147]) and to minimize scarring following acute injury or trauma (e.g., anti-TGF-β, transforming growth factor beta-3 (TGF-β3), 5-fluorouracil [148–150]). All this evidence adds further weight to the central role of fibroblasts during wound healing and the hypothesis that it is the dysfunctional role of wound fibroblasts that is key to chronic disease progression and persistence. However, the opportunities to test any therapies are limited as the disease-specific fibroblast phenotypes are evident at low passage/population doubling level but are eventually lost with serial culture. Therefore, the ability to generate immortalized human cells in the laboratory from a disease of interest as a target for therapies would be of considerable advantage in the study of human disease.

6.2 Immortalization of Disease-Specific Cells

From primary, chronic wound cell strains cell lines have now been generated via immortalization with a retrovirally introduced hTERT expression cassette. hTERT expression introduced into the primary cell cultures was sufficient to mediate an dramatic increase in growth potential (i.e., effectively immortalizing them) compared to their mock-transfected and non-transfected (primary) counterparts (Fig. 2). Crucially, immortalized DF-hTERT retained an impaired wound healing phenotype compared to patient-matched normal skin fibroblasts (NF-hTERT), a finding initially demonstrated for the primary cell strains [151]. This suggests (as has been demonstrated by others in relation to immortalization of various other, different cell types) that immortalization does not alter disease-specific responses. Crucially this means that there is no longer a need to have to seek new patient biopsy material, instead there is now potentially an inexhaustible supply of disease-specific cells that are available for the development of a reporter cell line.

The overall strategy for the development of such a human chronic wound cell-based reporter assay is set out in Fig. 3. Briefly, this involved microarray analysis of the patient-matched immortalized disease and normal fibroblasts to identify the expression of chronic wound-specific genes. Having verified this differential gene

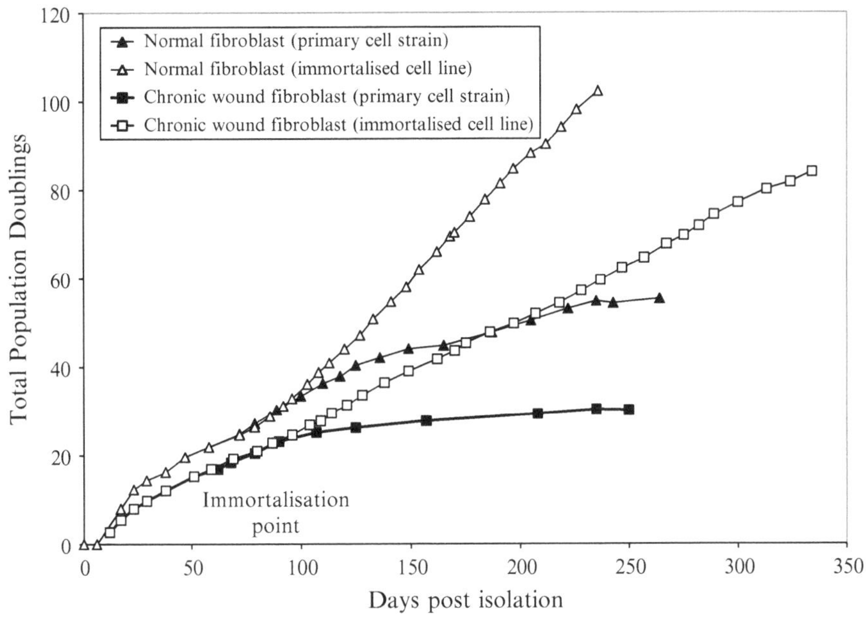

Fig. 2 Typical population doubling data for chronic wound and patient-matched normal skin fibroblasts (± immortalization with hTERT). The growth potential of chronic wound fibroblast cell strains is notably lower than for the normal fibroblasts, with chronic wound cells senescing (evidenced by the growth curve plateauing off) much earlier. However, post-immortalization, both the chronic wound and normal fibroblasts escape this senescence point and continue to proliferate for extended periods of time in culture

expression the promoters of these genes (not the genes themselves) were cloned and linked to fluorescent reporter molecules (specifically the destabilized variants of these fluorescent reporters such that any signal would not remain within the cells for long periods of time hence permitting real-time expression analysis of these disease genes). These constructs would then be transfected into chronic wound cells to assess their suitability as disease-specific reporters.

6.3 Identification of Disease-Specific Marker Genes

Affymetrix™ microarray analysis to investigate the stability of gene expression within the cell lines during extended periods of time in culture (10, 30, and 50 population doublings following the senescence of the parallel mock-infected cultures) demonstrated that, at a false discovery rate (FDR) level of 5 %, less than 0.1 % of genes were found to be significantly differentially regulated. Therefore, this stability of gene expression within the immortalized cell lines makes them a potentially, highly valuable resource for the development of a reporter cell line for commercial purposes (i.e., they can be substantially expanded for commercial scale up with little effect on their gene expression profiles).

Comparison of the array data for immortalized chronic wound fibroblasts with immortalized normal fibroblasts (5 % FDR cut off) demonstrated that of the 22,000 genes examined

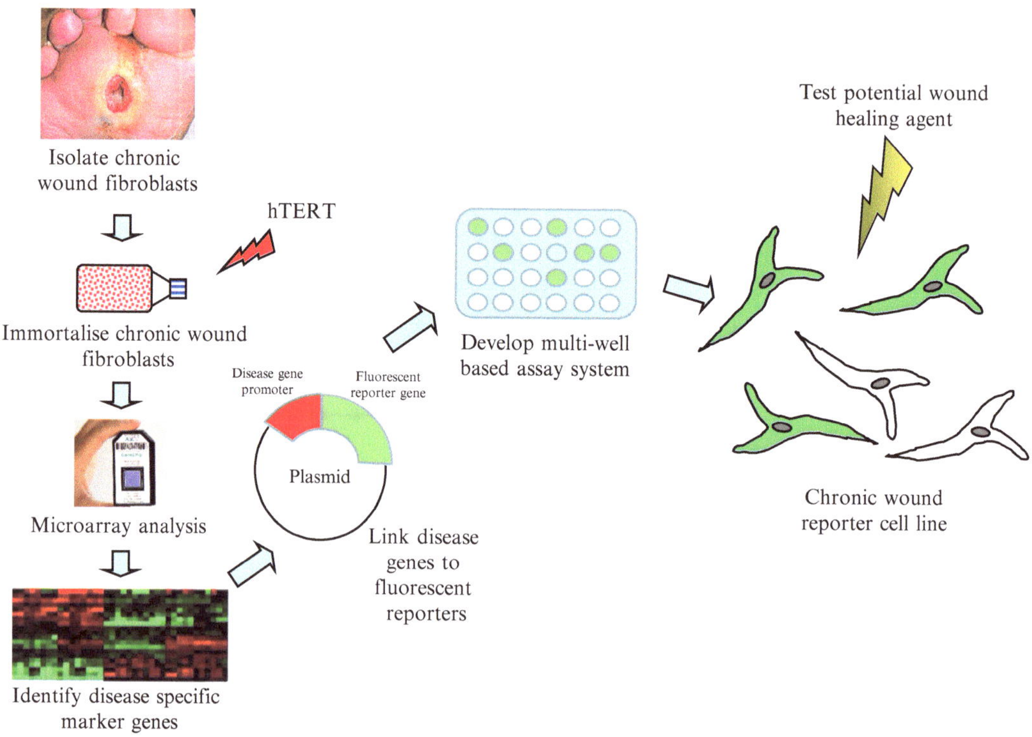

Fig. 3 The overall strategy for the development of a human chronic wound cell-based reporter assay. Chronic wound and patient-matched fibroblasts were isolated from patients and maintained in culture. These cells were then immortalized with human telomerase (hTERT; in order to provide a never-ending, consistent supply of disease cells for the development of the assay), RNA was extracted and *Affymetrix* microarray analysis undertaken to identify the expression of chronic wound-specific genes. Having verified this differential gene expression (qRT-PCR) the promoters of these genes (not the genes themselves) were cloned and linked to fluorescent reporter molecules (specifically the destabilized variants of these fluorescent reporters such that any signal would not remain within the cells for long periods of time hence permitting real-time expression/repression analysis of these disease genes). These constructs were then transfected into the chronic wound cell line to assess their suitability as disease-specific reporters

1,336 were altered in a disease-specific manner in response to a wound healing stimulus (Fig. 4). Further data mining and pathway analysis identified these genes to be involved in the following biological pathways: cell migration and motility, cell adhesion and ECM, inflammatory responses, cell division and cell cycle, angiogenesis, cytokine activity, growth factor activity, apoptosis and its regulation, transcription factors, and transcriptional regulation. This, together with additional manual data curation and validation via qRT-PCR, has enabled, for the first time, the identification of specific chronic wound disease marker genes which are stable in culture and which can be exploited in the development a reporter cell line.

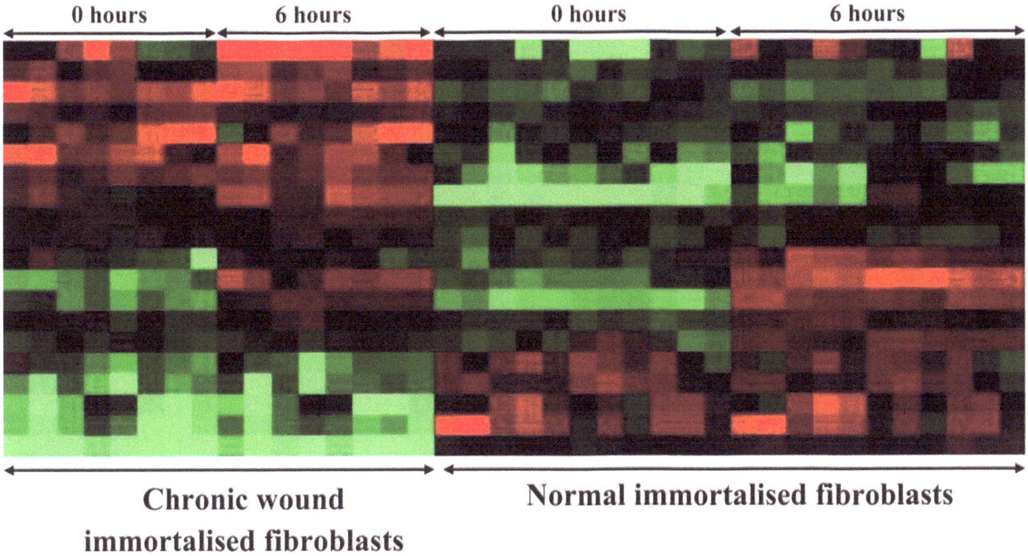

Fig. 4 A typical heat map demonstrating a selection of genes specifically up (*green*) or down (*red*) regulated consistently between the chronic wound and normal immortalized fibroblast cell lines ($n = 3$ patients) at quiescence (0 h) or after serum stimulation (6 h)

6.4 Generation of a Disease-Specific Reporter Cell Line

As the promoter/enhancer elements of the disease marker genes (rather than the genes themselves) were required (as these would drive the fluorescent reporter molecule), then these were identified (with the assistance of Dr. Llewellyn Rodderick, Babraham Institute, Cambridge, UK), genomic DNA was isolated from the bank of chronic wound fibroblasts and the promoters/enhancers of the genes of interest amplified by PCR, cloned into entry clone plasmids and sequenced. Similarly, the destabilized ZsGreen fluorescent reporter open reading frame from the ZsGreen1-DR plasmid (Clontech) was also cloned and sequenced. Utilizing the MultiSite Gateway® Cloning methodology (Invitrogen), a *disease gene-promoter* ZsGreen reporter construct was generated which was transfected into the immortalized chronic wound cell lines. Over a period of 24 h fluorescent time-lapse microscopy analysis demonstrated a significant level of fluorescence within the chronic wound cells demonstrating that a functional cell-based reporter system had been developed (Fig. 5). This system now has the potential to be further developed and validated as a real-time, high-throughput, human-based in vitro test system for (a) the prescreening of agents that could have a beneficial/detrimental effect on chronic wound healing and (b) reducing a significant amount of unnecessary and unwarranted animal experimentation.

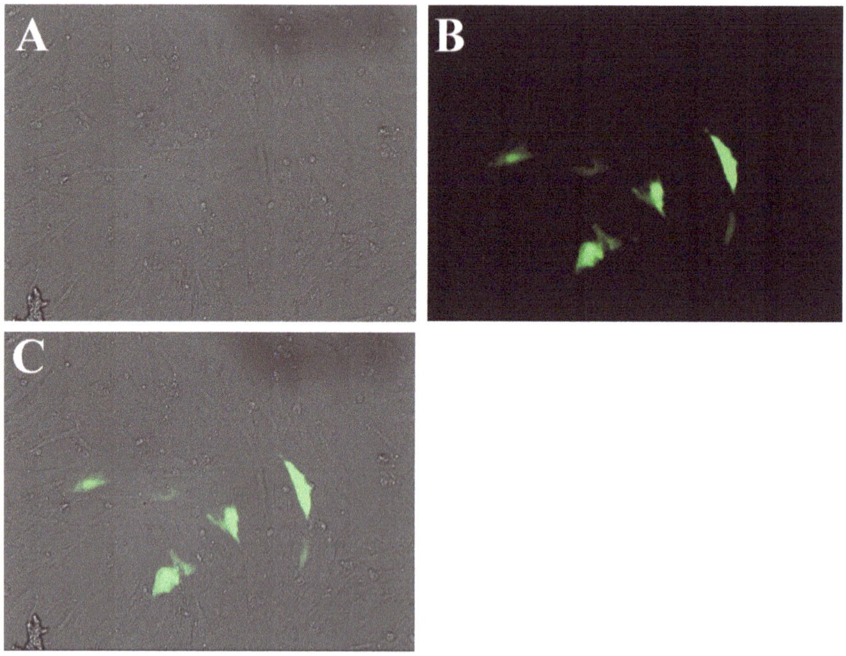

Fig. 5 Chronic wound fibroblasts demonstrating expression of a chronic disease-specific marker; (**a**) phase contrast image, (**b**) confocal laser scanning microscope image, (**c**) merged image. Scale bar = 50 μm

Acknowledgements

The authors would like to thank the National Centre for the Replacement, Refinement and Reduction of Animals in Research (NC3Rs), the Laboratory Animal Science Association (LASA) and the Dr Hadwen Trust for funding our research in this area. We would also like to thank Dr Llewellyn Rodderick (Babraham Institute, Cambridge, UK) for his assistance in our research and Professor Keith Harding (Wound Healing Research Unit, School of Medicine, Cardiff University, UK) for the provision of the clinical chronic wound samples and images.

References

1. Galiano RD, Michaels JT, Dobryansky M, Levine JP, Gurtner GC (2004) Quantitative and reproducible murine model of excisional wound healing. Wound Repair Regen 12:485–492
2. Wong VW, Sorkin M, Glotzbach JP, Longaker MT, Gurtner GC (2011) Surgical approaches to create murine models of human wound healing. J Biomed Biotechnol 2011:969618
3. Williams-Boyce PK, Daniel JC Jr (1986) Comparison of ear tissue regeneration in mammals. J Anat 149:55–63
4. Williams-Boyce PK, Daniel JC Jr (1980) Regeneration of rabbit ear tissue. J Exp Zool 212:243–253
5. Bos PK, van Osch GJVM, Frenz DA, Verhaar JAN, Verwoerd-Verhoef HL (2001) Growth factor expression in cartilage wound healing: temporal and spatial immunolocalization in a

rabbit auricular cartilage wound model. Osteoarthritis Cartilage 9:382–389
6. Kryger ZB, Sisco M, Roy NK, Lu L, Rosenberg D, Mustoe TA (2007) Temporal expression of the transforming growth factor-Beta pathway in the rabbit ear model of wound healing and scarring. J Am Coll Surg 205:78–88
7. Clark LD, Clark RK, Heber-Katz E (1998) A new murine model for mammalian wound repair and regeneration. Clin Immunol Immunopathol 88:35–45
8. Heber-Katz E (1999) The regenerating mouse ear. Semin Cell Dev Biol 10:415–419
9. Colwell AS, Thomas M, Krummel TM, Kong W, Longaker MTH, Lorenz P (2006) Skin wounds in the MRL/MPJ mouse heal with scar. Wound Repair Regen 14:81–90
10. Singer AJ, Taira BR, Anderson R, McClain SA, Rosenberg L (2010) The effects of rapid enzymatic debridement of deep partial-thickness burns with Debrase on wound reepithelialization in swine. J Burn Care Res 31:795–802
11. Agren MS (1996) Four alginate dressings in the treatment of partial thickness wounds: a comparative experimental study. Br J Plast Surg 49:129–134
12. Sanz GS, Santos HX, Izquierdo HA, Pascual PE, Bilbao de Aledo G, Hamann C (2000) Experimental model for local application of growth factors in skin re-epithelialisation. Scand J Plast Reconstr Surg Hand Surg 34:199–206
13. Breuing K, Kaplan S, Liu P, Onderdonk AB, Eriksson E (2003) Wound fluid bacterial levels exceed tissue bacterial counts in controlled porcine partial-thickness burn infections. Plast Reconstr Surg 111:781–788
14. Cuttle L, Kempf M, Phillips GE, Mill J, Hayes MT, Fraser JF, Wang XQ, Kimble RM (2006) A porcine deep dermal partial thickness burn model with hypertrophic scarring. Burns 32:806–820
15. Knabl JS, Bayer GS, Bauer WA, Schwendenwein I, Dado PF, Kucher C, Horvat R, Turkof E, Schossmann B, Meissl G (1999) Controlled partial skin thickness burns: an animal model for studies of burn-wound progression. Burns 25:229–235
16. Nanchahal J, Riches DJ (1982) The healing of suction blisters in pig skin. J Cutan Pathol 9:303–315
17. Rommain M, Brossard C, Piron MA, Smets P (1991) A skin suction blister model in hairless rats: application to the study of anti-inflammatory and immunomodulatory drugs. Int J Immunopharmacol 13:379–384
18. Levy JJ, von Rosen J, Gassmuller J, Kleine Kuhlmann R, Lange L (1995) Validation of an in vivo wound healing model for the quantification of pharmacological effects on epidermal regeneration. Dermatology 190:136–141
19. Leivo T, Kiistala U, Vesterinen M, Owaribe K, Burgeson RE, Virtanen I, Oikarinen A (2000) Re-epithelialization rate and protein expression in the suction-induced wound model: comparison between intact blisters, open wounds and calcipotriol-pretreated open wounds. Br J Dermatol 142:991–1002
20. Cribbs RK, Luquette MH, Besner GE (1998) A standardized model of partial thickness scald burns in mice. J Surg Res 80:69–74
21. Brans TA, Dutrieux RP, Hoekstra MJ, Kreis RW, du Pont JS (1994) Histopathological evaluation of scalds and contact burns in the pig model. Burns 20(Suppl 1):S48–S51
22. Becic F, Mornjakovic Z, Zulic I, Prasovic S, Mulabegovic N, Kapic E, Mijanovic M (2003) Animal model of thermal injuries. Bosn J Basic Med Sci 3:41–46
23. von Bulow S, Hartmann T, Fuchs PC, Schrimpf C, Pallua N (2005) Endothelial thrombomodulin (CD 141) in a rabbit burn model. Burns 31:459–464
24. Wang XQ, Kravchuk O, Winterford C, Kimble RM (2011) The correlation of in vivo burn scar contraction with the level of alpha-smooth muscle actin expression. Burns 37:1367–1377
25. Wang XQ, Kravchuk O, Kimble RM (2010) A retrospective review of burn dressings on a porcine burn model. Burns 36:680–687
26. Sullivan TP, Eaglstein WH, Davis SC, Mertz P (2001) The pig as a model for human wound healing. Wound Repair Regen 9:66–76
27. Lutz W, Sanderson W, Scherbov S (1997) Doubling of world population unlikely. Nature 387:803–805
28. Stephens P (2011) Dysfunctional wound healing in chronic wounds. In: Farrar D (ed) Advanced wound repair therapies. Woodhead Publishing Limited, Oxford, pp 1–38
29. Sen CK, Gordillo GM, Roy S, Kirsner R, Lambert L, Hunt TK, Gottrup F, Gurtner GC, Longaker MT (2009) Human skin wounds: a major and snowballing threat to public health and the economy. Wound Repair Regen 17:763–771
30. Inoue N, Nishikata S, Furuya E, Takita H, Kawamura M, Nishikaze O (1985) Streptozotocin diabetes: prolonged inflammatory response with delay in granuloma formation. Int J Tissue React 7:27–33

31. Lee JH, Yang SH, Oh JM, Lee MG (2010) Pharmacokinetics of drugs in rats with diabetes mellitus induced by alloxan or streptozocin: comparison with those in patients with type I diabetes mellitus. J Pharm Pharmacol 62:1–23
32. Kwon AH, Qiu Z, Hashimoto M, Yamamoto K, Kimura T (2008) Effects of medicinal mushroom (*Sparassis crispa*) on wound healing in streptozotocin-induced diabetic rats. Am J Surg 197:503–509
33. Lau TW, Sahota DS, Lau CH, Chan CM, Lam FC, Ho YY, Fung KP, Lau CB, Leung PC (2008) An in vivo investigation on the wound-healing effect of two medicinal herbs using an animal model with foot ulcer. Eur Surg Res 41:15–23
34. Morikawa T, Toyama T, Kudo N, Kawashima Y (2007) Reducing effect of matrix metalloproteinase inhibitors on serum triacylglycerol in streptozotocin-induced diabetic rats and Zucker fa/fa rats. Biol Pharm Bull 30:1461–1467
35. Rabelo SB, Villaverde AB, Nicolau R, Salgado MC, Melo M, Pacheco MT (2006) Comparison between wound healing in induced diabetic and nondiabetic rats after low-level laser therapy. Photomed Laser Surg 24:474–479
36. Velander P, Theopold C, Hirsch T, Bleiziffer O, Zuhaili B, Fossum M, Hoeller D, Gheerardyn R, Chen M, Visovatti S, Svensson H, Yao F, Eriksson E (2008) Impaired wound healing in an acute diabetic pig model and the effects of local hyperglycemia. Wound Repair Regen 16:288–293
37. Chen H, Charlat O, Tartaglia LA, Woolf EA, Weng X, Ellis SJ, Lakey ND, Culpepper J, Moore KJ, Breitbart RE, Duyk GM, Tepper RI, Morgenstern JP (1996) Evidence that the diabetes gene encodes the leptin receptor: identification of a mutation in the leptin receptor gene in db/db mice. Cell 84:491–495
38. Fang RC, Kryger ZB, Buck DW 2nd, De la Garza M, Galiano RD, Mustoe TA (2010) Limitations of the db/db mouse in translational wound healing research: is the NONcNZO10 polygenic mouse model superior? Wound Repair Regen 18:605–613
39. Scott JR, Tamura RN, Muangman P, Isik FF, Xie C, Gibran NS (2008) Topical substance P increases inflammatory cell density in genetically diabetic murine wounds. Wound Repair Regen 16:529–533
40. Trousdale RK, Jacobs S, Simhaee DA, Wu JK, Lustbader JW (2009) Wound closure and metabolic parameter variability in a db/db mouse model for diabetic ulcers. J Surg Res 151(1):100–107
41. Michaels J, Churgin SS, Blechman KM, Greives MR, Aarabi S, Galiano RD, Gurtner GC (2007) db/db mice exhibit severe wound-healing impairments compared with other murine diabetic strains in a silicone-splinted excisional wound model. Wound Repair Regen 15:665–670
42. Gottrup F, Agren MS, Karlsmark T (2000) Models for use in wound healing research: a survey focusing on in vitro and in vivo adult soft tissue. Wound Repair Regen 8:83–96
43. Galeano M, Deodato B, Altavilla D, Cucinotta D, Arsic N, Marini H, Torre V, Giacca M, Squadrito F (2003) Adeno-associated viral vector-mediated human vascular endothelial growth factor gene transfer stimulates angiogenesis and wound healing in the genetically diabetic mouse. Diabetologia 46:546–555
44. Kirchner LM, Meerbaum SO, Gruber BS, Knoll AK, Bulgrin J, Taylor RA, Schmidt SP (2003) Effects of vascular endothelial growth factor on wound closure rates in the genetically diabetic mouse model. Wound Repair Regen 11:127–131
45. Breitbart AS, Laser J, Parrett B, Porti D, Grant RT, Grande DA, Mason JM (2003) Accelerated diabetic wound healing using cultured dermal fibroblasts retrovirally transduced with the platelet-derived growth factor B gene. Ann Plast Surg 51:409–414
46. Galeano M, Torre V, Deodato B, Campo GM, Colonna M, Sturiale A, Squadrito F, Cavallari V, Cucinotta D, Buemi M, Altavilla D (2001) Raxofelast, a hydrophilic vitamin E-like antioxidant, stimulates wound healing in genetically diabetic mice. Surgery 129:467–477
47. Hamed S, Ullmann Y, Masoud M, Hellou E, Khamaysi Z, Teot L (2010) Topical erythropoietin promotes wound repair in diabetic rats. J Invest Dermatol 130:287–294
48. Altavilla D, Saitta A, Cucinotta D, Galeano M, Deodato B, Colonna M, Torre V, Russo G, Sardella A, Urna G, Campo GM, Cavallari V, Squadrito G, Squadrito F (2001) Inhibition of lipid peroxidation restores impaired vascular endothelial growth factor expression and stimulates wound healing and angiogenesis in the genetically diabetic mouse. Diabetes 50:667–674
49. Goova MT, Li J, Kislinger T, Qu W, Lu Y, Bucciarelli LG, Nowygrod S, Wolf BM, Caliste X, Yan SF, Stern DM, Schmidt AM (2001) Blockade of receptor for advanced glycation end-products restores effective wound healing in diabetic mice. Am J Pathol 159:513–525
50. Buemi M, Galeano M, Sturiale A, Ientile R, Crisafulli C, Parisi A, Catania M, Calapai G, Impala P, Aloisi C, Squadrito F, Altavilla D,

Bitto A, Tuccari G, Frisina N (2004) Recombinant human erythropoietin stimulates angiogenesis and healing of ischemic skin wounds. Shock 22:169–173

51. Gould LJ, Leong M, Sonstein J, Wilson S (2005) Optimization and validation of an ischemic wound model. Wound Repair Regen 13:576–582
52. Roy S, Biswas S, Khanna S, Gordillo G, Bergdall V, Green J, Marsh CB, Gould LJ, Sen CK (2009) Characterization of a preclinical model of chronic ischemic wound. Physiol Genomics 37:211–224
53. Wu L, Mustoe TA (1995) Effect of ischemia on growth factor enhancement of incisional wound healing. Surgery 117:570–576
54. Kamler M, Lehr HA, Barker JH, Saetzler RK, Galla TJ, Messmer K (1993) Impact of ischemia on tissue oxygenation and wound healing: intravital microscopic studies on the hairless mouse ear model. Eur Surg Res 25:30–37
55. Mogford JE, Liu WR, Reid R, Chiu C, Said H, Chen S, Harley CB, Mustoe TA (2006) Adenoviral human telomerase reverse transcriptase dramatically improves ischemic wound healing without detrimental immune response in an aged rabbit model. Hum Gene Ther 17:651–660
56. Mogford JE, Sisco M, Bonomo SR, Robinson AM, Mustoe TA (2004) Impact of aging on gene expression in a rat model of ischemic cutaneous wound healing1. J Surg Res 118:190–196
57. Chien S (2007) Ischemic rabbit ear model created by minimally invasive surgery. Wound Repair Regen 15:928–935
58. Kloeters O, Jia S-X, Roy N, Schultz GS, Leinfellner G, Mustoe TA (2007) Alteration of Smad3 signaling in ischemic rabbit dermal ulcer wounds. Wound Repair Regen 15:341–349
59. Manna V, Bem J, Marks R (1982) An animal model for chronic ulceration. Br J Dermatol 106:169–181
60. Pierce GF, Mustoe TA, Lingelbach J, Masakowski VR, Gramates P, Deuel TF (1989) Transforming growth factor beta reverses the glucocorticoid-induced wound-healing deficit in rats: possible regulation in macrophages by platelet-derived growth factor. Proc Natl Acad Sci USA 86:2229–2233
61. Aksoy B, Aksoy HM, Civas E, Ustun H, Atakan N (2009) A new experimental delayed wound healing model in rabbits. Eur J Dermatol 19:565–569
62. Singer AJ, Taira BR, Anderson R, McClain SA, Rosenberg L (2010) Does pressure matter in creating burns in a porcine model? J Burn Care Res 31:646–651
63. Rico RM, Ripamonti R, Burns AL, Gamelli RL, DiPietro LA (2002) The effect of sepsis on wound healing. J Surg Res 102:193–197
64. Gurjala AN, Geringer MR, Seth AK, Hong SJ, Smeltzer MS, Galiano RD, Leung KP, Mustoe TA (2011) Development of a novel, highly quantitative in vivo model for the study of biofilm-impaired cutaneous wound healing. Wound Repair Regen 19:400–410
65. Schierle CF, De la Garza M, Mustoe TA, Galiano RD (2009) Staphylococcal biofilms impair wound healing by delaying reepithelialization in a murine cutaneous wound model. Wound Repair Regen 17:354–359
66. Hinrichsen N, Birk-Sorensen L, Gottrup F, Hjortdal V (1998) Wound contraction in an experimental porcine model. Scand J Plast Reconstr Surg Hand Surg 32:243–248
67. Wright JB, Lam K, Buret AG, Olson ME, Burrell RE (2002) Early healing events in a porcine model of contaminated wounds: effects of nanocrystalline silver on matrix metalloproteinases, cell apoptosis, and healing. Wound Repair Regen 10:141–151
68. Jacobsen F, Fisahn C, Sorkin M, Thiele I, Hirsch T, Stricker I, Klaassen T, Roemer A, Fugmann B, Steinstraesser L (2011) Efficacy of topically delivered moxifloxacin against wound infection by *Pseudomonas aeruginosa* and methicillin-resistant *Staphylococcus aureus*. Antimicrob Agents Chemother 55:2325–2334
69. Bernatchez SF, Parks PJ, Grussing DM, Matalas SL, Nelson GS (1998) Histological characterization of a delayed wound healing model in pig. Wound Repair Regen 6:223–233
70. Hudson LG, Newkirk KM, Chandler HL, Choi C, Fossey SL, Parent AE, Kusewitt DF (2009) Cutaneous wound reepithelialization is compromised in mice lacking functional Slug (Snai2). J Dermatol Sci 56:19–26
71. AlDahlawi S, Eslami A, Hakkinen L, Larjava HS (2006) The alphavbeta6 integrin plays a role in compromised epidermal wound healing. Wound Repair Regen 14:289–297
72. Campbell L, Emmerson E, Davies F, Gilliver SC, Krust A, Chambon P, Ashcroft GS, Hardman MJ (2010) Estrogen promotes cutaneous wound healing via estrogen receptor beta independent of its antiinflammatory activities. J Exp Med 207:1825–1833
73. Rudolph KL, Chang S, Lee HW, Blasco M, Gottlieb GJ, Greider C, DePinho RA (1999) Longevity, stress response, and cancer in aging telomerase-deficient mice. Cell 96:701–712
74. Poujade M, Grasland-Mongrain E, Hertzog A, Jouanneau J, Chavrier P, Ladoux B, Buguin A, Silberzan P (2007) Collective

migration of an epithelial monolayer in response to a model wound. Proc Natl Acad Sci USA 104:15988–15993
75. Kole TP, Tseng Y, Jiang I, Katz JL, Wirtz D (2005) Intracellular mechanics of migrating fibroblasts. Mol Biol Cell 16:328–338
76. Kitano A, Saika S, Yamanaka O, Ikeda K, Okada Y, Shirai K, Reinach PS (2007) Emodin suppression of ocular surface inflammatory reaction. Invest Ophthalmol Vis Sci 48:5013–5022
77. Borensztajn K, Stiekema J, Nijmeijer S, Reitsma PH, Peppelenbosch MP, Spek CA (2008) Factor Xa stimulates proinflammatory and profibrotic responses in fibroblasts via protease-activated receptor-2 activation. Am J Pathol 172:309–320
78. Diaz G, Batetta B, Sanna F, Uda S, Reali C, Angius F, Melis M, Falchi AM (2008) Lipid droplet changes in proliferating and quiescent 3T3 fibroblasts. Histochem Cell Biol 129:611–621
79. Stevenson S, Taylor AH, Meskiri A, Sharpe DT, Thornton MJ (2008) Differing responses of human follicular and nonfollicular scalp cells in an in vitro wound healing assay: effects of estrogen on vascular endothelial growth factor secretion. Wound Repair Regen 16:243–253
80. Goldberg SR, Quirk GL, Sykes VW, McKinstry RP, Kordula T, Lanning DA (2008) Differential use of Erk1/2 and transforming growth factor beta pathways by mid- and late-gestational murine fibroblasts. J Pediatr Surg 43:971–976
81. Loryman C, Mansbridge J (2008) Inhibition of keratinocyte migration by lipopolysaccharide. Wound Repair Regen 16:45–51
82. Stephens P, Wall IB, Wilson MJ, Hill KE, Davies CE, Hill CM, Harding KG, Thomas DW (2003) Anaerobic cocci populating the deep tissues of chronic wounds impair cellular wound healing responses in vitro. Br J Dermatol 148:456–466
83. Wall IB, Moseley R, Baird DM, Kipling D, Giles P, Laffafian I, Price PE, Thomas DW, Stephens P (2008) Fibroblast dysfunction is a key factor in the non-healing of chronic venous leg ulcers. J Invest Dermatol 128:2526–2540
84. Hardwicke J, Moseley R, Stephens P, Harding K, Duncan R, Thomas DW (2010) Bioresponsive dextrin-rhEGF conjugates: in vitro evaluation in models relevant to its proposed use as a treatment for chronic wounds. Mol Pharm 7:699–707
85. Paul Ehrlich H, Sun B, Kainth KS, Kromah F (2006) Elucidating the mechanism of wound contraction: rapid versus sustained myosin ATPase activity in attached-delayed-released compared with free-floating fibroblast-populated collagen lattices. Wound Repair Regen 14:625–632
86. O'Gorman DB, Wu Y, Seney S, Zhu RD, Gan BS (2006) Wnt expression is not correlated with beta-catenin dysregulation in Dupuytren's disease. J Negat Results Biomed 5:13
87. Mukhopadhyay A, Tan EK, Khoo YT, Chan SY, Lim IJ, Phan TT (2005) Conditioned medium from keloid keratinocyte/keloid fibroblast coculture induces contraction of fibroblast-populated collagen lattices. Br J Dermatol 152:639–645
88. Bell E, Ivarsson B, Merrill C (1979) Production of a tissue-like structure by contraction of collagen lattices by human fibroblasts of different proliferative potential in vitro. Proc Natl Acad Sci USA 76:1274–1278
89. Buttle DJ, Ehrlich HP (1983) Comparative studies of collagen lattice contraction utilizing a normal and a transformed cell line. J Cell Physiol 116:159–166
90. Dallon JC, Ehrlich HP (2008) A review of fibroblast-populated collagen lattices. Wound Repair Regen 16:472–479
91. Ehrlich HP, Gabbiani G, Meda P (2000) Cell coupling modulates the contraction of fibroblast-populated collagen lattices. J Cell Physiol 184:86–92
92. Ehrlich HP, Rittenberg T (2000) Differences in the mechanism for high-versus moderate-density fibroblast-populated collagen lattice contraction. J Cell Physiol 185:432–439
93. Ehrlich HP, Wyler DJ (1983) Fibroblast contraction of collagen lattices in vitro: inhibition by chronic inflammatory cell mediators. J Cell Physiol 116:345–351
94. Nakagawa S, Pawelek P, Grinnell F (1989) Long-term culture of fibroblasts in contracted collagen gels: effects on cell growth and biosynthetic activity. J Invest Dermatol 93:792–798
95. Nakagawa S, Pawelek P, Grinnell F (1989) Extracellular matrix organization modulates fibroblast growth and growth factor responsiveness. Exp Cell Res 182:572–582
96. Sarber R, Hull B, Merrill C, Soranno T, Bell E (1981) Regulation of proliferation of fibroblasts of low and high population doubling levels grown in collagen lattices. Mech Ageing Dev 17:107–117
97. van Bockxmeer FM, Martin CE, Constable IJ (1984) Effect of cyclic AMP on cellular contractility and DNA synthesis in chorioretinal fibroblasts maintained in collagen matrices. Exp Cell Res 155:413–421
98. Stephens P, Davies KJ, Al-Khateeb T, Shepherd JP, Thomas DW (1996) A comparison of the

ability of intra-oral and extra-oral fibroblasts to stimulate extracellular matrix reorganization in a model of wound contraction. J Dent Res 75:1358–1364
99. Stephens P, Davies KJ, Occleston N, Pleass RD, Kon C, Daniels J, Khaw PT, Thomas DW (2001) Skin and oral fibroblasts exhibit phenotypic differences in extracellular matrix reorganization and matrix metalloproteinase activity. Br J Dermatol 144:229–237
100. Irwin CR, Myrillas T, Smyth M, Doogan J, Rice C, Schor SL (1998) Regulation of fibroblast-induced collagen gel contraction by interleukin-1beta. J Oral Pathol Med 27:255–259
101. Irwin CR, Picardo M, Ellis I, Sloan P, Grey A, McGurk M, Schor SL (1994) Inter- and intra-site heterogeneity in the expression of fetal-like phenotypic characteristics by gingival fibroblasts: potential significance for wound healing. J Cell Sci 107(Pt 5):1333–1346
102. Schor SL, Ellis I, Irwin CR, Banyard J, Seneviratne K, Dolman C, Gilbert AD, Chisholm DM (1996) Subpopulations of fetal-like gingival fibroblasts: characterisation and potential significance for wound healing and the progression of periodontal disease. Oral Dis 2:155–166
103. Cook H, Stephens P, Davies KJ, Harding KG, Thomas DW (2000) Defective extracellular matrix reorganization by chronic wound fibroblasts is associated with alterations in TIMP-1, TIMP-2, and MMP-2 Activity. J Invest Dermatol 115:225–233
104. Stephens P, Cook H, Hilton J, Jones CJ, Haughton MF, Wyllie FS, Skinner JW, Harding KG, Kipling D, Thomas DW (2003) An analysis of replicative senescence in dermal fibroblasts derived from chronic leg wounds predicts that telomerase therapy would fail to reverse their disease-specific cellular and proteolytic phenotype. Exp Cell Res 283:22–35
105. Stephens P, Wood EJ, Raxworthy MJ (1996) Development of a multilayered in vitro model for studying events associated with wound healing. Wound Repair Regen 4:393–401
106. Bell E, Ehrlich HP, Sher S, Merrill C, Sarber R, Hull B, Nakatsuji T, Church D, Buttle DJ (1981) Development and use of a living skin equivalent. Plast Reconstr Surg 67:386–392
107. Schneider RK, Neuss S, Stainforth R, Laddach N, Bovi M, Knuechel R, Perez-Bouza A (2008) Three-dimensional epidermis-like growth of human mesenchymal stem cells on dermal equivalents: contribution to tissue organization by adaptation of myofibroblastic phenotype and function. Differentiation 76:156–167
108. Kinsner A, Lesiak-Cyganowska E, Sladowski D (2001) In vitro reconstruction of full thickness human skin on a composite collagen material. Cell Tissue Bank 2:165–171
109. Wang TW, Huang YC, Sun JS, Lin FH (2003) Organotypic keratinocyte-fibroblast cocultures on a bilayer gelatin scaffold as a model of skin equivalent. Biomed Sci Instrum 39:523–528
110. Maas-Szabowski N, Stark HJ, Fusenig NE (2000) Keratinocyte growth regulation in defined organotypic cultures through IL-1-induced keratinocyte growth factor expression in resting fibroblasts. J Invest Dermatol 114:1075–1084
111. Stark HJ, Baur M, Breitkreutz D, Mirancea N, Fusenig NE (1999) Organotypic keratinocyte cocultures in defined medium with regular epidermal morphogenesis and differentiation. J Invest Dermatol 112:681–691
112. Auger FA, Berthod F, Moulin V, Pouliot R, Germain L (2004) Tissue-engineered skin substitutes: from in vitro constructs to in vivo applications. Biotechnol Appl Biochem 39:263–275
113. Prunieras M, Regnier M, Woodley D (1983) Methods for cultivation of keratinocytes with an air-liquid interface. J Invest Dermatol 81:28s–33s
114. Ouwehand K, Spiekstra SW, Waaijman T, Scheper RJ, de Gruijl TD, Gibbs S (2011) Technical advance: Langerhans cells derived from a human cell line in a full-thickness skin equivalent undergo allergen-induced maturation and migration. J Leukoc Biol 90:1027–1033
115. Ponec M, El Ghalbzouri A, Dijkman R, Kempenaar J, van der Pluijm G, Koolwijk P (2004) Endothelial network formed with human dermal microvascular endothelial cells in autologous multicellular skin substitutes. Angiogenesis 7:295–305
116. Monfort A, Soriano-Navarro M, Garcia-Verdugo JM, Izeta A (2012) Production of human tissue-engineered skin trilayer on a plasma-based hypodermis. J Tissue Eng Regen Med. doi:10.1002/term.548
117. Hoeller D, Huppertz B, Roos TC, Gutiérrez PP, Merk HF, Frank J, Jugert FK (2001) An improved and rapid method to construct skin equivalents from human hair follicles and fibroblasts. Exp Dermatol 10:264–271
118. Genever PG, Wood EJ, Cunliffe WJ (1993) The wounded dermal equivalent offers a simplified model for studying wound repair *in vitro*. Exp Dermatol 2:266–273
119. O'Leary R, Arrowsmith M, Wood EJ (2002) Characterization of the living skin equivalent as a model of cutaneous re-epithelialization. Cell Biochem Funct 20:129–141

120. Auxenfans C, Fradette J, Lequeux C, Germain L, Kinikoglu B, Bechetoille N, Braye F, Auger FA, Damour O (2009) Evolution of three dimensional skin equivalent models reconstructed in vitro by tissue engineering. Eur J Dermatol 19:107–113
121. Herman IM, Leung A (2009) Creation of human skin equivalents for the in vitro study of angiogenesis in wound healing. Methods Mol Biol 467:241–248
122. Rowden G, Lewis MG, Sheikh KM, Summerlin WT (1975) Long-term organ culture of human skin: an ultrastructural and immunochemical study. J Pathol 117:139–149
123. Holbrook KA, Hennings H (1983) Phenotypic expression of epidermal cells in vitro: a review. J Invest Dermatol 81:11s–24s
124. Hambrick GW Jr, Lamberg SI, Bloomberg R (1966) Observations on keratinization of human skin in vitro. J Invest Dermatol 47:541–550
125. Tammi R, Jansen CT, Santti R (1979) Histometric analysis of human skin in organ culture. J Invest Dermatol 73:138–140
126. Stoll SW, Kansra S, Elder JT (2003) Keratinocyte outgrowth from human skin explant cultures is dependent upon p38 signaling. Wound Repair Regen 11:346–353
127. Jameson J, Ugarte K, Chen N, Yachi P, Fuchs E, Boismenu R, Havran WL (2002) A role for skin gammadelta T cells in wound repair. Science 296:747–749
128. Onuma H, Mastui C, Morohashi M (2001) Quantitative analysis of the proliferation of epidermal cells using a human skin organ culture system and the effect of DbcAMP using markers of proliferation (BrdU, Ki-67, PCNA). Arch Dermatol Res 293:133–138
129. Kim LT, Wu J, Turnage RH (2001) FAK induction in keratinocytes in an in vitro model of reepithelialization. J Surg Res 96:167–172
130. Kratz G (1998) Modeling of wound healing processes in human skin using tissue culture. Microsc Res Tech 42:345–350
131. Moll I, Houdek P, Schmidt H, Moll R (1998) Characterization of epidermal wound healing in a human skin organ culture model: acceleration by transplanted keratinocytes. J Invest Dermatol 111:251–258
132. Greenwald DP, Gottlieb LJ, Mass DP, Shumway SM, Temaner M (1992) Full-thickness skin wound explants in tissue culture: a mechanical evaluation of healing. Plast Reconstr Surg 90:289–294
133. Toulon A, Breton L, Taylor KR, Tenenhaus M, Bhavsar D, Lanigan C, Rudolph R, Jameson J, Havran WL (2009) A role for human skin-resident T cells in wound healing. J Exp Med 206:743–750
134. Hayflick L, Moorhead PS (1961) The serial cultivation of human diploid cell strains. Exp Cell Res 25:585–621
135. Masters JR (2002) HeLa cells 50 years on: the good, the bad and the ugly. Nat Rev Cancer 2:315–319
136. Harley CB (1991) Telomere loss: mitotic clock or genetic time bomb? Mutat Res 256:271–282
137. Harley CB, Villeponteau B (1995) Telomeres and telomerase in aging and cancer. Curr Opin Genet Dev 5:249–255
138. Bodnar AG, Ouellette M, Frolkis M, Holt SE, Chiu CP, Morin GB, Harley CB, Shay JW, Lichtsteiner S, Wright WE (1998) Extension of life-span by introduction of telomerase into normal human cells. Science 279:349–352
139. Kipling D (2001) Telomeres, replicative senescence and human ageing. Maturitas 38:25–37
140. Jiang X-R, Jimenez G, Chang E, Frolkis M, Kusler B, Sage M, Beeche M, Bodnar AG, Wahl GM, Tlsty TD, Chiu C-P (1999) Telomerase expression in human somatic cells does not induce changes associated with a transformed phenotype. Nat Genet 21:111–114
141. Morales CP, Holt SE, Ouellette M, Kaur KJ, Yan Y, Wilson KS, White MA, Wright WE, Shay JW (1999) Absence of cancer-associated changes in human fibroblasts immortalized with telomerase. Nat Genet 21:115–118
142. Yang J, Chang E, Cherry AM, Bangs CD, Oei Y, Bodnar A, Bronstein A, Chiu C-P, Herron GS (1999) Human endothelial cell life extension by telomerase expression. J Biol Chem 274:26141–26148
143. Kampinga HH, Van Waarde-Verhagen MA, Van Assen-Bolt AJ, Nieuwenhuis B, Rodemann HP, Prowse KR, Linskens MH (2004) Reconstitution of active telomerase in primary human foreskin fibroblasts: effects on proliferative characteristics and response to ionizing radiation. Int J Radiat Biol 80:377–388
144. Meran S, Thomas D, Stephens P, Martin J, Bowen T, Phillips A, Steadman R (2007) Involvement of hyaluronan in regulation of fibroblast phenotype. J Biol Chem 282:25687–25697
145. Stephens P, Hiscox S, Cook H, Jiang WG, Zhiquiang W, Thomas DW (2001) Phenotypic variation in the production of bioactive hepatocyte growth factor/scatter factor by oral mucosal and skin fibroblasts. Wound Repair Regen 9:34–43
146. Wall SJ, Bevan D, Thomas DW, Harding KG, Edwards DR, Murphy G (2002) Differential expression of matrix metalloproteinases during impaired wound healing of the diabetes mouse. J Invest Dermatol 119:91–98

147. Chandler LA, Gu DL, Ma C, Gonzalez AM, Doukas J, Nguyen T, Pierce GF, Phillips ML (2000) Matrix-enabled gene transfer for cutaneous wound repair. Wound Repair Regen 8:473–479
148. Mead AL, Wong TT, Cordeiro MF, Anderson IK, Khaw PT (2003) Evaluation of anti-TGF-beta2 antibody as a new postoperative anti-scarring agent in glaucoma surgery. Invest Ophthalmol Vis Sci 44:3394–3401
149. Occleston NL, Daniels JT, Tarnuzzer RW, Sethi KK, Alexander RA, Bhattacharya SS, Schultz GS, Khaw PT (1997) Single exposures to antiproliferatives: long-term effects on ocular fibroblast wound-healing behavior. Invest Ophthalmol Vis Sci 38:1998–2007
150. So K, McGrouther DA, Bush JA, Durani P, Taylor L, Skotny G, Mason T, Metcalfe A, O'Kane S, Ferguson MW (2011) Avotermin for scar improvement following scar revision surgery: a randomized, double-blind, within-patient, placebo-controlled, phase II clinical trial. Plast Reconstr Surg 128:163–172
151. Stephens P (2010) Development of a cell-based diabetic wound assay. Altern Lab Anim 38(Suppl 1):45–48
152. Seitz O, Schürmann C, Hermes N, Müller E, Pfeilschifter J, Frank S, Goren I (2010) Wound healing in mice with high-fat diet- or ob gene-induced diabetes-obesity syndromes: a comparative study. Exp Diabetes Res. doi: 10.1155/2010/476969

Chapter 11

Novel Methods for the Investigation of Human Hypertrophic Scarring and Other Dermal Fibrosis

Dariush Honardoust, Peter Kwan, Moein Momtazi, Jie Ding, and Edward E. Tredget

Abstract

Hypertrophic scar (HTS) represents the dermal equivalent of fibroproliferative disorders that occur after injury involving the deep dermis while superficial wounds to the skin heal with minimal or no scarring. HTS is characterized by progressive deposition of collagen that occurs with high frequency in adult dermal wounds following traumatic or thermal injury. Increased levels of transforming growth factor-β1 (TGF-β1), decreased expression of small leucine-rich proteoglycans (SLRPs), and/or fibroblast subtypes may influence the development of HTS. The development of HTS is strongly influenced by the cellular and molecular properties of fibroblast subtypes, where cytokines such as fibrotic TGF-β_1 and CTGF as well as the expression of SLRPs, particularly decorin and fibromodulin, regulate collagen fibrillogenesis and the activity of TGF-β_1. Reduced anti-fibrotic molecules in the ECM of the deep dermis and the distinctive behavior of the fibroblasts in this region of the dermis which display increased sensitivity to TGF-β_1's biological activity contribute to the development of HTS following injury to the deep dermis. By comparing the cellular and molecular differences involved in deep and superficial wound healing in an experimental wound scratch model in humans that has both superficial and deep injuries within the same excisional model, our aim is to increase our understanding of how tissue repair following injury to the deep dermis can be changed to promote healing with a similar pattern to healing that occurs following superficial injury that results in no or minimal scarring. Studying the characteristics of superficial dermal injuries that heal with minimal scarring will help us identify therapeutic approaches for tissue engineering and wound healing. In addition, our ability to develop novel therapies for HTS is hampered by limitations in the available animal models used to study this disorder in vivo. We also describe a nude mouse model of transplanted human skin that develops a hypertrophic proliferative scar consistent morphologically and histologically with human HTS, which can be used to test novel treatment options for these dermal fibrotic conditions.

Key words Dermal fibroblasts, Wound healing, Hypertrophic scar, Nude mouse

1 Introduction

Hypertrophic scar (HTS) is a fibroproliferative disorder that occurs in human skin. Injury to the deep dermis often leads to HTS formation. Interestingly, superficial injuries often heal with minimal or no scarring. Cellular and molecular processes that lead to HTS

Robert G. Gourdie and Tereance A. Myers (eds.), *Wound Regeneration and Repair: Methods and Protocols*, Methods in Molecular Biology, vol. 1037, DOI 10.1007/978-1-62703-505-7_11, © Springer Science+Business Media New York 2013

development following injury to the deep dermis are not fully characterized. However, existing evidence indicates that heterogeneity of fibroblasts in terms of function, subtype, and population and differences in the level of growth factors between superficial and deep dermal wounds may influence HTS development after deep dermal injury. It has been reported that deep dermal fibroblasts have a higher expression of connective tissue growth factor, type I collagen, α-smooth muscle actin (α-SMA), and transforming growth factor-β_1 (TGF-β_1) when compared to cells from superficial layers [1]. Higher expression of TGF-β_1 results in increased number of α-SMA-expressing fibroblasts; excessive collagen production and abnormal extracellular matrix structure are considered the prominent features of the pathology of HTS. Fibroblasts are the key cells involved in wound healing and collagen fibrillogenesis. In scar-free regenerative healing such as in fetal wound healing, fibroblasts are dormant and show minimal proliferation and decreased extracellular matrix production; however, augmented fibroblast biological activity leads to increased fibril synthesis. As well, the down-regulation of collagen and remodelling contributes to fibrosis in the skin, resulting in HTS. At the cellular level, HTS is characterized by an accumulation of type I collagen comprising an increased population of fibroblasts, myofibroblasts, and endothelial cells [1, 2]. Myofibroblasts which are characterized by an increased expression of α-SMA and cytokines such as TGF-β_1 orchestrate extracellular matrix contraction in fibrosis [3]. The differentiation of fibroblasts to myofibroblasts is subject to TGF-β_1's biological activity. TGFβ_1 has been considered a key profibrotic mediator of different cell functions including cell migration, proliferation, differentiation, survival, and development of new blood vessels during wound healing. At the cell surface, the TGF-β ligands bind to TGF-β receptor II, which leads to cell signalling and increased bioactivity [4, 5]. Therefore, the expression of TGF-β receptor II during wound healing appears to be an important factor that may potentially influence TGF-β fibrogenic activity. Growing evidence indicates that differences between cellular and molecular constituents of papillary and deep reticular dermis can account for minimal scarring and HTS development, respectively, in these areas of the skin. For example, small leucine-rich proteoglycans (SLRPs), particularly decorin and fibromodulin, can influence scar formation by binding to and manipulating TGF-β_1 activity [6–8]. Furthermore, fibroblasts from deep and superficial layers of the skin may have distinctive differences in surface antigens which coordinate the structural organization of extracellular matrix and the interaction with different molecules [9]. One of the most relevant molecules of interest is fibroblast's surface antigen Thy-1 or CD90, an N-glycosylated, glycophosphatidylinositol (GPI)-anchored conserved cell surface protein. It has been shown that differential expression of Thy-1 affects myofibroblast differentiation by TGF-β_1 [10, 11]. Lack of Thy-1 [11] has been associated with an accumulation of collagen and increased tissue fibrosis [12]. To date the role of Thy-1 in the development of HTS has not been studied. Since Thy-1 is implicated in tissue fibrosis, studying its interaction with TGF-β_1 may reveal novel information to control dermal scarring. Taken together, compared to papillary fibroblasts, reduced levels of anti-fibrotic molecules, decorin, fibromodulin, Thy-1, and TGF-β_3 in deep dermis may determine the outcome of wound healing and increase HTS formation following injury that involves deeper layers of skin.

1.1 Linear Scratch Model

The procedure for creating the linear scratch in the wound scratch model has been explained elsewhere [13, 14]. The following briefly describes the steps in the procedure.

1.2 Fibroblasts from the Deep and Superficial Layers of the Skin

Dermal fibroblasts can be divided into distinct subpopulations with unique regenerative or fibrotic properties based on their depth [15–17]. Although we have previously described dividing the dermis into multiple layers [9], this requires a high level of technical skill in the use of dermatomes, which is typically difficult for non-surgeons to acquire. We have found that harvesting the most superficial and deepest portions of dermis for fibroblast extraction is more straightforward and is also the most experimentally relevant, and therefore we outline this process in detail. As with most techniques, there are multiple ways to isolate superficial and deep dermis. We present two options for each and highlight our preferred approach, which is designed to allow extraction of a wide variety of cells to maximize the yield from limited amounts of human tissue.

Human abdominoplasty specimens are collected under a health ethics review board-approved protocol along with relevant donor information including gender, age, smoking status, and medical comorbidities. Other tissues such as breast reduction specimens may be used, but in our practice these are unavailable since they are routinely sent for pathological examination. No such requirement exists for abdominoplasty specimens, and we have found that patients are generally willing to consent to donate their abdominoplasty tissue. As donors are not screened for blood-borne diseases (e.g., hepatitis C or human immunodeficiency virus), all of the collected tissues must be handled using universal precautions. Generally this includes appropriate personal protective equipment such as gloves, gowns, masks, and eye shields. All tissues are handled inside a biosafety cabinet, and all tissues are disposed of in appropriate biologic waste containers.

Specimens are collected and transported in a rapid, sterile, cooled fashion to the laboratory for processing. A dermatome is used to split the dermis into sequential layers of increasing depth. These individual dermal layers are further digested enzymatically or cut into small pieces for explantation and recovery of the fibroblasts. This allows one to harvest matched superficial and deep dermal fibroblasts for further experimentation.

1.3 A Nude Mouse Model of Human Hypertrophic Scar

Our understanding of HTS is hampered by limitations in the available animal models used to study this disorder. Here we describe a nude mouse model of HTS that can be used to create proliferative xenograft scars that have been shown to be morphologically and histologically consistent with human HTS [18]. Furthermore, this model can be used to develop as well as test the efficacy of novel treatment options for HTS.

The nude mouse model uses transplanted split-thickness human skin grafts to generate proliferative xenograft scars. First, human split-thickness skin is harvested from resected abdominal tissue from healthy females undergoing cosmetic abdominoplasty. The skin is then cut into 2.0×1.5 cm grafts that are sewn in a tie-over-bolus fashion to a full-thickness dorsal skin defect created on the backs of nude mice. Over time these wounds heal and become firm, thickened, and red scars that closely resemble human HTS.

In our initial experiments with this model both split- and full-thickness human grafts were transplanted onto nude mice [18]. Quantification of dermal thickness over time showed a greater percent increase in scar thickness over time using split-thickness human skin grafts [18]. For this reason in our subsequent studies we have chosen to use only split-thickness human skin grafts. Our current work with the nude mouse model supports our previously published morphological and histological observations. Red, thick, inelastic scars with an increase in dermal thickness, abnormal whorled collagen bundles, increased mast cell density, and an increase in α-SMA expression were observed following engraftment of split-thickness human skin onto the nude mouse. Our characterization of this model has been broadened to include immunohistochemical localization and quantification of SLRPs in proliferative xenograft scars including decorin and biglycan. These experiments demonstrate reciprocal changes in the expression of these proteins similar to those encountered in human HTS [19, 14]. These observations serve to further support the use of the nude mouse as practical and relevant research model for the study of human HTS.

2 Materials

2.1 Linear Wound Scratch Model

Antibodies

1. Polyclonal anti-TGF-β_1 (V; sc: 146), Santa Cruz Biotechnology.
2. Monoclonal anti-TGF-β_3 (Xn-12; sc: 80348), Santa Cruz Biotechnology.
3. Monoclonal anti-decorin (9XX; sc873896), Santa Cruz Biotechnology.

4. Polyclonal anti-biglycan (H-150; sc-33788), Santa Cruz Biotechnology.
5. Polyclonal anti-fibromodulin (H-50; sc-33772), Santa Cruz Biotechnology.
6. Polyclonal anti-lumican (H-90; sc-33785), Santa Cruz Biotechnology.
7. Polyclonal anti-TGF-β receptor type RII (C-16; sc-220), Santa Cruz Biotechnology.
8. Anti-CD90/Thy-1 from AbCam (Cambridge, MA, USA).
9. Monoclonal anti-procollagen (1912) from Chemicon International Inc.
10. Monoclonal anti-lymphocyte-specific protein-1 (LSP-1) from BD transduction.
11. Appropriate Alexa-conjugated anti-rat, anti-rabbit, or anti-mouse.
12. Polyclonal anti-β-actin from Sigma.
13. Anti-rabbit and anti-mouse IgG HRP-conjugated secondary antibodies.

2.2 Fibroblasts from the Deep and Superficial Layers of the Skin

All materials used should be autoclaved or otherwise appropriately sterilized prior to use to avoid contamination of harvested cells.

1. Sterile and waterproof transport containers.
2. Normal saline (0.9 %).
3. Dermatome with guards and blades.
4. Snaps or Kochers.
5. Non-toothed Adson forceps.
6. Scissors.
7. Mineral oil.
8. Dispase solution [16]: Dispase II (~1,000 Japanese units/mL or 1.67 Roche units/mL final concentration) in Dulbecco's modified Eagle medium.
9. Centrifuge tube (50 mL).
10. Petri dish.
11. Small-gauge needles (20 gauge).
12. Scalpel (#20 blade).
13. Tissue culture flask (75 cm^2).

2.3 A Nude Mouse Model of Human Hypertrophic Scar

Harvest of human split-thickness skin grafts

1. Padgett dermatome.
2. Sterile green towels.
3. Piercing towel clips.
4. Mineral oil.

5. Sterile gown.
6. Cap.
7. Sterile gloves.
8. Mask.
9. Eye protection.
10. Plastic carrier.
11. Single-use, number 20 disposable scalpel.
12. Conical tubes.
13. Normal saline.

Transplantation of human split-thickness skin grafts

1. 4–6-week-old male Bagg albino (BALB)/c-nu/nu nude mice.
2. 70 % ethanol.
3. Sterile green towels.
4. Heating pad.
5. Disposable paper ruler.
6. Sterile gown.
7. Cap.
8. Sterile gloves.
9. Mask.
10. Eye protection.
11. Isoflurane anesthetic.
12. Sterile gloves.
13. Tears Naturale® P.M.
14. Nair® hair remover.
15. Sharpie marker®.
16. Straight scissors.
17. Adson forceps.
18. Conical tubes.
19. Normal saline.
20. Human split-thickness skin grafts.
21. 4-0 braided silk suture, taper RB-1 needle.
22. Needle driver.
23. Non-toothed forceps.
24. Petrolatum 3 % bismuth tribromophenate gauze.
25. Single-use, sterile, 27 G syringes (1 mL).
26. Hydromorphone.

Harvest of normal skin and scar biopsies

1. 70 % ethanol.
2. Sterile green towels.
3. Heating pad.
4. Disposable paper ruler.
5. Sterile gown.
6. Cap.
7. Sterile gloves.
8. Mask.
9. Eye protection.
10. Straight scissors.
11. Adson forceps.
12. Conical tubes.
13. Normal saline.

Processing of normal skin and scar biopsies

1. 70 % ethanol.
2. Sterile gown.
3. Cap.
4. Sterile gloves.
5. Mask.
6. Eye protection.
7. Plastic carrier.
8. Sterile green towels.
9. Tissue cassettes.
10. Foam pads.
11. Embedding mold (22 × 22 × 20 mm).
12. Conical tubes.
13. Normal saline.
14. Single-use, number 20 disposable scalpel.
15. Adson forceps.
16. 10 % formalin.
17. Cryomatrix.

3 Methods

3.1 Creation of a Linear Wound Scratch Model

1. Following informed consent, burn patients received a standardized wound along the lines of relaxed skin tension, midway between the anterior superior iliac spine and the greater trochanter

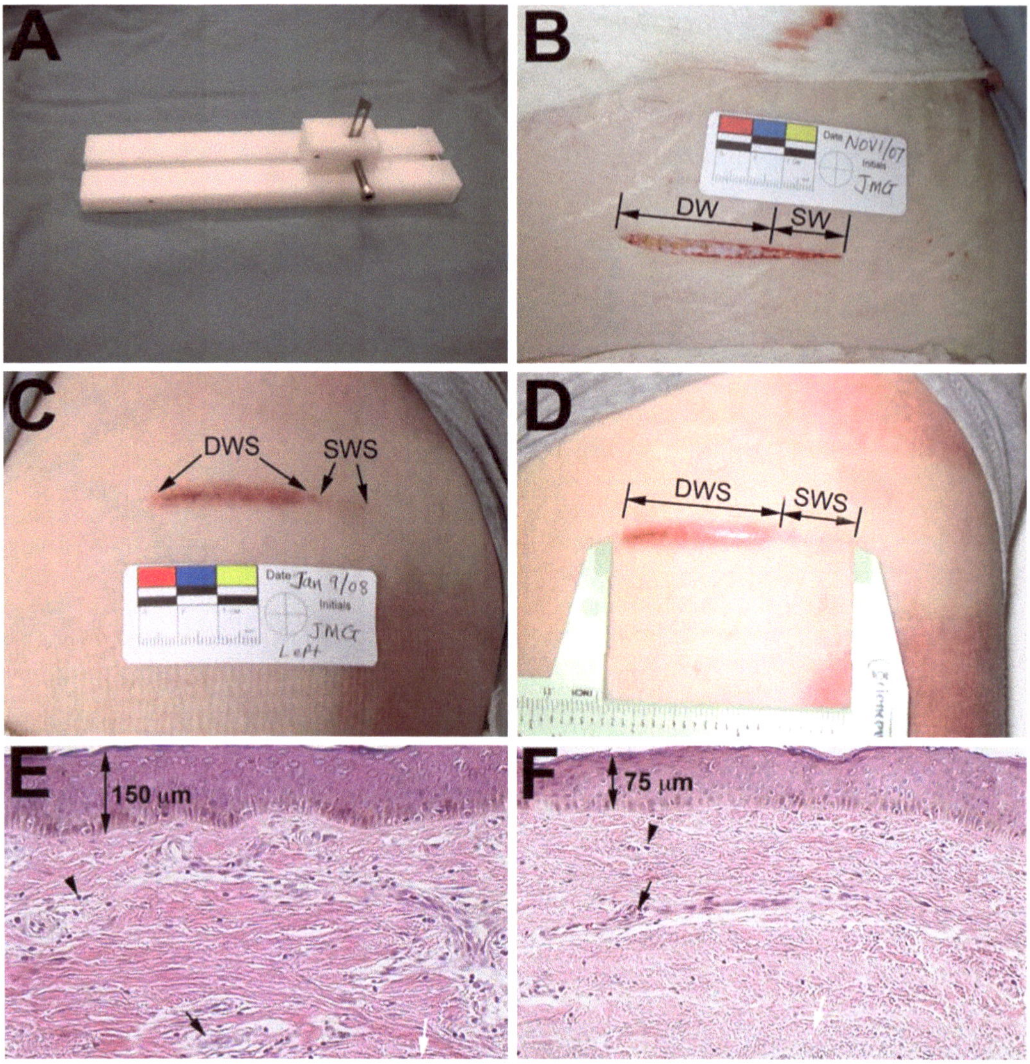

Fig. 1 Creation of deep and superficial scratch wounds and histological analysis of resulted scars. Jig used for the creation of scratch wound model (**a**). Wound created on the hip skin (**b**). Scratch wound 70 days post wounding (**c**). Deep and superficial wound scar (**d**). Deep wound scar tissue stained with H&E (**e**). Superficial wound scar tissue stained with H&E (**f**). *Double-head arrows* in (**e**) and (**f**) indicate average thickness of epithelium. *Arrowheads* point to cells. *Black arrows* point blood vessels. *White arrows* point collagen. *DW* deep wound, *SW* superficial wound, *DWS* deep wound scar, *SWS* superficial wound scar, *H&E* hematoxylin and eosin

under general anesthesia, under sterile conditions in an operating room at the time of split-thickness skin graft harvesting.

2. A specially designed jig using a No. 11 scalpel blade (Fig. 1a) allows the creation of two symmetrical wounds in the same subject 6 cm long, 0–0.75 mm deep at one end (superficial wound; SW) and 0.76–3 mm deep (deep wound; DW) at the other end above the superior border but outside the skin graft donor site (Fig. 1b) (*see* **Note 5**).

3. To confirm the depth of injury, a pilot study was performed on excised lower abdominal skin obtained from abdominoplasty patients.
4. The wound was dressed with an occlusive dressing (Opsite™) and examined and digitally photographed on days 14, 28, 60, and 120 post injury.
5. Scars were analyzed by using the revised Vancouver Burn Scar Score and Mexameter™ for erythema and pigmentation as previously described.
6. The entire scar was excised at day 70 post wounding and representative 4 mm diameter punch biopsies were taken from the SW scar (SWS) and the DW scar (DWS) regions for analysis before the wounds were closed with 5(O) monocril suture.

Immunofluorescence staining and confocal microscopy (*see* **Notes 6–11**, and **14–16**)

1. For immunofluorescence staining, paraffin-embedded tissue sections were deparaffinized in two washes of xylene for 5 min and hydrated gradually through 100, 90, 80, and 70 % alcohol for 5 min each.
2. For antigen unmasking, tissue sections were exposed to 0.05 % saponin in deionized water at room temperature for 30 min.
3. Tissue sections were washed three times with phosphate-buffered saline (PBS), and blocked with PBS containing bovine serum albumin (BSA; 10 mg/ml) and Triton X-100 (0.01 %) for 1 h at room temperature.
4. The tissue sections were incubated with the primary antibody diluted in PBS containing BSA (1 mg/ml) and Triton X-100 (0.01 %) overnight at 4 °C.
5. For the staining of cells, fibroblasts were permeabilized using 0.5 % Triton X-100 for 5 min before incubating with blocking buffer.
6. For cell surface receptor staining, the permeabilization step was not performed.
7. The tissue sections were washed and incubated with the appropriate secondary antibody for 1 h at room temperature before mounting using immuno-mount solution.
8. For double immunofluorescence staining, the tissue sections were incubated with the first primary antibodies above.
9. Sections were washed and incubated with the second primary antibody overnight in the cold followed by 1-h incubation with the appropriate Alexa-conjugated anti-rat or anti-mouse antibody.

10. Images were captured using Carl Ziess Laser Confocal Microscope equipped with LSM5 software.
11. Control immunostaining was performed by omitting the primary antibody incubation step resulting in negative staining results (data not shown).

Western blotting (*see* **Note 12**)

To determine protein expression levels of SLRPs and TGF-β receptor II in cultured fibroblasts the following steps were performed:

1. Cells were grown to 95 % confluency in DMEM/10 % FBS in 25 ml culture flasks.
2. After lysing the cells by adding 1 ml of lysis buffer (PBS, Triton X-100, EDTA, protease inhibitor cocktail) to each flask, the dissolved cells were transferred to 1.5 ml Eppendorf tubes.
3. Samples were left on ice for 30 min, vortexed intermittently, and then centrifuged at 21913 × *g* (rotor radius of 10 cm) for 15 min at 4 °C.
4. Supernatants were collected and the protein concentration was determined using a Bradford protein assay.
5. Equal amounts of solubilized protein from cultured fibroblasts were electrophoresed in 10 % SDS-polyacrylamide gels in 0.1 M Tris borate/SDS buffer.
6. The proteins were transferred to PVDF membranes.
7. Membranes were incubated with primary antibodies overnight at 4 °C, followed by 1-h incubation with appropriate horseradish peroxidase-conjugated antibody as previously described.
8. Antibody binding was probed using a chemiluminescence detection system
9. The blots were processed to be visualized by exposing membranes to Kodak X-ray film.
10. The blots were scanned and the band intensity was quantified using Image J software (http:/rsb.info.nih.gov/ij/).
11. Statistical analysis was performed by one-way ANOVA and Dunnett's multiple comparison test. Values of $p < 0.05$ were considered to be statistically significant.

To determine protein expression levels of SLRPs and TGF-β receptor II in cultured fibroblasts the following steps were performed:

Reverse transcription and real-time polymerase chain reaction (*see* **Note 12**)

1. In order to extract total RNA, the cultured fibroblasts were washed with PBS and solubilized in Trizol reagent.
2. Total RNA was extracted from the cell suspension using RNeasy Mini Kit following the manufacturer's instructions.

3. Samples were spectrophotometrically quantified ($\lambda = 260$ nm).
4. Next step was extraction of 0.5 μg of the RNA for first-strand cDNA synthesis using M-MLV (murine Moloney virus) reverse transcriptase by incubation at three different conditions, 25 °C for 10 min followed by 37 °C for 50 min and 70 °C for 15 min.
5. The resulting cDNA was used as a template for real-time polymerase chain reaction (RT-PCR) amplification of the TGF-β1 type II receptor gene.
6. RT-PCR was done using Power Sybr GreenTM PCR master mix in a total reaction volume of 25 μl containing 5 μl of a 1:10 dilution of cDNA product from the first-strand reaction and 1 μM of forward and reverse primers of TGF-β1 receptor type II gene on a StepOne Plus Real Time PCR system.
7. The amplification conditions included initial denaturation at 95 °C for 3 min followed by cycles of denaturation at 95 °C for 15 s and annealing and primer extension at 60 °C for 30 s.
8. Gene amplification was measured in terms of the cycle threshold (CT) value and the obtained CT value was normalized with the CT value of the housekeeping gene, hypoxanthine-guanine phosphoribosyltransferase (HPRT), and expressed as mean fold change ± standard error.
9. The primers for TGF-β receptor type II gene were AACCACCAGGGCATCCA (forward) and TCGTGGTCCCAGCACTCA (reverse).
10. The primers for HPRT were GACCAGTCAACAGGGGACA (forward) and ACACTTCGTGGGGTCCTTTT (reverse).

Flow cytometry analysis for Thy-1 expression (*see* **Note 13**)

1. Fibroblasts from L1 and L5 skin layers were grown in DMEM supplemented with 10 % FBS until they reached 95 % confluency.
2. The cells were harvested using 0.5 % EDTA for 10 min and fixed using 4 % paraformaldehyde.
3. Cells were incubated in 10 % BSA in PBS for 1 h at room temperature for blocking the nonspecific antigen-binding sites.
4. After washing with PBS, cells were incubated with FITC-conjugated anti-CD90/Thy1 at a concentration of 1 μg/ml on ice for 1 h.
5. Fluorescence was measured by flow cytometry.
6. The level of FITC fluorescence was measured through the FL-1 channel equipped with a 488 nm filter (42 nm band pass).
7. Data were acquired on 10,000 FITC-positive cells per sample with fluorescent signals at logarithmic gain.
8. Data was analyzed with CellQuest software.

3.2 Fibroblasts from the Deep and Superficial Layers of the Skin

Preparation of tissue

1. Specimens are collected from the surgical suite in sterile containers containing a volume of cold saline sufficient to keep the tissues moist during transport (*see* **Note 1**).
2. Containers are kept cool and specimens are brought to the laboratory as rapidly as possible and processed in a biosafety cabinet. In cases where this is not feasible, tissue may be stored for up to 24 h at 4 °C, although prolonged storage may affect cellular viability.
3. Tissues are thoroughly cleansed with saline prior to processing.

Harvesting of superficial dermal fibroblasts
Superficial dermis can be harvested in one of the two ways as outlined in Fig. 2. A dermatome can be used to either first remove the epidermis and then the superficial dermis (Sequential Dermatome Use to Separate Tissue) or remove the epidermis and superficial dermis in continuity and these can then be separated using dispase digestion (Enzyme Digestion to Separate Tissue). Following this step the epidermis may be discarded or further processed to extract matched keratinocytes.

Sequential dermatome use to separate tissue

1. Place abdominoplasty tissue with skin up and attach snaps or Kocher to corners of tissue such that even tension can be placed on the skin (*see* **Notes 2 and 3**).
2. Cover skin with a thin layer of mineral oil so that the dermatome will glide smoothly over the surface (*see* **Note 4**).
3. Set dermatome to thickness of 0.25 mm. The dermatome can be used on a small test area of skin to verify the correct thickness.
4. Remove epidermis using dermatome and discard.
5. Cover dermis with a thin layer of mineral oil so that the dermatome will glide smoothly over the surface.
6. Set dermatome to thickness of 0.25 mm.
7. Remove superficial dermal layer using dermatome and store in saline.

Enzyme digestion to separate tissue
This is our preferred technique as it allows simultaneous extraction of keratinocytes and superficial dermal fibroblasts, and also leads to fewer difficulties with fibroblast contamination of keratinocytes.

1. Place abdominoplasty tissue with skin up and attach snaps to corners of tissue such that even tension can be placed on the skin (*see* **Notes 2 and 3**).
2. Cover skin with a thin layer of mineral oil so that the dermatome will glide smoothly over the surface (*see* **Note 4**).

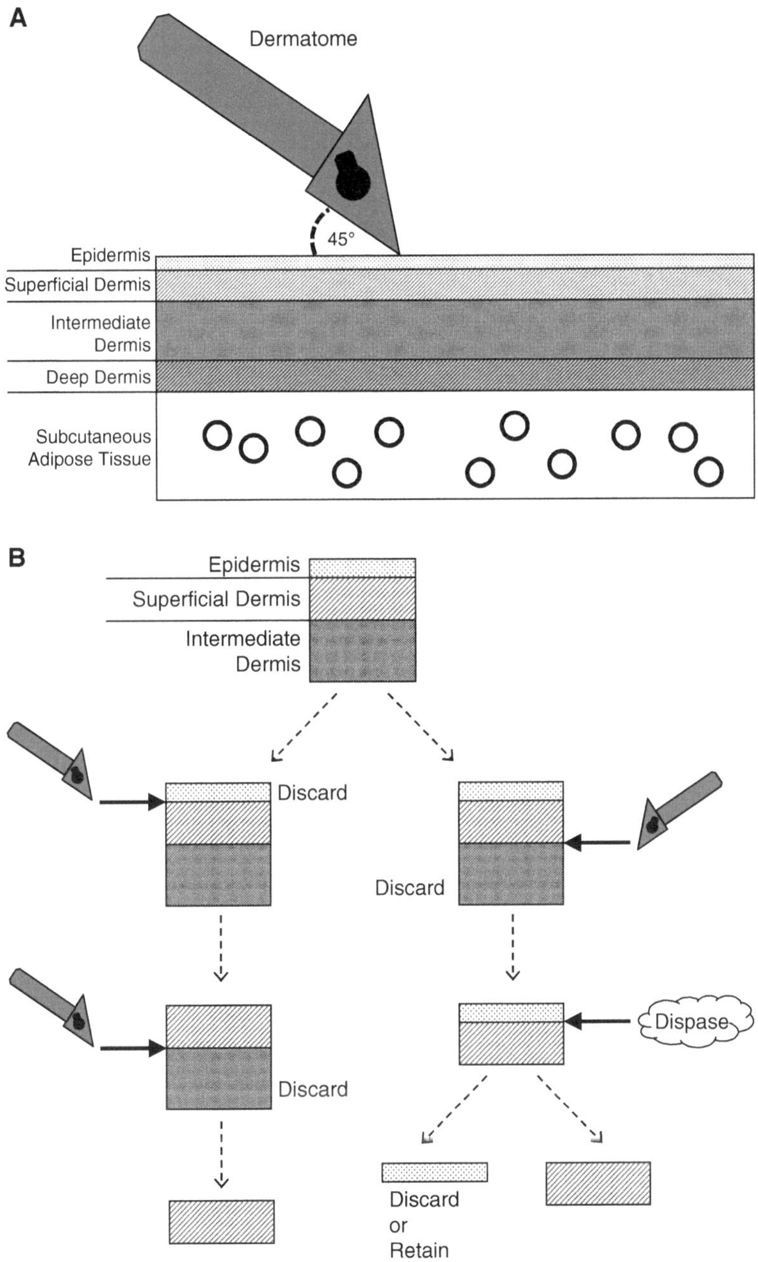

Fig. 2 (**a**) Abdominoplasty specimen showing tissue layers and dermatome orientation for harvesting. (Not to scale.) (**b**) Schematic representation of methods for harvesting superficial dermal tissue. (**c**) Schematic representation of methods for harvesting deep dermal tissue. (**d**) Photographs demonstrating (**d** and **e**) usage of a dermatome to harvest tissue, (**f**) epidermal and superficial dermal layers, (**g**) separation of subcutaneous adipose tissue from underside of dermis, and (**h**) harvesting of deep dermal tissue using a dermatome

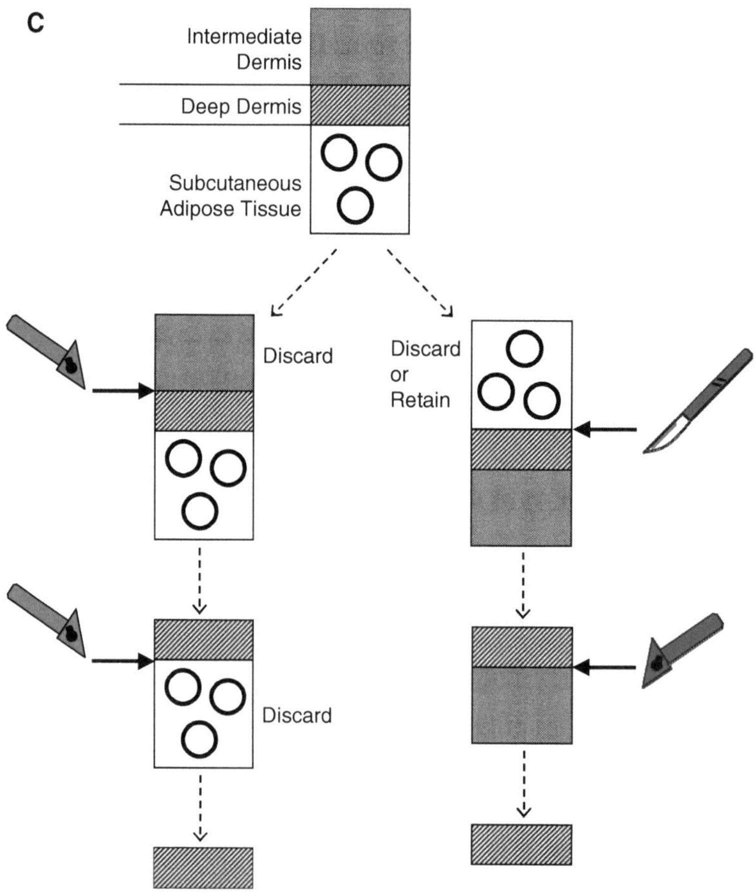

Fig. 2 (continued)

3. Set dermatome to thickness of 0.5 mm. The dermatome can be used on a small test area of skin to verify the correct thickness.
4. Remove epidermis and superficial dermis as contiguous sheet using dermatome. Cut sheet into strips approximately 2 cm wide.
5. Place harvested tissue in covering dispase digestion solution in a centrifuge tube and place on gentle agitator at 4 °C for 8 h or overnight.
6. Remove tissue from dispase solution, place epidermal side down in Petri dish, and spread out so that tissue is flat and epidermis is adherent to Petri dish.
7. Using small-gauge needles lift and pull on superficial dermis in a gentle scraping motion so that epidermis and superficial dermis separate.
8. Place epidermis and superficial dermis into separate containers of normal saline.

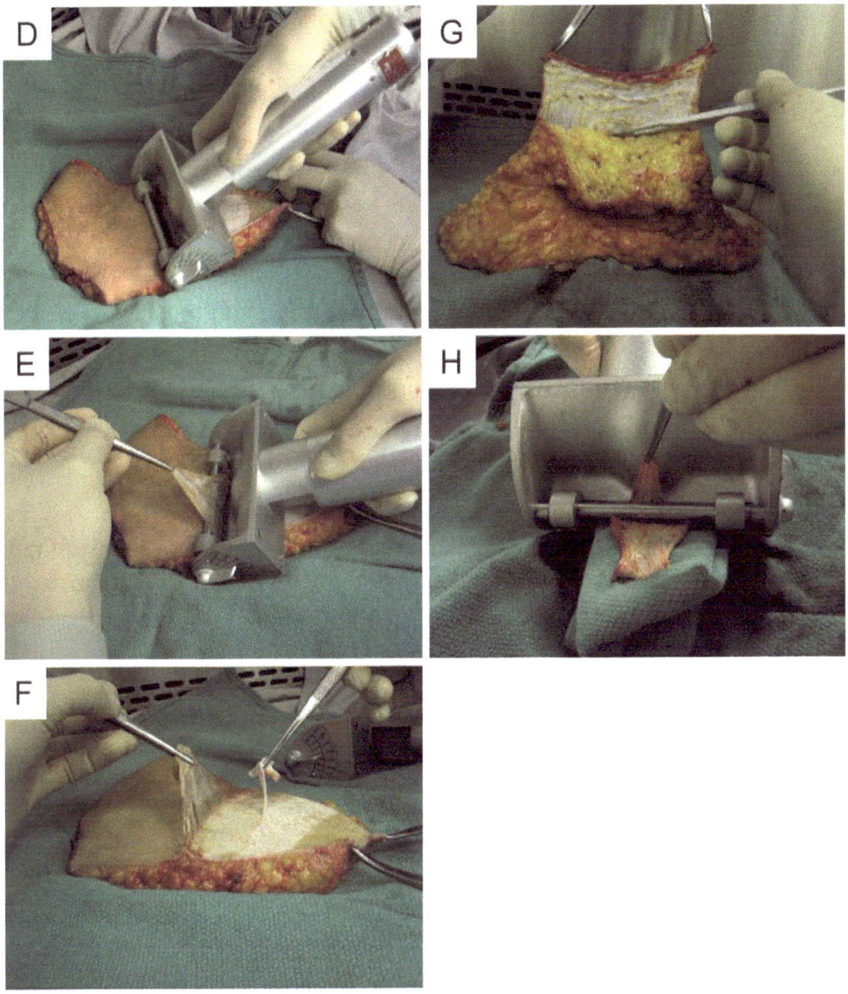

Fig. 2 (continued)

Harvesting of deep dermal fibroblasts
Deep dermis can be harvested in one of the two ways as outlined in Fig. 2. A dermatome can be used to remove either serial sections until the deep dermis is left (Sequential Dermatome Use to Separate Tissue) or the deep dermis from the underside of the skin once the subcutaneous adipose tissue has been removed (Separation of Adipose Tissue and Deep Dermis). Following this step the subcutaneous adipose tissue may be discarded or further processed to extract matched adipose-derived stem cells.

Separation of adipose tissue and deep dermis

This is our preferred technique as it allows simultaneous extraction of adipose-derived stem cells and deep dermal fibroblasts, and also ensures that deep dermal fibroblasts are from the deepest part of the dermis.

1. Attach snaps to one side of abdominoplasty tissue and suspend tissue.
2. Using scalpel incise tissue at junction between dermis and subcutaneous adipose tissue.
3. As subcutaneous adipose tissue drops away from dermis continue to separate tissue by sweeping scalpel along the plane of dissection.
4. Place dermal tissue upside down and attach snaps to corners of tissue such that even tension can be placed on the skin.
5. Cover dermis with a thin layer of mineral oil so that the dermatome will glide smoothly over the surface.
6. Set dermatome to thickness of 0.5 mm and remove the bottom layer of dermis for deep dermal fibroblasts.

Extraction of dermal fibroblasts from tissue

Fibroblasts can be extracted from dermal tissue using either of the two common methods: enzyme digestion or explantation.

Enzyme digestion of dermis for extraction of fibroblasts

While we do not outline the specific steps involved in enzyme digestion we have used this method with great success. However, it is important to note that this technique is significantly more expensive than explantation because of the requirement for the use of collagenase. In addition, there can be great variability between collagenase from various supply companies, and even between different batches from the same manufacturer, which creates a requirement for titration to determine the optimal concentration for digestion that simultaneously avoids inducing cellular death. Protocols for enzymatic digestion are widely available in series such as Methods in Molecular Biology and Nature Protocols.

Explantation of fibroblasts from dermis

The advantages of this method are lower material cost, and a more forgiving process that is less likely to induce cell death.

1. Cut dermal tissue into small pieces (<1 mm^3) and place into tissue culture flasks at a density of approximately 1 piece per cm^2.
2. Place flask in incubator for 30 min to allow tissue to adhere to flask surface and then gently add culture media without disturbing tissue attachment.
3. Replace flask in incubator for 3–5 days with media changes every 2 days to allow fibroblasts to migrate out of tissue. The degree of fibroblast migration can be assessed using an inverted light microscope.
4. Remove tissue and continue to subculture fibroblasts as per usual protocol.

3.3 A Nude Mouse Model of Human Hypertrophic Scar

Harvest of human split-thickness skin grafts

1. Obtain abdominoplasty tissue from surgery suite; store in normal saline (*see* **Note 17**).
2. Secure one-half of the resected abdominal tissue to a suture tray covered with a sterile green towel using piercing towel clips (*see* **Note 18**).
3. Apply mineral oil to the surface of the tissue section and the contact surfaces of the Padgett dermatome.
4. With the tissue section taut and the dermatome thickness set at 0.035 cm (14/1,000″) harvest a split-thickness skin graft using the 4″ (largest) guard.
5. Immerse the skin graft in normal saline and drape skin over a plastic carrier.
6. Using a prefabricated, 2.0 × 1.5 cm plastic template, cut out grafts from the harvested split-thickness skin graft using a number 20 scalpel (*see* **Note 19**).
7. Store the cut 2.0 × 1.5 cm split-thickness human skin grafts in a 50 mL conical tubes filled with normal saline and place on ice.

Transplantation of human split-thickness skin grafts

1. Prepare the fume hood by wiping the surface with 70 % ethanol and draping with sterile green towels (*see* **Notes 18, 20,** and **21**).
2. Set up a heating pad and drape with sterile green towel.
3. Secure paper ruler oriented vertically in the center of the operative area with adhesive tape.
4. Prepare a sufficient number of postoperative dressings consisting of 2 × 2 dry gauze folded in half and xeroform dressings cut to match the dry gauze size.
5. Prepare a sufficient number of cages for postoperative recovery (*see* **Note 22**)
6. Remove the animal from its cage and place it in the induction chamber using forceps (*see* **Note 18**).
7. Fill the chamber with a mixture of isoflurane anesthetic and oxygen until the animal is sufficiently induced, then remove the animal from the chamber, and transfer to a nose cone with the animal in the prone position (*see* **Note 23**).
8. Apply a small amount of Tears Naturale P.M. to each eye.
9. Spread a quarter-sized aliquot of commercial hair remover across the dorsal surface of the animal. Wait for 90–120 s prior to wiping of the cream.
10. Prepare the skin on the dorsal surface with iodine.
11. Use a 2.0 × 1.5 cm plastic template to mark the dorsal skin resection margins.

12. Use straight scissors to cut around the perimeter of the skin markings and sharply dissect, elevate, and excise a 2.0 × 1.5 cm full-thickness defect on the dorsal surface (see **Note 24**).
13. Suture the prepared 2.0 × 1.5 cm human split-thickness skin grafts into the dorsal defect using 4-0 braided silk suture on a taper RB-1 needle (*see* **Notes 25** and **26**).
14. Gently wash and dry the wound with normal saline and dry gauze.
15. Dress grafts with prepared non-adherent petrolatum (Xeroform™, Covidien, Mansfield, MA) and dry gauze tie-over-bolus dressing to ensure adherence of the graft to the wound bed (*see* **Note 27**).
16. Remove the animal from the nose cone and inject 0.02 mL of hydromorphone (hydromorphone HP 10 diluted to 0.05 mg/mL) subcutaneously using a 27 G syringe for post-operative pain management.
17. Wipe the animal's eyes clean and return it to the cage for recovery.
18. Monitor the animal carefully during the immediate postoperative period (*see* **Note 28**).
19. All sutures and dressings are removed 7 days following grafting (*see* **Note 29**).

Harvest of normal skin and scar biopsies

1. Repeat **steps 1–3** from Subheading 3.2 (*see* **Note 18**).
2. Euthanize animals at desired time points using a lethal dose of isoflurane anesthetic.
3. Normal skin and scar biopsies are harvested using sharp scissors and forceps lifted off the panniculus carnosus.
4. Immediately store skin samples in a 50 mL conical tube in normal saline.

Processing of normal skin and scar biopsies

1. Prepare the fume hood by wiping the surface with 70 % ethanol and draping with sterile green towels (*see* **Note 18**).
2. Prepare a suitable number of tissue cassettes with two foam pads inside each cassette. Appropriately pre-label all cassettes (*see* **Note 30**).
3. Prepare a suitable number of embedding molds (22 × 22 × 20 mm) by filling each half full with cryomatrix and freezing at −80 °C (*see* **Note 31**). Appropriately pre-label all embedding molds.
4. Divide freshly harvested normal skin and scar biopsy samples into thirds using a single-use, number 20 disposable scalpel.

5. Cut one-third of the sample into ~2 mm vertical strips (*see* **Note 32**).
6. Sandwich vertically cut skin samples between foam pads inside pre-labeled tissue cassettes (*see* **Note 33**).
7. Immerse cassettes in 10 % formalin for at least 24 h at room temperature prior to processing and paraffin embedding.
8. Cut paraffin-embedded sections to 5 μm and mount on glass slides. Store the slides at room temperature.
9. Cut another one-third of the sample into ~2 mm vertical strips.
10. Lay vertically cut strips flat inside embedding mold, cover with cryomatrix, and immediately freeze at −80 °C (*see* **Note 34**).
11. Cut OCT sections to 10 μm and mount on glass slides. Store at −20 °C.
12. Snap freeze the remaining one-third of the sample for future RNA and protein isolation and analysis. Store at −80 °C.

4 Notes

1. When choosing abdominoplasty tissue for processing we discard those portions with striae distensae (stretch marks) as this alters the dermal architecture and we feel that the resulting dermal fibroblast subpopulations are uncharacterized.
2. Maintaining tension on the skin while harvesting sections using the dermatome can be challenging and is often simplified by having one individual maintain tension while a second maneuvers the dermatome.
3. Even application of the dermatome along the entire width of the blade to the skin is essential to harvesting tissue with a uniform thickness, as outlined in Fig. 2. This can be facilitated by stretching the skin over a convex deformable object, such as a normal saline bag, and then running the dermatome over top.
4. Smooth gliding of the dermatome over the tissue is essential for consistency and uniform depth. This is made much simpler by generous use of mineral oil and reapplication before each pass of the dermatome is made.
5. The deep scratch wound scar shows features of hypertrophic scarring. The SW at one end of the wound scratch model was a partial/superficial thickness wound between 0.1 and 0.75 mm deep that regenerated with minimal scarring (SWS) (Fig. 1b–d). The DW at the other end was considered to be a full-thickness deep dermal wound between 0.76 and 3.0 mm that resulted in an erythmatous, elevated, inflamed scar (DWS) restricted to the site of injury and overall showed gross symptoms of HTS (Fig. 1b–d). Histological analysis of the scar

showed that compared to SWS (Fig. 1f), the DWS (Fig. 1e) developed thicker epithelium (double-head arrows), an accumulation of thick collagen bundle fibers (white arrow), hypercellularity (arrowhead), and an increased number of blood vessels per unit area (black arrow). Fibroblasts from superficial and deep layers of normal human dermis that represent papillary and reticular fibroblasts, respectively, were isolated from layer 1, 75 μm below the epithelium and layer 5 of the dermis, 100 μm above the dermal adipose tissue. Layer 1 is thought to be correlated with superficial wounds that heal with minimal scarring whereas, layer 5 involves deep dermal injuries that result in hypertrophic scarring.

6. Decorin is primarily localized to the extracellular matrix (Fig. 3a) while fibromodulin was associated with cells (Fig. 3d). Staining intensity of decorin and fibromodulin was noticeably lower in DWS (Fig. 3b and e, respectively) compared to SWS (Fig. 3a, d, respectively); however, decorin and fibromodulin were significantly reduced in DWS compared to SWS (Fig. 3c, f). In L5 fibroblasts, the immunoreactivity of decorin and fibromodulin (Fig. 3h, k, respectively) was considerably reduced compared to fibroblasts from superficial layers (L1) (Fig. 3g, j, respectively).
7. Compared to L1 fibroblasts, the protein expression of decorin and fibromodulin in L5 fibroblasts was significantly decreased (Fig. 3i, l). Continuous deposition and/or abnormal turnover of collagen or the ratio of collagen type I/type III was considered to be the main reason for HTS development [20]. Collagen-binding proteoglycans, SLRPs, decorin, biglycan, and fibromodulin can bind to and inhibit TGF-β_1 activity in vitro and in vivo [7, 8, 21, 22]. These results suggest a role for SLRPs in the reduction of scarring in superficial dermal wounds. Taken together, the findings may indicate that the decreased expression of decorin and fibromodulin especially in deep dermis can contribute to the development of HTS after deep dermal injury wounds.
8. TGF-β_1 appeared to be primarily cell associated in scar tissue and its immunoreactivity was significantly higher in DWS tissue compared to SWS (Fig. 4a–d, i). TGF-β3 localized to cells' extracellular matrix and was significantly down-regulate in DWS compared to SWS (Fig. 4e, f, i). We also saw TGF-β3 accumulated in the area below the basement membrane of the epithelium and in basal epithelial cells in SWS but it was absent from DWS tissue suggesting a novel function for TGF-β3 in basal epithelial cells (Fig. 4e, f, arrows).
9. In cultured fibroblasts, cells from L5 had significantly higher TGF-β1 immunoreactivity compared to L1 fibroblasts (Fig. 4c, d, j) whereas TGF-β_3 was significantly higher in L1 fibroblasts when compared to L5 fibroblasts (Fig. 4g, h, j). However,

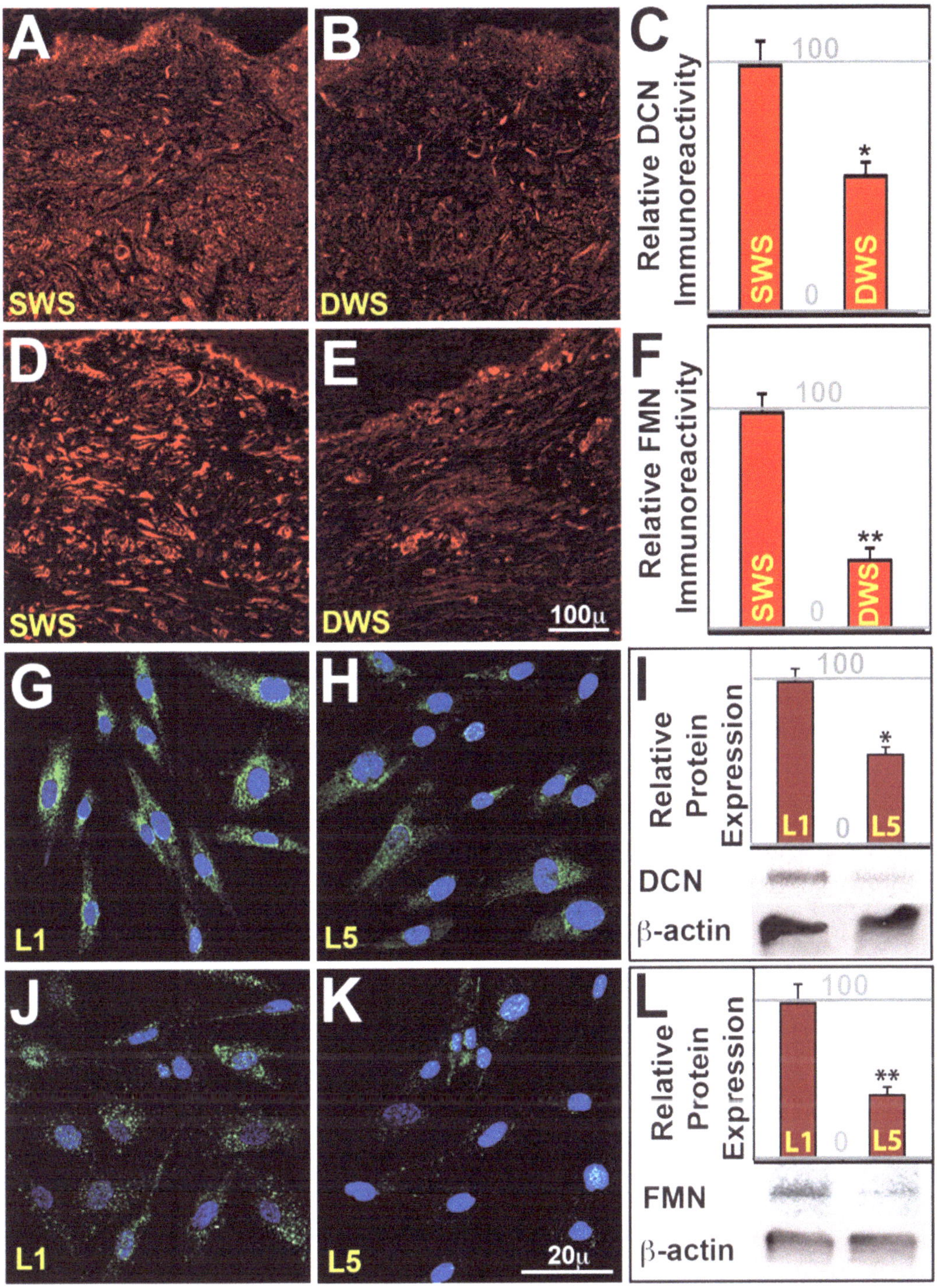

Fig. 3 Expression of SLRPs in scar and cultured dermal fibroblasts. Representative images immunostained for decorin and fibromodulin. Decorin staining in the scar tissues (**a** and **b**). Quantitation of relative decorin immunoreactivity in the scar tissues (**c**). Fibromodulin staining in the scar tissues (**d** and **e**). Quantitation of relative fibromodulin immunoreactivity in the scar tissues (**f**). Decorin staining in cultured dermal fibroblasts (**g** and **h**). Immunoblotting for decorin in cultured dermal fibroblasts (**i**). Fibromodulin staining in cultured dermal fibroblasts (**j** and **k**). Immunoblotting for fibromodulin in cultured dermal fibroblasts (**l**). *DCN* decorin, *FMN* fibromodulin, *DWS* deep wound scar, *SWS* superficial wound scar, *L1* layer 1 dermis, *L5* layer 5 dermis. 100 μm: Scale bar for (**a**), (**b**), (**d**), and (**e**). 20 μm: Scale bar for (**g**), (**h**), (**j**), and (**k**). *$P<0.05$, **$P<0.001$

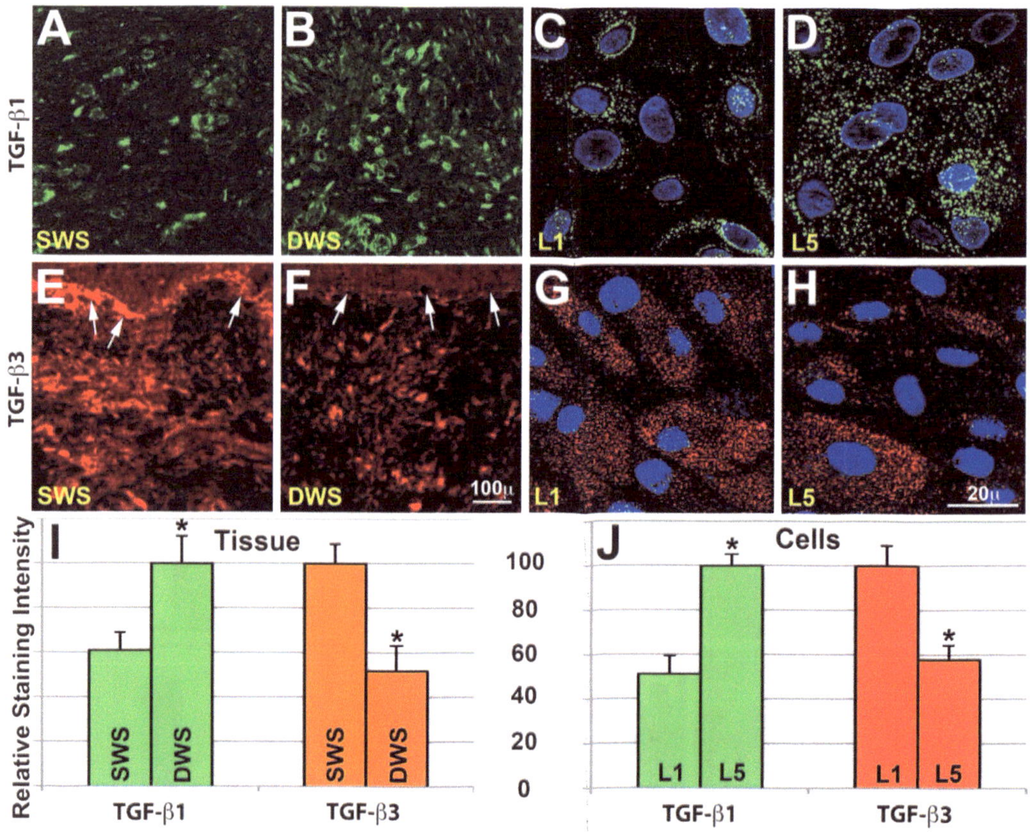

Fig. 4 Localization and immunoreactivity of TGF-βs in scar and cultured dermal fibroblasts. Representative images immunostained for TGF-β1 in *green* in SWS and DWS (**a** and **b**). Representative images immunostained for TGF-β1 in *green* in L1 and L5 cultured fibroblasts (**c** and **d**). Representative images immunostained for TGF-β3 in *red* in SWS and DWS (**e** and **f**). Representative images immunostained for TGF-β3 in *red* in L1 and L5 cultured fibroblasts (**g** and **h**). *Blue* indicates DAPI staining for nucleus. Relative staining intensity quantitation of TGF-βR11 and TGF-β3 in SWS and DWS (**i**). Relative staining intensity quantitation of TGF-β1 and TGF-β3 in L1 and L5 cultured fibroblasts (**j**). *Arrows* in (**e**) and (**f**) indicate epithelium. 100 μm: Scale bar for (**a**), (**b**), (**e**), and (**f**). 20 μm: Scale bar for (**c**), (**d**), (**g**), and (**h**). *SWS* superficial wound scar, *DWS* deep wound scar, *L1* layer 1 dermis, *L5* layer 5 dermis. *$P < 0.05$

differences in extracellular matrix constituents such as SLRPs and ratio of TGF-β_1/TGF-β_3 and differences in the fibroblast subtypes and population between superficial and deep dermal wounds may play a key role in the development of HTS following deep dermal injury. An increase in the activity and prolonged expression of TGF-β1 is considered to be a major factor that leads to the development of HTS.

10. TGF-β3 is a non-fibrotic cytokine that down-regulates TGF-β1 expression [23–25]. The balance between TGF-β1 and TGF-β3 may be an important regulator of scar formation. Interestingly, it has been shown that in a non-scarring gingival wound model, the ratio of TGF-β3 to TGF-β1 is higher at later wound healing time points [26]. As well, fibroblasts

from fetal tissue that heal without scarring exhibited an increased expression of TGF-β3 [27, 28]. In this study, the accumulation of TGF-β3 was considerably higher in superficial scars but TGF-β1 was lower. In contrast, TGF-β1 expression was pronounced in HTS whereas TGF-β3 was decreased. Thus, decreased TGF-β3 in the deeper areas of dermis may aggravate the outcome of wound healing.

11. In order to determine the expression of TGF-β3 affinity to TGF-β1, fibroblasts from different layers of dermis were analyzed using immunofluorescence staining, western blotting, and RT-PCR. The number of TGF-β1 receptor type II localized to the cell surface in L5 fibroblasts was prominently higher than that in L1 fibroblasts (Fig. 4a, b, arrows). Similarly, the immunoreactivity of TGF-β receptor type II in L5 fibroblasts was significantly higher compared to fibroblasts from L1 (Fig. 5c).
12. As compared to L1 fibroblasts, fibrotic L5 fibroblasts showed significantly higher levels of TGF-β receptor type II protein expression (Fig. 5d). Quantitative RT-PCR analysis also showed close to a threefold increase in the expression of TGF-BR11 receptor type II in L5 fibroblasts compared to L1 fibroblasts (Fig. 5e) suggesting that injuries to the deep dermis, which contains more TGF-β receptor type II-expressing fibroblasts, make it prone to TGF-β1 fibrogenic activity and can account, in part, for the development of HTS formation.
13. Cell surface antigen, Thy-1, has been reported to mediate differentiation of fibroblasts to myofibroblasts by TGF-β1 and fibroblasts that are Thy-1 negative are more responsive to TGF-β fibrogenic stimulation [10–12]. Two distinct Thy-1-positive (a and b) and Thy-1-negative (c and d) cultured fibroblasts from the superficial layer of dermis are shown in Fig. 6A. Control immunostaining resulted in no immunoreactivity (Fig. 6B). As verified by flow cytometry, L5 fibroblasts showed lower Thy-1 expression when compared to L1 fibroblasts (Fig. 6C). As well, immunofluorescence staining demonstrated that compared to superficial dermal fibroblasts from L1, the number of fibroblasts from the deep dermis (L5) that show Thy-1 immunoreactivity was notably lower (Fig. 6D–F). As such, Thy-1 is strongly expressed by superficial dermal fibroblasts whereas fibroblasts from deeper areas weakly express or lack expression of Thy-1. Therefore, development of HTS in the deep skin wounds could be associated with differences in the expression of Thy-1 by these two phenotypically different fibroblasts to produce excess collagen under the influence of TGF-β1.
14. Fibrocytes are circulating mesenchymal progenitor cells that are implicated in tissue response to injury, development of diseases characterized by chronic inflammation, excessive collagen deposition, and scar formation [29]. The inference of

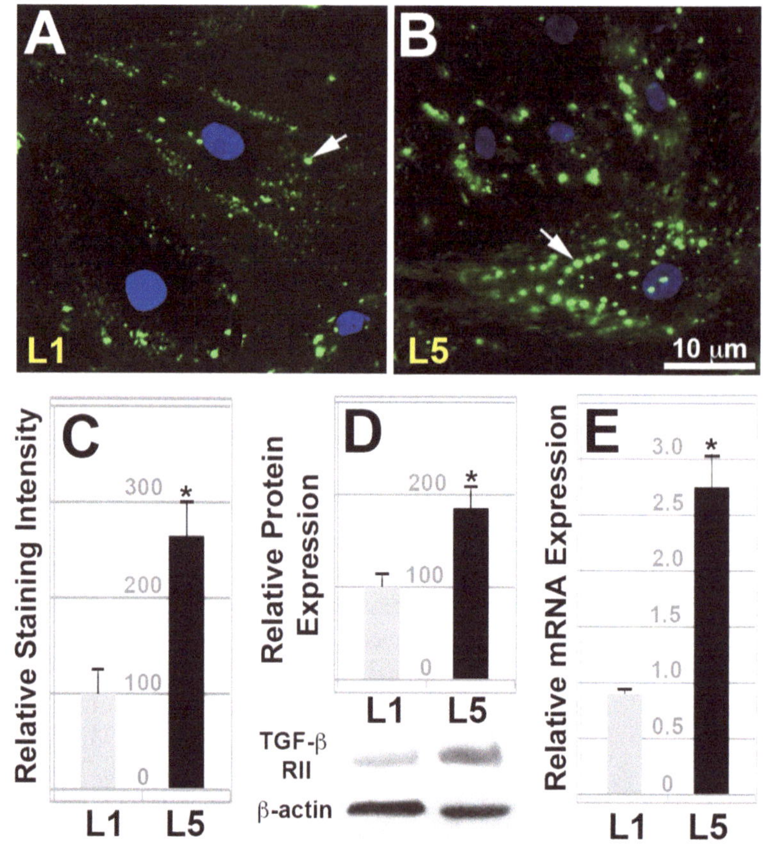

Fig. 5 Expression of TGF-βR11 receptor type II in cultured dermal fibroblasts. Representative images immunostained for TGF-β receptor type II in *green* in L1 and L5 cultured fibroblasts (**a** and **b**). *Arrows* point to TGF-β receptor type II staining at the cell surface. *Blue* indicates DAPI staining for nucleus. Relative staining intensity of TGF-βR11 receptor type II in L1 and L5 cultured fibroblasts (**c**). Relative protein expression of TGF-βR11 receptor type II measured from immunoblotting band intensity in L1 and L5 cultured fibroblasts (**d**). Relative mRNA expression of TGF-β receptor type II in L1 and L5 cultured fibroblasts (**e**). *L1* layer 1 dermis, *L5* layer 5 dermis, *RII* receptor type II. β-actin indicates loading control. *$P<0.05$

fibrocytes as a bone marrow supply of originator cells that persuade severe fibrosis mediated by TGF-β has been progressively more recognized as a general pathway essential in the HTS formation. Fibrocytes appear to be able to move to regional lymph nodes and wounds. Taken together, fibrocytes may play a role in the development of HTS [30, 31].

15. There is a difference in the population of fibrocytes between the superficial and deep scratch wound scars. Double immunofluorescence staining using known markers for fibrocytes, LSP-1, and procollagen indicated that in DWS the number of cells that showed co-localization of LSP-1 and procollagen was

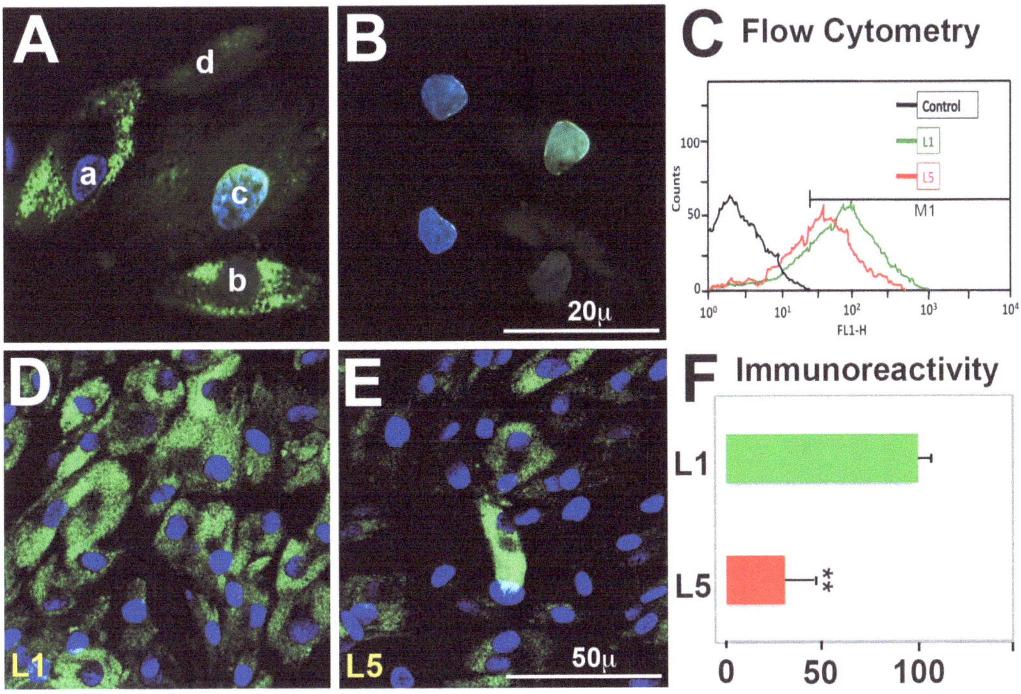

Fig. 6 Thy-1 expression in cultured dermal fibroblasts. Dermal fibroblasts stained positively for Thy-1 in *green* (**A**a, **A**b). Thy-1-negative dermal fibroblasts (**A**c, **A**d). *Blue* indicates DAPI staining for nucleus. Control immunostaining performed by omitting primary antibody incubation step (**B**). Flow cytometry analysis showing higher fluorescence intensity for Thy-1 staining in L1 fibroblasts (**C**). Representative images immunostained for Thy-1 in L1 and L5 cultured fibroblasts, respectively (**D**, **E**). Measurement and statistical analysis of Thy-1 immunoreactivity in L1 and L5 dermal fibroblasts (**F**). *Green* and *red* in (**C**) and (**F**) indicate L1 and L5 fibroblasts, respectively. 20 μm: Scale bar for (**A**) and (**B**). 50 μm: Scale bar for (**D**) and (**E**). *L1* layer 1 dermis, *L5* layer 5 dermis. *$P<0.05$. **$P<0.001$

considerably higher (Fig. 7d–f, g) compared to SWS (Fig. 7a–c, g) suggesting that fibrocytes may contribute to the development of HTS. Therefore, an increased number of fibrocytes in the DWS that showed the characteristics of HTS in our wound scratch model suggests a role for fibrocytes in promoting HTS following deep dermal injury.

16. Taken together, the findings in the present study suggest that transplantation of Thy-1+fibroblasts to the site of injury, treatment of wounds with TGF-β3, and/or targeted local overexpression of SLRPs, decorin or fibromodulin, during wound healing may be useful to ameliorate HTS formation or tissue fibrosis. Therefore, deep dermal injury can be directed to mimic superficial skin wound healing to regenerate without hypertrophic scarring. Further functional studies to elucidate the role of these anti-fibrotic molecules are warranted.
17. All human abdominal tissue samples were collected from female patients following informed consent.

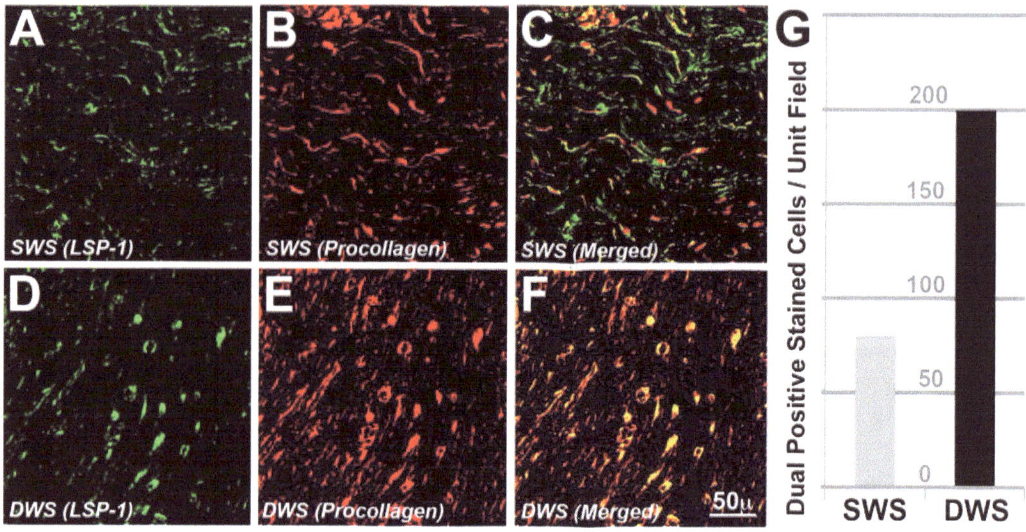

Fig. 7 Fibrocyte population in superficial and deep wound scar tissue. Comparison of fibrocyte population in superficial and deep wound scar tissue per unit field by double-immunofluorescence staining using antibodies against fibrocyte markers leukocyte-specific gene-1 protein (LSP-1) and procollagen. Superficial wound scar (SWS) (**a**–**c**). Deep wound scar (DWS) (**d**–**f**). Semiquantitative analysis of fibrocyte population in SWS and DWS by two separate individuals yielded similar results (**g**). *Yellow* in (**c**) and (**f**) indicates co-localization of LSP-1 and procollagen

18. All steps should be completed in a fume hood using strict sterile technique and with full personal protective equipment including a sterile gown, cap, sterile gloves, mask, and eye protection.
19. This is best achieved by placing the template over the harvested graft and cutting around it using a rolling motion with the belly of the scalpel. Take care to avoid harvesting skin containing striae and other skin abnormalities.
20. All animal experiments were performed using protocols approved by the University of Alberta Animal Care and Use Committee and in accordance with the standards of the Canadian Council on Animal Care.
21. Animals were housed and experiments were carried out in a virus antibody-free biocontainment facility. Animals were conditioned for two weeks prior to grafting.
22. Animals are housed individually from the time of grafting until they are euthanized. In our experience a minimum of five animals is needed per time point of interest.
23. Wait for 30–60 s after the animal is paralyzed and the respiration rate has decreased before transferring it from the induction chamber to the nose cone.
24. In removing the skin from the mouse for the creation of the excisional wound for grafting, take care to leave the underlying panniculus carnosus layer intact.

25. The four corners of the graft should be anchored first, followed by two additional sutures in between each corner stitch. Ensure that one tail of each of the four medial and four lateral sutures is left long to facilitate tying of the tie-over-bolus dressing.
26. In our experience at least five animals should be autografted with full-thickness nude mouse skin in order to provide suitable, negative controls representative of the normal morphologic and histologic changes that occur in uncomplicated wound healing.
27. The tails of the four medial and four lateral sutures should be tied to one another in the midline using a needle driver and non-toothed forceps. It is important to tie together the tails that are immediately across from one another, thereby creating four horizontal tie-over-bolus sutures.
28. Animals should awaken within approximately 3 min of being removed from the nosecone anesthetic. A return to normal feeding and grooming behavior can be expected within 10–15 min.
29. Our recommendation is to initially graft 1–2 animals when first attempting this experimental protocol. Following this it is not unreasonable to graft 5–10 animals in one sitting.
30. Pre-labeling cassettes with pencil is preferred to avoid running and smudging of labels that occur with handling after immersion in Zinc Formal Fixx™.
31. Pre-filling of embedding molds allows tissue samples to be sandwiched between cryomatrix, which facilitates cutting of OCT slides.
32. Cut scar biopsy samples into ~2 mm vertical strips such that the scar biopsy is flanked in the center by normal skin.
33. Carefully roll the thin, vertical strip of tissue onto the foam pad such that it is completely flat inside the tissue cassette.
34. Remove prefilled embedding molds containing cryomatrix from the freezer ~5 min prior to embedding samples to allow for some thawing of cryomatrix and ease of proper tissue orientation and placement within the mold.

Acknowledgements

The wound healing research in the author's laboratory is funded by the Canadian Institutes of Health Research, the Alberta Heritage Foundation for Medical Research, and Firefighters' Burn Trust Fund of the University of Alberta.

References

1. Wang J, Dodd C, Shankowsky HA et al (2008) Deep dermal fibroblasts contribute to hypertrophic scarring. Lab Invest 88:1278
2. Gallant CL, Olson ME, Hart DA (2004) Molecular, histologic and gross phenotype of skin wound healing in red Duroc pigs reveals an abnormal healing phenotype of hypercontracted, hyperpigmented scarring. Wound Repair Regen 12:305
3. Sappino AP, Masouye I, Saurat JH et al (1990) Smooth muscle differentiation in scleroderma fibroblastic cells. Am J Pathol 137:585
4. Wrana JL, Attisano L, Cárcamo J et al (1992) TGF-beta signals through a heteromeric protein kinase receptor complex. Cell 71:1003
5. Souchelnytskyi S, Rönnstrand L, Heldin CH et al (2001) Phosphorylation of Smad signalling proteins by receptor serine/threonine kinases. Methods Mol Biol 124:107
6. Hildebrand A, Romaris M, Rasmussen LM et al (1994) Interaction of the small interstitial proteoglycans biglycan, decorin and fibromodulin with transforming growth factor beta. Biochem J 302:527
7. Soo C, Hu FY, Zhang X et al (2000) Differential expression of fibromodulin, a transforming growth factor-beta modulator, in fetal skin development and scarless repair. Am J Pathol 157:423
8. Kolb M, Margetts PJ, Sime PJ et al (2001) Proteoglycans decorin and biglycan differentially modulate TGFβ-mediated fibrotic responses in the lung. Am J Physiol Lung Cell Mol Physiol 280:1327
9. Sorrell JM, Baber MA, Caplan AI (2004) Site-matched papillary and reticular human dermal fibroblasts differ in their release of specific growth factors/cytokines and in their interaction with keratinocytes. J Cell Physiol 200:134
10. Hagood JS, Lasky JA, Nesbitt JE et al (2001) Differential expression, surface binding, and response to connective tissue growth factor in lung fibroblast subpopulations. Chest 120:64
11. Zhou Y, Hagood JS, Murphy-Ullrich JE (2004) Thy-1 expression regulates the ability of rat lung fibroblasts to activate transforming growth factor-beta in response to fibrogenic stimuli. Am J Pathol 165:659
12. Hagood JS, Prabhakaran P, Kumbla P et al (2005) Loss of fibroblast Thy-1 expression correlates with lung fibrogenesis. Am J Pathol 167:365
13. Dunkin CS, Pleat JM, Gillespie PH et al (2007) Scarring occurs at a critical depth of skin injury: precise measurement in a graduated dermal scratch in human volunteers. Plast Reconstr Surg 119:1722
14. Honardoust D, Varkey M, Hori K et al (2011) Small leucine-rich proteoglycans, decorin and fibromodulin, are reduced in postburn hypertrophic scar. Wound Repair Regen 19:368–378
15. Ali-Bahar M, Bauer B, Tredget EE, Ghahary A (2004) Dermal fibroblasts from different layers of human skin are heterogeneous in expression of collagenase and types I and III procollagen mRNA. Wound Repair Regen 12(2):175–182
16. Kitano Y, Okada N (1983) Separation of the epidermal sheet by dispase. Br J Dermatol 108(5):555–560
17. Schönherr E, Beavan LA, Hausser H, Kresse H, Culp LA (1993) Differences in decorin expression by papillary and reticular fibroblasts in vivo and in vitro. Biochem J 290(Pt 3): 893–899
18. Mustoe TA, Cooter RD, Gold MH et al (2002) International clinical recommendations on scar management. Plast Reconstr Surg 110:560–571
19. Scott PG, Dodd CM, Tredget EE et al (1995) Immunohistochemical localization of the proteoglycans decorin, biglycan and versican and transforming growth factor-beta in human post-burn hypertrophic and mature scars. Histopathology 26:423
20. Oliveira GV, Hawkins HK, Chinkes D et al (2009) Hypertrophic versus non hypertrophic scars compared by immunohistochemistry and laser confocal microscopy: type I and III collagens. Int Wound J 6:445
21. Mohan RR, Gupta R, Mehan MK et al (2010) Decorin transfection suppresses profibrogenic genes and myofibroblast formation in human corneal fibroblasts. Exp Eye Res 91:238
22. Stoff A, Rivera AA, Mathis JM et al (2007) Effect of adenoviral mediated overexpression of fibromodulin on human dermal fibroblasts and scar formation in full-thickness incisional wounds. J Mol Med 85:481
23. Shah M, Foreman DM, Ferguson MW (1995) Neutralization of TGF-β 1 and TGF-β 2 or exogenous addition of TGF-β 3 to cutaneous rat wounds reduces scarring. J Cell Sci 108:985
24. Pohlers D, Brenmoehl J, Löffler I et al (2009) TGF-beta and fibrosis in different organs—molecular pathway imprints. Biochim Biophys Acta 1792:746
25. Bandyopadhyay B, Fan J, Guan S, Li Y et al (2006) A "traffic control" role for TGF-beta3: orchestrating dermal and epidermal cell motility during wound healing. J Cell Biol 172:1093
26. Honardoust D, Eslami A, Larjava H, Häkkinen L (2008) Localization of small leucine-rich proteoglycans and transforming growth factor-

beta in human oral mucosal wound healing. Wound Repair Regen 16:814
27. Colwell AS, Yun R, Krummel TM et al (2007) Keratinocytes modulate fetal and postnatal fibroblast transforming growth factor-beta and Smad expression in co-culture. Plast Reconstr Surg 119:1440
28. Goldberg SR, McKinstry RP, Sykes V et al (2007) Rapid closure of midgestational excisional wounds in a fetal mouse model is associated with altered transforming growth factor-beta isoform and receptor expression. J Pediatr Surg 42:966
29. Keeley EC, Mehrad B, Strieter RM (2009) The role of circulating mesenchymal progenitor cells (fibrocytes) in the pathogenesis of fibrotic disorders. Thromb Haemost 101:613
30. Yang L, Scott PG, Dodd C et al (2005) Identification of fibrocytes in postburn hypertrophic scar. Wound Repair Regen 13:398
31. Wang J, Jiao H, Stewart TL et al (2007) Improvement in postburn hypertrophic scar after treatment with IFN-alpha2b is associated with decreased fibrocytes. J Interferon Cytokine Res 27:921

Chapter 12

Study of the Human Chronic Wound Tissue: Addressing Logistic Barriers and Productive Use of Laser Capture Microdissection

Sashwati Roy and Chandan K. Sen

Abstract

Direct procurement of tissue samples from clinically presented chronic human wounds is a powerful approach to understand mechanism at play in an actual problem wound. While such approach suffers from limitations related to lack of reproducible conditions across wounds, something that we are used to in the laboratory while studying wounds on experimental animals, the direct study of human wound tissue helps recognize the right questions to ask in the laboratory. Going back and forth between human wound and experimental animal studies helps steer studies on experimental wounds in a clinically relevant direction. In this chapter, we describe critical factors that need to be considered prior to planning a study involving human wound samples. In addition, we describe an approach to capture wound hyperproliferative epithelium (HE) from chronic human wound biopsies using laser capture microdissection (LCM). LCM is a new technology applicable to a broad range of clinical research and represents a catalyst of sophisticated translational research.

Key words Chronic human wounds, Hyperproliferative epithelium, Laser capture microdissection

1 Introduction

Wound healing studies designed to address mechanisms or causality, specifically at a molecular level, lag far behind other medical conditions because even the most sophisticated animal wound model fails to capture the complexities associated with a clinically presented chronic wound. It is indeed difficult to control for the multitude of conditions that impact clinical wound healing outcomes. On the other hand, studies on experimental wounds are empowered to dissect underlying mechanisms in a well-controlled manner. The risk here is that one may deconvolute a complex mechanism only to discover that the mechanism is of little clinical relevance. Thus, it is imperative that the direction of experimental studies be supervised by findings of studies examining the clinically presented chronic human wound. This strategy should help unveil

Robert G. Gourdie and Tereance A. Myers (eds.), *Wound Regeneration and Repair: Methods and Protocols*, Methods in Molecular Biology, vol. 1037, DOI 10.1007/978-1-62703-505-7_12, © Springer Science+Business Media New York 2013

mechanisms underlying wound chronicity and those findings would be clinically relevant, therefore laying the foundation for clinically valuable solutions. In this chapter, we share the essence of what we have learnt as experimental wound biologists as we sought to undertake studies of the chronic human tissue. We hope that you will find this chapter useful to recognize and address some of the logistic as well as scientific barriers that are integral to the system.

1.1 Infrastructure

Availability of appropriate infrastructure is critical to the success of wound research studies involving chronic wound patients.

1.1.1 Wound Center: A Catalyst

Continuous access to chronic wound patients (as they visit over time) and their comprehensive medical records (etiology, comorbidity, medication, outcome, etc.) represents a critical barrier to performing patient-based wound healing research. At most health care facilities, wound care is not recognized as a specialized clinical discipline. In such setting, wound patients are treated across a wide range of clinics making tracking of such patients very difficult. Few medical centers, including ours, are fortunate to have a Comprehensive Wound Center that covers both outpatient as well as inpatient services. This not only helps us access chronic wound patients but also enables us to track them as they are discharged from inpatient facilities.

1.1.2 Multicenter Studies

Multicenter collaborations are useful in boosting the volume of patient enrollment. Lack of standardized wound care protocols among participating centers may lead to variation in data and sample collection, thus confounding results. Adoption of standardized protocols for routine wound care in participating centers is another key catalyst that drives patient-based wound research. At Ohio State, we are fortunate to have wound centers at five different locations in Columbus under the banner of our Comprehensive Wound Center. These clinics care for over 1,500 patient visits a month. Wound care practices across these locations are guided by standardized algorithms.

1.1.3 Personnel Training

As in most clinics, abrupt introduction of research into clinical operations is disruptive and alienates the clinical staff. The first steps in successful introduction of the research culture to a high-volume clinic is to make sure that the entire team of clinical staff understands that research and education are integral components of clinical care. This process requires extra effort of already over-burdened clinicians and thus must be appropriately incentivized. Next, personnel directed at clinical operations should not be additionally burdened with research tasks. Separate full-time effort must be budgeted and the clinics must be staffed with clinicians dedicated to the research mission. Such research staff must be trained to observe regulatory and compliance requirements. Topics to be addressed include the informed consent process, confidentiality, identifying and managing risks, protocol development, protocol adherence, and ethical conduct of research as approved by the IRB. Standardized DVD-based training is desirable. Appropriate oversight of such staff by designated principal investigator(s) is required. Furthermore, within the perimeters of the clinic, dedicated research space must be identified and made available for patient handling and sample collection. To keep the clinical team motivated, it is useful to add a research agenda item on each and every clinical operation meeting. This is a good venue to update the group on research progress and also to recognize the most productive clinicians who have helped recruit most subjects. This is also a good opportunity to find out which clinicians may want to voluntarily elevate their contribution to research to a point where they want coauthorship of publications. A few such cases soon get viral and become a big motivation factor for other clinicians—many of whom soon join in. Such transformation of culture in the clinics is a critical driver of effective research. Above all, it needs to be recognized that such change in work culture cannot be drilled bottom-up by passionate researchers who know a clinician or two at the wound clinic. Such transformation must be initiated with the complete buy in of top hospital leadership who need to be convinced to integrate clinical and research productivity as parts of one score card.

1.1.4 Institutional Review Board (IRB)

The design, conduct, and monitoring of a research activity involving human subjects is the responsibility of the principal investigator (PI) who provides a complete written description of the proposed research to an IRB and obtains appropriate approvals to conduct such research. No activity involving human subjects may begin until it has been reviewed and approved by an IRB. The primary responsibility of IRBs is to safeguard the rights and welfare of human research subjects. Therefore, a PI must provide enough information for the IRB to determine that human subjects will be adequately protected and that the research will be conducted in full compliance with federal regulations and policies in order to approve a research activity involving human subjects. For specific details on IRB approval, please consult your own institution. Typically, the following sections need to be addressed:

1. Introduction
2. Objectives
3. Study design and methods
4. Inclusion and exclusion criteria
5. Monitoring subjects and criteria for withdrawal of subjects from the study
6. Analysis of the study
7. Human subject protections
8. Rationale for subject selection and procedures for recruitment (include a targeted/planned enrollment table)

9. Rationale for the involvement of special classes of subjects, if any, such as fetuses, pregnant women, children, cognitively impaired individuals, prisoners or other institutionalized individuals, or others who are likely to be vulnerable
10. Evaluation of benefits and risks/discomforts
11. Adverse event reporting and data monitoring
12. Collection and storage of human specimens or data
13. Remuneration/compensation
14. Consent and assent processes and documents
15. References

Note. If the protocol involves subject enrollment at multiple sites, describe plans for ensuring appropriate IRB review and approval at each site.

1.1.5 Clinical–Basic Science Researcher Partnership

Obtaining tissue from human subjects with chronic wounds requires participation of not only physicians but also all clinical staff. It is therefore important to create and maintain a culture at the wound clinics/centers that supports research by (a) keeping an ongoing dialogue between research staff and all clinical staff on research-related logistics as well as findings; (b) minimizing disruption of clinical work flow because of research activity; (c) incentivizing (e.g., academic credits or monetary reimbursement) physicians to participate in research; and (d) developing a scorecard to monitor the impact of research on overall clinical revenues.

1.1.6 Wound Data Collection Tools

The high degree of variability in clinical conditions among chronic wound patients means that a large number of subjects must be screened and stratified based on their medical history to identify comparable cohorts of patients suitable for productive data analyses. An integrated data management platform that can access and intelligently mine electronic medical records as well as research data records is of extraordinary value. It is recommended that local bioinformatics expertise be engaged to provide support in this regard.

1.2 Laser Capture Microdissection: Spatially Resolved Molecular Analyses of the Wound Tissue Section

Biopsies collected from human cutaneous wounds presented in the clinic are highly heterogeneous in cellular composition. The small sample volume poses additional challenges in detailed molecular analyses. The composition of the tissue may vary from one collection to another complicating comparison of results derived from tissue homogenates. Thus, the utility of such tissue material is primarily limited to histological studies. Laser capture microdissection (LCM) is a powerful technique to collect specific single-cell or multicellular structures from within the tissue section. These tiny tissue samples lend themselves to subsequent molecular analyses [1]. In 2007, we reported the first evidence demonstrating LCM-based microdissection of blood vessels from human wound tissue sections [2]. The collected tissue elements were subjected to high-density microarray analysis and quantitative PCR-based validation of microarray data. Residual tissue in the single biopsy was used to perform immunohistochemistry aimed at validating microarray data. All of the above-said analyses were performed using no more than a single 3 mm punch biopsy from the chronic wound tissue. Comparison of results from blood vessels at the edge of chronic wound tissue with those of vessels from intact human skin demonstrated a striking contrast between the transcriptome of vessels collected from the two different locations [2]. LCM is applicable to a broad range of clinical research and represents a catalyst of sophisticated translational research. Here, we detail the approach to capture wound hyperproliferative epithelium (HE) from chronic human wound biopsies.

2 Materials

2.1 Chronic Wound Tissue Biopsy Collection

1. 3 mm Biopsy tool.
2. Optimum cutting temperature (OCT) compound.
3. Aluminum foil.
4. Dry ice.
5. Cryo-molds.
6. Styrofoam box to store dry ice (*see* **Notes 1–4**).

2.2 Cryosectioning

1. PEN membrane slides.
2. RNA Zap®.
3. Diethyl pyrocarbonate (DEPC)-treated water.

2.3 Rapid Staining: Visualize Tissue Architecture Yet Preserve RNA Integrity

Truncated hematoxylin and eosin staining [3–5].

1. Hematoxylin QS.
2. Eosin Y, 2 % in 95 % ethanol.
3. RNALater®.
4. Nuclease-free water, molecular biology-grade ethanol and xylene.
5. Mini-stainer jars.

2.4 Laser Capture Microdissection (LCM) or Laser Microdissection and Pressure Catapulting (LMPC)

1. PALM MicroBeam IV (Manufacturer: P.A.L.M. Microlaser Technologies GmbH, a company of the Carl Zeiss Group, Bernried, Germany).
2. Stained tissue sections on a PEN membrane slide (see sections for preparing stained tissues for LCM).

2.5 RNA Isolation [2, 4–6] (Fig. 4)

1. PicoPure RNA Isolation Kit.
2. NanoDrop system.
3. Agilent bioanalyzer.

2.6 Reverse Transcription and Quantitative Real-Time PCR [2, 4–6]

1. Arcturus® RiboAmp® HS PLUS cDNA Kit.
2. TaqMan® Gene Expression Assays.

3 Methods

3.1 Preparing the Wounds for Sampling

There is a lack of consensus on the correct way to prepare and sample a wound. We collect biopsies following routine marginal debridement of the wounds to avoid collection of dead necrotic tissue. Prior to biopsy collection, the wound bed is prepared with local antiseptics. Local anesthetics are administered according to patient needs. If the effects of specific anesthetics are being tested, it is necessary to avoid the use of local anesthetics.

3.1.1 Collection Procedure

Wound-edge tissue specimen is collected as 3 mm biopsy from the boundary (1 cm) between skin and open wound.

3.2 Sample Storage and Transport

Samples are immediately rinsed with ice-cold saline, patted dry on Kimwipe®, then placed in collection vial containing appropriate collection medium (e.g., OCT for cryosectioning), and snap frozen in liquid nitrogen as appropriate. The samples can be transported from clinic to laboratory in properly secured biohazard-safe containers under appropriate temperature conditions (e.g., in portable liquid nitrogen tanks). Shipments of any biological materials are regulated by the US Department of Transportation (DOT) and the International Air Transport Association (IATA). Additional regulations may apply while shipping of wound samples in dry ice.

3.3 Embedding of Sample in OCT Compound

Place dry ice in a thick-walled Styrofoam box. Place a cryo-mold on the dry-ice block. Add OCT compound to the cryo-mold. Place the biopsy in appropriate orientation to obtain cross section of the wound tissue. Add more OCT to cover the tissue. Let the OCT solidify on dry ice. Once solidified, the cryo-mold containing embedded tissue biopsies can be stored in liquid nitrogen vapor.

3.4 Cryosectioning

1. Mark side (non-glossy) of the slide that contains the membrane with pencil.
2. Place slide on clean paper towel.
3. Dip the membrane slide into RNA Zap® ten times.
4. Dip the membrane slide into DEPC water 10×.

5. Repeat with fresh DEPC water.
6. Flick the slide to remove excess water and let slide dry.
7. Take out the OCT-embedded tissue block and mount on a cutting chuck. Perform sectioning using standard cryosectioning equipment. We routinely use 10 μm thick sections for LCM captures of hyperproliferative epithelium from wound tissues.

3.5 Rapid Staining: Visualize Tissue Architecture Yet Preserve RNA Integrity

1. Thaw and dry frozen slides by placing them in a box containing silica gel for 5 min. This should be performed in a biosafety cabinet to minimize contamination. Do not allow the silica gel to come into direct contact with the sections.
2. Add RNALater® to cover the sections and incubate for 4 min.
3. Add a drop of hematoxylin QS on the section and incubate for 30 s.
4. Wash the slide in DEPC water by gently dipping the slide up and down five times.
5. Repeat wash with fresh DEPC water.
6. (*Optional step*): Add a drop of 2 % eosin Y in 95 % EtOH to the sections and incubate for 30 s.
7. Wash in 95 % EtOH by gently dipping the slide up and down; then let the slide sit in fresh 95 % EtOH for a period of 1 min.
8. Place the slide in 100 % EtOH for 1 min.
9. Place the slide in xylene for 1 min.
10. Shake off xylene and let the slide dry in the hood by placing it in a box containing silica gel for 2 min (*see* **Notes 5–7**).

3.6 Laser Capture Microdissection or Laser Microdissection and Pressure Catapulting

1. The PALM MicroBeam LMPC system consists of inverted microscope with a motorized stage and a "cold" nitrogen UV laser that is not harsh on DNA/RNA or proteins. For cell or tissue element capture, the laser is focused onto the tissue section by adjustment of objectives. The laser traverses the glass slide from below and cuts the section from beneath (Fig. 1).
2. The microscope stage and UV laser are controlled by a PC (through RoboSoftware™), and a video camera allows the tissue sections to be displayed on the PC screen. Viewing of tissue is possible with all objectives.
3. The captured tissue element is collected using Robomover Z, the robotic unit of PALM MicroBeam controlled by RoboSoftware™ with multipurpose functions (Fig. 2). It is a collection device with the adaptors for the versatile capture vials such as single microfuge caps, multitude of microfuge tubes, multicap strips, or plates. Different customized holders allow collection of samples from single or multiple object slides, LMPC-optimized slides, or culture dishes. The Robomover Z can process up to three slides simultaneously.

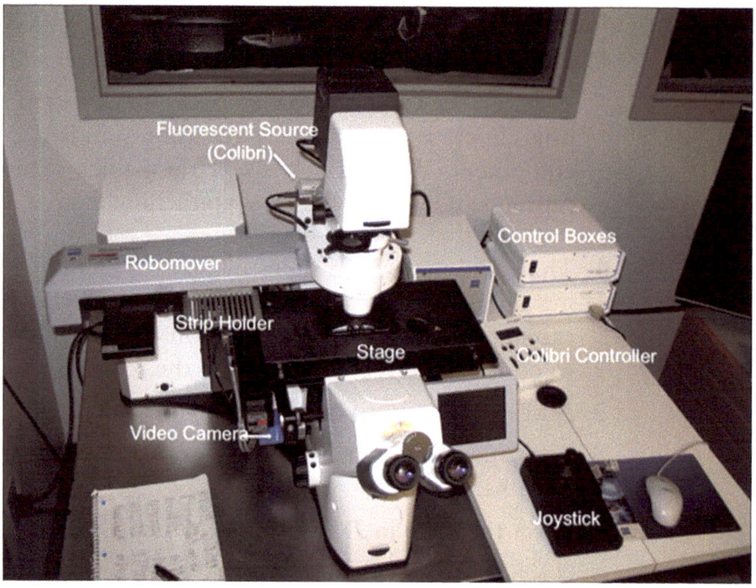

Fig. 1 Major components of the PALM MicroBeam IV system

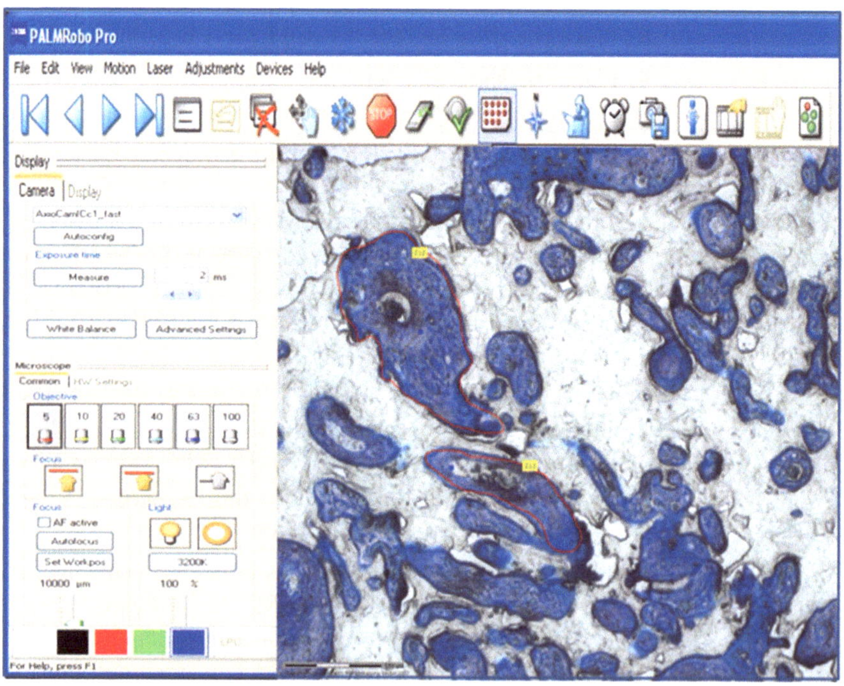

Fig. 2 A representative screen shot of RoboSoftware™ Pro software to show tissue elements marked and ready for capture

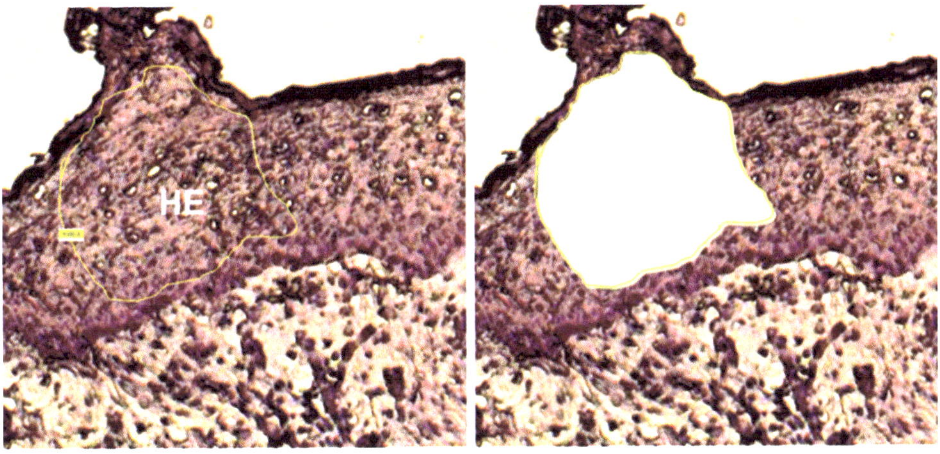

Fig. 3 LCM of human wound tissue hyperproliferative epithelium (HE). *Left*. Hematoxylin-stained HE area marked for capture. *Right*. The same area shown after capture

4. *Principles of LMPC*. In LMPC, the biological material is placed on a glass slide covered with a polyethylene naphthalate (PEN) membrane. The laser beam source is positioned and focused. A tightly focused laser beam is used to cut the PEN membrane and the tissue sample on it. A defocused laser beam is then used to "pressure catapult" the cut membrane with dissected cells of interest into a sample collector. Thus, cell retrieval is accomplished by LMPC approach against gravity. This is a noncontact method that prevents possible contamination by surrounding tissue as opposed to technologies that rely on gravity-driven "falling" of the dissected tissue element into the sample collector.
5. Mount the stained tissue-containing slide on the slide holder on motorized stage.
6. Load appropriate collection vial onto the Robomover. The collection vial should contain some collection solution. For gene expression studies, RNA extraction lysis buffer is appropriate.
7. Use appropriate objective to visualize the tissue and region of interest (ROI).
8. Mark the ROI utilizing the pen tool feature of the Robosoftware™.
9. Activate the laser to cut and catapult to the collection vial.
10. Collect images before and after the captures as well as of the collected tissue in vial (Fig. 3).
11. The laser focus, laser energy, and speed of cutting need to be optimized for specific type of tissues to obtain efficient tissue cutting and catapulting.

biopsy collection from human chronic wounds

↓

sample storage, sectioning & staining

↓

LCM/LMPC of wound hyperproliferative epithelium (HE)

↓

RNA extraction & RNA QC

↓

quantitative PCR for gene expression

Fig. 4 Workflow of gene expression analysis from LCM-captured human wound HE (*see* **Notes 8 and 9**)

3.7 RNA Isolation [2, 4–6] (Fig. 4)

1. Tissue elements are cut and captured in LCM collection vial containing RNA extraction buffer.
2. Perform total RNA isolation from captured tissue elements using a small sample RNA extraction kit, for example Picopure RNA isolation kit, following the manufacturer's instructions.
3. After isolation, the RNA quantity (Nanodrop) and quality (Bioanalyzer) are measured to determine the quantity as well as to ensure the integrity of the RNA. An RNA integrity number (RIN, 1 is worst quality vs. 10 is the best) above 5 is considered as acceptable RIN to proceed with gene expression analysis studies (*see* **Note 10**).

3.8 Reverse Transcription and Quantitative Real-Time PCR [2, 4–6]

1. The ARCTURUS® RiboAmp® HS PLUS cDNA Kit permits quantitative real-time PCR (qRT-PCR)-based gene expression analysis from small (as little as 100 pg) LCM captured of total RNA.
2. Perform reactions using instructions included with the kit.
3. The cDNA ready from reaction above can be used directly for qRT-PCR assay.
4. TaqMan® Gene Expression Assays are specifically designed to determine gene expression via qRT-PCR. The kit comes with detailed instructions to perform the assays.

4 Notes

1. It is essential to maintain the sample temperature below −20 °C during the entire sectioning process.
2. Before collection, the freshly cut section should be left for 5 min in cryosectioning equipment for appropriate cooling.
3. The slides containing tissue sections should be kept on dry ice immediately after they come out of cryosectioning equipment.
4. To get best results, cryosection no earlier than an hour before LCM.

5. Use molecular-grade ethanol, RNAse-free solutions, DEPC-treated water, and fresh xylene during staining.
6. Reduce time of staining as much as possible to minimize RNA degradation.
7. Avoid storage of stained tissue. The staining should be performed immediately before LCM.
8. It is recommended to optimize LCM conditions with a trial tissue to minimize loss of time.
9. For gene expression studies, the LCM process must not exceed 30 min to secure high-quality RNA integrity. We have reported a gradual loss in RNA stability in tissue sections as a function of time following staining [4].
10. The optimum tissue elements for gene expression studies vary with type of tissues. In general, 0.1–1 million square micron tissue elements provide sufficient quantity of RNA to perform gene expression studies.

Acknowledgments

Wound healing research in the authors' laboratory is funded by NIDDK R01 DK076566 (SR), and NIGMS GM069589 and GM 077185 (CKS).

References

1. Biswas S, Roy S, Sen CK (2010) Laser capture microdissection as a tool to address spatially resolved biological mechanisms in tissue injury and repair. In: Das DK (ed) Methods in redox signaling. Mary Ann Liebert, New York, p 264
2. Roy S, Patel D, Khanna S, Gordillo GM, Biswas S, Friedman A, Sen CK (2007) Transcriptome-wide analysis of blood vessels laser captured from human skin and chronic wound-edge tissue. Proc Natl Acad Sci U S A 104:14472–14477
3. Kuhn DE, Roy S, Radtke J, Gupta S, Sen CK (2006) Laser microdissection and pressure-catapulting technique to study gene expression in the reoxygenated myocardium. Am J Physiol Heart Circ Physiol 290:H2625–H2632
4. Kuhn DE, Roy S, Radtke J, Khanna S, Sen CK (2007) Laser microdissection and capture of pure cardiomyocytes and fibroblasts from infarcted heart regions: perceived hyperoxia induces p21 in peri-infarct myocytes. Am J Physiol Heart Circ Physiol 292:H1245–H1253
5. Roy S, Khanna S, Kuhn DE, Rink C, Williams WT, Zweier JL, Sen CK (2006) Transcriptome analysis of the ischemia-reperfused remodeling myocardium: Temporal changes in inflammation and extracellular matrix. Physiol Genomics 25:364–374
6. Roy S, Khanna S, Azad A, Schnitt R, He G, Weigert C, Ichijo H, Sen CK (2010) Fra-2 mediates oxygen-sensitive induction of transforming growth factor beta in cardiac fibroblasts. Cardiovasc Res 87:647–655

Chapter 13

The Wound Watch: An Objective Staging System for Wounds in the Diabetic (db/db) Mouse Model

G. Pietramaggiori, S. Scherer, and D.P. Orgill

Abstract

As in cancer biology, in wound healing there is a need for objective staging systems to decide for the best treatment and predictors of outcome. We developed in the diabetic (db/db) wound healing model, a staging system, the "wound watch," based on the quantification of angiogenesis and cell proliferation in open wounds. In chronic wounds, there is often a lack of cellular proliferation and angiogenesis that leads to impaired healing. The wound watch addresses this by quantifying the proliferative phase of wound healing in two dimensions (cellular division and angiogenesis). The results are plotted in a two-dimensional graph to monitor the course of healing and compare the response to different treatments.

Key words Wound staging, Angiogenesis, Cell proliferation, Wound healing, Diabetic (db/db) mouse model, ki-67, CD31

1 Introduction

One of the biggest limitations in the study and development of wound treatments is the lack of reliable models to quantify objectively the course of healing. Our wound healing staging system, developed to parallel wound healing and tumor biology, is based on the quantification of neovascularization and stromal cell proliferation in experimental diabetic (db/db) wounds [1]. The diabetic mouse, with its impaired angiogenesis and wound healing [2], and decreased production of growth factors [3, 4], serves as a reproducible surrogate for the study of the course of healing of recalcitrant wounds. Previous experiments have shown that diabetic mice have delayed wound contraction and hair regrowth, allowing for better wound dressing and closure analysis compared to wild-type mice [5]. We further developed the methods, which allowed for multiple studies on topics such as growth factors [6, 7], platelets [8, 9], cytokines [10], mechanical forces, and the VAC [11]. For the accurate staging of healing wounds, we selected two main markers for vascularization (CD31) and cell proliferation (Ki-67),

Robert G. Gourdie and Tereance A. Myers (eds.), *Wound Regeneration and Repair: Methods and Protocols*, Methods in Molecular Biology, vol. 1037, DOI 10.1007/978-1-62703-505-7_13, © Springer Science+Business Media New York 2013

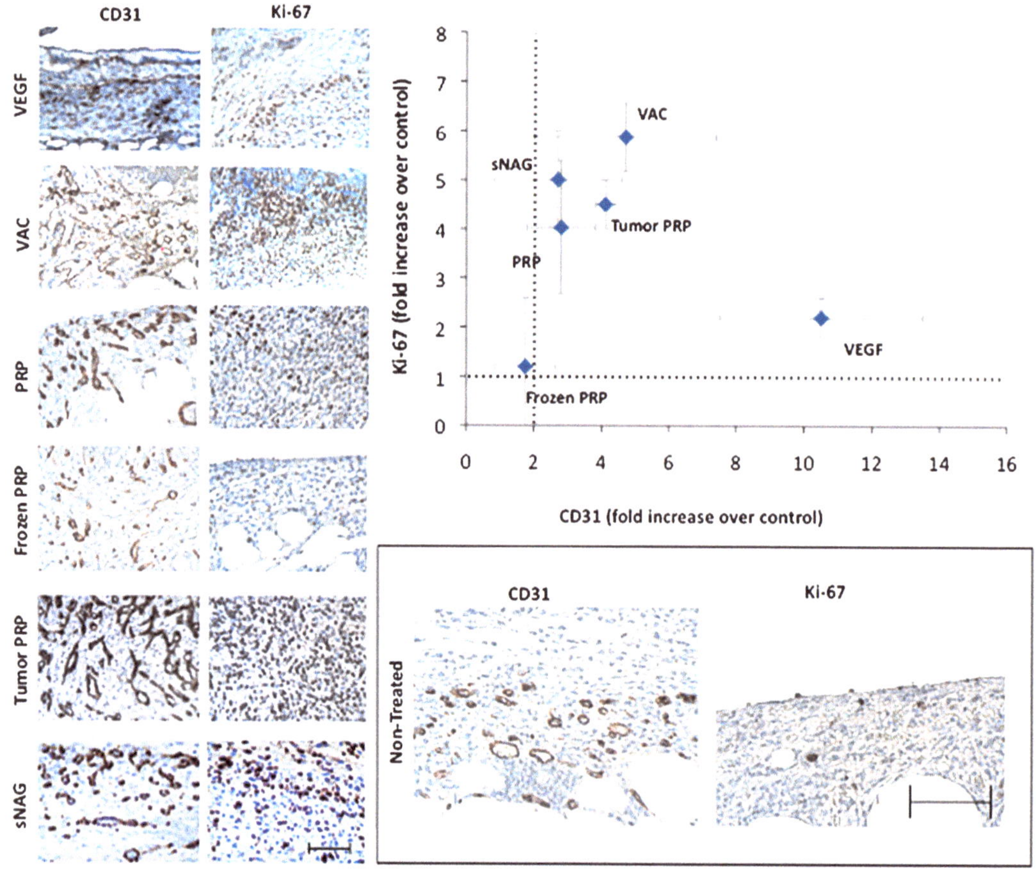

Fig. 1 Staining examples of CD31 and Ki67 for the wound watch: The *left column* shows CD31, while Ki-67 is in the *right column*. Each *row* corresponds to a different treatment. In the *box*: Control wounds, showing low levels of angiogenesis and cell proliferation (set to 1 in the plot). The quantification of the results is plotted in the wound watch staging system. Results are reported as mean ± SD. Error bar 50 μm

which are plotted together in the wound watch plot. This chapter describes the entire protocol to stage wounds using the wound watch staging system.

While the wound watch is used to stage the healing of a skin defect, the response to different treatments can be also evaluated. The following protocol allows for the high-quality staining condition, required in order to proceed with the quantification steps (Fig. 1). The level of angiogenesis and cell proliferation in control, non-treated wounds is set to 1 (Fig. 1). In this model, we selected day 10 for the analysis, corresponding to about 50 % closure in non-treated wounds. In this stage, we measured the largest differences between experimental treatments.

Although wound closure is the most widely accepted outcome to assess the effectiveness of a therapy, for particular wounds, the desired biological effect might be different. For example, in arterial wounds, angiogenesis might be preferred to cell proliferation. VEGF in our model increased vascularity by over tenfold while barely increasing cell proliferation by a factor of 2.2 (Fig. 1) [12]. On the opposite end of stimulation, we found scaffolds made of polymers (such as poly-*N*-Acetyl glucosamine, sNAG, Fig. 1), which allow for cell attachment: in this case cell proliferation reaches a 5-fold increase when compared to controls and 2.7-fold when compared to angiogenesis [13].

Mechanical forces, such as the vacuum-assisted closure device, when tested in our model, showed a strong increase of cell proliferation, with up to 5.9-fold increase over control wounds of Ki-67-positive cells (Fig. 1) [11]. Another commonly used strategy to stimulate wounds is fresh platelet-rich plasma (100 μl topically administered dose in our model), which increases cell proliferation by a factor of 4 and angiogenesis by a factor of 2.8 (Fig. 1) [9]. Interestingly, when PRP is frozen or lyophilized platelets lose some of the beneficial effects, with only a 1.7 increase in angiogenesis and 1.2 increase in cell proliferation, respectively [9], and [6] (Fig. 1). Interestingly, certain pathologic states, such as tumors, seem to stimulate the platelet ability to induce angiogenesis, as PRP derived from tumor-bearing mice stimulated angiogenesis and cell proliferation further than PRP from healthy animals, with 4.1 increase in angiogenesis and 4.5 in cell proliferation [8] (Fig. 1).

In conclusion, the wound watch staging system is a bi-dimensional (angiogenesis and cell proliferation) wound healing staging system, which can be used to compare the effects of several experimental treatments at the same time.

2 Materials

2.1 Mice

Homozygous genetically diabetic 8–12-week-old Lep/r—db/db male mice (strain C57BL/KsJ-Lepr^{db}) are used under an approved animal protocol in accredited facility.

2.2 Excisional Wounding

1. Anesthesia solution: Nembutal 60 mg/kg (Pentobarbital).
2. EtOH (70 %): Disinfection pads.
3. Single-use, nontoxic, non-pyrogenic syringes (3 ml).
4. Single-use, nontoxic, non-pyrogenic needles (0.5 × 25 mm).
5. Gauzes, examination gloves, 50 ml conical tubes.
6. Electric razor.
7. Depilatory cream.
8. Scissor and forceps.
9. Surgical marker.
10. Benzoin solution.
11. Tegaderm dressing.

2.3 Immunohistochemistry

1. Xylene.
2. Alcohol 100 %.
3. dH_2O.
4. Dulbecco's PBS.
5. H_2O_2.
6. Methanol.
7. Tris–H_2O solution.
8. Proteinase K.
9. Citrate solution.
10. TNB solution.
11. CD31 primary antibody.
12. Ki67 primary antibody.
13. CD31 secondary antibody.
14. Ki67 secondary antibody.
15. Tween-20.
16. Streptavidin–HRP.
17. Biotin.
18. DAB chromogen.
19. Hematoxylin.
20. PapPen.
21. Object slides.
22. Coverslips.
23. Slide wet chamber.

3 Methods

3.1 The Diabetic Mouse Model

The day before surgery, hair is clipped and depilated. On the day of the surgery, animals are weighed and anesthetized with 60 mg/kg Nembutal (Pentobarbital, freshly prepared). A dorsal 1.0 cm^2 area of skin and panniculus carnosus is excised and the wounds are photographed. All wounds are covered with a semi-occlusive polyurethane dressing. On postoperative day 10, the animals are euthanized and the wounds are photographed, excised, and fixed in 10 % neutral-buffered formalin solution.

3.2 Staining Instructions

Day 1

1. Label slides with a lead pencil either "Ki67" or "CD31."
2. Warm slides on a heat plate (for at least 30 min to 1 h, or even overnight), temp between 50 and 60 °C (never >60 °C to prevent denaturation) (*see* **Note 15**).

3. Remove paraffin using two washes of xylene for 5 min each.
4. Hydrate slides with a dehydration alcohol series, diluting dehydration alcohol in dH_2O. 100 % (2 min), 95 % (2 min), 85 % (1 min), and 70 % (1 min) (*see* **Note 16**).
5. Wash the slides in PBS three times for 10 min each.
6. Immerse slides in 30 % H_2O_2 solution (diluted in methanol), *in the dark*, for 20 min—methanol dilution: 25 ml of 30 % H_2O_2 in 225 ml methanol (*see* **Note 17**).
7. Wash the slides in PBS three times for 5 min each.
8. *CD-31* (surface antigen retrieval).
9. Prepare a water bath at 37 °C (*see* **Note 18**).
10. Prepare a [0.2 M] Tris–H_2O solution.
11. Place [0.2 M] Tris–H_2O solution in water bath to equalize temperature before adding proteinase and slides (*see* **Note 19**). Add 90 μl of Proteinase K for every 50 ml to Tris–H_2O solution (*see* **Note 20**), and place the slides in a water bath for 20 min.
12. Wash them in PBS three times for 5 min each.
13. *Ki67 (intracellular antigen retrieval).*
14. Immerse the slides in citrate (pH = 6. 10 mM) and microwave slides until solution begins to boil (breaking cell membranes to expose antigen).
15. STOP the microwave, and allow container to rest for 5–6 min.
16. Microwave the slides again until solution begins to boil.
17. STOP the microwave, and allow to *rest* until solution reaches room temperature for at least 20 min.
18. Wash the slides in PBS three times for 5 min each.

Antigen enhancing

19. Choose 1–2 sections per slide to stain. Wipe off other sections.
20. Dry the area around desired sections.
21. Draw a circle around the remaining sections with a marker.
22. Add 100–150 μl TNB (allow to defrost beforehand) to sections and leave in wet chamber for 30 min at room temperature (*see* **Note 21**).
23. Shake off the agent.
24. Replace slides in wet chamber, and pipette the primary antibody (1 μl:100 μl in TNB for CD31, 1:200 for Ki67), approximately 70–80 ml onto each section.
25. Leave in fridge overnight at 4 °C.

Day 2

1. Wash in PBS-Tween three times for 5 min each.
2. As you place slides in wet chamber, pipette secondary antibody (60–70 ml) onto each section, and then leave for 1 h at room temperature.
3. Biotinlated secondary antibody: 1 μl antibody:200 μl in TNB.
4. Wash in PBS-Tween, three times for 5 min each.

Signal Amplification

5. As you place slides in wet chamber, pipette streptavidin–HRP (60–70 μl, 1 μl:100 μl in TNB) onto each section, and then leave for 30 min at room temperature.
6. Wash in PBS-Tween, three times for 5 min each.
7. Replace slides in wet chamber. Pipette biotin on slides (1 μl:50 μl in amplification diluent), approximately 60–70 ml pipetted onto each section (*see* **Note 22**).
8. After 4 min, place slides in PBS-Tween, and wash three times for 5 min each.
9. Repeat the amplification with streptavidin–HRP: As you place slides in wet chamber, pipette streptavidin–HRP (60–70 ml) onto each section, and then leave for 30 min at room temperature.
10. Wash in PBS-Tween, three times for 5 min each.

Activation

11. Prepare chromogen agent: Dissolve ½ tablet of DAB in 5 ml PBS.
    To activate chromogen, add 1 μl 30 % H_2O_2 per 1 ml DAB/PBS solution. Once activated, this solution is only good for a maximum of 30 min (*see* **Note 23**).
12. Have two containers of dH_2O ready.
13. Bring slides (in PBS-Tween chamber), dH_2O containers, and activated DAB solution to light microscope.
14. Place slide on microscope under 10× magnification.
15. Activation is often instantaneous: Just put a few drops (approximately three drops) of activated DAB solution evenly onto the section, look under microscope, and STOP the reaction once the positive-staining structures are darkened (brown) by placing slide in dH_2O (*see* **Note 24**).

Counterstain

16. Fill one container with hematoxylin; make sure that there is no layer of film on top (*see* **Note 25**).
17. Fill another container with tap water.

18. Immerse slides in hematoxylin for 3 min.
19. Immerse slides in water for a few seconds, and change water immediately at least one time.
20. Leave slides in the container with constant water flow (not too strong) until water runs clear.
21. Fill one container with ammonium-H_2O as per above instructions.
22. After being in water, dip slides in ammonium-H_2O ten times.
23. Blue with properly diluted bluing agent for 1 min (PBS).
24. Put them back in tap water.

Montage

25. Reverse EtOH cascade = Dehydrate dehydration alcohol tissue with 70 % (1 min), 85 % (1 min), 100 % (1 min), and 100 % (3 min); make sure that slide-labeled area is covered with 100 %. There should be no H_2O molecules on the slide or labels of the slides.
26. Safe Clear or xylene to remove PapPen two times for 5 min.
27. Coverslip with permanent mount and let sit at room temp until dry (24 h).
28. To remove excess glue from coverslip, dip slides in xylene 2–3 times and allow them to dry.

3.3 Wound Watch Staging System

Three digital images of PECAM-1- and Ki-67-stained slides are captured for each wound sample, one in the middle and two on the edges of the wound bed. Pictures are viewed with Adobe Photoshop CS Software. Blood vessels (CD-31 positive) in each high-powered field are marked and counted. Ki-67-positive cells are expressed as a ratio of proliferating nuclei to total nuclei. Ratios were calculated between results obtained from the center of the lesions and from the edges of each treatment group to the control group (fixed to 1). Results from CD-31 are plotted in the *x* axis, while from Ki-67 in the *y*-axis. Fifteen samples are evaluated at 40× magnification for each experimental treatment.

4 Notes

1. To stain paraffin-embedded sections, the paraffin material must first be removed and then the sample must be rehydrated. Rehydration also reactivates enzymes that must be blocked with H_2O_2.
2. Prior fixation with formalin causes many proteins to cross-link preventing accessibility to surface antigens. These antigens are retrieved using a protease. When trying to detect an internal

antigen, the cell membrane is broken by microwaving the samples in a basic citrate releasing the internal antigen.

3. In order for the primary antibody to bind only to the antigen of interest, the samples are treated with an unspecific protein-based blocker *TNB* that acts like a blanket. Now, the antigen of interest is marked with its specific *primary antibody*. This antibody is a highly specific "rat anti-mouse antibody" made by injecting the mouse antigen into a rat.
4. A biotinylated anti-rat (mouse adsorbed) *secondary antibody* is then added (super-specific). The biotin unspecifically binds to negatively charged molecules.
5. Streptavidin–HRP is then added to amplify the signal. Streptavidin is also a negatively charged molecule which binds to biotin, while HRP is an enzyme. Further amplification is achieved by adding biotin, and then again adding S–HRP. Leaving biotin on the sample too long will cause unspecific binding to other less negative molecules, reducing the specificity of the signal.
6. Finally, DAB chromogen is added and acts like a substrate for HRP. Positive staining for the antigen will cause a brown color change (can be instantaneous). If the activated chromogen is left on the samples too long the background will become brown. Stop the reaction by immersing slides in dH_2O as soon as the positive cells have become brown, and minimal background has stained.
7. "Wet" chamber: Place wet towels along the bottom of a common slide box. Close lid. When done with staining, remove wet paper towels and hang slide boxes to dry to prevent mold.
8. 1 M PBS: Dilute 10 M PBS stock solution in dH_2O (100 ml 10 M PBS:1,000 ml dH_2O).
9. PBS-Tween: Add 0.5 ml Tween for every 1 l of 1 M PBS (Tween is a detergent).
10. Ammonium-H_2O: Dilute 0.5 ml Ammonium Hydroxide (Menconi) in 500 ml tap water.
11. 1 M Tris pH 8: Prepare both 1 M HCl Tris (acid) and 1 M base Tris (Basic) desolving the powder in dH_2O, with a ratio 1 l H_2O:1 weight in grams of the formula weight (FW, found on the box of the salts). Then mix the two solutions, adding acid to base until the needed pH is reached.
12. Citrate (pH 6.0): In 1,000 ml of dH_2O add 1.92 g of anhydrous citric acid. Adjust pH to 6.0 by adding NaOH. Then add 0.5 ml Tween 20 and mix. Store at room temperature.
13. Steps will vary in time especially when slide rack is lifted up and down in the solution. This gets the solutions moving over the tissue better, thus speeding up the desired interaction of the

applied step. For example the xylene steps can be cut down to 20 dips depending on the amount of paraffin left on the slide after baking.

14. Items to be placed in *ice* when not in immediate use:
    (a) TNB.
    (b) Pap green marker.
    (c) Primary antibody.
    (d) Secondary antibody.
    (e) Streptavidin–HRP (*light sensitive*).
    (f) Biotin.
    (g) DAB dissolved in PBS.
15. Keep the temperature between 50 and 60 °C (never >60 °C to prevent protein denaturation).
16. Change the alcohol solution every two immersions.
17. *In the dark*, for 20 min—methanol dilution: 25 ml of 30 % H_2O_2 in 225 ml methanol.
18. Start beforehand; it will take time to get the temperature right.
19. Dilute 1 M Tris (basic pH = 8.0) with dH_2O 1 ml:5 ml (50 ml Tris in 200 ml dH_2O).
20. Do this step just before immersion of slides.
21. Allow to defrost beforehand.
22. Biotin should remain on each slide for only 4 min.
23. Only activate 2–3 ml of this solution at a time.
24. Waiting too long to stop the reaction will cause the background to stain as well.
25. Hematoxylin can be reused, but if there is a top film, filter the hematoxylin and only use for an additional few times.

References

1. Pietramaggiori G, Scherer SS, Mathews JC et al (2008) Healing modulation induced by freeze-dried platelet-rich plasma and micronized allogenic dermis in a diabetic wound model. Wound Repair Regen 16:218–225
2. Greenhalgh DG (2005) Models of wound healing. J Burn Care Rehabil 26:293–305
3. Beer HD, Longaker MT, Werner S (1997) Reduced expression of PDGF and PDGF receptors during impaired wound healing. J Invest Dermatol 109:132–138
4. Lauer G, Sollberg S, Cole M et al (2000) Expression and proteolysis of vascular endothelial growth factor is increased in chronic wounds. J Invest Dermatol 115:12–18
5. Scherer S, Pietramaggiori G, Mathews J, Chan R, Fiorina P, Orgill DP (2008) Wound healing kinetics of genetically diabetic mouse. Wounds 20(1):18–28
6. Pietramaggiori G, Kaipainen A, Czeczuga JM, Wagner CT, Orgill DP (2006) Freeze-dried platelet-rich plasma shows beneficial healing properties in chronic wounds. Wound Repair Regen 14:573–580
7. Chan RK, Liu PH, Pietramaggiori G, Ibrahim SI, Hechtman HB, Orgill DP (2006) Effect of

recombinant platelet-derived growth factor (Regranex) on wound closure in genetically diabetic mice. J Burn Care Res 27:202–205
8. Pietramaggiori G, Scherer SS, Cervi D, Klement G, Orgill DP (2008) Tumors stimulate platelet delivery of angiogenic factors in vivo: an unexpected benefit. Am J Pathol 173:1609–1616
9. Pietramaggiori G, Scherer SS, Mathews JC et al (2010) Quiescent platelets stimulate angiogenesis and diabetic wound repair. J Surg Res 160:169–177
10. Fiorina P, Pietramaggiori G, Scherer SS et al (2010) The mobilization and effect of endogenous bone marrow progenitor cells in diabetic wound healing. Cell Transplant 19:1369–1381
11. Scherer SS, Pietramaggiori G, Mathews JC, Prsa MJ, Huang S, Orgill DP (2008) The mechanism of action of the vacuum-assisted closure device. Plast Reconstr Surg 122:786–797
12. Pietramaggiori G, Kaipainen A, Ho D et al (2007) Trehalose lyophilized platelets for wound healing. Wound Repair Regen 15:213–220
13. Scherer SS, Pietramaggiori G, Matthews J et al (2009) Poly-N-acetyl glucosamine nanofibers: a new bioactive material to enhance diabetic wound healing by cell migration and angiogenesis. Ann Surg 250:322–330

Chapter 14

Human Ex Vivo Wound Healing Model

Stojadinovic Olivera and Marjana Tomic-Canic

Abstract

Wound healing is a spatially and temporally regulated process that progresses through sequential, yet overlapping phases and aims to restore barrier breach. To study this complex process scientists use various in vivo and in vitro models. Here we provide step-by-step instructions on how to perform and employ an ex vivo wound healing model to assess epithelization during wound healing in human skin.

Key words Epithelization, Ex vivo, Healing model, Human, Acute wounds

1 Introduction

Wound healing is a multifaceted process that involves different cell types, complex signaling network, and advances through orderly and timely different, yet overlapping phases [1]. However, this process can be influenced by various factors that may impair the healing response, resulting in a healing-impaired wound that fails to proceed through the usual progression. In addition to profound suffering of patients, healing-impaired wounds represent a major area of unmet clinical needs, leading to significant morbidities and mortality.

Various in vitro and in vivo models have been used to study wound healing process. Studies in many animal models of wound healing have examined cellular and molecular changes and their resultant delay in wound healing; however, our field is still lacking an adequate wound healing model that will be readily translatable to a human setting. As a consequence of this limited understanding, current treatments are often ineffective and very scarce.

Studying different phases of wound healing is dependent on the use of models. Although by using animal wound healing models different wound healing phases can be assessed simultaneously, differences in the skin morphology and physiology between, for example, rodents and humans make wound healing studies difficult to translate to human settings. Here we describe an ex vivo

Robert G. Gourdie and Tereance A. Myers (eds.), *Wound Regeneration and Repair: Methods and Protocols*, Methods in Molecular Biology, vol. 1037, DOI 10.1007/978-1-62703-505-7_14, © Springer Science+Business Media New York 2013

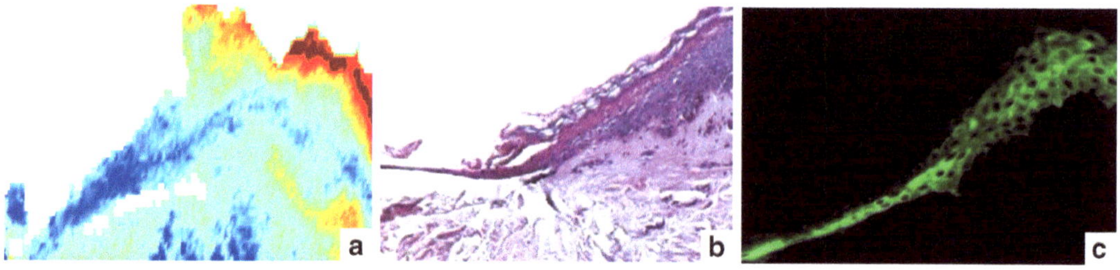

Fig. 1 Images of infrared (**a**), H& E-stained (**b**), keratin 17-stained (**c**) ex vivo wound edge specimen

human wound healing model adopted and further developed in our laboratory to study wound healing in human skin [2–5]. Comparative analyses between acute human wounds and ex vivo wound models showed comparable expression patterns for multiple genes involved in wound healing and epithelization phase, supporting human ex vivo model as reliable and functional in assessing human epidermal healing [5–8]. Human ex vivo wound model has been successfully used to assess epithelization rate after topical treatments [9–12]. Recently, we utilized this model to assess re-epithelization over time using infrared (IR) and confocal Raman spectroscopic imaging analysis [13] (Fig. 1). Our data revealed fine spectral differences that corresponded to distinct spatial distributions and we correlated these findings with keratin expression using immunostainings. Therefore human ex vivo wound healing model can be used to acquire detailed molecular structure information from many proteins involved in the wound healing process.

An advantage of this model is that the epithelization process can be assessed histologically at different time points. The skin specimen is typically obtained from reduction surgeries and brought to the laboratory within an hour to 2 h. Institutional Review Board (IRB) approvals are necessary but are straightforward to obtain, given that this is a discarded and de-identified tissue and, as such, may qualify for an exempt. Importantly, a large number of wounds can be obtained from the same donor. One limitation of this experimental model is lack of blood supply. Despite that, ex vivo model provides a very useful tool for preclinical testing, to study behavior of particular cell type, e.g., keratinocytes, during wound healing process and the epithelization phase in particular.

2 Materials

2.1 Ex Vivo Wounding

1. Sterile phosphate-buffered saline (1× PBS).
2. Red DMEM.
3. Antibiotic–antimicotic.
4. Fetal bovine serum.

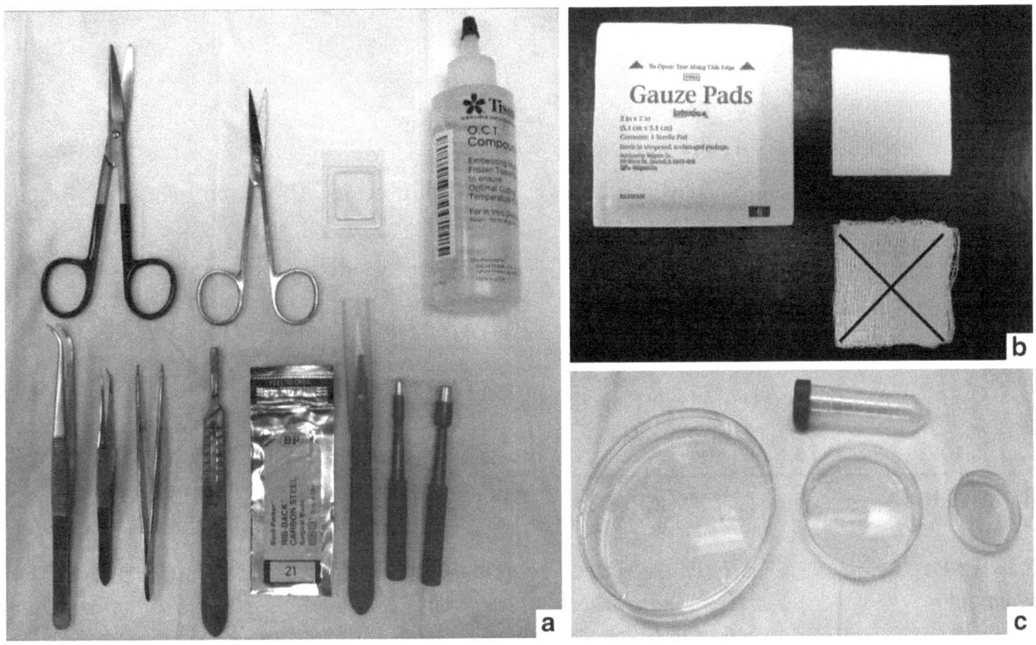

Fig. 2 Equipment and plasticware required for ex vivo wounding. Scissors, embedding mold, OCT media, curved- and flat-tip forceps, scalpels, biopsy punches (**a**). Gauze pads, *crossed* is non-recommended type of gauze pad (**b**). Falcon tube and tissue culture dishes (**c**)

5. Sterile gauze pads, highly absorbent (5.1 cm × 5.1 cm) (Fig. 2b).
6. Scissors and forceps (Fig. 2a).
7. 50 ml Falcon tubes.
8. Biopsy punches (3 mm and 6 mm).
9. 150 mm × 25 mm Cell culture dish.
10. 100 mm × 20 mm Cell culture dish.
11. 60 mm × 20 mm Cell culture dish.
12. Surgical gloves.
13. Disposable scalpel.

2.2 Embedding Ex Vivo Wounds for Histology

1. Tissue Tek®, Cryomold® Standard, disposable vinyl specimen molds (25 mm × 20 mm × 5 mm) (Fig. 2a).
2. Tissue Tek, OCT Compound (Fig. 2a).
3. Forceps.
4. Dry ice.

2.3 Cutting OCT-Embedded Ex Vivo Wounds Using Cryostat

1. Box of dry ice.
2. Paintbrush.
3. VWR low profile coated microtome blades.

2.4 H&E Staining

1. Microscope slides.
2. Acetone.
3. 1× PBS.
4. Hematoxylin.
5. Acid ethanol (1 % HCl diluted in 70 % ethanol).
6. Eosin.
7. Ethanol.
8. Xylene.
9. Coverslips.
10. Mounting media.

3 Methods

3.1 Ex Vivo Wounding

All the steps need to be carried under the tissue culture hood and all the equipment needed for ex vivo wounding must be sterilized before use!

1. Prepare ex vivo media using red DMEM supplemented with 1 % antibiotic–antimycotic and 10 % fetal bovine serum.
2. In a sterile tissue culture dish place sterile gauze pad and add 5 ml of media.
3. Using scalpel cut piece of skin received from reduction surgery to a 10 cm × 10 cm piece and place it in a Falcon tube filled with sterile 1× PBS. Cap the Falcon tube securely and shake the tube containing tissue to wash of the blood from skin.
4. Repeat washing step three times or until excess blood has been washed off the tissue.
5. Place skin, with epidermis facing up, into a 100 mm tissue culture dish and tap dry epidermis using sterile gauze pad.
6. Once the skin is dry, transfer the piece of skin to a clean tissue culture dish with epidermis facing down. Using scissors and forceps to hold the skin, cut off underlying fat tissue until dermis is reached. A skin specimen cleaned from fat should contain only dermis and epidermis (*see* **Note 1**).
7. Wash the specimen with the sterile 1× PBS three times (*see* **Note 2**).
8. Place skin specimen, with dermis facing down, into a clean tissue culture plate. Touch the skin surface with a 3 mm biopsy punch and gently punch the epidermis and the upper dermis by rotating a biopsy punch to create a wound bed (*see* **Note 3**).
9. Using forceps to pull out excised epidermis and upper dermis, excise 3 mm biopsy punch to create a central wound (*see* **Note 4**).

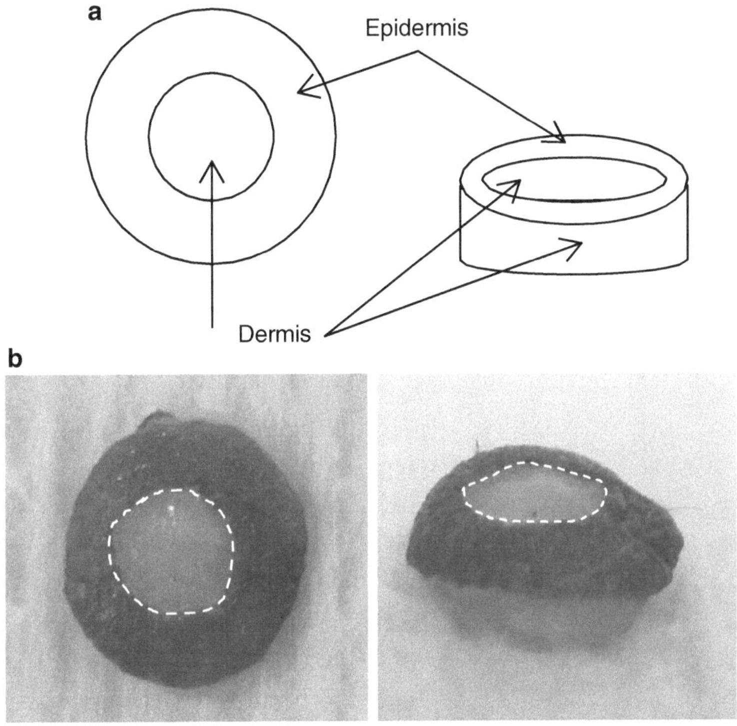

Fig. 3 Scheme of a doughnut-shaped ex vivo wound as seen from above and from a side (**a**). Human ex vivo wound. *White dotted circles* trace inner wound edge and demarcate wound bed (**b**)

10. Position a 6 mm biopsy punch containing the recently created 3 mm wound positioned in the center of the punch. Rotating the 6 mm biopsy punch, punch through the full-thickness skin to create a doughnut-shaped ex vivo wound (Fig. 3a, b).
11. Gently, without touching the wound bed place ex vivo wound on the top of the sterile gauze pad containing 5 ml of media and spread it out to lay flat so that the dermis is touching the gauze with the media completely. Avoid air pockets between the dermis and the air–liquid interface (*see* **Note 5**).
12. Apply topical treatment by pipetting it into the wound bed.
13. Cover the dish and place it at 37 °C in a humidified atmosphere of 5 % CO_2.
14. Change media every other day and reapply treatments every day (*see* **Note 6**).

3.2 Embedding Ex Vivo Wounds for Histology

1. Label the Tissue Tek disposable tissue specimen molds.
2. Fill in the tissue pocket with the Tissue Tek, OCT Compound.
3. Using tweezers detach ex vivo wound by stridently removing the wound from the underlying gauze (*see* **Note 7**).

4. Slowly by holding tissue specimen mold filled with OCT media on a dry ice submerge the ex vivo wound so that epidermis is parallel and faces the label. Prevent folding by holding in place until tissue is frozen.
5. Store labeled tissue at –80 °C until use.

3.3 Cutting OCT-Embedded Ex Vivo Wounds Using Cryostat

1. Place OCT-embedded ex vivo wounds in cryostat or –20 °C refrigerator to equilibrate the temperature for at least 5 min.
2. Place the chuck into the hole of a fast-freezing rail and cover it with OCT. Remove wound specimen from embedding mold, place it on a chuck coated with OCT, and allow OCT media to freeze. Use a metal blade to chip away any OCT around the chuck in order to fit it securely into a cryostat head.
3. Lock the handwheel brake when working with the specimen holder and the knife carrier.
4. Insert chuck with frozen tissue into the specimen clamping, lock it in place, and set the section thickness (*see* **Note 8**).
5. Insert the knife into the knife carrier from the side and tighten the clamping screws (*see* **Note 9**).
6. To cut sections, move the tissue towards the blade until thin sections can be cut. Use paintbrush to guide the cut section (*see* **Note 10**).
7. Begin by cutting 20 μm sections to reach to the specimen. Once you reach the wound bed tissue, cut sections at 5–8 μm and place section on slide by gently touching cut tissue with the slide (*see* **Note 11**).
8. Observe under microscope to assess the quality of the section (*see* **Note 12**).
9. Observe sections under microscope and label the slide.
10. Dry slides with sections at room temperature overnight and proceed with H&E staining.

3.4 H&E Staining

1. To fix the tissue place the slides with up to 8 μm thick tissue sections in a staining jar filled with cold (4 °C) acetone for 1 min at room temperature.
2. To wash slides submerge them in 1× PBS three times, 5 min each.
3. Submerge slides in ddH_2O for 5 min and blot access water from slide before submerging into hematoxylin.
4. Transfer slides to hematoxylin and keep them submerged for 2 min.
5. Rinse slides with ddH_2O for 3 min and then under the tap water for 15 min.
6. Dip slides in acid ethanol to destain.

7. Wash slides under the tap water for 5 min.
8. Rinse slides in ddH_2O for 2 min. Blot excess water from slides before transferring into eosin.
9. Submerge slides in eosin for 45 s.
10. Submerge slides in three changes of 95 % ETOH, 5 min each.
11. Submerge slides in three changes of 100 % ETOH, 5 min each. Blot excess ethanol before moving slides into a xylene.
12. Submerge slides in three changes of xylene, 5 min each.
13. Mount the sections using permanent mounting media and coverslip them avoiding any bubbles (*see* **Note 13**).

3.5 Quantification of Epithelization

1. To quantify epithelization use appropriate software (*see* **Note 14**).
2. Determine the location of wound edges and drag a measuring tool horizontally from one to the other wound edge to measure the size of the total wound bed (*see* **Note 15**). If the entire wound bed cannot be observed on the monitor, take multiple measurements until the entire wound bed is measured and record measurements under the appropriate sample label.
3. Next, measure migrating epithelial tongue (the distance keratinocytes migrated) from both wound edges and record measurement.
4. To determine the percent of epithelization divide a total distance measured as a total length of migrated epithelial tongue visible from both wound edges by length of the initial wound bed and multiply by 100 %.

4 Notes

1. Start cleaning the fat tissue by cutting it out with scissors, but do not cut too deep to injure the dermis. Once the fat tissue is cut out, level the dermis using scissors by cutting the small pieces of fat that were left attached to the dermis.
2. Washing cleaned specimen is extremely important in order to avoid lipid contamination. We noticed the presence of a lipid droplet in the wound bed and a skin surface while using infrared and confocal Raman spectroscopic imaging to acquire detailed molecular structure information from the diverse proteins and their subclasses during ex vivo wound healing process.
3. This is a fundamental step in creating the wound. It is important that the punch is sharp. The biopsy punch should be pressed down gently touching the epidermis. A few rotations of a punch while applying a slight pressure to the epidermis are sufficient to create a wound through epidermis and upper dermis (Fig. 3a, b).

4. In our experience the best way to separate epidermis and upper dermis from the rest of the skin in order to create a wound bed is to use pointed forceps and slowly pull out the 3 mm biopsy punch created. Once the 3 mm biopsy punch is pulled up allowing visualization of the dermis below, cut out the tissue by placing scissors parallel to the surface of the skin (parallel to the wound bed) just below epidermal-dermal junction. The greatest variability in the wound depth comes from this step and it is extremely important to keep wound depth as constant as possible. We recommend one person to perform wounding experiments for any given skin specimen.
5. It is important to separate wounds from each other in order to avoid additional tissue injury during the process of wound detachment from the gauze pad. We recommend specific gauze pads (Fig. 2b) to circumvent a tight dermis attachment and tear of the newly formed epithelium during separation process.
6. Perform this step under the sterile conditions. Put plate under the hood and use Pasteur pipette to aspirate media out from the gauze pad and pipette in fresh media. Transferring ex vivo wounds from one dish to another with a new gauze pad and fresh media can lead to injury of the migrating epithelial tongue and compromise the healing outcome. In our experience, keeping the wounds on the same gauze pad throughout the duration of the experiment yields the best results. Treatments should be re-applied daily by pipetting 25 μl of the treatment directly to the wound bed without touching the skin surface. In general at least three wounds per condition, per time point, should be made and experiment should be repeated at least three times using different skin specimens. To observe an effect of tested topical treatment we recommend assessing wound epithelization 4 days after wounding given that control, untreated wound would fully epithelializate by day 7.
7. To remove ex vivo wound from the gauze pad we recommend using curved-tip forceps (Fig. 2a). One should start detaching the wound from the underlying gauze pad by lifting the outer wound edge on one side of the ex vivo wound. Once the skin is lifted hold the skin between the curved tips and using one sharp move detach ex vivo wound by pulling it up and sideways.
8. The orientation of the specimen in relation to the cutting edge is important to achieve nice sections. The chuck has to be parallel in relation to the knife. Position the specimen so that epidermis faces either left or right.
9. Cutting area of the knife should be changed if it is not usable anymore by moving the knife to the left or the right side so that unused area of the knife can come in contact with the specimen.

10. To obtain continuous sections use a smooth cutting stroke. Stopping in the middle of the block causes tears and shreds the section.
11. To quantify epithelization it is necessary to place the section without any wrinkling on the side of the slide with the white area where can be written on to label the slide.
12. One will know that the wound bed was reached by visualizing indentation first macroscopically by looking at the cut tissue and confirming it by microscopical observation. We strongly recommend that after each section is placed onto the slide, a slide should be observed under the microscope. This way one can clearly see and distinguish the wound bed from the wound edges where the full thickness of epidermis is evident.
13. The mounting of sections on microscope slides is important for successful observation of the epithelization. A mounting media placed between coverslip and slide should spread by surface tension. To exclude bubbles that can interfere with visualization use forceps with tooth and press gently to smooth away bubbles from the section.
14. Any microscope with software that allows for precise measurements can be used. We recommend NIS-Elements BR 3.10 software.
15. To determine the wound edge we suggest that one first observes the whole section. Indentation into the dermis from biopsy punch along with the discontinuation of stratum corneum layer demarcates the wound edge and should be used as a starting point to measure wound bed length. At least ten, but one can use more, different wound sections derived from different parts of the wound should be measured and their average should represent the healing outcome for that particular specimen.

References

1. Barrientos S, Stojadinovic O, Golinko MS, Brem H, Tomic-Canic M (2008) Growth factors and cytokines in wound healing. Wound Repair Regen 16:585–601
2. Kratz G (1998) Modeling of wound healing processes in human skin using tissue culture. Microsc Res Tech 42:345–350
3. Lee B, Vouthounis C, Stojadinovic O, Brem H, Im M, Tomic-Canic M (2005) From an enhanceosome to a repressosome: molecular antagonism between glucocorticoids and EGF leads to inhibition of wound healing. J Mol Biol 345:1083–1097
4. Stojadinovic O, Brem H, Vouthounis C, Lee B, Fallon J, Stallcup M, Merchant A, Galiano RD, Tomic-Canic M (2005) Molecular pathogenesis of chronic wounds: the role of beta-catenin and c-myc in the inhibition of epithelialization and wound healing. Am J Pathol 167:59–69
5. Pastar I, Stojadinovic O, Krzyzanowska A, Barrientos S, Stuelten C, Zimmerman K, Blumenberg M, Brem H, Tomic-Canic M (2010) Attenuation of TGFbeta signaling pathway in chronic venous ulcers. Mol Med 16:92–101
6. Pullar CE, Grahn JC, Liu W, Isseroff RR (2006) Beta2-adrenergic receptor activation delays wound healing. FASEB J 20:76–86
7. Roupe KM, Nybo M, Sjobring U, Alberius P, Schmidtchen A, Sorensen OE (2009) Injury is a major inducer of epidermal innate immune responses during wound healing. J Invest Dermatol 130:1167–1177

8. Heilborn JD, Nilsson MF, Kratz G, Weber G, Sorensen O, Borregaard N, Stahle-Backdahl M (2003) The cathelicidin anti-microbial peptide LL-37 is involved in re-epithelialization of human skin wounds and is lacking in chronic ulcer epithelium. J Invest Dermatol 120: 379–389
9. Tomic-Canic M, Mamber SW, Stojadinovic O, Lee B, Radoja N, McMichael J (2007) Streptolysin O enhances keratinocyte migration and proliferation and promotes skin organ culture wound healing in vitro. Wound Repair Regen 15:71–79
10. Vukelic S, Stojadinovic O, Pastar I, Rabach M, Krzyzanowska A, Lebrun E, Davis SC, Resnik S, Brem H, Tomic-Canic M (2011) Cortisol synthesis in epidermis is induced by IL-1 and tissue injury. J Biol Chem 286:10265–10275
11. Vukelic S, Stojadinovic O, Pastar I, Vouthounis C, Krzyzanowska A, Das S, Samuels HH, Tomic-Canic M (2010) Farnesyl pyrophosphate inhibits epithelialization and wound healing through the glucocorticoid receptor. J Biol Chem 285:1980–1988
12. Sivamani RK, Pullar CE, Manabat-Hidalgo CG, Rocke DM, Carlsen RC, Greenhalgh DG, Isseroff RR (2009) Stress-mediated increases in systemic and local epinephrine impair skin wound healing: potential new indication for beta blockers. PLoS Med 6:e12
13. Chan KL, Zhang G, Tomic-Canic M, Stojadinovic O, Lee B, Flach CR, Mendelsohn R (2008) A coordinated approach to cutaneous wound healing: vibrational microscopy and molecular biology. J Cell Mol Med 12(5B): 2145–2154

Chapter 15

Murine Models of Human Wound Healing

Jerry S. Chen, Michael T. Longaker, and Geoffrey C. Gurtner

Abstract

In vivo wound healing experiments remain the most predictive models for studying human wound healing, allowing an accurate representation of the complete wound healing environment including various cell types, environmental cues, and paracrine interactions. Small animals are economical, easy to maintain, and allow researchers to take advantage of the numerous transgenic strains that have been developed to investigate the specific mechanisms involved in wound healing and regeneration. Here we describe three reproducible murine wound healing models that recapitulate the human wound healing process.

Key words Wound healing, Mouse model, Excisional wound, Ischemic wound, Pressure ulcer

1 Introduction

1.1 Splinted Excisional Wound

Chronic non-healing wounds are a morbid condition and place a huge financial burden on the health system. Non-healing ulcers in diabetics alone account for almost two-thirds of all non-traumatic amputations in the United States [1]. Excisional wound models are commonly employed to recapitulate this disease process in animal models. We have utilized a splinted excisional wound model which prevents wound margin contracture. Wound contracture by the panniculus carnosus is the primary mode of murine wound healing as opposed to granulation tissue formation and reepithelialization in humans [2, 3]. In addition to preventing wound contracture, the silicone splint provides a constant reference to use for calculating wound size when assessing wound closure. The wound bed can also be easily accessed to apply topical or subcutaneous agents to study wound healing modulation.

1.2 Ischemia Reperfusion Model

Pressure ulcers commonly affect the elderly population and patients who are debilitated due to spinal cord or traumatic brain injury and cause significant morbidity and even mortality. They are increasingly becoming a complication of acute hospitalizations as well with an 80 % increased incidence in the United States between

Robert G. Gourdie and Tereance A. Myers (eds.), *Wound Regeneration and Repair: Methods and Protocols*, Methods in Molecular Biology, vol. 1037, DOI 10.1007/978-1-62703-505-7_15, © Springer Science+Business Media New York 2013

1993 and 2006. The treatment of these wounds costs up to $11 billion per year in the United States [4]. The pathogenesis of pressure ulcers is thought to be mediated by cycles of ischemia followed by reperfusion injury. Wasserman et al. implemented a model using a metal disk implanted below the mouse gluteal muscle [5]. A magnet is placed on the skin to cause ischemia in the tissue between the magnet and metal disk and is cycled on and off. This model is able to reproduce up to a stage IV pressure ulcer with muscle necrosis. This approach however requires invasive methods and leaves a foreign material within the wound, both of which can confound wound characteristics. A simple noninvasive method was described by Stadler et al. which can achieve a stage III pressure ulcer [6]. Two ceramic magnetic disks are placed to "pinch" the dorsal skin and apply 50 mmHg of pressure which has been shown to decrease blood flow by 80 % [7]. This is followed by removal of the magnets to allow for a period of reperfusion injury. Wounds can be followed to analyze wound closure rate and tissue is easily harvested for histology, RNA, and protein analysis.

1.3 Ischemic Flap Model

Neovascularization is critical in the wound healing process. Recent studies have suggested that this process consists of both angiogenesis (sprouting of new vessels from existing endothelium) and vasculogenesis (creating of de novo vessels by progenitor cells). Neovascularization is dependent on numerous factors including ischemic signaling and mobilization and migration of progenitor cells. We have employed a three-sided full-thickness peninsular flap on the dorsum of the mouse with an impermeable silicone sheet beneath. This prevents neovascularization from below the wound leaving only the flap pedicle as a vascular source. A reproducible ischemic gradient is created within the flap allowing for analysis of differential ischemic signaling.

2 Materials

2.1 Splinted Excisional Wound

Fabrication of Silicone Splints

1. 0.5 mm thick silicone sheet.
2. 10 and 16 mm skin biopsy punches.
3. 70 % EtOH.
4. 50 mL polypropylene conical tube.

Excisional Wounding

1. Anesthesia.
2. Electric hair trimmer.
3. Depilatory cream.
4. Gauze.

5. 70 % EtOH.
6. 6 mm Skin biopsy punch.
7. Cyanoacrylate adhesive (super glue).
8. Forceps.
9. Scissors.
10. Needle driver.
11. 6-0 nylon suture.
12. Mastisol.
13. Occlusive dressing (Tegaderm).

Digital Analysis of Wound Closure

1. Digital camera.
2. ImageJ software.

Harvesting of Tissue for Analysis

1. Forceps.
2. Scissors.
3. Tissue capsule pads.
4. Biopsy cassette.
5. Cryovials.
6. Liquid nitrogen.

2.2 Ischemia Reperfusion Model

1. 12 mm diameter × 5 mm thick magnetic ceramic disk.
2. Anesthesia.
3. Electric hair trimmer.
4. Depilatory cream.
5. Gauze.
6. 70 % EtOH.

2.3 Ischemic Flap Model

1. 0.133 mm thick medical grade silicone sheet.
2. Anesthesia.
3. Electric hair trimmer.
4. Depilatory cream.
5. Gauze.
6. 70 % EtOH.
7. Template (10 mm × 25 mm).
8. Forceps.
9. #10 scalpel.
10. Microsurgical scissors.

11. Needle driver.
12. 6-0 nylon suture.
13. Phosphate-buffered saline.

3 Methods

3.1 Splinted Excisional Wound

Fabrication of Silicone Splints

1. Punch out the desired number of silicone splints using 16 mm biopsy punch (Fig. 1a). Each animal requires two splints.
2. Using the 10 mm biopsy punch, remove the inner diameter from each silicone disk to form a ring (Fig. 1b).
3. Place splints in 70 % EtOH in a 50 ml conical tube to keep disinfected until ready for use.

Excisional Wounding

1. Induce adequate anesthesia through either intraperitoneal or inhaled anesthesia.
2. Shave back of anesthetized mouse with electric hair trimmer. Be careful not to induce any trauma with razor teeth.
3. Apply depilatory cream to shaved skin and let sit for 2 min.
4. Wipe off cream and hair with gauze. Prepared skin should be completely bare.
5. Wipe the shaved back with 70 % EtOH.
6. Using a 6 mm biopsy punch, mark out the location of excisional wounds by pressing down gently and tracing with a fine marking pen. Bilateral wounds should be equidistant from the midline and spaced on either side of the dorsum (Fig. 1c).
7. Lift the skin away from the dorsum to incise and perform full-thickness excisional wound through the panniculus carnosus carefully with scissors (Fig. 1d).
8. Remove the skin flap by sharply dissecting the skin from the wound bed and cutting any connective tissue.
9. Apply a thin layer of glue fixative to one side of the silicone splint and affix the splint concentrically around the wound (Fig. 1e).
10. Secure each splint with eight evenly spaced interrupted stitches (Fig. 1f, *see* **Note 1**).
11. Apply Mastisol to intact skin surrounding the silicone splints and the splints themselves. Place occlusive dressing over wound to protect the splint from scratching or chewing by the animal.

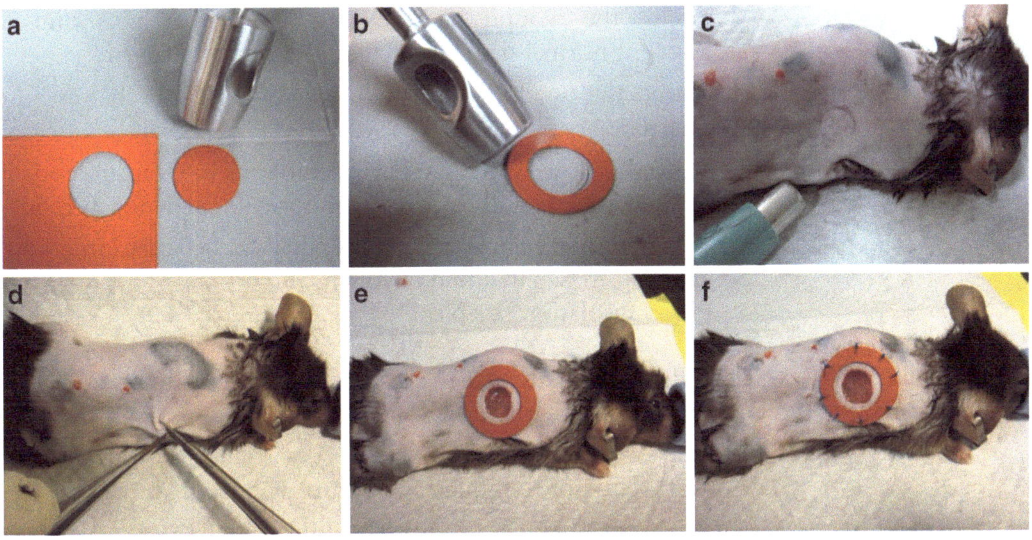

Fig. 1 Splinted excisional wound model. (**a**) The 16 mm punch is used to cut out the desired number of splints. (**b**) The 10 mm punch is then placed inside each silicone circle to remove the inner diameter and create the silicone splint. (**c**) A 6 mm punch is used to mark the location of the excisional wound. (**d**) Forceps retract the skin outward and scissors are used to sharply excise a circular piece of skin down through the panniculus carnosus. (**e**) The silicone splint is glued concentrically around the excisional wound. (**f**) Eight interrupted stitches reinforce the splint to the intact tissue surrounding the wound to prevent wound edge contracture

12. Return mouse into individual cage and observe to ensure recovery from anesthesia and surgery (*see* **Note 2**).

Digital Analysis of Wound Closure

1. Obtain digital pictures of the wounds on the day of surgery and every other day thereafter (*see* **Note 3**).
2. Open the image of each wound in ImageJ software.
3. Draw a line of the inner diameter using the straight measuring tool.
4. Set scale under the analyze menu in the toolbar by entering known distance as 16 mm.
5. Measure the area of each wound by tracing the perimeter of the wound using the freehand tracing tool and selecting measure under the analyze menu (*see* **Note 4**).
6. After obtaining all wound areas, calculate the size of the wound at each time point as a percentage of the wound size on day 1 (*see* **Note 5**).

Harvesting of Tissue for Analysis

1. Determine which animals will be sacrificed for each time point (*see* **Note 6**).
2. Euthanize animal by CO_2 and confirm by cervical dislocation (*see* **Note 7**).

3. Excise the wound bed keeping approximately 2 mm of adjacent normal tissue to include the wound margin.
4. Bisect the wound and then cut one of the halves into quarters.
5. Place the larger specimen between foam tissue capsule pads in a tissue cassette and fix in 4 % paraformaldehyde for fixation.
6. The two smaller quartered specimens can be stored in cryovials, snap-frozen in liquid nitrogen, and processed for RNA and protein isolation (*see* **Note 8**).

3.2 Ischemia Reperfusion Model

1. Induce adequate anesthesia through either intraperitoneal or inhaled anesthesia.
2. Shave back of anesthetized mouse with electric hair trimmer. Be careful not to induce any trauma with razor teeth.
3. Apply depilatory cream to shaved skin and let sit for 2 min.
4. Wipe off cream and hair with gauze. Prepared skin should be completely bare.
5. Wipe the shaved back with 70 % EtOH.
6. Draw two outlines of the magnets, each 5 mm from the midline (Fig. 2a, *see* **Note 9**).
7. Place the magnets on the designated areas so that a fold of skin is pinched between the magnets (Fig. 2b, *see* **Note 10**).
8. Return mouse to an individual cage.
9. Keep magnets on for 12 h of ischemia time.
10. After 12 h remove the magnets from each mouse (*see* **Note 11**).
11. Repeat 24-h ischemia–reperfusion cycle for a total of three cycles (*see* **Note 12**).
12. Monitor mice daily; ulcers will become visible after 3–4 days. Ulcers reach maximum depth at 10 days (Fig. 2c).

3.3 Ischemic Flap

1. Induce adequate anesthesia through either intraperitoneal or inhaled anesthesia.
2. Shave back of anesthetized mouse with electric hair trimmer. Be careful not to induce any trauma with razor teeth.
3. Apply depilatory cream to shaved skin and let sit for 2 min.
4. Wipe off cream and hair with gauze. Prepared skin should be completely bare.
5. Wipe the shaved back with 70 % EtOH.
6. Using a template, mark off a 10 mm × 25 mm longitudinal rectangle on the dorsal midline. The cranial edge of the rectangle should be at the level of the shoulder blades (Fig. 3a).
7. Use scalpel to incise the skin along the caudal edge.

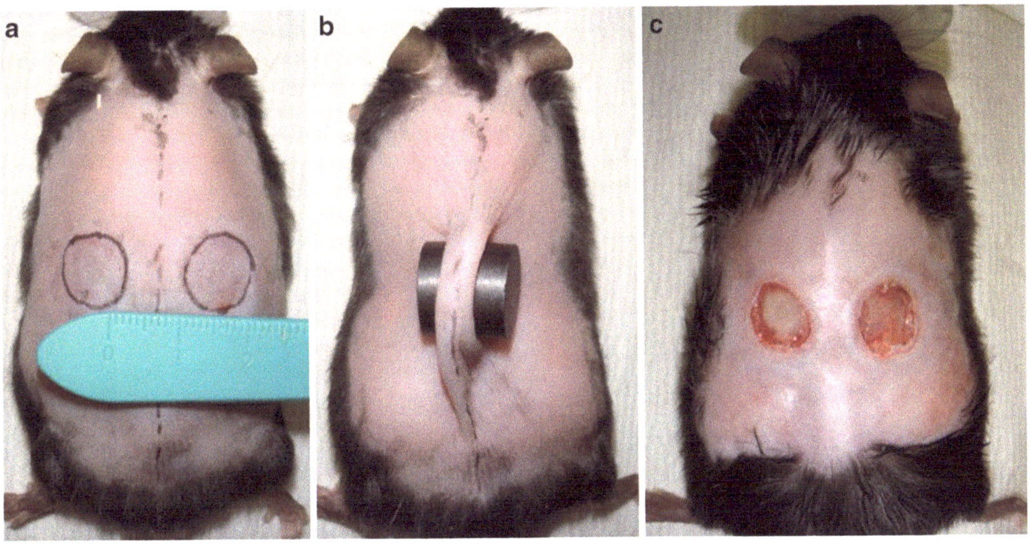

Fig. 2 Ischemia reperfusion model. (**a**) Use a template to mark the position of the magnets half way between the fore and hindlimbs. Note a 1 cm skin bridge between the magnet placement. (**b**) Pinch fold of skin and place magnets around skin fold. (**c**) Representative animal on day 5 displaying bilateral stage 3 ulcers

8. Use microsurgical scissors to extend the incisions along the caudal edge and along the longitudinal edges. The cranial edge will remain intact as the pedicle.
9. Raise a full-thickness peninsular flap by sharply dissecting any connective tissue between the skin flap and underlying muscle (Fig. 3b).
10. Place a 10 mm×25 mm silicone sheet within the wound between the muscle and skin (Fig. 3c, *see* **Note 13**).
11. Keep the wound bed moist by periodically wetting with PBS.
12. Suture the flap using interrupted 6-0 nylon stitches (Fig. 3d, *see* **Note 14**).
13. Return mouse to individual cage.
14. Monitor the mouse to ensure recovery from anesthesia and surgery.
15. Monitor wounds daily.

4 Notes

4.1 Splinted Excisional Wound

1. Sutures should be tied with minimal tension as suture material can be easily cut through the silicone rings if too tight. Splints may need to be resutured or replaced especially in wild-type mice as they tend to be more active and are able to bite at the splint.

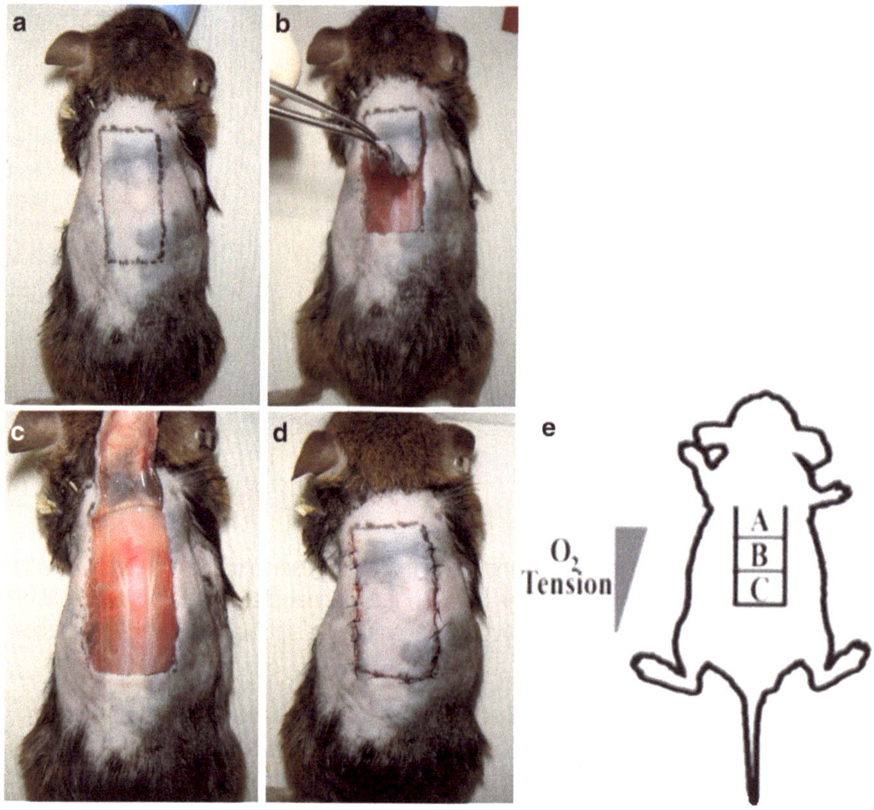

Fig. 3 Ischemic flap model. (**a**) Mark a 10 mm × 25 mm rectangle with the long dimension in the cranial–caudal axis. The cranial edge of the rectangle should be between the scapulae. (**b**) Raise a full-thickness flap leaving the cranial pedicle intact. (**c**) Place the impermeable silicone sheet flat on the wound and lay the flap down over the sheet. Note the superficial perforators that can be seen originating from the pedicle. (**d**) Suture the flap down using interrupted stitches. (**e**) Schematic of the O_2 gradient produced with the ischemic flap model

2. Daily monitoring should take place to both evaluate the wound and replace the dressing should it start to come off. Diabetic mice often are unable to reach the splint; however daily monitoring should still occur.
3. Individual cages are necessary as fighting may lead to damaged splints and other mice will chew through the silicone splints.
4. We obtain the best results using the macro function on most digital cameras and avoiding the use of flash. Care should be taken to make sure that images are obtained holding the camera in a plane parallel to the wound.
5. The scale must be set individually for each image as photographs are likely to be taken at various distances.
6. The wound area measured on day 1 is used as the original wound area as opposed to day 0 because the excisional wound tends to transiently widen immediately after surgery.

7. In our experience, usual time points to evaluate the acute phases of wound healing are days 0, 3, 5, 7, and 14. A minimum of three mice (six wounds) should be allocated to each time point to increase statistical validity.
8. Confirmation of death by cervical dislocation is necessary; however it should be performed carefully to prevent disruption of the wound which would affect histological results.

   In our laboratory, common histological assays include measuring the epithelial gap and immunostaining for CD31 to measure vessel density in each tissue sample. RNA and protein are evaluated for the presence of multiple vasculogenic genes including VEGF, EPO, SDF-1α, and HIF-1α. It is not within the scope of this chapter to detail the methods for these assays; however standard techniques are adequate to provide results.

4.2 Ischemia Reperfusion Model

9. Magnets should be placed equidistant from the forelegs and hindlegs to minimize tension on the skin bridge and reduce chances of skin necrosis. The original method by Stadler describes a 5 mm skin bridge. In our experience, especially with mice with impaired wound healing (diabetic, aged, etc.), this bridge is too narrow and leads to ischemia and skin necrosis. A 1 cm bridge provides adequate perfusion to the bridge.
10. The skin fold should include epidermis, dermis, subcutaneous fat, panniculus carnosus, and subcutaneous loose connective tissue layer (hypodermis), but not muscle.
11. The skin may remain creased after removing the magnets. The skin may need to be stretched back out in this case to insure proper reperfusion.
12. In our experience of using leptin receptor-deficient diabetic (*db/db*) mice, three 6-h ischemia/6-h reperfusion cycles provide adequate trauma to induce pressure ulcers.

4.3 Ischemic Flap

13. The oxygen tension gradient produced by this model has been validated with direct oxygen tension measurements. The cranial 1/3rd of the flap exhibits oxygen tensions of 22 mmHg proximally and 18 mmHg distally. The middle 1/3rd of the flap exhibits oxygen tensions of 18 mmHg proximally and 11 mmHg distally. The caudal 1/3rd of the flap exhibits oxygen tensions of 11 mmHg proximally and 4 mmHg distally.
14. Start by suturing down the two caudal corners. Eight stitches for each longitudinal edge and five stitches for the caudal edge are adequate for closure. The devascularized skin flap is very delicate and care should be taken to minimize manipulation. Always use a new needle for each animal and avoid handling the flap with forceps when possible.

References

1. Ramsey SD, Newton K, Blough D, McCulloch DK, Sandhu N, Reiber GE, Wagner EH (1999) Incidence, outcomes, and cost of foot ulcers in patients with diabetes. Diabetes Care 22:382–387
2. Davidson JM (1998) Animal models for wound repair. Arch Dermatol Res 290(Suppl):S1–S11
3. Galiano RD, Michaels J, Dobryansky M, Levine JP, Gurtner GC (2004) Quantitative and reproducible murine model of excisional wound healing. Wound Repair Regen 12:485–492
4. Reddy M, Gill SS, Rochon PA (2006) Preventing pressure ulcers: a systematic review. JAMA 296:974–984
5. Wassermann E, van Griensven M, Gstaltner K, Oehlinger W, Schrei K, Redl H (2009) A chronic pressure ulcer model in the nude mouse. Wound Repair Regen 17:480–484
6. Stadler I, Zhang RY, Oskoui P, Whittaker MS, Lanzafame RJ (2004) Development of a simple, noninvasive, clinically relevant model of pressure ulcers in the mouse. J Invest Surg 17:221–227
7. Peirce SM, Skalak TC, Rodeheaver GT (2000) Ischemia–reperfusion injury in chronic pressure ulcer formation: a skin model in the rat. Wound Repair Regen 8:68–76

Part IV

Cardiovascular Wound Healing Models

Chapter 16

A Corneal Scarring Model

Daniel J. Gibson and Gregory S. Schultz

Abstract

Corneal opacification (i.e., haze) following a non-denaturing acute injury to the cornea is a process which takes about 5 days to manifest itself, indicating that it is the consequence of cellular and molecular biological processes. In order to obtain a better understanding of the haze development process, and to test candidate anti-haze therapies, we use a corneal scarring model whereby we create an excimer laser wound in the center of rabbit corneas. The primary data generated by this model are (1) changes in corneal thickness with time; (2) wound closure rates, or re-epithelialization; (3) changes in the location and density of corneal sub-epithelial haze; and (4) molecular and histological changes leading up to, during, and following the formation of haze.

While the use of excimer lasers to generate consistent wounds in rabbit corneas is not a novel protocol for the study of corneal haze, the photographic technique presented here for the more objective recording and quantification of corneal haze is. At present, a qualitative, semiquantitative, grading system is employed whereby the amount of iris detail discernible through the scar is assigned a value between 0 and 4. Such a system makes direct comparisons amongst reported anti-haze trials nearly impossible. Furthermore, the additional "geographic" detail provided by the image provides a new layer of information about the formation of haze and the ability to troubleshoot dosing regimens. Altogether, with the information present herein, we believe that the study of corneal haze formation and the ability to compare and contrast candidate therapies are both greatly improved.

Key words Cornea, Haze, Fibrosis, Model, Excimer, Refractive surgery

1 Introduction

The center of any experiment is a measurement regimen, because it is the contrasting of the measurements made at various times and/or under a variety of conditions that gives rise to the new knowledge about the system being investigated. In studying the cornea's response to acute injury, or an intervention's capacity to improve healing outcomes, a measurement regimen which consists of ensuring the consistency of wounding, measuring the side effects of the intervention, and finally measuring both the macroscopic effects and molecular and cellular effects of the intervention is necessary to gain insight into the tolerability and efficacy of an intervention.

Robert G. Gourdie and Tereance A. Myers (eds.), *Wound Regeneration and Repair: Methods and Protocols*, Methods in Molecular Biology, vol. 1037, DOI 10.1007/978-1-62703-505-7_16, © Springer Science+Business Media New York 2013

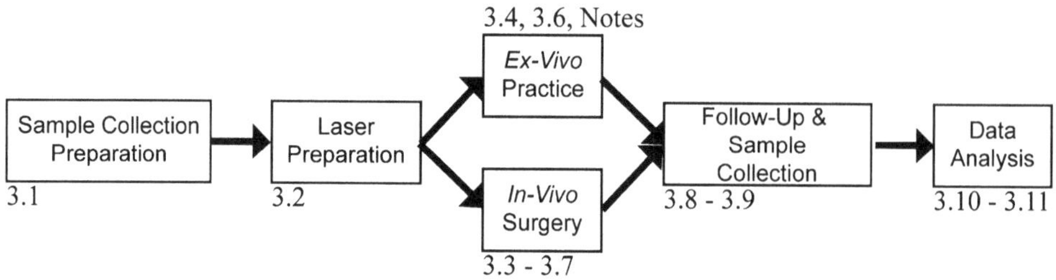

Fig. 1 General workflow schematic. Most of the steps will be the same for practice as they are for surgery. The divergence in the middle depicts how practicing with nonviable tissue (essential) can fit in with the rest of the workflow. The *numerals* refer to the sections that cover the procedures for that portion of the flowchart

This same regimen, sans intervention, can be used to measure the wound healing process per se, as well. While the measurements presented herein are presented as being done for each experiment, they need not be. At times, the measurements themselves, or the frequency of prolonged observations, can interfere with the experiment. In these cases, experiments which, for example, establish the intervention's side effects can be done separately from those done to establish the intervention's effects on healing.

The protocol described herein relies upon an excimer laser surgery model, which can model most wounds arising due to mechanical disruption of the tissue. It will not model thermal or chemical burns which opacify the cornea via a different, unrelated mechanism. Acute wounds to the cornea caused by laser surgery or due to surgical incisions or traumatic lacerations give rise to a light-reflecting scar which begins to be present about 5–7 days following surgery. The use of in vivo confocal microscopy has revealed that wound healing cells themselves are the light-reflecting entities which give rise to the corneal opacifying, lattice-like, "haze" [1, 2]. In contrast, chemical and thermal burns give rise to an immediate and homogenous opacification which is due to denaturation and modification of the stromal extracellular matrix.

While the excimer laser presented here is the most reproducible means of creating an acute wound in the cornea, two alternative techniques exist, which while less reproducible are both more affordable than an excimer laser and have in our experience been approved as IACUC equivalents (e.g., pain, suffering, and analgesia). These alternatives will only be mentioned in passing with a brief explanation of the known benefits and shortcomings of the techniques. Citations to publications making use of these techniques will be provided for those who are interested in employing these more economical alternatives. Aside from the surgery itself, the remaining workflow and measurements made are both applicable for all three wounding techniques.

The general protocol schematized in Fig. 1 consists of preparing for sample collection, preparing the laser, anesthetizing the model animal, making initial pre-surgical measurements, performing the surgery, making postsurgical measurements, applying the experimental treatment (if any), making posttreatment measurements, making follow-up measurements, euthanizing the animals at prescribed time points, and harvesting the cornea for a variety of downstream molecular biological and histological analyses. The final steps include data analysis of re-epithelialization, edema, and corneal haze. Analysis of molecular biological and histological data varies from analyte to analyte and is left with the investigator to determine which methods are best for his or her investigation. Overall, this protocol is presented as it is done in vivo. However, notes on how to obtain surgical practice with ex vivo enucleated globes (rabbits) are also given. We cannot stress enough the importance of practicing this protocol, from start to finish, with ex vivo tissue prior to beginning in vivo experimentation.

The presentation of this protocol will be primarily centered on the measurement regimen, and it will be described as it is done using rabbits. We also provide side notes with non-comprehensive insights into the use of mice and rats. With this protocol, an individual or a group can become acquainted with, and/or adept at, the study of corneal wound healing.

2 Materials

2.1 Equipment

2.1.1 Presurgery

1. A corneal ultrasonic pachymeter.
2. Animal restraint.
3. Certified calibrated isoflurane evaporator.
4. A fresh isoflurane scavenger.
5. Nose cone suitable for the animal model used.
6. Eyelid speculum.
7. Eye proptoser: Two wooden shafted swabs and a twist-tie or a metal wire of similar malleability (Fig. 2).
8. Digital SLR camera.
9. Dedicated macrophotography lens capable of 1:1 or greater reproduction (*see* **Note 1**).
10. Dedicated macrophotography flash.
11. A refractive surgery excimer laser (*see* **Notes 2** and **3**).

2.1.2 Tissue Excision

1. Corneal button punch block.
2. 8.0 mm Biopsy punches (*see* **Note 4**).

2.2 Drugs

1. Isoflurane.
2. Ophthalmic tetracaine or proparacaine eye drops.
3. Meloxicam, oral suspension.

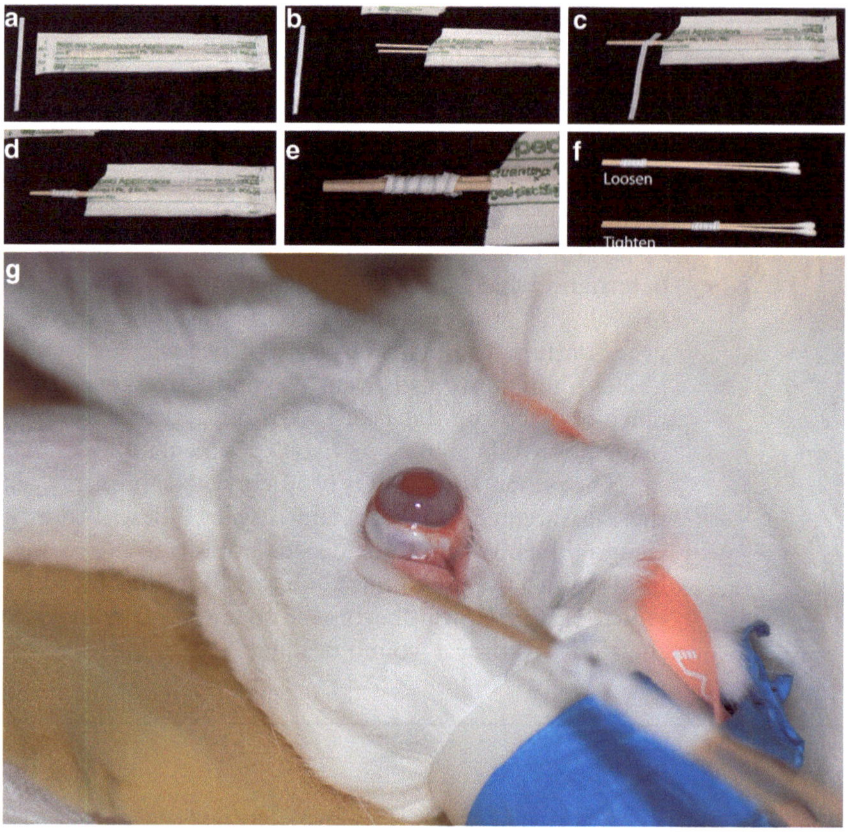

Fig. 2 Constructing a disposable rabbit eye proptoser. (**a**) One-half of a twist-tie and a pack with two sterile or clean wooden-shafted cotton swabs. (**b**) Open the end of the swab package exposing the non-swab ends. (**c**) Begin wrapping the twist-tie around the butt of the swabs. (**d**) The finished swab wrapping, and (**e**) detail of the finished swab wrapping. (**f**) The twist-tie wrapping should be able to slide up and down the shaft. Slide the twist-tie towards the swabs to tighten the pair once in place below the equator of the proptosed eye. (**g**) The proptoser in place on an anesthetized rabbit. The swabs should not be reused, but the twist-tie can. Slide the twist-tie off of the butt and can be slided onto a fresh pair of swabs

4. Buprenorphine, injectable.
5. Tropicamide.
6. Phenylephrine.

2.3 Reagents

1. Phosphate-buffered saline (PBS): 137 mM NaCl, 2.7 mM KCl, 10 mM $Na_2HPO_4{\cdot}2\ H_2O$, 1.76 mM KH_2PO_4, pH 7.4.
2. Sodium fluorescein solution: PBS saturated with sodium fluorescein, pH 7.4.
3. 4 % Paraformaldehyde in PBS ("fixative"): One volume of 39 % paraformaldehyde, nine volumes of PBS.
4. RNA later.
5. 70 % Ethanol: Three parts dH_2O, seven parts 100 % ethanol.
6. Antihistamine (*see* **Note 5**).

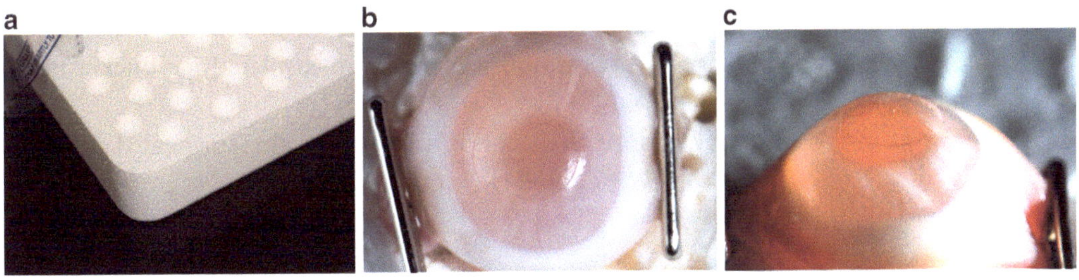

Fig. 3 Immobilizing the ex vivo globe for practice surgery. (**a**) A polystyrene 15 ml tube tray. (**b**) An ex vivo rabbit eye immobilized on the polystyrene tray using two dissection "t"-pins penetrating the residual conjunctiva and muscle tissue. Note that the ablated region is not centered on the pupil; this is why we practice. (**c**) Another ablated eye viewed side-on to see the detail of the sharp wound border provided by PTK

2.4 Biologics and Tissue

1. Frozen ex vivo rabbit eyes from an abattoir.
2. Alternative: Rat or mouse cadavers from colleagues.
   (a) There is a constant stream of mice without the appropriate genotype which are euthanized and typically suitable for practice once euthanized.

2.5 Analysis

1. Image analysis software.
2. Data analysis software.

2.6 Consumables and Miscellaneous

1. PPE.
2. Scalpels.
3. Tubes.
4. Empty 15 ml tube, polystyrene packaging (Fig. 3a).

3 Methods

In order to ensure the competence of the individual performing the surgery, practice with ex vivo globes is absolutely essential. Using whole rabbit eyes obtained from an abattoir and a mounting system depicted in Fig. 3, this protocol can and should be practiced. By performing this protocol with ex vivo tissues, the investigator can be sure to gain the experience necessary to conduct most of the essential steps of the actual surgery prior to operating on live rabbits, and thereby dramatically reduce the chance of costly errors.

3.1 Sample Collection Preparation

1. Prepare vessels and buffer for fixing or extracting samples depending upon the planned experiment.
2. Pre-label each vessel for each sample.
   (a) Make several extra, unlabeled vessels just in case.
3. Add the appropriate amount of fixative or extraction buffer, if any, for each sample.

4. Prechill the vessels at 4 °C if possible.
5. Place the vessels on wet ice in an insulated ice bucket in the surgical suite.
6. Make more than enough copies of the surgical data sheets (*see* **Note 6**, Fig. 4a).

a

Rabbit/Rat/Mouse#:________ ExpDay#:________ Date/Time:________

Treatment

Right	**Left**
________ BeamDiameter(mm)	________ BeamDiameter(mm)
________ TxDeliverable(μl)	________ TxDeliverable(μl)

WoundMeasurements

Pachymetry

	Measurement	Reading 1	2	3	4	5
LEFT	Pre-Ablation					
	Post-Ablation					
	Post-Treatment					
	Misc					
	Misc2					
RIGHT	Measurement					
	Pre-Ablation					
	Post-Ablation					
	Post-Treatment					
	Misc					
	Misc2					

MacroImages

FrameNumbers

	LEFT	RIGHT
Initial		
PostWound Post		
Tx Fluroscein		
Stained		

Notes:

Fig. 4 Data collection templates to facilitate organized data collection. (**a**) Surgical template and (**b**) follow-up template

b

Rabbit/Rat/Mouse#: ____________ ExpDay#: ________ Date/Time: ____________

FollowupScreening

	Right Y	Right N	Left Y	Left N
Photosensitive?	☐	☐	☐	☐
Weaping?	☐	☐	☐	☐
HazeDetected?	☐	☐	☐	☐

	Right Y	Right N	Left Y	Left N
Inflammed?	☐	☐	☐	☐
WoundClosed?	☐	☐	☐	☐

Lesion(mmxmm): Left: ________ Right: ________

Pachymetry

Eye	Reading 1	2	3	4	5
Left					
Right					

Haze

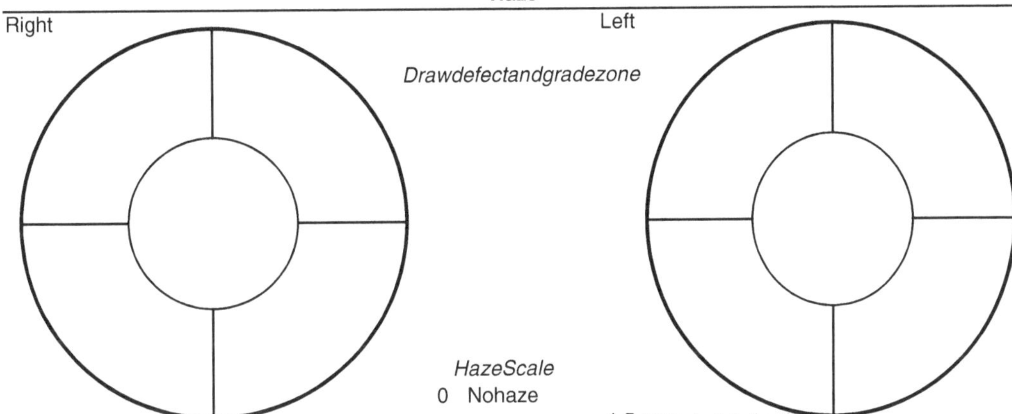

HazeScale

0 Nohaze

1.5 Mild, butdistincthaze

1 Well-defineddiffusehaze

2 Irisdetailedobscured

3 Anteriorchamberobscured

MacroImages

FrameNumbers

	Right	Left
HazeGrading		
Re-Epithelialization		
Misc.		

AdditionalNotes: ____________________________

Fig. 4 (continued)

7. Bring a vessel with enough 70 % ethanol to clean forceps, scissors, etc. in between each sample (*see* **Notes** 7 and **8**).
8. Ensure that the ultrasonic pachymeter is in the surgical suite and ready to use.
9. Ensure that the digital camera, flash, and filters are in the surgical suite and ready to use.
10. Bring the sodium fluorescein solution to the surgical suite.

3.2 Laser Preparation

1. Prepare and calibrate the excimer laser for use according to the manufacturer's protocol.
2. Set the laser to phototherapeutic keratectomy (PTK) mode (*see* **Note 9**).
3. For rabbits, set the beam diameter at 6.0 mm with no transition zone (*see* **Notes 10** and **11**).
4. We do not pre-remove the epithelium with a scalpel, and as such we set the ablation depth to at least 125 μm, with approximately 50 μm being the epithelium (*see* **Note 12**).
5. The laser should be ready, but not yet armed.

3.3 In Vivo Pre-surgical Preparation

1. Restrain the animal.
2. Anesthetize the corneas with a single drop of tetracaine (preferred) or proparacaine.
3. Place the anesthetic nose cone on the rabbit.
4. Induce general anesthesia by setting the oxygen flow rate to 1.0 L/min and the isoflurane to 3.5–4.0 %.
5. Once induced, the isoflurane can be reduced to 3.25–3.5 %.
6. Remove the rabbit from the restraint and place it on its side.
7. Proptose the eye using the instrument depicted in Fig. 2 (*see* **Note 13**).

3.4 Pre-surgical Measurements

The measurements made before, during, and after surgery should be the same for both the ex vivo practice and in vivo experiments.

3.4.1 Photograph the Cornea (See Note 14)

1. The following camera settings will be used for haze measurement (Fig. 5a and Table 1).
2. If the camera has an automatic ISO mode, disable it and set the ISO to the lowest setting (typically ISO 100 or 200).
3. Set the camera's color setting to "Normal" or "Standard" (*see* **Note 15**).
4. Set the exposure compensation to "0."
5. Set the exposure mode on the camera to "Manual."
6. Set the exposure mode on the flash to "Manual."
7. Set the camera's shutter speed to at least 1/250th of a second (i.e., "250").
8. Set the aperture to f/16 or f/18.
9. Set the flash to 1/8th power. If there is more than one flash, ensure that they are all set to the same power.
10. Prefocus the lens to the closest focusing distance (i.e., highest magnification).
11. Center the cornea in the center of the view finder and move slowly back and forth until the cornea is in focus.

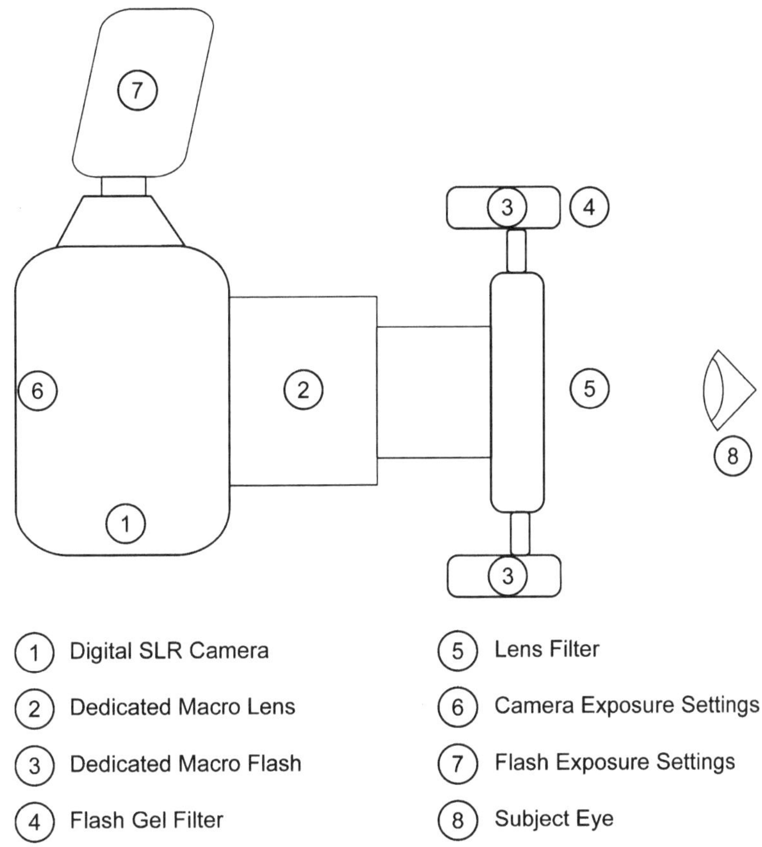

Fig. 5 Camera components for macrophotography-based data collection

Table 1
General camera/flash settings for three common imaging situations

Type of image	Camera/flash exposure	Camera/flash filters
General image	Program auto/auto	n/a
Haze measurement	Manual/manual	n/a
Re-epithelialization	Program auto/auto	Cobalt blue/deep yellow

12. Take the picture.
13. Check the image on the camera to see how the exposure settings worked, whether the focus was good, and whether the flash reflection overlapped with the central region of interest.
14. Check the histogram (if available) for saturation.
15. If the image is underexposed (too dark), increase the flash power to 1/6th.
16. If the image is overexposed (too light), decrease the flash power to 1/10th.

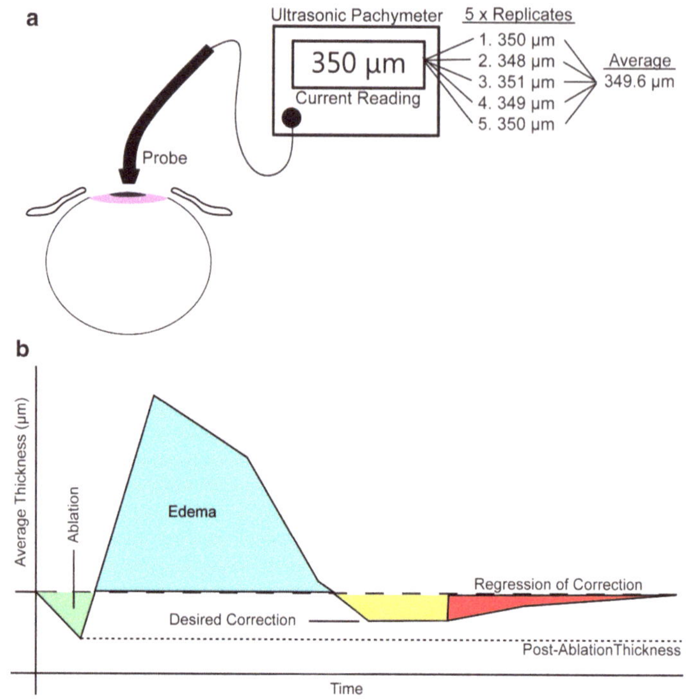

Fig. 6 Pachymetry. (**a**) Schematic of measuring central thickness. We typically use five replicate measurements each time the corneal thickness is measured. (**b**) Fluctuations in corneal thickness with time and what they correlate to. The *dashed line* represents the initial thickness of the cornea. The thickness decreases due to the excimer-based tissue removal. The cornea then swells (edema). Once the cornea is re-epithelialized, the edema is gone, but the epithelium is once again contributing to the total corneal thickness, thus preventing a return to the post-ablation thickness. To varying extents, the ablated stroma regenerates itself and can result in a return to the pre-ablation refractive power (regression)

17. Retake the image.
18. Recheck the image.
19. Readjust the flash power if necessary.
20. Once a good setting is found, use it for all further haze-related images.
21. Once a good image is acquired, record all of the frame numbers (good and bad) of the images taken on the surgical sheet.

3.4.2 Corneal Thickness

1. Measure the cornea's central thickness via ultrasonic pachymetry by lightly touching the central cornea with the probe (Fig. 6a).
2. Perform five replicate measurements (*see* **Note 16**).
3. Record the pachymetry measurements on the surgical sheet.

3.5 Surgery

1. While we do not manually debride the epithelium, if you choose to do so, now would be the time to do so (*see* **Note 17**).
2. Center the eye beneath the laser.
3. Focus the laser ensuring that you are focusing on the corneal surface and not some structure within the eye.
4. Arm the laser.
5. Perform the ablation, and periodically stop and refocus as necessary.

3.6 Postsurgical Measurements

1. Immediately after ablation, remeasure the thickness via pachymetry with five replicate measurements.
2. Record the measurements on the surgical sheet.
3. Rephotograph the cornea as done before.
4. Record the frame numbers on the surgical sheet.
5. Place the blue filters on the flash and the deep yellow filter on the lens of the camera (Table 1).
6. Apply fluorescein solution to the cornea.
7. Rinse away excess fluorescein with saline.
8. Change both the camera in the programmed auto mode ("P") and the flash into auto exposure mode.
9. Immediately photograph the fluorescein-stained wound.
10. Remove the proptoser.

For experiments testing the efficacy of an intervention, we suggest that both eyes of a rabbit be used in paired analysis with one eye receiving the intervention and the other either placebo or no intervention. This suggestion is due to the observation that the extent of scarring appears to be more consistent between the eyes of a given rabbit than amongst the eyes of all rabbits. If this route is chosen, at this point the rabbit can be flipped over and the preceding steps repeated on the opposing eye.

3.7 Postsurgical Wrap-Up

1. Turn off the isoflurane, but allow the oxygen to continue flowing.
2. Periodically pet/stir the rabbit.
3. Once it demonstrates some degree of coming up from the anesthesia, attempt to administer the oral suspension of meloxicam.
4. Once the meloxicam is imbibed, replace the rabbit in its transport container and/or its cage.

3.8 Follow-Up

In the follow-up period, the rabbits will likely require analgesia. We prefer meloxicam due to its 24-h effect. The rabbits will likely require analgesia until the cornea is completely re-epithelialized. Additional analgesia might be required if the cornea spontaneously ulcerates (*see* **Note 18**).

3.8.1 Follow-Up Measurements

1. Make more than enough copies of the follow-up data sheets (Fig. 4b).
2. Prior to restraint, dilate the pupils of all animals undergoing measurement with a single drop each of tropicamide and phenylephrine in each eye.
3. Once the pupils are dilated, restrain an animal.
4. Anesthetize the corneas with a single drop of tetracaine (preferred) or proparacaine.
5. Place the anesthetic nose cone on the rabbit.
6. Induce general anesthesia by setting the oxygen flow rate to 1.0 m^3/min and the isoflurane to 3.5–4.0 %.
7. Once induced, the isoflurane can be reduced to 3.25–3.5 %.
8. Remove the rabbit from the restraint and place it on its side.
9. Proptose the eye using the instrument depicted in Fig. 2.
10. Set the camera and flash to the previously determined manual settings.
11. Photograph the center of the cornea.
12. Record the frame numbers on the follow-up sheet.
13. Measure the central cornea's thickness via pachymetry.
14. Record values on the follow-up sheet.
15. Prepare the camera for fluorescein photography as before.
16. Rinse the cornea with fluorescein solution.
17. Rinse away excess fluorescein with PBS or BSS.
18. Photograph the fluorescein-stained cornea as before.
19. Record the frame numbers on the follow-up sheet.
20. Flip the rabbit over and repeat and record the measurements for the other eye.

3.9 Terminal Sample Collection

The final stage is the harvesting of tissue for either histological or molecular analysis. The time points and particular molecular analyses performed will depend upon the investigator's aims. Here, we describe the tissue excision and process techniques that we use for a variety of analyses. Two distinct protocols are presented for either molecular (ELISA, Western blot, etc.) or histological (gross histology, immunohistochemistry, etc.) analysis which primarily differ in that the tissues will either be homogenized or preserved. For either process, be sure to euthanize the rabbit using a humane and IACUC-approved method prior to tissue excision.

3.10 Molecular Analysis

This method is suitable for both protein and nucleic acid analysis with the primary point of variation being the composition of the extraction buffer. We tend to wait until all tissues are collected prior to further tissue homogenization and extraction.

1. Excise the cornea with a stromal rim attached for handling.
2. Place the excised cornea face down in the punch block.
3. Using a new biopsy punch, punch out a portion of the cornea centered on the wound. Absolute precision is not necessary, but the biopsy cut should not intersect with the wound.
4. Transfer the corneal button to an empty labeled 1.5 ml microcentrifuge tube.

3.10.1 RNA Analysis

1. If RNA is going to be analyzed, add a sufficient volume of an RNA-stabilizing buffer (e.g., RNA later) to the tube.
2. Store the tube at −20 °C.
3. Wait until all samples are collected prior to proceeding.

3.10.2 Protein Analysis

1. Close the tube lid, and then pierce the center of the lid with an 18G needle (*see* **Note 19**).
2. Immediately snap freeze the tissue in liquid nitrogen.
3. Leave the tube in the liquid nitrogen until it can be transferred to a −80 °C freezer or a liquid nitrogen storage vessel.
4. Once all of the corneas for all of the time points have been collected, then in a single batch process and homogenize the samples.

3.10.3 Tissue Homogenization

1. Transport all of the samples to a suitable workspace, maintaining them at the temperature they were stored at.
2. Take one sample at a time, and rapidly thaw the sample if necessary with a 37 °C water bath.
3. Once thawed, place the button on a clean cutting block.
4. With a fresh razor blade, dice the cornea.
5. Transfer the diced corneal material back into the storage tube and add extraction buffer (*see* **Note 20**).
6. With a clean fresh microcentrifuge tube dounce, sheer the corneal materials in the centrifuge tube, occasionally repositioning the dounce to ensure that all materials have been sufficiently sheered.
7. Carefully remove the dounce, and scrape any attached material on the edge of the tube so that it remains in the tube.
8. Vortex the sample to ensure that all materials are submerged in the extraction buffer.
9. Place the tube in wet ice.
10. Repeat **steps 1–9** until all samples are processed before proceeding.
11. Prepare an ~300 ml bath of iced saline in a plastic vessel.
12. Bring the samples and bath to a probe ultrasonicator.

13. Ensure that the probe is clean.
14. With the base of the tube well submerged in the iced saline, and the probe well submerged in the sample, ultrasonicate the sample with five pulses for about 5 s each (*see* **Note 21**).
15. Place the sample back in wet ice.
16. Clean the probe.
17. Repeat **steps 13–16** until all samples are processed before proceeding.
18. Pellet the insoluble materials in a microcentrifuge.
19. Transfer the supernatant to a new labeled tube.
20. We retain all materials until the experiment is completely finished.
21. Snap freeze all materials, and store at −80 °C in liquid nitrogen until needed.

3.10.4 Histology

1. Bathe the corneal surface with fixative prior to excision.
2. With a sharp pointed scalpel, push the point into the sclera near the cornea (~1/8" away from the cornea, *see* **Note 22**).
3. While withdrawing the scalpel, apply slight pressure in the cutting direction to further open the initial cut.
4. Using curved fine-serrated scissors, with the curve matching the curve of the corneal/scleral interface, continue expanding the incision (*see* **Note 23**).
5. A pair of toothed forceps can be used to provide additional traction or to control globe position if needed.
6. Fill a well in a 12-well cell culture dish with fixative.
7. Place the cornea face down into the fixative and fill the posterior corneal "cup" with fixative.
8. Repeat **steps 1–7** for all eyes before proceeding.
9. Move the 12-well plate(s) into a well-vented hood.
10. Fix the corneas like this for about 2 h.
11. Transfer the corneas into individually labeled vessels with fresh fixative and continue the fixation for 18–24 h at 4 °C.
12. Decant the fixative, and rinse the cornea and vessel with 70 % ethanol.
13. Add enough 70 % ethanol to completely cover the cornea.
14. The corneas can be stored like this until ready to proceed for further histological processing.

3.11 Data Analysis

3.11.1 Pachymetry

In all cases, the amount of edema or tissue replacement is calculated daily by the difference between the corneal thickness on that day and the immediate post-wounding thickness.

1. Transcribe the pachymetry measurements from the surgery and follow-up sheets into your data analysis program of choice.
2. Average the five replicate measurements for each measurement made.
3. Use only the averaged values from here on.
4. Calculate the percent difference from the ablated thickness for each measurement:

$$\%\ \text{Difference} = \frac{\text{Current thickness} - \text{ablated thickness}}{\text{Ablated thickness}} \quad (1)$$

5. Perform the appropriate statistical test for the experiment you designed.
6. For the paired analysis we suggest, perform a paired Student's *t*-test, pairing the opposing eyes of a given rabbit, for each time point.
7. Choose a two-tailed analysis if there is not an anticipated effect, or a one-tail analysis if a particular effect is expected a priori.
8. Under normal circumstances, differences in thicknesses will represent differences in edema during re-epithelialization, and tissue replacement for time points thereafter (*see* Fig. 6b and **Note 24**).

3.11.2 Re-epithelialization

Re-epithelialization can be measured in two ways: "clinically" (Fig. 7a) and "technically" (Fig. 7b). We consider the first day that an eye does not stain with fluorescein to be the "clinical" re-epithelialization rate, while we use digital photogrammetry to measure the actual fluorescein-positive area to calculate a "technical" rate of closure in terms of a change in area with respect to time. Given that both can be done with the same data, and the simplicity of both metrics, we choose to report both. We choose this because the technical rate gives insight into the molecular biology of wound healing, while the clinical rate conveys whether or not the difference is clinically relevant.

1. Open the image in a program capable of photographic measurement (Fig. 7b).
2. Choose the selection tool which selects regions of contiguous color (e.g., Photoshop's "Quick Selection Tool").
3. Select the fluorescein-stained region in the image.
4. Record the measurement (see Note 25).
5. Repeat steps 1–4 for all images prior to proceeding.
6. Export the measurements to your data analysis program of choice.
7. Convert the area as measured in pixels into spatial terms (i.e., mm2) using the value determined for your camera (see Notes 26 and 27).

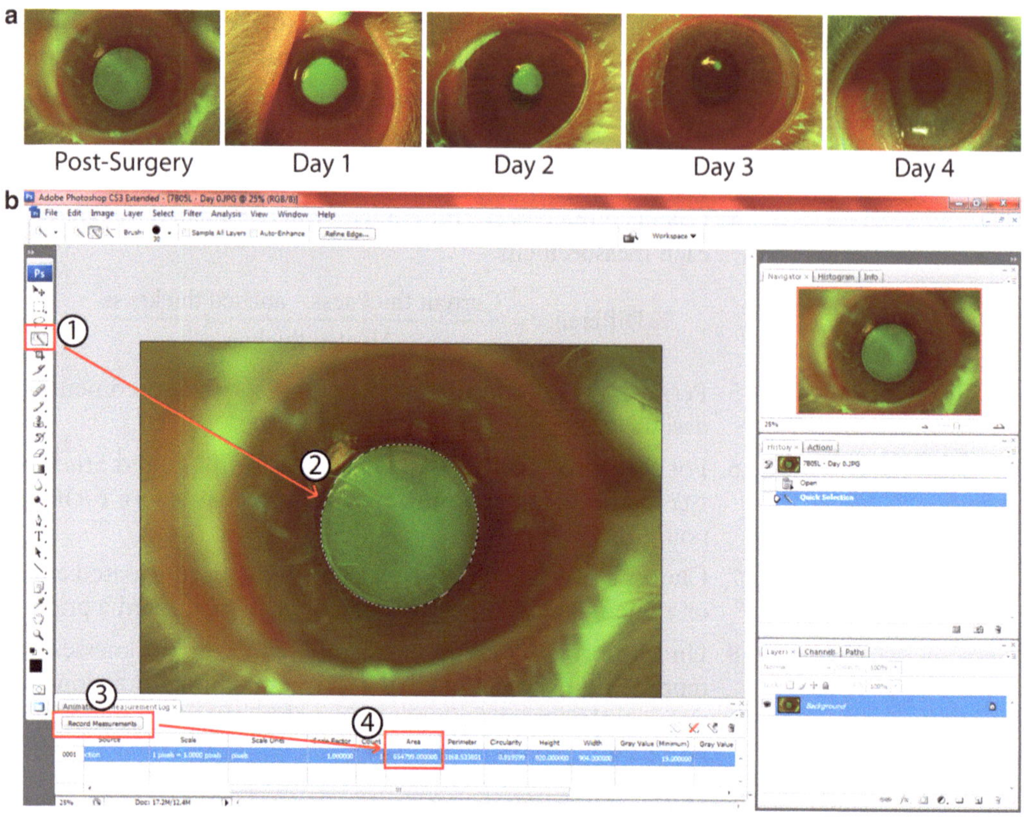

Fig. 7 (**a**) An example wound closure image series. (**b**) How to quantify wound area using a fluorescent macrophotograph. (**b**1–**b**2) Using Photoshop CS3 Extended as an example, use the "Quick Selection Tool" which quickly selects continuous regions of similar colors. (**b**3) Push the "Record Measurements" button. (**b**4) Scroll the table over to obtain the total number of pixels selected. Depending on the reproduction ratio (magnification) and the camera used (pixel dimensions), this number can be used to determine the actual wound area. We strongly suggest using a dedicated macro lens capable of 1:1 reproduction. Doing so enables the investigator to easily quantify the area using the manufacturer's published pixel dimensions to calculate the area per pixel which can then be used to convert pixels into spatial area units (e.g., mm^2)

8. Perform the appropriate statistical test for the experiment you designed.
9. For the paired analysis we suggest, perform a paired Student's t-test, pairing the opposing eyes of a given rabbit, for each time point.
10. Choose a two-tailed analysis if there is not an anticipated effect, or a one-tail analysis if a particular effect is expected a priori.
11. For the clinical analysis, perform the same paired analysis using the day number that the cornea had no apparent fluorescein staining.

3.11.3 Haze

The first set of images will be created with an unwounded eye and a piece of filter paper, which will serve as a pseudo-wound. These initial images will be used to create a filter which will be

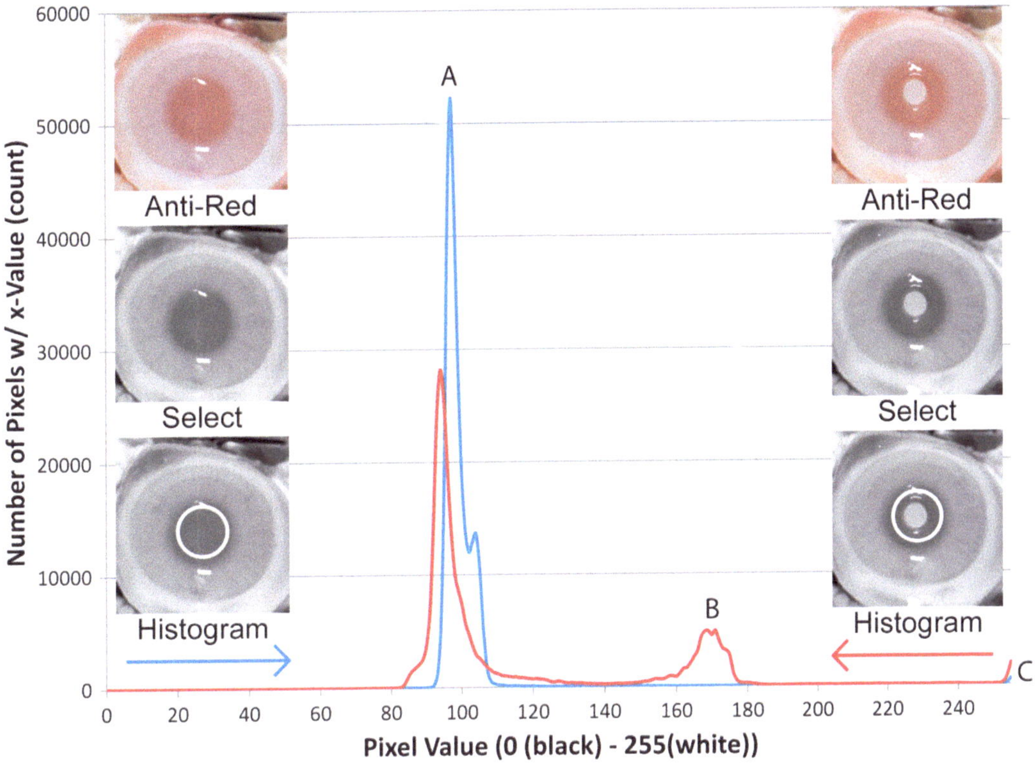

Fig. 8 Haze analysis using an ex vivo model which employs a 4 mm circle of filter paper to mimic the worst scar possible. (**a**) Background, (**b**) haze (filter paper in this example), and (**c**) flash reflections. By choosing a cutoff value between (**a**, **b**) and (**b**, **c**), the investigator can create a "band-pass" filter which prevents reflections from the camera and flash from affecting the haze score

applied to all further images of corneal haze in order to remove the background and any reflections (i.e., the flash) on the corneas (Fig. 8).

1. Determine the band pass
   (a) Dilate the pupils of a rabbit.
   (b) Wait for dilation.
   (c) Anesthetize a rabbit as before.
   (d) Proptose as before.
   (e) Photograph the eye for haze (i.e., manual settings).
   (f) Place a 4–6 mm biopsy-punched filter paper in the center of the eye.
   (g) Photograph again.
   (h) Restore the rabbit as before.
2. Convert all of the images to grayscale using the data only in the blue channel.
3. Open the grayscale images from the +/− filter paper rabbit.
4. Manually create a circular selection by typing in the dimensions.

5. A selection region should remain within the margins of the dilated pupil of the plain eye, and it should encompass the filter paper in the pseudo-wounded eye.
6. Save the dimensions of the selection circle as they will be used for all samples.
7. Record the measurement with histogram.
8. Use the same selection circle size to select the filter paper pseudo-wound.
9. Import the histograms into a charting utility.
10. Compare the histograms to identify the pixel intensities which correlate with the background (Fig. 7a), the wound (Fig. 7b), and the flash reflections (Fig. 7c).
11. The pixel values between the background and wound (e.g., <80 in Fig. 7) and the wound and flash reflections (e.g., >168 in Fig. 7) will serve as a "band-pass" filter for quantifying haze (*see* **Note 28**).
12. Select and record the values and histograms for the remaining experiment wound images.
13. Import the experiment histograms into a data analysis program (e.g., Microsoft Excel).
14. For the pixel values within the band pass, multiply the pixel value with the number of pixels measured with that value (i.e., the histogram data).
15. Sum all of these values within the band pass to obtain an unweighted haze score (*see* **Note 29**).
16. Perform the appropriate statistical test on the haze scores for the experiment you designed.
17. For the paired analysis we suggest, perform a paired Student's *t*-test, pairing the haze scores from opposing eyes of a given rabbit, for each time point.
18. Choose a two-tailed analysis if there is not an anticipated effect, or a one-tail analysis if a particular effect is expected a priori.

3.11.4 Molecular and Histological Analysis

Both molecular and histological analysis will vary greatly depending upon the experimental design and the analyte(s) of interest. The only suggestion we will make is to continue to perform paired analysis using opposing eyes.

4 Notes

1. If a dedicated macro lens is not available, or the reproduction ratio is not obtainable, be sure to include a ruler or an object of known length within the image to empirically determine the pixel-to-area conversion value.

2. A week prior to in vivo experimentation, ensure that all laser supplies (i.e., gasses) are adequate for the experiment and that the laser is in good working order. Performing a practice surgery on an ex vivo globe is a good way to do this, but performing a laser calibration which typically entails firing the laser should suffice. All of this will enable ordering additional supplies (i.e., gasses) if necessary.
3. Alternatives to the excimer laser include lamellar keratectomy and corneal abrasion with a diamond-dusted burr [3–14]. Considering that all three techniques are done on an outpatient basis with similar level of analgesia, we have successfully obtained IACUC approval to use these methods interchangeably with the lower technology techniques as backup methods in the event of laser equipment problems.
4. The diameter should exceed the wound size.
5. If anyone who is going to be present for the surgery has allergies that might be triggered by rabbit hair or dander, be sure that they take any medications they need well in advance of the surgery. They may also want to consider acquiring a robust respirator mask.
6. Data collection templates: I cannot stress the value of these enough. Make them, have them, and use them.
7. Since the sample collection is being done either on nonviable tissue or following euthanasia for the in vivo experiments, the instruments do not need to be sterilized unless the tissues are collected for tissue/organ culturing.
8. Use a fresh scalpel for each tissue in all cases to reduce cross-contamination and to ensure the best cut possible.
9. We choose PTK due to the nice sharp stromal wound margin it creates which provides a sharp contrast between the wound and surrounding tissue.
10. Some lasers request three different diameters which together enable one to control the shape of the laser cut profile. Choose the diameters such that the tissue volume removed is a cylinder. For the Nidek EC-5000, all three diameters should be 6.0 mm.
11. For mice use a 1.0 mm beam diameter and for rats use 4.0 mm.
12. If you choose to manually debride the epithelium, adjust the ablation depth accordingly.
13. Alternatively, an eyelid speculum can be used, but we have better results with this custom proptoser.
14. Practicing the camera placement and focus is important. Focusing on a viable cornea is much like focusing on a fingerprint left on a glass surface; accordingly such can be used for focusing and camera handling practice. The day prior to surgery,

be sure that the batteries are charged and extra batteries (for the flash) are readily available. Also, be sure that the memory card has sufficient space for the experiment (2–4 GB cards are fine for most, but will be dependent upon the camera).

15. Cameras typically have a variety of settings which affect the color and tone curves which are used for a variety of photography situations. Examples include a mode with high color saturation for nature or flower photography, or mode with more muted colors for photographing people. For photographing corneas, we typically use the "Normal" or "Standard" (or others; the name is manufacturer dependent) color mode.
16. If the corneal surface is too dry, the pachymeter will not be able to measure the corneal thickness. Slightly moisten the tip of the probe or the corneal surface to remedy this.
17. If you manually debride the epithelium, repeat the photography and pachymetry steps to measure differences due to the debridement.
18. Within the context of this model, we have found that frequent observation of the cornea can lead to spontaneous sloughing of the yet-to-be stabilized epithelium. This problem might be addressable with anti-drying gels, but we have yet to validate this hypothesis.
19. Pierce the center of the tube with an 18 G needle or dissection "t"-pin to provide a vent. Failure to do this can lead to the tube opening "explosively" upon thawing and can result in loss of sample.
20. We use 600 μl for protein extraction from an 8.0 mm punch. If too little is used we have had our sample extracts congeal in the tube (i.e., stromal collagen forming a gelatin within the tube). For RNA, use the volume suggested by the manufacturer of your kit of choice.
21. The ultrasonication routine will be dependent upon the extractability and stability of the investigator's analyte(s) of interest.
22. Be sure to go in at an angle which is parallel with the iris to avoid coming into contact with the iris which will bleed (contamination) and interfere with excising the cornea.
23. The fine serrations provide grip/traction and vastly improve the controllability of the cut.
24. The corneal epithelium is highly unstable during healing and can spontaneously "slough" off in places forming ulcers. These corneas will begin to swell via edema again until the ulcer is once again closed. These eyes should be removed from any further study as the ulcer represents another wounding event.

25. Some programs enable a pixel-to-unit of measurement conversion factor prior to making measurements. This enables the program to immediately report the measurement in relevant terms. If the pixel-to-area conversion factor is known, you can save yourself some time by entering it into the software prior to starting.
26. At a 1:1 reproduction ratio, the dimensions of a pixel will be the camera's sensor's width (in millimeter) divided by the total width of the image in pixels by the sensor's height (in millimeter) divided by the image's height in pixels. Alternatively, some manufacturers publish the pixel size of their sensors. This process is highly dependent upon using the macrophotography settings suggested (i.e., 1:1 reproduction ratio). If another reproduction ratio is used, the pixel-to-area value must be scaled accordingly (i.e., 1:2 -> double the area).
27. Use the image analysis software to determine the pixel-to-area conversion factor by measuring your empirical standard if you included in your image.
28. These values will vary somewhat from camera to camera and by the exposure conditions used, but should be consistent from eye to eye with the same camera and camera/lens/flash exposure settings.
29. The use of a weighting scheme is currently being investigated, but would include giving more intense pixels (i.e., worse haze), a higher per-pixel score.

References

1. Moller-Pedersen T (2004) Keratocyte reflectivity and corneal haze. Exp Eye Res 78:553
2. Dawson DG et al (2005) Ex vivo confocal microscopy of human LASIK corneas with histologic and ultrastructural correlation. Ophthalmology 112:634
3. Sridhar MS, Rapuano CJ, Cosar CB, Cohen EJ, Laibson PR (2002) Phototherapeutic keratectomy versus diamond burr polishing of Bowman's membrane in the treatment of recurrent corneal erosions associated with anterior basement membrane dystrophy. Ophthalmology 109:674
4. Reidy JJ et al (1996) Comparison of corneal epithelial wound healing after photorefractive and lamellar keratectomy. J Refract Surg 12:352
5. Tuft SJ, Zabel RW, Marshall J (1989) Corneal repair following keratectomy. A comparison between conventional surgery and laser photoablation. Invest Ophthalmol Vis Sci 30:1769
6. Szerenyi KD, Campos M, McDonnell PJ (1994) Prostaglandin E2 production after lamellar keratectomy and photorefractive keratectomy. J Refract Corneal Surg 10:413
7. Bokosky JE, Meyer RF, Sugar A (1985) Surgical treatment of calcific band keratopathy. Ophthalmic Surg 16:645
8. Lance SE, Capone A Jr, SundarRaj N, Roat MI, Thoft RA (1988) Diamond burring and surgical keratectomy. Morphologic comparison in the rabbit. Arch Ophthalmol 106:830
9. Malta JB, Soong HK (2008) Diamond burr superficial keratectomy in the treatment of visually-significant anterior corneal lesions. Arq Bras Oftalmol 71:415
10. Ramamurthi S, Rahman MQ, Dutton GN, Ramaesh K (2006) Pathogenesis, clinical features and management of recurrent corneal erosions. Eye (Lond) 20:635
11. Schipper I (2003) Diamond burr polishing. Ophthalmology 110:1855, author reply 1855
12. Soong HK, Farjo Q, Meyer RF, Sugar A (2002) Diamond burr superficial keratectomy for recurrent corneal erosions. Br J Ophthalmol 86:296

13. Tzelikis PF, Rapuano CJ, Hammersmith KM, Laibson PR, Cohen EJ (2005) Diamond burr treatment of poor vision from anterior basement membrane dystrophy. Am J Ophthalmol 140:308

14. Wong VW, Chi SC, Lam DS (2009) Diamond burr polishing for recurrent corneal erosions: results from a prospective randomized controlled trial. Cornea 28:152

Chapter 17

A Novel and Efficient Model of Coronary Artery Ligation in the Mouse

Erhe Gao and Walter J. Koch

Abstract

Coronary artery ligation to induce myocardial infarction (MI) and ischemia/reperfusion (I/R) injury in mice is typically performed by an invasive and time-consuming approach that requires ventilation and a full thoracotomy (classical method), often resulting in extensive tissue damage and high mortality. Here, we describe a novel and rapid surgical method to induce MI that does not require ventilation. This method is much more efficient and safer than the classical method of MI and I/R injury.

Key words Animal models of human disease, Myocardial infarction, Ischemia reperfusion, Surgical efficiency, Mouse

1 Introduction

The increase in the availability of various types of genetically engineered mice has brought about the blooming need for more efficient ways to induce myocardial damage for both molecular mechanistic studies as well as the testing of potentially therapeutic interventions. Currently, two of the most common models used by researchers to induce ischemic cardiac damage are permanent left main descending coronary artery (LCA) occlusion to induce an MI and the temporary coronary artery occlusion to induce I/R injury [1, 2]. The MI model is usually used to investigate myocardial changes such as remodeling that occur over an extended period of time, whereas the I/R model is generally used to examine the short-term consequences of ischemic injury. Since the first attempt of coronary ligation in mouse was presented by Johns and Olson in 1954, where a ventilation-based thoracotomy was introduced [3], a variety of surgical manipulations have been made to induce the cardiac ischemic event [4, 5]. However, LCA ligation remains the most commonly practiced ischemic injury [2, 6, 7]. This continues to utilize methodology requiring ventilation and full opening of the chest (referred to below as the classical method).

Robert G. Gourdie and Tereance A. Myers (eds.), *Wound Regeneration and Repair: Methods and Protocols*, Methods in Molecular Biology, vol. 1037, DOI 10.1007/978-1-62703-505-7_17, © Springer Science+Business Media New York 2013

This technique can cause extensive tissue damage and high surgical related death and can also be quite time consuming for most surgeons [8–11].

Over the last few years we have developed a new approach of coronary artery ligation in mice without the need of ventilation. This method can be applied to not only LCA ligation and the MI procedure [7, 12–22] but also I/R injury [23–37] as well as gene or cell therapy procedures to the heart [13, 38–41]. Compared to the above-mentioned classical method this novel method is much more efficient and safer for mice while not compromising the extent of cardiac injury [15]. In this chapter, we focus on the technical details of this procedure.

2 Materials

2.1 Anesthesia and Analgesia

1. Isoflurane anesthesia delivery system as shown in Fig. 1.
2. Buprenorphine.

2.2 Surgical Instruments: Five Instruments Are Needed as Shown in Fig. 2

1. A needle holder for skin suturing.
2. Needle holder for suturing mouse heart coronary artery.
3. Hartman mosquito hemostatic forceps.
4. Surgical scissors.
5. Dressing forceps.

2.3 Surgical Accessories

1. Dry bead sterilizer (Germinator 500).
2. Surgical light.

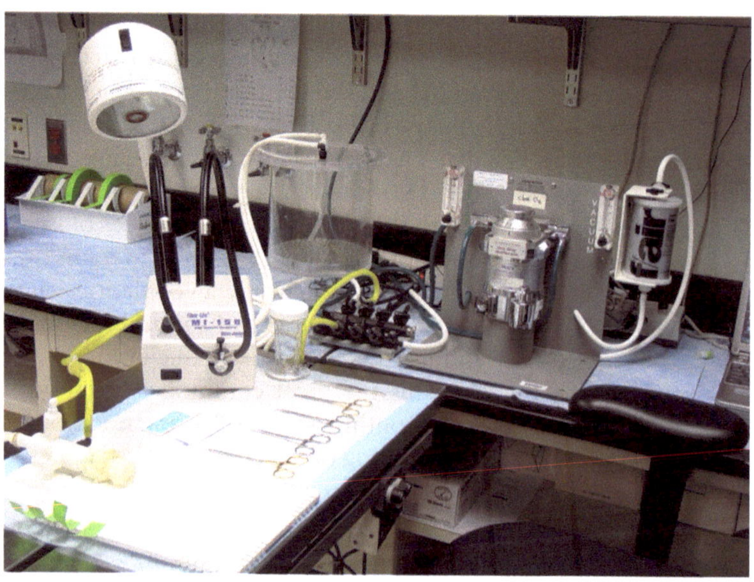

Fig. 1 Isoflurane anesthesia delivery system

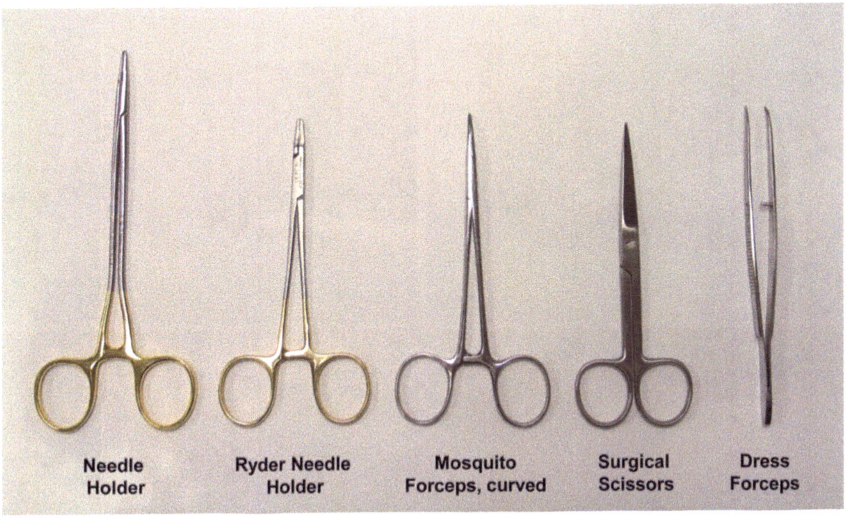

Fig. 2 Five surgical instruments: A needle holder for skin suturing, needle holder for suturing mouse heart coronary artery, Hartman mosquito hemostatic forceps, surgical scissors, and dressing forceps

3. Surgical board (Fig. 1).
4. Heating blank.
5. 4/0 silk suture.
6. 6-0 silk suture.
7. Taps (Fig. 1).
8. Betadine and alcohol pads.
9. Examination gloves.

3 Methods

3.1 Permanent Coronary Artery Occlusion Without Ventilation

1. Sterilize surgical instruments with a dry bead sterilizer (Germinator 500).
2. Mouse (generally 2–3 months of age or at least 18 g body weight) is anesthetized with 2–3 % isoflurane inhalation in an inducing chamber.
3. Once anesthetized, the mouse is removed from the inducing chamber to the surgical board, immobilized with tape, and continuously anesthetized with 2 % isoflurane via coaxial breathing apparatus but not ventilated.
4. Remove the fur with a standard depilatory (e.g., Nair) and clean the skin with water and then betadine and alcohol pads. In order to perform this procedure more efficiently the step of fur-removing could be done earlier.
5. A small skin cut (1.2 cm long) is made over the left chest with the scissors and a purse suture is made as shown in Fig. 3a, b.

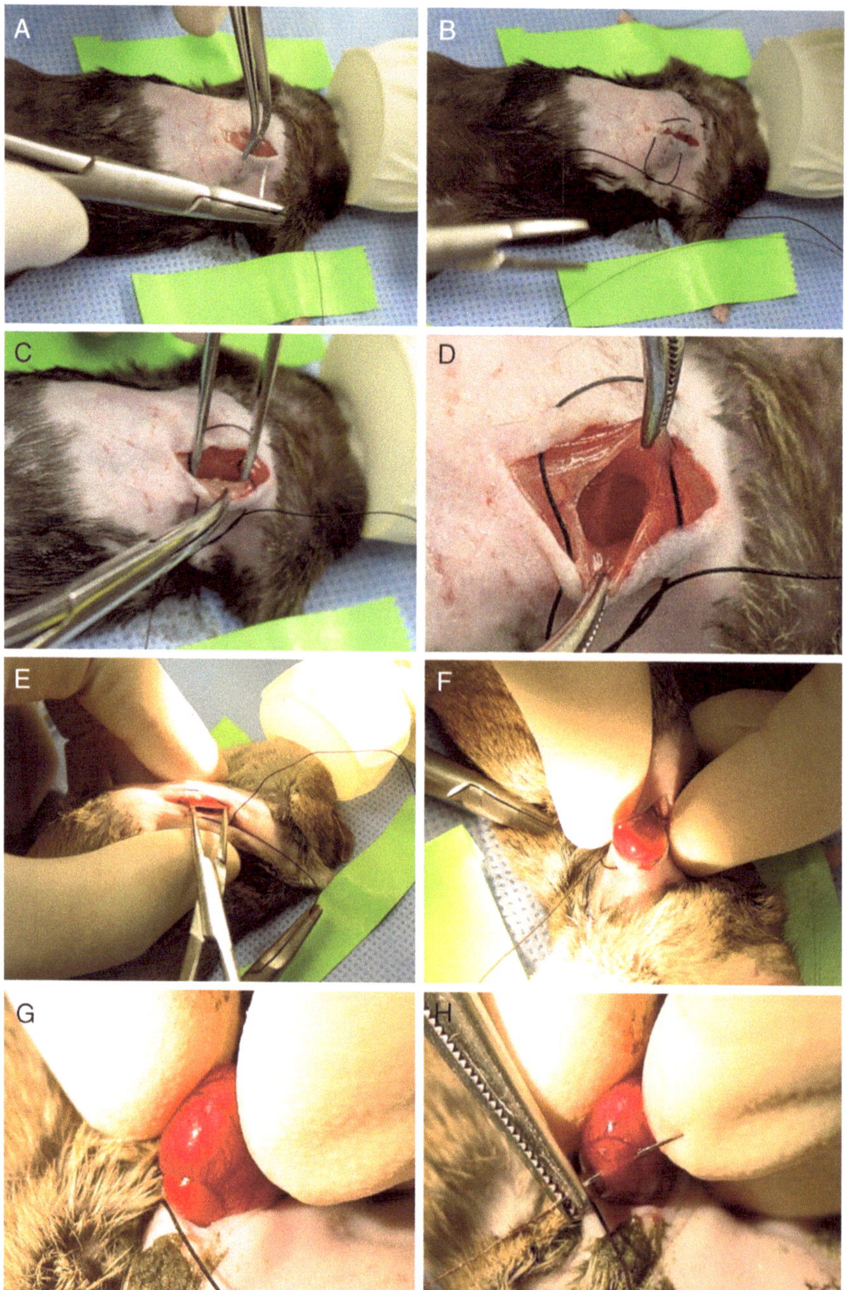

Fig. 3 Permanent coronary artery occlusion without ventilation: (**a**, **b**) A small skin cut (1.2 cm long) is made over the left chest with the scissors making a purse suture. (**c**, **d**) Dissection and retraction of the pectoral major and minor muscle to expose the fourth intercostal space. (**e**) Externalize (pop out) the heart. (**f**) Heart "popped out" through the hole. (**g**–**l**) Ligation of the main descending left ventricle coronary artery (LCA also called LAD). (**m**) Placing the heart back. (**n**) Evacuation of air out of the thoracic cavity. (**o**, **p**) Closing the skin incision

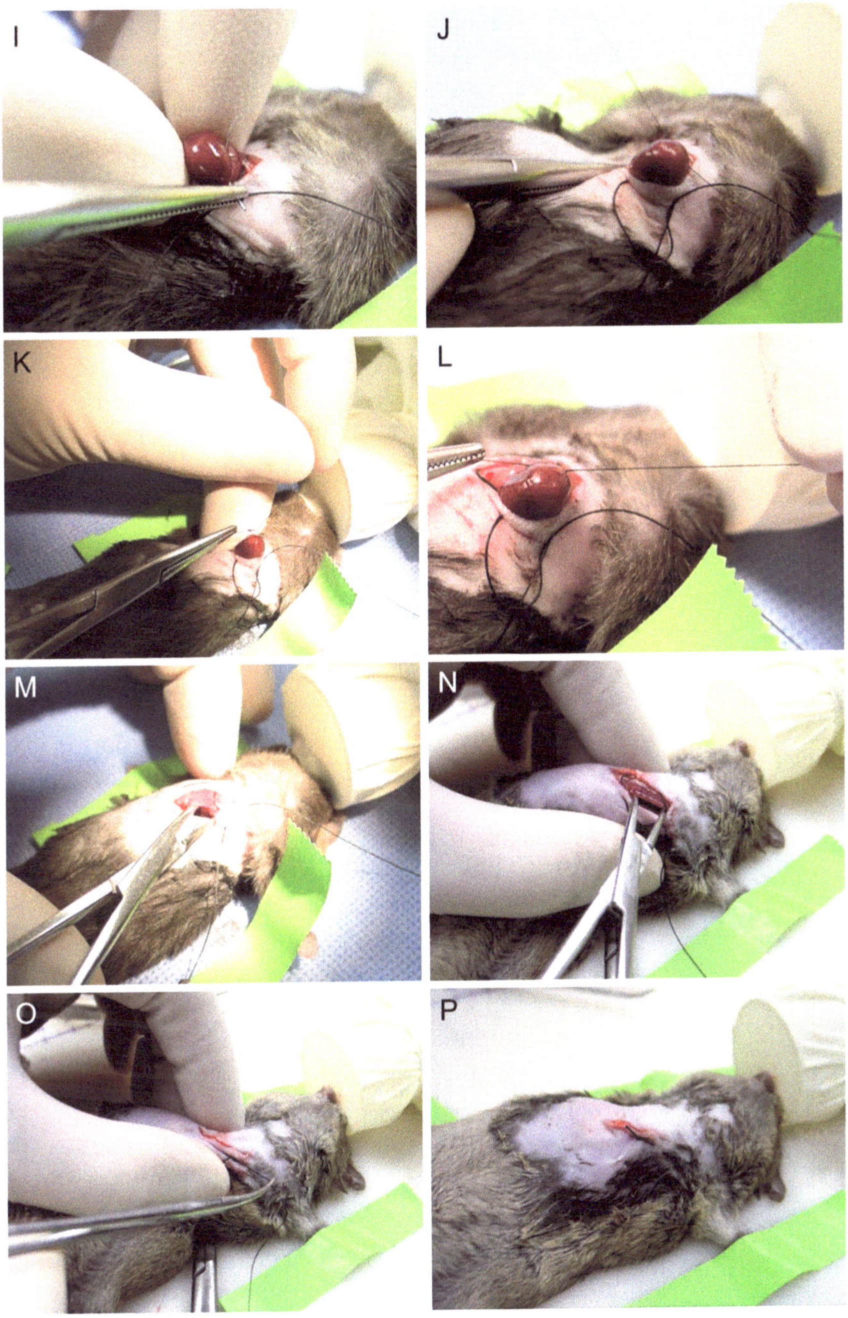

Fig. 3 (continued)

6. Dissection and retraction of the pectoral major and minor muscle are done to expose the fourth intercostal space (Fig. 3c, d). For this step, first hold the hemostatic forceps in left hand to grasp the pectoral major, and the mosquito clamp in right hand to underneath dissect the pectoral major and expose the pecto-

ral minor muscle, then using the mosquito clamp (right hand) to grasp the pectoral minor, and use forceps (left hand) to go underneath and dissect the pectoral minor (Fig. 3c) exposing the intercostal space (Fig 3d).

7. Externalize (pop out) the heart. This is the most difficult part for this procedure. First, using left-hand finger and thumb to hold the mouse chest, right hand to hold the mosquito forceps, and then using the forceps punch a small hole at the fourth intercostal space to penetrate the intercostal muscle, the pleural membrane, and the pericardium. Hold the clamp slightly open with the right hand (Fig. 3e) and using the left-hand finger, press slightly down the chest; the heart is then smoothly and gently "popped out" through the hole as shown in Fig. 3f.
8. Ligation of the main descending left ventricle coronary artery (LCA also called LAD): LCA is located, sutured, and ligated at a site about 2–3 mm from its origin using a 6-0 silk suture as shown in Figs. 3g–l and 4. The ligation is deemed successful when the anterior wall of the LV turns pale or proved by the typical ischemic change in ECG (S-T segment elevation, Fig. 5). Suturing the LCA is also a critical step of this procedure; typically, once the heart is popped out, using the left-hand digital finger and thumb to hold the heart, find the LCA (*see* Subheading 4), and using right hand to grab the needle holder toothed with a previously prepared 6-0 suture, suture the LCA (Fig. 3I); pull the needle suture with the left digital finger and thumb and at the same time use the needle holder to put pressure on the right side of the chest to prevent the heart pull back (Fig. 3j); and then replace the needle holder with the right middle finger to press the right side of the chest (Fig. 3k) and make the knot (Fig. 3k, l).
9. After ligation, the heart is immediately placed back into the intrathoracic space with the help of the clamp (Fig. 3m) followed by manual evacuation of air via left hand (Fig. 3n, o) and closure of the skin incision by means of the previously placed purse-string suture (Fig. 3p).
10. The mouse is then allowed to breathe room air and monitored on a heating blanket during the recovery period, which is generally complete within 3–5 min. No artificial respiratory aid is required during the recovery time.
11. The sham group undergoes the same surgical procedures except that the LCA is not occluded.
12. One dose of buprenorphine (0.1 mg/kg) is administered subcutaneously (s.c.) immediately after the incision is closed.
13. Clean surgical tools with PBS and alcohol.

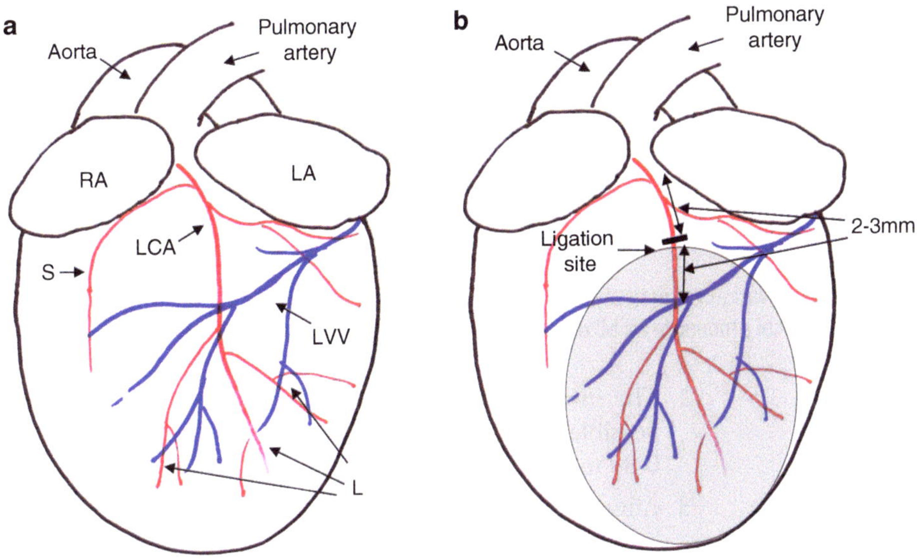

Fig. 4 Representative graph of artery and vein distribution in mouse heart (**a**) and left coronary artery ligation site in mouse heart (**b**). *LA* left atrial, *RA* right atrial, *LCA* left coronary artery, *LVV* left ventricular vein, *S* septal branch of LCA, *L* left ventricle branches of LCA

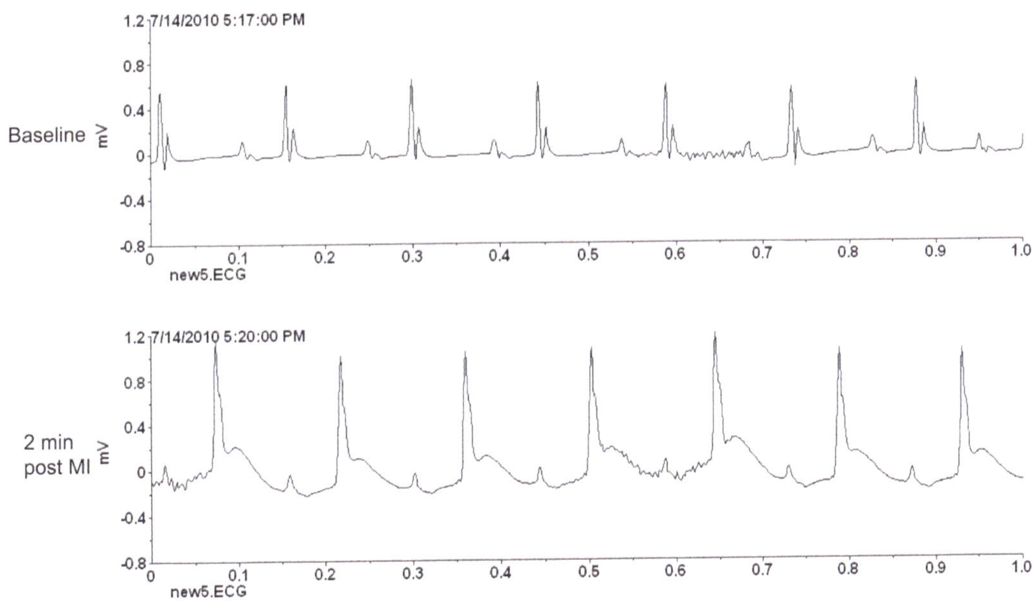

Fig. 5 Representative electric cardiogram recorded at baseline and 2 min after left coronary artery ligation in mouse

3.2 Method of Induction of Myocardial I/R Injury Without Ventilation

1. This I/R injury procedure in mice is essentially the same as the procedure for inducing MI except that a slipknot is tied around the LCA 2–3 mm from its origin with a 6-0 silk suture as shown in Fig. 6a. The heart is then quickly placed back into the thoracic space followed by manual evacuation of air and the skin closing (Fig. 6b, c).

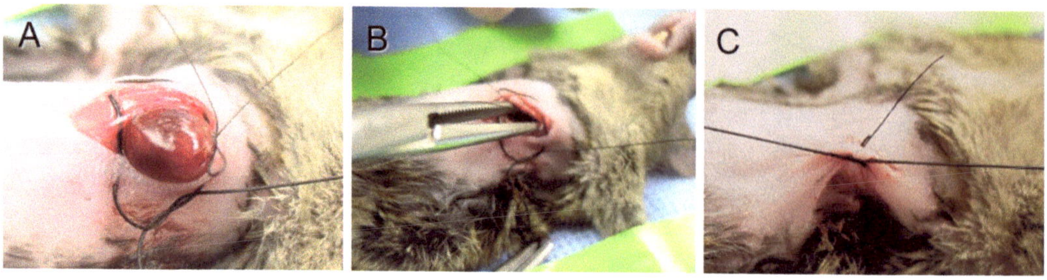

Fig. 6 Photographs of various stages of the novel, rapid I/R injury model in mice. Mice were subjected to I/R using the same surgical procedure as MI except that a slipknot was made to occlude LCA (**a**–**c**)

2. The internal needle end of slipknot suture is cut as short as possible and the other end of the suture is approximately 0.8 cm long and remains outside of the chest (Fig. 6c).
3. After 30 min of ischemia, the slipknot is released by pulling the long end of slipknot suture smoothly and gently until a feeling of release is sensed at which time the myocardium begins reperfusion. This outside-the-skin suture knot releasing method should only be attempted by the experienced surgeon.
4. Alternatively, the mouse can be re-anesthetized with 2 % isoflurane inhalation, the chest reopened, and the slipknot released by pulling the long end of slipknot suture smoothly and gently, and then following manual evacuation of the pneumothorax and chest closure.

All animals should be monitored after the surgery and should be given one dose (0.1 mg/kg) of buprenorphine within 6 h post surgery with another dose administered the following morning. No further analgesia is needed after that.

4 Notes

1. The procedure of murine model of MI or I/R injury without ventilation is technically demanding and challenging. Critical to the success of the new model is keeping the time that the heart is outside the body to a minimum. Based on our experience, which now totals over 40,000 cases in mice, the time allowed for heart externalization should be no more than 30 s to limit global hypoxia. This will give the best prognostic result. Therefore practice is the key to success.
2. The key step in this heart "pop out" procedure is to externalize the heart from the chest, which is the most difficult thing for the beginner.
   (a) Technically, the two hands should be working together; once the chest is open, left-hand digit finger and thumb

should push down and the middle finger should be on the right side of mouse chest and push toward left side. Once the heart is popping out, the left middle figure should keep pressure a little bit to prevent the heart go back to the chest.

(b) The performer should avoid pushing too hard to cause lung damage.

(c) If the anesthesia is too deep the heart is barely beating and it will be even more difficult to externalize the heart. The perfect technique actually uses the force of the beating heart to externalize, essentially "popping" itself out, and the fingers guide it.

(d) For practice purposes, prior to perfecting this surgical model we recommend ventilating the mouse before and after popping the heart.

3. Localization of the LCA is another difficult part in this procedure and should be based on earlier studies on mouse coronary circulation by Ahn et al. [6] and Michael et al. [8]. Once the heart is properly externalized (popped out) the LCA can be visualized as being light red in color (Fig 3 g). The artery originates at the aorta and goes down to base of LV on the left side of the atria. It intercrosses with the ventricular vein at about the middle part of LV toward the apex (Fig. 4a). The suture for the coronary artery ligation is placed between the inner upper part of the left atria and the intersection of the artery/vein and at about the middle part of that segment of LCA (2–3 mm from aural and the intersection of artery/vein (Fig. 4b)). Since the trace of LCA has less variation at this part, ligation at this level will induce about 40 % LV myocardial infarction in MI model (Fig. 7a) and 33 % infarction in ischemic area (also called area at risk, AAR) in I/R model at 24 h post surgery (Fig. 7b) [15, 27, 31]. MI at this level also causes significant LV functional decline and structural remodeling (Fig. 8) [15, 16, 18, 27].

4. One of the advantages of this new procedure is that the pectoral muscles remain intact. This is very important because intact pectoral muscles are necessary to cover the hole (the mini thoracotomy) after the heart has been placed back to the thoracic space and eliminate the repairing suture of the muscle.

5. Bleeding is one of the major factors that contribute to the early surgical related deaths in this "heart exteriorization" MI and I/R mice model. Based on our experience, there are five common reasons that account for bleeding.

(a) First, bleeding can arise because of heart puncture. The heart puncture happens when the clamp punches too deep

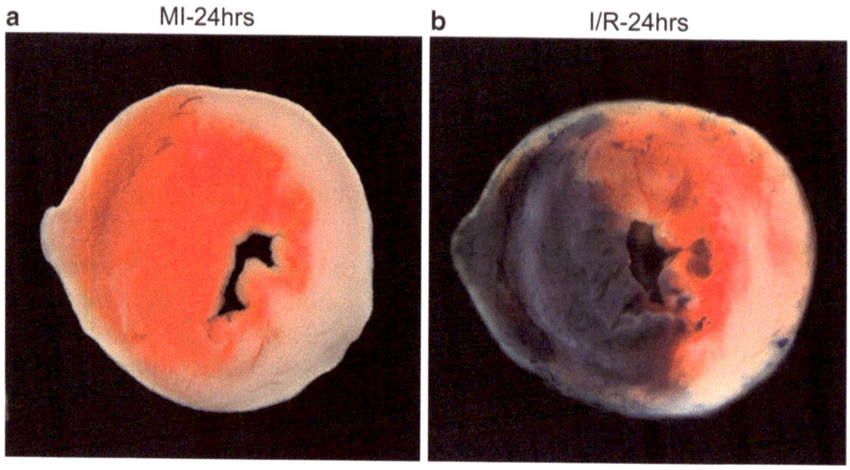

Fig. 7 Representative pictures of myocardial infarction from ischemic mouse hearts. (**a**) Heart section stained with TTC from MI model at 24 h post MI. (**b**) Heart tissue section stained with Evan's blue/TTC from I/R model at 24 h post I/R. The *white* color indicated a dead tissue, the *red* color indicated a viable tissue, and the *blue* color shows that the area is not at risk

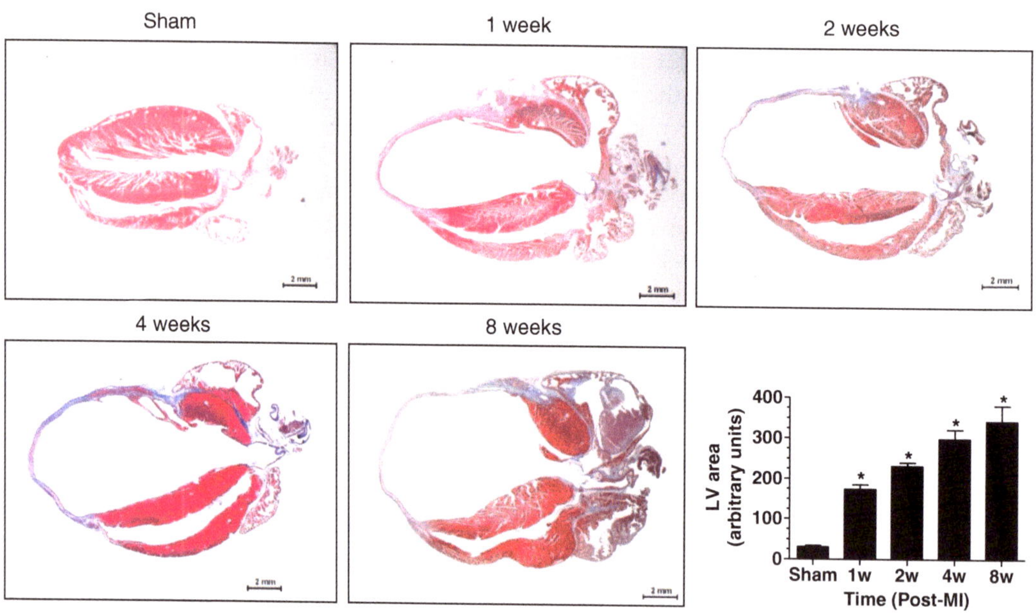

Fig. 8 Representative Masson's trichrome-stained tissue sections from sham and MI groups at 1, 2, 4, and 8 weeks post MI. The bar graph shows the left ventricle area and indicates a structure remodeling change

and damages the left atrium (through the third intercostal space) or ventricle (through the fourth intercostal space). The bleeding is profuse and the mouse usually dies on the table or within few hours post surgery. To avoid this problem, adjust the deepness of the clamp and choose the correct intercostal space.

(b) Secondly, rib fracture and costal artery tear can cause bleeding, which occurs most often when the surgeon tries to spread the opening between the fourth or the fifth intercostal space too wide against the ribs. The bleeding from this type of damage is usually not severe and stops spontaneously in approximately 50 % of cases. To avoid this problem, use the widest intercostal space (fourth to fifth space) and be cautious to spread the ribs only as far as necessary.

(c) A third cause of bleeding is from the damage of the internal mammary artery. The internal mammary artery goes perpendicularly to the ribs and about 1–2 mm apart from the sternum. The damage of this artery happens when the surgeon tries to open the hole too widely along the ribs with the clamp, or the clamp entry point is too close to the sternum. Once damaged, the bleeding is hard to stop. To avoid this, use the widest intercostal space, which is usually 5 mm away from the sternum, and do not open too wide.

(d) A fourth cause of bleeding is tearing of pulmonary vessels. Bleeding from a torn pulmonary vessel is hard to stop and is usually fatal. This damage occurs when the heart is exteriorized and pulled too far out of the chest. The key to prevent this is to try not to externalize the whole heart; in most cases, the right atria should be in the thoracic space and left atria is partially in the thoracic cage.

(e) Finally, bleeding can occur when suturing the LCA. Bleeding could be the result of either tearing of the artery directly or wide suturing that may go through the septa and right ventricle (RV). The wall of the RV is thin and bleeding from the hole made when the wide suture is tied is hard to stop. The animal may survive the MI procedure itself but dies usually within first 24 h. MI or I/R without ventilation is technically a challenging procedure but with practice can be learned and a perfect surgery is bloodless.

6. Pneumothorax is another contributor to early postsurgical death and occurs when the operator forgets to either displace air before closing the suture or forgets to keep the "hole" open with a mosquito hemostatic forceps when manually evacuating air out of the thoracic cavity.

References

1. Guo Y, Wu WJ, Qiu Y, Tang XL, Yang Z, Bolli R (1998) Demonstration of an early and a late phase of ischemic preconditioning in mice. Am J Physiol 275(4 Pt 2):H1375–H1387
2. Klocke R, Tian W, Kuhlmann MT, Nikol S (2007) Surgical animal models of heart failure related to coronary heart disease. Cardiovasc Res 74(1):29–38
3. Johns TN, Olson BJ (1954) Experimental myocardial infarction. I. A method of coronary occlusion in small animals. Ann Surg 140(5): 675–682

4. van den Bos EJ, Mees BM, de Waard MC, de Crom R, Duncker DJ (2005) A novel model of cryoinjury-induced myocardial infarction in the mouse: a comparison with coronary artery ligation. Am J Physiol Heart Circ Physiol 289(3):H1291–H1300
5. Tarnavski O, McMullen JR, Schinke M, Nie Q, Kong S, Izumo S (2004) Mouse cardiac surgery: comprehensive techniques for the generation of mouse models of human diseases and their application for genomic studies. Physiol Genomics 16(3):349–360
6. Ahn D, Cheng L, Moon C, Spurgeon H, Lakatta EG, Talan MI (2004) Induction of myocardial infarcts of a predictable size and location by branch pattern probability-assisted coronary ligation in C57BL/6 mice. Am J Physiol Heart Circ Physiol 286(3):H1201–H1207
7. Most P, Seifert H, Gao E, Funakoshi H, Volkers M, Heierhorst J et al (2006) Cardiac S100A1 protein levels determine contractile performance and propensity toward heart failure after myocardial infarction. Circulation 114(12):1258–1268
8. Michael LH, Entman ML, Hartley CJ, Youker KA, Zhu J, Hall SR et al (1995) Myocardial ischemia and reperfusion: a murine model. Am J Physiol 269(6 Pt 2):H2147–H2154
9. Patten RD, Aronovitz MJ, Deras-Mejia L, Pandian NG, Hanak GG, Smith JJ et al (1998) Ventricular remodeling in a mouse model of myocardial infarction. Am J Physiol 274(5 Pt 2):H1812–H1820
10. Lutgens E, Daemen MJ, de Muinck ED, Debets J, Leenders P, Smits JF (1999) Chronic myocardial infarction in the mouse: cardiac structural and functional changes. Cardiovasc Res 41(3):586–593
11. Gehrmann J, Frantz S, Maguire CT, Vargas M, Ducharme A, Wakimoto H et al (2001) Electrophysiological characterization of murine myocardial ischemia and infarction. Basic Res Cardiol 96(3):237–250
12. Tsuda T, Gao E, Evangelisti L, Markova D, Ma X, Chu ML (2003) Post-ischemic myocardial fibrosis occurs independent of hemodynamic changes. Cardiovasc Res 59(4): 926–933
13. Botta R, Gao E, Stassi G, Bonci D, Pelosi E, Zwas D et al (2004) Heart infarct in NOD-SCID mice: therapeutic vasculogenesis by transplantation of human CD34+ cells and low dose CD34+KDR+ cells. FASEB J 18(12):1392–1394
14. Raake PW, Vinge LE, Gao E, Boucher M, Rengo G, Chen X et al (2008) G protein-coupled receptor kinase 2 ablation in cardiac myocytes before or after myocardial infarction prevents heart failure. Circ Res 103(4): 413–422
15. Gao E, Lei YH, Shang X, Huang ZM, Zuo L, Boucher M et al (2010) A novel and efficient model of coronary artery ligation and myocardial infarction in the mouse. Circ Res 107(12):1445–1453
16. Woulfe KC, Gao E, Lal H, Harris D, Fan Q, Vagnozzi R et al (2010) Glycogen synthase kinase-3beta regulates post-myocardial infarction remodeling and stress-induced cardiomyocyte proliferation in vivo. Circ Res 106(10):1635–1645
17. Zhang H, Chen X, Gao E, MacDonnell SM, Wang W, Kolpakov M et al (2010) Increasing cardiac contractility after myocardial infarction exacerbates cardiac injury and pump dysfunction. Circ Res 107(6):800–809
18. Zhou J, Lal H, Chen X, Shang X, Song J, Li Y et al (2010) GSK-3alpha directly regulates beta-adrenergic signaling and the response of the heart to hemodynamic stress in mice. J Clin Invest 120(7):2280–2291
19. Ciccarelli M, Chuprun JK, Rengo G, Gao E, Wei Z, Peroutka RJ et al (2011) G protein-coupled receptor kinase 2 activity impairs cardiac glucose uptake and promotes insulin resistance after myocardial ischemia. Circulation 123(18):1953–1962
20. Lal H, Zhou J, Ahmad F, Zaka R, Vagnozzi RJ, Decaul M et al (2012) Glycogen synthase kinase-3alpha limits ischemic injury, cardiac rupture, post-myocardial infarction remodeling and death. Circulation 125(1):65–75
21. Tsuda T, Wu J, Gao E, Joyce J, Markova D, Dong H et al (2012) Loss of fibulin-2 protects against progressive ventricular dysfunction after myocardial infarction. J Mol Cell Cardiol 52(1):273–282
22. Zhang H, Makarewich C, Kubo H, Wang W, Duran J, Li Y et al (2012) Hyperphosphorylation of the cardiac ryanodine receptor at serine 2808 is not involved in cardiac dysfunction after myocardial infarction. Circ Res 110: 831–840
23. Liu HR, Gao F, Tao L, Yan WL, Gao E, Christopher TA et al (2004) Antiapoptotic mechanisms of benidipine in the ischemic/reperfused heart. Br J Pharmacol 142(4): 627–634
24. Gao F, Tao L, Yan W, Gao E, Liu HR, Lopez BL et al (2004) Early anti-apoptosis treatment reduces myocardial infarct size after a prolonged reperfusion. Apoptosis 9(5):553–559
25. Liu HR, Gao E, Hu A, Tao L, Qu Y, Most P et al (2005) Role of Omi/HtrA2 in apoptotic cell death after myocardial ischemia and reperfusion. Circulation 111(1):90–96
26. Tao L, Jiao X, Gao E, Lau WB, Yuan Y, Lopez B et al (2006) Nitrative inactivation of thioredoxin-1 and its role in postischemic myocardial apoptosis. Circulation 114(13):1395–1402

27. Gao E, Boucher M, Chuprun JK, Zhou RH, Eckhart AD, Koch WJ (2007) Darbepoetin alfa, a long-acting erythropoietin analog, offers novel and delayed cardioprotection for the ischemic heart. Am J Physiol Heart Circ Physiol 293(1):H60–H68
28. Tao L, Gao E, Jiao X, Yuan Y, Li S, Christopher TA et al (2007) Adiponectin cardioprotection after myocardial ischemia/reperfusion involves the reduction of oxidative/nitrative stress. Circulation 115(11):1408–1416
29. Zhang H, Tao L, Jiao X, Gao E, Lopez BL, Christopher TA et al (2007) Nitrative thioredoxin inactivation as a cause of enhanced myocardial ischemia/reperfusion injury in the aging heart. Free Radic Biol Med 43(1):39–47
30. Boucher M, Pesant S, Lei YH, Nanton N, Most P, Eckhart AD et al (2008) Simultaneous administration of insulin-like growth factor-1 and darbepoetin alfa protects the rat myocardium against myocardial infarction and enhances angiogenesis. Clin Transl Sci 1(1):13–20
31. DeGeorge BR Jr, Gao E, Boucher M, Vinge LE, Martini JS, Raake PW et al (2008) Targeted inhibition of cardiomyocyte Gi signaling enhances susceptibility to apoptotic cell death in response to ischemic stress. Circulation 117(11):1378–1387
32. Hu A, Jiao X, Gao E, Li Y, Sharifi-Azad S, Grunwald Z et al (2008) Tonic beta-adrenergic drive provokes proinflammatory and proapoptotic changes in aging mouse heart. Rejuvenation Res 11(1):215–226
33. Jiao XY, Gao E, Yuan Y, Wang Y, Lau WB, Koch W et al (2009) INO-4885 [5,10,15,20-tetra[N-(benzyl-4'-carboxylate)-2-pyridinium]-21H,23H-porphine iron(III) chloride], a peroxynitrite decomposition catalyst, protects the heart against reperfusion injury in mice. J Pharmacol Exp Ther 328(3):777–784
34. Liu HR, Tao L, Gao E, Qu Y, Lau WB, Lopez BL et al (2009) Rosiglitazone inhibits hypercholesterolaemia-induced myeloperoxidase upregulation–a novel mechanism for the cardioprotective effects of PPAR agonists. Cardiovasc Res 81(2):344–352
35. Brinks H, Boucher M, Gao E, Chuprun JK, Pesant S, Raake PW et al (2010) Level of G protein-coupled receptor kinase-2 determines myocardial ischemia/reperfusion injury via pro- and anti-apoptotic mechanisms. Circ Res 107(9):1140–1149
36. Yi W, Sun Y, Gao E, Wei X, Lau WB, Zheng Q et al (2011) Reduced cardioprotective action of adiponectin in high-fat diet-induced type II diabetic mice and its underlying mechanisms. Antioxid Redox Signal 15(7):1779–1788
37. Kerkela R, Boucher M, Zaka R, Gao E, Harris D, Piuhola J et al (2011) Cytosolic phospholipase A(2)alpha protects against ischemia/reperfusion injury in the heart. Clin Transl Sci 4(4):236–242
38. Degeorge BR Jr, Rosenberg M, Eckstein V, Gao E, Herzog N, Katus HA et al (2008) BMP-2 and FGF-2 synergistically facilitate adoption of a cardiac phenotype in somatic bone marrow c-kit+/Sca-1+ stem cells. Clin Transl Sci 1(2):116–125
39. Liang H, Hou H, Yi W, Yang G, Gu C, Lau WB et al (2011) Increased expression of pigment epithelium-derived factor in aged mesenchymal stem cells impairs their therapeutic efficacy for attenuating myocardial infarction injury. Eur Heart J. 2011 May 23. [Epub ahead of print]
40. Wang Y, Lau WB, Gao E, Tao L, Yuan Y, Li R et al (2010) Cardiomyocyte-derived adiponectin is biologically active in protecting against myocardial ischemia-reperfusion injury. Am J Physiol Endocrinol Metab 298(3): E663–E670
41. Pu J, Yuan A, Shan P, Gao E, Wang X, Wang Y et al (2012) Cardiomyocyte-expressed farnesoid-X-receptor is a novel apoptosis mediator and contributes to myocardial ischaemia/reperfusion injury. Eur Heart J 2012 Feb 3. [Epub ahead of print]

Chapter 18

Cardiac Wound Healing Post-myocardial Infarction: A Novel Method to Target Extracellular Matrix Remodeling in the Left Ventricle

Rogelio Zamilpa, Jianhua Zhang, Ying Ann Chiao, Lisandra E. de Castro Brás, Ganesh V. Halade, Yonggang Ma, Sander O. Hacker, and Merry L. Lindsey

Abstract

Myocardial infarction (MI) is a leading cause of death worldwide. Permanent ligation of the left anterior descending coronary artery (LAD) is a commonly used surgical model to study post-MI effects in mice. LAD occlusion induces a robust wound healing response that includes extracellular matrix (ECM) remodeling. This chapter provides a detailed guide on the surgical procedure to permanently ligate the LAD. Additionally, we describe a prototype method to enrich cardiac tissue for ECM, which allows one to focus on ECM remodeling in the left ventricle following surgically induced MI in mice.

Key words Myocardial infarction, Cardiac wound healing, Mice, Extracellular matrix, Matrix metalloproteinases, Inflammation, Decellularization

1 Introduction

Over the last 40 years, successful cardiovascular research has led to increases in 30-day post-myocardial infarction (MI) survival rates of 60 % in the 1970s to current rates of >90 %. Based on this achievement, the current therapeutic challenge has shifted from immediate survival issues to ways to improve the long-term survival of patients undergoing cardiac infarct healing [1, 2]. Cardiac infarct healing in the left ventricle (LV) is referred to as LV remodeling and includes changes in LV size, shape, and function [3–5].

The extent of LV remodeling is dependent on the severity of the following component events that occur after the infarct: (1) myocyte death, (2) inflammatory response, (3) granulation tissue deposition to form the scar, and (4) granulation tissue remodeling [6, 7]. Throughout LV remodeling, the critical balance between extracellular matrix (ECM) degradation and

Robert G. Gourdie and Tereance A. Myers (eds.), *Wound Regeneration and Repair: Methods and Protocols*, Methods in Molecular Biology, vol. 1037, DOI 10.1007/978-1-62703-505-7_18, © Springer Science+Business Media New York 2013

deposition influences ventricular function and patient survival [5, 8]. Excessive ECM degradation predisposes the LV to aneurysm formation and ventricular rupture, while excessive ECM deposition may result in arrhythmias and congestive heart failure [8]. In this chapter, we describe an established murine MI model followed by a protocol to enrich the ECM content during the LV remodeling process post MI.

The surgically induced MI model presented here consists of permanent occlusion of the LAD at the site where it emerges from under the left atrium. This chronic infarct model evokes an anterolateral, apical LV infarct with an average MI size that equals 44 ± 2 % ($n = 25$) of the total LV muscle mass by day 7 post MI in C57BL/6 wild-type mice of both genders [9–12]. The mortality associated with the MI procedure ranges from 37 to 50 % over the first 7 days and is primarily due to ventricular rupture, sudden cardiac death, or acute heart failure, with mortality being greater for male mice [13]. This model is suitable for cardiac pathophysiology studies of MI responses, as indicated by the significantly decreased ejection fraction at day 7 post MI (64 ± 2 % for controls versus 18 ± 2 % post MI, $p < 0.05$) [11]. In addition to inducing progressive wall thinning, the permanent LAD occlusion model induces an increase in LV dilation. Further, this model is a convenient tool to investigate biochemical, cellular, and molecular responses to MI.

The advantage of using the mouse permanent occlusion MI model is that the availability of genetically modified mice makes this a practical model to investigate the role of proteins of interest during the LV remodeling process. A constraint of using this animal model is that the mouse infarct and non-infarct LV tissue sizes provide limited amounts of material. Therefore, the experimental design and execution must be carefully planned in order to completely test the hypothesis under investigation. It is important to account for the mortality rate when determining the number of animals required for the study, in order to have an appropriate sample size.

The quality control for infarct confirmation in our permanent occlusion MI model is based on visual inspection of the LV for blanching, electrocardiogram (EKG) assessment for ST segment elevation, and imaging by two-dimensional echocardiography. For every mouse, baseline echocardiograms of the long and short axes should be recorded and analyzed before surgery to assure that the animals have normal LV function. During the surgery, an EKG is used to continuously monitor the animal. Post surgery, LV function is assessed by echocardiography after 3 h and serial echocardiograms are recorded for up to 4 weeks post MI. At the time of sacrifice, the LV is dissected, sectioned into three slices (apex, midcavity, and base), and stained using 1 % 2,3,5-triphenyltetrazolium chloride (TTC). Viable myocardium stains red, which facilitates infarct sizing [14].

The decellularization protocol that follows the permanent LAD occlusion surgery is a perfusion-free version of the procedure described for rats [15]. In addition to decellularizing the heart to remove cellular constituents, the glycosaminoglycans, collagen type I, collagen type III, laminin, and fibronectin that form the scaffold are preserved [15]. The ECM enrichment process described below yields a fully functional three-dimensional scaffold that facilitates investigation of ECM responses post MI [15].

2 Materials

2.1 Mouse Intubation and Coronary Artery Ligation

All animal procedures should be conducted according to the *Guide for the Care and Use of Laboratory Animals* and need to be reviewed and approved by the appropriate institutional animal care and use committee.

1. Paper towels and examination gloves.
2. Glass bead sterilizer.
3. Surgical instruments (Fig. 1).
4. Oxygen cylinder equipped with regulator.
5. Isoflurane.
6. Induction chamber for anesthesia.
7. Mouse Surgery Board (Fig. 2A).
8. Light source (Fig. 2B) and surgical microscope (Fig. 2C).
9. Mouse ventilator equipped with a mouse nose cone (Fig. 2D).
10. Anesthetic vaporizer (Fig. 2E).
11. Surgical tape.
12. Hair remover lotion (Nair®).
13. Cotton-tipped applicators.
14. EtOH (70 % [v/v] solution in H_2O).
15. Povidone-Iodine Prep Solution USP (Betadine).
16. EKG recording system (iWorx Software) and computer (Fig. 2F).
17. IV catheter 20 ga × 1.
18. Sterile sutures.

    3–0 AROSurgical™ (Black Polyamide Monofilament) sterile suture.

    6–0 Ethicon Prolene (Polypropylene) sterile suture.

    8–0 AROSurgical™ (Black Polyamide Monofilament) sterile suture.
19. Gauze sponges (100 % cotton, 4 in. × 4 in., 8-ply).

a

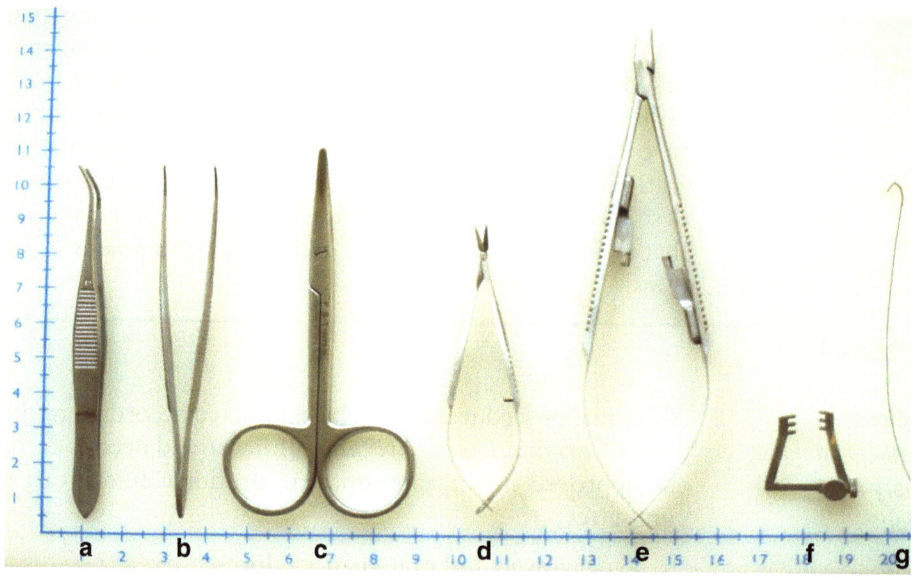

b

Instrument	Company	Catalog #
(a) blunt-end, curved forceps	Fine Science Tools	11054 -10
(b) extra-fine, curved forceps	Fine Science Tools	11151-10
(c) blunt-ended, curved scissors	Fine Science Tools	14041-10
(d) Vannas-Tubingen angled, scissors	Fine Science Tools	15005-08
(e) Castroviejo needle holder	Biomedical Research Instruments	20-1340
(f) Mini-Goldstein retractor	Fine Science Tools	17002-02
(g) Modified 3-0 Sterile Suture	ARO Surgical	001-SP24A03N-45

Fig. 1 (**a**) Instruments recommended for the surgical procedure. (**b**) Instrument names and catalogue numbers

20. 0.9 % Sodium Chloride Irrigation Solution USP.
21. Buprenorphine hydrochloride.

2.2 Extracellular Matrix Decellularization and Proteomic Analysis

1. 1 % Sodium dodecyl sulfate [w/v] in distilled, deionized water.
2. Distilled, deionized water.
3. 1× Complete Mini Protease Inhibitor cocktail with 1 mM EDTA.
4. Protein extraction reagent type 4.
5. 5 mL Round-bottom tubes.
6. 1.5 mL tube.
7. Gyro Twister™ 3D Shaker.
8. Homogenizer.

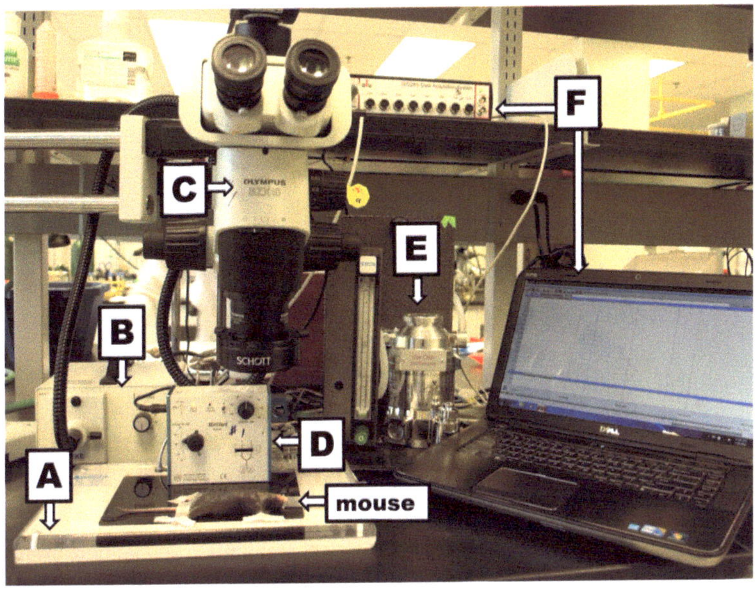

Fig. 2 Surgical equipment and setup include (A) surgical board, (B) microscope light source, (C) microscope, (D) rodent ventilator, (E) vaporizer, and (F) EKG system with computer

3 Methods

3.1 Mouse Intubation and Coronary Artery Ligation

Pre-surgical Preparation

1. Sterilize surgical instruments using the glass bead sterilizer for 12–15 s per instrument.
2. Place the mouse in the induction chamber.
3. Adjust vaporizer to instill a 2 % isoflurane in 100 % oxygen mix into induction chamber.
4. Induce anesthesia: 30-s to 1-min exposure or until mouse is unconscious.
5. Move mouse to a warm surgical board and place anesthetic nose cone over the face of the mouse.
6. Administer inhalant anesthetic to maintain an unconscious state, i.e., 2 % isoflurane and 100 % oxygen mix via ventilator. Adjust stroke volume and ventilation rate according to the manufacturer's recommendations.
7. Place the mouse in a supine position; tape its paws and base of the tail to the surgical board (Fig. 3).
8. Using a cotton-tipped applicator, gently apply Nair® to the thorax and neck area to remove hair.
9. Sanitize the neck and thorax with a gauze containing 70 % EtOH followed by a gauze containing betadine.

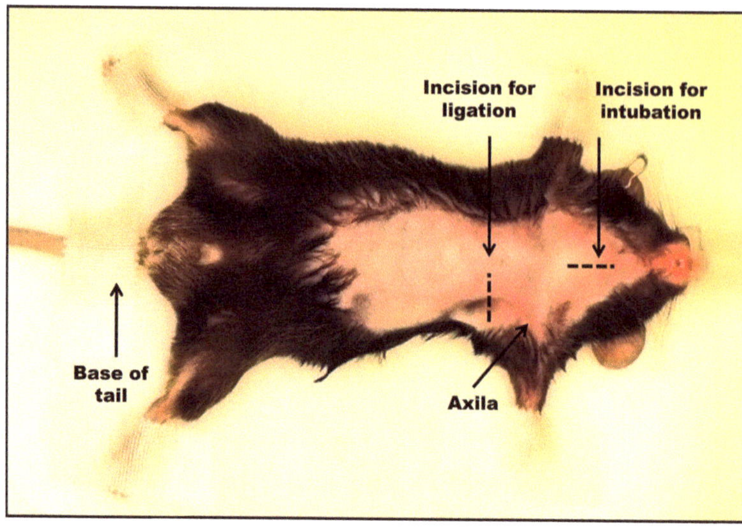

Fig. 3 Pre-surgical preparation and immobilization of mouse. The mouse is in the supine position and immobilized at paws and base of tail, under anesthesia delivered through a mouse nose cone. Area to be shaved and surgical incisions are indicated by the *broken lines*

Intubation Procedure

1. Insert needle electrodes into the subcutaneous space of the forelimbs and hindlimbs to monitor the animal's EKG. Record baseline EKG.
2. Make a 1 cm longitudinal incision on the ventral midline of the neck (Fig. 4a) exposing the submandibular glands; use blunt dissection to expose the sternothyroid muscles (*see* Fig. 4b and **Note 1**).
3. Retract the sternothyroid muscles laterally to expose the ventral aspect of the trachea (Fig. 4c).
4. Before intubation, isoflurane percentage may be increased for a brief period in order to increase the level of anesthesia and facilitate intubation.
5. The nose cone is removed; a 10–15 cm piece of string is placed around the upper incisors and gentle traction is applied cranially so that the airway is open and straight.
6. To intubate, lift the tongue with forceps and guide 20 ga IV catheter into trachea with IV catheter needle as support while monitoring with microscope to ensure clear passage. Advance the catheter until the tip is visible within the trachea through the neck incision (Fig. 4d). The tip of the catheter can be colored with a black marker to facilitate visualization within the trachea.
7. Remove the support needle and connect the anesthesia tubing from the nose cone to the IV catheter. Once intubated, isoflurane percentage should be adjusted to 2 % and the stroke volume and stroke/min adjusted based on the size of the animal.

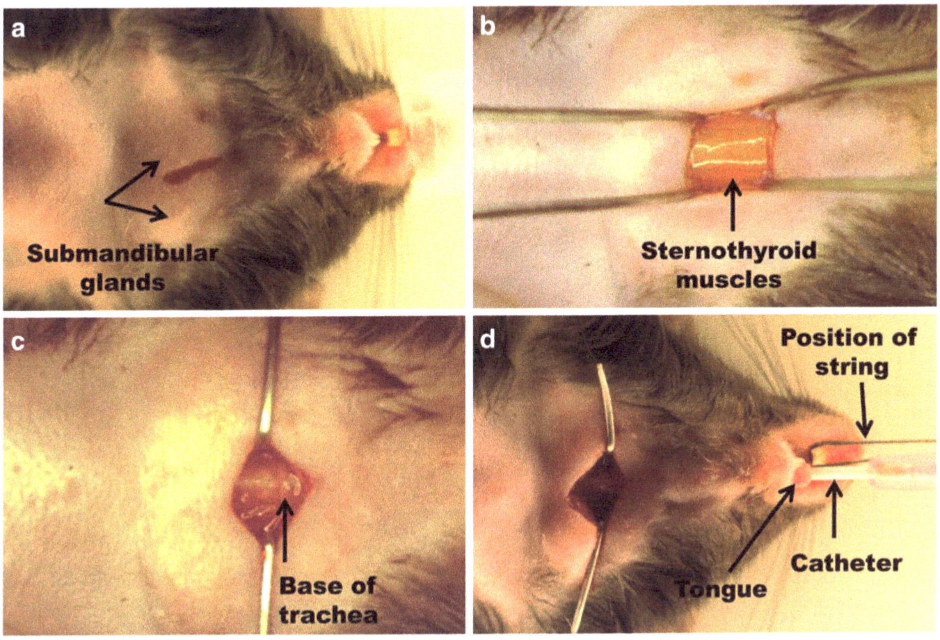

Fig. 4 Intubation procedure. (**a**) The incision for intubation is made using the submandibular glands as a landmark. (**b**) Beneath submandibular glands the sternothyroid muscles cover the trachea. (**c**) Retraction of the sternothyroid muscles allows for visualization of the base of trachea. (**d**) The catheter is properly guided using the incision located in the neck area

Ligation Procedure

1. Cut a 1 cm × 1 cm strip of gauze and place in saline.
2. Make a 1–1.5 cm transverse incision at mid-thorax starting left of ventral midline and extending laterally, parallel to the ribs and approximately 2 cm below the left axilla.
3. Use blunt dissection to expose the left pectoralis major muscles. Place a 3–0 retention suture around the muscles and retract toward the right shoulder of the mouse in order to expose the rib cage (Fig. 5a, b).
4. Place additional 3–0 retention suture around the left rectus thoracis and serratus ventralis muscles. Retract these muscles toward the left side of the mouse and affix the retention suture with tape to the surgical board as close to the body of the mouse as possible (Fig. 5a, b).
5. Bluntly dissect the muscle striations at the intercostal space between the third and fourth ribs to make a 0.25 cm incision that penetrates into the thoracic cavity (Fig. 5b).
6. Place the strip of gauze into the thoracic incision and gently push down toward lungs (Fig. 5c). It is important to avoid damage to the lungs.

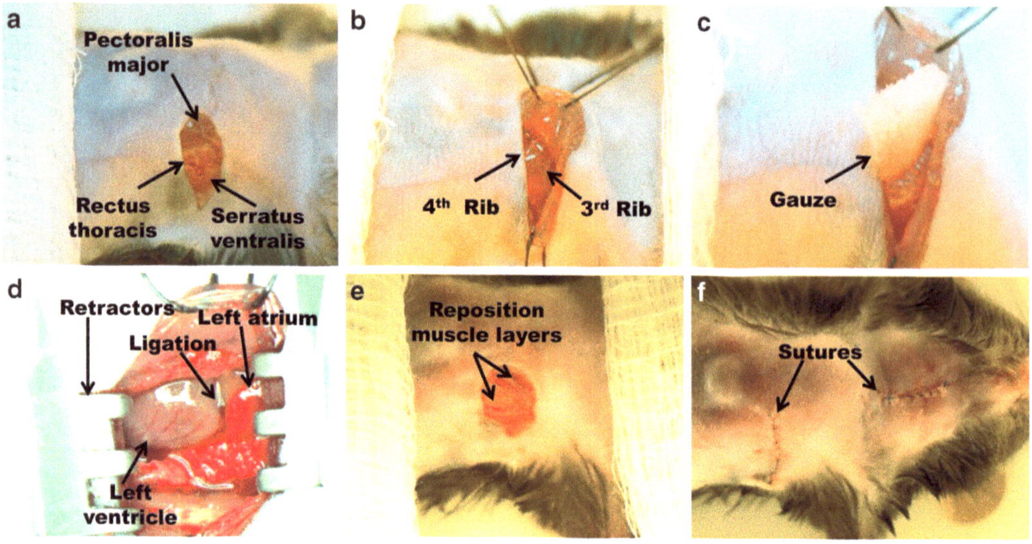

Fig. 5 Ligation procedure. (**a**) Incision and identification of the muscle layers covering the rib cage. (**b**) Retraction of muscle layers and location of the third intercostal space. (**c**) Position of the gauze to protect the lungs and assist with the intercostal incision. (**d**) Position of LAD ligation with respect to left atrium. (**e**) Relocation of muscle layers and (**f**) suturing the skin incisions

7. Slowly enlarge the thoracic incision to approximately 1 cm long using blunt dissection technique applied medially and laterally while using the gauze to protect the lungs.
8. Apply the Mini-Goldstein retractor to the incision and gently spread ribs in order to clearly visualize the left atrium and left ventricle (Fig. 5d).
9. Use fine forceps and blunt-ended curved scissors to remove pericardium and provide access to the left ventricular free wall for identification of the left anterior descending coronary artery (LAD). Identifying the artery is required for successful occlusion. For references on the murine coronary artery anatomy, please see refs. 9, 16.
10. Using a Castroviejo needle holder, guide an 8–0 suture under the LAD 1 mm distal to the left atrium and ligate securely with a square knot.
11. Confirm a discrete blanched region on the LV under microscope and ST segment elevation on the EKG.
12. Remove the rib retractor.
13. Place one 6–0 suture to encircle the outer edges of the separated ribs, remove the gauze, and secure the knot so that the ribs are repositioned in the chest and the incision is closed. Remember to double-check that the gauze was removed prior to closing the chest.
14. Remove the retention sutures and reposition the retracted muscle layers of the chest (Fig. 5e).

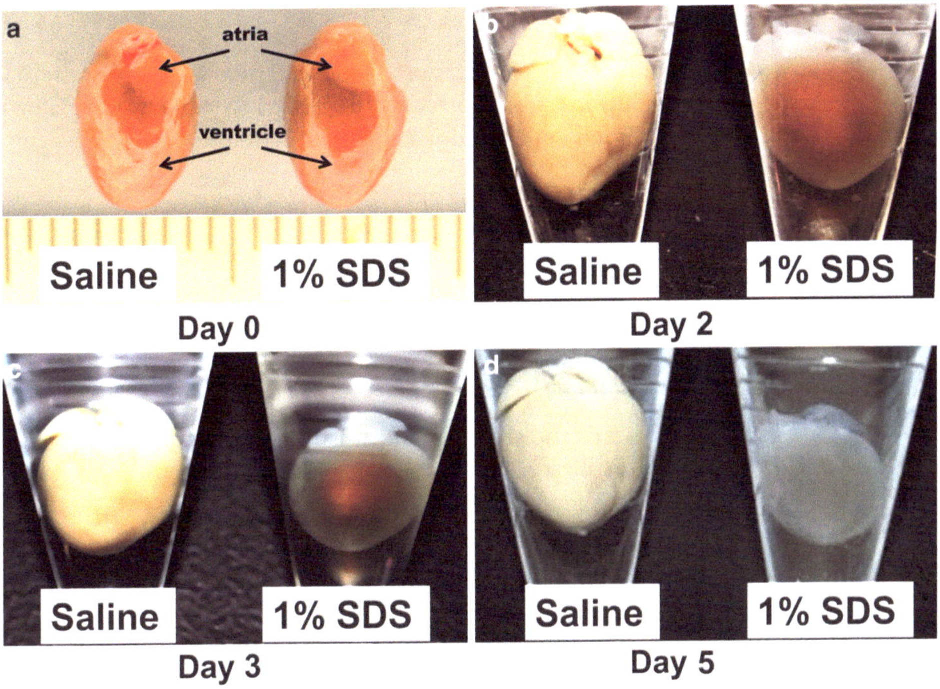

Fig. 6 Decellularization process of the whole heart. (**a**) Representative day 0 images of freshly dissected heart ventricle and atria in saline (*left*) and 1 % SDS (*right*). (**b–d**) Representative images of (**b**) day 2, (**c**) day 3, and (**d**) day 5 whole heart in saline (*left*) and 1 % SDS (*right*)

15. Close the thoracic and neck incisions using 6–0 suture (*see* Fig. 5f and **Note 2**).
16. Administer Buprenex (0.1 mg/kg IP).
17. Remove all the tape, and turn off isoflurane but allow the oxygen to continue to flow.
18. Move the mouse to the prone position and extubate. After extubation, spontaneous breathing should begin immediately.
19. Remove the EKG needle electrodes.
20. Place animal in a 37 °C incubator to recover (*see* **Notes 3** and **4**).

Day 0 (Fig. 6a)

1. Dissect the LV and separate remote from infarct tissue, based on TTC staining.
2. Record infarct and remote wet tissue weight separately.
3. Incubate remote and infarct tissue separately in a round-bottom tube containing 4mL of distilled H_2O with 1× protease inhibitor for 30 min with gentle shaking.
4. Decant the solution and replace with 1 % SDS with 1× protease inhibitor cocktail at room temperature for 24 h with gentle shaking.

Day 1

1. Collect 1 mL of the supernatant, discard the remaining solution, and replace with fresh 1 % SDS with 1× protease inhibitor cocktail for 24 h with gentle shaking.

Day 2 (Fig. 6b), Day 3 (Fig. 6c), and Day 4

1. Each day, discard the solution and replace with fresh 1 % SDS with 1× protease inhibitor cocktail at room temperature for 24 h with gentle shaking.

Day 5 (Fig. 6d)

1. Discard the solution and wash decellularized tissue with 2 mL of distilled H_2O with 1× protease inhibitor for 5 min with gentle shaking (*see* **Note 5**).
2. Discard the solution and wash with 4 mL of distilled H_2O with 1× protease inhibitor for 24 h with gentle shaking.
3. Discard the solution and transfer decellularized tissue to 1.5 mL tube and add Sigma reagent 4 with 1× protease inhibitor cocktail at 5 μL for every 1 mg of wet tissue weight.
4. Homogenize the tissue for 5 s, four times with 1-min intervals.
5. Sonicate the homogenate for 5 s, four times. Incubate the sample at 30 °C for 1 h.
6. Store at −80 °C.
7. Rewarm the sample to 30 °C for 1 h before determining protein concentration assay.

4 Notes

1. Blunt dissection is the primary technique used in this surgical procedure. The only steps that require the use of surgical scissors are **step 2** during intubation and **steps 2**, 7, and **9** during the ligation procedure.
2. When performed correctly, no bleeding should occur throughout the entire procedure. However, if blood loss occurs, saline can be injected intraperitoneally.
3. Average time for the entire procedure is 25–30 min.
4. Representative baseline and post-MI EKGs, echocardiograms, and TTC-stained images are shown in Fig. 7.

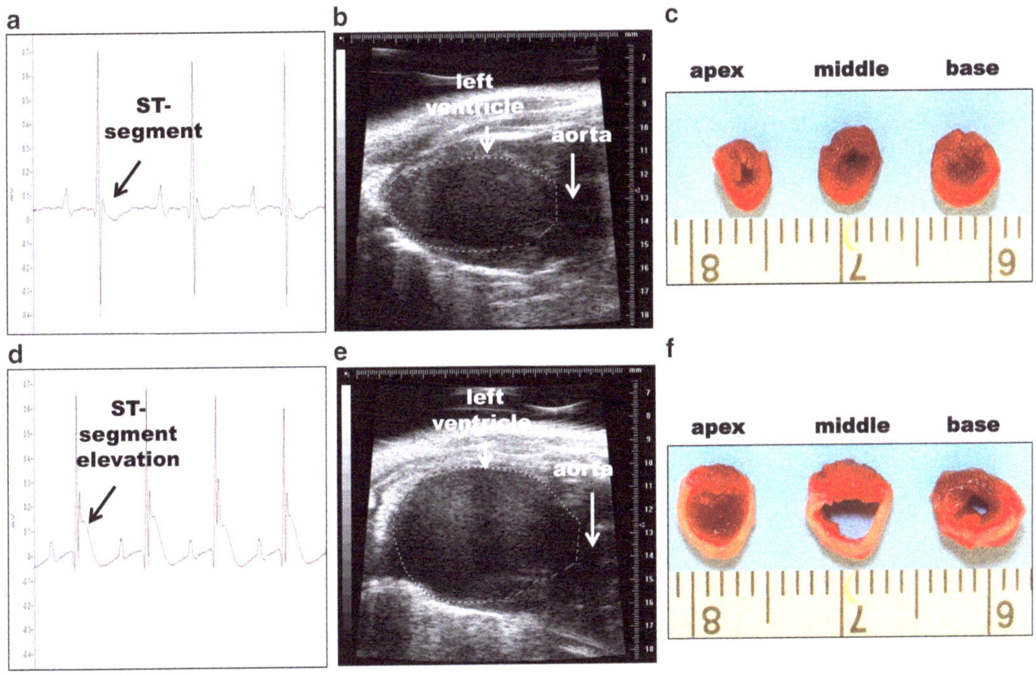

Fig. 7 Control (**a–c**) and post-MI (**d–f**) EKGs, echocardiograms, and TTC-stained images. (**a**) Murine EKG, (**b**) long-axis echocardiogram, and (**c**) TTC-stained left ventricle in control mice. (**d**) 15-min post-MI EKG, (**e**) day 7 post-MI long-axis echocardiogram, and (**f**) matching day 7 TTC-stained left ventricle

5. Decellularized tissue appears translucent in color. If the tissue still appears yellow in the center after 5 days, add fresh 1 % SDS with 1× protease inhibitor cocktail and continue to incubate at room temperature for an additional 24 h.

Acknowledgements

We acknowledge funding from AHA (09POST2150178) to RZ, NCCAM (1K99AT006704-01) to GH, and NHLBI HHSN 268201000036C (N01-HV-00244) for the UTHSCSA Cardiovascular Proteomics Center and R01 HL-075360, the Max and Minnie Tomerlin Voelcker Fund, and the Veteran's Administration (Merit) to MLL.

References

1. Krumholz HM, Wang Y, Chen J, Drye EE, Spertus JA, Ross JS, Curtis JP, Nallamothu BK, Lichtman JH, Havranek EP, Masoudi FA, Radford MJ, Han LF, Rapp MT, Straube BM, Normand SL (2009) Reduction in acute myocardial infarction mortality in the united states: risk-standardized mortality rates from 1995–2006. JAMA 302:767–773
2. Lloyd-Jones D, Adams RJ, Brown TM, Carnethon M, Dai S, De Simone G, Ferguson TB, Ford E, Furie K, Gillespie C, Go A, Greenlund K, Haase N, Hailpern S, Ho PM, Howard V, Kissela B, Kittner S, Lackland D, Lisabeth L, Marelli A, McDermott MM, Meigs J, Mozaffarian D, Mussolino M, Nichol G, Roger VL, Rosamond W, Sacco R, Sorlie P, Stafford R, Thom T, Wasserthiel-Smoller S, Wong ND, Wylie-Rosett J (2010) Executive summary: heart disease and stroke statistics–2010 update: a report from the American Heart Association. Circulation 121:948–954
3. Pfeffer MA, Pfeffer JM (1987) Ventricular enlargement and reduced survival after myocardial infarction. Circulation 75:IV93–IV97
4. Pfeffer MA, Braunwald E (1990) Ventricular remodeling after myocardial infarction. Experimental observations and clinical implications. Circulation 81:1161–1172
5. Sutton MG, Sharpe N (2000) Left ventricular remodeling after myocardial infarction: pathophysiology and therapy. Circulation 101: 2981–2988
6. Blankesteijn WM, Creemers E, Lutgens E, Cleutjens JP, Daemen MJ, Smits JF (2001) Dynamics of cardiac wound healing following myocardial infarction: observations in genetically altered mice. Acta Physiol Scand 173: 75–82
7. Lindsey ML, Zamilpa R (2012) Temporal and spatial expression of matrix metalloproteinases and tissue inhibitors of metalloproteinases following myocardial infarction. Cardiovasc Ther 30(1):31–41
8. Zamilpa R, Lindsey ML (2010) Extracellular matrix turnover and signaling during cardiac remodeling following MI: causes and consequences. J Mol Cell Cardiol 48:558–563
9. Salto-Tellez M, Yung Lim S, El-Oakley RM, Tang TP, ZA AL, Lim SK (2004) Myocardial infarction in the c57bl/6j mouse: a quantifiable and highly reproducible experimental model. Cardiovasc Pathol 13:91–97
10. Wang J, Bo H, Meng X, Wu Y, Bao Y, Li Y (2006) A simple and fast experimental model of myocardial infarction in the mouse. Tex Heart Inst J 33:290–293
11. Zamilpa R, Kanakia R, Jt C, Dai Q, Escobar GP, Martinez H, Jimenez F, Ahuja SS, Lindsey ML (2011) Cc chemokine receptor 5 deletion impairs macrophage activation and induces adverse remodeling following myocardial infarction. Am J Physiol Heart Circ Physiol 300:H1418–H1426
12. Kumar D, Hacker TA, Buck J, Whitesell LF, Kaji EH, Douglas PS, Kamp TJ (2005) Distinct mouse coronary anatomy and myocardial infarction consequent to ligation. Coron Artery Dis 16:41–44
13. Klocke R, Tian W, Kuhlmann MT, Nikol S (2007) Surgical animal models of heart failure related to coronary heart disease. Cardiovasc Res 74:29–38
14. Pfeffer MA, Pfeffer JM, Fishbein MC, Fletcher PJ, Spadaro J, Kloner RA, Braunwald E (1979) Myocardial infarct size and ventricular function in rats. Circ Res 44:503–512
15. Ott HC, Matthiesen TS, Goh SK, Black LD, Kren SM, Netoff TI, Taylor DA (2008) Perfusion-decellularized matrix: using nature's platform to engineer a bioartificial heart. Nat Med 14:213–221
16. Michael LH, Entman ML, Hartley CJ, Youker KA, Zhu J, Hall SR, Hawkins HK, Berens K, Ballantyne CM (1995) Myocardial ischemia and reperfusion: a murine model. Am J Physiol 269:H2147–H2154

Chapter 19

Injury Models to Study Cardiac Remodeling in the Mouse: Myocardial Infarction and Ischemia–Reperfusion

Daniel J. Luther, Charles K. Thodeti, and J. Gary Meszaros

Abstract

Deep tissue wound healing requires a complex sequence of several factors working in unison to repair the organ at risk. Myocardial infarction (MI) is particularly complex due to several local and systemic factors mediating the repair process within the heart. The wound healing process during this time is critical—the cardiac myocytes are at risk of apoptotic cell death, autophagy, and necrosis. During the early remodeling period, the fibroblasts and myofibroblasts play critical roles in infarct scar formation, a process that is greatly influenced by a robust inflammatory response. Construction of the infarct scar is a "necessary evil" that helps to limit expansion of the infarction; however, the collagen and matrix deposition will often spread to the healthy areas of the heart, causing reactive fibrosis in areas remote from the original damage. This chapter outlines in detail the procedures for two myocardial infarction injury models as well as how to quantify the size of the experimentally induced injury. These procedures are critical to the development of in vivo approaches to study myocardial injury, particularly for use in knockout and transgenic mice.

Key words Myocardium, Ischemia, Cardiac remodeling, Fibrosis, Extracellular matrix, Collagen

1 Introduction

Myocardial infarction (MI) continues to be the leading clinical cause of heart failure and mortality. Depending on the severity of MI, the heart may undergo a complexity of structural changes known as ventricular remodeling that begins with compensatory hypertrophy; however, decompensation ensues leading to chamber dilation, wall thinning, reparative fibrosis in the infarcted region, and reactive fibrosis in remote regions [1–5]. At the cellular level, cardiac fibroblasts and myofibroblasts are the key mediators of remodeling by producing several types of collagen, the vast majority being the fibrillar type I collagen; however, other fibrillar and non-fibrillar collagens are also synthesized during injury repair. Both of these cell types can synthesize and secrete collagen;

Robert G. Gourdie and Tereance A. Myers (eds.), *Wound Regeneration and Repair: Methods and Protocols*, Methods in Molecular Biology, vol. 1037, DOI 10.1007/978-1-62703-505-7_19, © Springer Science+Business Media New York 2013

however, the myofibroblasts are hypersecretory and considered as specialized wound healing cells.

The majority of the initial studies on cardiac remodeling and fibrosis were based on in vitro models using cardiac fibroblasts isolated from rat hearts. This work was important to characterize the fibroblast phenotype and to identify the hormonal and mechanical mediators of cardiac fibroblast function. This in vitro work was viewed as being largely descriptive by several investigators and with the movement towards translational research, it was clear that those who studied reparative wound healing in the heart would have to develop in vivo models of cardiac injury and repair.

Our laboratory also performed our initial studies on isolated cardiac fibroblasts using an in vitro cell–matrix model that initially involved comparisons of how these cells responded to various extracellular matrix substrates. We first compared several types of purified collagens as substrates for the fibroblasts, and measured which types of collagen could alter cardiac function in terms of proliferation and myofibroblast differentiation. We found that type I and III collagen both induced proliferation; however, the most interesting observation was that type VI collagen was capable of inducing myofibroblast differentiation [6]. We speculated that this non-fibrillar collagen, which was previously thought to be of minor importance in the heart, may possibly play a larger role in the cardiac remodeling process by mediating the fibroblast-to-myofibroblast transition during injury. To test this, we developed the MI mouse model in our laboratory which can be extended to any future mouse models that investigators may want to pursue.

We used the collagen VI knockout mouse originally developed and described by Bonaldo and colleagues as a model for Bethlem myopathy [7, 8]. These mice suffer from early apoptosis of the skeletal myocytes as they age, causing progressive weakness and limited life span. Given this, we predicted that similar loss of function would occur in the hearts of these animals, particularly after induction of MI injury. Surprisingly, we have discovered that the lack of type VI collagen actually *improves* cardiac remodeling and function following MI, results that were completely unexpected. Our work on this knockout model is ongoing as we search for the critical mechanisms leading to the improved remodeling and cardioprotection. Importantly, we believe that this potentially novel role for collagen VI in myocardial remodeling would not have been uncovered without applying the MI injury model to this knockout model and in vivo models of MI are indispensable tools for studying the role of collagens as well as other key proteins in the in vivo setting. In this chapter, we describe in detail protocols for mouse models of MI and ischemia–reperfusion (I–R) injury along with the assessment of infarct size. These methodologies can provide a means to bridge in vitro studies to the whole animal and create new avenues and targets for translational studies in cardiac remodeling.

2 Materials

2.1 Mouse Model of Myocardial Infarction

1. Drugs:
   (a) *Anesthetic: Nembutal*® Sodium Solution (sodium pentobarbital, 50 mg/mL). Immediately prior to use, dilute Nembutal solution in 0.9 % saline to achieve 20 mg/mL concentration. *Dosage*: 70 mg/kg, intraperitoneal (i.p.) injection.
   (b) Atropine sulfate: *Dosage*: 0.04 mg/kg, intramuscular (i.m.) injection.
   (c) *Ketofen*® (Ketoprofen): *Dosage*: 3 mg/kg, subcutaneous (s.c.) injection.
   (d) 1 % Lidocaine HCl.
   (e) 2 % Lidocaine HCl Oral Topical Solution (Viscous).
   (f) Penject (Penicillin G w/ Procaine): *Dosage*: 44,000 units/kg, s.c.
2. Surgical tools:
   (a) Two curved forceps.
   (b) Scalpel w/no. 10 blade.
   (c) Strabismus scissors—curved/blunt–blunt/11.5 cm.
   (d) Spring scissors—slightly curved/sharp/15 mm cutting edge.
   (e) Olsen-Hegar needle holder with scissors—12 cm w/lock.
   (f) Castroviejo micro needle holder—straight/smooth/9 cm w/lock.
   (g) COOK eye speculum (used as rib spreaders).
   (h) Small cauterizer.
3. Suture:
   (a) 6–0 Ethilon®, nylon suture w/reverse cutting (C-2) needle.
   (b) 6–0 Vicryl®, polyglactin suture w/taper (RB-1) needle.
   (c) 8–0 Ethilon®, nylon suture w/taper (BV130-5) needle.
4. MiniVent Type 845 mouse ventilator.
5. ECG apparatus—PowerLab® 4/25T
6. *Betadine*® solution (10 % providone-iodine).
7. *Nair*® hair removal gel.
8. Intubation tube (PE-60 tubing approximately 45–50 mm long).
9. 5 mL Syringe w/blunted 23 G needle fitted w/PE-50 tubing (chest tube).

10. Rectal temperature probe.
11. Surgical lamp (120 W).
12. Heating lamp (60 W).
13. 70 % (v/v) EtOH in H_2O.
14. Lactated Ringer's solution.
15. Sterilized H_2O.
16. Surgical gauze, tape, and cotton-tipped applicators.

2.2 Ischemia/Reperfusion Surgery

1. *See* Subheading 2.1, **items 1–16**.
2. 3–4 mm piece of PE-90 tubing (used as snare for LAD ligation).
3. Timer.

2.3 Whole-Animal Perfusion (Gravity Fed)

1. *Anesthetic*. Fatal-Plus® Solution (pentobarbital sodium solution, 390 mg/mL): *Dosage*: 1 mL/10 lbs (*as indicated on bottle*).
2. Heparinized normal saline [heparin sodium, 10 U/mL of 0.9 % NaCl in H_2O (DEPC-H2O if for RNA in situ)].
3. *Fixative*. 4 % Paraformaldehyde, made immediately prior to use.
4. Phosphate-buffered saline (PBS).
5. 70 % (v/v) EtOH in H_2O.
6. Perfusion apparatus:
   (a) 60 cc Syringe housing (no plunger) (used as vessel for holding fixative—alternative vessels can be substituted, i.e., Empty Saline I.V. Bag).
   (b) Tygon® Formula 3350 silicone tubing (I.D.: 1/16 in. (1.6 mm); wall thickness: 1/32 in. (0.8 mm); O.D.: 1/8 in. (3.2 mm).
   (c) Male-to-male Luer fitting.
   (d) 2-way stopcock.
   (e) Two 23 g 1″ needles (one needle can be broken off from Luer stub adapter with a pair of hemostats—care must be taken not to pinch needle closed when removing).
   (f) 12″ PE-50 tubing.
7. Ring stand (200 cm minimum height).
8. 5–10 mL vials w/lid for samples.
9. Surgical scissors.
10. Curved forceps.
11. Hemostat.
12. Absorbent pad.
13. Surgical gauze, tape, and cotton-tipped applicators.

2.4 Infarct Size Assessment: TTC Method

1. *Anesthetic*. Fatal-Plus® Solution (pentobarbital sodium solution, 390 mg/mL): *Dosage*: 1 mL/10 lbs (*as indicated on bottle*).
2. Heparinized normal saline (heparin sodium, 10 U/mL of 0.9 % NaCl in H_2O).
3. 2, 3, 5-triphenyltetrazolium chloride (TTC) (1 % w/v in PBS).
4. 2 % Evan's blue dye.
5. 4 % Paraformaldehyde, made immediately prior to use.
6. PBS.
7. 70 % (v/v) EtOH in H_2O.
8. Perfusion apparatus (*see* Subheading 2.3) 23 G needle should be blunted for this procedure.
9. Shaking water bath (37 °C).
10. Acrylic heart matrix (mouse, 1.0 mm section width).
11. Western blotting glass spacer slide (1.0 mm spacing) and flat glass cover slide.
12. Four small binder clips.
13. Micro-centrifuge tubes (1.7 mL).
14. 12-Well plate.
15. Razor blades.
16. 5–0 silk suture.
17. Surgical scissors.
18. Small, fine spring scissors (2 mm cutting edge).
19. Curved forceps.
20. Dumont #5 fine forceps.
21. Hemostats.
22. Absorbent pad.
23. Surgical gauze, tape, and cotton-tipped applicators.

3 Methods

3.1 Myocardial Infarction Surgery

1. Prepare fresh aliquot of diluted Nembutal® for anesthesia.
2. Once animal is weighed and drug doses are calculated, administer atropine sulfate (i.m.) as a preanesthetic.
3. After 5 min, gently grasp mouse by the tail between two fingers and with the same hand grab the scruff on the back of the neck. Invert mouse and inject anesthesia approximately 1/3 of the way up the abdomen into the peritoneal cavity.
4. Isolate mouse in separate cage under a heating lamp (60 W) approximately 18–24″ away from source. Allow animal to rest

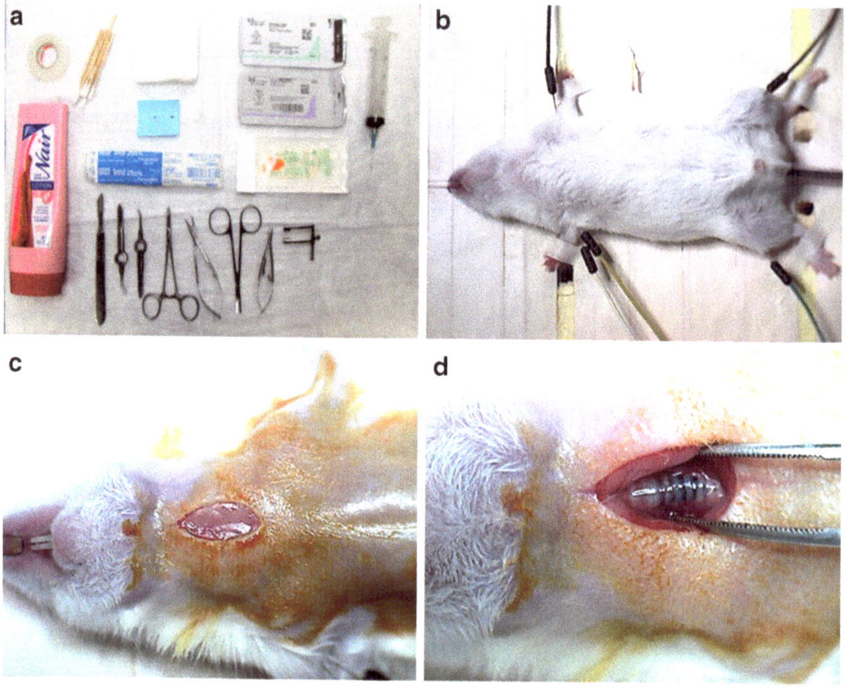

Fig. 1 (**a**) Surgical materials and instruments for MI and I/R surgery. (**b**) Secure animal on a surgical platform in the supine position. Insert rectal temperature probe and ECG limb leads. (**c**) Remove hair from the neck and chest of animal and prep skin with Betadine® solution. Make a small (<1 cm) incision in the neck of the mouse. (**d**) Blunt dissection of the submaxillary gland and underlying muscles with forceps to expose the trachea. *Black* tip of intubation tube (PE-60 tubing) is shown inside the trachea

undisturbed for approximately 5–10 min. After 10 min check reflexes by toe-pinch method (*see* **Note 1**).

5. Once animal is at an adequate plane of anesthesia, remove from cage and place in the supine position onto a surgical table. Secure limbs using surgical tape. Secure the head by hooking the incisors with a loop of suture, pulling taut and securing the other end with a tape or a pin (Fig. 1b).
6. Insert rectal temperature probe and ECG limb electrodes. Temperature should be maintained between 36 and 37 °C. A heat lamp may be necessary in addition to the surgical lamp in order to maintain the desired temperature range (*see* **Note 2**).
7. Apply a dime sized amount of Nair® to the chest and neck of the animal and rub into the hair. Wait for 1 min and remove hair with moistened gauze.
8. Apply Betadine® solution to the center of chest and work outwards.
9. Using the scalpel, make a small <1 cm incision over the neck along the sagittal plane of the animal. Blunt dissect the

underlying submaxillary gland and sternohyoid muscle along the natural divisions of each to visualize the trachea (Fig. 1c).

10. To intubate the animal, first, lubricate the beveled tip of the intubation tube (PE-60 tubing) with 2 % lidocaine oral gel. Next, with one hand, grasp the tongue and gently lift and move to the side as you insert intubation tube at a slightly upward angle with the other hand. Once in the oral cavity you can release the tongue and with sterile forceps open the window to the trachea so that you can visualize the intubation tube once inside the trachea. Gently insert the intubation tube into the trachea until the tip is visible through the neck incision (Fig. 1d) (*see* **Notes 3–4**).
11. Once intubated, check for normal breathing patterns to ensure that airway and tube are not obstructed. When breathing is stable, connect to the ventilator and begin to ventilate with room air (*see* **Notes 5–6**).
12. Using 6–0 nylon suture, close neck incision.
13. While paying close attention to the intubation, carefully rotate mouse onto its right side and prop with gauze pillow secured with tape.
14. Locate the axilla and tip of sternum. Make a transverse skin incision on the left side of animal halfway between these points extending from the sternum to the lateral side of chest (Fig. 2a).
15. Dissect pectoral muscles to visualize ribs. Begin to perform left thoracotomy between the fourth and fifth rib using sharp spring scissors. Once incision is begun, gently grasp fourth rib with forceps and lift away from internal organs. Before penetrating thoracic cavity switch to blunt-tipped Strabismus scissors to avoid damaging underlying organs and make 1 cm incision in intercostal muscle (Fig. 2b) (*see* **Note 7**).
16. Insert COOK eye speculum into incision and gently spread ribs to visualize heart. Cauterize any bleeding vessels if necessary (Fig. 2c).
17. Using forceps, blunt dissect the pericardial sac and clear pericardium away from anterior wall of heart (Fig. 2d).
18. Identify the left atrium. The LAD runs from underneath the left atrium towards the apex of the heart, branching approximately at the mid papillary level. Using an 8–0 nylon suture and Castroviejo micro needle holders pass suture underneath the LAD approximately 1–2 mm down from the left atrium towards the apex, trying to avoid penetrating into the ventricular chamber (Fig. 3a) (*see* **Note 8**).
19. Securely tie suture to induce infarction. Blanching of the apex should be immediately apparent (Fig. 3b). Monitor the ECG for changes, particularly ST segment elevation to confirm infarction.

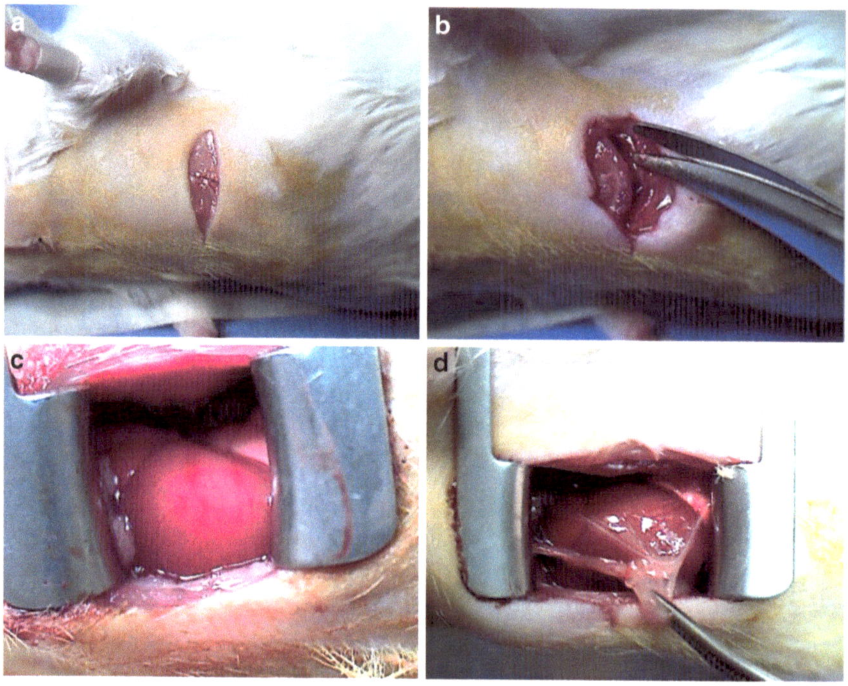

Fig. 2 (**a**) Once intubated, position animal on its side (*right* side down) and secure the *left* forelimb of the animal above its head with surgical tape. Using a scalpel, make a skin incision halfway down the *left* rib cage from the sternum to the lateral chest wall. (**b**) Perform a left thoracotomy at the fourth intercostal space using scissors. (**c**) Insert the COOK eye speculum into the rib incision and gently open the chest. (**d**) Blunt dissection of the pericardial sac using forceps

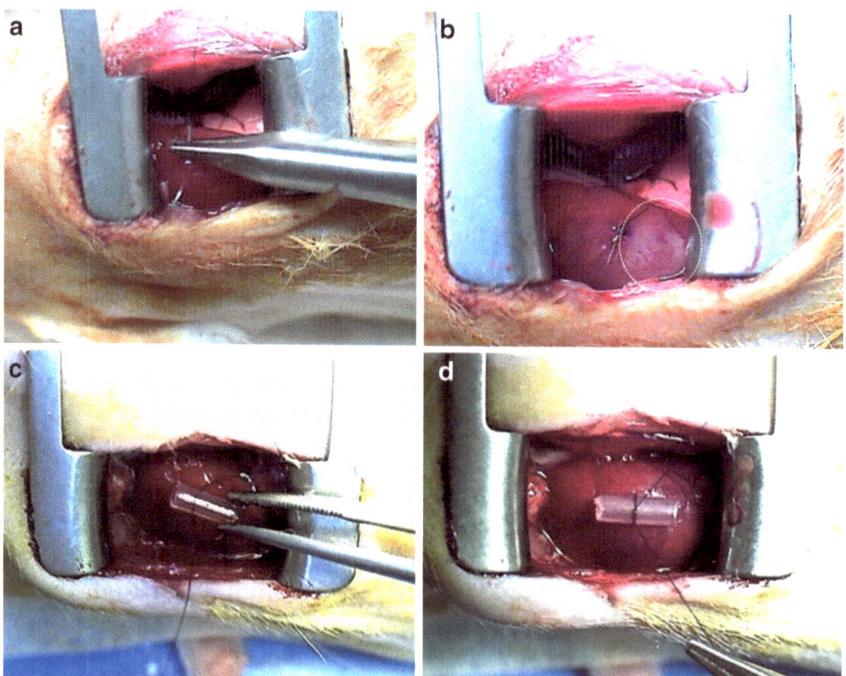

Fig. 3 (**a**, **b**) Pass an 8–0 nylon suture underneath the LAD and for MI procedure, tie the suture to permanently occlude the LAD, and induce MI (ischemic region outlined in *white*, above). (**c**) For I/R surgery, before tying the suture, insert small (3–4 mm) PE-90 tube between the myocardium and the suture to act as a snare. (**d**) Secure the tubing in place by tying the suture to induce ischemia

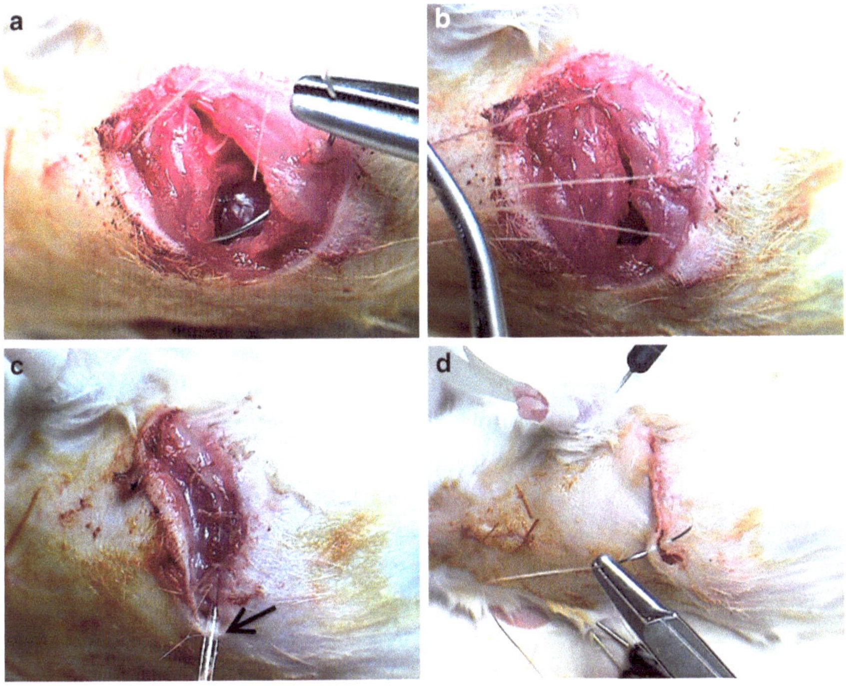

Fig. 4 (**a**, **b**) To close the chest incision, use an interrupted suturing pattern and place 6–0 Vicryl® sutures together through both sides of the ribs and the chest muscle layer. (**c**) Before tying the last suture, gently insert the chest tube (*arrow*) and secure by tying the remaining suture. (**d**) Once negative pressure inside the thoracic cavity is restored, remove the chest tube, check for leaks in the chest wall, and close skin incision with 6–0 nylon suture

20. Once confirmed, place 1–2 drops of 1 % lidocaine directly onto the heart and remove COOK eye speculum.
21. With extreme care, gently insert closed forceps between left lung lobe and rib cage and lay the left lung lobe onto the heart to avoid puncturing when closing the ribs.
22. Using a 6–0 Vicryl suture, place sutures through both sides of the ribs and the chest muscle layer together using an interrupted suturing pattern (Fig. 4a, b). Do not tie the sutures until all sutures are in place. This allows for visualization of visceral organs to avoid puncturing while placing the sutures.
23. Once sutures are in place, begin to tie sutures (starting laterally and working medially towards sternum). Before tying the last suture, insert the chest tube between the last suture and sternum, and secure by tying the last suture (Fig. 4c) (*see* **Note 9**).
24. Draw back on the syringe connected to chest tube until adequate negative pressure is achieved. Maintain negative pressure with chest tube for 1–2 min. A small amount of blood, if present in the cavity, may become visible in tube (*see* **Note 10**). Expand (sigh) lungs to remove additional air in the thoracic

cavity before removing chest tube. This can be accomplished simply by blocking the exhaust on the ventilator for three cycles. DO NOT OVERINFLATE LUNGS.

25. Once negative pressure is reestablished in the chest cavity, gently remove the chest tube. Use the pair of forceps to pinch muscle and skin layer around chest tube as it is slowly removed.
26. Using a 6–0 nylon suture, close skin incision (Fig. 4d).
27. Reapply Betadine® solution to incision sites.
28. Administer analgesic (Ketofen®) and penicillin via s.c. injection in an area away from surgical site. A bolus s.c. injection of lactated Ringer's solution is also suggested for fluid replacement (1 % of animal's weight in grams; i.e., 0.3 mL for a 30 g mouse).
29. Check for reflexes via toe-pinch method. Once animal is responding, attempt to remove animal from ventilator, leaving the intubation tube in place. Observe for chest movement to ensure breathing. If breathing does not start then put back on ventilator until animal regains breathing reflex.
30. Once off ventilator, place animal in an isolated recovery area under heat lamps to maintain body temperature until fully recovered and the mouse is able to maintain body temperature on its own (*see* **Note 11**).
31. After 24 h, additional analgesic should be administered to animal.

3.2 Ischemia/Reperfusion Surgery

1. (*See* Subheading 3.1, **steps 1–16**.)
2. Identify the left atrium and the LAD as previously described (*see* Subheading 3.1, **step 17**). Using an 8–0 nylon suture and Castroviejo micro needle holders pass suture underneath the LAD approximately 2 mm down from the left atrium towards the apex, trying to avoid penetrating into the ventricular chamber (*see* **Note 12**). Begin to make a surgeon's knot but do not pull tight completely. Before tightening, insert the suture snare (PE-90 tubing) between the myocardium and suture and then secure knot (Fig. 3c, d) (*see* **Note 13**). At this point, start the timer to begin measuring the period of ischemia.
3. Be sure that tubing is not resting against atria or other internal organs that may be damaged as it will vibrate while the heart is beating. Observe apex for blanching and ST-elevation on ECG. Loosen and close COOK eye speculum and partially close chest. Cover incision with moistened gauze and monitor animal's vital signs during length of ischemia. The underlying visceral organs, muscles, and skin may need to be moistened with saline during this time if they become dry (*see* **Note 14**).
4. Once the desired length of ischemia has passed, expand the eye speculum and reopen the chest. Gently remove the snare so

that slack is created in the suture. Insert small spring scissors into the slack created and cut the suture. This will allow the distal portions of the LAD and myocardium to reperfuse. Look for color to return to the apex of heart and resolution of the ST elevation in the ECG to ensure that reperfusion has occurred.

5. Once reperfusion is confirmed, place 1–2 drops of 1 % lidocaine onto the heart and remove COOK eye speculum.
6. Close chest and skin incision as previously described (*see* Subheading 3.1, **steps 20–29**).

3.3 Whole-Animal Perfusion (Gravity Fed)

1. Inject 0.5 mL of heparinized saline i.p. Allow animal to rest for approximately 10 min for the heparin to take effect.
2. Set up perfusion apparatus on ring stand or I.V. pole approximately 125–150 cm above the animal. The perfusion apparatus is made using a 60 cc syringe with 2-way stopcock. Silicone tubing is then attached to the stopcock and fitted with male-to-male Luer stub fitting at the other end. The Luer fitting is then connected to a 23 G syringe needle fitted with PE-50 tubing and another 23 G needle for a cannula (Fig. 5a) (*see* **Note 15**).
3. Fill syringe housing with 3–5 mL of heparinized saline. Turn stopcock to "on" position and allow small amount of saline to drip from perfusion needle to remove all air from lines. Turn stopcock to "off" position once air is removed.
4. Weigh mouse and calculate dose for Fatal-Plus® Solution. Inject anesthetic i.p.
5. Secure mouse in the supine position on an absorbent pad using adhesive surgical tape. Wet the chest and abdominal hair with 70 % EtOH to sterilize the area and prevent spreading of hair.
6. Work quickly here, using surgical scissors and forceps to grasp abdominal skin, and begin to make an abdominal incision into the cavity just below the sternum.
7. Using scissors make an incision in the diaphragm from lateral wall to lateral wall to enter the thoracic cavity. Avoid cutting any visceral organs. Cut ribs on both lateral sides of rib cage towards the shoulders.
8. Grasp the sternum with hemostats and retract rib cage towards the head of the animal to visualize the heart. Using forceps, gently grasp the free wall of the right atrium and insert the perfusion needle into the apex of the left ventricle, penetrating the chamber for intraventricular administration of fixative (Fig. 5b).
9. Use scissors to snip the right atrium to allow blood and perfusate to escape.

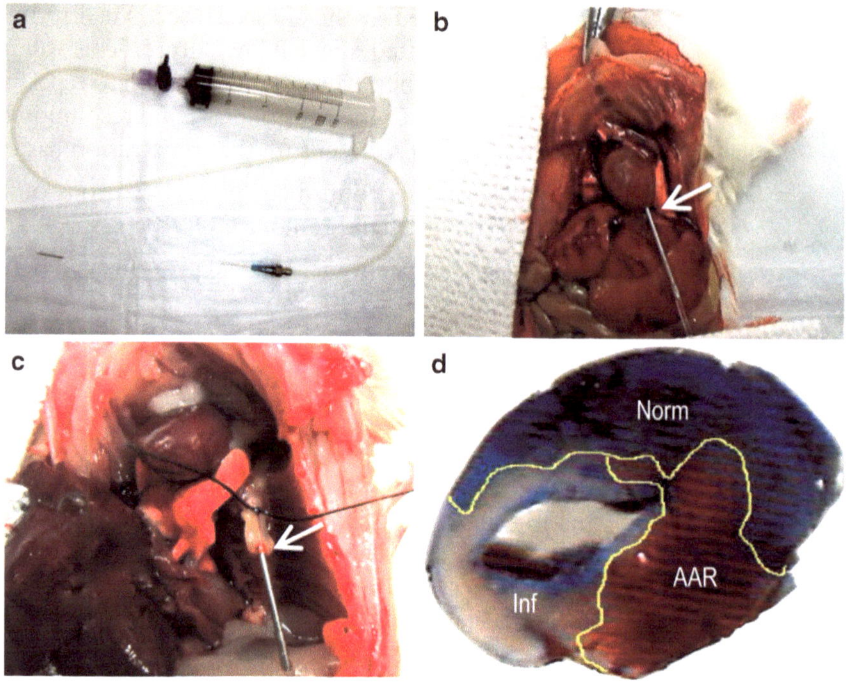

Fig. 5 (**a**) Perfusion apparatus used for whole-animal fixation protocol and perfusion of Evan's blue dye in infarct size assessment procedure. (**b**) For whole-animal fixation, insert the 23 G needle of the perfusion device through the apex of the heart (*arrow*) and into the left ventricle. (**c**) For perfusion of Evan's blue dye, cannulate the thoracic aorta (*arrow*) with a *blunted* 23 G needle and tie with suture to secure in place. (**d**) Heart section after TTC staining for infarct size assessment illustrating the normal/non-ischemic zone (*blue*), the AAR region (*red*), and the infarcted zone (*white*)

10. Turn stopcock to "on" position to start perfusion of saline in order to flush blood from circulation. Color should begin to fade from heart and liver during perfusion. Also note that perfusate from right atrium should become clear as blood is removed from the animal.
11. As the syringe approaches empty (approximately <0.5 mL), fill syringe with 10 mL of freshly prepared 4 % PFA. Avoid introducing any air bubbles into the system. Allow the entire volume of PFA to perfuse (approximately 10 min) (*see* **Notes 16–17**).
12. Once perfusion is complete, turn off system and remove needle from heart. Isolate and collect heart and any other tissues of interest.
13. Once tissue is trimmed and cleaned, it is recommended to place hearts in specimen vials and submerse with 4 % PFA for a minimum of 2 h to overnight at 4 °C.

3.4 Infarct Size Assessment (TTC Method)

1. Make fresh 1 % TTC solution in amber glass bottle or a similar vessel protected from light. Warm TTC to 37 °C prior to use.
2. Inject animal with 0.5 mL of heparinized saline via i.p. injection. Allow animal to rest for approximately 10 min for the heparin to take effect.
3. Prepare perfusion apparatus and animal as described above in Subheading 3.3, **steps 2–5**.
4. Once animal is prepared, grasp the skin just below the sternum and make an incision into the abdominal cavity. Reposition the liver by pulling it outside of the cavity to enhance visualization of the thoracic and abdominal aorta.
5. Make an incision in the diaphragm from lateral wall to lateral wall to enter the thoracic cavity. Avoid cutting any visceral organs. Cut ribs on both lateral sides of rib cage towards the shoulders. Clamp the sternum with hemostats and retractrib cage towards the head of the animal.
6. Identify the aorta, which runs parallel to the vertebral column in the posterior mediastinal cavity. Once identified, using care not to puncture any visceral organs or vessels, use the Dumont #5 fine forceps to pass a 8–10 cm length of suture underneath the thoracic aorta approximately halfway down its length. Begin to tie the suture, but do not tighten the knot completely. Knot should be loose enough to be able to pass a 23 G needle inside of the aorta through this knot.
7. If animal has undergone MI (permanent ligation) then proceed to next step. However, if animal underwent ischemia/reperfusion protocol then it is necessary at this point to re-occlude the LAD. If suture has been left in from the procedure it can be easily identifiable and can be retied. An additional suture may be passed underneath the LAD at the same exact location and tied to occlude if unable to retie the original suture.
8. Use small, fine spring scissors to make a small incision in the thoracic aorta just proximal to the suture knot. With the aid of the Dumont #5a fine forceps, insert the *blunted* 23 G needle attached to the perfusion apparatus into the aorta in a retrograde direction towards the heart. Secure needle by tightening the suture knot (Fig. 5c).
9. Once secured, using the spring scissor make a small nick in the right atrium to allow perfusate to drain.
10. Turn the stopcock to the "on" position and begin to perfuse heart retrogradely with heparinized saline. Once flushed, add 3–5 mL of 2 % Evan's blue dye to syringe housing and begin to perfuse. Immediate color change should be noticeable throughout the animal. However, the ischemic region of the

heart (antero-apical portion of heart if LAD occlusion) should not change color due to the occluded vessel preventing perfusion with Evan's blue dye (*see* **Note 18**).

11. Once heart has been adequately perfused and desired intensity of staining has been achieved, turn off perfusion apparatus and remove cannula from aorta. Excise heart and immediately rinse in PBS to flush the remaining dye from chamber. Once flushed, remove heart and dry and wrap in small piece of plastic wrap. To aid in cutting, place heart in freezer for 10 min (optional).
12. Once frozen, remove heart and place in heart matrix. Starting from the base of the heart, insert razor blades in 2 mm intervals towards the apex. Do not cut the heart until all blades are in place. Once all blades are in place, simultaneously section heart with all blades.
13. Remove section from matrix and place in labeled (e.g., base, middle, apex) Eppendorf tubes and fill tubes with warmed (37 °C) 1 % TTC solution. Insert tubes into foam holder and place in shaking water bath (37 °C) for 15–20 min. This will stain the viable regions of the ischemic myocardium brick red and necrotic tissues white. Normal zones without ischemia will remain blue.
14. Next, remove samples and place in appropriately labeled wells of 12-well plate. Submerge sections in freshly prepared 4 % PFA. Place 12-well plate onto lab shaker plate and incubate for 15–20 min at room temperature. This step will help bleach the white necrotic regions and enhance the contrast of colors between zones.
15. Remove from PFA and blot dry using gauze. Align sections serially (keeping orientation similar and in order from base to apex) on western spacer plates. Once aligned, sandwich sections by clamping a glass plate on top of spacer plate using small binder clips on all sides. By adding this slight pressure to sections, zones become much more distinguishable for measure.
16. Using a macroscopic imaging device, image both sides of each section at a magnification that allows for clear identification and measure of normal zone (blue), area at risk (AAR; red), and infarcted zone (white) (Fig. 5d). Assess images by planimetry method using ImageJ software (NIH, Bethesda, Maryland, USA) or other imaging software.
17. Infarct size should be described relative to AAR to standardize for any variability in size of ischemic region due to any inconsistencies in technique or animal variability. This is most commonly expressed as an infarct-to-AAR ratio.

4 Notes

1. It is important to leave the animal undisturbed after anesthetizing. If repeatedly checked for reflexes or if in an environment where ambient noise is high, animal may never reach the plane of anesthesia necessary to begin the procedures described above.
2. Body temperature can have a profound effect on infarct size. Therefore, monitoring and maintenance of body temperature within the range of 36–37 °C during the infarction procedure are essential for consistent infarct size.
3. The intubation steps have proven to be the most difficult portion of the procedure. It may take multiple practice procedures before becoming adjusted to the delicacy of the tissues as well as the angle of approach and light force necessary to guide the intubation tube into the trachea. Larger mice (>35 g) are often easier to intubate and are very suitable for practice. This may also be taken into consideration when choosing the desired age and strain of mice for experimental models if you are experiencing difficulties intubating. A technique that has worked for our lab has been to enter the oral cavity parallel to the surgical platform until reaching the epiglottis while visualizing the entrance to the trachea through the window created by the neck incision. Once at the epiglottis, a slight upward angle of the tube is often necessary to enter the trachea and avoid slipping into the esophagus, which is common. Sometimes a gentle side-to-side rotation of the tube during this process aids in getting the intubation tube beyond the epiglottis. It is highly recommended to bevel and blunt the tip of the PE-60 tubing. Be sure that edges are not sharp to avoid puncturing of the trachea. Also, coloring the tip of the intubation tube black with permanent marker is suggested to help visualize the depth of the tube through neck incision (Fig. 1b).
4. Once intubation is in place, a suture may be placed through the cheek of the animal and used to secure the intubation tube and prevent possibility of extubation. Also, the neck incision may be left open until the end of the procedure to aid in monitoring of intubation depth and re-intubation if the tube moves during the procedure.
5. The tidal volume and ventilation rates are calculated from the following equation (as provided by Harvard Apparatus):

   $V_t = 6.2 \times M_b^{1.01}$ where V_t is tidal volume (mL) and M_b is animal body mass (kg).

   $BPM = 53.5 \times M_b^{-0.26}$ where BPM is ventilation rate (breaths/minute).

6. Our protocol describes the ventilation of animals during surgery with ambient room air. However, other groups also describe the use of room air mixed with 100 % oxygen at a flow of 2 L/min. Initially, our lab utilized this mixture of ambient air supplemented with oxygen; however, we have found that using solely room air has greatly improved our survival rates.
7. When performing the left thoracotomy in infarction procedures, care should be taken when cutting the intercostal muscle close to the sternum. The internal mammary artery runs within close proximity (≈1 mm) and parallel to the sternum. Although usually not fatal, if cut large amounts of bleeding could occur.
8. Consistency of the placement of LAD occlusion is critical to minimize variability in infarct size. Identifying landmarks of the heart (e.g., left atrium, pulmonary artery, coronary sulcus) can help in consistent placement of ligature. Due to this, it is recommended to have MI groups of $n \geq 8$ to account for variability.
9. It is critical that the rib closure is adequate to maintain negative pressure inside the thoracic cavity. The presence of a pneumothorax will cause lungs to remain collapsed and will most likely result in death of the animal. The site of chest tube insertion is the most common location of any leaks. Therefore, careful placement of the chest tube and the suture used to secure both the ribs and the chest tube is essential. It is an option to make a small 1 mm incision through the pectoral muscles and the fifth intercostal space adjacent to thoracotomy incision and insert the chest tube at this point. Then secure the chest tube in place using a purse stitch through the chest muscle and around the tubing. Drip lidocaine solution onto incision to ensure that chest is tightly closed. If lidocaine is pulled into the chest cavity, then an additional suture may be needed near the leak and negative pressure must then be again reestablished.
10. When making the chest tube it is suggested to add additional openings at the distal end of PE-tubing. This will allow for continued suction if one opening becomes clogged by clots or connective tissue inside the chest.
11. While the animal is resting in the recovery cage, it may be placed inside of a large clear plastic bag loosely tied closed while being supplied with 100 % oxygen through a line passed through the opening (i.e., oxygen tent). Creating this oxygen-rich environment during the initial recovery period may help in survivability but should be utilized consistently between groups to further minimize variability in surgical outcomes between groups.

12. When placing the 8–0 suture used for the LAD occlusion in the ischemia/reperfusion surgery, it is an option to place a second suture in the exact location of the first during this time (however, left untied) to be later tied at the time of sacrifice to re-ligate the LAD during Evan's blue perfusion in infarct size assessment. This may help in accuracy of re-ligation to ensure proper Evan's blue staining, consistent with actual ischemic zone during the time of surgery.
13. Care must be taken when tying the ligature to occlude the LAD so as not to sever the artery. Reperfusion will not occur if severed.
14. If necessary, during longer surgical procedures (ischemia/reperfusion surgery) an additional dose of anesthetic (Nembutal®) may be needed to keep animal at the necessary plane of anesthesia. If so, dosage given should be 30 mg/kg i.p.
15. If desired, a perfusion pump can be utilized instead of a gravity-based perfusion setup. Just replace the perfusion apparatus with the pump, calculate appropriate perfusion rate (dependent on syringe size), and follow the same surgical steps described previously (*see* Subheading 3.3).
16. Depending on the ultimate use of tissue samples, the whole animal perfusion protocol can be adjusted to suit the needs of the investigation. This includes substituting different fixatives for PFA that are more compatible with specific histological/immunological techniques. For example, for electron microscopy studies, PFA can be substituted with a solution of 1.5 % glutaraldehyde/2.0 % paraformaldehyde to adequately fix tissues.
17. During the whole-animal fixation procedure, look for signs of a successful perfusion and fixation which include blanching of heart and liver, skeletal muscle twitching, curling of the tail, and stiffening of limbs.
18. During the perfusion of Evan's blue dye in the infarct assessment, it is important that the dye perfusate does not come into contact with the epicardial surface of the myocardium. This will complicate the distinguishability between zones, particularly the white infarcted regions. Laying the animal on its side once the cannula is secured in place and using gauze to absorb perfusate may help prevent this. It is also optional to excise the heart, leaving the aorta intact for cannulation and hanging the heart during the perfusion.

Common applications. The post-MI time period can vary from hours to days to weeks, depending upon the nature of the study. The histological assessments can also vary by need: selection of cryosections or paraffin sections is dictated by the outcomes to

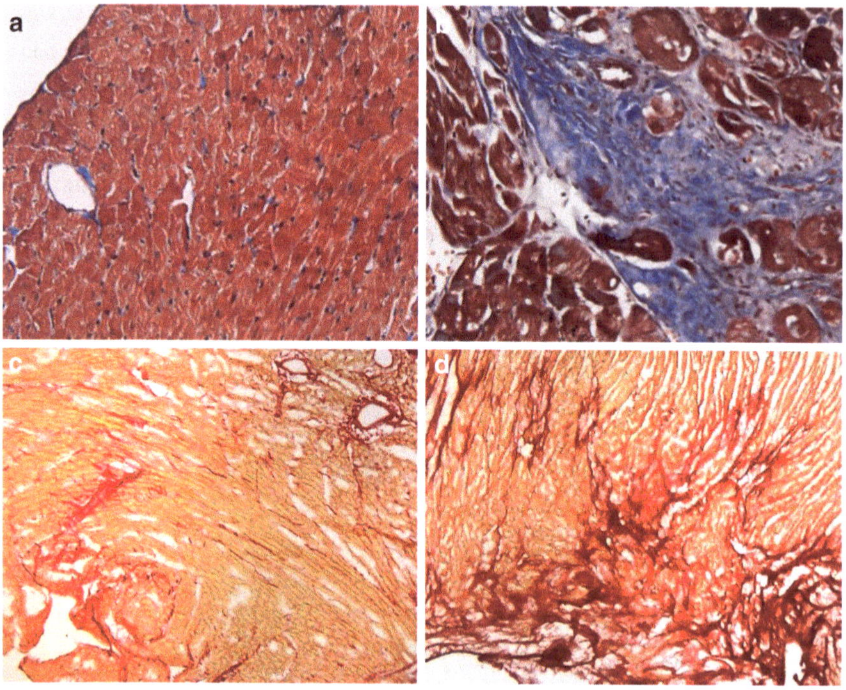

Fig. 6 Masson's trichrome staining of the non-infarcted myocardium (**a**) and infarcted regions (**b**) illustrating collagen deposition (*blue*) during scar formation. (**c**, **d**) PSR staining of non-infarcted and infarcted myocardium, respectively, depicting the occurrence of fibrosis

be evaluated. Fig. 6 depicts examples of typical types of staining we perform for fibrosis and infarct scar assessment, Masson's trichrome (Fig. 6a, b) and picrosirius red (Fig. 6c, d).

References

1. Shamhart PE, Meszaros JG (2010) Non-fibrillar collagens: key mediators of post-infarction cardiac remodeling? J Mol Cell Cardiol 48:530–537
2. Lindsey ML, Mann DL, Entman ML, Spinale FG (2003) Extracellular matrix remodeling following myocardial injury. Ann Med 35:316–326
3. Sun Y, Weber KT (2002) Infarct scar: a dynamic tissue. Cardiovasc Res 46:250–256
4. Wei S, Chow LT, Shum IO, Qin L, Sanderson JE (1999) Left and right ventricular collagen type I/III ratios and remodeling post-myocardial infarction. J Card Fail 5:117–126
5. Opie LH, Commerford PJ, Gersh BJ, Pfeffer MA (2006) Controversies in ventricular remodeling. Lancet 367(9507):356–367
6. Naugle JE, Olson ER, Zhang X, Mase SE, Pilati CF, Maron MB, Folkesson HG, Horne WI, Doane KJ, Meszaros JG (2006) Type VI collagen induces cardiac myofibroblast differentiation: implications for postinfarction remodeling. Am J Physiol Heart Circ Physiol 290:H323–H330
7. Bonaldo P, Braghetta P, Zanetti M, Piccolo S, Volpin D, Bressan GM (1998) Collagen VI deficiency induces early onset myopathy in the mouse: an animal model for Bethlem myopathy. Hum Mol Genet 7:2135–2140
8. Irwin WA, Bergamin N, Sabatelli P, Reggiani C, Megighian A, Merlini L, Braghetta P, Columbaro M, Volpin D, Bressan GM, Bernardi P, Bonaldo P (2003) Mitochondrial dysfunction and apoptosis in myopathic mice with collagen VI deficiency. Nat Genet 35:367–371

Chapter 20

Cryoinjury Models of the Adult and Neonatal Mouse Heart for Studies of Scarring and Regeneration

Erik G. Strungs, Emily L. Ongstad, Michael P. O'Quinn, Joseph A. Palatinus, L. Jane Jourdan, and Robert G. Gourdie

Abstract

A major limitation in studies of the injured heart is animal-to-animal variability in wound size resulting from commonly used techniques such as left anterior descending coronary artery ligation. This variability can make standard errors sufficiently large that mean separation between treatment and control groups can be difficult without replicating numbers (*n*) of animals in groups by excessive amounts. Here, we describe the materials and protocol necessary for delivering a standardized non-transmural cryoinjury to the left ventricle of an adult mouse heart that may in part obviate the issue of injury variance between animals. As reported previously, this cryoinjury model generates a necrotic wound to the ventricle of consistent size and shape that resolves into a scar of uniform size, shape, and organization. The cryo-model also provides an extended injury border zone that exhibits classic markers of remodeling found in surviving cardiac tissue at the edge of a myocardial infarction, including connexin43 (Cx43) lateralization. In a further extension of the method, we describe how we have adapted the model to deliver a cryoinjury to the apex of the heart of neonatal mice—a modification that may be useful for studies of myocardial regeneration in mammals.

Key words Cryoinjury, Mouse, Scarring, Regeneration, Myocardial infarction, Cardiac wound healing

1 Introduction

Animal models of cardiac injury facilitate the study of wound healing processes and the evaluation of therapies that may improve or hasten this response. One of the most widely used models of cardiac injury is the left anterior descending artery (LAD) ligation model [1–4]. LAD ligation mimics myocardial infarction (MI) seen in ischemic heart disease by artificially occluding a major coronary artery, causing ischemia in a portion of the myocardium and recapitulating aspects of the characteristic pathologic progression found subsequent to MI, beginning with necrosis, and ending with the formation of a mature scar.

Robert G. Gourdie and Tereance A. Myers (eds.), *Wound Regeneration and Repair: Methods and Protocols*, Methods in Molecular Biology, vol. 1037, DOI 10.1007/978-1-62703-505-7_20, © Springer Science+Business Media New York 2013

LAD ligation is a reliable and useful method of cardiac wound healing, but it has limitations [1]. First, there is consideration of the extent to which ligating a coronary artery in a healthy animal actually models ischemic heart disease in humans. Perhaps of greater importance is that coronary artery distribution and arterial supply are not consistent between individuals within the same species. Such normal anatomic variations can result in variability of the injury size observed within a group of animals. This variability becomes particularly important in studies of the wound healing response, where measurement of injury size is a key parameter for discerning differences between experimental groups. The efficiency of using LAD ligation is also limited by higher postoperative mortality, as compared to other myocardial injury models [5].

In earlier work, we reported that a peptide based on the carboxyl terminus of the gap junction (GJ) protein Cx43 (αCT1) inhibited remodeling of GJs in cultured cells [6], as well as beneficially affecting the progression of healing of skin wounds [7, 8]. In follow-up studies, we sought to determine whether αCT1 had similar effects on GJ remodeling and recovery from injury to the mouse heart [9]. In initial experiments, it was found that coronary arterial ligation was problematic in our hands owing to difficulties in achieving a repeatable injury to the left ventricle (LV). To circumvent this, a cryoinjury model that provided a wound on the LV of the mouse heart of uniform size and geometry was developed. The method was based on one described by Van den Bos and coworkers who used a liquid nitrogen-cooled cryoprobe [5]. We modified their protocol to include probe prechilling and a non-transmural injury of LV, as opposed to the more severe transmural injury generated by this group (Fig. 1). We did so as non-transmural injuries provide extended IBZs—a tissue of particular interest in our experiments because of its putative causal role in reentrant arrhythmia mechanisms.

In this chapter, we describe the materials and protocol that were used to deliver a standardized non-transmural cryoinjury to the LV of an adult mouse. In addition to providing an injury and extended IBZ of uniform size and shape, this approach has further advantages over LAD of being technically straightforward, providing high postoperative survival and generating healed scars of relatively consistent volume and organization. As we have also reported, mouse hearts receiving cryoinjury demonstrated loss of mechanical LV function as assessed by echocardiography, slowed

Fig. 1 (continued) scar immunolabeled for Mlc2a—normally expressed in the embryonic ventricle (*gray*). Dapi nuclear signal (*dark gray*). *Inset.* Cx43 immunolabeled sister section at IBZ. Note lateralized Cx43 in IBZ. *Inset. Dashed lines* represent border of injury/scar. Scale (**c**) = 500 μm, (**d**) = 25 μm, (**c**, **d**) inset = 10 μm

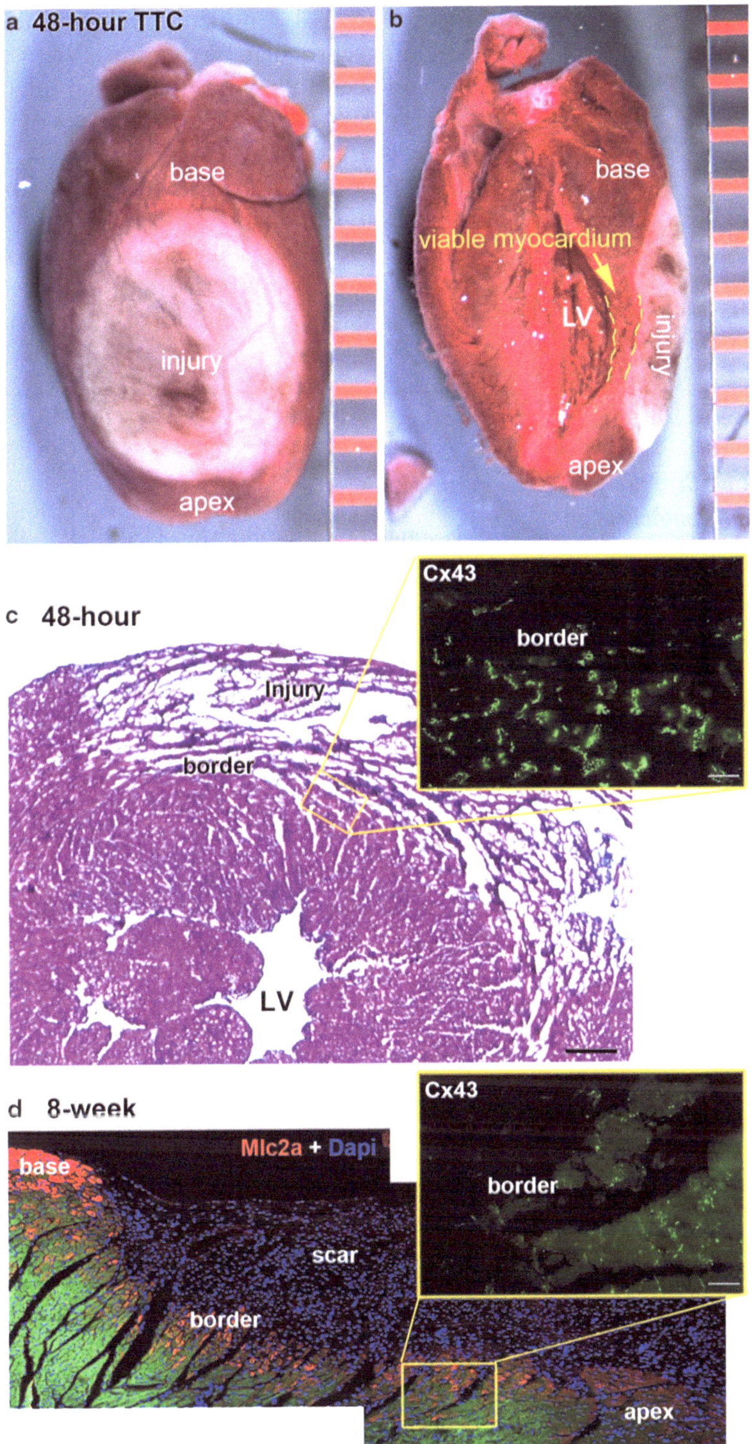

Fig. 1 Cryoinjury of the LV of an adult mouse heart. (**a**, **b**) Whole mount and cross section of a TTC-stained heart 48-h after 5-s exposure to cryoprobe. (**c**) Montage of H&E section through LV 48 h after standard 5-s exposure. *Inset*. Cx43 immunolabeled (*gray*) sister section at IBZ. (**d**) Montage of section from 8-week LV

action potential conduction velocity, and increased propensity to develop ventricular arrhythmias consistent with pathological changes seen following myocardial infarction [9, 10]. Moreover, the cryo-IBZ exhibits classic markers of remodeling seen in surviving myocardium at the edge of the MI, including Cx43 lateralization, interspersion of scar tissue with myocardial tissue, and reexpression of markers normally only found in the embryonic ventricle [9, 11]. In a further extension of the method, we describe how we have adapted cryoinjury to deliver an injury to the apex of the neonatal heart useful for studies of myocardial regeneration.

2 Materials

2.1 Cryoinjury of the Left Ventricle of an Adult Mouse Heart

Anesthesia

1. Anesthetic: Isoflurane, USP.
2. 2 % Lidocaine solution, USP.
3. Oxygen cylinder and regulator.
4. Compact anesthesia vaporizer system, Harvard Apparatus.
5. MiniVent Type 845 mouse ventilator, Harvard Apparatus.

Surgical Preparation and Procedure

1. Cork surgery board.
2. Umbilical tape.
3. Surgical stereomicroscope (Wild M3Z) and adjustable (snake) light source.
4. Sterile gauze, tape, and cotton-tipped applicators.
5. Nair® hair removal product.
6. Betadine® solution (10 % providone-iodine).
7. 70 % (v/v) EtOH in H_2O.
8. Surgical gloves.
9. Liquid nitrogen.
10. 16-G angiocatheter (needle bevel removed).
11. Brymill Cryo-Gun apparatus with 3 mm circular, flat copper cryoprobe.

Surgical Tools (Sterilized)

1. Small surgical scissors.
2. Small surgical retractor.
3. 2× small surgical forceps.
4. Needle holder.

Suture

1. 4–0 Silk suture.
2. Gluture® skin glue.

Postoperative Care

1. Heating pad.

2.2 Cryoinjury of the Ventricular Apex of a Neonatal Mouse Heart

Anesthesia

1. Ice water bath.
2. Latex barrier.

Surgical Preparation and Procedure

1. Surgical stereomicroscope (Wild M3Z) and adjustable (snake) light source.
2. Sterile gauze, tape, and cotton-tipped applicators.
3. Betadine® solution (10 % providone-iodine).
4. 70 % (v/v) EtOH in H_2O.
5. Surgical gloves.
6. Liquid nitrogen.
7. Brymill Cryo-Gun apparatus with 1 mm circular, flat copper cryoprobe.

Surgical Tools (Sterilized)

1. Small surgical scissors.
2. 2× Small surgical forceps.
3. Needle holder.

Suture

1. 6–0 Prolene suture.
2. Gluture® skin glue.

Postoperative Care

1. Heating pad.

3 Methods

3.1 Cryoinjury of the Left Ventricle of an Adult Mouse Heart

Intubation and Anesthesia

1. Begin airflow to the anesthesia induction chamber. Use a mixture of 100 % oxygen and 2 % isoflurane and a flow rate of 0.8–1.0 L/min.

2. Place animal in the induction chamber, and wait until all voluntary motion ceases, approximately 5 min.
3. Remove the mouse from the induction chamber and place in a supine position on the table with the nose of the mouse slightly overhanging the edge of the table. Run the umbilical tape behind the upper middle two incisors and secure the umbilical tape to the edge of the table using tape.
4. Adjust the arms of the adjustable "snake" light so they overlie the neck of the mouse. This will illuminate the upper airway and facilitate visualization of the subtle opening and closing of the vocal folds as the mouse breathes.
5. Use a cotton swab to gently brush the tongue out of the mouse's mouth. Using the opposite end of the same cotton swab, using gentle pressure, press the tongue against the floor of the mouth and hold the mouth open.
6. Place a small amount of 2 % lidocaine on another cotton swab, and apply to the vocal folds as a local anesthetic.
7. With the blunted catheter introducer still in place, guide the catheter past the vocal folds and the epiglottis, and into the trachea.
8. Remove the catheter introducer from the catheter and attach the catheter to the tubing running from the ventilator. Switch the valves of the ventilation system to redirect airflow from the induction chamber to the ventilator. Turn on the ventilator, using a stroke volume of 260 μL and a stroke rate of 350 breaths/min. If the intubation catheter has been properly placed in the trachea, the chest wall will rise and fall in synchrony with the ventilator; if the intubation catheter has been erroneously placed in the esophagus, there will be no rhythmic rising and falling of the chest wall.

Surgical Procedure

1. Position the mouse supine on the cork board on the operating table, taking care to not remove the intubation catheter. Tape the mouse's fore- and hindlimbs to the operating table to prevent any muscular movements during the course of the operation.
2. Apply a hair removal product, such as Nair®, on the left side of the anterior chest wall. Wipe away after about a minute to remove the hair from the surgical area.
3. Apply Betadine followed by 70 % ethanol to the now-hairless area of skin to sterilize the surgical area.
4. Make a transverse skin incision using scissors. Palpation of the point of maximum impact of the apex against the chest wall provides a useful landmark for the location of the skin incision.

5. Using a pair of small forceps, bluntly dissect between the layers of skeletal muscle overlying the chest wall. In most cases, the muscle layers can be divided and individually retracted without the need for transection.
6. Use blunt dissection to open the chest wall by transecting the intercostal muscles between the fourth and fifth ribs. Retract the ribs. Blunt dissection during this step will prevent transection of the internal mammary artery, which will run along the medial edge of the resulting thoracotomy. Following this step, the surgeon should be able to visualize the heart through the pericardial sac.
7. Gently open the pericardial sac using blunt dissection. Move the thymus off of the surface of the heart if it is overlying the area of the ventricle to be injured.
8. After the proposed injury site is totally visualized and free from overlying obstructions, prechill the cryoprobe for 10 s using the Brymill Cryo-Gun apparatus—prefilled with liquid nitrogen (**Note 1**).
9. Cease chilling the cryoprobe and apply the cryoprobe to the surface of the heart for 5 s (**Notes 2** and **3**).
10. Gently remove cryoprobe, making sure not to tug on the heart or disrupt the epicardium in the process. Excess moisture on the heart could cause the cryoprobe to stick to the surface, potentially creating a traumatic injury.
11. Apply gentle pressure to the chest wall to remove excess air in the chest cavity. Close the chest wall incision using 4–0 silk suture with two or three sutures tied in an interrupted fashion. Be sure to include the superior and inferior ribs within this closure, as the intercostal muscle and parietal pleura will tear under the force of the sutures. Close any transected skeletal muscle layers using 4–0 silk suture in a running continuous fashion. Close the skin incision using Gluture skin glue with standard techniques.

Postoperative Care

1. Remove tape restraints from the mouse's limbs and turn the isoflurane to 0 % on the gas mixer, leaving the oxygen flow and ventilator running.
2. Allow the mouse to pull itself off of the intubation tube as it ascends from anesthesia, rather than removing it immediately following incision closure.
3. Place the mouse on a heating pad until it regains sufficient movement to be placed into its cage.

3.2 Cryoinjury of the Ventricular Apex of a Neonatal Mouse Heart

Rather than forming scar tissue as observed in the adult mouse heart, the ventricular apex of newborn mouse hearts has been reported to possess a transient regenerative potential similar to that seen in newts

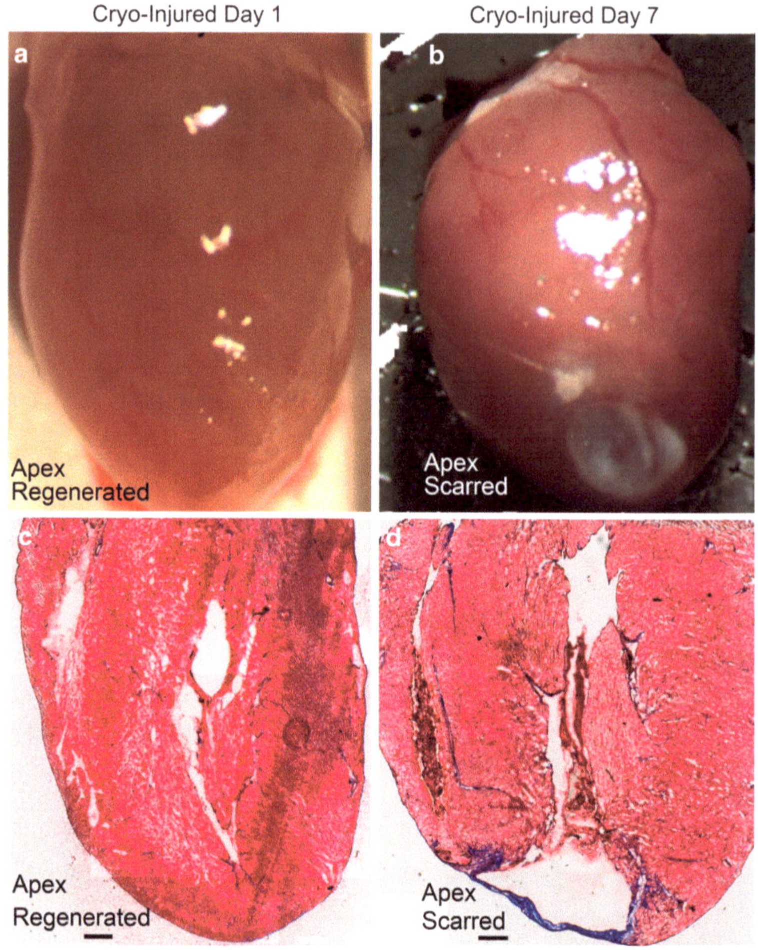

Fig. 2 Cryoinjury of the ventricular apex of neonatal mouse heart. (**a**, **b**) 21-day-old postnatal mouse hearts cryoinjured at either 1 (**a**) or 7 (**b**) postnatal days. Note only the 21-day heart injured 7 days after birth has a visible scar. (**c**, **d**) H&E sections through 21-day-old postnatal hearts cryoinjured at either 1 (**c**) or 7 (**d**) postnatal days. Note the heart cryoinjured at 1 day has no visible scar and fully regenerated myocardium at its ventricular apex, whereas the heart injured at 7 days displays transmural scar tissue (*black*). Scale 100 μm

and zebrafish [12]. The earlier study in neonatal mouse used partial surgical resection as an injury model. Below we describe an approach to reproducing a similar regenerative outcome by adapting the cryo-injury approach we describe for adult hearts for 1 day neonates (Fig. 2). Cryoinjury has the advantage of reducing perioperative mortality by preventing excess bleeding from surgical breach of the ventricular wall and leaving a scaffold of extracellular matrix in place for repair processes to proceed upon. This approach may also enable improved resolution of steps in the wound healing process including inflammation, granulation tissue deposition and scar differentiation, remodeling, and resorption compared to what is observed following a mechanical injury of the ventricle.

Animal Preparation and Anesthesia

1. Remove litter of mice from mother's cage and place in a new, clean cage.
2. Induce anesthesia by inducing hypothermia. Using a latex barrier to protect the mouse's skin, submerge the mouse in an ice water bath until an anesthetic state is reached (**Note 4**). This may occur at any point between 4 and 20 min, so close observation is required.
3. Place the mouse in a supine position on the surgical table and clean the skin using Betadine solution followed by 70 % ethanol (**Note 5**).

Surgical Procedure

1. Make a horizontal incision in the dermis using a small scissors from the midline to the left axilla, about 1 cm in length.
2. Gently separate the ribs by inserting the closed forceps through the intercostal muscles and slowly opening the forceps such that the ribs separate.
3. Apply gentle pressure to the dorsum of the mouse, such that the heart extrudes through the opening in the rib cage with the apex leading.
4. Prechill the cryoprobe for 10 s and apply the probe to the surface of the heart's apex for 5 s (**Note 6**). Remove pressure from the dorsum of the mouse and allow the heart to reenter the chest cavity.
5. Close the chest cavity using one or two discontinuous sutures of 6–0 prolene. Close the skin incision using Gluture skin glue.

Postoperative Care

1. Clean any blood or topical antiseptic from the skin of the mouse.
2. Allow the mouse to recover from hypothermic anesthetic state by placing on a heating pad. After the mouse resumes voluntary movement, place the mouse into a new, clean cage until operations on the entire litter are complete.
3. Return the postoperative mice to the mother's cage as a complete group (**Note 7**).

4 Notes

1. Gentle rotation of the exposed heart such that the cryoprobe is in contact with the left ventricular surface of the heart may be required.
2. We used a 5-s exposure to the cryoprobe to generate a non-transmural injury in adult mice. We noted that the blanched frozen area of the heart took just over 19 s to recover a pink reperfused appearance if cryoinjury had been successful. Shorter or longer periods of recovery generally indicated an unsuccessful injury that was either too little or too extensively injured. Such hearts were discarded.
3. The size of the injury and the extent of its transmurality can also be controlled by adjusting the length of time the cryoprobe is in contact with the surface of the heart.
4. A latex surgical glove serves as an effective and convenient barrier between the mouse's skin and the ice water bath during hypothermia induction of neonates.
5. Time to anesthetic induction of neonates can vary widely from animal to animal. It is, therefore, extremely important to closely monitor each animal independently as overcooling can result in mortality. After induction of anesthesia, the surgeon oftentimes has less than 5 min before the mouse begins to awaken, so the procedure must be performed expediently.
6. It is important for the surface of the heart to be dry at the time of application of the cryoprobe. If the surgeon encounters the issue of the cryoprobe adhering to the surface of the heart during the wounding process, it is likely that there is excess moisture on the surface of the heart.
7. Postoperative maternal cannibalization is a major problem with neonates, but several measures can be taken to limit morbidity and mortality. Precondition to the scents acquired during the procedure by having the surgeon handle the mother daily prior to the birth of the pups and place a small gauze containing a few drops of Betadine solution in the cage. Remove and replace the entire litter before and after the operation, respectively. Careful cleaning of all traces of blood from the animals' skin can also reduce this form of mortality.

References

1. Klocke R, Tian W, Kuhlmann MT, Nikol S (2007) Surgical animal models of heart failure related to coronary heart disease. Cardiovasc Res 74(1):29–38
2. Tarnavski O (2009) Mouse surgical models in cardiovascular research. Methods Mol Biol 573:115–137
3. Madeddu P, Emanueli C, Spillmann F, Meloni M, Bouby N, Richer C, Alhenc-Gelas F, Van Weel V, Eefting D, Quax PH, Hu Y, Xu Q, Hemdahl AL, van Golde J, Huijberts M, de Lussanet Q, Struijker Boudier H, Couffinhal T, Duplaa C, Chimenti S, Staszewsky L, Latini R, Baumans V, Levy BI (2006) Murine models

of myocardial and limb ischemia: diagnostic end-points and relevance to clinical problems. Vascul Pharmacol 45(5):281–301
4. Zamilpa R, Zhang J, Chiao Y-A, de Castro Bras L, Halade G, Ma Y, Hacker SO, Lindsey ML (2013) Cardiac wound healing post-myocardial infarction: a novel method to target extracellular matrix remodeling in the left ventricle. Wound Regeneration and Repair: Methods and Protocols. Eds. Gourdie RG and Myers T
5. van den Bos EJ, Mees BM, de Waard MC, de Crom R, Duncker DJ (2005) A novel model of cryoinjury-induced myocardial infarction in the mouse: a comparison with coronary artery ligation. Am J Physiol Heart Circ Physiol 289:H1291–H1300
6. Hunter AW, Barker RJ, Zhu C, Gourdie RG (2005) Zonula occludens-1 alters connexin43 gap junction size and organization by influencing channel accretion. Mol Biol Cell 16(12):5686–5698, Epub 2005 Sep 29
7. Ghatnekar GS, O'Quinn MP, Jourdan LJ, Gurjarpadhye AA, Draughn RL, Gourdie RG (2009) Connexin43 carboxyl-terminal peptides reduce scar progenitor and promote regenerative healing following skin wounding. Regen Med 4(2):205–223
8. Rhett JM, Ghatnekar GS, Palatinus JA, O'Quinn M, Yost MJ, Gourdie RG (2008) Novel therapies for scar reduction and regenerative healing of skin wounds. Trends Biotechnol 26(4):173–180
9. O'Quinn MP, Palatinus JA, Harris BS, Hewett KW, Gourdie RG (2011) A peptide mimetic of the connexin43 carboxyl terminus reduces gap junction remodeling and induced arrhythmia following ventricular injury. Circ Res 108(6):704–715
10. Ongstad EL, O'Quinn MP, Ghatnekar G, Yost MJ, Gourdie RG (2013) A connexin43 mimetic peptide promotes regenerative healing and improves mechanical properties in skin and heart. Advances in Wound Care 2(2):55–62
11. Rémond MC, Iaffaldano G, O'Quinn MP, Mezentseva NV, Garcia V, Harris BS, Gourdie RG, Eisenberg CA, Eisenberg LM (2011) GATA6 reporter gene reveals myocardial phenotypic heterogeneity that is related to variations in gap junction coupling. Am J Physiol Heart Circ Physiol 301(5):H1952–H1964
12. Porrello ER, Mahmoud AI, Simpson E, Hill JA, Richardson JA, Olson EN, Sadek HA (2011) Transient regenerative potential of the neonatal mouse heart. Science 331(6020):1078–1080

Chapter 21

Targeting Wnt Signaling to Improve Wound Healing After Myocardial Infarction

Evangelos P. Daskalopoulos, Ben J.A. Janssen, and W. Matthijs Blankesteijn

Abstract

Myocardial infarction is one of the major causes of left ventricular dilatation, frequently leading to heart failure. In the last decade, the wound healing process that takes place in the infarct area after infarction has been recognized as a novel therapeutic target to attenuate left ventricular dilatation and preserve an adequate cardiac function. In this chapter, we discuss the role of Wnt signaling in the wound healing process after infarction, with a specific focus on its modulating effect on myofibroblast characteristics.

Key words Wnt, Wound healing, Myofibroblast, Heart failure

1 Introduction

Following myocardial infarction (MI), a plethora of cellular and noncellular mechanisms is activated, in order to overcome myocardial injury. These mechanisms lead to the formation of fibrotic tissue in the infarcted area of the myocardium, which is crucial for the maintenance of the architectural and functional integrity of the heart. However, structural and functional changes also occur in areas remotely from the infarct area, leading gradually to ventricular remodeling and eventually to heart failure (HF) [1, 2]. Heart failure is a worldwide major public health issue with high prevalence, high morbidity, and a substantial burden for health care systems [3].

Lately, the interest for the myofibroblast and its role in the wound healing of the heart tissue after an infarction is growing. Myofibroblasts are cells with contractile properties that appear shortly after MI, produce collagen, communicate with myocytes and non-myocyte cells, and are the major mediators of the fibrotic process and infarct healing [4]. Nevertheless, they are also implicated in the fibrosis of non-infarcted myocardial areas, which leads

Robert G. Gourdie and Tereance A. Myers (eds.), *Wound Regeneration and Repair: Methods and Protocols*, Methods in Molecular Biology, vol. 1037, DOI 10.1007/978-1-62703-505-7_21, © Springer Science+Business Media New York 2013

to adverse remodeling and HF. Hence, factors and mechanisms that can alter the numbers and functioning of these cells can have notable effects on the structure, morphology, and performance of heart tissue [5].

The purpose of this chapter is to discuss the basic concepts of the myofibroblast actions during the cardiac wound healing process and the factors that influence them. Moreover, we give an overview of the myofibroblast involvement in wound healing after MI, the pathological effect of ventricular remodeling and its crucial role in the development of HF, as well as the current pharmacological treatment on these cells. Furthermore, a part of this chapter is dedicated to describing the general principles of the Wnt/Frizzled (Wnt/Fzd)-signaling pathway and provide an insight on the role it plays in the wound healing following an infarct. Although, the Wnt/Fzd-signaling pathway has several effects on the cardiovascular system, it is beyond the scope of this chapter to address any of these that are not involved in post-MI wound healing.

2 Myocardial Infarction, Wound Healing, Adverse Remodeling, and Heart Failure

2.1 Epidemiology of MI and HF

MI is one of the major causes of mortality and morbidity in the Western world. According to a report produced by the WHO in the dawn of 2011, cardiovascular disease is the most common cause of death worldwide. It was estimated that in the year 2004, coronary heart disease caused approximately 7.2 million deaths [6]. In the USA alone, the overall prevalence of MI in the population is 7.9 million [7]. The main cause of MI is partial or total occlusion of an epicardial coronary artery by a thrombus or an atheromatous plaque that ruptures or erodes [8]. Blood flow to the affected area of the myocardium is insufficient—or might even be completely blocked—leading to cardiomyocyte death due to ischemia [1]. The magnitude of the myocardial damage depends on the location of the coronary occlusion, on its severity and the time taken for the artery to be reperfused by itself, with the use of fibrinolytics or by Percutaneous Coronary Intervention (PCI). Moreover, several pharmacological agents can be used during the acute phase of MI, in order to manage the deleterious effects of the ischemic injury. Examples are antiplatelet and antithrombotic treatment, glycoprotein IIb/IIIa inhibitors, direct thrombin inhibitors, Factor X inhibitors, etc. [8].

Myocardial injury that follows MI, promotes left ventricular remodeling events that continue for months or years and eventually lead to the development of heart failure which is one of the most disabling, deadly, and costly clinical syndromes for the Western societies [3]. Worldwide, there are more than 23 million people suffering due to HF, while in the USA alone, the prevalence is above 5.8 million [9]. As documented in the latest American Heart Association Report, mortality for patients with HF is still very high and about one in two patients diagnosed with HF will die within 5 years. The financial burden for the health care systems is enormous. In 2007, about 11 % of the total deaths (277,193 deaths) in the USA were attributed to HF [7]. More than 40 % of patients firstly diagnosed with HF will return to the hospital within the first 6 months, with each readmission costing more than US $7,000 per patient to the US health care system [10]. The total cost due to HF-related admissions in the USA hospitals was US $39 billion per year [9], while in the UK the NICE 2010 report revealed that the cost for community-based drug therapy of HF in the NHS was approximately £129 million per year [11].

2.2 The Wound Healing Process After MI

Cardiac wound healing following MI is a complex and dynamic process that involves a myriad of chemical molecules, cells, and mechanisms that lead to the establishment of a fibrous scar. Cardiac wound healing can be divided into four distinct phases. The first phase of wound healing starts straight after the MI and is characterized by the death of cardiomyocytes. Adult cardiomyocytes are terminally differentiated cells, which means that substitution of dead myocytes after MI is very limited (if possible at all) [12]. Myocyte death occurs via necrosis (cell-swelling) and/or apoptosis (cell-shrinkage), with apoptosis reaching a peak 6–8 h after MI in humans and *Rattus norvegicus* [1].

The acute inflammatory response (12–16 h post-MI) signifies the second phase of wound healing [13]. The complement system is activated and several cytokines and chemokines are released (e.g., tumor necrosis factor-α [TNF-α], interleukin-1β [IL-1β], interleukin-6 [IL-6], interleukin-8 [IL-8], and interleukin-18 [IL-18]) [14, 15]. Neutrophilic granulocytes infiltrate the area and remove dead myocytes, while at the same time attracting lymphocytes, macrophages, and plasma cells [1, 14].

The third phase of wound healing in the infarcted myocardium is characterized by granulation tissue deposition. This is done first in the border zone and then in the center of the infarct [1]. The main factor initiating this phase is the fibrogenetic mediator cytokine, transforming growth factor-β (TGF-β). The main function of its many roles is to stimulate fibroblast-to-myofibroblast differentiation [16]. Myofibroblasts differentiate, proliferate, and start surrounding the infarcted area and depositing collagen (types I and III) which eventually with the addition of several other factors and proteins will make up the extracellular matrix (ECM). The ECM is a 3D mesh that is the scaffolding for myocytes and non-myocytes which maintain the cardiac architecture and function [15, 17]. Along with collagen deposition, it has been proven that collagen degradation is also activated and is mediated via metalloproteinases (MMPs) [18]. Lastly, neovascularization of the infarct area is an integral part of the third phase of wound healing. New blood vessels are formed and aim to restore the perfusion of the injured area [14, 19]. It is worth noting that studies in rats have shown that coronary flow is restored completely 7 days post-MI. The third phase lasts for 2–3 weeks in mice, but it is longer in the human myocardium (up to 2 months) [1].

Finally, the fourth stage of cardiac tissue healing is characterized by the maturation of granulation tissue and the formation of a stabilized scar. The main attributes of this phase are disappearance of the inflammatory cells—mostly via apoptosis—and cross-linking of the ECM [15]. Although myofibroblast numbers start decreasing about 7 days post-MI [15], a study by Willems et al. has demonstrated that human myofibroblasts might persist in the infarct area for up to 17 years post-MI [20]. Poorly healed infarcts usually are shown to lose their myofibroblasts very early in the healing process, hence the presence (and persistence) of myofibroblasts in the infarct area for a long time is crucial [2]. An overview of the four phases of the wound healing following MI is given in Table 1 [1, 21].

It has to be noted that the wound healing process in the myocardium is accelerated in smaller animals like rats and mice, in comparison to humans. Hence one should be careful when extrapolating from animal studies and drawing conclusions for cardiac wound healing in humans [1]. Moreover, the presence of myofibroblasts seems to be directly connected to infarct healing. In a recent study by our group, marked deviations between the cardiac remodeling of five mouse strains were demonstrated. High numbers of myofibroblasts were correlated to higher ejection fraction (EF) and lower end diastolic volume (EDV) measurements. Hence, it was demonstrated that myofibroblasts are key players in wound healing and that their presence prevents infarct dilatation and attenuates the development of HF [22].

2.3 Adverse Remodeling of the Myocardium and HF Development

The post-MI wound healing process of the myocardium is characterized by drastic molecular, cellular, structural, and functional alternations in the infarct area [1]. The healing process is associated with the activation of the renin–angiotensin–aldosterone system (RAAS), the adrenoceptor signaling pathways, the release of a wide range of cytokines and chemokines, and other chemical molecules, as well as the mobilization of various cell types [21, 23]. Cardiomyocyte death and the presence of cross-linked ECM in the void spaces lead to depressed pumping capacity and increased ventricular chamber volume, in order to sustain a normal cardiac output [23].

Per contra, the fibrotic changes are not confined to the infarct site, but also occur in areas that lie remote from the infarct area (non-infarcted left ventricle [LV] or even right ventricle [RV]) [1]. The fibrotic tissue spreads into the non-infarcted areas of the heart and is characterized by deposition of ECM with abnormal cross-linking. The main effect of the ECM deposition in the unaffected

Table 1
Overview of the wound healing phases of the infarcted heart in humans

Phase 1	*Cardiomyocyte death*	Immediately after MI until fourth day post-MI
	Necrosis	
	Apoptosis	
Phase 2	*Acute inflammatory reaction*	6 h post-MI until fifth to sixth day post-MI
	Complement activation	
	Release of cytokines and chemokines	
	Recruitment of granulocytes, lymphocytes, macrophages	
	Phagocytosis	
Phase 3	*Granulation tissue formation*	Second/third day until first month post-MI
	Myofibroblast differentiation/migration/proliferation	
	Collagen secretion and deposition	
	ECM degradation by MMPs (mostly in the infarcted area)	
	Neovascularization	
Phase 4	*Scar maturation/cardiac remodeling*	Third week until months or years post-MI
	Collagen cross-linking	
	Disappearance of cells (except myofibroblasts)	

MI myocardial infarction, *ECM* extracellular matrix, *MMPs* matrix metalloproteinases

myocardium is the development of muscle stiffness [24]. The stiff cardiac muscle needs to work harder in order to pump adequate amounts of blood, causing cardiomyocytes of the non-infarcted sites to become hypertrophic [25]. This eventually leads to a decline in their contractile capacity and worsening of the cardiac performance [23]. If LV hypertrophy, ventricular dilatation, and fibrotic myocardial stiffness are left untreated, eventually the heart becomes unable to fulfill its primary function (pumping of blood) and HF develops [2, 3, 23, 26].

3 The Role of Fibroblasts and Myofibroblasts During Cardiac Wound Healing

3.1 Fibroblasts and the Origins of Myofibroblasts

Fibroblasts are cells producing connective tissue and can be found in a variety of tissues, in all vertebrates. They are flat, spindle-shaped cells with an elliptical-shaped nucleus [27], a large Golgi apparatus, and an extensive rough endoplasmic reticulum [17]. Fibroblasts show great variability regarding their phenotype and it

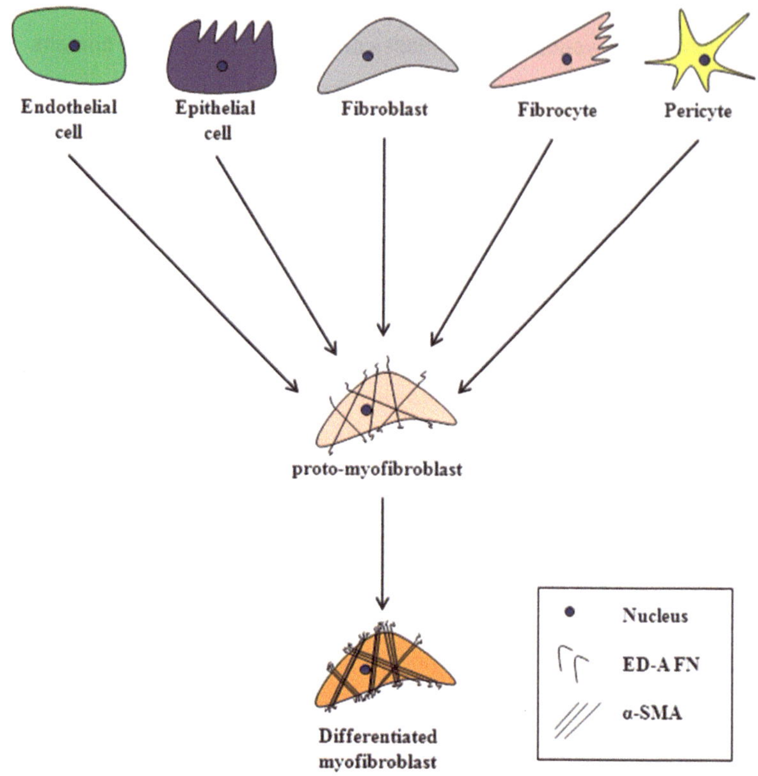

Fig. 1 The origins of the myofibroblast. A variety of cells, under the influence of specific factors, stimulate the differentiation of fibroblasts to the activated myofibroblasts. Myofibroblasts originate mostly from local fibroblasts, but also from endothelial cells (endothelial–mesenchymal transition), epithelial cells (epithelial–mesenchymal transition), fibrocytes (mesenchymal progenitors), or pericytes. These cells differentiate into an intermediate form (proto-myofibroblast) and under further stimulation by mechanical stress, TGF-β1 or FN ED-A the intermediate form differentiates to the activated myofibroblast. *Abbreviations*: *TGF-β1* transforming growth factor β1, *FN ED-A* fibronectin extra-domain A

is characteristic that they can be heterogeneous even within the same anatomical site (e.g., in different parts of the heart muscle), depending on the condition (normal or pathological) [28, 29].

Under physiological conditions only fibroblasts are present in the heart. Several stimuli (e.g., ischemia or mechanical stress) can however lead to their differentiation into the activated myofibroblasts that are normally not found in the healthy adult heart [30]. Myofibroblasts may originate from preexisting locally present fibroblasts that spontaneously differentiate into their activated counterparts under the influence of local factors (e.g., TGF-β). Moreover, a wide range of cells (Fig. 1) can differentiate into myofibroblasts under the pressure of various stimuli. Myofibroblasts can arise from epithelial [31] or endothelial cells [32], fibrocytes

[33], or pericytes [34]. Nevertheless, the exact mechanisms by which specific cells differentiate into the myofibroblast phenotype still remains unclear.

3.2 Differentiation of Activated Myofibroblasts and Their Function in the Heart

Cardiac fibroblasts are responsible for maintaining the architecture and the functional integrity of the healthy heart by controlling homeostasis of the ECM. This is done by regulating the secretion of ECM, the activity of matrix MMPs that metabolize collagen and the activity of the tissue inhibitors of metalloproteinases (TIMPs) which inhibit the activity of MMPs [4]. In case of a myocardial injury, drastic changes occur at the molecular level and fibroblasts differentiate into the proto-myofibroblast phenotype and then to the activated myofibroblasts [2, 35]. The main difference of the proto-myofibroblast from the fibroblast phenotype is the enhanced expression of fibronectin (mainly the splice variant with the extra-domain A [ED-A] module); however, the proto-myofibroblast does not express α-smooth muscle actin (α-SMA) [36]. The latter is detected only in fully differentiated myofibroblasts, and it is crucial for the contractile properties of myofibroblasts [37]. Moreover, differentiated myofibroblasts possess hyper-mature focal adhesions and have increased amounts of ED-A fibronectin (ED-A FN) which also plays a key role in the functionality of the myofibroblast [38].

Myofibroblasts are responsible for "fine-tuning" collagen turnover during the wound healing and remodeling that follows MI [4]. They migrate to the infarct area (Fig. 2) under the influence of various factors and produce pro-collagen (a collagen precursor), which is transformed into the active forms (collagen I and III) by pro-collagen proteinases [39]. Collagen and several other components (fibronectin, proteoglycan, elastin, matricellular proteins, cytokines, growth factors, proteases, etc.) cross-link and form a complex matrix, the ECM [40–42]. ECM is deposited first in the border zone (the site between the infarcted and the non-infarcted area) and then in the infarct [1]. It serves the vital role of filling empty spaces that have appeared after the death of cardiomyocytes and forms the fibrotic scar which heals the wound caused after ischemic injury. Furthermore, the ECM serves as the network for communication and transmission of signals between myocytes and other heart cells, both in the infarcted and the unaffected myocardium [40]. The actions and functions of the differentiated cardiac myofibroblast are summarized in Fig. 3.

3.3 Factors Modulating the Cardiac Myofibroblasts

Myofibroblasts can be stimulated by a broad spectrum of factors that range from chemical mediators, proteins, and receptors to mechanical stretch and electrophysiological changes.

3.3.1 Chemical Stimulation

As mentioned above, several endogenous compounds are released at the site of injury following MI. These compounds stimulate

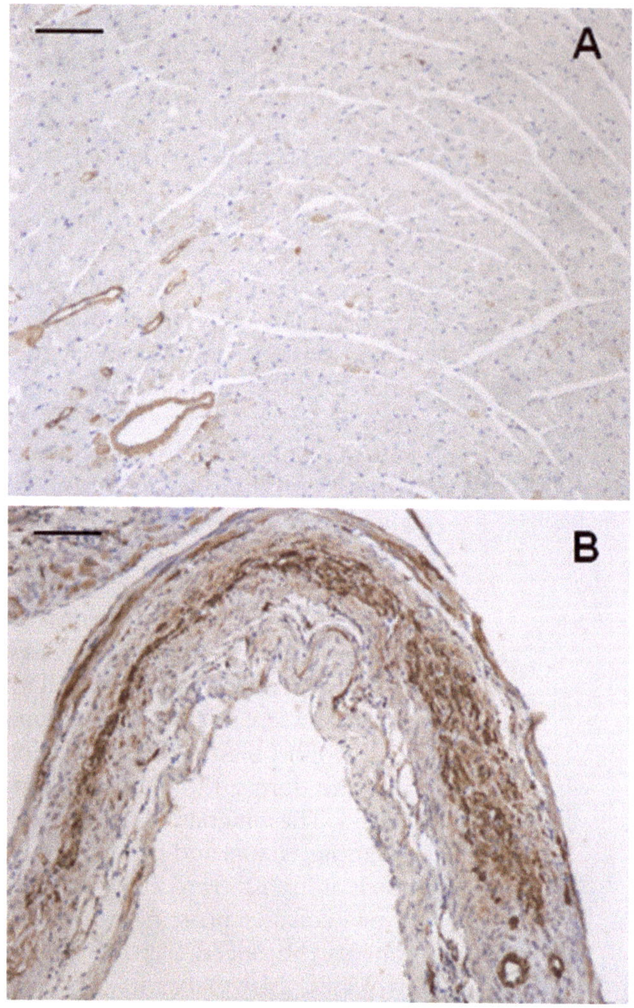

Fig. 2 Microscopic images of mouse myocardium. Staining with α-SMA antibody (Sigma) correlates to α-SMA expression by myofibroblasts. Image (**a**) corresponds to non-infarcted myocardium, while image (**b**) is a section from an infarcted area (14 days post-MI). Note that the staining observed in image (**a**) is due to α-SMA staining of smooth muscle cells that line the blood vessels. Scale bar equals 100 μm. *Abbreviations*: *MI* myocardial infarction, *α-SMA* α-smooth muscle actin

the myofibroblasts in order to differentiate, proliferate, and migrate into the infarct site. Examples include vasoactive peptides, growth factors, and several cytokines. TGF-β is a pleiotropic growth factor of vital importance for several cellular functions. Three TGF-β forms exist, with TGF-β1 being the most important. It acts on its receptors (activin receptor-like kinase 1 [ALK1] and activin receptor-like kinase 5 [ALK5]) which in turn activate the Smad proteins. Smads form complexes and translocate into the nucleus in order to induce target gene expression [43, 44].

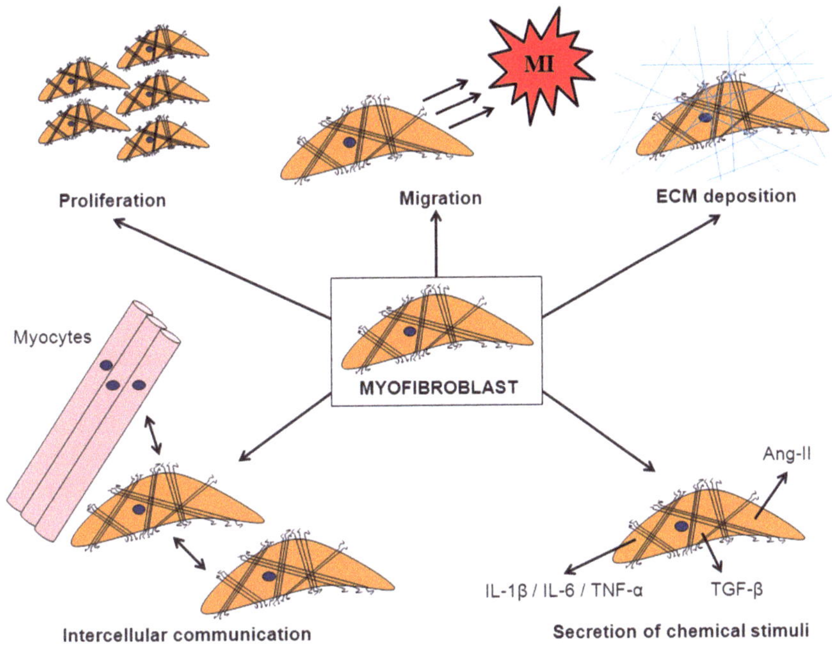

Fig. 3 Overview of the most important functions of the differentiated myofibroblast. The activated myofibroblast can proliferate, migrate to the site of the infarction, secrete and deposit collagen and ECM proteins, communicate with other cells such as myocytes or non-myocyte cells, and secrete a plethora of chemical stimuli (Ang-II, TGF-β1, TNF-α, interleukins IL-1β/IL-6). Due to the wide range of its functions, the myofibroblasts is a key factor during the healing process after MI. *Abbreviations*: *ECM* extracellular matrix, *Ang-II* angiotensin-II, *TGF-β1* transforming growth factor β1, *TNF-α* tumor necrosis factor α, *IL-1β/IL-6* interleukin-1β/interleukin-6

In the heart, TGF-β modulates proliferation and migration of fibroblasts and stimulates fibroblast-to-myofibroblast differentiation after MI [38, 43–45]. Further, it induces collagen secretion, suppresses MMPs and induces TIMPs, leading to a net increase of the ECM deposition [44] and hence playing a major role in the fibrosis that follows MI.

It has been well established that RAAS is activated following MI and that this is one of the major factors determining cardiac remodeling [46]. Angiotensin-converting enzyme (ACE) and the angiotensin-II type 1 receptors (AT1-receptors) are expressed by myofibroblasts during fibrotic repair after MI [30]. Ang-II acts by stimulating the proliferation and differentiation of fibroblasts and modulates ECM deposition. This modulatory effect appears to be mediated via AT1-receptors [16, 47]. On the other hand, there is a hypothesis for a cross-talking mechanism between TGF-β1 and Ang-II, which is supported by the observation that drugs inhibiting Ang-II signaling (ACE inhibitors and AT1-receptor blockers) after MI, are shown to deplete TGF-β levels [48]. Lately, the beneficial role of ACE2 and Ang (1–7)—a peptide with opposite effects to Ang-II—has been receiving interest as a modulator of cardiac remodeling [49] with effects mediated via MMPs [50].

A broad spectrum of cytokines, such as TNF-α and interleukins modulate cardiac fibroblasts. TNF-α is a pro-inflammatory cytokine expressed in the cardiac fibroblast. It suppresses ECM deposition in rat cardiac fibroblasts [51] and this is probably via a modulation of MMPs [52]. Interleukin-6 (IL-6) is involved in cell growth, differentiation, and apoptosis and has been shown to be present in cardiac fibroblasts [53]. Just like TNF-α, IL-6 suppresses collagen deposition and secretion [51]. It has also been shown that its receptor (soluble IL-6 receptor) is of major importance for myofibroblast differentiation as well as for collagen deposition [54]. IL-1β suppresses the proliferation of rat cardiac fibroblasts [55] and it has a direct negative effect on fibrosis by inhibiting collagen deposition and enhancing MMP activity [51]. The other member of the IL-1 family, IL-1α has been shown to modulate the expression of various MMPs and TIMP-1 in cultured myofibroblasts from human hearts [56]. Lastly, IL-17 and IL-18 also seem to play minor roles in the migration of cardiac fibroblasts and collagen deposition [57, 58].

Osteopontin (OPN) is a protein (also known as cytokine Eta-1) [59] that is involved in the wound healing process. The expression of OPN in mice is induced in the infarct area as well as in areas remote from the infarct, and this appears to modulate differentiation and activity of myofibroblasts [60]. Trueblood et al. demonstrated substantial changes in the collagen secretion and the MMP activity, in knockout mice for the OPN ($OPN^{-/-}$). These mice showed increased LV dilatation and depressed heart function measurements [61], indicating that OPN plays a crucial role in the remodeling after myocardium injury.

Fibronectin (FN) is one of the main components of the ECM. When an extra-domain A module (ED-A) is incorporated into the fibronectin structure, this leads to the formation of the ED-A fibronectin (ED-A FN) splice variant. This splice variant has enhanced adhesive abilities and is a critical factor in the cross-linking between fibronectin and various cardiac cells as well as the rest of the ECM components. This is crucial for TGF-β-mediated fibroblast-to-myofibroblast differentiation [62, 63]. ED-A FN is induced in cardiac myofibroblasts in vitro as well as in healed infarcts in vivo [64], hence its role in the wound healing process is decisive.

Moreover, hyaluronan (HN) is a glucosamine that forms polymeric chains and associates with cardiac cells via CD44 [65]. In an early study on an experimental rat MI model, the increased presence of HN in the infarct area and the non-infarcted myocardium was observed [66]. Infarcts of CD44-knockout mice demonstrate suppressed collagen secretion and reduced migration and proliferation of fibroblasts [67]. The inhibition of fibroblast HN leads to a suppressed α-SMA expression during differentiation of myofibroblasts [68], implying that the CD44-mediated HN is essential for maintenance of the myofibroblast phenotype and infarct healing.

Lastly, the peroxisome proliferator-activated receptor γ (PPAR-γ) belongs to the nuclear hormone receptor superfamily. PPAR members exert various important functions with a special focus on the lipid regulation and the metabolism of glucose [69]. PPAR-γ not only regulates glucose levels, but has also been proven to be expressed in cardiac myofibroblasts [70]. PPAR-γ is mostly expressed in cardiomyocytes under physiological conditions, but its expression is increased in cardiac fibroblasts of the infarct area, 3 weeks post-MI [71], hence it seems to be a modulator of the wound healing process, although the PPAR-γ effects on fibroblast migration and differentiation are still not clear-cut [4].

3.3.2 Mechanical and Electrophysiological Stimulation

Following MI, mechanical stretch stimulates the myofibroblasts, in order to increase the expression of collagen fibers I and III [72]. The orientation of collagen fibers is closely connected to the mechanical stretch applied and is a major determinant of the resistance of the myocardial scar [41]. A scar that is inadequately healed cannot withstand the large forces applied during systole and this may lead to life-threatening rupturing of the myocardium [73].

During wound healing in the border zone, ECM, and myofibroblasts lie between cardiomyocytes and this creates gaps between cardiomyocytes. Myofibroblasts form gap junctions with myocytes and other myofibroblasts or even with other heart cells. These gap junctions are composed of connexins (Cx43 and Cx45) which appear to directly modulate the proliferation of cardiac fibroblasts. Increased levels of Cx43 expression correlate to reduced proliferation of myofibroblasts and vice versa [74]. Nevertheless, the formation of gap junctions can lead to prevention of impulse propagation, which may induce conductivity problems and consequently cause life-threatening arrhythmias [75].

4 Current Therapy for Heart Failure and Novel Insights

The latest research shows that life expectancy in the Western world population is constantly rising [76]. Due to improved medical care, survival of patients suffering MI has increased greatly and it is estimated that the prevalence of heart failure will rise [77]. The pharmacological arsenal for the treatment of heart failure has evolved dramatically in the last 40 years and comprises a wide range of drug classes, such as diuretics, digoxin, hydralazine, nitrates, angiotensin-converting enzyme inhibitors (ACEI), angiotensin-receptor blockers (ARB), β-blockers, mineralocorticoid-receptor antagonists, and statins. These are used as monotherapies or in combinations depending on the specific needs of patients, possible comorbidities, etc. [78, 79]. Drug classes such as the ones affecting RAAS (ACEI, ARB, mineralocorticoid-receptor antagonists), β-blockers, and statins act mainly outside of the infarct area. The latest (2009) ACC/AHA guidelines for the treatment of patients following an acute MI or patients with HF are recommending the use of ACEI or ARB therapy, with aldosterone blockers and β-blockers following as second-line treatment [80].

4.1 Targeting the RAAS

The RAAS system has been established as a key player in normal physiology of the heart as well as wound healing following MI. Hence, the targeting of its components is the gold standard in treatment for HF patients that have suffered MI. Three main drug classes will be discussed: the ACEI, the ARB, and the mineralocorticoid-receptor antagonists.

4.1.1 ACEI and ARB

The main molecule that acts as the effector of the RAAS system is Ang-II. Ang-II is secreted de novo by myofibroblasts and modulates the deposition of collagen, thus affecting fibrosis of the myocardium (both infarcted and non-infarcted). This promotes adverse remodeling and eventually leads to HF [81]. Several experimental studies with ACEI or ARB in animals have proven to be inhibitory for fibrosis and have shown beneficial effects on reducing mortality and morbidity [82, 83]. The favorable effects of these drug classes are achieved via maturation of scar tissue in the infarct border zone, which prevents ECM from being deposited in areas that are far from the infarct [84]. Their effect on collagen might be more complex. The ACEI seem to use five different mechanisms by which they modulate collagen formation and deposition: by decreasing Ang-II production, inhibiting Ang-(1–7) degradation, suppressing MMP activity, amplifying the action of bradykinin, and inhibiting the *N*-acetyl-seryl-aspartyl-lysyl-proline (Ac-SDKP) degradation. On the other hand, the ARB exhibit their anti-fibrotic effects via the following mechanisms: decreasing Ang-II actions on AT-1 receptors, inducing Ang-(1–7) actions, inhibiting MMP activity, increasing the bradykinin effects, and activating AT-2 receptor activity [85]. A large number of randomized controlled trials (RCTs) in human patients using ACEI, ARB or a combination of the two, have demonstrated that suppression of Ang-II inhibits cardiac adverse remodeling and reduces overall mortality and morbidity following MI [86–89]. These two drug classes are thus the standard therapy in the pharmacological HF treatment of patients post-MI.

4.1.2 Mineralocorticoid-Receptor Antagonists

Aldosterone also plays an important role in the RAAS system. Several experimental groups have demonstrated that aldosterone directly affects cardiac fibroblasts and leads to cardiac fibrosis. Elevated levels of aldosterone have been shown to increase proliferation and induce collagen synthesis in animal [90, 91] and human fibroblasts [92]. In a quite recent study in rats, high doses of aldosterone given intravenously (I/V) induced collagen levels in the LV, while intrapericardial administration had no effect.

On the other hand, when lower doses of aldosterone were used, it was the intrapericardial administration that elevated collagen secretion in the LV, while I/V administration showed no significant effects regarding collagen secretion and deposition [93]. Two RCTs have shown that the use of aldosterone antagonists can be beneficial in reducing mortality and morbidity of patients with HF. In the RALES study [94] spironolactone (an aldosterone receptor blocker) was utilized in combination with the standard therapy (ACEI plus a loop diuretic), while in the EPHESUS study [95] eplerenone (a more selective aldosterone blocker) was coadministered with ACEI and loop diuretics. These two RCTs demonstrated a survival benefit and a decreased morbidity in patients with HF. Furthermore a study by Zannad et al. showed that the addition of spironolactone to standard therapy suppressed the levels of serum markers of cardiac fibrosis in HF patients, which is translated into a direct clinical benefit [96].

4.1.3 Renin Inhibitors

Lastly, the direct renin inhibitor Aliskiren has been shown to suppress de novo Ang-II synthesis [97] and could potentially have a place in the treatment of HF post-MI. It is not however, licensed yet for use in the treatment of HF [98].

4.2 Targeting β-Adrenoceptors

The drug class of β-adrenoceptor blockers is also one of the main tools for pharmacotherapy in the treatment of HF after MI and they have been shown to slow down the progression of HF by attenuating ventricular remodeling [99]. Following MI, cardiomyocytes are lost and the blood pumping ability of the heart muscle is compromised. Hence, in order to sustain the organs with an adequate blood supply there is a need for an increase in the cardiac output, [100]. This is done via the secretion of catecholamines that activate the β-adrenoceptor pathway and increase cardiac output [101].

On the other hand, sustained increase in β-adrenoceptor activity can cause adverse effects, especially in a patient with comorbidities, so the use of β-blockers is encouraged. This drug class can have several beneficial effects such as reducing heart rate, contractility of the myocardium and the blood pressure, limiting cardiac remodeling, and decreasing the risk for another MI [85]. β1-Selective antagonists are routinely used in the treatment of HF [102]; however, in the COMET study it was demonstrated that carvedilol, a nonselective agent which blocks both β1- and β2-adrenoceptors, is more beneficial compared to β1-selective agents (metoprolol) [103]. Cardiomyocytes express predominantly the β1-adrenoceptor, while the β2-adrenoceptor is the receptor type mainly expressed in cardiac fibroblasts [104]. β2-Antagonism is found to inhibit fibroblast proliferation [105] and modulate the secretion of collagen [106]. Nevertheless, it is not clear yet how the β-adrenoceptor signaling pathways affect myofibroblast differentiation and how they exert their effects on the ECM [4].

4.3 Statins

Finally, the HMG-CoA reductase inhibitors (also known as statins) have been receiving more and more attention lately as part of the treatment of HF in patients after MI. Various research groups have established that several members of the statins drug class can affect both fibroblasts and myofibroblasts. Nattel et al. utilized canine cardiac fibroblasts to show that simvastatin decreased the fibroblast α-SMA expression and could be counteracted by the administration of TGF-β. This was an indication that simvastatin can suppress the differentiation of fibroblasts to myofibroblasts, an action that would be beneficial in the remote area from the infarct [107]. Moreover, the proliferation of myofibroblasts was suppressed in mice by pravastatin [108] and by simvastatin in rats [109] and humans [110], while the deposition of collagen was found to be decreased after atorvastatin administration in rat cardiac fibroblasts [111]. Hence, there is evidence that apart from their well-established beneficial effect on the lowering of the blood cholesterol, statins offer an all-around cardiovascular protective role. They not only act on cardiac fibroblasts via a variety of mechanisms but also have pleiotropic effects on cardiomyocytes. Thus, they modulate myocardial remodeling and can be regarded as indispensible tools in the pharmacological treatment of patients following MI [112].

4.4 Novel Pharmacological Tools

Moreover, several novel drug therapies are under way targeting myofibroblasts and aiming to prevent heart failure development after MI. Some of these challenging therapies involve NADPH oxidase 4 (NOX4) [113, 114], TGF-β [44, 115], microRNAs [116, 117], ET-1 [118, 119], myocardin-related transcription factor-A (MRTF-A) [120], and others. Nevertheless, a detailed analysis of these novel opportunities is beyond the scope of this chapter. In Subheading 5 we discuss the challenging area of Wnt/Fzd-signaling pathway and its key role in the wound healing process following MI.

5 The Wnt/Fzd-Signaling Pathway and Its Implication in the Wound Healing Process

5.1 Wnt Signaling

Proteins belonging to the Wnt/Fzd signaling pathway have been identified as key factors in a variety of systems. These glycoproteins play major roles in many animal developmental processes, while at the same time they are implicated in cell adhesion, cell polarity, cell migration and proliferation, and the self-renewal of stem cells. Moreover, Wnt family proteins are involved in the progression of cancer (colon, ovarian, prostate, breast, and hepatocellular), cardiovascular diseases, osteoporosis, etc.; hence interest in them has been growing lately. Nineteen different Wnt proteins have been described so far in mammals and they are divided into two subfamilies, based on their transforming ability of the mouse mammary epithelial cell line. The Wnt-1 subfamily includes Wnt-1, -3,

-3a, and -7a, while the Wnt-5a subfamily comprises of Wnt-2, -4, -5a, -5b, -6, -7b, and -11 [121]. Frizzled (Fzd) are the main receptors that are activated by the Wnt proteins. They are seven transmembrane receptors (7-TM) and carry a large N-terminal cysteine-rich extracellular domain, which is of key importance for binding of the Wnt protein to the Fzd receptor. In mammals, ten members of the Fzd family have been characterized so far [122]. The low-density lipoprotein receptor-related protein (LRP) family has also been identified as an important factor of the Wnt/Fzd-signaling pathway, where it acts as a co-receptor [123]. The intracellular proteins Disheveled (Dvl) also play a crucial role in the signaling pathway [124].

The Wnt/Fzd-signaling pathway is involved in various, high complexity mechanisms that transduce signals in the cell. These signal transduction systems are divided into "canonical" (Wnt/β-catenin pathway) and "noncanonical" (independent of β-catenin) signaling pathways. The noncanonical pathway is comprised of two subdivisions, one involving Ca^{2+} that activates protein kinase C (PKC) and Ca-calmodulin kinase II (CaCKII) and another one that utilizes Jun-N-kinase (JNK) [121]. The canonical pathway is the most commonly studied and involves β-catenin which is a crucial transcription coactivator controlling cell adhesion and structure. β-Catenin is known to stimulate a wide range of genes that are involved in the hypertrophic process, such as c-jun, c-fos, c-myc, and cyclin D1. This process is mediated via the T-cell factor/lymphocyte enhancer factor (Tcf/Lef). The Wnt/β-catenin signaling pathway comprises a number of components that interact with each other and eventually activate target gene expression [125]. In this complex system, glycogen synthase kinase 3β (GSK-3β) is a key player [126]. More specifically, when the endogenous ligand Wnt is present, it binds to the Fzd receptor and the co-receptor LRP. The interaction between these three molecules activate Dvl, which in turn induces the formation of a complex comprising of the scaffolding protein Axin, GSK-3β, the adenomatous poliposis coli gene product (APC), and casein kinase 1 (CK1). Formation of this Axin complex prevents the phosphorylation of β-catenin, hence the latter is stabilized and translocates to the nucleus where it forms a complex with Tcf/Lef and induces the expression of target genes. When the Wnt ligand is not present (not activating the Fzd receptor), the Axin complex is unaffected and it acts by degrading β-catenin. Hence, the translocation of the latter to the nucleus and the induction of target gene expression is prevented [123, 127, 128].

5.2 Wnt Signaling in Cardiac Development and in Disease

The *Wnt* gene was first discovered by Nusse and Varmus in the early 1980s and focus was given to its involvement in tumors [129]. Since then, the Wnt/Fzd-signaling pathway has been established as a crucial player in the cardiovascular system and specifically in all developmental phases of the vertebrate heart. McMahon and McMahon demonstrated that the mouse *irp* gene (which was later renamed to *Wnt-2*) is expressed in heart tissue [130]. The Wnt/Fzd pathway (both canonical and noncanonical) has been proven to affect cardiogenesis, where the canonical pathway acts in an inhibitory way. Wnt signaling inhibition has further been implicated in malformation of heart valves and great arteries, while it also plays a major role in the differentiation of embryonic cardiomyocytes involving formation of Purkinje fibers. The importance of both canonical and noncanonical Wnt signaling pathways for the normal development of the heart is well-established. Apart from its role in the normal cardiac development, a range of cardiac pathological states has been associated with the Wnt-signaling pathway. Cardiac hypertrophy and wound healing following MI, along with changes in blood vessels have been shown to be regulated in varying degrees by the Wnt/Fzd-signaling pathway [131].

5.3 Effects of Wnt/Fzd Pathway on Cardiac Remodeling

Several studies have been undertaken on the role of Wnt/Fzd-signaling pathway on wound healing following MI and the effect that the manipulation of this signaling system might have in the development of HF.

5.3.1 Use of sFRPs

One area that has been studied is the soluble frizzled related proteins (sFRPs). These proteins can bind and inactivate both Wnt and Fzd [132]. Early studies have shown that the expression of sFRP-3 and -4 in the cardiomyocytes of failing hearts is elevated and that soluble β-catenin is reduced, implying a direct effect of the antagonists sFRP-3 and -4 on the canonical signaling pathway [133]. Barandon et al., using coronary artery ligation model in mice, underscored the beneficial role of the Wnt/Fzd pathway blockade. The infarct size of mice overexpressing FzdA (a homologue of sFRP-1) was found to be reduced and cardiac function measurements were greatly improved, compared to wild-type mice [134]. Later studies demonstrated that sFRP-1 can have a direct effect on cells involved in the inflammatory process following MI. sFRP-1 overexpression in leukocytes caused a reduction of IL-6 (pro-inflammatory cytokine) expression and an increase of IL-10 (antiinflammatory cytokine) expression. These events lead to improvement in hemodynamic measurements and a reduction of the infarct size, signifying the importance of fine balance between anti- and pro-inflammatory factors in wound healing [132]. Antagonism of sFRP-2 was shown to enhance the pro-collagen C-proteinase (pCP) activity (which is responsible for converting pro-collagen precursors to various ECM components) and to have a beneficial effect on cardiac performance compared to control [135]. Recently, the group of He et al. demonstrated that sFRP2 reduces collagen deposition, prevents wall thinning and improves cardiac function measurements in a rat MI model [136]. Furthermore, work on stem cells has suggested a beneficial role of the sFRPs, with sFRP-2 substantially suppressing the infarct size and greatly improving cardiac function measurements post-MI [137].

5.3.2 Targeting the Fzd Receptor

As mentioned earlier, the presence of myofibroblasts in the infarct area following MI has been shown to prevent the dilatation of the infarct and in this way plays a major role in the preservation of cardiac function. Previous studies undertaken in our laboratory have made an association of the Fzd-2 receptor with the actions of myofibroblasts after MI and have highlighted its importance during the healing process post-MI [138]. A study by Aisagbonhi et al. showed that the majority of the myofibroblasts present in the infarcted area are of endothelial–mesenchymal transition (EMT) origin and that this transition is mediated by the canonical Wnt signaling, playing an important role in the fibrosis of the infarct. One interesting conclusion of this study was that the β-catenin pathway does not seem to play a major role in the early stages following MI but functions mostly during the formation of the granulation tissue [139]. In an in vitro study by our group, immortalized cardiac fibroblast cells that were named cardiac fibroblasts immortalized with telomerase (CFIT) were utilized and manipulations of the Wnt/Fzd pathway were undertaken. It was proven that the Wnt/Fzd-signaling pathway has an inductive effect on myofibroblast migration and differentiation but not on myofibroblast proliferation. The Wnt/Fzd-signaling effect on CFIT migrations seemed to be mediated via the noncanonical pathway, while the alternations on the CFIT differentiation were mostly a result of the canonical (β-catenin) pathway. When various combinations of Wnt ligands (Wnt3a or Wnt5a) and Fzd receptors (rFzd-1 or rFzd-2) were used, totally opposite effects were observed, demonstrating the high complexity of the Wnt/Fzd-signaling pathway [140].

In another study, the role of peptide fragments of Wnt5a in the regulation of Wnt/Fzd pathway was investigated. Three different peptides (UM206, UM207, and UM208) acting as antagonists of the Fzd receptor were developed. As UM206 had the highest receptor affinity, it was the one studied in more depth. It was found that UM206 antagonized the effects of both Wnt3a and Wnt5a on myofibroblasts. UM206 was also administered in vivo in a murine MI model. When UM206 was given immediately after the MI induction or 2 weeks post-MI for a total of 5 weeks, it increased myofibroblast numbers in the infarct area, suppressed collagen secretion and the expansion of the fibrosis in areas remote from the infarct. In addition, it dramatically improved cardiac function measurements (EF, EDV) and mortality due to heart failure was completely prevented, compared to controls (Fig. 4) [141]. This emerging data has strengthened the hypothesis that myofibroblasts can be targeted via blockade of Wnt/Fzd signaling, in the hunt for a new therapeutic strategy to prevent adverse cardiac remodeling and heart failure post-MI.

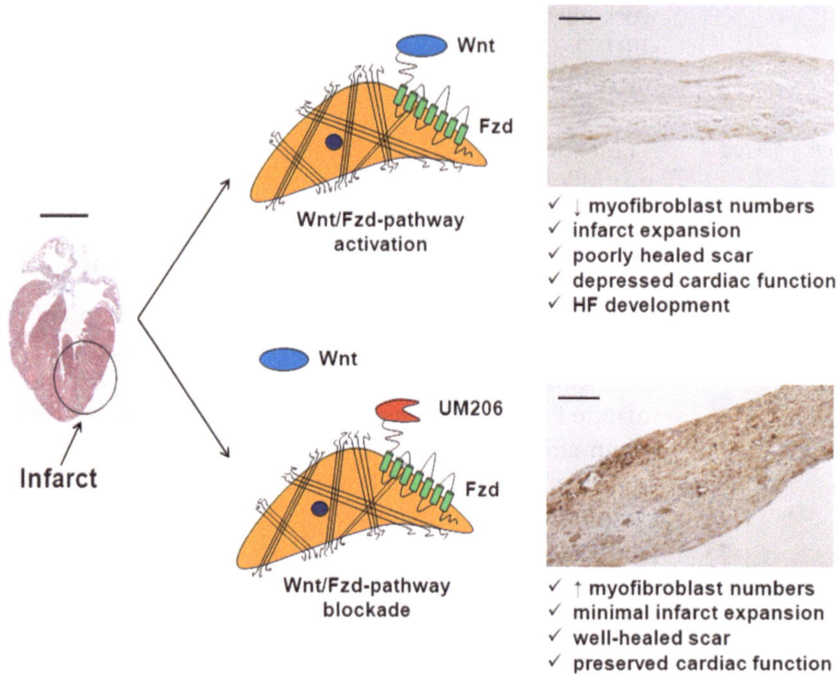

Fig. 4 The involvement of Wnt/Fzd-signaling pathway during cardiac remodeling (5 weeks post-MI). The blockade of Fzd1/2-receptors with UM206 after MI induction increases myofibroblast numbers, leads to a well-healed scar and reduced infarct expansion, and preserves cardiac function. Treatment with saline (control group) permits the activation of the Wnt/Fzd pathway, which causes reduced myofibroblast presence, expansion of infarct area, poor healing of the infarct scar, worsening of cardiac function markers, and HF. Images are taken from mouse hearts, 5 weeks post-MI. Scale bar on the mouse heart (*left side*) section equals 3 mm. Scale bars on the heart section images (*right side*) equal 100 μm. *Abbreviations*: *MI* myocardial infarction, *Fzd1/2-receptors* frizzled1/2-receptors, *HF* heart failure

5.3.3 Targeting Dvl

An early in vivo study by Chen et al. showed that Dvl-1 mRNA is present at the border zone of the myocardium on day 1 post-MI. The mRNA levels of Dvl-1 in rats that have suffered MI are induced on day 4 post-MI compared to sham and reach a peak on day 7. Dvl-1 mRNA is not detected in the non-infarcted myocardium or in the hearts of the sham group. This study also made a connection between the Wnt/Fzd/Dvl system activation and the effect that it might have on myofibroblast migration and proliferation [142]. Furthermore, transgenic mice overexpressing Dvl showed severe ventricular hypertrophy with a reduction of EF. These mice died prematurely (i.e., before reaching 6 months of age). Both the canonical and noncanonical signaling pathways seem to be implicated in these effects [143]. More research is still necessary on the Dvl-level, to ascertain its actual beneficial effects as a target for the prevention of the adverse remodeling.

6 Conclusions

In this review, we have made an effort to give an overview of the main characteristics of the cardiac myofibroblast, the factors that affect its function and its major role in the wound healing process following MI, the remodeling and the development of HF. Myofibroblasts are becoming the center of attention with regards to cardiac wound healing. They are the cells producing and depositing collagen that leads not only to the formation of the scar in the infarct area but also plays a role in the fibrosis of the non-infarcted area and the adverse remodeling of the heart. Hence, the myofibroblast impact on the development of LV dysfunction and HF seems to be established. The exact underlying factors regulating myofibroblasts during the wound healing process after MI however are not yet fully understood and more research is needed in this area.

The ultimate goal of modern cardiovascular pharmacology is to devise strategies that will target three separate areas after MI: (1) inhibit death of the cardiomyocytes, (2) improve myocardial wound healing, and (3) limit the adverse cardiac remodeling [144]. Currently available drug therapies are quite efficacious in reducing mortality and morbidity; however, they have not been successful in reducing dilatation or reversing the adverse remodeling once begun. They may also cause significant side effects, which lead to noncompliant patients and reduced therapeutic efficacy. It is time for new pharmacological treatments to be discovered.

Optimal drug therapy would lead to a well-healed compacted scar, prevent fibrosis in areas that lie in the non-infarcted areas, and have minor side effects. Several novel and promising pharmacological tools are under development in order to target the myofibroblast itself. The emerging data is showing that the Wnt/Fzd-signaling pathway is a key player in the regulation of the myofibroblast and that its manipulation could prevent adverse cardiac remodeling and the development of HF post-MI. Still, the clinical advantage and possible side effects are yet to be addressed.

Acknowledgments

The financial support for this research line by the Department of Economic Affairs of The Netherlands (BSIK grant 03033) is gratefully acknowledged.

References

1. Cleutjens JP, Blankesteijn WM, Daemen MJ, Smits JF (1999) The infarcted myocardium: simply dead tissue, or a lively target for therapeutic interventions. Cardiovasc Res 44: 232–241
2. van den Borne SW, Diez J, Blankesteijn WM, Verjans J, Hofstra L, Narula J (2010) Myocardial remodeling after infarction: the role of myofibroblasts. Nat Rev Cardiol 7:30–37
3. McMurray JJ, Pfeffer MA (2005) Heart failure. Lancet 365:1877–1889
4. Porter KE, Turner NA (2009) Cardiac fibroblasts: at the heart of myocardial remodeling. Pharmacol Ther 123:255–278
5. Baudino TA, Carver W, Giles W, Borg TK (2006) Cardiac fibroblasts: friend or foe? Am J Physiol Heart Circ Physiol 291: H1015–H1026
6. WHO (2011) WHO Fact Sheet No.317 - Cardiovascular diseases (CVDs)
7. Roger VL, Go AS, Lloyd-Jones DM, Adams RJ, Berry JD, Brown TM, Carnethon MR, Dai S, de Simone G, Ford ES, Fox CS, Fullerton HJ, Gillespie C, Greenlund KJ, Hailpern SM, Heit JA, Ho PM, Howard VJ, Kissela BM, Kittner SJ, Lackland DT, Lichtman JH, Lisabeth LD, Makuc DM, Marcus GM, Marelli A, Matchar DB, McDermott MM, Meigs JB, Moy CS, Mozaffarian D, Mussolino ME, Nichol G, Paynter NP, Rosamond WD, Sorlie PD, Stafford RS, Turan TN, Turner MB, Wong ND, Wylie-Rosett J (2011) Heart disease and stroke statistics–2011 update: a report from the American Heart Association. Circulation 123:e18–e209
8. White HD, Chew DP (2008) Acute myocardial infarction. Lancet 372:570–584
9. Bui AL, Horwich TB, Fonarow GC (2011) Epidemiology and risk profile of heart failure. Nat Rev Cardiol 8:30–41
10. Wexler DJ, Chen J, Smith GL, Radford MJ, Yaari S, Bradford WD, Krumholz HM (2001) Predictors of costs of caring for elderly patients discharged with heart failure. Am Heart J 142:350–357
11. NICE (2010) Chronic Heart Failure, National clinical guideline for diagnosis and management in primary and secondary care (National Institute for Health and Clinical Excellence Clinical Guideline No 108)
12. Soonpaa MH, Daud AI, Koh GY, Klug MG, Kim KK, Wang H, Field LJ (1995) Potential approaches for myocardial regeneration. Ann N Y Acad Sci 752:446–454
13. Buja LM, Vela D (2008) Cardiomyocyte death and renewal in the normal and diseased heart. Cardiovasc Pathol 17:349–374
14. Dewald O, Ren G, Duerr GD, Zoerlein M, Klemm C, Gersch C, Tincey S, Michael LH, Entman ML, Frangogiannis NG (2004) Of mice and dogs: species-specific differences in the inflammatory response following myocardial infarction. Am J Pathol 164:665–677
15. Dobaczewski M, Bujak M, Zymek P, Ren G, Entman ML, Frangogiannis NG (2006) Extracellular matrix remodeling in canine and mouse myocardial infarcts. Cell Tissue Res 324:475–488
16. Rosenkranz S (2004) TGF-beta1 and angiotensin networking in cardiac remodeling. Cardiovasc Res 63:423–432
17. Souders CA, Bowers SL, Baudino TA (2009) Cardiac fibroblast: the renaissance cell. Circ Res 105:1164–1176
18. Tyagi SC, Campbell SE, Reddy HK, Tjahja E, Voelker DJ (1996) Matrix metalloproteinase activity expression in infarcted, noninfarcted and dilated cardiomyopathic human hearts. Mol Cell Biochem 155:13–21
19. van der Laan AM, Piek JJ, van Royen N (2009) Targeting angiogenesis to restore the microcirculation after reperfused MI. Nat Rev Cardiol 6:515–523
20. Willems IE, Havenith MG, De Mey JG, Daemen MJ (1994) The alpha-smooth muscle actin-positive cells in healing human myocardial scars. Am J Pathol 145:868–875
21. Frantz S, Bauersachs J, Ertl G (2009) Post-infarct remodelling: contribution of wound healing and inflammation. Cardiovasc Res 81:474–481
22. van den Borne SW, van de Schans VA, Strzelecka AE, Vervoort-Peters HT, Lijnen PM, Cleutjens JP, Smits JF, Daemen MJ, Janssen BJ, Blankesteijn WM (2009) Mouse strain determines the outcome of wound healing after myocardial infarction. Cardiovasc Res 84:273–282
23. Konstam MA, Kramer DG, Patel AR, Maron MS, Udelson JE (2011) Left ventricular remodeling in heart failure: current concepts in clinical significance and assessment. JACC Cardiovasc Imaging 4:98–108
24. Beltrami CA, Finato N, Rocco M, Feruglio GA, Puricelli C, Cigola E, Quaini F, Sonnenblick EH, Olivetti G, Anversa P (1994) Structural basis of end-stage failure in ischemic cardiomyopathy in humans. Circulation 89:151–163
25. Olivetti G, Melissari M, Balbi T, Quaini F, Sonnenblick EH, Anversa P (1994) Myocyte nuclear and possible cellular hyperplasia contribute to ventricular remodeling in the hypertrophic senescent heart in humans. J Am Coll Cardiol 24:140–149
26. Frangogiannis NG (2006) Targeting the inflammatory response in healing myocardial infarcts. Curr Med Chem 13:1877–1893

27. Camelliti P, Borg TK, Kohl P (2005) Structural and functional characterisation of cardiac fibroblasts. Cardiovasc Res 65:40–51
28. Fries KM, Blieden T, Looney RJ, Sempowski GD, Silvera MR, Willis RA, Phipps RP (1994) Evidence of fibroblast heterogeneity and the role of fibroblast subpopulations in fibrosis. Clin Immunol Immunopathol 72:283–292
29. Sorrell JM, Caplan AI (2004) Fibroblast heterogeneity: more than skin deep. J Cell Sci 117:667–675
30. Sun Y, Weber KT (1996) Angiotensin converting enzyme and myofibroblasts during tissue repair in the rat heart. J Mol Cell Cardiol 28:851–858
31. Potts JD, Runyan RB (1989) Epithelial-mesenchymal cell transformation in the embryonic heart can be mediated, in part, by transforming growth factor beta. Dev Biol 134:392–401
32. Zeisberg EM, Tarnavski O, Zeisberg M, Dorfman AL, McMullen JR, Gustafsson E, Chandraker A, Yuan X, Pu WT, Roberts AB, Neilson EG, Sayegh MH, Izumo S, Kalluri R (2007) Endothelial-to-mesenchymal transition contributes to cardiac fibrosis. Nat Med 13:952–961
33. Strieter RM, Keeley EC, Burdick MD, Mehrad B (2009) The role of circulating mesenchymal progenitor cells, fibrocytes, in promoting pulmonary fibrosis. Trans Am Clin Climatol Assoc 120:49–59
34. Diaz-Flores L, Gutierrez R, Madrid JF, Varela H, Valladares F, Acosta E, Martin-Vasallo P, Diaz-Flores L Jr (2009) Pericytes. Morphofunction, interactions and pathology in a quiescent and activated mesenchymal cell niche. Histol Histopathol 24:909–969
35. Baum J, Duffy HS (2011) Fibroblasts and myofibroblasts: what are we talking about? J Cardiovasc Pharmacol 57:376–379
36. Meran S, Steadman R (2011) Fibroblasts and myofibroblasts in renal fibrosis. Int J Exp Pathol 92:158–167
37. Hinz B, Celetta G, Tomasek JJ, Gabbiani G, Chaponnier C (2001) Alpha-smooth muscle actin expression upregulates fibroblast contractile activity. Mol Biol Cell 12:2730–2741
38. Tomasek JJ, Gabbiani G, Hinz B, Chaponnier C, Brown RA (2002) Myofibroblasts and mechano-regulation of connective tissue remodelling. Nat Rev Mol Cell Biol 3: 349–363
39. Prockop DJ, Kivirikko KI (1995) Collagens: molecular biology, diseases, and potentials for therapy. Annu Rev Biochem 64:403–434
40. Holmes JW, Borg TK, Covell JW (2005) Structure and mechanics of healing myocardial infarcts. Annu Rev Biomed Eng 7: 223–253
41. Fomovsky GM, Thomopoulos S, Holmes JW (2010) Contribution of extracellular matrix to the mechanical properties of the heart. J Mol Cell Cardiol 48:490–496
42. Matsui Y, Morimoto J, Uede T (2010) Role of matricellular proteins in cardiac tissue remodeling after myocardial infarction. World J Biol Chem 1:69–80
43. Hinz B (2007) Formation and function of the myofibroblast during tissue repair. J Invest Dermatol 127:526–537
44. Dobaczewski M, Chen W, Frangogiannis NG (2011) Transforming growth factor (TGF)-beta signaling in cardiac remodeling. J Mol Cell Cardiol 51:600–606
45. Bujak M, Frangogiannis NG (2007) The role of TGF-beta signaling in myocardial infarction and cardiac remodeling. Cardiovasc Res 74:184–195
46. Sun Y (2010) Intracardiac renin-angiotensin system and myocardial repair/remodeling following infarction. J Mol Cell Cardiol 48:483–489
47. Lijnen PJ, Petrov VV, Fagard RH (2001) Angiotensin II-induced stimulation of collagen secretion and production in cardiac fibroblasts is mediated via angiotensin II subtype 1 receptors. J Renin Angiotensin Aldosterone Syst 2:117–122
48. Yu CM, Tipoe GL, Wing-Hon Lai K, Lau CP (2001) Effects of combination of angiotensin-converting enzyme inhibitor and angiotensin receptor antagonist on inflammatory cellular infiltration and myocardial interstitial fibrosis after acute myocardial infarction. J Am Coll Cardiol 38:1207–1215
49. Lin CS, Pan CH, Wen CH, Yang TH, Kuan TC (2010) Regulation of angiotensin converting enzyme II by angiotensin peptides in human cardiofibroblasts. Peptides 31:1334–1340
50. Kassiri Z, Zhong J, Guo D, Basu R, Wang X, Liu PP, Scholey JW, Penninger JM, Oudit GY (2009) Loss of angiotensin-converting enzyme 2 accelerates maladaptive left ventricular remodeling in response to myocardial infarction. Circ Heart Fail 2:446–455
51. Siwik DA, Chang DL, Colucci WS (2000) Interleukin-1beta and tumor necrosis factor-alpha decrease collagen synthesis and increase matrix metalloproteinase activity in cardiac fibroblasts in vitro. Circ Res 86:1259–1265
52. Porter KE, Turner NA, O'Regan DJ, Ball SG (2004) Tumor necrosis factor alpha induces human atrial myofibroblast proliferation, invasion and MMP-9 secretion: inhibition by simvastatin. Cardiovasc Res 64:507–515
53. Banerjee I, Fuseler JW, Intwala AR, Baudino TA (2009) IL-6 loss causes ventricular dysfunction, fibrosis, reduced capillary density,

and dramatically alters the cell populations of the developing and adult heart. Am J Physiol Heart Circ Physiol 296:H1694–H1704
54. Dixon IM (2010) The soluble interleukin 6 receptor takes its place in the pantheon of interleukin 6 signaling proteins: phenoconversion of cardiac fibroblasts to myofibroblasts. Hypertension 56:193–195
55. Palmer JN, Hartogensis WE, Patten M, Fortuin FD, Long CS (1995) Interleukin-1 beta induces cardiac myocyte growth but inhibits cardiac fibroblast proliferation in culture. J Clin Invest 95:2555–2564
56. Turner NA, Warburton P, O'Regan DJ, Ball SG, Porter KE (2010) Modulatory effect of interleukin-1alpha on expression of structural matrix proteins, MMPs and TIMPs in human cardiac myofibroblasts: role of p38 MAP kinase. Matrix Biol 29:613–620
57. Cortez DM, Feldman MD, Mummidi S, Valente AJ, Steffensen B, Vincenti M, Barnes JL, Chandrasekar B (2007) IL-17 stimulates MMP-1 expression in primary human cardiac fibroblasts via p38 MAPK- and ERK1/2-dependent C/EBP-beta, NF-kappaB, and AP-1 activation. Am J Physiol Heart Circ Physiol 293:H3356–H3365
58. Fix C, Bingham K, Carver W (2011) Effects of interleukin-18 on cardiac fibroblast function and gene expression. Cytokine 53: 19–28
59. Singh M, Foster CR, Dalal S, Singh K (2010) Osteopontin: role in extracellular matrix deposition and myocardial remodeling post-MI. J Mol Cell Cardiol 48:538–543
60. Lenga Y, Koh A, Perera AS, McCulloch CA, Sodek J, Zohar R (2008) Osteopontin expression is required for myofibroblast differentiation. Circ Res 102:319–327
61. Trueblood NA, Xie Z, Communal C, Sam F, Ngoy S, Liaw L, Jenkins AW, Wang J, Sawyer DB, Bing OH, Apstein CS, Colucci WS, Singh K (2001) Exaggerated left ventricular dilation and reduced collagen deposition after myocardial infarction in mice lacking osteopontin. Circ Res 88:1080–1087
62. Manabe R, Ohe N, Maeda T, Fukuda T, Sekiguchi K (1997) Modulation of cell-adhesive activity of fibronectin by the alternatively spliced EDA segment. J Cell Biol 139:295–307
63. Serini G, Bochaton-Piallat ML, Ropraz P, Geinoz A, Borsi L, Zardi L, Gabbiani G (1998) The fibronectin domain ED-A is crucial for myofibroblastic phenotype induction by transforming growth factor-beta1. J Cell Biol 142:873–881
64. Santiago JJ, Dangerfield AL, Rattan SG, Bathe KL, Cunnington RH, Raizman JE, Bedosky KM, Freed DH, Kardami E, Dixon IM (2010) Cardiac fibroblast to myofibroblast differentiation in vivo and in vitro: expression of focal adhesion components in neonatal and adult rat ventricular myofibroblasts. Dev Dyn 239:1573–1584
65. Ponta H, Sherman L, Herrlich PA (2003) CD44: from adhesion molecules to signalling regulators. Nat Rev Mol Cell Biol 4:33–45
66. Waldenstrom A, Martinussen HJ, Gerdin B, Hallgren R (1991) Accumulation of hyaluronan and tissue edema in experimental myocardial infarction. J Clin Invest 88: 1622–1628
67. Huebener P, Abou-Khamis T, Zymek P, Bujak M, Ying X, Chatila K, Haudek S, Thakker G, Frangogiannis NG (2008) CD44 is critically involved in infarct healing by regulating the inflammatory and fibrotic response. J Immunol 180:2625–2633
68. Webber J, Jenkins RH, Meran S, Phillips A, Steadman R (2009) Modulation of TGFbeta1-dependent myofibroblast differentiation by hyaluronan. Am J Pathol 175: 148–160
69. Huss JM, Kelly DP (2004) Nuclear receptor signaling and cardiac energetics. Circ Res 95:568–578
70. Chintalgattu V, Harris GS, Akula SM, Katwa LC (2007) PPAR-gamma agonists induce the expression of VEGF and its receptors in cultured cardiac myofibroblasts. Cardiovasc Res 74:140–150
71. Fliegner D, Westermann D, Riad A, Schubert C, Becher E, Fielitz J, Tschope C, Regitz-Zagrosek V (2008) Up-regulation of PPARgamma in myocardial infarction. Eur J Heart Fail 10:30–38
72. Husse B, Briest W, Homagk L, Isenberg G, Gekle M (2007) Cyclical mechanical stretch modulates expression of collagen I and collagen III by PKC and tyrosine kinase in cardiac fibroblasts. Am J Physiol Regul Integr Comp Physiol 293:R1898–R1907
73. van den Borne SW, Narula J, Voncken JW, Lijnen PM, Vervoort-Peters HT, Dahlmans VE, Smits JF, Daemen MJ, Blankesteijn WM (2008) Defective intercellular adhesion complex in myocardium predisposes to infarct rupture in humans. J Am Coll Cardiol 51: 2184–2192
74. Zhang Y, Kanter EM, Laing JG, Aprhys C, Johns DC, Kardami E, Yamada KA (2008) Connexin43 expression levels influence intercellular coupling and cell proliferation of native murine cardiac fibroblasts. Cell Commun Adhes 15:289–303
75. Pellman J, Lyon RC, Sheikh F (2010) Extracellular matrix remodeling in atrial fibrosis: mechanisms and implications in atrial fibrillation. J Mol Cell Cardiol 48:461–467

76. Leon DA (2011) Trends in European life expectancy: a salutary view. Int J Epidemiol 40:271–277
77. Owan TE, Hodge DO, Herges RM, Jacobsen SJ, Roger VL, Redfield MM (2006) Trends in prevalence and outcome of heart failure with preserved ejection fraction. N Engl J Med 355:251–259
78. Krum H, Teerlink JR (2011) Medical therapy for chronic heart failure. Lancet 378:713–721
79. Minicucci MF, Azevedo PS, Polegato BF, Paiva SA, Zornoff LA (2011) Heart failure after myocardial infarction: clinical implications and treatment. Clin Cardiol 34:410–414
80. Hunt SA, Abraham WT, Chin MH, Feldman AM, Francis GS, Ganiats TG, Jessup M, Konstam MA, Mancini DM, Michl K, Oates JA, Rahko PS, Silver MA, Stevenson LW, Yancy CW (2009) 2009 Focused update incorporated into the ACC/AHA 2005 Guidelines for the diagnosis and management of heart failure in adults. A Report of the American College of Cardiology Foundation/ American Heart Association Task Force on Practice Guidelines Developed in Collaboration With the International Society for Heart and Lung Transplantation. J Am Coll Cardiol 53:e1–e90
81. Cleutjens JP, Verluyten MJ, Smiths JF, Daemen MJ (1995) Collagen remodeling after myocardial infarction in the rat heart. Am J Pathol 147:325–338
82. Schieffer B, Wirger A, Meybrunn M, Seitz S, Holtz J, Riede UN, Drexler H (1994) Comparative effects of chronic angiotensin-converting enzyme inhibition and angiotensin II type 1 receptor blockade on cardiac remodeling after myocardial infarction in the rat. Circulation 89:2273–2282
83. Taylor K, Patten RD, Smith JJ, Aronovitz MJ, Wight J, Salomon RN, Konstam MA (1998) Divergent effects of angiotensin-converting enzyme inhibition and angiotensin II-receptor antagonism on myocardial cellular proliferation and collagen deposition after myocardial infarction in rats. J Cardiovasc Pharmacol 31:654–660
84. van den Borne SW, Isobe S, Verjans JW, Petrov A, Lovhaug D, Li P, Zandbergen HR, Ni Y, Frederik P, Zhou J, Arbo B, Rogstad A, Cuthbertson A, Chettibi S, Reutelingsperger C, Blankesteijn WM, Smits JF, Daemen MJ, Zannad F, Vannan MA, Narula N, Pitt B, Hofstra L, Narula J (2008) Molecular imaging of interstitial alterations in remodeling myocardium after myocardial infarction. J Am Coll Cardiol 52:2017–2028
85. Jugdutt BI (2008) Pleiotropic effects of cardiac drugs on healing post-MI. The good, bad, and ugly. Heart Fail Rev 13:439–452
86. Kober L, Torp-Pedersen C, Carlsen JE, Bagger H, Eliasen P, Lyngborg K, Videbaek J, Cole DS, Auclert L, Pauly NC (1995) A clinical trial of the angiotensin-converting-enzyme inhibitor trandolapril in patients with left ventricular dysfunction after myocardial infarction. Trandolapril Cardiac Evaluation (TRACE) Study Group. N Engl J Med 333:1670–1676
87. McKelvie RS, Yusuf S, Pericak D, Avezum A, Burns RJ, Probstfield J, Tsuyuki RT, White M, Rouleau J, Latini R, Maggioni A, Young J, Pogue J (1999) Comparison of candesartan, enalapril, and their combination in congestive heart failure: randomized evaluation of strategies for left ventricular dysfunction (RESOLVD) pilot study. The RESOLVD Pilot Study Investigators. Circulation 100:1056–1064
88. McMurray JJ, Ostergren J, Swedberg K, Granger CB, Held P, Michelson EL, Olofsson B, Yusuf S, Pfeffer MA (2003) Effects of candesartan in patients with chronic heart failure and reduced left-ventricular systolic function taking angiotensin-converting-enzyme inhibitors: the CHARM-Added trial. Lancet 362:767–771
89. Pfeffer MA, McMurray JJ, Velazquez EJ, Rouleau JL, Kober L, Maggioni AP, Solomon SD, Swedberg K, Van de Werf F, White H, Leimberger JD, Henis M, Edwards S, Zelenkofske S, Sellers MA, Califf RM (2003) Valsartan, captopril, or both in myocardial infarction complicated by heart failure, left ventricular dysfunction, or both. N Engl J Med 349:1893–1906
90. Brilla CG, Zhou G, Matsubara L, Weber KT (1994) Collagen metabolism in cultured adult rat cardiac fibroblasts: response to angiotensin II and aldosterone. J Mol Cell Cardiol 26:809–820
91. Brilla CG, Maisch B, Zhou G, Weber KT (1995) Hormonal regulation of cardiac fibroblast function. Eur Heart J 16(Suppl C): 45–50
92. Neumann S, Huse K, Semrau R, Diegeler A, Gebhardt R, Buniatian GH, Scholz GH (2002) Aldosterone and D-glucose stimulate the proliferation of human cardiac myofibroblasts in vitro. Hypertension 39:756–760
93. Minnaard-Huiban M, Hermans JJ, Essen H, Bitsch N, Smits JF (2008) Comparison of the effects of intrapericardial and intravenous aldosterone infusions on left ventricular fibrosis in rats. Eur J Heart Fail 10:1166–1171
94. Pitt B, Zannad F, Remme WJ, Cody R, Castaigne A, Perez A, Palensky J, Wittes J (1999) The effect of spironolactone on morbidity and mortality in patients with severe heart failure Randomized aldactone evaluation study investigators. N Engl J Med 341: 709–717

95. Pitt B, Remme W, Zannad F, Neaton J, Martinez F, Roniker B, Bittman R, Hurley S, Kleiman J, Gatlin M (2003) Eplerenone, a selective aldosterone blocker, in patients with left ventricular dysfunction after myocardial infarction. N Engl J Med 348:1309–1321
96. Zannad F, Alla F, Dousset B, Perez A, Pitt B (2000) Limitation of excessive extracellular matrix turnover may contribute to survival benefit of spironolactone therapy in patients with congestive heart failure: insights from the randomized aldactone evaluation study (RALES). Rales Investigators. Circulation 102:2700–2706
97. Singh VP, Baker KM, Kumar R (2008) Activation of the intracellular renin-angiotensin system in cardiac fibroblasts by high glucose: role in extracellular matrix production. Am J Physiol Heart Circ Physiol 294:H1675–H1684
98. Novartis (2011) Highlights of prescribing information - Tekturna® (aliskiren) Tablets, Oral (http://www.pharma.us.novartis.com/product/pi/pdf/tekturna.pdf). Accessed 9 Mar 2011
99. Klapholz M (2009) Beta-blocker use for the stages of heart failure. Mayo Clin Proc 84:718–729
100. Lefkowitz RJ, Rockman HA, Koch WJ (2000) Catecholamines, cardiac beta-adrenergic receptors, and heart failure. Circulation 101:1634–1637
101. Post SR, Hammond HK, Insel PA (1999) Beta-adrenergic receptors and receptor signaling in heart failure. Annu Rev Pharmacol Toxicol 39:343–360
102. de Groote P, Ennezat PV, Mouquet F (2007) Bisoprolol in the treatment of chronic heart failure. Vasc Health Risk Manag 3:431–439
103. Kveiborg B, Major-Petersen A, Christiansen B, Torp-Pedersen C (2007) Carvedilol in the treatment of chronic heart failure: lessons from the Carvedilol Or Metoprolol European Trial. Vasc Health Risk Manag 3:31–37
104. Meszaros JG, Gonzalez AM, Endo-Mochizuki Y, Villegas S, Villarreal F, Brunton LL (2000) Identification of G protein-coupled signaling pathways in cardiac fibroblasts: cross talk between G(q) and G(s). Am J Physiol Cell Physiol 278:C154–C162
105. Turner NA, Porter KE, Smith WH, White HL, Ball SG, Balmforth AJ (2003) Chronic beta2-adrenergic receptor stimulation increases proliferation of human cardiac fibroblasts via an autocrine mechanism. Cardiovasc Res 57:784–792
106. Ostrom RS, Naugle JE, Hase M, Gregorian C, Swaney JS, Insel PA, Brunton LL, Meszaros JG (2003) Angiotensin II enhances adenylyl cyclase signaling via Ca2+/calmodulin. Gq-Gs cross-talk regulates collagen production in cardiac fibroblasts. J Biol Chem 278:24461–24468
107. Shiroshita-Takeshita A, Brundel BJ, Burstein B, Leung TK, Mitamura H, Ogawa S, Nattel S (2007) Effects of simvastatin on the development of the atrial fibrillation substrate in dogs with congestive heart failure. Cardiovasc Res 74:75–84
108. Chen J, Mehta JL (2006) Angiotensin II-mediated oxidative stress and procollagen-1 expression in cardiac fibroblasts: blockade by pravastatin and pioglitazone. Am J Physiol Heart Circ Physiol 291:H1738–H1745
109. He YP, Zhao LY, Zheng QS, Liu SW, Zhao XY, Lu XL, Niu XL, Li X (2008) Involvement of ERK and AKT signaling in the growth effect of arginine vasopressin on adult rat cardiac fibroblast and the modulation by simvastatin. Mol Cell Biochem 317:33–41
110. Porter KE, Turner NA, O'Regan DJ, Balmforth AJ, Ball SG (2004) Simvastatin reduces human atrial myofibroblast proliferation independently of cholesterol lowering via inhibition of RhoA. Cardiovasc Res 61:745–755
111. Martin J, Denver R, Bailey M, Krum H (2005) In vitro inhibitory effects of atorvastatin on cardiac fibroblasts: implications for ventricular remodelling. Clin Exp Pharmacol Physiol 32:697–701
112. Porter KE, Turner NA (2011) Statins and myocardial remodelling: cell and molecular pathways. Expert Rev Mol Med 13:e22
113. Cucoranu I, Clempus R, Dikalova A, Phelan PJ, Ariyan S, Dikalov S, Sorescu D (2005) NAD(P)H oxidase 4 mediates transforming growth factor-beta1-induced differentiation of cardiac fibroblasts into myofibroblasts. Circ Res 97:900–907
114. Hecker L, Vittal R, Jones T, Jagirdar R, Luckhardt TR, Horowitz JC, Pennathur S, Martinez FJ, Thannickal VJ (2009) NADPH oxidase-4 mediates myofibroblast activation and fibrogenic responses to lung injury. Nat Med 15:1077–1081
115. Tan SM, Zhang Y, Connelly KA, Gilbert RE, Kelly DJ (2010) Targeted inhibition of activin receptor-like kinase 5 signaling attenuates cardiac dysfunction following myocardial infarction. Am J Physiol Heart Circ Physiol 298:H1415–H1425
116. Thum T, Gross C, Fiedler J, Fischer T, Kissler S, Bussen M, Galuppo P, Just S, Rottbauer W, Frantz S, Castoldi M, Soutschek J, Koteliansky V, Rosenwald A, Basson MA, Licht JD, Pena JT, Rouhanifard SH, Muckenthaler MU, Tuschl T, Martin GR, Bauersachs J, Engelhardt S (2008) MicroRNA-21 contributes to myocardial disease by stimulating MAP kinase signalling in fibroblasts. Nature 456:980–984

117. van Rooij E, Sutherland LB, Thatcher JE, DiMaio JM, Naseem RH, Marshall WS, Hill JA, Olson EN (2008) Dysregulation of microRNAs after myocardial infarction reveals a role of miR-29 in cardiac fibrosis. Proc Natl Acad Sci U S A 105:13027–13032
118. Peng HJ, Dai DZ, Ji H, Dai Y (2010) The separate roles of endothelin receptors participate in remodeling of matrix metalloproteinase and connexin 43 of cardiac fibroblasts in maladaptive response to isoproterenol. Eur J Pharmacol 634:101–106
119. Rodriguez-Pascual F, Busnadiego O, Lagares D, Lamas S (2011) Role of endothelin in the cardiovascular system. Pharmacol Res 63:463–472
120. Small EM, Thatcher JE, Sutherland LB, Kinoshita H, Gerard RD, Richardson JA, Dimaio JM, Sadek H, Kuwahara K, Olson EN (2010) Myocardin-related transcription factor-a controls myofibroblast activation and fibrosis in response to myocardial infarction. Circ Res 107:294–304
121. Kikuchi A, Yamamoto H, Kishida S (2007) Multiplicity of the interactions of Wnt proteins and their receptors. Cell Signal 19:659–671
122. Schulte G, Bryja V (2007) The frizzled family of unconventional G-protein-coupled receptors. Trends Pharmacol Sci 28:518–525
123. Cadigan KM, Liu YI (2006) Wnt signaling: complexity at the surface. J Cell Sci 119: 395–402
124. Krasnow RE, Wong LL, Adler PN (1995) Dishevelled is a component of the frizzled signaling pathway in Drosophila. Development 121:4095–4102
125. Reya T, Clevers H (2005) Wnt signalling in stem cells and cancer. Nature 434:843–850
126. Blankesteijn WM, van de Schans VA, ter Horst P, Smits JF (2008) The Wnt/frizzled/GSK-3 beta pathway: a novel therapeutic target for cardiac hypertrophy. Trends Pharmacol Sci 29:175–180
127. van Gijn ME, Daemen MJ, Smits JF, Blankesteijn WM (2002) The wnt-frizzled cascade in cardiovascular disease. Cardiovasc Res 55:16–24
128. MacDonald BT, Tamai K, He X (2009) Wnt/beta-catenin signaling: components, mechanisms, and diseases. Dev Cell 17:9–26
129. Nusse R, Varmus HE (1982) Many tumors induced by the mouse mammary tumor virus contain a provirus integrated in the same region of the host genome. Cell 31:99–109
130. McMahon JA, McMahon AP (1989) Nucleotide sequence, chromosomal localization and developmental expression of the mouse int-1-related gene. Development 107:643–650
131. van de Schans VA, Smits JF, Blankesteijn WM (2008) The Wnt/frizzled pathway in cardiovascular development and disease: friend or foe? Eur J Pharmacol 585:338–345
132. Barandon L, Casassus F, Leroux L, Moreau C, Allieres C, Lamaziere JM, Dufourcq P, Couffinhal T, Duplaa C (2011) sFRP-1 improves postinfarction scar formation through a modulation of inflammatory response. Arterioscler Thromb Vasc Biol 31:e80–e87
133. Schumann H, Holtz J, Zerkowski HR, Hatzfeld M (2000) Expression of secreted frizzled related proteins 3 and 4 in human ventricular myocardium correlates with apoptosis related gene expression. Cardiovasc Res 45:720–728
134. Barandon L, Couffinhal T, Ezan J, Dufourcq P, Costet P, Alzieu P, Leroux L, Moreau C, Dare D, Duplaa C (2003) Reduction of infarct size and prevention of cardiac rupture in transgenic mice overexpressing FrzA. Circulation 108:2282–2289
135. Kobayashi K, Luo M, Zhang Y, Wilkes DC, Ge G, Grieskamp T, Yamada C, Liu TC, Huang G, Basson CT, Kispert A, Greenspan DS, Sato TN (2009) Secreted Frizzled-related protein 2 is a procollagen C proteinase enhancer with a role in fibrosis associated with myocardial infarction. Nat Cell Biol 11:46–55
136. He W, Zhang L, Ni A, Zhang Z, Mirotsou M, Mao L, Pratt RE, Dzau VJ (2010) Exogenously administered secreted frizzled related protein 2 (Sfrp2) reduces fibrosis and improves cardiac function in a rat model of myocardial infarction. Proc Natl Acad Sci U S A 107:21110–21115
137. Mirotsou M, Zhang Z, Deb A, Zhang L, Gnecchi M, Noiseux N, Mu H, Pachori A, Dzau V (2007) Secreted frizzled related protein 2 (Sfrp2) is the key Akt-mesenchymal stem cell-released paracrine factor mediating myocardial survival and repair. Proc Natl Acad Sci U S A 104:1643–1648
138. Blankesteijn WM, Essers-Janssen YP, Verluyten MJ, Daemen MJ, Smits JF (1997) A homologue of Drosophila tissue polarity gene frizzled is expressed in migrating myofibroblasts in the infarcted rat heart. Nat Med 3:541–544
139. Aisagbonhi O, Rai M, Ryzhov S, Atria N, Feoktistov I, Hatzopoulos AK (2011) Experimental myocardial infarction triggers canonical Wnt signaling and endothelial-to-mesenchymal transition. Dis Model Mech 4:469–483
140. Laeremans H, Rensen SS, Ottenheijm HC, Smits JF, Blankesteijn WM (2010) Wnt/frizzled signalling modulates the migration

and differentiation of immortalized cardiac fibroblasts. Cardiovasc Res 87:514–523
141. Laeremans H, Hackeng TM, van Zandvoort MA, Thijssen VL, Janssen BJ, Ottenheijm HC, Smits JF, Blankesteijn WM (2011) Blocking of Frizzled signaling with a homologous peptide fragment of Wnt3a/Wnt5a reduces infarct expansion and prevents the development of heart failure after myocardial infarction. Circulation 124:1626–1635
142. Chen L, Wu Q, Guo F, Xia B, Zuo J (2004) Expression of Dishevelled-1 in wound healing after acute myocardial infarction: possible involvement in myofibroblast proliferation and migration. J Cell Mol Med 8:257–264
143. Malekar P, Hagenmueller M, Anyanwu A, Buss S, Streit MR, Weiss CS, Wolf D, Riffel J, Bauer A, Katus HA, Hardt SE (2010) Wnt signaling is critical for maladaptive cardiac hypertrophy and accelerates myocardial remodeling. Hypertension 55:939–945
144. Vilahur G, Juan-Babot O, Pena E, Onate B, Casani L, Badimon L (2011) Molecular and cellular mechanisms involved in cardiac remodeling after acute myocardial infarction. J Mol Cell Cardiol 50:522–533

Chapter 22

Vascular Connexins in Restenosis After Balloon Injury

Sandrine Morel and Brenda R. Kwak

Abstract

Atherosclerosis is an arterial progressive disease characterized by accumulation of lipids, macrophages, T lymphocytes, and smooth muscle cells in large- and medium-sized arteries. Erosion and rupture of the atherosclerotic plaque may induce myocardial infarction and cerebrovascular accidents that are responsible for a large percentage of sudden death. Atherosclerosis is often treated by angioplasty generally followed by stent implantation. Although angioplasty and stent implantation are necessary for the survival of the patient, they induce a trauma in the vessel wall that favors a vascular reaction called restenosis and the associated de-endothelialization increases the risk of thrombosis. To study mechanisms involved in restenosis and thrombus formation, animal models have been developed. In this chapter, we describe the experimental model of balloon injury adapted for mice and apply it to study the role of Cx43 in this process. Connexins are members of a large family of transmembrane proteins that allow exchange of ions and small metabolites between cytosol and extracellular space or between neighboring cells. Connexins are important in vascular physiology, they support radial and longitudinal cell-to-cell communication in the vascular wall, and have been shown to modulate vascular pathologies such as atherosclerosis and hypertension. We also describe the various connexin-specific tools, for example, transgenic mice, blocking peptides, antisense, and siRNA, and their value in obtaining insight into the role of Cx43 in restenosis and thrombus formation after vascular injury.

Key words Connexins, Cx43, Atherosclerosis, Myocardial infarct, Restenosis, Thrombus

1 Introduction

Atherosclerosis, an inflammatory disease of large- and medium-sized arteries, involves the formation of intimal lesions that are characterized by a dysfunctional endothelium, lipid accumulation, inflammation, cell death, and fibrosis. Approximately 75 % of acute coronary events and 60 % of recently symptomatic carotid artery disease are caused by disruption of an atherosclerotic plaque [1, 2]. This vascular disease is often treated by angioplasty and stenting, but clinical studies have shown that treatment efficacy may be reduced by thrombus formation or restenosis at the site of the intervention. Development of animal models to study cellular mechanisms involved in these deleterious processes is of high

Robert G. Gourdie and Tereance A. Myers (eds.), *Wound Regeneration and Repair: Methods and Protocols*, Methods in Molecular Biology, vol. 1037, DOI 10.1007/978-1-62703-505-7_22, © Springer Science+Business Media New York 2013

interest to discover implicated proteins that might represent novel therapeutical targets. In this chapter, we describe the tools developed for the study of vascular injury in mice and summarize results showing a role for connexins in restenosis after injury.

1.1 Atherosclerotic Plaque Development and Its Consequences

Normal arterial wall is composed by three layers: the intima, the media, and the adventitia. The tunica intima, the most internal layer of the wall, corresponds to a monolayer of endothelial cells (ECs) surrounded by minimal connective tissue and by the internal elastic lamina. The tunica media is composed of several layers of smooth muscle cells (SMCs) and extracellular matrix (ECM) surrounded by the external elastic lamina. The external part of the blood vessel wall is the tunica adventitia which is composed of connective tissue, microvessels, and nerve endings. Development of atherosclerotic plaques is a progressive pathology that takes place in the intima of large- and medium-sized arteries. The initiating step of this disease is an endothelial dysfunction [3] often induced by cardiovascular risk factors such as hypercholesterolemia, hyperglycemia, hypertension, free radicals, and infectious microorganisms. This activation of ECs leads to an increase in the expression of different cell adhesion molecules and to the secretion of chemoattractants which induce the recruitment of monocytes, T lymphocytes, and platelets [4, 5]. Monocytes transmigrate between ECs to infiltrate into the arterial intima where they mature into macrophages. These intimal macrophages accumulate lipids and transform into foam cells creating the earliest atherosclerotic lesion called fatty streak. Later, this original atherosclerotic lesion might become covered by SMCs that migrate from the media to the intima attracted by cytokines, chemokines, and growth factors. In the intima, SMCs proliferate and secrete ECM components that participate in the formation of a strong fibrous cap. In the center of the atherosclerotic plaque, foam cells die and release lipids that form the necrotic core of this lesion. A deleterious consequence of atherosclerosis might be luminal thrombotic occlusion due to the superficial erosion of the endothelial monolayer or to the rupture of the plaques' fibrous cap [6, 7]. Plaque erosion leads to the activation of platelets and in consequence to their adhesion into the endothelium. The rupture of the fibrous cap of the atherosclerotic plaque allows contact between the blood and the constituents of the center of the plaque. As a consequence, platelets and leukocytes will become activated by thrombogenic tissue factor produced in the lesion leading to thrombus formation in the arterial lumen [8]. Plaque erosion and plaque rupture induce myocardial infarctions and cerebrovascular accidents that are responsible for a large percentage of sudden death [9]. This clinical complication of atherosclerosis is often treated by angioplasty consisting of balloon dilation of the occluded vessel nowadays generally followed by stent implantation.

1.2 Angioplasty and Restenosis

Angioplasty is an effective way to reopen arteries blocked by a thrombus onto a ruptured or eroded atherosclerotic plaque. The procedure consists of the introduction of a catheter in the occluded artery. Once inserted, a balloon positioned at the tip of the catheter is inflated to compress the plaque and reopen the artery. This technique is now often associated with the implantation of a stent at the site of the occlusion to keep the artery open and to improve patient outcome [10]. Although angioplasty and stent implantation are necessary for the survival of the patient, they induce a trauma in the vessel wall that favors a vascular reaction called restenosis. Moreover, they induce de-endothelialization with increased risk of early thrombosis (shortly after the intervention). Restenosis may be described as the re-occlusion of the artery at the site of the angioplasty or stent implantation due to tissue growth at the site of the intervention, and it occurs generally within 6 months after the initial procedure [11]. This exaggerated response to injury involves the recruitment and infiltration of leukocytes into the damaged site and a surge in cytokines and growth factors. Moreover, medial SMCs undergo a phenotypic modulation from a contractile to a synthetic phenotype, proliferate, and migrate towards the intima. Together, these events induce the formation of the neointima. Drug-eluting stents help to prevent restenosis by inhibiting neointimal hyperplasia. Unfortunately, they also delay re-endothelialization, which increases the period of time during which the stent remains thrombogenic leading to late in-stent thrombosis [12]. The molecular pathways implicated in all these phenomena remain poorly understood. Models of balloon injury have been developed in animals, and more interestingly in hypercholesterolemic mice [13], that allowed for detailed investigation of molecules involved in the restenotic process such as connexins [14].

1.3 Connexin Expression and Functions in Healthy Vessels

Connexins (Cx) are a family of transmembrane proteins that form channels to allow exchange between extracellular space and cytoplasm (hemichannels, also called connexons) or between two neighboring cells (gap junction channels) (for review *see* ref. 15). Hemichannels and gap junction channels allow the passage of ions and small molecules (up to ~1,000 Da). Under normal conditions, connexons are in a closed configuration and their opening is induced by different stimuli such as removal of extracellular calcium, hypoxic or ischemic stress, mechanical stimulation, and dephosphorylation [16, 17]. This way, connexons are involved in the regulation of cell volume, in paracrine or autocrine signalling, and in activation of survival pathways [18]. Gap junction channels are crucial for fast coordinated activities such as conduction of the action potential in the heart or transmission of neuronal signals at electrical synapses, but are also involved in the regulation of slower physiological processes such as cell growth and development.

Four connexins are expressed in the healthy vascular wall: Cx37, Cx40, Cx43, and Cx45. Their expression is not uniform in all blood vessels and varies with vascular territory and species. Usually, Cx37 and Cx40 are co-expressed in ECs, whereas Cx43 and Cx45 are present in SMCs (for review *see* ref. 19). Nevertheless, Cx37 and Cx40 are also found in SMCs of small elastic or resistance arteries or during development [20, 21], and Cx43 is expressed in ECs at branch points of arteries [22]. The vascular tone is under the control of paracrine molecules such as nitric oxide and prostaglandins [23, 24]; however, gap junctional intercellular communication (GJIC) plays also an important role in the control of vascular function [25, 26]. Homomeric (one connexin isoform) and heteromeric (multiple connexin isoforms) channels and homocellular and heterocellular gap junctions are found in the vascular wall between ECs, between SMCs, and in some cases between ECs and SMCs (myoendothelial junctions) allowing for the passage of transverse and longitudinal signals in the vessel wall [19, 27, 28]. This way, connexins provide, for example, the molecular basis for ascending dilations (i.e., conducted dilations) in arterioles that are required for substantial increases in blood flow during exercise [29, 30]. Thus, connexins have a central role in vascular physiology but also in vascular pathology such as hypertension, atherosclerosis, or restenosis.

1.4 Connexins in Restenosis

As described above, atherosclerotic plaque rupture/erosion and associated thrombus formation may lead to the occlusion of an artery. To allow the survival of the patient, angioplasty is performed to reopen the artery, but this surgical procedure induces a trauma that may favor, at long term, the re-occlusion of the artery. Proliferation and migration are two cellular processes highly implicated in restenosis. Several proteins, including connexins, have been described to regulate these processes.

In 1997, Yeh and colleagues have first described a possible involvement of connexins in restenosis after balloon catheter injury in rat carotid artery [31]. They have carefully followed Cx43 expression in the media and neointima during 14 days after ballooning, and they showed an up-regulation of Cx43 between medial and intimal SMCs with a higher expression in neointima. Moreover, the size of the gap junction was larger between SMCs in the intima than between SMCs of the media. Restenosis in rat carotid artery is mainly due to migration and proliferation of SMCs and to a lesser extend to leukocyte infiltration [31]. In a more inflammatory animal model of restenosis, i.e., the rabbit aorto-iliac bifurcation, Polacek and colleagues demonstrated that the expression of Cx43 after balloon injury is equivalent between intimal and medial SMCs [32]. Hypercholesterolemic diet given to these animals, submitted to mechanical injury or not, induced Cx43 up-regulation in macrophages, while Cx43 in medial SMCs was down-regulated. In addition, balloon injury increased Cx40 mRNA and protein expression in neointimal SMCs in this model [33]. Interestingly, experiments in the presence of 3-hydroxy-3-methylglutaryl-coenzyme A (HMG-CoA) reductase inhibitors (statins), which have anti-inflammatory and anti-proliferative properties, induced a reduction of neointimal formation and reduction of Cx43 and Cx40 expression after balloon injury in rabbit iliac artery [33]. To investigate the role of Cx43 in neointima formation, we have performed carotid balloon distension injury in hypercholesterolemic $Cx43^{+/+}LDLR^{-/-}$ and in $Cx43^{+/-}LDLR^{-/-}$ mice [14]. We have shown that reduction of Cx43 expression is associated with the limitation of neointima formation, and in vitro and in vivo experiments suggested that this reduction is due to a decrease of SMC migration and proliferation. In contrast with the rat carotid model, macrophages appear to have an additional important role in neointima formation in the mouse hypercholesterolemic model. Indeed, macrophage infiltration is lower in the carotid intima of $Cx43^{+/-}LDLR^{-/-}$ mice than in $Cx43^{+/+}LDLR^{-/-}$ mice submitted to balloon injury, and in vitro experiments showed that macrophages from $Cx43^{+/-}LDLR^{-/-}$ mice have reduced migration capacity compared with macrophages of $Cx43^{+/+}LDLR^{-/-}$ mice. Interestingly, in vitro experiments also showed that migration of SMCs is higher in the presence of conditioned medium obtained from $Cx43^{+/+}LDLR^{-/-}$ than $Cx43^{+/-}LDLR^{-/-}$ macrophages. This data suggest that macrophages from $Cx43^{+/-}LDLR^{-/-}$ mice secrete fewer chemoattractant factors than control macrophages resulting in vivo in the reduction of SMCs migration towards the injury site after balloon distention. To study the specific role of Cx43 in SMCs, Liao and colleagues have generated mice in which the deletion of Cx43 gene is confined to SMCs, and they have performed carotid artery denudation using a guide wire [34]. Seven days after surgery, the neointima in these mice is larger than the neointima in control mice, and is formed mostly by SMCs and only occasionally by ECs or leukocytes. The differences between these and our results have not yet been explained, but are likely induced by the difference in mouse models used (deletion of Cx43 in SMCs vs. half of Cx43 in all cells), but might also derive from different methodology (balloon injury with hypercholesterolemia vs. wire injury).

Two types of SMCs have been described in porcine coronary arteries: spindle-shaped SMCs (S-SMCs) which present the classical "hills-and-valleys" growth pattern and that are differentiated cells and rhomboid SMCs (R-SMCs) that have high proliferative, migratory, and proteolytic activities [35]. Similar SMC populations have been described in the rat aorta [36]. Interestingly, R-SMCs are present in higher concentration than S-SMCs in the intimal thickening induced by stent implantation in porcine coronary arteries suggesting their role in restenosis [35]. In vitro, it has been shown

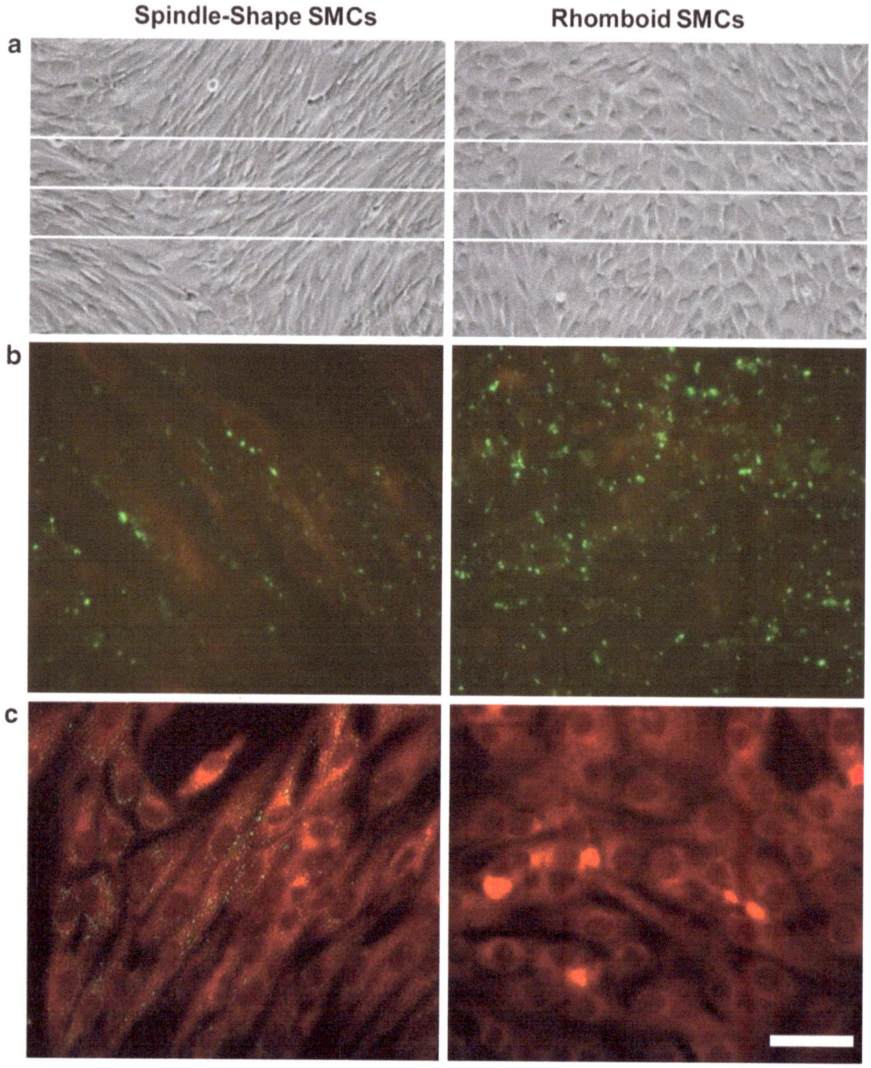

Fig. 1 Expression of Cx43 and Cx40 in Spindle-Shape and Rhomboid SMCs. (**a**) Phase-contrast microphotographs showing morphological features of Spindle-Shape- and Rhomboid-SMCs isolated from the normal media of porcine coronary artery. (**b**) Immunofluorescence staining (in *green*) for Cx43. (**c**) Immunofluorescence staining (in *green*) for Cx40. Cells were counterstained with Evans blue (in *red*). Scale bars: 120 μm (*top images*); 75 μm (*middle* and *bottom images*)

that S-SMCs expressed Cx43 and Cx40, whereas R-SMCs expressed only Cx43 [37] (Fig. 1). In vivo, Cx43 expression is strongly up-regulated in the intimal thickening after stent implantation, whereas Cx40 is absent, which is in accordance with the increased presence of R-SMCs [37]. It appeared that Cx43 expression and channel function are regulating the migration process of R-SMCs. Indeed, PDGF-BB-induced migration of R-SMCs is highly reduced in the presence of Cx43 antisense or Cx43 blocking peptide [37]. Furthermore, Cx43 antisense prevented PDGF-BB-induced deleterious phenotypic changes of porcine S-SMCs to R-SMCs [37]. Interestingly, down-regulation of Cx43 expression in rat aortic SMCs inhibited also their proliferation [38].

The roles of connexins in migration and proliferation of vascular cells and for vascular repair not yet clearly understood. Studies on different cell types towards the role of connexins as modulator of migration or proliferation suggest that connexins play a role in these processes by both channel-dependent (hemichannel or gap junction channel) or channel-independent effects on interactions with cytoskeletal proteins, junctional proteins, or enzymes (for reviews *see* ref. 39, 40). Moreover, restenosis after vascular surgery could also be influenced by the formation of a thrombus at the site of the injury. It has recently been described that Cx37 is expressed in platelets and is able to form gap junction channels between platelets that influence platelet activity and aggregation [41]. In vivo experiments confirmed that Cx37 plays a restraining role on thrombus propensity [41]. One can easily imagine that increased platelet activation at the site of injury results in increase PDGF-BB secretion, which has then additional effects on SMCs.

All of the data described above suggest that connexins, and more specifically Cx43, are implicated in restenosis after vascular injury. Moreover, the experiments show the potential for targeting Cx43 in reduction of restenosis. Finally, the potential for genetic manipulation in mice has made this species particularly attractive to identify the key molecules involved in restenosis.

2 Materials

2.1 Transgenic Mice for the Study of Atherosclerosis

In vivo studies towards atherosclerosis are currently performed by the use of two well-characterized mouse models: the apolipoprotein E ($ApoE^{-/-}$) knockout mice and the low-density lipoprotein-receptor knockout mice ($LDLR^{-/-}$). $ApoE^{-/-}$ mice present high plasma cholesterol concentration (400 500 mg/dL) and develop spontaneously foam cell-rich depositions throughout the arterial tree [42]. $LDLR^{-/-}$ mice have a lower increase of plasma cholesterol level (175–225 mg/dL) and they develop only minimal atherosclerotic lesions in aortic roots when fed a normal chow diet [43]. Upon feeding these mouse models with a high-cholesterol diet, they both rapidly develop advanced lesions throughout the vascular tree and they present systemic inflammation. Like in human, atherosclerosis in mice develops in regions of the vasculature subjected to low or oscillatory wall shear stress [44]. Predilection sites in the mouse are the aortic root, the lesser curvature of the aortic arch and branch points of the brachiocephalic, left carotid, and subclavian arteries. Of note, retrovalvular lesions in the aortic root either stop abruptly at the orifice of the common coronary artery or extend a short distance into the arterial trunks but, in contrast to the human coronary arteries, the first segment and the first branch of all major coronary arteries are usually protected from the disease [45]. $LDLR^{-/-}$ or $ApoE^{-/-}$ mice can be interbred with other knockout mice to study the effects of those specific molecules.

2.2 Transgenic Mice for the Study of Connexin Functions

Twenty connexins have been identified in mice, and many transgenic mice in which a connexin is deleted or modified have been created [46]. An important problem with mice knockout for vascular connexins is that these deletions are often lethal. For example, $Cx43^{-/-}$ and $Cx45^{-/-}$ die in utero or shortly after birth. $Cx40^{-/-}$ mice are viable but sometimes develop arrhythmias [47, 48] and they are hypertensive [49]. $Cx37^{-/-}$ mice are viable and their heart function is normal, but they recently appeared to have altered vasculogenesis, arteriogenesis, and angiogenesis [50, 51]. Moreover, females are infertile [50]. Connexin gene deletion in mice is frequently used to study disease processes; however, the absence of one connexin may lead to a decreased or increased expression of another connexin. In the context of myocardial disease, several transgenic mice, in which Cx43 expression or function was modified, have been used. At first, heterozygous Cx43 mice, which express 50 % of the normal Cx43 level, have been extensively used to study, for example, the effects of myocardial infarction and development of atherosclerosis [52, 53]. During the last years, it has become increasingly evident that the C-terminal (CT) part of the Cx43 is crucial for Cx43 channel function. Consequently, new transgenic mice have been generated in which the total expression of Cx43 is not changed but its function was altered. Maass and colleagues generated transgenic mice in which they deleted the C-terminal 125 amino acid residues of Cx43 (mice designated as Cx43K258stop mice) [54]. Unfortunately, these mice die shortly after birth due to a defective epidermal barrier. To counterbalance that, they have bred transgenic mice harboring one *Cx43K258stop* and one *Cx43knockout* allele. These mice do not suffer from a defective epidermal barrier and 50 % of them survive to adulthood [54]. They have shown that this deletion increased infarct size and the development of ventricular tachyarrhythmias after acute myocardial infarction [55]. In these mice with truncated CT domain the number, the size, and the localization of cardiac gap junction plaques was altered [56]. Otherwise, the phosphorylation status of the CT domain of Cx43 has been extensively described to affect the expression, trafficking, assembly, degradation, open probability, and single channel conductance of Cx43 gap junction channels in in vitro studies (for review *see* ref. 57). In this context, two new strains of genetically knock-in mice have been recently generated and characterized: the S3E mice in which the serines 325/328/330 were replaced with phosphomimetic glutamic acids, and the S3A mice in which the serines 325/328/330 were replaced by nonphosphorylatable alanines [58]. The authors show that the S3E mice were resistant to acute and chronic pathological gap junction remodeling and present a lower susceptibility to ventricular arrhythmias. In contrast, the S3A mice present deleterious cardiac gap junction formation and function, electric remodeling and higher susceptibility to arrhythmias.

In the context of vascular injury it appeared important to develop models in which connexin expression is modified specifically in one cell type (ECs, SMCs, or inflammatory cells) to circumvent deleterious effects of ubiquitous Cx43 deletion cq. alteration or of hypertension associated with ubiquitous Cx40 deletion. For example, the Cre–loxP system under the control of the Tie2 promoter has been used to create mice in which Cx40 or Cx43 have been deleted from the endothelium only [59–61]. Moreover, by using the same type of Cre–loxP system, mice in which Cx43 is deleted in SMCs only (smooth muscle myosin heavy chain promoter) have been generated [34, 61]. Finally, bone marrow transplantation has been performed and the role of Cx43 in the immune system of adult mice has been described in these $Cx43^{-/-}$, $Cx43^{+/-}$, and $Cx43^{+/+}$ chimeras [62].

As mice are the best known species for genetic manipulation, adaptation of the model of balloon injury for mice appeared to be of high interest to identify the key molecules involved in restenosis [13, 14, 63].

2.3 Balloon Catheter

The balloon catheter (Jomed, Beringen, Switzerland) is manufactured with a balloon length of around 5 mm and a diameter ranging from 0.63 mm at 4 bars to 0.83 mm at 16 bars. The balloon size is matched to the weight of the mice. The balloon is expanded using a water-filled inflation device with digital display and stiffened by a guidewire with a diameter of 0.15 mm.

2.4 Blocking Peptides

Different compounds such as heptanol, octanol, 18β-glycyrrhetinic acid, carbonoxolone, or oleamide are known to inhibit GJIC, but their actions are nonspecific [64–66]. In an attempt to be more specific for gap junction channels (vs. other membrane channels) and to be even more specific within the connexin family of proteins, various types of blocking peptides have been generated [67–69]. The nowadays most used blocking peptides for cardiovascular connexins, Gap26 and Gap27, correspond, respectively, to the sequence of the first and the second extracellular loop of various connexins. These blocking peptides have been used in studies towards gap junctional communication between ECs and SMCs and between ECs and macrophages [70–72]. They have a rapid and reversible mode of action, and they are nontoxic for the cells (for review *see* ref. 73). All derivates of these blocking peptides are more-or-less specific for Cx37, Cx40, and Cx43 (^{43}Gap26, ^{37,40}Gap26, ^{40}Gap27, ^{37,43}Gap27). They act on rodent and human cells, thus reflecting the high degree of amino acid conservation in the extracellular loops. In an atherosclerosis study, we have used these blocking peptides to prove the implication of Cx37 hemichannels in the adhesive property of macrophages [74]. In another study onto the phenotypic modulation of SMCs, we have shown that rhomboid SMCs express higher levels of Cx43 than spindle-shaped SMCS and that a Cx43 blocking peptide reduced PDGF-BB-induced migration of rhomboid SMCs [37]. Furthermore, reducing conductivity of Cx43 channels with ^{43}Gap26 decreased the adhesion of neutrophils to ECs in vitro and reduced neutrophil recruitment in a mouse model of acute lung inflammation in vivo [75].

Another type of blocking peptide has been developed, it corresponds to a 25-amino acid peptide comprised of a cell-permeabilization sequence linked to the last nine amino acids of the C-terminal part of the Cx43 and is called αCT1 [76]. This peptide binds to the Postsynaptic density/Disks-large/ZO-1 (PDZ2) domain of Zonula occludens (ZO)-1. This peptide favors healing rate after skin excisional wounds and the generation of more normal-looking histology skin in comparison with control [77]. This improvement of wound healing is associated with the reduction of the area of granulation tissue, the better regeneration of cutaneous structure and function, the inhibition of neutrophil infiltration, and the increased of gap junction size [78]. In a model of cryo-injury of the left ventricle of mouse heart, it has been shown that the αCT1 peptide limits the development of arrhythmias in response to programmed stimulation and increases conduction velocities [79]. These effects are associated with the stabilization of the Cx43 in intercalated disks in the injury border zone and with the increase of the phosphorylation of Cx43 on Ser368.

2.5 Connexin Antisense or siRNA

The function of connexins in cells or tissues has also been investigated using siRNA or antisense oligonucleotides. In experiments towards skin repair, a Cx43 antisense has been prepared in a 30 % Pluronic F-127 gel and used in combination with various types of skin lesion (incision or thermal injury). A single topical application of Cx43 antisense gel onto the skin lesion induced a transient Cx43 knockdown at the wound site. This reduction of expression of Cx43 is associated with an increase in the rate of wound closure and a reduction of inflammation [80, 81]. Indeed, decreasing Cx43 expression improves migration and proliferation of fibroblasts [82]. With respect to the inflammatory response, it has been shown that the peak of neutrophil infiltration 2 days after the injury is significantly reduced by the application of Cx43 antisense and that the number of macrophages 7 days after the injury is also significantly reduced [82]. At long term, the use of Cx43 antisense reduced the extent of granulation tissue deposition and favors a smaller and less distorted scar [80, 81]. This acceleration of skin wound healing was also associated with increased early angiogenesis in animals treated with Cx43 antisense [82]. This antisense has also been tested in models of spinal cord injury and authors have shown that it reduces inflammation and improves functional recovery [83]. In the context of vascular injury, we have tested Cx43 antisense on SMCs differentiation and migration, two processes implicated in restenosis after ballooning injury [37]. We have shown that Cx43 antisense reduced Cx43 expression and channel function in rhomboid SMCs, which is associated with the reduction of PDGF-BB-induced migration of rhomboid SMCs. Moreover, the PDGF-BB-induced change of spindle-shaped SMCs towards a rhomboid phenotype was prevented by Cx43 antisense [37]. More recently, it has been shown that shock wave therapy attenuates inflammatory responses and proliferation of neointima and SMCs following denudation by balloon catheter, and that these effects are associated with limitation of Cx43 expression [84]. Only few in vivo studies have described a protective effect of connexin knockdown by siRNA. For example, in a model of corneal wound healing, knockdown of Cx43 by a single injection of siRNA, improves wound closure by promoting endothelial proliferation [85].

3 Methods

3.1 Balloon Injury Model in Mice

The model of carotid balloon distension injury in hypercholesterolemic mice has been firstly published by Matter and colleagues [13], and we have then used it to study the implication of Cx43 in neointima formation [14].

3.1.1 Protocol for Carotid Balloon Distension Injury

1. To induce vascular inflammation, ApoE$^{-/-}$ or LDLR$^{-/-}$ mice are feed a cholesterol-rich diet during several days or weeks before balloon injury depending on the level of inflammation required. Moreover, mice receive aspirin (16 mg/kg/day) during 7 days before the surgery.
2. Mice are anesthetized by intraperitoneal injection of Ketamine (75–95 mg/kg) and Xylazine (4–8 mg/kg).
3. The hair is removed from the neck region and the left anterior cervical triangle will be accessed by a sagittal anterior neck incision.
4. The muscles around the carotid artery are drawn aside to distinguish the Left Common Carotid Artery (LCCA), the Left Internal Carotid Artery (LICA), and the Left External Carotid Artery (LECA) (Fig. 2a).

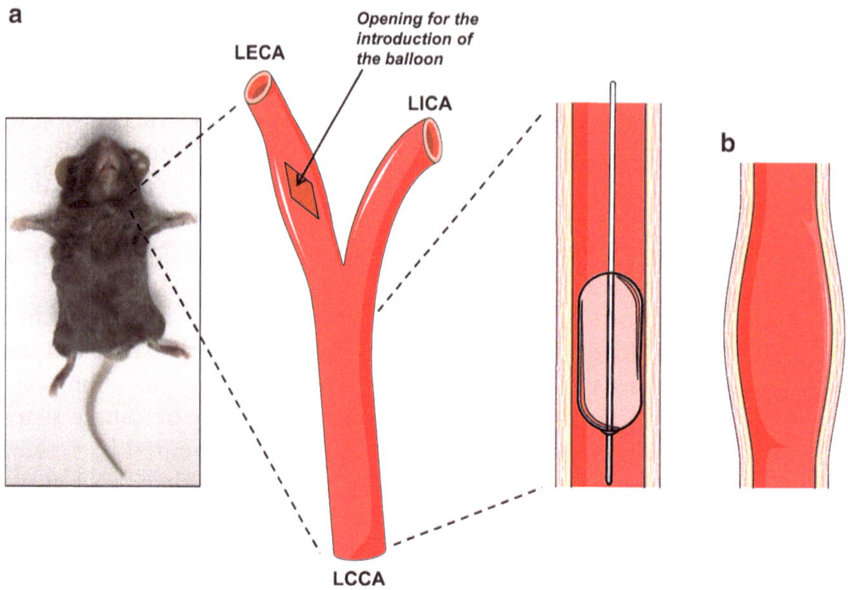

Fig. 2 Carotid balloon distension injury in mice. (**a**) Schematic representation of the left carotid artery bifurcation and of the site for balloon distension; *LECA* left external carotid artery, *LICA* left internal carotid artery, *LCCA* left common carotid artery. (**b**) Schematic representation of the left proximal part of the LCCA after balloon distension

5. To perform balloon distension injury in mice, the LECA is ligated distally and clamps are placed on the LICA and mid-LCCA (Fig. 2a). The balloon catheter is introduced through an arteriotomy on the proximal LECA, and after removing the clamp on the common artery, the catheter is advanced to the proximal non-dissected common carotid artery (Fig. 2a). The balloon is distended for 40 s with a pressure adapted to the weight of the animals.
6. After balloon distension (Fig. 2b), the catheter is withdrawn, the proximal LECA is ligated and the clamp on the LICA is removed to restore blood flow through the LCCA and the LICA.
7. Lidocaine (2 %) is applied onto operation field and the wound is closed. Aspirin treatment is continued until the end of the experiment to prevent acute vessel closure.
8. Analysis of initial apoptosis may be done after 24 h, of inflammatory infiltration after 4–7 days after balloon injury and analysis of restenosis and re-endothelialization after 14 days.

3.1.2 Protocol for Histochemical Analysis of Restenosis and Re-endothelialization

1. At the desired time point, mice are anesthetized and the thorax is opened to have access to the heart to allow the perfusion of the mouse with NaCl 0.9 % at a pressure of 100 mmHg.
2. Once all blood is eliminated from the vessels, the mouse is perfused with 4 % paraformaldehyde for 8 min. The right common

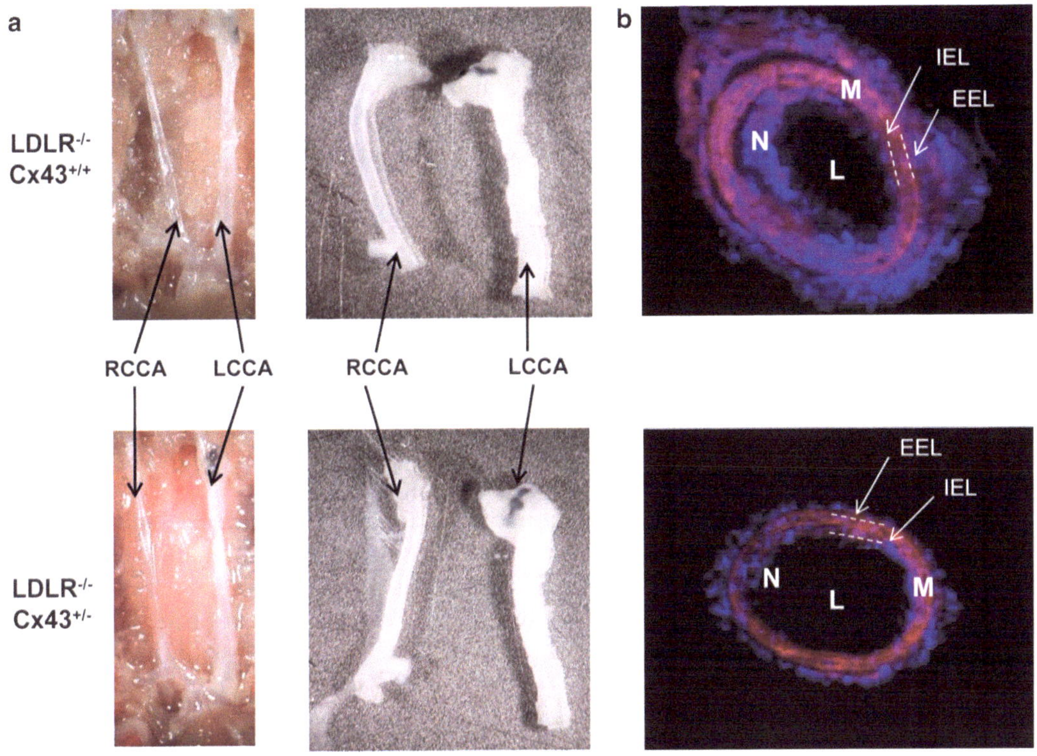

Fig. 3 Reduced neointima formation in LDLR$^{-/-}$Cx43$^{+/-}$ mice. (**a**) Pictures of right common carotid artery (RCCA) and left common carotid artery (LCCA) of LDLR$^{-/-}$Cx43$^{+/+}$ and LDLR$^{-/-}$Cx43$^{+/-}$ mice 14 days after balloon injury. (**b**) Cross sections of carotid arteries 14 days after balloon distension injury were stained with DAPI to visualize medial and neointimal nuclei (*blue*). Elastic laminas were counterstained with Evans Blue (*red*). Magnification 100× (L indicates lumen; M, media; N, neointima; IEL, inner elastic lamina; and EEL, external elastic lamina)

carotid artery (RCCA, untouched artery, internal control) and the injured LCCA are excised and the adventicia carefully removed (Fig. 3a).

3. The blood vessels are postfixed for an additional 2 h in 4 % paraformaldehyde and then immersed overnight in 30 % Sucrose.

4. Thereafter, the LCCA is cut into two parts at the level of the balloon distension and embedded in OCT compound and snap-frozen.

5. Tissues are stored at −80 °C. Serial cryosections (5 μm) are obtained from the middle portion of the injured LCCA and from the untouched RCCA.

6. Nuclei and elastic lamina are stained with 4′-6′-diamidino-2-phenylindole (DAPI) and Evans blue (0.3 %), respectively, to quantify lumen area and total number of intimal and medial cells (Fig. 3b).

7. Immunostaining may be performed on these cryosections to visualize SMCs, ECs, and inflammatory cells with antibodies against smooth muscle actin, von Willebrand factor, and CD68, for example, as, well as with antibodies against the proteins of interest such as connexins [14].

4 Notes

This chapter provides an overview of the role of connexins in restenosis after vascular injury. The development of tools to study restenosis in mice has allowed us to better understand the mechanisms implicated in this devastating process. Until now, our attention has been mostly focused on proliferation and migration of SMCs, but re-endothelialization of the injured vessels is also of high interest to limit re-occlusion of the artery by thrombosis and definitely needs to be better studied. The better understanding of the role of connexins in restenosis, re-endothelialization, and thrombus propensity will hopefully lead to the development of new therapeutic strategies.

Acknowledgments

We thank Drs. Christos Chadjichristos, Marie-Luce Bochaton-Piallat, and Christian Matter for their valuable help with preparing the data shown in Figs. 1 and 3, as well as for helpful discussions. This work was supported by the Swiss National Science Foundation (310030–143343 to B.R.K.), the Foundations Novartis (to B.R.K.), Prevot (to S.M.), and Bangerter-Rhyner (to S.M.).

References

1. Kolodgie FD, Narula J, Yuan C, Burke AP, Finn AV, Virmani R (2007) Elimination of neoangiogenesis for plaque stabilization: is there a role for local drug therapy? J Am Coll Cardiol 49:2093–2101
2. Yla-Herttuala S, Bentzon JF, Daemen M, Falk E, Garcia-Garcia HM, Herrmann J, Hoefer I, Jukema JW, Krams R, Kwak BR, Marx N, Naruszewicz M, Newby A, Pasterkamp G, Serruys PW, Waltenberger J, Weber C, Tokgozoglu L (2011) Stabilisation of atherosclerotic plaques. Position paper of the European Society of Cardiology (ESC) Working Group on atherosclerosis and vascular biology. Thromb Haemost 106:1–19
3. Ross R (1995) Cell biology of atherosclerosis. Annu Rev Physiol 57:791–804
4. Weber C, Noels H (2011) Atherosclerosis: current pathogenesis and therapeutic options. Nat Med 17:1410–1422
5. Libby P, Ridker PM, Hansson GK (2009) Inflammation in atherosclerosis: from pathophysiology to practice. J Am Coll Cardiol 54:2129–2138
6. Virmani R, Kolodgie FD, Burke AP, Farb A, Schwartz SM (2000) Lessons from sudden coronary death: a comprehensive morphological classification scheme for atherosclerotic lesions. Arterioscler Thromb Vasc Biol 20:1262–1275
7. Libby P (2009) Molecular and cellular mechanisms of the thrombotic complications of atherosclerosis. J Lipid Res 50:S352–S357

8. Jackson SP (2011) Arterial thrombosis-insidious, unpredictable and deadly. Nat Med 17:1423–1436
9. Libby P (2008) The molecular mechanisms of the thrombotic complications of atherosclerosis. J Intern Med 263:517–527
10. Dangas G, Kuepper F (2002) Cardiology patient page. Restenosis: repeat narrowing of a coronary artery: prevention and treatment. Circulation 105:2586–2587
11. Serruys PW, Luijten HE, Beatt KJ, Geuskens R, de Feyter PJ, van den Brand M, Reiber JH, ten Katen HJ, van Es GA, Hugenholtz PG (1988) Incidence of restenosis after successful coronary angioplasty: a time-related phenomenon. A quantitative angiographic study in 342 consecutive patients at 1, 2, 3, and 4 months. Circulation 77:361–371
12. Newsome LT, Kutcher MA, Royster RL (2008) Coronary artery stents: Part I. Evolution of percutaneous coronary intervention. Anesth Analg 107:552–569
13. Matter CM, Ma L, von Lukowicz T, Meier P, Lohmann C, Zhang D, Kilic U, Hofmann E, Ha SW, Hersberger M, Hermann DM, Luscher TF (2006) Increased balloon-induced inflammation, proliferation, and neointima formation in apolipoprotein E (ApoE) knockout mice. Stroke 37:2625–2632
14. Chadjichristos CE, Matter CM, Roth I, Sutter E, Pelli G, Luscher TF, Chanson M, Kwak BR (2006) Reduced connexin43 expression limits neointima formation after balloon distension injury in hypercholesterolemic mice. Circulation 113:2835–2843
15. Morel S, Kwak BR (2012) Roles of connexins in atherosclerosis and ischemia-reperfusion injury. Curr Pharm Biotechnol 13(1):17–26
16. John S, Cesario D, Weiss JN (2003) Gap junctional hemichannels in the heart. Acta Physiol Scand 179:23–31
17. Derouette JP, Desplantez T, Wong CW, Roth I, Kwak BR, Weingart R (2009) Functional differences between human Cx37 polymorphic hemichannels. J Mol Cell Cardiol 46(4):499–507
18. Goodenough DA, Paul DL (2003) Beyond the gap: functions of unpaired connexon channels. Nat Rev Mol Cell Biol 4:285–294
19. Brisset AC, Isakson BE, Kwak BR (2009) Connexins in vascular physiology and pathology. Antioxid Redox Signal 11:267–282
20. Little TL, Beyer EC, Duling BR (1995) Connexin 43 and connexin 40 gap junctional proteins are present in arteriolar smooth muscle and endothelium in vivo. Am J Physiol 268:H729–H739
21. Haefliger JA, Polikar R, Schnyder G, Burdet M, Sutter E, Pexieder T, Nicod P, Meda P (2000) Connexin37 in normal and pathological development of mouse heart and great arteries. Dev Dyn 218:331–344
22. Gabriels JE, Paul DL (1998) Connexin43 is highly localized to sites of disturbed flow in rat aortic endothelium but connexin37 and connexin40 are more uniformly distributed. Circ Res 83:636–643
23. Villar IC, Francis S, Webb A, Hobbs AJ, Ahluwalia A (2006) Novel aspects of endothelium-dependent regulation of vascular tone. Kidney Int 70:840–853
24. Bellien J, Thuillez C, Joannides R (2008) Contribution of endothelium-derived hyperpolarizing factors to the regulation of vascular tone in humans. Fundam Clin Pharmacol 22:363–377
25. de Wit C, Boettcher M, Schmidt VJ (2008) Signaling across myoendothelial gap junctions–fact or fiction? Cell Commun Adhes 15:231–245
26. Sandow SL, Haddock RE, Hill CE, Chadha PS, Kerr PM, Welsh DG, Plane F (2009) What's where and why at a vascular myoendothelial microdomain signalling complex. Clin Exp Pharmacol Physiol 36:67–76
27. Haefliger JA, Nicod P, Meda P (2004) Contribution of connexins to the function of the vascular wall. Cardiovasc Res 62:345–356
28. de Wit C, Hoepfl B, Wolfle SE (2006) Endothelial mediators and communication through vascular gap junctions. Biol Chem 387:3–9
29. De Wit C (2004) Connexins pave the way for vascular communication. News Physiol Sci 19:148–153
30. Segal SS (2005) Regulation of blood flow in the microcirculation. Microcirculation 12:33–45
31. Yeh HI, Lupu F, Dupont E, Severs NJ (1997) Upregulation of connexin43 gap junctions between smooth muscle cells after balloon catheter injury in the rat carotid artery. Arterioscler Thromb Vasc Biol 17:3174–3184
32. Polacek D, Bech F, McKinsey JF, Davies PF (1997) Connexin43 gene expression in the rabbit arterial wall: effects of hypercholesterolemia, balloon injury and their combination. J Vasc Res 34:19–30
33. Wang L, Chen J, Sun Y, Zhang F, Zhu J, Hu S, Wang DH (2005) Regulation of connexin expression after balloon injury: possible mechanisms for antiproliferative effect of statins. Am J Hypertens 18:1146–1153
34. Liao Y, Regan CP, Manabe I, Owens GK, Day KH, Damon DN, Duling BR (2007) Smooth muscle-targeted knockout of connexin43 enhances neointimal formation in response to vascular injury. Arterioscler Thromb Vasc Biol 27:1037–1042
35. Hao H, Ropraz P, Verin V, Camenzind E, Geinoz A, Pepper MS, Gabbiani G, Bochaton-Piallat ML

(2002) Heterogeneity of smooth muscle cell populations cultured from pig coronary artery. Arterioscler Thromb Vasc Biol 22:1093–1099
36. Bochaton-Piallat ML, Ropraz P, Gabbiani F, Gabbiani G (1996) Phenotypic heterogeneity of rat arterial smooth muscle cell clones. Implications for the development of experimental intimal thickening. Arterioscler Thromb Vasc Biol 16:815–820
37. Chadjichristos CE, Morel S, Derouette JP, Sutter E, Roth I, Brisset AC, Bochaton-Piallat ML, Kwak BR (2008) Targeting connexin 43 prevents platelet-derived growth factor-BB-induced phenotypic change in porcine coronary artery smooth muscle cells. Circ Res 102:653–660
38. Song M, Yu X, Cui X, Zhu G, Zhao G, Chen J, Huang L (2009) Blockade of connexin 43 hemichannels reduces neointima formation after vascular injury by inhibiting proliferation and phenotypic modulation of smooth muscle cells. Exp Biol Med (Maywood) 234:1192–1200
39. Dbouk HA, Mroue RM, El-Sabban ME, Talhouk RS (2009) Connexins: a myriad of functions extending beyond assembly of gap junction channels. Cell Commun Signal 7:4
40. Vinken M, Decrock E, De Vuyst E, Ponsaerts R, D'Hondt C, Bultynck G, Ceelen L, Vanhaecke T, Leybaert L, Rogiers V (2011) Connexins: sensors and regulators of cell cycling. Biochim Biophys Acta 1815:13–25
41. Angelillo-Scherrer A, Fontana P, Burnier L, Roth I, Sugamele R, Brisset A, Morel S, Nolli S, Sutter E, Chassot A, Capron C, Borgel D, Saller F, Chanson M, Kwak BR (2011) Connexin 37 limits thrombus propensity by downregulating platelet reactivity. Circulation 124:930–939
42. Zhang SH, Reddick RL, Piedrahita JA, Maeda N (1992) Spontaneous hypercholesterolemia and arterial lesions in mice lacking apolipoprotein E. Science 258:468–471
43. Ishibashi S, Brown MS, Goldstein JL, Gerard RD, Hammer RE, Herz J (1993) Hypercholesterolemia in low density lipoprotein receptor knockout mice and its reversal by adenovirus-mediated gene delivery. J Clin Invest 92:883–893
44. Suo J, Ferrara DE, Sorescu D, Guldberg RE, Taylor WR, Giddens DP (2007) Hemodynamic shear stresses in mouse aortas: implications for atherogenesis. Arterioscler Thromb Vasc Biol 27:346–351
45. Hu W, Polinsky P, Sadoun E, Rosenfeld ME, Schwartz SM (2005) Atherosclerotic lesions in the common coronary arteries of ApoE knockout mice. Cardiovasc Pathol 14:120–125
46. Dobrowolski R, Willecke K (2009) Connexin-caused genetic diseases and corresponding mouse models. Antioxid Redox Signal 11:283–295
47. Simon AM, Goodenough DA, Paul DL (1998) Mice lacking connexin40 have cardiac conduction abnormalities characteristic of atrioventricular block and bundle branch block. Curr Biol 8:295–298
48. Kirchhoff S, Nelles E, Hagendorff A, Kruger O, Traub O, Willecke K (1998) Reduced cardiac conduction velocity and predisposition to arrhythmias in connexin40-deficient mice. Curr Biol 8:299–302
49. de Wit C, Roos F, Bolz SS, Pohl U (2003) Lack of vascular connexin 40 is associated with hypertension and irregular arteriolar vasomotion. Physiol Genomics 13:169–177
50. Simon AM, Goodenough DA, Li E, Paul DL (1997) Female infertility in mice lacking connexin 37. Nature 385:525–529
51. Fang JS, Angelov SN, Simon AM, Burt JM (2011) Cx37 deletion enhances vascular growth and facilitates ischemic limb recovery. Am J Physiol Heart Circ Physiol 301:H1872–H1881
52. Guerrero PA, Schuessler RB, Davis LM, Beyer EC, Johnson CM, Yamada KA, Saffitz JE (1997) Slow ventricular conduction in mice heterozygous for a connexin43 null mutation. J Clin Invest 99:1991–1998
53. Kwak BR, Veillard N, Pelli G, Mulhaupt F, James RW, Chanson M, Mach F (2003) Reduced connexin43 expression inhibits atherosclerotic lesion formation in low-density lipoprotein receptor-deficient mice. Circulation 107:1033–1039
54. Maass K, Ghanem A, Kim JS, Saathoff M, Urschel S, Kirfel G, Grummer R, Kretz M, Lewalter T, Tiemann K, Winterhager E, Herzog V, Willecke K (2004) Defective epidermal barrier in neonatal mice lacking the C-terminal region of connexin43. Mol Biol Cell 15:4597–4608
55. Maass K, Chase SE, Lin X, Delmar M (2009) Cx43 CT domain influences infarct size and susceptibility to ventricular tachyarrhythmias in acute myocardial infarction. Cardiovasc Res 84:361–367
56. Maass K, Shibayama J, Chase SE, Willecke K, Delmar M (2007) C-terminal truncation of connexin43 changes number, size, and localization of cardiac gap junction plaques. Circ Res 101:1283–1291
57. Solan JL, Lampe PD (2009) Connexin43 phosphorylation: structural changes and biological effects. Biochem J 419:261–272
58. Remo BF, Qu J, Volpicelli FM, Giovannone S, Shin D, Lader J, Liu FY, Zhang J, Lent DS,

Morley GE, Fishman GI (2011) Phosphatase-resistant gap junctions inhibit pathological remodeling and prevent arrhythmias. Circ Res 108:1459–1466

59. Theis M, de Wit C, Schlaeger TM, Eckardt D, Kruger O, Doring B, Risau W, Deutsch U, Pohl U, Willecke K (2001) Endothelium-specific replacement of the connexin43 coding region by a lacZ reporter gene. Genesis 29:1–13
60. Chadjichristos CE, Scheckenbach KE, van Veen TA, Richani Sarieddine MZ, de Wit C, Yang Z, Roth I, Bacchetta M, Viswambharan H, Foglia B, Dudez T, van Kempen MJ, Coenjaerts FE, Miquerol L, Deutsch U, Jongsma HJ, Chanson M, Kwak BR (2010) Endothelial-specific deletion of connexin40 promotes atherosclerosis by increasing CD73-dependent leukocyte adhesion. Circulation 121:123–131
61. Regan CP, Manabe I, Owens GK (2000) Development of a smooth muscle-targeted cre recombinase mouse reveals novel insights regarding smooth muscle myosin heavy chain promoter regulation. Circ Res 87:363–369
62. Nguyen TD, Taffet SM (2009) A model system to study Connexin 43 in the immune system. Mol Immunol 46:2938–2946
63. Matter CM, Chadjichristos CE, Meier P, von Lukowicz T, Lohmann C, Schuler PK, Zhang D, Odermatt B, Hofmann E, Brunner T, Kwak BR, Luscher TF (2006) Role of endogenous Fas (CD95/Apo-1) ligand in balloon-induced apoptosis, inflammation, and neointima formation. Circulation 113:1879–1887
64. Takens-Kwak BR, Jongsma HJ, Rook MB, Van Ginneken AC (1992) Mechanism of heptanol-induced uncoupling of cardiac gap junctions: a perforated patch-clamp study. Am J Physiol 262:C1531–C1538
65. Guan X, Cravatt BF, Ehring GR, Hall JE, Boger DL, Lerner RA, Gilula NB (1997) The sleep-inducing lipid oleamide deconvolutes gap junction communication and calcium wave transmission in glial cells. J Cell Biol 139:1785–1792
66. Guo Y, Martinez-Williams C, Gilbert KA, Rannels DE (1999) Inhibition of gap junction communication in alveolar epithelial cells by 18alpha-glycyrrhetinic acid. Am J Physiol 276:L1018–L1026
67. Kwak BR, Jongsma HJ (1999) Selective inhibition of gap junction channel activity by synthetic peptides. J Physiol 516(3):679–685
68. Warner A, Clements DK, Parikh S, Evans WH, DeHaan RL (1995) Specific motifs in the external loops of connexin proteins can determine gap junction formation between chick heart myocytes. J Physiol 488(3):721–728
69. Chaytor AT, Evans WH, Griffith TM (1997) Peptides homologous to extracellular loop motifs of connexin 43 reversibly abolish rhythmic contractile activity in rabbit arteries. J Physiol 503(1):99–110
70. Griffith TM (2004) Endothelium-dependent smooth muscle hyperpolarization: do gap junctions provide a unifying hypothesis? Br J Pharmacol 141:881–903
71. Zahler S, Hoffmann A, Gloe T, Pohl U (2003) Gap-junctional coupling between neutrophils and endothelial cells: a novel modulator of transendothelial migration. J Leukoc Biol 73:118–126
72. Isakson BE, Duling BR (2005) Heterocellular contact at the myoendothelial junction influences gap junction organization. Circ Res 97:44–51
73. Evans WH, Leybaert L (2007) Mimetic peptides as blockers of connexin channel-facilitated intercellular communication. Cell Commun Adhes 14:265–273
74. Wong CW, Christen T, Roth I, Chadjichristos CE, Derouette JP, Foglia BF, Chanson M, Goodenough DA, Kwak BR (2006) Connexin37 protects against atherosclerosis by regulating monocyte adhesion. Nat Med 12:950–954
75. Sarieddine MZ, Scheckenbach KL, Foglia B, Maass K, GarciaI KBR, Chanson M (2009) Connexin43 modulates neutrophil recruitment to the lung. J Cell Mol Med 13:4560–4570
76. Hunter AW, Barker RJ, Zhu C, Gourdie RG (2005) Zonula occludens-1 alters connexin43 gap junction size and organization by influencing channel accretion. Mol Biol Cell 16:5686–5698
77. Gourdie RG, Ghatnekar GS, O'Quinn M, Rhett MJ, Barker RJ, Zhu C, Jourdan J, Hunter AW (2006) The unstoppable connexin43 carboxyl-terminus: new roles in gap junction organization and wound healing. Ann N Y Acad Sci 1080:49–62
78. Ghatnekar GS, O'Quinn MP, Jourdan LJ, Gurjarpadhye AA, Draughn RL, Gourdie RG (2009) Connexin43 carboxyl-terminal peptides reduce scar progenitor and promote regenerative healing following skin wounding. Regen Med 4:205–223
79. O'Quinn MP, Palatinus JA, Harris BS, Hewett KW, Gourdie RG (2011) A peptide mimetic of the connexin43 carboxyl terminus reduces gap junction remodeling and induced arrhythmia following ventricular injury. Circ Res 108:704–715
80. Qiu C, Coutinho P, Frank S, Franke S, Law LY, Martin P, Green CR, Becker DL (2003) Targeting connexin43 expression accelerates

the rate of wound repair. Curr Biol 13:1697–1703

81. Coutinho P, Qiu C, Frank S, Wang CM, Brown T, Green CR, Becker DL (2005) Limiting burn extension by transient inhibition of Connexin43 expression at the site of injury. Br J Plast Surg 58:658–667
82. Mori R, Power KT, Wang CM, Martin P, Becker DL (2006) Acute downregulation of connexin43 at wound sites leads to a reduced inflammatory response, enhanced keratinocyte proliferation and wound fibroblast migration. J Cell Sci 119:5193–5203
83. Cronin M, Anderson PN, Cook JE, Green CR, Becker DL (2008) Blocking connexin43 expression reduces inflammation and improves functional recovery after spinal cord injury. Mol Cell Neurosci 39:152–160
84. Shao PL, Chiu CC, Yuen CM, Chua S, Chang LT, Sheu JJ, Sun CK, Wu CJ, Wang CJ, Yip HK (2010) Shock wave therapy effectively attenuates inflammation in rat carotid artery following endothelial denudation by balloon catheter. Cardiology 115:130–144
85. Nakano Y, Oyamada M, Dai P, Nakagami T, Kinoshita S, Takamatsu T (2008) Connexin43 knockdown accelerates wound healing but inhibits mesenchymal transition after corneal endothelial injury in vivo. Invest Ophthalmol Vis Sci 49:93–104

Part V

Non Mammalian and Mammalian Models of Regenerative Repair

Chapter 23

Gain-of-Function Assays in the Axolotl (*Ambystoma mexicanum*) to Identify Signaling Pathways That Induce and Regulate Limb Regeneration

Jangwoo Lee, Cristian Aguilar, and David Gardiner

Abstract

The adult salamander has been studied as a model for regeneration of complex tissues for many decades. Only recently with the development of gain-of-function assays for regeneration, has it been possible to screen for and assay the function of the multitude of signaling factors that have been identified in studies of embryonic development and tumorigenesis. Given the conservation of function of these regulatory pathways controlling growth and pattern formation, it is now possible to use the functional assays in the salamander to test the ability of endogenous as well as small-molecule signaling factors to induce a regenerative response.

Key words Axolotl, Urodele, Regeneration, Fibroblasts, Accessory Limb Model, Excisional Regeneration Model

1 Introduction

Although regenerative abilities are widespread and essential to the maintenance of tissue homeostasis, they are largely limited to individual cell types within complex tissues. Many of the cell types within the human arm can regenerate (e.g., muscle, nerves, bone, and blood vessels); however, the limb itself cannot regenerate when amputated. In contrast, the urodele amphibians (salamanders and newts) have the remarkable ability to regenerate complex tissues, and thus have historically been the model organisms of choice for regeneration studies.

In spite of many decades of effort studying regeneration in salamanders, we are only now beginning to take advantage of modern techniques in molecular genetics to discover the critical

Jangwoo Lee and Cristian Aguilar have contributed equally.

Robert G. Gourdie and Tereance A. Myers (eds.), *Wound Regeneration and Repair: Methods and Protocols*, Methods in Molecular Biology, vol. 1037, DOI 10.1007/978-1-62703-505-7_23, © Springer Science+Business Media New York 2013

gene regulatory networks that regulate regeneration. Most studies have been limited in terms of describing the anatomy and patterns of gene expression associated with the regeneration process. Attempts to understand the function of candidate genes are limited by the lack of gain-of-function models for regeneration. It is paradoxical that it is experimentally challenging to test function in an animal that can regenerate perfectly. Although it is possible to test for the inhibition of regeneration (loss-of-function), it is hard to think of ways to induce a regenerative response in an animal that already regenerates. To address this issue of a need for gain-of-function assays, we began several years ago to develop and validate experimental models using the axolotl (*Ambystoma mexicanum*) to develop assays for the signaling pathways that control limb regeneration.

Initially, we focused on how to identify the signals that induce blastema formation and subsequent reformation of an entire limb. We took advantage of previous reports in the literature indicating that ectopic limbs could be induced to form from wounds on the side of the limb [1, 2]. Building on the insight that this response is dependent on signals from a nerve combined with interactions between dermal fibroblasts, we were able to optimize and validate the Accessory Limb Model (ALM) as an assay for the events that are induced by amputation and lead to regeneration of the entire limb [1, 2] (Fig. 1a, b). We subsequently developed the Excisional Regeneration Model (ERM) in the axolotl to identify the steps and signals leading to the regeneration of skeletal defects along the proximal–distal limb axis, which are common injuries that can lead to amputation given our limited ability to repair them [3] (Fig. 1c, e). In both these models, the strategy is to start with the axolotl that we know can regenerate all the limb structures perfectly when the limb is amputated. We then create wounds that in the absence of additional signals do not regenerate. We then can deliver and test candidate factors to these wounds so as to induce a regenerative response as a gain-of-function assay.

The ALM is based on the ability to induce an entire, ectopic limb to form from a wound on the side of the arm of an axolotl (Fig. 1a, b). From the earlier literature, it was evident that if all the necessary signals are provided, a limb is formed de novo; however, if one or more of these signals is not provided the regenerative response is impaired [1]. If a skin wound is made but no additional signals are provided (Fig. 3c), the wound becomes re-epithelialized and the dermis regenerates without forming a scar. If, however, the brachial nerve is surgically deviated to the site of the skin wound, dermal cells adjacent to the wound are induced to migrate to the site of the deviated nerve where they dedifferentiate to form an ectopic blastema [1] (Fig. 1a and 3e). The induced ectopic blastema is equivalent to the blastema formed in response to amputation [2], and thus the ALM is a model for studying the induction

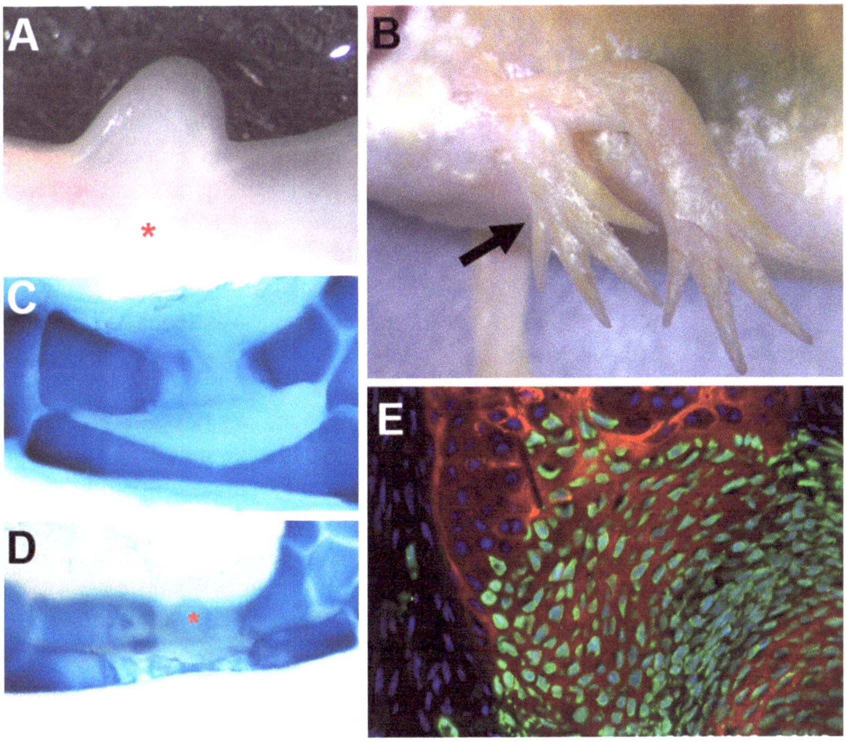

Fig. 1 Regeneration of an entire new limb or a radial defect in the axolotl. (**a**) In response to signals from the wound epithelium and a deviated nerve (*red asterisk*), dedifferentiated blastema cells accumulate on the side of the arm. (**b**) If in addition to the wound and nerve signals, cells from the side of the limb that is opposite to the wound are grafted into the wound (*see* also Fig. 3a, b), a normally patterned ectopic limb is induced to form de novo (*arrow*). (**c**) Surgical removal of the mid-diaphyseal region of the radius results in a defect (critical size defect, CSD) that is not regenerated. (**d**) Regeneration of a CSD can be induced by a variety of experimental manipulations, including the implantation of gelatin microcarrier beads soaked in BMP2 (*red asterisk*). (**e**) Grafted blastema cells (grafted cells are *green* from a GFP donor animal) regenerate the CSD and differentiate into chondrocytes that integrate into the cut ends of the host defect (*red* is immunostaining for Type II Collagen)

of dedifferentiation and blastema formation. The important advantage of the ALM, in contrast to amputation, is that there is very little damage to the stump tissues, and thus extraneous signaling events associated with the trauma of amputation are not induced.

Although nerve-associated signals are necessary and sufficient to induce blastema formation, additional signals are required to induce a limb de novo. As has been appreciated for decades, the genesis of an entire new limb requires the interaction of cells from opposite sides of the limb [4–6]. These cells are provided in the ALM by the grafting of skin from the side of the limb that is opposite the side on which the wound is created (e.g., a graft of posterior skin to an anterior wound; Fig. 3a, b, f). In response to signaling from both a deviated nerve and connective tissue cells grafted from the opposite side of the limb, a new and perfectly patterned limb can be induced to form at a high frequency [1] (Fig. 1b).

Results from ALM experiments have provided a number of insights into the mechanisms of limb regeneration. The fundamental contribution of the ALM is the demonstration that regeneration is a stepwise process (e.g., wounding, nerve signaling, and cell–cell interactions). Experimentally it is important that the ALM allows us to study each step individually, unlike the situation with an amputation in which all the steps are activated. Secondly, the ALM demonstrates the critical role of dermal fibroblasts in controlling growth and pattern formation during regeneration. This function had been established previously from studies based on the limb amputation model [4, 7]. With the ALM, the only tissues that are damaged are the severed nerve and the wounded/grafted dermal connective tissue. The interactions between the graft and host dermal fibroblasts in response to nerve signals subsequently generate the signals that activate and recruit the ingrowth and patterns of the other limb tissues (e.g., blood vessels, nerves, and myoprogenitor cells derived from satellite cells). Finally, the ALM can be used as an assay to test the function of candidate molecules and factors that can be delivered by a number of techniques (e.g., electroporation, microinjection of viral vectors, and implantation of microcarrier beads that have been soaked in growth factors). Given the ability to manipulate the steps in the limb regeneration cascade, it is possible to utilize the ALM to test for function in the control of cell migration, proliferation, and dedifferentiation leading to blastema and limb formation.

Although many injuries result in limb amputation, most involve the loss of tissues at an intermediate level along the proximal–distal limb axis. In such cases, the surgical challenge is to repair the damage and salvage as much of the remaining limb as possible so as to avoid amputation of the remaining limb tissues. Thus the ability to enhance a regenerative response would be an important advancement in treating such injuries. Although limb amputations in salamanders have been studied for decades, little is known about the regenerative response to injuries that result in structural defects in the limb. What is known is that typically such defects are not regenerated unless the entire limb is amputated [3, 8]. As with the ALM, this lack of a regenerative response in an animal that can regenerate perfectly provides the opportunity to develop a gain-of-function assay for regeneration.

The ERM involves the surgical removal of a central segment of the radius in the zeugopod. This mid-diaphyseal deletion fails to regenerate beyond a critical size limit (critical size defect, CSD) despite the immense regenerative ability of the axolotl [3, 9] (Fig. 1c). It is important to note that this response to injury in the axolotl is equivalent to that in the human. Mesenchymal progenitors are activated and form a soft callus structure made of fibrocartilage around the cut ends of the skeletal element, and without treatment or intervention, the gap created by the surgery will persist. However, in the axolotl the addition of multipotent blastema cells generated by dedifferentiation after limb amputation provides cells with the ability to be reprogrammed, leading to the regeneration of the excised skeletal element [3]. Taking these results together, the ERM creates an injury environment that is permissive and instructive for a regenerative response, but does not induce dedifferentiation so as to provide a source of regeneration-competent cells.

By creating an assay in which regeneration is not the endogenous response, we can begin to identify the molecular factors for inducing regeneration in a non-regenerating wound. As mentioned previously, grafting of dissociated blastema cells into the wound site results in a regenerated skeletal element that is integrated into the existing host bone (Fig. 1e). Other cell types can be grafted into the injury with various results. For example, dermal fibroblasts, the progenitors of the blastema, form dense connective tissue when grafted but do not repair the defect. However, the presence of a nerve deviated to the wound site promotes dedifferentiation of surrounding cells, as well as the grafted dermal cells, leading to regeneration of the skeletal element. Therapeutically introducing factors that induce regeneration in situ could also induce dedifferentiation, and the efficacy of candidate factors can be assayed based on the extent of CSD regeneration that occurs. One method of introducing factors involves the formation of microcarrier beads encapsulating growth factors or small molecules for delivery to the wound site (Fig. 1d and 4a, d). Alternatively, cells can be isolated, treated in culture, and subsequently grafted into the wound site. The extent of dedifferentiation induced during ex vivo culture of the donor cells (e.g., autologous dermal fibroblasts) can then be assayed by the ERM and quantified by the degree of regeneration that results.

Both the ALM and the ERM provide experimental models to test the function of specific signaling pathways involved in the regulation of limb regeneration. The key element of both is that they are gain-of-function assays that identify the progressive steps and the signals that allow for progression to the next step. Combined with the power of computational biology to identify the regulatory networks that are activated in response to these injuries in the axolotl, it will be possible to discover how each step of successful regeneration progresses from one to the next. While the ability of these animals to regenerate entire organs seems extraordinary, the mechanisms of regeneration involve many basic biological processes regulated by conserved signaling pathways [4]. Thus, we can expect to stimulate regeneration in a human by the same mechanisms used by the axolotl, and thus to enhance our regenerative responses to both acute injury and the accumulation of chronic damage associated with aging.

2 Materials

Methods for both surgical procedures are largely overlapping, and the following materials are often used for both (Fig. 2).

2.1 Induction and Maintenance of Anesthesia

1. Stock Holtfreter's salt solution: To make 2 L of stock Holtfreter's salt solution, add 320 g of NaCl, 28 g of $CaCl_2$, two tablespoons of $MgSO_4$-$7H_2O$, and one teaspoon of KCl.

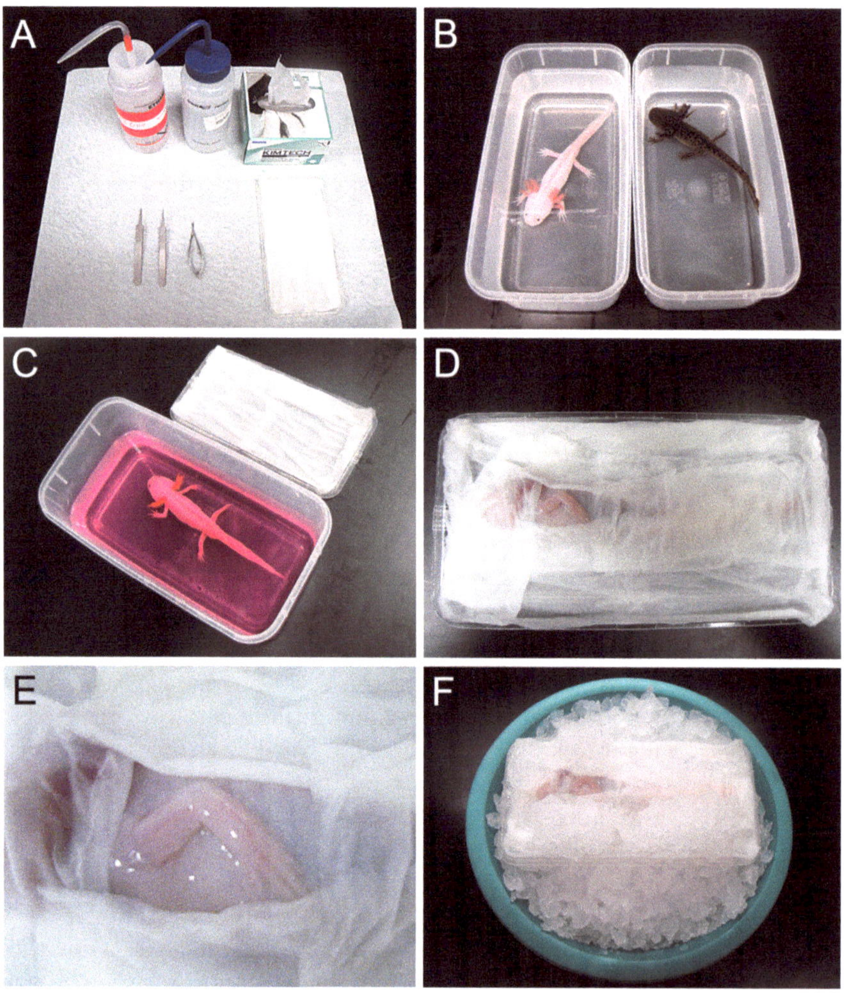

Fig. 2 Surgical and postsurgical handling of the axolotl (*Ambystoma mexicanum*). (**a**) Surgical supplies. Clockwise from upper left-hand corner: 70 % ethanol, 40 % Holtfreter's solution, Kimwipes™, surgery plate, microscissors, and microforceps. (**b**) White (*left*) and wild-type (*right*) axolotls in 40 % Holtfreter's salt solution. (**c**) The white axolotl is being anesthetizing in MS222 solution. MS22 solution contains phenol red for monitoring pH change during anesthesia, so the solution is *pink color* (*see* **Note 1**). (**d**) The animal is placed on the surgery plate and covered with moist Kimwipes™. (**e**) Surgery area (upper arm) is exposed from the moist Kimwipes™. (**f**) For recovery, the animal on the surgery plate is placed on ice

2. Holtfreter's salt solution, 40 %: Add 250 mL of stock Holtfreter's salt solution to 5 gal of deionized water.
3. Stock MS222 solution, 20× (*see* **Note 1**): Add 20 g of MS222 (ethyl 3-aminobenzoate methanesulfonate salt, Sigma) to 1 L of deionized water. Store the solution at 4 °C and avoid light.
4. Working MS222 solution, 1×: Add 50 mL of 20× MS222 stock solution, 1 mL of Tris–HCl (pH 7.4), and 0.5 mL of 0.5 % phenol red solution (Sigma) to 900 mL of 40 % Holtfreter's salt solution. Adjust the pH to 7.4 with NaOH. Adjust the volume to 1 L with 40 % Holtfreter's salt solution. Store the solution at 4 °C.
5. Plastic containers large enough to immerse an animal in the MS222 anesthetic solution (Fig. 2b)

2.2 Preparation of Microcarrier Beads for Delivery of Experimental Agents

2.2.1 Sol–Gel Beads

1. Tetramethyl orthosilicate.
2. Methanol, 100 %.
3. HCl, 0.1 N.
4. Glacial acetic acid, 0.1 N.
5. Growth factor/small molecule of interest.

2.2.2 Gelatin Microspheres

1. Type B gelatin.
2. Olive oil.
3. Acetone, 100 %.
4. 2-Propanol, 100 %.
5. Glutaraldehyde solution: Combine 100 mL H_2O, 100 μL Tween 20, and 36 μL of glutaraldehyde. Stir on stir plate for 2 min or until solution is homogenous.
6. 10 mM glycine.
7. Growth factor of interest.

2.3 Preparation of Blastema Cell Grafts

2.3.1 Collagen Clot Delivery

1. Type I collagen (rat tail), 3.6 mg/mL.
2. L-15 culture media, 600 %.
3. Fetal bovine serum.
4. Sodium bicarbonate, 7.5 % w/v.

2.3.2 Fibrin Clot Delivery

1. Fibrinogen, 25 mg/mL in 60 % DMEM (Dulbecco's Modified Eagle's Medium).
2. Thrombin, 200 U/mL in 60 % DMEM.

2.4 Surgical Procedures

1. Lab Tissues [e.g., Kimwipes™ (Kimtech)].
2. Holtfreter's solution, 40 % in squeeze bottle.
3. Two microforceps, Dumont #5 stainless steel (0.10 × 0.06 mm tips)—World Precision Instruments.

4. Microscissors, World Precision Instruments Noyes scissors (15 mm blades), WPI Vannas scissors (5 mm blades, 0.1 mm tips).
5. Ethanol, 70 %.
6. Plastic plate or tray (slightly larger than the animal) to place the animal on that will keep liquids contained during the surgery (Fig. 2d).
7. Ice and ice bucket.

2.5 Sample Collection and Processing for Histology/Immunohistochemistry/In Situ Hybridization

2.5.1 Whole-Mount Staining

1. Paraformaldehyde (PFA), 4 %.
2. Alcian blue, 7.5 mg/mL in acid alcohol (3:1 EtOH:Glacial acetic acid).
3. EtOH, 100 %.
4. KOH, 0.5 %.
5. KOH, 4 %.
6. Alizarin red (1:4 0.1 % w/v Alizarin red in EtOH:0.5 % KOH).
7. Glycerol.

2.5.2 Embedding for Cryosectioning

1. Paraformaldehyde, 4 %.
2. EDTA (Ethylenediaminetetraacetic acid), 10 %.
3. Sucrose, 30 %.
4. Tissue Tek® cryomold® intermediate, disposable vinyl specimen molds.
5. Tissue Tek®, O.C.T. compound, embedding medium for frozen tissue specimens.
6. Forceps.
7. Liquid nitrogen.

2.5.3 Histological Staining

1. Alcian blue, 7.5 mg/mL in acid alcohol (3:1 EtOH:Glacial acetic acid).
2. Ehrlich's hematoxylin.
3. Eosin, 0.25 % w/v in 70 % EtOH + 0.4 % v/v Glacial acetic acid.
4. EtOH, 70 %.
5. EtOH, 95 %.
6. EtOH, 100 %.
7. Cytoseal™ 60 mounting media.

3 Methods

3.1 Preparation of Surgical Area

1. Organize a surgical station that is equipped with a dissecting microscope and fiber-optics illumination.
2. Disinfect the surgical station with 70 % ethanol.

3. Sterilize the surgical instruments (*see* **Note 2**) with a glass bead sterilizer (e.g., Steri 250, Inotech).

3.2 Induction and Maintenance of Anesthesia

1. Anesthetize axolotl by immersion in 1× MS222 solution (*see* **Note 1**) until it is no longer responsive to a toe pinch (about 20–30 min for larger animals; less time for smaller animals).
2. Prepare a surgery plate that has a moist (saturated with 40 % Holtfreter's solution) Kimwipe™ on it.
3. After the axolotl has been anesthetized, put the animal on the surgery plate.
4. Cover the animal with moist Kimwipes™ to prevent dehydration of skin. Tease away the paper to expose the surgery area while keeping the rest of the animal covered (*see* **Note 3**).

3.3 Accessory Limb Model

3.3.1 Wounding

1. Expose the upper arm region by teasing back the moist Kimwipes™.
2. Make a full-thickness skin wound on the mid-anterior region with a rectangular incision (1.0–1.5 mm × 2.0–3.0 mm) using microscissors (Fig. 3c).
3. Remove the excised skin using fine forceps.

3.3.2 Nerve Deviation

1. To obtain a nerve fiber for surgical deviation to the wound site, lift the arm to expose the posterior-ventral side, and make an incision extending from the shoulder to the elbow (Fig. 3d).
2. Locate the brachial nerve, and transect it with microscissors at the level of the elbow.
3. Dissect the transected brachial nerve from the surrounding connective tissue being careful to not damage the nerve fiber.
4. To reroute the transected brachial nerve to the skin wound at the anterior site, use microforceps to grab the tip of the transected nerve and guide it beneath the skin to the skin wound (Fig. 3e).
5. After rerouting the nerve to the anterior skin wound, close the posterior-ventral incision using microforceps and allow the edges to heal into place (*see* **Note 4**).

3.3.3 Skin Grafting

1. Carefully turn the animal over to access the posterior side of contralateral arm (avoid having the deviated nerve retracting away from the anterior wound site).
2. Excise a square piece of skin (~1 mm on each side) from the posterior side of the arm using the technique for making the anterior skin wound (*see* **Note 5**).
3. Turn the animal over so as to access the nerve-deviated anterior wound.

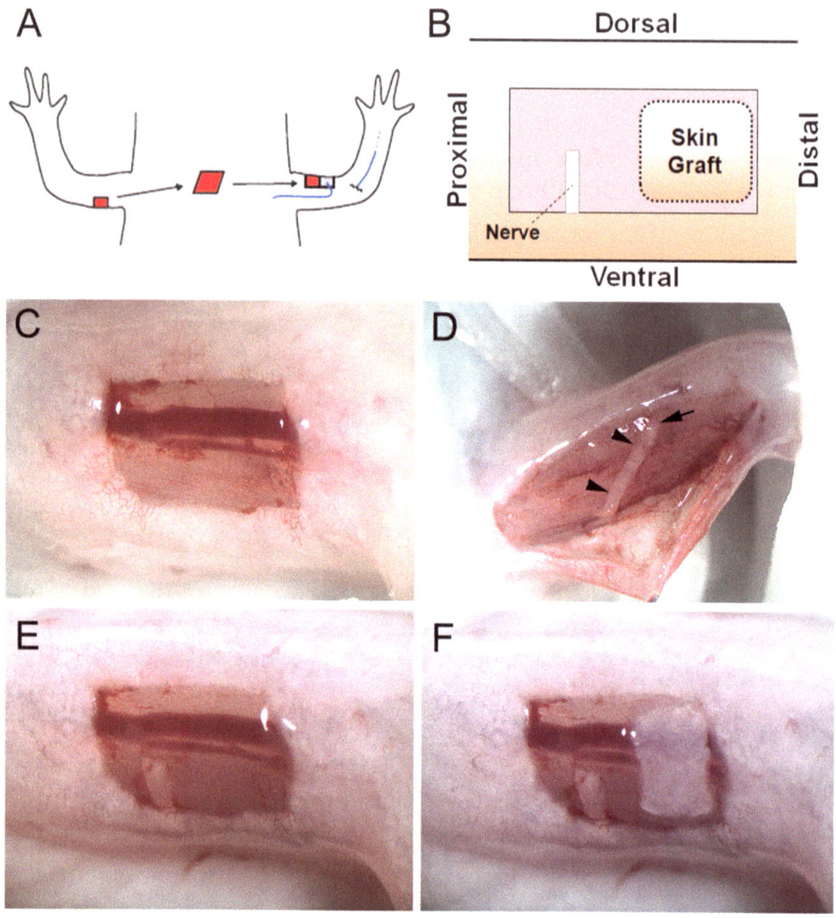

Fig. 3 Wounding, nerve deviation, and skin grafting to induce an ectopic limb (Accessory Limb Model). (**a**) Summary of surgical procedure for induction of ectopic limb. (**b**) A diagram that shows the relative positions of the wound, deviated nerve, and skin graft after a completed surgery. (**c**) Wounding on the anterior side of the upper arm showing the major blood vessel and nerve running from proximal (*left*) to distal (*right*) along the anterior side of the limb. (**d**) Incision on the posterior side of the arm, exposing the dissected brachial nerve (*arrowheads*) and the tip of the nerve that has been severed distally (*arrow*). (**e**) Deviated nerve on the anterior skin wound. (**f**) The posterior skin from the contralateral arm is grafted on the skin wound adjacent to the deviated nerve as in (**b**)

4. Trim the end of the deviated brachial nerve using microscissors and position the cut end so that it lies within the anterior wound (*see* **Note 6**; Fig. 3e). After trimming, the nerve tip should be positioned vertically in the middle and horizontally one-third on either side.
5. Place the posterior skin within the anterior skin wound, next to the end of the deviated nerve (Fig. 3f). It is important that the tip of the nerve does not touch the edge of either the grafted skin or the wound (*see* **Note 7**).
6. Place the animal on the ice to let the wound heal for 2 h (Fig. 2f).

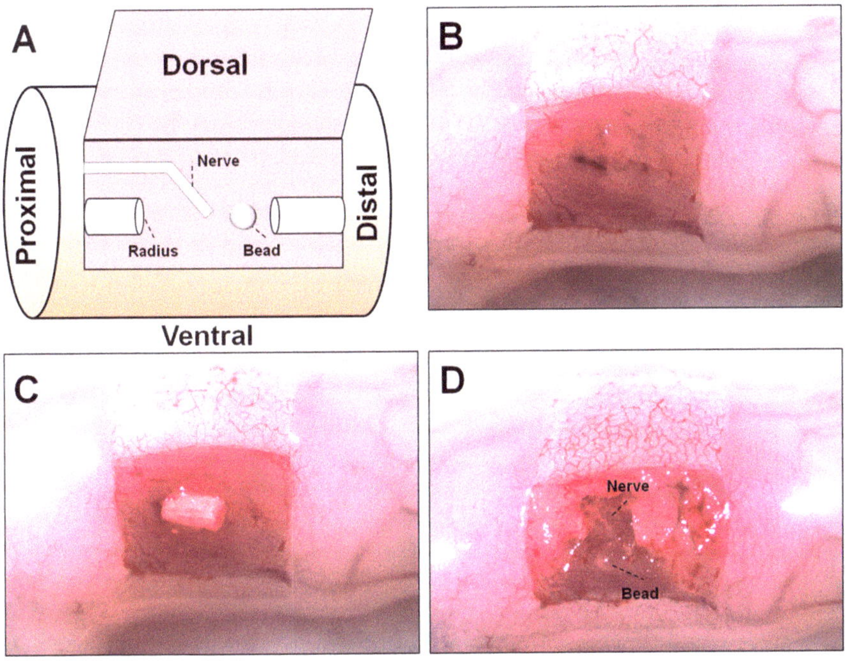

Fig. 4 Excision of the diaphyseal region of the radius to create a critical size defect in the excisional regeneration model (ERM). (**a**) Schematic representing a completed excision with a deviated nerve and microcarrier bead graft. Placement of the bead is subject to investigator's aim and discretion. (**b**) Skin flap window created by three incisions on the anterior side of the zeugopod. The underlying muscle has been reflected to allow access to the radius. Note the minimal damage to surrounding tissue. (**c**) 2 mm mid-diaphyseal segment of the radius after excision. (**d**) A deep wound surgery with a deviated nerve and microcarrier bead graft. Mesenchymal tissue has been removed from the top of the wound to below the radius. The cut ends of the radius are visible with the deviated nerve placed between them

7. Place the animal back in a housing container with 40 % Holtfreter's salt solution.

3.4 Excisional Regeneration Model

Both types of ERM wounds (simple and deep w/nerve deviation) are host sites into which microcarrier beads and/or cells can be grafted. Beads and cells are grafted into the surgically created cavity between the cut ends of the radius. Grafting is done before closing the skin flap in the case of the simple wound, or after nerve deviation in the case of the deep wound (there is no skin flap and the wound heals by reepithelialization from the surrounding epidermal keratinocytes).

3.4.1 Simple Wound

1. Anesthetize an 8–10 cm axolotl in MS222 solution for 25 min.
2. Prepare the animal for surgery, laying the axolotl such that the subject limb is easily accessible (*see* **Note 3**; Fig. 2d, e).
3. Using microscissors (5 mm blades, 0.1 mm tips), create a skin flap on the dorsal zeugopod by making three incisions in the shape of a rectangle. The attached side of the rectangle should be the posterior side (Fig. 4a, b).

4. Reflect the flap back so as to expose the underlying muscle tissue. Using forceps, expose the radius by carefully reflecting the muscle fibers, taking care to not sever major blood vessels or damage the muscle itself (*see* **Note 8**).
5. Excise a 2 mm mid-diaphyseal segment of the radius using microscissors (15 mm blades) (Fig. 4c).
6. Return the displaced muscle fibers to their original position.
7. Close the skin flap, ensuring contact on all sides with the adjacent, uninjured skin. Suturing is not necessary (*see* **Note 4**).
8. Place Kimwipes™ to cover the animal completely except for the wound site created by the surgery. Soak Kimwipes™ in 40 % Holtfreter's solution.
9. Place the animal on the ice to let the wound heal for 2 h (Fig. 2f) before returning it to the housing container with 40 % Holtfreter's salt solution.

3.4.2 Deep Wound and Nerve Deviation

1. Begin preparation as described above.
2. Using microscissors (5 mm blades, 0.1 mm tips), create a skin window on the dorsal zeugopod by making four incisions in the shape of a rectangle.
3. Surgically remove the underlying muscle tissue until the radius is exposed.
4. Excise a 2 mm mid-diaphyseal segment of the radius using microscissors (15 mm blades).
5. Continue to remove muscle tissue until the pair of nerves running between the radius and ulna is exposed (*see* **Note 9**).
6. Sever the nerves distally (at the carpal level) and position the cut ends in the center of the deep wound (Fig. 4d)
7. Place Kimwipes™ to cover the animal completely except for the wound site created by the surgery. Soak Kimwipes™ in 40 % Holtfreter's solution.
8. Place the animal on the ice to let the wound heal for 2 h (Fig. 2f) before returning it to the housing container with 40 % Holtfreter's salt solution.

3.5 Preparation of Microcarrier Beads

3.5.1 Sol–Gel Beads [10]

The following protocol describes the procedure for making 100 μL total volume of sol–gel solution. For larger volumes, the amounts specified can be scaled up. All work should be carried out in the hood unless otherwise specified.

1. In a microcentrifuge tube combine 37.6 μL H_2O, 31.6 μL tetramethyl orthosilicate (TMOS), and 8.5 μL methanol. TMOS is extremely hazardous (*see* **Note 10**).
2. Add 0.1 N HCl until the pH of the solution is below 2. Approximately 2 μL should be sufficient.

3. In a separate tube, prepare a mixture of the growth factor/small molecule of interest in 0.1 N acetic acid. The final volume of this mixture should be 20 μL. Use the appropriate amount of growth factor/small molecule and add 0.1 N acetic acid until the volume is 20 μL.
4. Add the resultant growth factor/small molecule solution to the TMOS mixture (*see* **Note 11**).
5. Dispense the solution into molds of a desired shape. Beads must be small for grafting into the ERM. The simplest method of bead formation is to dispense 1 μL volumes onto a sheet of parafilm.
6. Allow the beads to harden for 3 days at room temperature in a hood.
7. At this stage, the beads can be removed from the hood. Dry the beads at 37 °C overnight.

3.5.2 Gelatin Microsphere Beads [11] (See ***Note 11****)*

1. Add 5 g type B gelatin to 50 mL H_2O. Microwave until completely dissolved.
2. In a 500 mL beaker, heat 250 mL olive oil to 40 °C on a hot plate/stirrer. Add a stir bar.
3. Slowly pour the gelatin solution into the olive oil while stirring. Continue stirring for 5 min. The speed of stirring will affect the size of the beads produced. Faster speeds will produce smaller beads.
4. Transfer the beaker to a cool stir plate (no heat) and continue stirring for 30 min at room temperature.
5. Add 100 mL acetone to the beaker and increase the stir speed to the highest possible. The acetone will fix the beads. Remove any large aggregates that form during the stirring process. Keep the beaker covered to prevent contamination.
6. The solution will have separated into two layers. Remove the top layer and wash the beads with 75 mL of 2-propanol.
7. Transfer the mixture to two 50-mL conical tubes and centrifuge at 2,000 × *g* for 5 min at 4 °C.
8. Decant the supernatant and add 20 mL of glutaraldehyde solution. Immediately resuspend the pellet and pour into a 200 mL beaker.
9. Stir the mixture for approximately 2 min or until most of the beads are separated. Remove large aggregates and incubate covered for 12 h at 4 °C.
10. Transfer the contents of the beaker to a 50-mL conical tube and centrifuge at 2,000 × *g* for 2 min at 4 °C.
11. Decant the supernatant and wash the beads in 50 mL H_2O. Centrifuge at 2,000 × *g* for 2 min at 4 °C.

12. Decant the supernatant and transfer the beads to a beaker containing 100 mL of a 10 mM glycine solution. Stir covered for 1 h at 37 °C.
13. Transfer the contents of the beaker to two 50-mL conical tubes and centrifuge at 2,000 × *g* for 2 min.
14. Wash beads 2×: Decant supernatant and add 50 mL H_2O. Resuspend and centrifuge at 2,000 × *g* for 2 min.
15. Decant the supernatant and transfer the beads to a large Petri dish. Remove any large aggregates and dry the beads at 4 °C overnight.
16. In order to add a growth factor of interest, prepare the desired concentration of a growth factor solution. Soak beads of the desired size (usually 300–500 μM) in the growth factor solution overnight at 4 °C.

3.6 Preparation of Blastema Cell Grafts

3.6.1 Collagen Clot Delivery: 10 μL Volume Graft

1. In a sterile microcentrifuge tube, combine: 27 μL type I collagen, 5 μL L-15 culture media (600 %), 2.5 μL FBS, 13.5 μL autoclaved H_2O, and 3 μL $NaHCO_3$ (7.5 % w/v). Mix well.
2. Transfer 10 μL of the collagen solution into a tube containing a cell pellet of approximately 5×10^4 cells. Mix gently and place the mixture into a 35 mm tissue culture dish to clot.
3. Once clotted, graft the collagen clot into the gap created by excising a mid-diaphyseal segment of the radius.

3.6.2 Fibrin Clot Delivery

1. Centrifuge approximately 7.5×10^5 cells into a loose pellet at 200 × *g* for 5 min.
2. Decant the supernatant and resuspend in 150 μL fibrinogen solution. Centrifuge into a loose pellet at 200 × *g* for 5 min.
3. Decant the supernatant without disturbing the pellet.
4. Add 1.0 μL of thrombin to the pellet and spin at low speed (20 × *g* for 1 min).
5. Once the clot is set, you can trim the clot to the appropriate size/shape and graft into the gap created by excising a mid-diaphyseal segment of the radius.

3.7 Sample Collection and Processing

3.7.1 Whole-Mount Staining

1. Fix collected tissue in 4 % PFA overnight at room temperature.
2. Rinse the sample twice with H_2O.
3. Stain the sample with alcian blue for 24 h at room temperature.
4. Wash the sample twice with acid alcohol for 5 min.
5. Place the sample in 100 % EtOH overnight.
6. Transfer the sample to 0.5 % KOH solution for destaining. Once the sample sinks to the bottom of the vessel, replace the solution with fresh 0.5 % KOH and leave overnight.

7. Stain the sample with alizarin red for 24 h.
8. Destain in 4 % KOH, replacing the solution with fresh 4 % KOH periodically.
9. Once the destaining is complete, clear the sample in 50 % glycerol. After the sample sinks, transfer to 80 % glycerol. Once the sample sinks again, transfer to 100 % glycerol. This process will take several days.

3.7.2 Embedding for Cryosectioning

1. Fix collected tissue in 4 % PFA overnight at room temperature.
2. Replace the 4 % PFA with 10 % w/v EDTA and incubate overnight at room temperature. This step is necessary for samples containing bone.
3. Replace the 10 % EDTA with 30 % w/v sucrose and incubate overnight at 4 °C with gentle shaking. Proceed to the next step when the sample sinks to the bottom of the vessel.
4. Fill a cryomold halfway with Tissue Tek®, O.C.T. compound and add the tissue sample. The sample should be submerged and in contact with the bottom of the mold.
5. Incubate the sample at 4 °C for at least 4 h in order to remove bubbles.
6. Add liquid nitrogen to a Styrofoam cooler. Place the cooler on a slight incline so that the liquid nitrogen is present on one side only. Place the cryomold on the elevated side of the cooler. Avoid direct contact with the liquid nitrogen. Cover the cooler.
7. Once frozen, store the cryomold at −20 °C or colder.

3.7.3 Histological Staining

1. Rehydrate sectioned tissue in H_2O for 3–5 min.
2. Using a coplin jar, stain the slides with alcian blue for 30–45 min. The stain intensity depends upon the length of time spent in the staining solution.
3. Rinse the slides in H_2O.
4. Stain the slides in Ehrlich's hematoxylin for 5 min.
5. Place the slides under running tap water for 10 min. The water stream should not come into direct contact with the slides. Place the slides in a slide holder, and place into a water-tight container. The slides should face away from the stream, so as to not disturb the sections.
6. Rinse the slides in 70 % EtOH.
7. Stain the slides in Eosin for 2 min.
8. Rinse the slides in H_2O.
9. Rinse the slides in 70 % EtOH.
10. Wash in 95 % EtOH for 30 s.

11. Wash in 100 % EtOH twice for 3 min.
12. Allow the slides to dry before mounting a coverslip using Cytoseal™ 60 mounting media.

4 Notes

1. MS222 is acidic, and if not buffered appropriately (or if the pH of the solution changes over time during storage) the animals lose the ability to efficiently transport oxygen in their blood and will die when anesthetized. Addition of phenol red prevents the inadvertent use of MS222 that is at the wrong pH.
2. It is important to protect the mucous layer covering the skin of the animal (e.g., do not use antiseptic solutions to wash the surface of the skin prior to making an incision). Mucous in amphibians has antimicrobial activity, and therefore the surgical incisions do not get infected so long as the mucous layer is present. Similarly, autoclaving the surgical instruments is not effective since once you make the first incision, the instrument is contaminated. Use of a hot-bead sterilizer allows you to re-sterilize as necessary throughout the surgical procedures.
3. During surgery and subsequent recovery, keep the axolotl moist by periodically squirting 40 % Holtfreter's solution over the length of the animal. Axolotls have both lungs and gills, and therefore tolerate being out of the water; however, their skin does not have the water barrier function necessary to prevent dehydration during extended periods out of the water.
4. Generally it is not necessary to suture the edges of the wounds or the skin grafts. The edges of the skin adhere to the underlying connective tissues and reepithelialization is rapid in these animals, occurring within 4–6 h post surgery [12].
5. Pay attention to the orientation of skin grafts to avoid flipping them over (dermis side up) during transferring and grafting.
6. The brachial nerve is white, surrounded by a fibrous bundle, and can be distinguished easily from the adjacent blood vessels. It is very important not to damage either the blood vessels or the brachial nerve fiber during surgery. To avoid damage to the nerve fiber, lift only the tip of the nerve, which will then be trimmed off after being rerouted to the anterior skin wound.
7. The protocol for deviating the nerve is based on the assumption that posterior skin graft is going to be made. For experiments that involve induction of only an ectopic blastema (wound with deviated nerve but no posterior skin graft), the end of the nerve can be placed anywhere within the wound bed. Signals from the deviated nerve attract migrating cells from the periphery such that the ectopic blastema will form symmetrically over the cut end of the nerve.

8. Be sure to avoid severing major blood vessels or muscle tissue. In the event that a blood vessel is severed, control the bleeding before proceeding. Using a dry Kimwipe™, gently blot the wound until bleeding subsides. Remove any blood clots using forceps or a stream of 40 % Holtfreter's solution.
9. After removing the muscle tissue, keep the deep wound dry by regularly absorbing excess Holtfreter's solution with a dry Kimwipe™. This will aid the surgery by increasing visibility of the nerve.
10. TMOS is extremely hazardous and volatile. Lab coat, safety goggles, and double nitrile gloves are required. When creating the sol–gel solution, begin with the water in the microcentrifuge tube, and then add the TMOS. This will minimize the amount of TMOS exposure. Once TMOS and water are combined, the TMOS is hydrolyzed and safe.
11. Sol–gel beads are clear and may be hard to visualize. You may add a dye in order to increase their visibility.

Acknowledgments

We wish to thank Dr. Susan Bryant (UC Irvine), Dr. Tetsuya Endo (Aichi Gakuin University), and Dr. Akira Sato (Okayama University) for their insights and efforts in developing these two surgical models.

References

1. Endo T, Bryant SV, Gardiner DM (2004) A stepwise model system for limb regeneration. Dev Biol 270:135–145
2. Satoh A, Gardiner DM, Bryant SV, Endo T (2007) Nerve-induced ectopic limb blastemas in the Axolotl are equivalent to amputation-induced blastemas. Dev Biol 312:231–244
3. Satoh A, Cummings GM, Bryant SV, Gardiner DM (2010) Neurotrophic regulation of fibroblast dedifferentiation during limb skeletal regeneration in the axolotl (*Ambystoma mexicanum*). Dev Biol 337:444–457
4. Bryant SV, Endo T, Gardiner DM (2002) Vertebrate limb regeneration and the origin of limb stem cells. Int J Dev Biol 46:887–896
5. Bryant SV, French V, Bryant PJ (1981) Distal regeneration and symmetry. Science 212: 993–1002
6. French V, Bryant PJ, Bryant SV (1976) Pattern regulation in epimorphic fields. Science 193: 969–981
7. Muneoka K, Fox W, Bryant SV (1986) Cellular contribution from dermis and cartilage to the regenerating limb blastema in axolotls. Dev Biol 116:256–260
8. Goss RJ (1969) Principles of regeneration. Academic, New York
9. Hutchison C, Pilote M, Roy S (2007) The axolotl limb: a model for bone development, regeneration and fracture healing. Bone 40:45–56
10. Santos EM, Radin S, Ducheyne P (1999) Sol-gel derived carrier for the controlled release of proteins. Biomaterials 20:1695–1700
11. Tabata Y, Hijikata S, Muniruzzaman M, Ikada Y (1999) Neovascularization effect of biodegradable gelatin microspheres incorporating basic fibroblast growth factor. J Biomater Sci Polym Ed 10:79–94
12. Ferris DR, Satoh A, Mandefro B, Cummings GM, Gardiner DM, Rugg EL (2010) Ex vivo generation of a functional and regenerative wound epithelium from axolotl (*Ambystoma mexicanum*) skin. Dev Growth Differ 52:715–724

Chapter 24

The Mouse Digit Tip: From Wound Healing to Regeneration

Jennifer Simkin, Manjong Han, Ling Yu, Mingquan Yan, and Ken Muneoka

Abstract

A challenge to the study of regeneration is determining at what point the processes of wound healing and regeneration diverge. The mouse displays level-specific regeneration responses. An amputation through the distal third of the terminal phalanx will prompt a regeneration response and result in a new digit tip that mimics the morphology of the lost digit tip. Conversely, an amputation through the distal third of the intermediate phalanx initiates a wound healing and scarring response. The mouse, therefore, provides a model for studying the transition between wound healing and regeneration in the same animal. This chapter details the methods used in the study of mammalian digit regeneration, including a method to introduce exogenous protein into the mouse digit amputation model via microcarrier beads and methods for analysis of bone regeneration.

Key words Digit regeneration, Amputation model, Scarring response

1 Introduction

In some animals, injury prompts complete regrowth of complex tissues to perfectly replace the lost structures. Classical models of this response include tail, eye, and limb regeneration in many salamanders, claw regeneration in the crawfish, and head and tail regeneration of the hydra [1–3]. In contrast, an injury in mammals generally cues a wound healing response that ultimately ends with a scar and incomplete structures. Arm and leg amputations in humans terminate in a bone stump covered by scar tissue. However, data suggest that humans do retain a regenerative potential. For several decades, surgeons have described the regeneration of human digit tips following amputation injury [4]. In 1972, surgeons observed the regrowth of fingertips in amputation wounds that were cleaned but not sutured [5]. This conservative treatment of amputation injuries led to an increase in the observation of fingertip regeneration in both children and adults [5, 6]. It has been

Robert G. Gourdie and Tereance A. Myers (eds.), *Wound Regeneration and Repair: Methods and Protocols*, Methods in Molecular Biology, vol. 1037, DOI 10.1007/978-1-62703-505-7_24, © Springer Science+Business Media New York 2013

established that while some amputation injuries would mount a rebuilding response, others culminate in a wound healing and scarring process [7]. Ultimately, the response seen in the human digit, whether a scar or a new digit tip will form, depends upon the level of amputation injury.

Amputation proximal to the nail bed results in a wound healing response, the permanent loss of amputated structures, and scar formation [7]. In contrast, an amputation in the distal third of the distal phalanx, transecting the nail, bone, and connective tissue will trigger a regenerative response that results in a digit tip similar in structure and function to the lost fingertip. The newly regenerated digit tip is a complex structure including bone, nail, dermis, and epidermis, suggesting that this process is more than just a regrowth response but most likely includes tissue-specific interactions [8]. These clinical descriptions of human digit tip regeneration coupled with the regenerative capacity of other vertebrates, such as the salamander, suggest that humans have retained the ability to regenerate. Figuring out how to unlock this inherent potential in humans requires a practical model of regeneration.

The mouse adult and neonatal digit, like a human fingertip, display level-specific regeneration responses [9] (Fig. 1). The distal one-third of phalangeal element 3 (P3), including bone, surrounding soft connective tissue, nail, and vasculature can regrow in a pattern that mimics the lost digit tip [8, 10]. Conversely, proximal level amputations through the distal one-third of the second phalangeal element (P2) result in wound healing and scar formation. This juxtaposition of regeneration-competent and regeneration-incompetent tissue environments provides a model in which we can perform both loss-of-function and gain-of-function studies in the same animal to determine the fundamental difference between wound healing and regeneration.

To study what signaling pathways are necessary for regeneration, either the neonatal (postnatal day 3 mouse) or the adult (8 week old) amputation models are used in experiments to block the regenerative response. Both models have advantages and disadvantages. Neonatal bone is still developing and not as dense as adult bone which facilitates ease of processing for histology and in situ hybridization. In the adult at 8 weeks, growth plates have completed ossification and bone elongation. While this makes the adult bone more difficult to process, the adult regenerative response can easily be tracked in vivo using μCT or X-ray analysis.

This chapter focuses on level-specific amputations in the mouse model and on specific methods for inducing a regenerative response by manipulating growth factor signaling. Also detailed here are methods for tracking and evaluating the regenerative process.

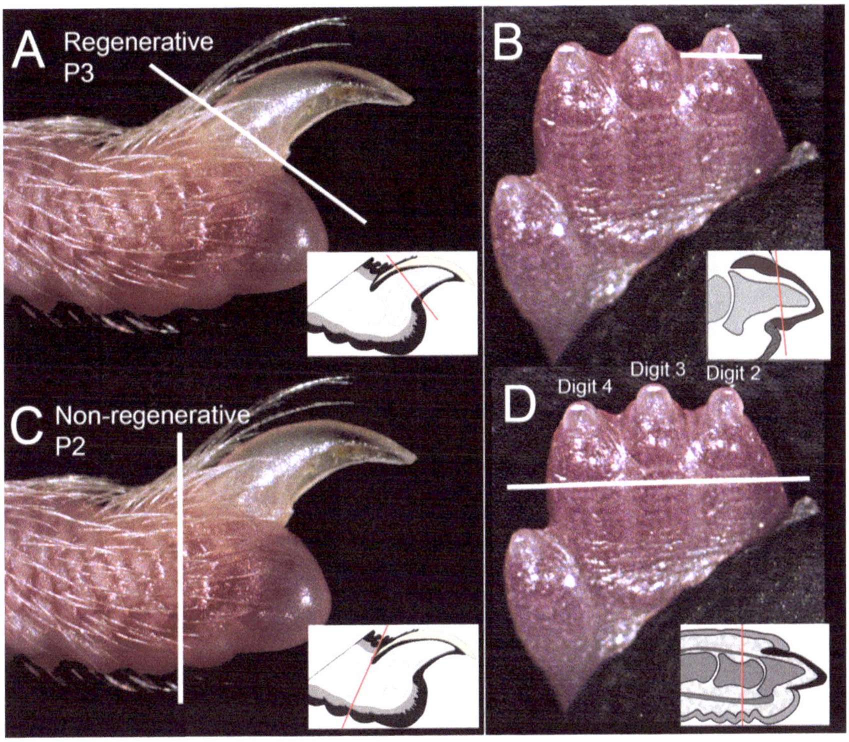

Fig. 1 Amputation planes in the adult (**a**, **c**) and neonatal digits (**b**, **d**). An amputation (*white lines*) across the distal third of the third phalangeal element (P3) will mount a regeneration response in both adults (**a**, lateral view) and neonates (**b**, ventral view). The amputation cuts through bone, connective tissue, epidermis, and nail (*inset* diagrams). An amputation through the distal third of the second phalangeal element (P2) in both adults (**c**, lateral view) and neonates (**d**, ventral view) will result in scar formation and missing structures. Note that digits 2, 3, and 4 are still fused in the neonate (**b**, **d**).

1.1 Stages of the Amputation Responses

Following amputation of the mouse digit, the various tissue responses, depending upon amputation level, can be monitored histologically by looking for several key stages (Figs. 2 and 3).

1. Inflammation—The initial response following any amputation, whether through adult or neonate P2 or P3, is similar to the wound healing response observed in full-thickness skin injuries [11]: blood vessels open, the soft tissue swells, a fibrin clot forms, the epidermis thickens at the wound site, and inflammatory cells infiltrate (Fig. 2 #1–3 and 3 #1–2).
2. Histolysis—Osteoclasts become evident at the amputated bone lining, both in the marrow and on the periosteum [12]. In the adult P3 amputation, the bone is completely severed by this degradation response at about 7–9 days post amputation (DPA), exposing the marrow to the surrounding tissue (Fig. 2 #4, 6). Degradation of the skeletal template can be seen in the neonate amputations, as well, but the response is not as easily

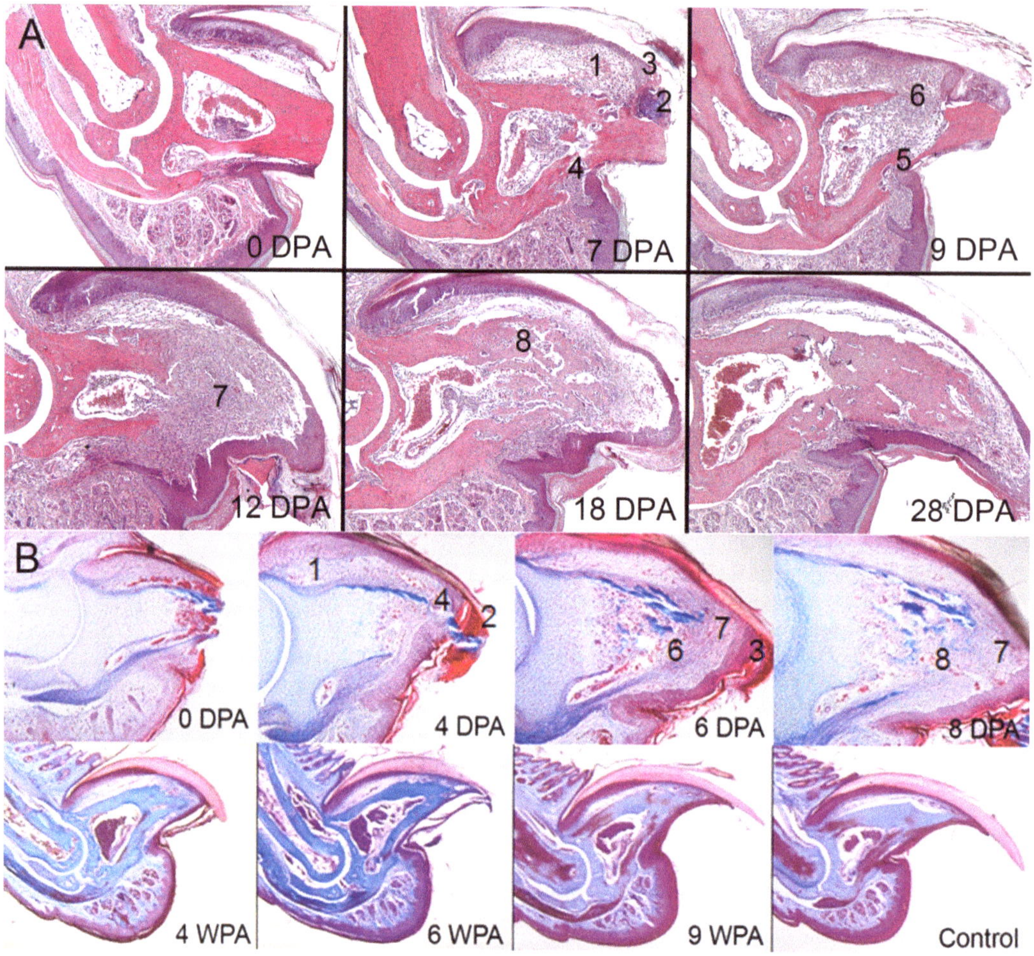

Fig. 2 Histological timeline of events during adult (**a**) and neonate (**b**) digit regeneration following a distal P3 amputation. (**a**) Adult timeline visualized by H&E stain. By 7 DPA digit shows (*1*) fragmented connective tissue surrounding bone, (*2*) a fibrin clot forming over cut area, (*3*) thickening epidermis to form wound epidermis, (*4*) bone degradation. By 9 DPA digit shows (*5*) epidermis migrating between degrading bone stump to eject bone, (*6*) bone marrow cavity open to surrounding tissue and hypercellular. At 12 DPA, (*7*) the blastema is more evident. At 18 DPA, (*8*) new trabecular bone has begun to form and at 28 DPA the digit has regained its original shape but continues to be remodeled. (**b**) Neonate timeline visualized by Mallory Trichrome stain: By 4 DPA, the digit is visibly swollen and the extracellular matrix of the connective tissue is fragmented (*1*). A fibrin clot forms over the distal tip of the amputation wound (*2*) and bone resorption is evident (*4*) just as in the adult digit. The epidermis begins to thicken at this stage as well, but can be easily visualized at 6 DPA when re-epithelialization is mostly complete (*3*). At 6 DPA, the marrow cavity, not yet completely formed in neonates, is open to the surrounding connective tissue (*6*) and a highly proliferative population of mesenchymal cells can be seen in front of the bone stump (*7*). At 8 DPA, bone begins to reform (*8*) as the blastema continues to move distally (*7*). The neonatal digit continues developing and at 4, 6, and 9 WPA the digit mimics an unamputated control digit. DPA 4, 6 and unamputated control micrographs reprinted with permissions [7]. *DPA* days post amputation, *WPA* weeks post amputation

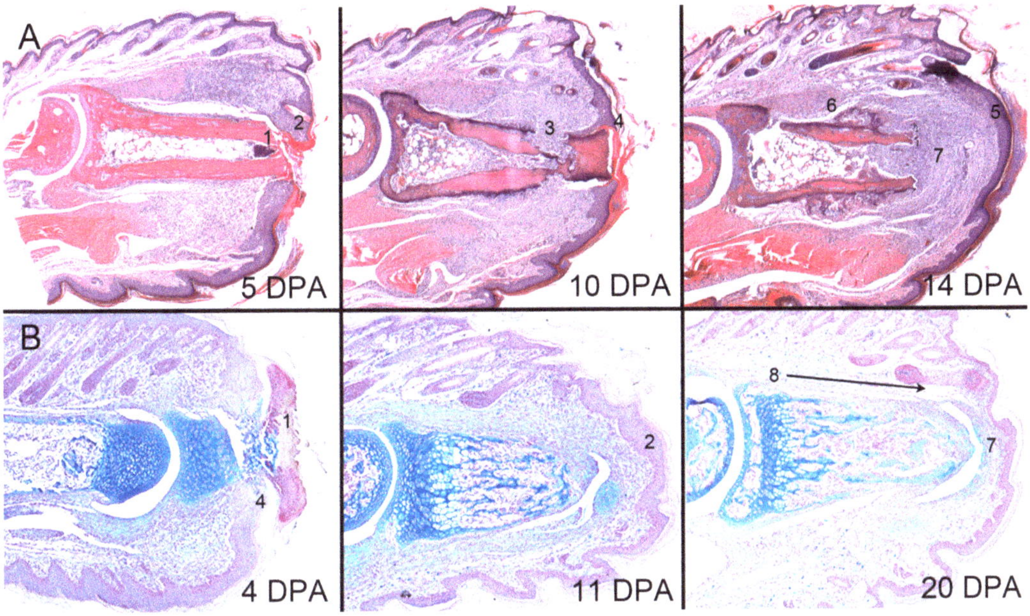

Fig. 3 Histological timeline of events following a non-regenerating P2 amputation of an adult (**a**) and neonatal (**b**) mouse. (**a**) Adult wound healing process shown by H&E. Following a P2 amputation, by as early as 2 DPA but still visible at 5 DPA, a fibrin clot caps the open bone marrow (*1*). The epidermis begins to thicken (*2*). Bone degradation peaks around 9 DPA (*3*) as the wound epidermis migrates between the cut surfaces of the bone (*4*). By 14 DPA the epidermis has completed closure of the wound (*5*), a cartilaginous callus can be seen ossifying annularly along the periosteal surface (*6*) and a dense collagen scar can be seen forming in front of the bone stump (*7*). (**b**) Neonatal wound healing process shown by Mallory's Trichrome. In the neonate, a P2 amputation results in a wound healing response characterized by tissue swelling, fibrin clot formation (*1*) and epidermal thickening (*2*). Re-epithelialization occurs between 4 and 6 DPA as the wound epidermis migrates across the amputated bone stump beneath the fibrin clot (*4*). A dense collagen scar forms at the amputated tip of the bone stump (*7*), but unlike in the adult, the neonatal bone continues to elongate (*8*) as the growth plate at the proximal end of P2 remains active

visualized (Fig. 2 #4 neonate). P2 amputations also demonstrate bone histolysis (Fig. 3 #3); however, bone degradation is not as consistent and depends upon the exact level of amputation.

3. Epidermal closure—The P2 and P3 level amputation responses diverge at this stage. Following a P3 amputation, the epidermis migrates between the degrading bone surfaces to close the wound and expel the self-severed bone (Fig. 2 #5). The wound epidermis is completely closed at about 6–8 DPA for the neonate model [7] and 9–10 DPA for the adult [12]. In contrast, following a P2 amputation, the wound epidermis migrates across the top of the bone between the fibrin clot and bone stump, completing re-epithelialization between 4 and 6 DPA in neonates, 8–13 DPA in adults (Fig. 3 #4).

4. Subsequent to the complete closure of the epidermis in a P2 amputation, fibroblastic cells at the distal tip of the bone stump deposit collagen fibrils, capping the bone in both the neonate and adult (Fig. 3 #7). In the adult, a chondrocytic callus forms annularly along P2 (Fig. 3 #6). This callus eventually ossifies creating a P2 bone that is wider and shorter in length than the original amputation level. In neonates, the bone, although capped by a fibrotic scar, continues to elongate due to the active growth plate at the proximal end of P2 (Fig. 3 #8). Neither in the adult nor in the neonate does P2 ever reach its original length or shape. This step concludes the P2 amputation response. In contrast, a P3 amputation response, in both models, continues through the next stage of the regenerative process: blastema formation.
5. Blastema formation—At 8 DPA for neonate, 12 DPA for adult, a highly proliferative, undifferentiated collection of cells can be observed at the distal end of the bone stump. In adults, this blastema is contiguous with the bone marrow and surrounding soft connective tissue of the amputated bone (Fig. 2 #7). In neonates, the bone marrow cavity is not yet delineated, but the blastema persists at the tip of the skeletal stump. This population of cells will contribute to the final stage of regeneration [7].
6. Redifferentiation—Histological staining at 8 days post amputation in the neonate, 12–14 DPA for adult, indicates the beginning of bone rebuilding [7, 12]. The newly forming bone is trabecular in appearance and highly vascularized (Fig. 2 #8). Bone continues to elongate until general morphology and patterning are restored at approximately 14 DPA in neonate, 28 DPA in adult. Following this, appositional growth and remodeling continues beyond 21 DPA for neonate, 28 DPA for adult. During this time, nail and surrounding soft connective tissue are also regrowing.

1.2 Manipulating Signaling Pathways

Studying the P3 regenerative response in transgenic models allows us to dissect pathways that are essential for regeneration and has carved out a prominent role for the BMP cascades. For example, the *Msx1* mutant embryo shows regeneration defects that can be rescued by exogenous addition of BMP4 [13]. To better study this pathway, the BMP inhibitor, NOGGIN, was found to inhibit regeneration when introduced to the P3 amputation wound using an Affi-gel agarose microcarrier bead. Alternatively, exogenous introduction of BMP2 or BMP7 to a normally non-regenerating amputation wound trigger a regeneration response [14].

Growth factor delivery using Affi-gel agarose beads with Cibacron Blue have been used for many years in embryonic development studies [15–17]. Cibacron Blue dye has been referred to as a "universal pseudoaffinity ligand" because this dye is able to bind many different water soluble proteins, enzymes, and small molecules. The Cibacron Blue molecule is composed of a combination of aromatic (nonpolar) and sulfonate (ionic) groups enabling this molecule to bind a wide variety of proteins [18].

The protein release dynamics of this bead system are quick, with adsorbed protein generally being depleted within the first 24 h of implantation. This quick release of protein allows for tight temporal control over delivery [19, 20]. The agarose beads survive tissue processing and staining, allowing easy visualization of cell interaction with the beads. One limitation of this system is not knowing exact protein concentrations that are carried by the beads and subsequently released into the animal, although we have used microcarrier beads to elicit a dose-dependent migratory response in vivo [21]. Delivering BMPs with microcarrier beads can influence proliferation and differentiation in vivo long after the protein is exhausted, thus suggesting that a single burst of activity is sufficient to dramatically change downstream signaling pathways in vivo [14].

Our lab has had success in translating these techniques to postnatal and adult mammals with the goal of manipulating signaling pathways that are important in regeneration.

2 Materials

2.1 Animals

Mice: CD#1 day-3-postnate, CD#1 8-week-old female adult (*see* **Note 1**).

2.2 Anesthesia

1. 4 °C incubation chamber: Ice, towels (Neonates).
2. 0.1 mg Ketamine/xylazine per gram mouse body weight (Adults).
3. Isoflurane.

2.3 Tools for Amputations

1. Iridectomy scissors: Vannas 7 mm straight blades (Neonates).
2. Scalpels: Miltex4-411 #10 blade (Adults).
3. Dermabond.
4. Dissection microscope.

2.4 Tools for Bead Implantations

1. Cibacron Blue cross-linked microcarrier beads: BioRad Affi-Gel Blue Gel beads, 150 μm in diameter for neonate models and 400 μm in diameter for adult models.
2. 0.01 M Phosphate-buffered saline (PBS): 0.138 M NaCl, 0.0027 M KCl pH 7.4, 0.1 % BSA.
3. Recombinant BMP2 reconstituted in PBS with 0.1 % BSA and 4 mM HCl.

4. Tungsten needle: electrolytically sharpened [22].
5. Forceps: World Precision Instruments, #5 Super Fine.

2.5 Analysis

1. High resolution micro Computed Tomography (μCT) scanner.
2. Calcein: diluted to 7.0 mg/kg mouse body weight in 2 % $NaHCO_3$, pH 7.4.
3. Alizarin Complexone (ACO): dilute to 50 mg/kg mouse body weight in 2 % $NaHCO_3$, pH 7.4.
4. Glycerol 100 %.

2.6 Tissue Collection, Fixation, Sectioning

1. 4 % Paraformaldehyde in PBS.
2. Decalcification solution: 10 % Formaldehyde, 8 % formic acid, 1 % methanol (Surgipath Decal Solution I).
3. Xylene and ethanols: 70, 95, 100 %.
4. Paraffin: 56 °C melting temperature.
5. Microtome.
6. Sta-on slide coater: Mix of gelatin and chromium potassium sulfate.
7. Slides: Superfrost Plus (Fischer), 25 × 75 × 1.0 mm glass.

2.7 Immunohistochemistry with In Situ Hybridization

1. RNAse-free PBT:PBS with 0.1 % Tween 20.
2. Proteinase K (1 μg/mL).
3. 2× Salt sodium citrate (SSC): 175.3 g NaCl, 88.2 g sodium citrate pH 4.5, dilute 1:10 with sterile H_2O.
4. Digoxigenin-labeled *Collagen X* RNA probe.
5. Probe hybridization buffer: 50 % Formamide, 5× SSC pH 4.5, 1 % SDS, 50 mg/mL tRNA, 50 mg/mL Heparin in sterile H_2O.
6. Buffer 1: Formamide, 20× SSC pH 4.5, 10 % SDS, sterile H_2O preheat to 70 °C.
7. Buffer 2: Formamide, 20× SSC pH 4.5, sterile H_2O preheat to 65 °C.
8. Tris-buffered saline solution (TBS): 0.05 M Tris–HCl, 0.138 M NaCl, 0.0027 M KCl, pH 8.0 with 0.05 % Tween 20.
9. Protein block, Serum-free solution.
10. Anti-digoxigenin antibody conjugated to horse radish peroxidase (HRP) (0.75 U/μL).
11. In situ blocking solution: 1.8 % Blocking reagent (Roche), 10 % heat inactivated sheep serum (HISS), 80 % PBT heat to 70 °C, let cool to room temp.
12. Tyramide conjugated to Alexa-Fluor 488.

13. Primary antibody: example MouseIgG1 anti-COLLAGENII.
14. Secondary antibody: example Goat anti-mouse IgG1 conjugated to Alexa-Fluor 568.
15. Antibody Diluent.
16. DAPI: 4′,6-diamidino-2-phenylindole.
17. Prolong Gold Antifade coverslip mounting medium.
18. Micro Cover Glass: 24 mm wide × 60 mm long × 0.19 mm thick.

3 Methods

3.1 Amputations

3.1.1 Neonate "Distal" P3 Amputation

1. Anesthetize the neonate by placing it in a 4 °C incubation chamber for about 1 min.
2. Hold the neonate's left hindpaw between your thumb and forefinger under a dissection microscope and apply slight pressure to fully extend the neonate's digits (Fig. 1b). Digits 2, 3, and 4 are fused below the P2–P3 joint at this stage of development.
3. Place scissors against base of the nail where skin meets nail on the ventral side of the paw (Fig. 4a).
4. Perform complete amputation without cutting through skin, so that only nail, underlying connective tissue, and developing bone are cut. Ensure cut is parallel to the base of the nail. Cut only digits 2 and 4, leaving digit 3 as an unamputated control (Fig. 1b).

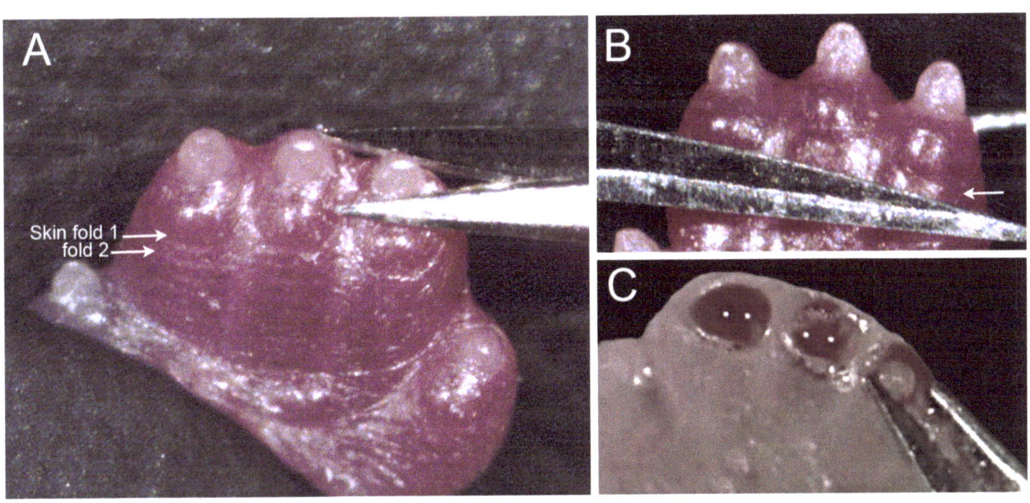

Fig. 4 Amputation procedure neonate. For a distal amputation, scissors are placed against the base of the nail (**a**). For a proximal amputation, the first cut (**b**) is made just distal to the second ventral skin fold (see *arrows* in **a**) across all three digits exposing the P2–P3 interphalangeal joint. *Arrow* in (**b**) shows the level the scissors will cut at. The second cut trims the head of the joint off P2 (**c**)

3.1.2 Neonate "Proximal" P2 Amputation

1. Anesthetize the neonate by placing it in a 4 °C incubation chamber for about 1 min.
2. Hold neonate hindpaw between your thumb and forefinger. Place slight pressure on the paw so that the digits 2, 3, and 4 are fully extended. Proximal to the P2–P3 joint the digits are still fused at this stage of development and digit 3 is slightly longer than digits 2 and 4 (Fig. 1b). Because of this, all three digits are cut at once. However, digit 3 is excluded from the experimental group because the amputation level is not comparable due to the length of digit 3.
3. Use the skin folds on the ventral side of the digits as landmarks. Perform amputation slightly distal to second skin fold (Fig. 4b).
4. The amputation exposes the distal end of P2. Wipe blood away with cotton swab to visualize the head of the P2 joint.
5. On digits 2 and 4, make a second cut to trim the distal end of the P2 bone (Fig. 4c). This will remove the head of the joint and expose the marrow cavity.

3.1.3 Adult "Distal" P3

1. To anesthetize mouse, inject 0.1 mg ketamine/xylazine per gram of mouse body weight intraperitoneally.
2. Using a dissection microscope, separate the digits of the hindpaw with forceps. Tape the paw and the lateral side of each digit to a plastic plate so that digits 2, 3, and 4 are fully extended and can be easily moved around under the microscope.
3. Place scalpel against base of the nail where the skin fold meets the nail (Fig. 1a). The cut angle should be parallel to the base line of the nail. Amputate through the digit in one motion, cleanly separating the tip of the digit from the base.
4. Perform complete amputation without cutting skin or fat pad, so that only nail, underlying connective tissue, and bone are cut and marrow is opened. Cut only digits 2 and 4 leaving digit 3 as an unamputated control.

3.1.4 Adult "Proximal" P2 Amputation

1. To anesthetize mouse, inject 0.1 mg ketamine/xylazine per gram of mouse body weight intraperitoneally.
2. Stem blood flow to the hind limb by tying a rubber band at the ankle. Tape the adult mouse paw on a plastic plate and apply slight pressure to fully extend the digits.
3. Using the ventral skin folds as landmarks, place the scalpel against the second skin fold. The cut angle should be perpendicular to the ventral line of the digit (Fig. 1c).
4. Amputate and completely bisect the digit with one motion. Cut only digits 2 and 4 leaving digit 3 as the unamputated control.
5. Apply a drop of Dermabond with a 1 mL micropipette and release the rubber band.

3.2 Bead Implants

1. In a 0.5 mL centrifuge tube, collect 40–60 beads, size 150 μm (neonate) or 400 μm (adult) in diameter and add PBS solution with 0.1 % BSA.
2. Centrifuge the tube at 10 K rpm for 3 min then wick the PBS solution out of the tube with a rolled up Kimwipe. Allow the remaining PBS solution to evaporate for 2–3 h by placing the tube on a 37 °C heating plate under a hood.
3. Add 2 μL of 0.5 μg/μL recombinant BMP2 protein in PBS with 0.1 % BSA and 4 mM HCl. Scrape beads off sides of tube into solution with a micropipette tip. Soak in protein solution for 2–3 h at 4 °C.

3.2.1 Into a Distal Amputation

4. For neonate and adult, first make a small hole in the base of the nail with the tungsten needle. The hole should be on the dorsal side of the digit (Fig. 5b).
5. Spear a bead with the tungsten needle and allow the bead to air dry for 2–3 min (Fig. 5a). Push the bead through the primed hole and let the needle sit for 2–5 s as the bead re-expands.
6. Remove needle and macroscopically check for bead position. The bead should sit distal to the amputated bone stump (Fig. 5d).

3.2.2 Into a Proximal Amputation

7. For neonates, start by removing the fibrin clot from previously amputated digits. Separate each digit with forceps. For adults, start by priming the digit with a small hole. This is done by inserting the tungsten needle directly underneath the scar tissue distal to the amputated bone stump.
8. With an electrolytically sharpened tungsten needle, spear and lift a bead out of solution (Fig. 5a). Allow the bead to air dry for 2–3 min. The bead should shrink to about one-half its original size (Fig. 5a).
9. Push the needle with the bead beneath the skin between the wound epidermis and the amputated bone stump (Fig. 5c). Hold the needle in place for 2–5 s. The bead will rehydrate and expand. This will allow the tungsten needle to be removed from the digit leaving the bead in place.
10. Remove needle and macroscopically check for bead position. The bead should sit beneath the wound epidermis distal to the bone stump (Fig. 5e). For adults, apply Dermabond to stop bleeding.

3.3 Quantification of Bone Formation

The measurement and analysis of digit regeneration focuses on bone for a couple reasons: First, bone growth and volume are easy to quantify without sacrificing the mouse and second, bones show patterns that can also be easily quantified. The following two procedures allow for bone elongation, pattern change, and volume analyses.

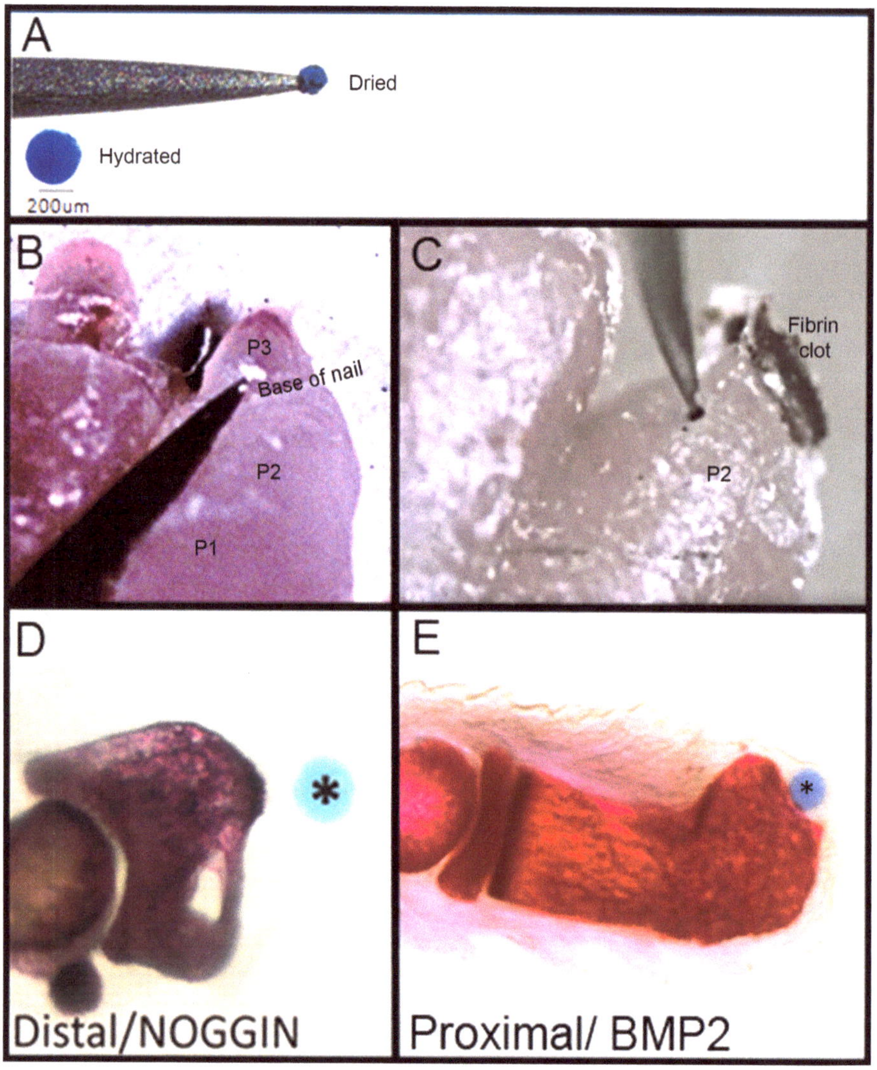

Fig. 5 Bead placement. An electrically sharpened tungsten needle can be used to pierce the *Cibacron Blue* cross-linked agarose beads. When allowed to air dry, the beads shrink to about 1/2 their original size (**a**). To implant a bead into a P3 amputation, the needle is pushed through the dorsal side of P3 at the base of the nail (**b**). To implant a bead into a P2 amputation, the needle is pushed between the wound epidermis and the P2 bone stump (**c**). The final bead placement for a P3 amputation can be visualized in (**d**, *asterisk*) and the final bead placement for a P2 amputation can be viewed in (**e**, *asterisk*). (**d**) Reprinted with permissions [14]

3.3.1 Calcein Incorporation

1. First amputate digits and implant beads.
2. Inject calcein (CCN) intraperitoneally (7.0 mg/kg of body weight) at a determined time point. The fluorochrome chelates calcium ions at the surface of newly mineralized bone. Incorporation produces a green fluorescent band along the bone at the time of injection. The width of the band is determined by how fast the tissue mineralizes and how long the fluorochrome stays in circulation [23].

3. Inject Alizarin Complexone intraperitoneally (50 mg/kg) five prior to sacrifice. This red fluorescent fluorochrome can be used in conjunction with calcein to identify all bone, both existing and newly formed [24].
4. Sacrifice the mouse and cut the digit through the mid-point of the first phalangeal element (P1). Dissect the bone from the skin and surrounding tissue in 100 % glycerol.
5. Image the sample in 100 % glycerol under a fluorescent stereoscope. New bone can be measured starting at the calcein-labeled band to the distal tip of the Alizarin Complexone stained digit (Fig. 6b).

3.3.2 μCT

1. At desired time point, induce anesthesia in the mouse with 3 % isoflurane, 1 L/min flow rate. Maintain sedation with 2 % isoflurane pumped at 1 L/min into the μCT imaging chamber.
2. Position mouse on imaging chamber stage so that ventral pad of paw is pointing toward the ceiling. Tape each digit to the stage so that digits are fully extended and immobilized.
3. Adult bone images are acquired using the following parameters: 1,000 projections per 180°, voxel resolution of 10.5 μm^3, X-ray energy at 55 kV, and intensity at 145 μA, integration time at 380 ms with continuous rotation at high resolution (*see* **Note 2** for neonatal settings).
4. DICOM images are segmented with the BoneJ "Optimize Threshold" plugin for ImageJ (version 1.45b). Detailed volume analysis can be carried out with the BoneJ "Volume Fraction" plugin for ImageJ [25], and 3D images are created with the ImageJ 3D viewer [26] (Fig 6a).

3.4 Quantification of Cellular Response Using Double Staining Detection of Protein and mRNA

Cell type and signaling pathways involved in the wound healing/regeneration process are determined using immunohistochemistry and in situ hybridization to detect specific cell markers and RNA transcripts. Digoxigenin-conjugated RNA probes are generated and used in conjunction with an anti-digoxygenin antibody conjugated to HRP. A tyramide-Alexa-Fluor conjugate is used as a substrate for HRP. A primary antibody against the cell surface protein of choice is also included. Alexa-Fluor secondary antibodies are used for detection. The result is a dual fluorescent stain for a specific RNA and a specific protein.

3.4.1 Tissue Collection, Fixation, Sectioning

1. Sacrifice the mouse at the appropriate time point and collect the digit tips by harvesting tissue just proximal to the P1–P2 interphalangeal joint.
2. Place tissue in RNAse free 4 % paraformaldehyde at 4 °C for 24–48 h. Decalcification for 6–8 h in a solution of formic acid with EDTA is required for mice older than postnatal day 10.

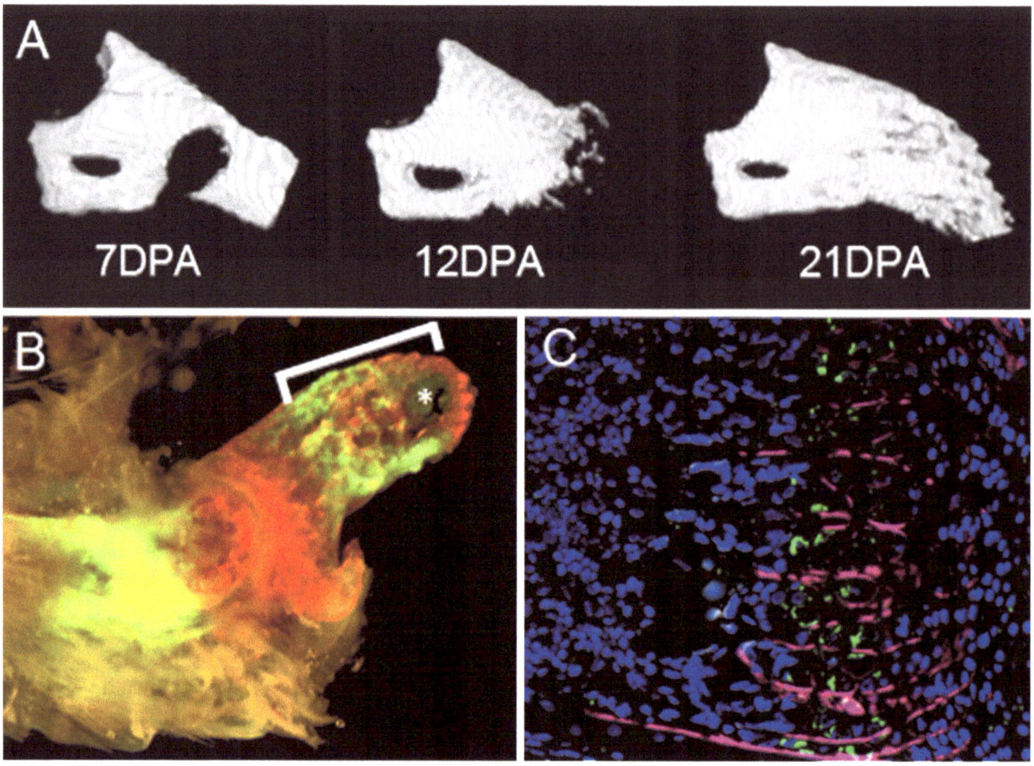

Fig. 6 Bone growth and analysis. The regeneration response can be tracked following bead implantation with μCT scans (**a**), calcein/alizarin complexone incorporation (**b**), or combined RNA in situ hybridization/immunohistochemistry (**c**). (**a**) μCT shows the bone degradation and regrowth that occurs following a P3 amputation in an adult digit. (**b**) Calcein is incorporated into newly forming bone at the time of injection. Unlabeled bone distal to this band of calcein can be identified as new bone growth and measured (*white bracket*) to quantify growth in response to the bead implant (*asterisk*). (**c**) Double staining of protein and mRNA allows for a more in depth analysis of digit regeneration. An RNA probe against *Collagen X* (*green*) combined with an antibody against COLLAGEN II (*red*) helps identify the growth plate and allows for easy tracking during the wound healing response of P2. *Blue* = DAPI

Incubate digits in the decalcifying solution on a nutator to reduce the formation of calcium crystals.

3. Wash the tissue in clean water and dehydrate using a series of graded ethanols. Clear and permeabilize the tissue in xylene, and let soak in liquid paraffin for 4–6 h.
4. Embed the digit lateral side down in paraffin. Prepare a water bath with 2 % Sta-On solution. Collect serial sagittal sections of the digits at 4–5 μm each. Let the sections float on the water bath at 42 °C for 1 min to allow the tissue to flatten and coat in Sta-On.
5. Place tissue sections on positively charged glass slides and dry horizontally at 37 °C.

3.4.2 Slide Preparation

1. Deparaffinize tissues in three washes of Xylenes and rehydrate through a series of graded ethanols ending in RNAse-free deionized water (*see* **Note 3**).
2. Rinse slides in RNAse-free PBS with 0.1 % Tween 20 (PBT).
3. Perform an antigen retrieval step by incubating the slides with 1 μg/mL Proteinase K in PBT for 10 min at room temperature.
4. Wash off Proteinase K in three changes of PBT for 5 min each.
5. Refix tissue in RNase-free 4 % PFA for 20–30 min. Follow this fixation with a wash in three changes of PBT for 5 min each.
6. Denature nucleic acids in the tissue by incubating slides in 2× SSC pH 4.5 at room temperature for 5 min.
7. Dehydrate the slides to 100 % alcohol and let the slides air dry under a hood for 1 h.

3.4.3 RNA In Situ Hybridization

1. Block nonspecific binding sites with in situ blocking solution for 1 h.
2. Perform hybridization using digoxygenin-labeled RNA probes: 1 μg/mL of probe in 100 μl of buffer. Cover slides with clean coverslips to prevent drying and let incubate in a humidified chamber overnight at 70 °C.
3. Wash off any non-bound probe in Buffer 1 at 70 °C for 30 min. Repeat three times. Wash the slides in Buffer 2 at 65 °C for 30 min. Repeat three times. Finally, wash slides in TBST pH 7.5 at room temperature for 5 min. Repeat wash three times.

3.4.4 Immunohistochemistry

1. Block nonspecific antibody binding sites by incubating the slides in a protein block solution from Dako for 1 h.
2. Add primary antibody in Antibody Diluent and incubate overnight at 4 °C (*see* **Note 4**).
3. Wash off unbound antibody by running the slides with PBT pH 7.4 for 5 min. Repeat three times.
4. Add anti-digoxigenin antibody conjugated to HRP (*see* **Note 5**) and incubate for 1 h.
5. Apply Tyramide—Alexa-Fluor 488 conjugate as substrate for HRP. Wash three times using PBT for 5 min.
6. Apply 4 μg/mL secondary antibody-Alexa-Fluor 488 conjugate in Antibody Diluent and let slides incubate for 45 min. Wash off unbound antibody with three changes of PBT for 5 min.
7. Counterstain the nuclei with DAPI diluted to 2 μg/mL in 1× PBS for 5 min. Wash briefly in dH_2O.
8. Vacuum dry and mount coverslips with ProLong Gold Antifade Reagent.

4 Notes

1. The timeline of events after an amputation will differ between mouse strains. We have found C57/Black6 mice complete the regeneration process more quickly than CD#1 mice. We have also seen differences in timing between male and female mice.
2. At postnatal day 8, the neonatal mouse has bone dense enough for detection using μCT. However, because the neonatal skeleton is less dense than the adult, lower X-ray energy is used to better capture the soft tissue. Therefore neonatal parameters for the μCT are as follows: 1,000 projections per 180°, voxel resolution of 10.5 μm^3, X-ray energy at 45 kV and intensity at 88 μA, integration time at 380 ms with continuous rotation at high resolution.
3. The high temperatures of the hybridization steps can cause bone tissue to come off the slides. To prevent this from happening, slides can be baked at 60 °C for 45 min followed by 37 °C for 48 h before deparaffinization.
4. The concentration of the primary antibody that should be added will depend on the exact antibody. For Mouse IgG1 anti-COLLAGEN II, 0.2 μg/mL is used. A secondary antibody specific for mouse IgG1 and the serum-free blocking step are crucial to minimize background staining against endogenous mouse antibodies.
5. The anti-digoxigenin antibody can be made in a protein block solution as follows: Add 3 mg of mouse embryo powder to 1 mL 2 % blocking reagent in PBT. Incubate this solution for 30 min at 70 °C, vortex the solution for 10 min, then store on ice. Add 100 μL of 10 % HISS, 3 μL of anti-digoxigenin-HRP (0.75 U/μL) and shake for 2 h at 4 °C. Centrifuge this solution at 14 K for 10 min at 4 °C. Mix the supernatant with 5.4 mL 2 % blocking reagent and 0.6 mL 10 % HISS.

References

1. Dinsmore CE (ed) (1991) A history of regeneration research: milestones in the evolution of a science. Cambridge University Press, Cambridge
2. Tanaka EM, Reddien PW (2011) The cellular basis for animal regeneration. Dev Cell 21:172–185
3. Stocum DL, Cameron JA (2011) Looking proximally and distally: 100 years of limb regeneration and beyond. Dev Dyn 240: 943–968
4. McKim LH (1932) Regeneration of the distal phalanx. Can Med Assoc J 26:549–550
5. Douglas BS (1972) Conservative management of guillotine amputation of the finger in children. Aust Paediatr J 8:86–89
6. Allen MJ (1980) Conservative management of finger tip injuries in adults. Hand 12:257–265
7. Han M, Yang X, Lee J, Allan CH, Muneoka K (2008) Development and regeneration of the neonatal digit tip in mice. Dev Biol 315: 125–135
8. Reginelli AD, Wang YQ, Sassoon D, Muneoka K (1995) Digit tip regeneration correlates with regions of Msx1 (Hox 7) expression in fetal and newborn mice. Development 121: 1065–1076
9. Neufeld DA, Zhao W (1995) Bone regrowth after digit tip amputation in mice is equivalent in adults and neonates. Wound Repair Regen 3:461–466

10. Borgens RB (1982) Mice regrow the tips of their foretoes. Science 217:747–750
11. Clark RA (ed) (1995) The molecular and cellular biology of wound repair, 2nd edn. Plenum Press, New York
12. Fernando WA, Leininger E, Simkin J, Li N, Malcom CA, Sathyamoorthi S, Han M, Muneoka K (2011) Wound healing and blastema formation in regenerating digit tips of adult mice. Dev Biol 350:301–310
13. Han M, Yang X, Farrington JE, Muneoka K (2003) Digit regeneration is regulated by Msx1 and BMP4 in fetal mice. Development 130:5123–5132
14. Yu L, Han M, Yan M, Lee EC, Lee J, Muneoka K (2010) BMP signaling induces digit regeneration in neonatal mice. Development 137: 551–559
15. Ros MA, Lopez-Martinez A, Simandl BK, Rodriguez C, Izpisua Belmonte JC, Dahn R, Fallon JF (1996) The limb field mesoderm determines initial limb bud anteroposterior asymmetry and budding independent of sonic hedgehog or apical ectodermal gene expressions. Development 122:2319–2330
16. Ganan Y, Macias D, Duterque-Coquillaud M, Ros MA, Hurle JM (1996) Role of TGF beta s and BMPs as signals controlling the position of the digits and the areas of interdigital cell death in the developing chick limb autopod. Development 122:2349–2357
17. Ferguson CM, Schwarz EM, Puzas JE, Zuscik MJ, Drissi H, O'Keefe RJ (2004) Transforming growth factor-beta1 induced alteration of skeletal morphogenesis in vivo. J Orthop Res 22:687–696
18. Subramanian S (1984) Dye-ligand affinity chromatography: the interaction of Cibacron Blue F3GA with proteins and enzymes. CRC Crit Rev Biochem 16:169–205
19. Fallon JF, Lopez A, Ros MA, Savage MP, Olwin BB, Simandl BK (1994) FGF-2: apical ectodermal ridge growth signal for chick limb development. Science 264:104–107
20. Schreiber AB, Winkler ME, Derynck R (1986) Transforming growth factor-alpha: a more potent angiogenic mediator than epidermal growth factor. Science 232:1250–1253
21. Li S, Muneoka K (1999) Cell migration and chick limb development: chemotactic action of FGF-4 and the AER. Dev Biol 211:335–347
22. Brady J (1965) A simple technique for making very fine, durable dissecting needles by sharpening tungsten wire electrolytically. Bull World Health Organ 32:143–144
23. Erben RG (2003) Bone-labeling techniques. In: An YH, Martin KL (eds) Handbook of histology methods for bone and cartilage. Humana Press Inc., Totowa, pp 99–117
24. Neufeld DA, Mohammad KS (2000) Fluorescent bone viewed through toenails of living animals: a method to observe bone regrowth. Biotech Histochem 75:259–263
25. Doube M, Klosowski MM, Arganda-Carreras I, Cordelieres FP, Dougherty RP, Jackson JS, Schmid B, Hutchinson JR, Shefelbine SJ (2010) BoneJ: free and extensible bone image analysis in ImageJ. Bone 47:1076–1079
26. Schmid B, Schindelin J, Cardona A, Longair M, Heisenberg M (2010) A high-level 3D visualization API for Java and ImageJ. BMC Bioinformatics 11:274

Chapter 25

Lower Vibrissa Follicle Amputation: A Mammalian Model of Regeneration

James M. Waters and Allison J. Cowin

Abstract

Regeneration of damaged or lost cells, tissues, and organs continues to fascinate and intrigue researchers with the lure of creating beneficial therapeutics for use in wound healing and regenerative medicine. However, unlike many other animals, wound healing in mammalian species typically proceeds via imperfect repair rather than authentic regeneration of tissues. Here, we describe a model of mammalian regeneration which can be used by researchers to investigate conditions that permit renewal of lost tissue and identify potential barriers to mammalian regeneration. The methods describe the surgical procedures for amputation of the lower third of the whisker follicle (vibrissa) in the mouse, as well as subsequent isolation and processing of the regenerating follicles for analysis.

Key words Regeneration, Wound healing, Vibrissa, Hair follicle

1 Introduction

Regeneration of damaged or lost tissues is an impressive trait of many lower order animals including vertebrates such as fish [1] and axolotls [2]. Significant insights into the process and mechanisms of regeneration across diverse species have been largely achieved using these traditional models of regeneration [3]. However, mammals are notoriously poor at regenerating tissues following injury. The predominating route of mammalian wound healing is via tissue repair, resulting in scar formation with impaired function at the site of injury. Typically, little if any regeneration of lost anatomical structures occurs. However, in specialized cases, mammals are capable of limited regeneration. Many researchers have focused on these exceptional cases attempting to elucidate the factors promoting regeneration rather than repair. Further insights into the regenerative process and significant barriers to regeneration in mammals will undoubtedly have important implications for the future of wound healing and regenerative medicine.

Robert G. Gourdie and Tereance A. Myers (eds.), *Wound Regeneration and Repair: Methods and Protocols*, Methods in Molecular Biology, vol. 1037, DOI 10.1007/978-1-62703-505-7_25,

The hair follicle, paradoxically a defining characteristic of mammals, is one such example of a mammalian organ capable of regeneration following partial amputation. Whisker follicles (vibrissae) of rats and mice as well as human hair follicles have remarkable regenerative abilities. Following microsurgical removal of up to one-third of the active lower follicle, regrowth of all amputated components essential for normal hair production occurs [4–8]. The regeneration of amputated hair follicles proceeds through a number of sequential but overlapping phases, which can be categorized into four stages: (1) the acute wound response, (2) cellular and extracellular reorganization and remodeling, (3) neogenesis of new hair follicle structures, and (4) differentiation and production of the new hair fiber. This model of regeneration encompasses multiple processes from the hyperproliferative acute wound response to cellular and extracellular remodeling, neogenesis of lost structures that mimics embryological development, and finally normal differentiation and cycling of the regenerated follicle, making it a comprehensive model of wound healing and regeneration.

This chapter aims to outline the materials and methods for performing lower follicle amputations and subsequent isolation of partial or fully regenerated follicles for further analysis. The methods summarized in this chapter specifically relate to amputation of lower vibrissae in mice. However, these protocols can be simply adapted for other rodent species including rats. We have described the process for mice due to the many advantages of working with mice, including the availability of numerous transgenic mouse models which may be utilized to further elucidate the processes of regeneration.

Mice of comparable age, weight, sex, and genetic background should be used when feasible to minimize confounding factors and maximize reproducibility of independent experiments. The amputation protocol outlined in this chapter begins with incisions being made with fine scissors posterior to the whisker pads of anesthetized mice, creating a free flap which is then reflected to reveal the deep structures of the vibrissae. The connective tissue is then carefully removed from around the end bulbs of vibrissae using fine forceps, and the lower third of the follicle containing essential hair-forming structures is transected using microscissors. Hair fibers are then plucked from the follicle, the whisker pad is sutured back into place, and the mice are allowed to recover in normal housing conditions with general health monitoring. Mice typically cope well following the procedure with normal feeding and habits quickly resumed.

Mice can be killed at desired time points to coincide with the chosen stage of the regeneration process. We have described the protocol for isolating experimental follicles and embedding them for histological and immunological analysis. However, other methods for analyzing the regeneration process could be utilized, including immunoblotting or various gene expression techniques.

Even though lower follicle amputation has been around for more than four decades, it has been underutilized as a model for studying mammalian regeneration. Currently, only one study has utilized transgenic animals to assess the influence of modified gene expression on regeneration of vibrissae [9]. That study looked at the influence of Flightless I (FLII) on regeneration and found that in contrast to its negative influence on wound healing [10], it positively regulated regeneration. Our hope is that more researchers will use these methods to investigate mammalian regeneration and increase the body of knowledge available to current and future researchers.

2 Materials

Appropriate care and adequate personal protective equipment should be used when preparing, using, and disposing of all materials and solutions. Disposal of all waste materials should follow relevant regulations and guidelines. Many of the materials required for amputation of vibrissae end bulb, the subsequent isolation of experimental follicles, and embedding of experimental follicles are shown in Fig. 1.

2.1 Amputation of Vibrissae End Bulbs

1. Anesthetic agents: Combination of medetomidine (Dormitor®, stock concentration of 1 mg/mL) and ketamine (stock concentration of 100 mg/mL) to be given at a final dose of 1 and 75 μg/g body weight, respectively. Immediately prior to use, add 200 μL of Dormitor® and 150 μL of ketamine to 650 μL of sterile saline solution (0.9 % w/v), and mix carefully by inversion. This 1 mL solution will suffice for six 30 g mice.
2. Single-use syringes with needles.
3. Iodine solution (Riodine™), Povidone-Iodine solution (10 % w/v) equivalent to 1 % available iodine.
4. Dissecting stereomicroscope.
5. Light source.
6. Electric heat pad.
7. Disposable protective surgical mat.
8. Surgical instruments:
   (a) Forceps.
   (b) Iris scissors.
   (c) Hemostat clamp.
   (d) Microforceps (pattern #4 and/or #5).
   (e) Microdissecting scissors.
9. Sterile saline solution for injection (0.9 % w/v).

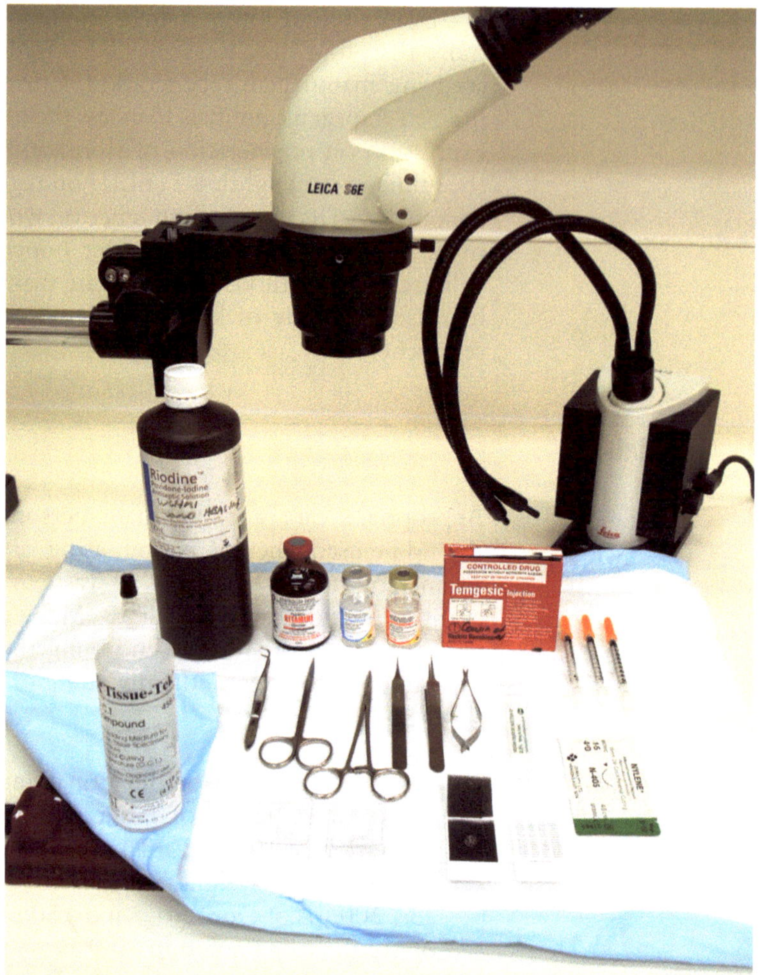

Fig. 1 Instruments and equipment required for amputation of follicle end bulbs, and subsequent isolation and embedding of experimental follicles. From *left* to *right* (*Front row*): Electric heat pad, disposable protective surgical mat, Tissue-Tek® O.C.T. embedding compound, Tissue-Tek® cryomolds (small), tissue biopsy cassettes (small) with foam inserts (shown with a drop of saline on the foam insert). (*Second row*): Forceps, iris scissors, hemostat clamp, microforceps (pattern #4 and #5), microdissecting scissors, sterile saline solution for injection (0.9 % w/v), sutures. (*Third row*): Iodine solution, ketamine, Dormitor®, Antisedan®, Temgesic®, single-use syringes with needles. (*Back row*): Dissecting stereomicroscope, light source

10. Sutures.
11. Anesthetic reversal agent: Atipamezole (Antisedan®, stock concentration of 5 mg/mL) to be given at a final dose of 1 μg/g body weight. Immediately prior to use, add 50 μL of Antisedan® to 950 μL of sterile saline solution (0.9 % w/v),

and mix carefully by inversion. This 1 mL solution will suffice for six 30 g mice.

12. Analgesic: Buprenorphine (Temgesic®, stock concentration of 300 μg/mL) to be given at a final dose of 0.05 μg/g body weight. Immediately prior to use, add 42 μL of Temgesic® to 958 μL of sterile saline solution (0.9 % w/v), and mix carefully by inversion. This 1 mL solution will suffice for six 30 g mice.

2.2 Isolation of Treated Follicles

1. Carbon dioxide.
2. Ethanol (70 % v/v).
3. Dissecting stereomicroscope.
4. Light source.
5. Surgical instruments:
   (a) Forceps.
   (b) Iris scissors.
   (c) Hemostat clamp.
   (d) Microforceps (pattern #4 and/or #5).
6. Sterile saline solution (0.9 % w/v).

2.3 Embedding Treated Follicles for Histology and Immunological Analysis

Paraffin embedding

1. Tissue biopsy cassettes, small, with foam inserts.
2. Sterile saline solution (0.9 % w/v).
3. 10 % buffered formalin.
4. Dehydration solvents (ethanol—70, 90, 95, and 100 %).
5. Clearing agent (xylene).
6. Paraffin wax (molten).
7. Wax embedding mold.
8. Dissecting stereomicroscope.
9. Light source.
10. Microforceps.

Cryogenic embedding

1. Tissue Tek® cryomold®, small, disposable specimen molds.
2. Embedding medium for frozen tissue specimens, Tissue Tek® O.C.T. compound.
3. Dissecting stereomicroscope.
4. Light source.
5. Microforceps.
6. Liquid nitrogen, dry ice, or –80 °C freezer.

3 Methods

All procedures to be performed involving animals must first be approved by relevant animal welfare, ethics, and biosafety regulatory bodies.

3.1 Amputation of Vibrissae End Bulbs

1. Weigh and check general health of all mice.
2. Administer the appropriate dose (5 μL per gram body weight of pre-diluted working solution; *see* Subheading 2.1, **item 1**) of anesthetic by intraperitoneal injection. Return animal to cage in isolation while the anesthetic takes effect.
3. Once the mouse is fully anesthetized, place on its side on the heated pad with a single whisker pad in the dissecting microscope field of view (*see* **Note 1**).
4. Using the dissecting microscope, locate the two or the three most posterior columns of follicles, then identify, and record follicles which are in the anagen phase of the hair cycle (*see* **Notes 2** and **3**, Fig. 2a, b).
5. Swab the whisker pad with iodine solution taking care to avoid delicate tissues including the eye and nasal cavities, remove excess, and allow them to dry (*see* **Note 4**).
6. Using the iris scissors, make an incision parallel but posterior to the most posterior column of follicles (column A and Fig. 2c). Then slightly extend the opening at right angles to the first incision, ventral to the column A follicles.
7. Using the hemostat clamp reflect the whisker pad to expose the deepest surface (*see* **Note 5**, Fig. 2d).
8. While the whisker pad is reflected, keep the internal tissues hydrated by occasional irrigation with saline.
9. Carefully loosen the connective tissue surrounding each follicle of interest using the pointed tips of the fine forceps (No. 4 and No. 5) to expose the end bulb and lower regions of the follicles (*see* **Note 6**, Fig. 2e).
10. Amputate the lower section of the follicles which includes the end bulb. Make the transverse cut using microscissors below the site of innervation, leaving the nerve intact (*see* **Notes 7** and **8**, Fig. 2f, i–k).
11. Once the lower portions of all target follicles have been amputated, return the whisker pad flap to its original position. Pluck both the growing and club fibers of amputated follicles (*see* **Note 9**, Fig. 2g, k).
12. Suture the whisker pad flap back into place, taking care to avoid placing the suture needle too close to the follicles (Fig. 2h).

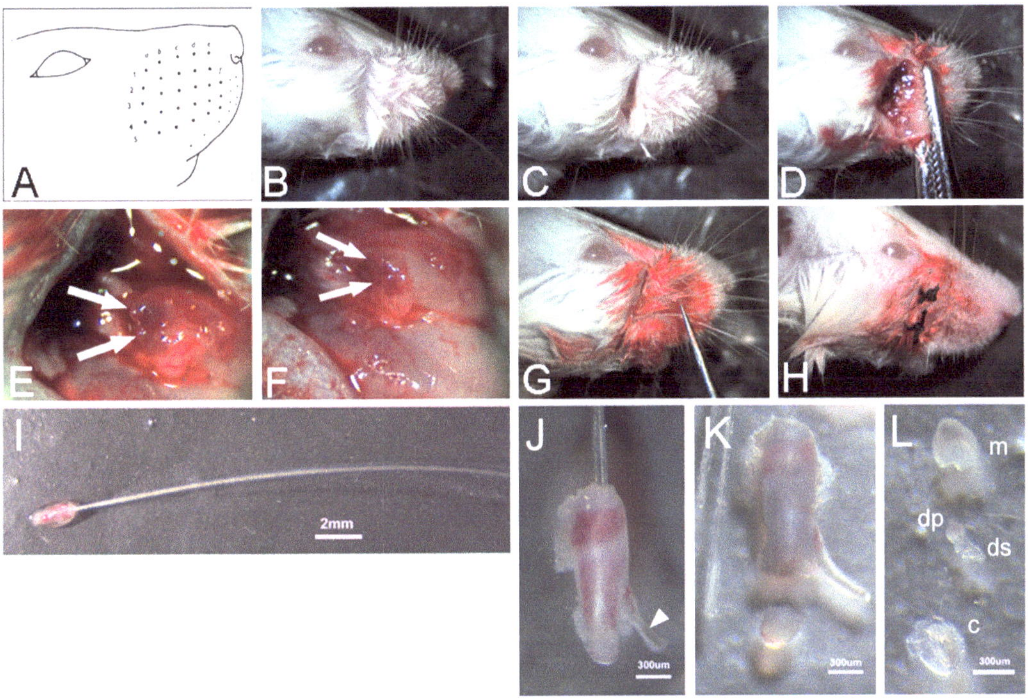

Fig. 2 Procedure for amputation of follicle end bulbs. (**a** and **b**) Appropriate follicles are identified and their coordinate recorded. (**c**) An incision is made posterior to the whisker pad, parallel to the first column of follicles. (**d**) The whisker pad is reflected to reveal follicle end bulbs (**e**, *arrows*). (**f**) End bulbs are amputated below the nerve using microscissors. (**g**) The whisker pad is replaced and the hair fibers plucked from wounded follicles, before closing the surgical wound (**h**). (**i** and **j**) An isolated mid-anagen mouse vibrissa follicle and nerve (*arrowhead*) are shown to clearly demonstrate the correct site for amputation. (**k**) Amputated end bulb and plucked fibers. (**l**) Amputated end bulb components include epithelial matrix (m), dermal papilla (dp), and attached dermal sheath (ds), and the collagen capsule (c)

13. Administer the appropriate dose (4 μL per gram body weight of pre-diluted working solution; *see* Subheading 2.1, **item 11**) of anesthetic reversal agent and place the animal in a suitable environment to recover.

    Appropriate doses: 4 μL per gram body weight of pre-diluted working solution (Subheading 2.1, **item 12**) of analgesic should be administered immediately post surgery and subsequently if required.

3.2 Isolation of Treated Follicles

The reconstitution of lower follicles is a complex progression of multiple overlapping phases of regeneration. A guide to the stages of regeneration has been previously described in rat vibrissae [7], which in the authors experience correlate well with mice vibrissae. Briefly, the four phases and approximate timing for the regeneration process consist of the following: stage 1—an acute wound healing response lasting up to 6 days post amputation, stage 2—remodelling of cellular and extracellular components of the lower follicle (days 7–10 post amputation), stage 3—development of new follicle structures including the dermal papilla (days 11–16 post amputation), and stage 4—differentiation and growth of the new hair fiber (from day 17 onwards). Examples of histological, longitudinal sections of regenerating follicles in each of these stages are shown in Fig. 3. These time points should be used as a guide only, as variation in timing of the stages of regeneration is expected.

1. Prepare and label the tissue biopsy cassettes and/or cryomolds prior to follicle isolation (*see* **Notes 12** and/or **14**) to avoid delays once the follicles have been isolated.
2. Kill mice painlessly by carbon dioxide asphyxiation followed by cervical dislocation.
3. Using the dissecting microscope, locate the follicles which were previously amputated and record the presence or the absence of new hair fiber(s) visible externally (*see* **Note 10**).
4. Swab the whisker pad with 70 % ethanol, remove excess, and allow to dry.
5. Using the iris scissors, make an incision parallel but posterior to the original incision. Then slightly extend the opening at right angles to the first incision, ventral to the original incision (*see* **Note 11**).
6. Using the hemostat clamp reflect the whisker pad to expose the deepest surface.
7. While the whisker pad is reflected, keep the internal tissues hydrated by occasional irrigation with saline.
8. Carefully loosen the connective tissue surrounding a follicle of interest using fine forceps (No. 4 and No. 5) to isolate and

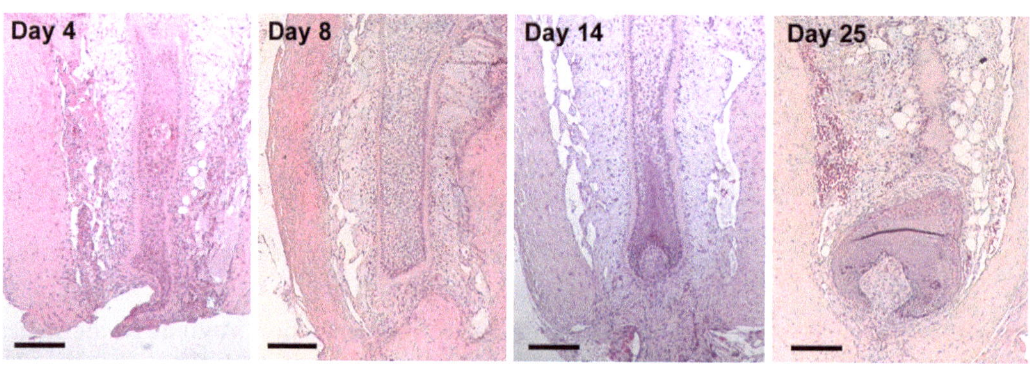

Fig. 3 Stages of vibrissa regeneration. Stage 1 (day 4)—acute wound healing response with epithelial outgrowth below the site of amputation and downward movement of dermal sheath cells. Stage 2 (day 8)—remodelling of cellular and extracellular components of the lower follicle with regression of the epithelial outgrowth. Stage 3 (day 14)—development of new follicle structures including the dermal papilla. Stage 4 (day 25)—differentiation and growth of the new hair fiber. Scale bars = 50 μm

expose the entire follicle, including the site of attachment to the interfollicular skin (infundibulum).

9. Remove the entire follicle by using iris scissors to cut around the circumference of the follicle opening. Alternatively, use the fine forceps (No. 4) to pinch the most superficial region of the follicle and remove by plucking the tissue from the surrounding skin. Although quicker and easier, plucking whole follicles in this way may result in damage to the infundibulum and nearby structures.
10. Proceed without delay to embedding treated follicles.

    Repeat **steps 7–10** for all target follicles in the whisker pad.

3.3 Embedding Treated Follicles for Histology and Immunological Analysis

The analysis of regenerating and regenerated follicles will be guided by the interests and preferences of individual researchers. Specific details regarding the most favorable conditions for sectioning, post-sectioning fixation, antigen retrieval, antibody concentrations, etc. should be determined by end users. Below are some general methods for fixation, orienting, and embedding of follicles in either paraffin or cryogenic material.

Paraffin embedding

1. Place isolated follicles in tissue processing cassettes lined with foam inserts (*see* **Note 12**).
2. Fix tissue in 10 % buffered formalin solution for up to 24 h.
3. After fixation, transfer cassettes to 70 % ethanol. Cassettes can be stored temporarily at this point or processed immediately.
4. Proceed with tissue processing by dehydrating the tissue by serial incubation in increasing concentrations of ethanol, followed by clearing agent (xylene), and then infiltration with molten paraffin (*see* **Note 13**).
5. Using the dissecting microscope and a heated work surface, position the follicle in the embedding mold using fine forceps in an appropriate orientation for longitudinal sectioning. If orientation for transverse sectioning is desired, transfer the mold to a non-heated surface until the paraffin begins to solidify, and then place the follicle on its end into the solidifying paraffin using fine forceps until it is retained without assistance.
6. Carefully transfer to a refrigerated work surface to solidify the base of the paraffin block containing the follicle and quickly place the identifying component of the plastic biopsy cassette into the mold before the paraffin has completely solidified.
7. Tissue samples embedding in paraffin can be stored indefinitely at room temperature.

Cryogenic embedding

1. Add a small amount of Tissue-Tek® O.C.T. compound to the Tissue-Tek® cryomold (*see* **Note 14**).
2. Place the freshly isolated follicle in the center of the cryomold.
3. Fill the cryomold with Tissue-Tek® O.C.T. compound.
4. Using the dissecting microscope and fine forceps, position the follicle in the cryomold in an appropriate orientation for sectioning.
5. Carefully transfer the cryomold to the surface of a small amount of liquid nitrogen or a piece of dry ice to snap freeze the O.C.T. embedded sample (*see* **Note 15**).
6. Store frozen samples at −80 °C until required for cryosectioning.

4 Notes

1. Only one whisker pad per animal should be used. This places a reduced burden on each animal and allows the intact whisker pad to be used as an untreated control. Right-handed operators may find that using the left whisker pad allows for efficient retraction using their left hand while their dominant hand remains available for isolation and amputation of follicle end bulbs.
2. A convenient identification system for whisker follicles has been described by Oliver [4], which labels columns of follicles running dorsoventrally alphabetically, starting with the most posterior (column A) and continuing anteriorly. It also labels the rows of follicles running along the antero-posterior axis numerically, starting with the most dorsal (row 1) and continuing ventrally (Fig. 2a).
3. Anagen follicles can be easily identified by the presence of two fibers emerging from a single follicle. The longer of the two is the terminal or club fiber, and the shorter fiber is the actively growing (anagen) fiber. Mid-anagen follicles, which have the growing fiber between one- and two-thirds of the club fiber length, should be chosen for amputation to standardize experiments.
4. Do not substitute 70 % ethanol for iodine solution at this point as accidental exposure of the delicate tissues including those in the nasal cavity can cause serious irritation and even breathing difficulties for the animal.
5. If effective retraction is not achieved at this point you can slightly extend the initial incision to allow increased exposure

of the follicles. However, care should be taken to keep the surgical wound to the minimum required.

6. Mechanical manipulation of the external fibers with forceps will result in visible mobility of the internal follicle structure and can be used to confirm the position of target follicles.
7. Amputating the follicle above the site of innervation will *not* result in regeneration of the lower follicle structures. A simple wound repair process will occur and no new hair fibers will result. This type of non-regenerating amputation can be used as a control to contrast with regenerating follicles.
8. Amputated lower follicles can be fixed and sectioned for histological confirmation of complete amputation of all end bulb structures including the apex of the dermal papilla.
9. Fibers should slide easily out of the follicle with little or no resistance when plucked. If resistance is felt when plucking, check that the correct follicle has been amputated by manipulation of the fibers (*see* **Note 6**).
10. Regenerated fibers are unlikely to be visible externally within the first 3 weeks following amputation.
11. Larger incisions than those used during the amputation surgery are possible at this point to help with subsequent reflection of the flap and to avoid excessive stretching of the skin and follicles.
12. Placing a droplet of saline on the foam inserts prior to follicle isolation can help the placement of the follicle into the cassette and prevent desiccation of the tissue. The foam inserts are essential for preventing small tissues from exiting the cassette during processing.
13. Automatic tissue processors can be programmed using the following schedule as a guide: one-hour incubations in 70 % ethanol twice, 90 % ethanol, 95 % ethanol, 100 % ethanol twice, xylene twice, and then finally in molten paraffin twice. Do not keep the tissue in molten paraffin for too long as the tissues will become hard and brittle when sectioning. Other schedules may work equally well but should be evaluated by end users.
14. Care should be taken to avoid introducing air bubbles to the cryomold when dispensing O.C.T. compound as this can lead to cracking of the sample upon freezing and cause difficulties when cryosectioning.
15. A stable, flat surface within a −80 °C freezer is a reasonable alternative to liquid nitrogen or dry ice as the small size of individual samples means that they will still freeze reasonably quickly.

References

1. Whitehead GG, Makino S, Lien CL, Keating MT (2005) fgf20 is essential for initiating zebrafish fin regeneration. Science 310:1957–1960
2. Kragl M, Knapp D, Nacu E, Khattak S, Maden M, Epperlein HH, Tanaka EM (2009) Cells keep a memory of their tissue origin during axolotl limb regeneration. Nature 460:60–U69
3. Alvarado AS, Tsonis PA (2006) Bridging the regeneration gap: genetic insights from diverse animal models. Nat Rev Genet 7:873–884
4. Oliver RF (1966) Histological studies of whisker regeneration in hooded rat. J Embryol Exp Morphol 16:231–244
5. Oliver RF (1966) Whisker growth after removal of dermal papilla and lengths of follicle in hooded rat. J Embryol Exp Morphol 15: 331–347
6. Ibrahim L, Wright EA (1982) A quantitative study of hair-growth using mouse and rat vibrissal follicles. 1. Dermal papilla volume determines hair volume. J Embryol Exp Morphol 72:209–224
7. Jahoda CAB, Horne KA, Mauger A, Bard S, Sengel P (1992) Cellular and extracellular involvement in the regeneration of the rat lower vibrissa follicle. Development 114:887–897
8. Jahoda CAB, Oliver RF, Reynolds AJ, Forrester JC, Horne KA (1996) Human hair follicle regeneration following amputation and grafting into the nude mouse. J Invest Dermatol 107:804–807
9. Waters JM, Lindo JE, Arkell RM, Cowin AJ (2011) Regeneration of hair follicles is modulated by flightless I (Flii) in a rodent vibrissa model. J Invest Dermatol 131:838–847
10. Cowin AJ, Adams DH, Strudwick XL, Chan H, Hooper JA, Sander GR, Rayner TE, Matthaei KI, Powell BC, Campbell HD (2007) Flightless I deficiency enhances wound repair by increasing cell migration and proliferation. J Pathol 211:572–581

Chapter 26

Using *Drosophila* Larvae to Study Epidermal Wound Closure and Inflammation

Sirisha Burra, Yan Wang, Amanda R. Brock, and Michael J. Galko

Abstract

This methods chapter describes two methods for creating epithelial wounds in *Drosophila* larvae: pinch and puncture wounding. It also covers protocols for visualizing epithelial wounds, either in a dissected whole mount preparation or, using transgenic reporter larvae, in a live whole mount preparation. Finally, useful transgenic lines for live genetic screening of genes required for wound closure or inflammation are described.

Key words Drosophila, Genetic screening, Wound healing, Inflammation, Transgenic reporters

1 Introduction

Traditionally, wound healing has been studied using clinical patient samples or vertebrate models where the tissue architecture of the skin and dermis resembles more closely that of humans [1]. Further, these models have other substantial advantages, including similar phases of repair and diversity of wound-responsive cell types (especially in the immune system that mediates the wound-induced inflammatory response). Despite these advantages, rapid and unbiased discovery of new wound closure genes remains a challenge in vertebrate systems due to the cost and time involved in performing large-scale genetic screens with complicated physiological readouts [2].

Due to its substantial advantages for genetic analysis [3], in recent years a number of groups have begun to use the fruit fly, *Drosophila melanogaster*, to model various phases of both embryonic [4, 5] and postembryonic wound healing [6–9]. Here we focus on the use of *Drosophila* larvae to genetically dissect wound-induced epidermal cell migration and blood cell recruitment. Although the wormlike larvae are a provisional stage of the *Drosophila* life cycle (sandwiched between embryogenesis and the metamorphic events

Robert G. Gourdie and Tereance A. Myers (eds.), *Wound Regeneration and Repair: Methods and Protocols*, Methods in Molecular Biology, vol. 1037, DOI 10.1007/978-1-62703-505-7_26, © Springer Science+Business Media New York 2013

that shape the adult fly) they contain fully differentiated epidermal and innate immune cells and should thus be regarded as a useful model of postembryonic wound repair.

The *Drosophila* larval epidermis is composed of a monolayer of terminally differentiated epidermal cells that are functionally equivalent to vertebrate keratinocytes in that they form the barrier epidermis and secrete a cuticle that is molecularly distinct from, but functionally similar to, the stratum corneum of vertebrate skin. These epidermal cells are separated from the underlying hemolymph (blood) of the larval open circulatory system by a basal lamina. This hemolymph contains three major types of blood cells that comprise a robust innate immune system [10]. Each blood cell type is wound responsive [6, 7, 11, 12]. Although the tissue architecture of fly larvae is substantially different from vertebrates (there is no dermis, the epidermis is a non-proliferative monolayer, and there are fewer blood cell types and no lymphocyte-mediated adaptive immunity) the *Drosophila* system trades a somewhat less familiar set of wound-responsive cells for power and ease of genetic dissection. Furthermore, the migration of epidermal cells to close the wound gap bears significant similarities to the same process observed during re-epithelialization in vertebrates.

Here, we describe two methods for creating physical wounds to the epidermis of *Drosophila* larvae: pinch and puncture wounding (for schematic of wounding and subsequent visualization, *see* Fig. 1a). Pinch wounding creates a gap in the epidermal sheet but leaves the overlying cuticle barrier intact, thus resulting in a sterile wound. These wounds provoke a robust inflammatory response and heal over a 24-h period through a process of directed cell migration (Fig. 1b). Puncture wounding, by contrast, breaches both the cuticle and epidermis. Although the puncture wounds are smaller and heal faster (~8 h), they are not sterile and result in a robust coagulation response and epidermal cell–cell fusion (Fig. 1c) at the puncture site. In practice, pinch wounds are substantially more useful for studying both epidermal cell migration and blood cell recruitment since there is no melanized scab to obscure one's view of the epidermis. In addition to these wounding procedures we also describe genetic tools and mounting techniques appropriate for visualizing epidermal tissue damage and recruitment of circulating blood cells in both live animals and in dissected, fixed, and immunostained whole mounts. *See* Table 1 for summary of reporters and antibodies useful for visualizing epidermal wounds in *Drosophila* larvae.

The advantage of immunostaining is that it affords a clearer visualization of the epidermal morphology and is applicable to virtually any mutant genotype. Immunostaining, however, is a fairly tedious method involving manual dissection of the larval epidermis and a multiday staining procedure. Therefore, it would be difficult to employ this technique for high- or medium-throughput

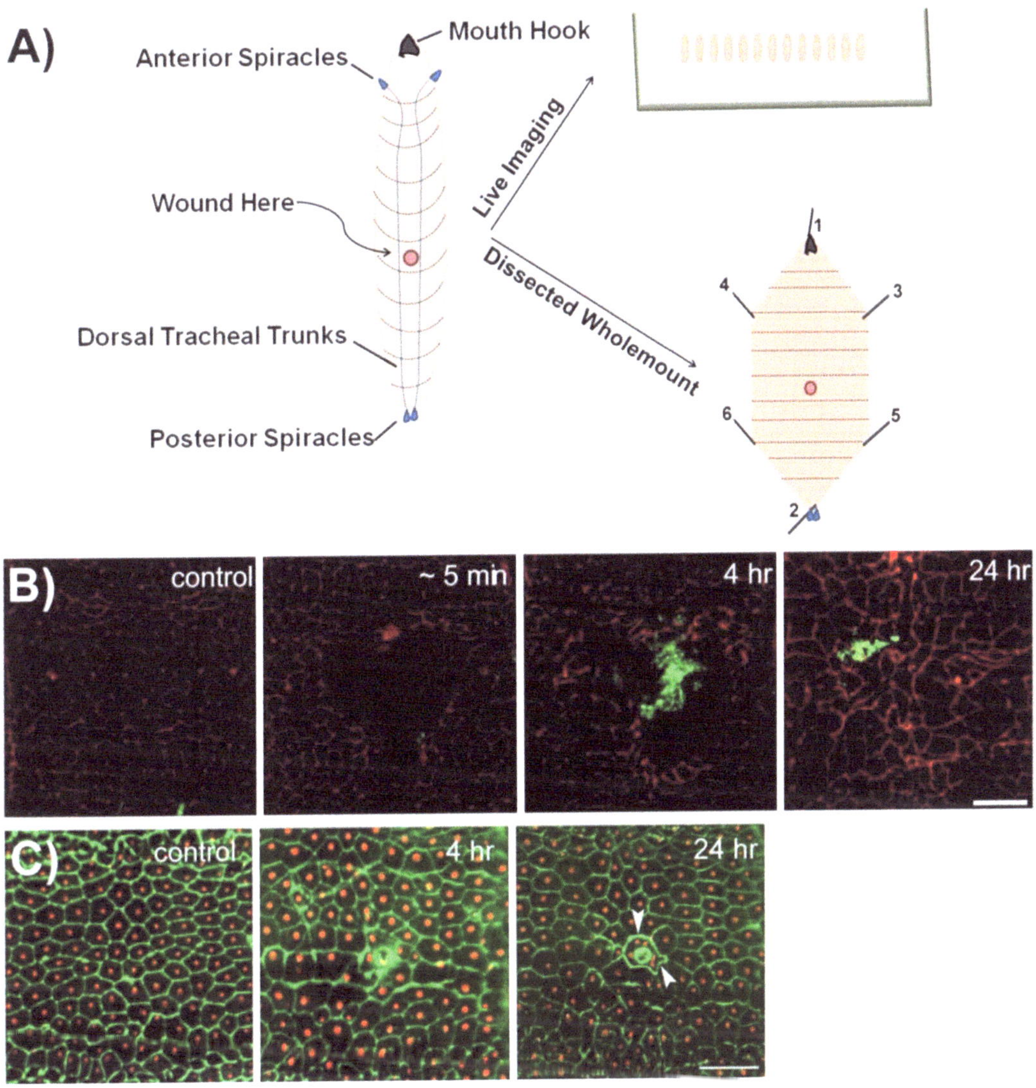

Fig. 1 Two methods for epidermal wounding of *Drosophila* larvae. (**a**) Schematic representation of wounding and mounting of *Drosophila* larvae. The left illustration depicts the dorsal side of the *Drosophila* larvae, where the dorsal tracheal trunks connecting the anterior and posterior spiracles are clearly visible. Typically, the A4 segment of the *Drosophila* larva is chosen for wounding. The epidermal wounds can be visualized either by live imaging (*upper right*) when reporter lines are used or by dissecting the larval epidermis and immunostaining for appropriate markers (*lower right*). Numbers on dissected larva indicate the order in which pins should be inserted during dissections (*see* Subheading 3.6). (**b**) Dissected epidermal whole mounts of control or pinch-wounded larvae stained for Fasciclin III (*red*) at the indicated times after wounding. Larvae are of genotype *w;;Pxn-Gal4, UAS-GFP/+*. The GFP-expressing hemocytes are stained with anti-GFP (*green*) antibody and the rapid accumulation of hemocytes at the wound site can be seen. At 24 h after wounding, the wound gap is completely closed as observed through Fasciclin III staining. Scale bar 100 μm (adapted from Babcock et al. [17], Copyright © 2008 by The National Academy of Sciences of the USA). (**c**) Dissected epidermal whole mounts of control or puncture-wounded *w; e22c-Gal4, UAS DsRed2-Nuc/+* larvae stained with anti-Fasciclin III (*green*). Scab formation and epidermal cell–cell fusion around the scab (*arrowhead*) can be seen. Scale bar 100 μm

Table 1
Transgenic lines and antibodies used to label blood cells and epidermis in Drosophila larvae

Tissue of interest	Antibodies	Direct labels	*Gal4/UAS* labels
Blood cells (plasmatocytes)	α-Pl[a]		*Pxn-Gal4>GFP*[e] *HmlΔ-Gal4>GFP*[f]
Epidermis	α-Fas III[b]	*Fas III-GFP*[c] *Nrg-GFP*[d]	*A58-Reporter*[g] *e22c-Reporter*[g]
Both			*Nrg-GFP* *Pxn-Gal4,* *UAS-GFP*[h]

[a]Kurucz E et al., Nimrod a putative phagocytosis receptor with EGF repeats in *Drosophila* plasmatocytes. Curr Biol. 2007 Apr 3;17(7):649–54
[b]Patel NH et al., Characterization and cloning of fasciclin III: a glycoprotein expressed on a subset of neurons and axon pathways in *Drosophila*. Cell. 1987 Mar 27;48(6):975–88
[c]Hudson AM et al., Mononuclear muscle cells in *Drosophila* ovaries revealed by GFP protein traps. Dev Biol. 2008 Feb 15;314(2):329–40
[d]Morin X et al., A protein trap strategy to detect GFP-tagged proteins expressed from their endogenousloci in Drosophila. Proc Natl Acad Sci U S A. 2001 Dec 18;98(26): 15050–5
[e]Stramer B et al., Live imaging of wound inflammation in Drosophila embryos reveals key roles for small GTPases during in vivo cell migration. J Cell Biol. 2005 Feb 14;168(4):567–73
[f]Chew SK et al., The apical caspase drone govems programmed and unprogrammed cell death in Drosophila. Dev Cell. 2004 Dec; 7(6):897–907
[g]Lesch C et al., A targeted UAS-RNAi screen in Drosophila larvae identifies wound closure genes regulating distinct cellular processes. Genetics. 2010 Nov;186(3):943–57
[h]Babcock DT et al., Circulating blood cells function as a surveillance system for damaged tissue in Drosophila larvae. Proc Natl Acad Sci U S A. 2008 Jul 22;105(29):10017–22

genetic screening. An alternative approach is to use wound reporter lines to accelerate and simplify wound visualization. These reporters mostly employ the *Gal4/UAS* system for tissue-specific ectopic transgene expression [13] (Fig. 2a) to express membrane- and nuclear-localized fluorescent proteins in either the epidermis or the blood cells (Table 1) and can be used to quickly visualize wound morphology in live larvae (Fig. 2b, c). Due to the rapidity of visualization and the fact that they can be combined with near-whole-genome screening technologies [14, 15], the wound reporter lines can be used for medium-throughput genetic screening. The trade-off is that the visualization of wound morphology is not as detailed in the live preparation. Regardless of the visualization method used, the larvae can be analyzed 24 h after epidermal wounding (when the epidermal wound gaps are typically closed in controls) or can be analyzed at intermediate time points to observe the dynamics of cell migration/elongation as closure progresses.

We expect that these detailed protocols for both wounding and wound visualization will be useful for vertebrate wound healing researchers who wish to quickly test the function of orthologs of their favorite genes (or interactors with these genes) in the *Drosophila* system. Such a strategy allows a rapid assessment of whether the gene is important, either for epidermal wound closure

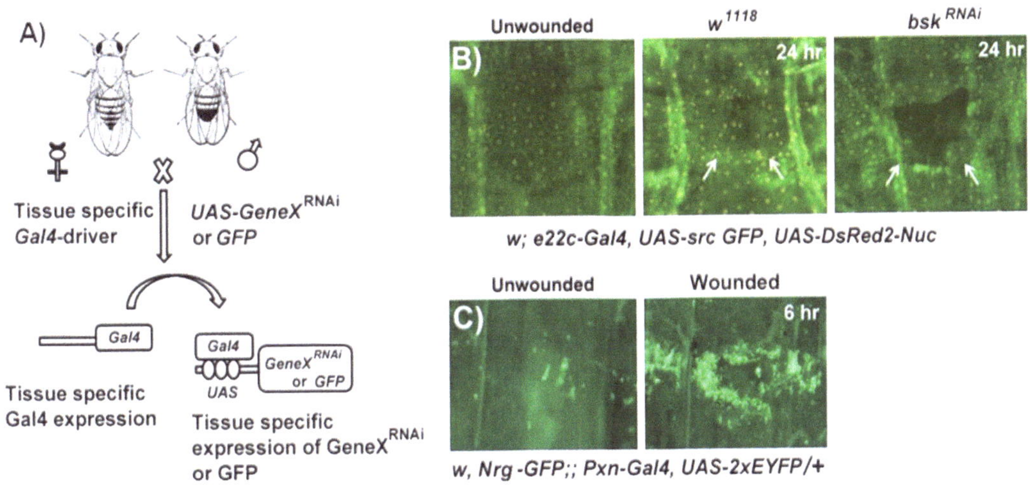

Fig. 2 Application of the *Gal4-UAS* system for live imaging of wound responses and genetic screening. (**a**) A schematic overview of the *Gal4-UAS* system in *Drosophila*. When flies (females shown here) carrying a tissue-specific *Gal4* driver are crossed to flies (males shown here) with a transgene under UAS control (for example, *UAS-GeneX*^{RNAi} or *UAS-GFP*), progeny containing both genetic elements are produced leading to expression of the *UAS-transgene* in the tissues that express Gal4. (**b**) Live whole mounts of the pinch-wounded larval progeny obtained after mating *w; e22c-Gal4, UAS src-GFP, UAS DsRed2-Nuc/+ (e22c-reporter)* flies with either *w*^{1118} (*middle panel*) or *UAS-bsk*^{RNAi} (*right panel*: to silence jun N-terminal kinase (JNK)) flies. The unwounded *w; e22c-Gal4, UAS src-GFP, UAS DsRed2-Nuc/+* larvae are used as controls (*left panel*). Twenty-four hours after wounding the *w*^{1118} larvae show complete wound gap closure, but disorganized epidermal tissue (indicated by *arrows*) compared to the unwounded control. The larvae with epidermal JNK knockdown show open wounds indicated by the *dark gap* in the epidermal sheet (*arrows*). (**c**) Live larval whole mounts of the reporter line *w, Nrg-GFP;; PxnGal4, UAS-2xEYFP/+* can be used to study the wound-induced inflammatory response, where a robust accumulation of hemocytes (*green*) labeled with YFP at the wound site can be seen 6 h after pinch wounding (*right panel*, compare to control)

or recruitment of blood cells, without the time and expense commitment of making a mouse knockout or transgenic model. A positive result in *Drosophila* larvae could provide the extra motivation to commit to a more clinically relevant model.

2 Materials

2.1 Fly Husbandry

1. Cornmeal dextrose media (for media preparation, refer to the Bloomington *Drosophila* Research Center [http://flystocks.bio.indiana.edu/Fly_Work/media-recipes/dextrosefood.htm], Indiana University) (*see* **Note 1**).
2. Plastic vials and cotton plugs.
3. 25 °C incubator, preferably with humidity control to ~50 %.

2.2 Pinch and Puncture Wounding

1. Diethyl ether.
2. Coplin jar (Fig. 3a).

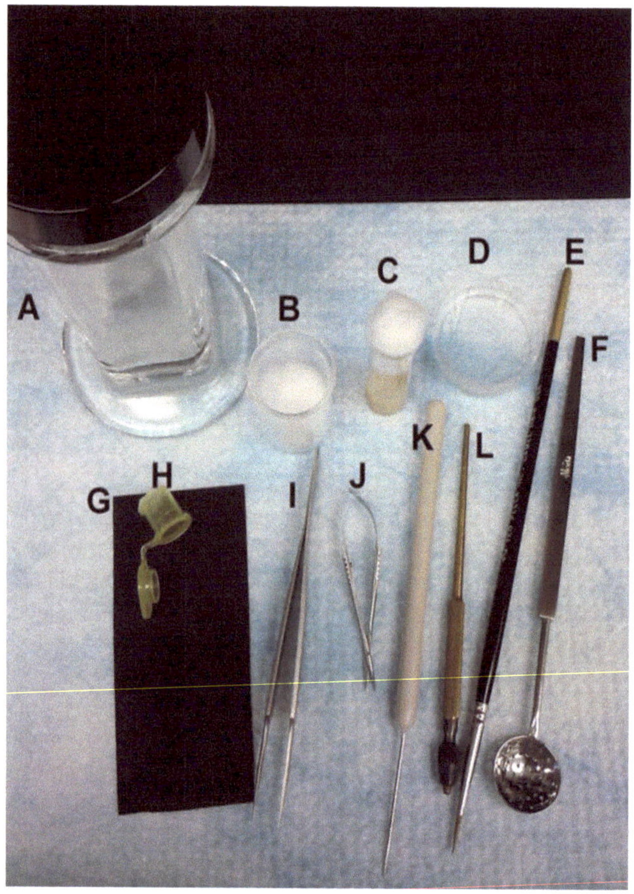

Fig. 3 The instruments used for epidermal wounding and dissection of *Drosophila* larvae. (**a**, **b**) Coplin jar (**a**) and 10 ml beaker with a cotton ball (**b**, for etherization). (**c**) Wounding vial. (**d**) Sylgard plate with pins for dissections. (**e**) Paintbrush. (**f**) Perforated spoon. (**g**) Vinyl sheet (wounding pad). (**h**) Etherization cage, cut from a 2 ml microcentrifuge tube. (**i**) Forceps. (**j**) Dissecting scissors. (**k**) Needle to soften the fly food. (**l**) Pin holder

3. 10 ml glass beaker, stuffed with a cotton ball (Fig. 3b).
4. Wounding vials: 1.0 dram glass vials with 1.0 ml of cornmeal dextrose media and sealed with a small cotton plug (Fig. 3c).
5. Paintbrush (Fig. 3e).
6. Perforated steel spoon (Fig. 3f).
7. Wounding pad (a thin vinyl sheet, preferably black to contrast the white larvae) (Fig. 3g).
8. 2.0 ml Microcentrifuge tubes (Fig. 3h).
9. Nitex nylon mesh (pore size: 630 μm).
10. Dissecting stereomicroscope with brightfield gooseneck illumination (Zeiss Stemi 2000 microscope or equivalent).

11. Nitex Nylon mesh (pore size: 100 μm).
12. *For Pinch wounding*: Blunted dissecting forceps (Fig. 3i).
13. *For puncture wounding*: 0.1 mm steel needle held in needle or pin holder (Fig. 3l).

2.3 Live Scoring of Epidermal Morphology or Inflammation After Pinch or Puncture Wounding

1. Epidermal reporter line chosen from among options in Table 1: *w^{1118};; UAS-src-GFP, UASDsRed2-Nuc, A58-Gal4/TM6B* (*A58-reporter*) or *w^{1118};e22c-Gal4, UAS-src-GFP, UAS-DsRed2Nuc/CyO* (*e22c-reporter*) or blood cell-specific reporter line: *w, Nrg-GFP;; Pxn-Gal4, UAS2xEYFP*.
2. 70 % Glycerol.
3. Micro glass slides (3in. × 1 in.) and coverslips (24 × 60 mm). Double-sided clear tape.
4. Stereomicroscope equipped with dual filter capable of simultaneous visualization of GFP and RFP channels (Leica MZ16FA or equivalent) and color digital camera (Leica DFC300 FX or equivalent).

2.4 Dissection of Larval Epidermis

1. 35× 10 mm plastic petri dishes.
2. Sylgard elastomer base and curing agent (Dow Corning, Catalog no. 3555-47-3 and 68988-896, respectively).
3. 35× 10 mm petri dishes partially filled with polymerized Sylgard (preparation of Sylgard plates is described in detail methods Subheading 3.5 and [16]) (Fig. 3d).
4. 0.1 mm thick dissecting pins.
5. Dissecting scissors (Fig. 3j).
6. Forceps.
7. Phosphate buffer saline (pH 7.2) (PBS).
8. 3.7 % formaldehyde made in PBS.

2.5 Immunostaining and Mounting of Dissected Epidermis

1. PBS.
2. Goat serum. Inactivate the goat serum by incubating at 55 °C for 1 h.
3. Tween-20.
4. Anti-Fasciclin III antibody (Developmental Studies Hybridoma Bank, Catalog no. 7G10).
5. Fluorescein (FITC)-conjugated goat anti-mouse IgG.
6. BD Clay Adams Nutator.
7. Micro glass slides.
8. Coverslips.
9. Vectashield™ mounting medium.

3 Methods

3.1 Fly Husbandry and Harvesting Larvae

1. Culture appropriate wound reporter lines on standard cornmeal dextrose fly media at 25 °C. For genetic screening, cross the wound reporter lines to an appropriate *UAS-RNAi* line targeting your gene of interest (experimental condition) or to a w^{1118} control strain (closure should proceed normally) (*see* **Notes 1** and **2**).
2. Transfer the flies onto fresh food every other day (*see* **Note 3**).
3. Collect early-L3-stage larvae after 5 days of incubation at 25 °C. With a metal spatula, scoop out the soft food to a nylon mesh (630 μm pore size) and rinse gently but thoroughly with running tap water. Very little food should remain when done.
4. Using forceps, quickly select the green fluorescent larvae bearing the reporter using a fluorescence stereomicroscope (*see* **Note 4**).

3.2 Pinch and Puncture Wounding

Here we describe two techniques to physically wound larvae. Pinch wounding is recommended for studying the cell migration events [15, 17] (Fig. 1b) and puncture wounding is best suited for studying hemolymph coagulation and wound-induced epidermal cell–cell fusion [6] (Fig. 1c). Prepare a small etherization cage to anesthetize the larvae using a 2.0 ml microcentrifuge tube. Briefly, sever the tube with a razor blade above the taper. Save the part with the cap and melt the cut edge of the tube with a hot metal spatula. While the melted edge is still hot, attach a small square of nylon mesh (100 μm pore size) to the melted region. This will allow access to ether fumes but prevent the larvae from escaping.

1. Transfer rinsed L3 larvae into the cap of the etherization cage (*see* **Note 5**). To expose to ether, place a 10 ml glass beaker containing a cotton ball soaked in ~1.0 ml of diethyl ether in a Coplin jar. This should be done in a fume hood. Place the etherization cage on the cotton ball and close the lid of the Coplin jar. Wait for 2.5 min.
2. Remove etherization chamber from Coplin jar, open the cap, and then gently rinse larvae into a 35 × 10 mm petri dish using a water squirt bottle.
3. Using forceps, gently place up to 14 larvae onto the wounding pad with their dorsal sides up (*see* **Note 6** and Fig. 1a).
4. *Pinch wounding*: While observing under the dissecting microscope, pinch wound each larva by grabbing the cuticle and epidermis with forceps and holding for 10 s (*see* **Notes 7–9**). The pinch wounds should be made in the smooth cuticle at the dorsal midline between the third and the fourth segmental boundaries marked by dorsal hairs (Fig. 1b).

5. *Puncture wounding*: Pierce the larval cuticle with a 0.1 mm steel needle at the dorsal midline between the third and the fourth dorsal segmental boundaries. Placing paper towels underneath the wounding pad will cushion both the larva and the pin during piercing (*see* **Note 10** and Fig. 1c).

3.3 Recovery

1. After wounding, use a water squirt bottle to gently rinse the larvae off the pad and into a petri dish filled with water.
2. Gently transfer the larvae into wounding vials using forceps (*see* **Note 11**).
3. Incubate the larvae at 25 °C for the desired length of time. In control lines, wound closure is completed by 24 h.

3.4 Mounting for Live Imaging

1. After recovery for the desired length of time, use forceps to carefully remove the wounded larvae from the food within the wounding vial and rinse with water.
2. Anesthetize the larvae using ether for 5 min (see the above protocol).
3. Wash the larvae into a 35× 10 mm petri dish using a tap water squirt bottle.
4. Prepare the micro glass slide for mounting the larvae by sticking a length of double-sided tape sufficient to mount the number of larvae chosen for analysis.
5. Place the etherized larvae on the glass side adjacent to the tape with the dorsal side facing up (*see* **Note 12**). Now, hold the larva with forceps near the posterior spiracles and attach the mouth hook to the tape. Gently stretch the larva without breaking the cuticle and attach the posterior spiracle to the tape as well. Flatten the larva to the extent possible to obtain better quality images.
6. Add approximately 100 μl of 70 % glycerol onto the coverslip and spread it in a line sufficient to cover the larvae. Take the slide with the larvae affixed to it and invert it onto the coverslip to cover the larvae uniformly in glycerol (*see* **Note 13**).

3.5 Dissections and Immunostaining of Dissected Samples

Dissection of the larval epidermis allows for a clearer visualization of the epidermal wound morphology. *See* Fig. 1a for layout of a properly dissected larva. In addition to anti-Fasciclin III which labels epidermal cell membranes, whole mounts can be immunostained for other cellular markers if desired.

1. Prepare polymerized Sylgard plates by mixing 1 part of elastomer with 10 parts of base. Pour ~3 ml of the mixture into a 35 × 10 mm petri dish and allow the gel to set by incubating the plates at room temperature for 2 days.

2. Gently transfer the larvae from wounding vials and rinse with water or PBS. Transfer a larva to a Sylgard plate containing enough PBS to cover the surface. Using one pair of forceps, grab and stabilize the larva. Using a second pair of forceps, grab a 0.1 mm pin and pin the larva to the soft Sylgard surface. The pin should pierce the anterior of the larva near the mouth hooks and the ventral epidermis should be facing up.
3. Grab the posterior of the larva and pin near the posterior spiracles while gently stretching the larva along the anteroposterior axis.
4. Using dissecting scissors make a small incision at the midbody along the anteroposterior axis and cut the ventral epidermis/cuticle both anteriorly and posteriorly towards the head and tail pins. Take care not to nick the underlying dorsal epidermis while cutting.
5. Using one pair of forceps, stretch the right-hand anterior epidermal flap just shy of breaking. Using a second pair of forceps, grab an insect pin and pin the right epidermal flap to the Sylgard. Similarly, pin the left-hand anterior flap, the right-hand posterior flap, and the left-hand posterior flap in that order (Fig. 1a). These steps of the dissection will take practice to perfect (*see* **Note 14**).
6. Eviscerate by gently grabbing the internal organs at the posterior and gently pulling upward and anteriorward. If one grabs between the epidermis and the tracheal dorsal trunks, one can usually pull up all of the viscera (gut, fat bodies, malpighian tubules, and associated trachea) at once. Use a second pair of forceps to sever the connection between the pulled organs and the anterior dorsal epidermis (avoid poking the dorsal epidermis). Residual internal organs such as brain and fat body should be gently picked off the epidermis by forceps.
7. Fix in a fume hood for 1 h at room temperature by replacing the PBS with 3.7 % formaldehyde in PBS.
8. Using forceps gently unpin each larva and transfer the epidermis to a 1.5 ml microcentrifuge tube containing PBS (*see* **Note 15**).

3.6 Immunostaining of Dissected Larval Epidermis with Anti-Fasciclin III

The epidermal wounds can be studied by immunostaining the larval epidermis for markers such as Fasciclin III. The antibody stains epidermal cell membranes (Fig. 1b, c) and the pinch and puncture wounds can be clearly visualized. The staining protocol described here can be adapted to any marker of interest.

1. Block fixed larvae for at least 1 h at room temperature in PHT by rotating on an Adams nutator, but any equivalent will do.
2. Dilute anti-Fasciclin III 1:50 in PHT, which is PBS containing 5 % heat-inactivated normal goat serum and 0.3 % Tween-20.

3. Replace block solution with sufficient diluted primary antibody to cover all of the larvae being stained.
4. Incubate on a nutator overnight at 4 °C.
5. Remove and save the primary antibody and wash the larvae at least six times for at least 30 min with 1.0 ml PHT.
6. Replace final wash with diluted secondary antibody: Goat anti-mouse IgG conjugated to the fluor of your choice. Incubate overnight at 4 °C.
7. Replace secondary antibody with PHT and wash again as in **step 5**.
8. Add a drop of Vectashield™ to a glass microscope slide and place the epidermal whole mount into the drop. Slowly mount a coverslip onto the drop without creating air bubbles. Seal the coverslip to the glass slide using nail polish.

3.7 Microscopy

1. Live or dissected larvae can be viewed with an appropriate fluorescent filter on a Leica MZ16FA stereomicroscope using Planapo 1.6× objective. Capture images using a color digital camera (Leica DFC300 FX or equivalent) and Image-Pro AMS v5.1 software (Media Cybernetics or equivalent).

4 Notes

1. Please refer to the Bloomington Stock Center Web site or to Ralph Greenspan's book "Flypushing" for basics of fly husbandry.
2. We recommend 25 °C for *Drosophila* culture. Note that the *Gal4/UAS* system and *Drosophila* in general are temperature sensitive. Decreasing the temperature will give a longer generation time and lower transgene expression while increasing the temperature will give a shorter generation time and higher transgene expression although fecundity may be compromised. This is useful to keep in mind when encountering a transgene that is lethal at 25 °C or gives no phenotype. Lower temperatures may promote viability and higher temperatures may reveal a phenotype.
3. It is only necessary to brood the flies if the cross is going slowly (low number of embryo/larval progeny) and until a sufficient population of larvae is obtained for wounding.
4. After sorting the fluorescent larvae, it is advisable to leave the larvae on small moist pads of fly food in a 35 mm × 10 mm petri dish. This practice avoids starvation of larvae before wounding.

5. Larvae should be transferred to the etherization cage with as little accompanying water as possible since water will hinder the desired anesthetization.
6. The dorsal side is marked by the presence of the two prominent dorsal tracheal trunks and the absence of dark denticles that mark the segmental boundaries on the ventral side.
7. A good pair of dissecting forceps is very important for pinch wounding. The tips should be filed blunt and the edges made smooth so that the forceps do not puncture the larval cuticle. Forceps can be blunted by filing the tips with a sanding stone. Tips should be rounded when done but blades should still be flat and close flush.
8. Standardize the forceps used for pinching by checking for the wound morphology immediately after wounding. Occasional refiling or replacement of the wounding forceps may be necessary if the initial wound morphology drifts from the ideal that is a well-rounded wound that is centered in a single body segment (*see* Fig. 1a, b).
9. The amount of pressure applied while pinch wounding should be enough to create a gap in the epidermis but not so much as to puncture the cuticle and thus create a scab. The proper amount of pressure can be determined through practice.
10. During puncture wounding, it is typical to poke the needle through both the dorsal and ventral cuticle. This is because the cuticle is firm but the interior of the larva is filled with liquid hemolymph. Either the dorsal entry or ventral exit wound can be used for further analysis although the investigator has more control over the location of the dorsal wound.
11. A maximum of 14 larvae should be transferred to a single wounding vial to avoid overcrowding and excess competition for food. Likewise, a minimum of 5 or 6 larvae should be in each vial so that they can assist each other in working up the food. Pre-mashing the food (with the needle shown in Fig. 3k) before transferring the larvae to the wounding vial helps in quick revival of the larvae.
12. After transferring the etherized larvae onto the slide from water, there will be a small drop of water around the larvae. Remove this water droplet using a kimwipe before mounting onto the tape as they will stick better.
13. Live-mounted larvae should be observed immediately after mounting and photographed quickly. Occasionally they will wake up and wriggle free from the tape. This can be minimized by chilling them for approximately 30 min at 4 °C.
14. When pinning out the larva during dissection it is imperative that the pins be placed symmetrically to ensure that there are

no stretch lines skewing across the epidermis (*see* Fig. 1a). The ventral cut edges should be parallel to each other, and the two anterior and two posterior pins should be on the same line perpendicular to the anteroposterior midline. Finally, all flaps should be fully stretched and form an angle of 45° with respect to the anteroposterior midline.

15. It is a good idea to use a separate pair of forceps for dissecting and for unpinning to avoid damage to the tips. In general, since the forceps are used to grab and place the pins, the tips will only last in good condition for up to a few months even if good care is taken.

References

1. Grose R, Werner S (2004) Wound-healing studies in transgenic and knockout mice. Mol Biotechnol 28(2):147–166
2. Liu J, Johnson K, Li J, Piamonte V, Steffy BM, Hsieh MH, Ng N, Zhang J, Walker JR, Ding S, Muneoka K, Wu X, Glynne R, Schultz PG (2011) Regenerative phenotype in mice with a point mutation in transforming growth factor β type I receptor (TGFBR1). Proc Natl Acad Sci 108(35):14560–14565
3. Bier E (2005) Drosophila, the golden bug, emerges as a tool for human genetics. Nat Rev Genet 6(1):9–23
4. Wood W, Jacinto A, Grose R, Woolner S, Gale J, Wilson C, Martin P (2002) Wound healing recapitulates morphogenesis in Drosophila embryos. Nat Cell Biol 4(11):907–912
5. Stramer B, Wood W, Galko MJ, Redd MJ, Jacinto A, Parkhurst SM, Martin P (2005) Live imaging of wound inflammation in Drosophila embryos reveals key roles for small GTPases during in vivo cell migration. J Cell Biol 168(4):567–573
6. Galko MJ, Krasnow MA (2004) Cellular and Genetic Analysis of Wound Healing in *Drosophila* Larvae. PLoS Biol 2(8):e239
7. Babcock DT, Brock AR, Fish GS, Wang Y, Perrin L, Krasnow MA, Galko MJ (2008) Circulating blood cells function as a surveillance system for damaged tissue in Drosophila larvae. Proc Natl Acad Sci 105(29):10017–10022
8. Bosch M, Serras F, Martín-Blanco E, Baguñà J (2005) JNK signaling pathway required for wound healing in regenerating Drosophila wing imaginal discs. Dev Biol 280(1):73–86
9. Rämet M, Lanot R, Zachary D, Manfruelli P (2002) JNK Signaling Pathway Is Required for Efficient Wound Healing in Drosophila. Dev Biol 241(1):145–156
10. Evans CJ, Hartenstein V, Banerjee U (2003) Thicker Than Blood: Conserved Mechanisms in Drosophila and Vertebrate Hematopoiesis. Dev Cell 5(5):673–690
11. Bidla G, Dushay MS, Theopold U (2007) Crystal cell rupture after injury in Drosophila requires the JNK pathway, small GTPases and the TNF homolog Eiger. J Cell Sci 120(7):1209–1215
12. Márkus R, Kurucz É, Rus F, Andó I (2005) Sterile wounding is a minimal and sufficient trigger for a cellular immune response in Drosophila melanogaster. Immunol Lett 101(1):108–111
13. Duffy JB (2002) GAL4 system in drosophila: A fly geneticist's swiss army knife. Genesis 34(1–2):1–15
14. Dietzl G, Chen D, Schnorrer F, Su K-C, Barinova Y, Fellner M, Gasser B, Kinsey K, Oppel S, Scheiblauer S, Couto A, Marra V, Keleman K, Dickson BJ (2007) A genome-wide transgenic RNAi library for conditional gene inactivation in Drosophila. Nature 448(7150):151–156
15. Lesch C, Jo J, Wu Y, Fish GS, Galko MJ (2010) A Targeted UAS-RNAi Screen in Drosophila Larvae Identifies Wound Closure Genes Regulating Distinct Cellular Processes. Genetics 186(3):943–957
16. Brent JR, Werner KM, McCabe BD (2009) Drosophila Larval NMJ Dissection. J Vis Exp (24):e1107.
17. Wu Y, Brock AR, Wang Y, Fujitani K, Ueda R, Galko MJ (2009) A Blood-Borne PDGF/VEGF-like Ligand Initiates Wound-Induced Epidermal Cell Migration in Drosophila Larvae. Curr Biol 19(17):1473–1477

Chapter 27

Zebrafish Cardiac Injury and Regeneration Models: A Noninvasive and Invasive In Vivo Model of Cardiac Regeneration

Michael S. Dickover, Ruilin Zhang, Peidong Han, and Neil C. Chi

Abstract

Despite current treatment regimens, heart failure still remains one of the leading causes of morbidity and mortality in the world due to failure to adequately replace lost ventricular myocardium from ischemia-induced infarct. Although adult mammalian ventricular cardiomyocytes have a limited capacity to divide, this proliferation is insufficient to overcome the significant loss of myocardium from ventricular injury. However, lower vertebrates, such as the zebrafish and newt, have the remarkable capacity to fully regenerate their hearts after severe injury. Thus, there is great interest in studying these animal model systems to discover new regenerative approaches that might be applied to injured mammalian hearts. To this end, the zebrafish has been utilized more recently to gain additional mechanistic insight into cardiac regeneration because of its genetic tractability. Here, we describe two cardiac injury methods, a mechanical and a genetic injury model, for studying cardiac regeneration in the zebrafish.

Key words Heart regeneration, Zebrafish, Resection, Genetic ablation, Ischemia, Nitroreductase

1 Introduction

Despite current treatment regimens, heart failure still remains one of the leading causes of morbidity and mortality in the world due to failure to adequately replace lost ventricular myocardium from ischemia-induced infarct. Although adult mammalian ventricular cardiomyocytes have a limited capacity to divide, this proliferation is insufficient to overcome the significant loss of myocardium from ventricular injury [1–3]. As a result, the damaged human heart generates fibrous scar tissue and pathologically remodels itself in an attempt to unsuccessfully compensate for reduced cardiac function and altered cardiac architecture, subsequently leading to heart failure. Thus, developing innovative approaches to produce new healthy cardiomyocytes in order to replace damaged heart tissue may provide novel regenerative cellular therapies for treating heart failure.

Robert G. Gourdie and Tereance A. Myers (eds.), *Wound Regeneration and Repair: Methods and Protocols*, Methods in Molecular Biology, vol. 1037, DOI 10.1007/978-1-62703-505-7_27, © Springer Science+Business Media New York 2013

Because of their robust capacity to regenerate a wide variety of tissues, including limb, brain, spinal cord, and retina, lower vertebrates, such as amphibians and fish, have been classically used to study organ regeneration [4]. Many of these regeneration studies were initially performed in newts and Xenopus [5–12]; however, because of its genetic tractability, the zebrafish has been utilized more recently to gain additional mechanistic insight into vertebrate organ regeneration [13–19]. As a consequence, researchers have taken advantage of these lower vertebrate models to further investigate cardiac regeneration with the hopes of translating these findings to the regeneration of mammalian hearts. Towards this end, similar cardiac injury and regeneration models used in lower vertebrate systems have recently been employed in mouse cardiac models [20]. These studies reveal that early postnatal mouse hearts, like zebrafish hearts, possess the ability to regenerate from the proliferation of existing cardiomyocytes after ventricular apex resection. However, unlike zebrafish cardiac regeneration, this mouse cardiac regenerative potential appears to significantly diminish 7 days after birth. Consequently, there is intense interest to investigate cardiac regeneration in these different animal model systems in order to identify new regenerative cellular strategies for repair of damaged adult human hearts.

Various cardiac injury techniques have been utilized to study the process of cardiac regeneration in multiple animal models. The most common and well-established method has been the mechanical resection procedure [8, 13]. In this cardiac injury model, an incision is created in the chest/ventral surface of the animal to expose the heart for mechanical ventricular apical transection with a pair of scissors (Fig. 1). Using this technique, several groups have reproducibly observed in zebrafish and newts that after resection of 15–20 % of the ventricular apex, adult and juvenile hearts are able to generate sufficient new myocardium within several weeks to fully replace the transected apex [13, 14, 17, 18, 21]. Because of its reliability, this cardiac injury model has been recently used in the mouse to investigate the cardiac regenerative potential in mammalian hearts [20]. Moreover, in order to simulate myocardial infarction-like cardiac injuries, researchers have also used cryocauterization in both mouse and zebrafish hearts to study cardiac regeneration [19, 22]. This technique involves contacting the cardiac ventricle with a copper filament cooled in liquid nitrogen, thereby causing necrotic death of the surrounding cells. Similar to myocardial infarction, this injury results in fibrotic scar tissue formation. Eventually, the scar tissue is resorbed to allow for integration of newly generated cardiomyocytes to the damaged myocardium. Although these invasive techniques can create a robust and fairly reproducible injury to the heart, they are

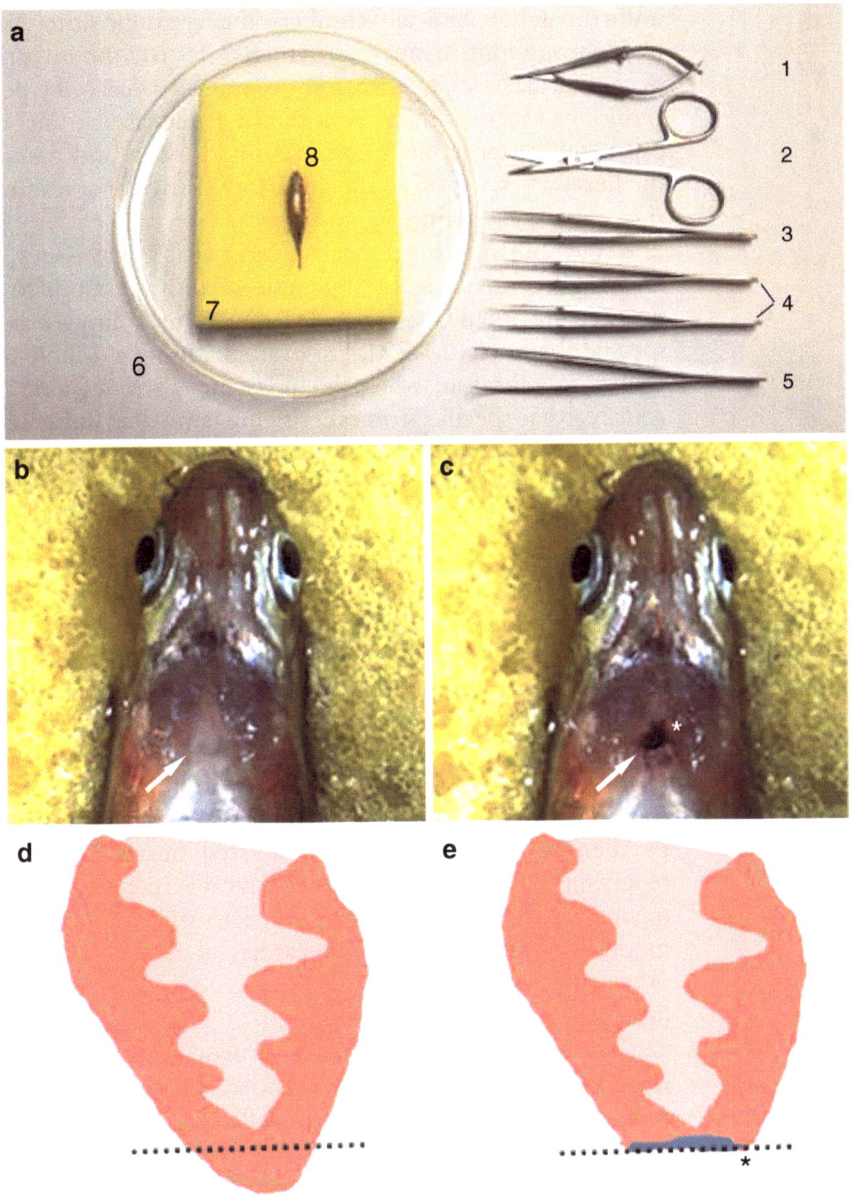

Fig. 1 Surgical transection of the zebrafish ventricular apex. Six month old adult zebrafish is mounted in sponge ventral side up for incision of chest wall and transection of cardiac ventricle apex. (**a**) Materials required to surgically transect the ventricular apex include: (1) Vannas scissors, (2) Iris scissors, (3) #5 Dumont tweezers, (4) a pair of #55 Dumont tweezers, (5) WPI inc. Economy tweezers #7, (6) petri dish, and (7) sponge with slit cut into the top in order to hold (8) anesthetized zebrafish. Anesthetized 6 month old adult zebrafish mounted ventral side up (**b**) prior to surgical resection and (**c**) after resection. Cartoon schematic of a section of the ventricle (**d**) before and (**e**) after resection. *Dashed line* marks the plane of resection. *White arrows* point to the heart. *Asterisks mark* scarring after resection. Note blood clot that forms immediately after resection in **c**. Head is to the top and the tail is to the bottom in panels (**a**–**c**)

unfortunately technically cumbersome, are difficult to employ in a high-throughput manner, structurally destroy the integrity of the overall cardiac architecture, remove other cardiac cell types beyond cardiomyocytes, such as endothelial, endocardial, and epicardial cells, and can potentially damage other organs and tissues besides the heart when exposing the heart for ventricular apical transection (i.e., skin, skeletal muscle, and bone).

To circumvent these mechanical cardiac injury issues, researchers have recently developed genetic ablation methods to noninvasively create organ/tissue-specific injuries, including cardiomyocyte specific ablations [23–32]. For example, using a double transgenic system in zebrafish, which includes the *Tg(cmlc2:CreER)* line—*cmlc2* cardiac specific promoter driving tamoxifen inducible *CreER*, and *Tg(bactin2:loxp-mCherry-STOP-loxp-DTA)* line—a ubiquitously expressing *diphtheria toxin A chain* (*DTA*) switch line, cardiac-specific expression of DTA, which blocks protein synthesis, can be induced with tamoxifen treatment in order to ablate cardiomyocytes [23]. As a result, high dose tamoxifen injections into 6 month old zebrafish harboring both transgenes can specifically and reproducibly destroy over 60–80 % of the ventricular myocardium, causing heart failure and in some cases death. In contrast to the ventricular apical transection model where these hearts cannot fully regenerate after more than a 20–25 % cardiac resection [13], the DTA-genetically ablated ventricles were able to regenerate after ablation of over 60 % of ventricular cardiomyocytes [23]. Furthermore, these genetically ablated hearts also significantly regenerated faster than the mechanically resected hearts. These regenerative differences may be due to the mechanical resection model severely disrupting the cardiac architecture and removing other cardiac cell types, which may be necessary for more efficient myocardial regeneration [33].

Though the DTA-inducible genetic cardiac ablation shows great potential as a noninvasive cardiac injury model, this method can be technically and genetically difficult because intercrossing of two different transgenic lines is required to generate fish for this cardiac injury model. Thus, another promising genetic cell ablation model, which only requires one transgenic line and utilizes the bacterial nitroreductase (NTR) enzyme, has been used recently to specifically destroy tissues/organs in both mouse and zebrafish after treatment with the prodrug, metronidazole (MTZ) [30–32, 34]. In this technique, tissue-specific NTR expression in vertebrate cells is innocuous; however, MTZ exposure to these NTR-expressing cells leads to the conversion of MTZ into a toxic compound, which then effectively ablates these cells through induction of apoptosis. Using the *cmlc2*, cardiac specific promoter, to drive expression of NTR in zebrafish hearts—*Tg(cmlc2:NTR)*, cardiomyocytes can be reliably ablated after MTZ treatment [30]. Moreover, to study the recovery and regeneration of the damaged heart, the MTZ can be effectively washed away to discontinue the ablations [35]. Overall, these genetic ablation techniques are noninvasive, rapid, and easy to carry out, thus permitting researchers in the future to more easily execute high-throughput studies, such as chemical and forward genetic screens, of cardiac regeneration.

Overall, the surgical and genetic cardiac injury models are both useful and complementary tools to study different aspects of cardiac regeneration. Because of their genetic tractability (i.e., transgenics, mutants, morpholino knockdowns) and inherent regenerative capacity [36], the zebrafish has gained significant popularity for performing both cardiac injury models in order to dissect the cellular and molecular mechanisms of cardiac regeneration. In particular, zebrafish transgenic tools have been recently created to investigate critical signaling pathways that may be required for cardiac regeneration [14, 37] as well as to genetically fate map cells to identify the source of the cardiac regenerative tissue [17, 18]. Thus, in this chapter, we describe protocols for performing both the surgical resection and the NTR-mediated genetic ablation cardiac injury models in the zebrafish heart. In the future, we anticipate that these protocols may be applicable in other animal models. Already, the mechanical cardiac injury model has been successfully used in mice, and recent mouse gene trap screens using NTR have been performed [31], which may yield cardiac specific NTR expressing lines that can be used for mouse NTR-mediated cardiac genetic ablation in the future.

2 Materials

2.1 Surgical Resection of the Zebrafish Heart

1. Iris Scissors.
2. Vannas Scissors.
3. Dumont Tweezers #5.
4. Dumont Tweezers #55.
5. Economy Tweezers #7.
6. Petri Dish.
7. Double Sided Tape.
8. Sponge.

2.2 Transgenic Fish Lines

The *Tg(cmlc2:mCherry-NTR)* zebrafish line, which expresses the mCherry-nitroreductase fusion protein in the myocardium, was used to genetically ablate cardiomyocytes after MTZ treatment (Fig. 2). In order to more easily observe the cardiac ablation, this line was crossed into the *Tg(kdrl:GFP)* reporter line (Fig. 2), which expresses green fluorescent protein (GFP) in the endocardium and vasculature [38]. During cardiac ablation, the cardiomyocytes are specifically destroyed and can be observed by the reduction of cherry red cardiomyocytes, whereas the endocardium marked in GFP remains intact (Figs. 2 and 3).

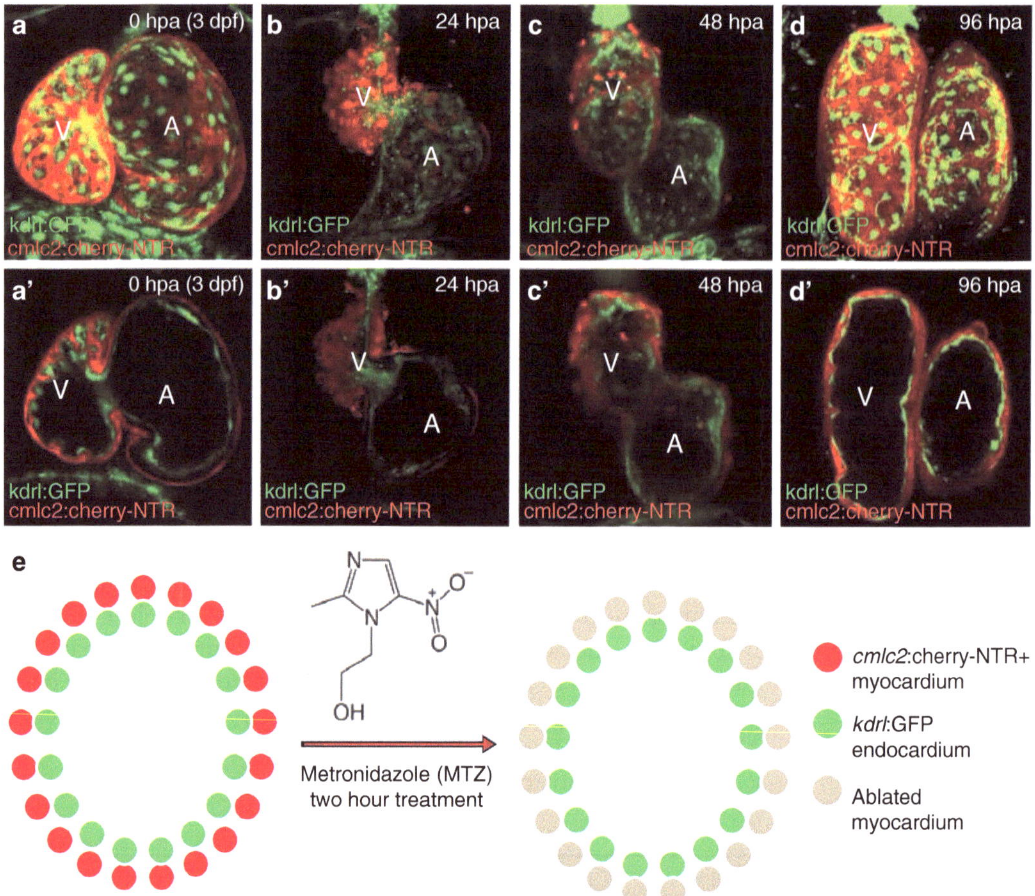

Fig. 2 In vivo cardiac genetic ablation and regeneration model. 3 dpf (days post-fertilization) *Tg(cmlc2:mCherry-NTR; kdrl:GFP)* larvae are treated with 5 mM metronidazole (MTZ) for 2 h. Live confocal imaging of ablated and regenerating heart at (**a**, **a**′) 0 hpa (hour post-ablation); (**b**, **b**′) 24 hpa; (**c**, **c**′) 48 hpa; and (**d**, **d**′) 96 hpa. **a**, **b**, **c**, **d** are full confocal image projections of the hearts and **a**′, **b**′, **c**′, **d**′ are representative confocal optical section of the **a**, **b**, **c**, **d** heart. (**e**) Schematic representation of myocardial genetic ablation strategy. *cmlc2:mcherry-NTR* (*red*) and *kdrl:GFP* (*green*) are expressed in the myocardium and endocardium, respectively. By 24 hpa (**b**, **b**′), much of the myocardium (*red*) is specifically ablated, whereas the endocardium (*green*) remains intact. Regeneration of cardiomyocytes begins ~48 hpf (**c**, **c**′) and continues to 96 hpa (**d**, **d**′) when the heart has fully regenerated. A, atrium; V, ventricle

2.3 Reagents for Cardiomyocyte Ablation and Cell Death Detection

1. Egg water for zebrafish embryos: 0.3 g Instant Ocean Sea Salt, 0.075 g $CaSO_4$, 1 L distilled water.
2. 20× Phenylthiourea (PTU) stock solution: 0.06 % (w/v) 1-phenyl-2thiourea (Sigma P7629) dissolved in egg water.
3. Dimethyl sulfoxide (DMSO) control solution: 0.2 % (v/v) DMSO (Sigma D8418) in egg water.
4. 10 mM metronidazole (MTZ) solution: 1.712 g metronidazole, 2 mL DMSO, 1 L egg water (*see* **Note 1**).

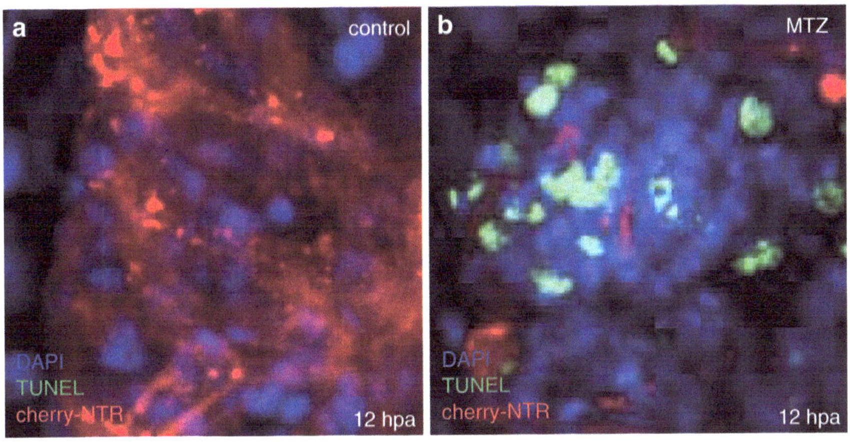

Fig. 3 Cell death in *Tg(cmlc2:mCherry-NTR)* ablated hearts. TUNEL staining (*green*) was performed in 3.5 dpf *Tg(cmlc2:mCherry-NTR)* larvae 12 h after being treated with (**a**) DMSO or (**b**) 5 mM metronidazole (MTZ) for 2 h. TUNEL positive staining, which indicates apoptosis, has increased significantly in ablated cardiac ventricle (**b**) compared to DMSO-treated age-matched control (**a**, control). *Red—cmlc2:mcherry-NTR*; *Green*—TUNEL; *Blue*—DAPI/nuclei

5. 10× Phosphate buffered saline (PBS) solution: 80 g NaCl, 2 g KCl, 14.4 g Na_2HPO_4, 2.4 g KH_2PO_4, 1 L distilled water, pH adjusted to 7.4 with 5 M NaOH.
6. Paraformaldehyde (PFA): 4 % (w/v) paraformaldehyde in 1× PBS.
7. PBST: 0.5 % (v/v) Triton X-100 in 1× PBS.
8. In situ cell death detection kit.

2.4 Microscopy for Regeneration Analysis

1. 20× Tricaine stock solution: 4 g Ethyl 3-aminobenzoate methanesulfate salt, 21 mL 1 M Tris–HCl (pH 9), 1 L distilled water, pH adjusted to 7.0 with 12 M HCl.
2. Low melting point agarose: 1 % (w/v) SeaPlaque Agarose dissolved in egg water. Microwave to dissolve, aliquot in 1.5 or 2 mL microcentrifuge tubes and keep at 37 °C.
3. Glass bottom culture dish.
4. Stereomicroscope.
5. Confocal microscope.

3 Methods

3.1 Surgical Resection of the Zebrafish Heart

1. Affix sponge to a petri dish using double-sided tape (Fig. 1a).
2. Using a razor blade, cut a slot into the sponge big enough to hold an adult zebrafish (Fig. 1a).

3. Anesthetize zebrafish (>3 months old) with tricaine solution and place ventral side up into the sponge (Fig. 1a–c).
4. Determine location of the heart by observing motion under the skin.
5. Observing under a stereomicroscope, make an incision in the skin above the heart using iris scissors. This will expose the silver-colored pericardial sac.
6. Open the pericardial sac using #55 tweezers (Fig. 1a4).
7. Using #5 tweezers (Fig. 1a3), which are not too sharp, gently pull up the apex of the ventricle to expose it.
8. Using Vannas scissors (Fig. 1a1), remove approximately 20 % of the ventricular apex.
9. Once the wound has clotted and stopped bleeding, the fish can be returned to a tank (Fig. 1c).
10. Use a plastic transfer pipette to gently push water across the gills of the fish until it is able to swim freely on its own (generally 3–5 min).
11. Allow the fish to recover for several days (full recovery is usually between 30 and 60 days). The chest incision normally heals without problems within 24–48 h, with more than 80 % of resected fish surviving following a procedure.

3.2 MTZ-dependent Ablation and Cell Death Detection

1. Collect eggs from a pair mating of *Tg(cmlc2:mCherry-NTR)* and *Tg(kdrl:GFP)* fish. Incubate the eggs at 28 °C in egg water until they reach the desired stage for ablation (Fig. 2). To maintain optical clarity of the zebrafish embryo or larvae, add PTU after gastrulation to prevent pigmentation of the fish. Cardiac ablation can be performed as soon as *cmlc2* is expressed (16–18 hpf). However, in order to investigate that cardiac recovery is due to cardiac regeneration rather than normal cardiac development, we have specifically ablated at 3–4 days post-fertilization (dpf), when cardiogenesis has completed and the proliferation rate of the cardiomyocytes has dramatically decreased [39].
2. Before ablation, sort the larvae into two groups: (1) *cmlc2:mCherry-NTR*$^+$ siblings; 0.2 % DMSO (negative control), and (2) *cmlc2:mCherry-NTR*$^+$ siblings; +MTZ (experimental) (Fig. 2). Use a minimum of 30 larvae per group in order to account for any variability.
3. Remove as much egg water as possible and add 5 mM MTZ solution or 0.2 % DMSO control solution at 28 °C. Incubate at 28 °C in the dark for 2 h (*see* **Note 2**).
4. Remove the solution at the end of the incubation. Wash four to five times with egg water and replace with fresh egg water + PTU. Check under the microscope over the next few hours for

any indication of cell death, such as decreased levels of mCherry fluorescence, decreased ventricular function, or blood pooling in the heart (Fig. 2b, b′ and c, c′).

5. In addition to these indicators of myocardial dysfunction and cardiomyocyte ablation as outlined in point 4, TUNEL (Terminal deoxynucleotidyl transferase dUTP nick end labeling) can be used to assess and confirm cell death (Fig. 3). Thus, following cardiac ablation, larvae can be briefly fixed with 4 % PFA solution, washed three times with 1× PBS, and permeabilized with PBST for 2 h or longer at room temperature. Next, 50 μL of staining solution is added, and the larvae are incubated at 37 °C for 1 h. The samples are then examined under the microscope (*see* **Note 3**).

3.3 Post-ablation Regeneration Analysis

1. The optical clarity of the zebrafish embryo/larvae is particularly useful to monitor heart regeneration in vivo in zebrafish (Fig. 2c, c′, d, d′) and is maintained in the zebrafish larvae by preventing the pigmentation of the zebrafish using PTU (*see* Subheading 3.2, **step 1**). In order to image the zebrafish cardiac regeneration in vivo, transfer fish to a 1.5 mL Eppendorf tube that contains 37 °C pre-warmed liquid 1 % low melting point agarose with 3–4 drops of tricaine. Transfer the anesthetized larvae with low melt agarose using a pipette onto a Mat-Tek glass bottom dish (*see* Subheading 2.4, **item 3**). Use a long flat Eppendorf gel loading pipette tip to position the larvae so that the entire heart can be observed. This must be done quickly before the agarose solidifies (*see* **Note 4**). Once the fish is positioned and agarose solidifies, the larvae can then be imaged under the microscope (*see* **Note 5**). Afterwards, the agarose can be broken apart with forceps to free the larvae for transfer back to egg water. Regeneration normally occurs by 3–4 days post-ablation.

4 Notes

1. Shake vigorously until MTZ is dissolved. The MTZ may be more difficult to completely dissolve in a higher stock solution concentration. MTZ may be toxic at high concentrations or after prolonged exposure, so use gloves when preparing MTZ solution.
2. The concentration and incubation times with MTZ solution should be experimentally determined based on the age of the fish and the degree of damage desired. For example, while 2 h 5 mM MTZ treatments at 3 dpf is sufficient to cause substantial damage to the heart, it may require 6–8 h at 5 dpf to achieve the same degree of damage.

3. For better penetration, open the pericardial sac with forceps before permeabilization or proteinase K digest followed by a 4 % PFA re-fixation.
4. Do not transfer too many fish at the same time otherwise the agarose will solidify before all the fish are positioned. If working with an inverted microscope, position 3–4 dpf larvae with the heads down close to the glass bottom slide and tilt the tail up 45°. For 5 dpf or later, rotate the body to the side as well.
5. For time-lapse imaging, use 0.5 % low melt agarose and as little agarose as possible to allow for better diffusion of egg water to the fish. Fill the dish with egg water and change it every day to keep the fish alive.

References

1. Hsieh PC, Segers VF, Davis ME, MacGillivray C, Gannon J, Molkentin JD, Robbins J, Lee RT (2007) Evidence from a genetic fate-mapping study that stem cells refresh adult mammalian cardiomyocytes after injury. Nat Med 13:970–974
2. Bergmann O, Bhardwaj RD, Bernard S, Zdunek S, Barnabe-Heider F, Walsh S, Zupicich J, Alkass K, Buchholz BA, Druid H, Jovinge S, Frisen J (2009) Evidence for cardiomyocyte renewal in humans. Science 324:98–102
3. Bersell K, Arab S, Haring B, Kuhn B (2009) Neuregulin1/ErbB4 signaling induces cardiomyocyte proliferation and repair of heart injury. Cell 138:257–270
4. Major RJ, Poss KD (2007) Zebrafish heart regeneration as a model for cardiac tissue repair. Drug Discov Today Dis Models 4: 219–225
5. Mochii M, Taniguchi Y, Shikata I (2007) Tail regeneration in the Xenopus tadpole. Dev Growth Differ 49:155–161
6. Lin G, Chen Y, Slack JM (2007) Regeneration of neural crest derivatives in the Xenopus tadpole tail. BMC Dev Biol 7:56
7. Oberpriller JO, Oberpriller JC, Aafedt BC (1987) Changes in binucleation and cellular dimensions of rat left atrial myocytes after induced left ventricular infarction. Am J Anat 179:285–290
8. Oberpriller JO, Oberpriller JC (1974) Response of the adult newt ventricle to injury. J Exp Zool 187:249–253
9. Flink IL (2002) Cell cycle reentry of ventricular and atrial cardiomyocytes and cells within the epicardium following amputation of the ventricular apex in the axolotl, Amblystoma mexicanum: confocal microscopic immunofluorescent image analysis of bromodeoxyuridine-labeled nuclei. Anat Embryol (Berl) 205:235–244
10. Endo T, Yoshino J, Kado K, Tochinai S (2007) Brain regeneration in anuran amphibians. Dev Growth Differ 49:121–129
11. Mescher AL, Neff AW (2006) Limb regeneration in amphibians: immunological considerations. ScientificWorldJournal 6(Suppl 1): 1–11
12. Kragl M, Knapp D, Nacu E, Khattak S, Maden M, Epperlein HH, Tanaka EM (2009) Cells keep a memory of their tissue origin during axolotl limb regeneration. Nature 460:60–65
13. Poss KD, Wilson LG, Keating MT (2002) Heart regeneration in zebrafish. Science 298:2188–2190
14. Lepilina A, Coon AN, Kikuchi K, Holdway JE, Roberts RW, Burns CG, Poss KD (2006) A dynamic epicardial injury response supports progenitor cell activity during zebrafish heart regeneration. Cell 127:607–619
15. Poss KD (2007) Getting to the heart of regeneration in zebrafish. Semin Cell Dev Biol 18:36–45
16. Poss KD, Shen J, Nechiporuk A, McMahon G, Thisse B, Thisse C, Keating MT (2000) Roles for Fgf signaling during zebrafish fin regeneration. Dev Biol 222:347–358
17. Jopling C, Sleep E, Raya M, Marti M, Raya A, Belmonte JC (2010) Zebrafish heart regeneration occurs by cardiomyocyte dedifferentiation and proliferation. Nature 464:606–609
18. Kikuchi K, Holdway JE, Werdich AA, Anderson RM, Fang Y, Egnaczyk GF, Evans T, Macrae CA, Stainier DY, Poss KD (2010) Primary contribution to zebrafish heart regeneration by gata4(+) cardiomyocytes. Nature 464: 601–605

19. Gonzalez-Rosa JM, Martin V, Peralta M, Torres M, Mercader N (2011) Extensive scar formation and regression during heart regeneration after cryoinjury in zebrafish. Development 138:1663–1674
20. Porrello ER, Mahmoud AI, Simpson E, Hill JA, Richardson JA, Olson EN, Sadek HA (2011) Transient regenerative potential of the neonatal mouse heart. Science 331:1078–1080
21. Lien CL, Schebesta M, Makino S, Weber GJ, Keating MT (2006) Gene expression analysis of zebrafish heart regeneration. PLoS Biol 4:e260
22. Robey TE, Murry CE (2008) Absence of regeneration in the MRL/MpJ mouse heart following infarction or cryoinjury. Cardiovasc Pathol 17:6–13
23. Wang J, Panakova D, Kikuchi K, Holdway JE, Gemberling M, Burris JS, Singh SP, Dickson AL, Lin YF, Sabeh MK, Werdich AA, Yelon D, Macrae CA, Poss KD (2011) The regenerative capacity of zebrafish reverses cardiac failure caused by genetic cardiomyocyte depletion. Development 138:3421–3430
24. Akazawa H, Komazaki S, Shimomura H, Terasaki F, Zou Y, Takano H, Nagai T, Komuro I (2004) Diphtheria toxin-induced autophagic cardiomyocyte death plays a pathogenic role in mouse model of heart failure. J Biol Chem 279:41095–41103
25. Breitman ML, Clapoff S, Rossant J, Tsui LC, Glode LM, Maxwell IH, Bernstein A (1987) Genetic ablation: targeted expression of a toxin gene causes microphthalmia in transgenic mice. Science 238:1563–1565
26. Brockschnieder D, Lappe-Siefke C, Goebbels S, Boesl MR, Nave KA, Riethmacher D (2004) Cell depletion due to diphtheria toxin fragment A after Cre-mediated recombination. Mol Cell Biol 24:7636–7642
27. Lee P, Morley G, Huang Q, Fischer A, Seiler S, Horner JW, Factor S, Vaidya D, Jalife J, Fishman GI (1998) Conditional lineage ablation to model human diseases. Proc Natl Acad Sci USA 95:11371–11376
28. Saito M, Iwawaki T, Taya C, Yonekawa H, Noda M, Inui Y, Mekada E, Kimata Y, Tsuru A, Kohno K (2001) Diphtheria toxin receptor-mediated conditional and targeted cell ablation in transgenic mice. Nat Biotechnol 19:746–750
29. Stanger BZ, Tanaka AJ, Melton DA (2007) Organ size is limited by the number of embryonic progenitor cells in the pancreas but not the liver. Nature 445:886–891
30. Curado S, Anderson RM, Jungblut B, Mumm J, Schroeter E, Stainier DY (2007) Conditional targeted cell ablation in zebrafish: a new tool for regeneration studies. Dev Dyn 236: 1025–1035
31. Medico E, Gambarotta G, Gentile A, Comoglio PM, Soriano P (2001) A gene trap vector system for identifying transcriptionally responsive genes. Nat Biotechnol 19:579–582
32. Pisharath H, Rhee JM, Swanson MA, Leach SD, Parsons MJ (2007) Targeted ablation of beta cells in the embryonic zebrafish pancreas using E. coli nitroreductase. Mech Dev 124: 218–229
33. Laflamme MA, Murry CE (2011) Heart regeneration. Nature 473:326–335
34. Davison JM, Akitake CM, Goll MG, Rhee JM, Gosse N, Baier H, Halpern ME, Leach SD, Parsons MJ (2007) Transactivation from Gal4-VP16 transgenic insertions for tissue-specific cell labeling and ablation in zebrafish. Dev Biol 304:811–824
35. Anderson RM, Bosch JA, Goll MG, Hesselson D, Dong PD, Shin D, Chi NC, Shin CH, Schlegel A, Halpern M, Stainier DY (2009) Loss of Dnmt1 catalytic activity reveals multiple roles for DNA methylation during pancreas development and regeneration. Dev Biol 334:213–223
36. Beis D, Stainier DY (2006) In vivo cell biology: following the zebrafish trend. Trends Cell Biol 16:105–112
37. Stoick-Cooper CL, Weidinger G, Riehle KJ, Hubbert C, Major MB, Fausto N, Moon RT (2007) Distinct Wnt signaling pathways have opposing roles in appendage regeneration. Development 134:479–489
38. Jin SW, Beis D, Mitchell T, Chen JN, Stainier DY (2005) Cellular and molecular analyses of vascular tube and lumen formation in zebrafish. Development 132:5199–5209
39. de Pater E, Clijsters L, Marques SR, Lin YF, Garavito-Aguilar ZV, Yelon D, Bakkers J (2009) Distinct phases of cardiomyocyte differentiation regulate growth of the zebrafish heart. Development 136:1633–1641

Part VI

Wound Healing Cell Biology

Chapter 28

Quantifying Alterations in Cell Migration: Tracking Fluorescently-Tagged Migrating Cells by FACs and Live-Imaging

Rachael Z. Murray

Abstract

Cell migration is fundamental to many different physiological processes including embryonic development, inflammation and wound healing. Given the range and importance cell migration plays a number of assays have been developed to measure different aspects of cell migration. Here we describe two different methods to analyze cell migration. The first method analyzes the migration of fluorescently tagged cells using Boyden chambers and FACs and the second looks at migration properties using time-lapse microscopy.

Key words Cell migration, Live-imaging, Boyden chamber, Fluorescence

1 Introduction

1.1 Cell Migration and Wound Healing Assays

The skin serves as a protective barrier to the environment and loss of its integrity leads to a typical wound healing response. Cell migration is a key component of many of the stages of this repair response, particularly in the inflammatory stage, where many different immune cells migrate into the site of injury, and in the reepithelialization stage of repair. Like wound healing, migration is also a highly dynamic and complex process that can be separated into distinct steps including; the extension of membrane towards the direction of movement, formation of adhesive contacts at the leading edge, deadhesion at the rear of the cell, and cytoskeletal contraction to propel the cell forward [1]. Cells can migrate as a sheet of cells, such as the movement of keratinocyte sheets during the reepithelialization of the wound, or as individual cells, for instance the infiltration of sites of injury by macrophages [1–4]. Migrating cells are often drawn toward a gradient of chemoattractive factors and the surfaces they must migrate over may differ depending on where in the body they are targeted.

Given the complexity of migration, the number of different migratory cell types, and the fundamental role that it plays in many

Robert G. Gourdie and Tereance A. Myers (eds.), *Wound Regeneration and Repair: Methods and Protocols*, Methods in Molecular Biology, vol. 1037, DOI 10.1007/978-1-62703-505-7_28, © Springer Science+Business Media New York 2013

different physiological processes, it is not surprising that a variety of assays have been developed to look at the different aspects of cell migration. It is near impossible to find an assay to analyze the migratory properties of cells that can encompass all of the different factors, such as migration of individual and sheets of cells, and so the choice of assay should be determined by the questions you want to answer. For example, to determine whether your protein of interest can alter reepithelialization then the basic scratch wound assay may be the one for you [5], while determining whether your protein of interest can act as a chemoattractant or your cells are altered in their ability to migrate towards a chemoattractant use of the Boyden chamber or a derivative of this chamber may be your best choice [6].

Often the tools we have at our disposal dictate how we can best analyze the effect of our protein of interest on migration and for many who lack a transgenic or knockout mouse the transient expression of these proteins in cells is the way to go. Proteins can also be stably expressed in cell lines but for some proteins overexpressing or loss can be detrimental to the cells or altering their expression may lead to the up regulation of other family members over time that can compensate for loss of function, as is the case with many of the proteins may laboratory works with. For this reason we often transiently transfect our gene of interest and then measure its effect on cell migration. Unfortunately the transient transfection of your gene of interest does not lead to its expression in every cell often making subtle changes harder to detect. We show here how using FACs we have modify the conventional Boyden chamber to analyzes alterations in only the cells expressing fluorescently tagged wild-type and mutant protein of interest. Additionally, we show how to image live cells expressing a GFP-tagged protein of interest and analyze the data to look at alterations in speed and directionality.

1.2 Boyden Chamber and its Modified Derivatives

The Boyden chamber is one of the most common assays used to measure cell migration. These chambers are often used to study migration towards a higher concentration of a soluble factor (chemotaxis) or for migration over a gradient of extracellular matrix (ECM) bound chemoattractants (haptotaxis) [7, 8]. These properties play key roles during a number of physiological processes such as the inflammatory phase of repair, where immune cells are guided towards to site of injury by these gradients, and in the migration of keratinocytes and fibroblasts over the ECM. Many commercial Boyden chamber inserts are available making the procedure relatively easy. This method is also less time consuming than scratch wound assays, where cells must be grown to confluence prior to "wounding."

The assay generally involves two medium filled chambers separated by a microporous membrane, the lower one enclosing media with a chemoattractant in it and the upper one containing the cells of interest. Cells in the upper chamber are able to migrate through the pores of the membrane towards the lower chemoattractant filled chamber. The time for this to occur is cell type, as well as chemoattractant, dependent. These wells can also be coated in individual ECM components or Matrigel™, which is a gelatinous mixture that resembles the complex ECM environment found in many tissues, prior to the start of the assay. Alternatively, wells can be coated in a layer of cells; for example, the plating of immune cells on to a confluent layer of endothelial cells is often used to study the passage of immune cells through the blood vessel wall. The typical method involves removing the cells from the upper chamber, fixing the membrane, staining with a dye such as hematoxylin and the number of cells that have migrated to the lower side of the membrane can then be counted. We provide a quick and simple method using FACs analysis to determine the effect of transiently overexpressing a GFP-tagged protein on the migration of cells towards a chemoattractant using a commercially available Boyden chamber (Corning's Transwells) [9].

When selecting your choice of commercial Boyden chamber the membrane pore size you choose is important and is very much cell size dependent. The pore size should allow for active transmigration through the membrane; for instance, many immune cells are small and a 3–5 μm pore is most suitable, while the larger epithelial and fibroblasts may require the larger 8 μm pore size. Several different companies provide disposable tissue culture plastic inserts, such as Corning's Transwell inserts, with these membranes in them. The inserts are available in a range of sizes to fit different tissue culture wells, allowing you to reduce the number of cells if necessary, which is particularly important when working with a limited number of primary cells or small amounts of chemoattractant. For information on insert sizes and the membrane pore sizes available, as well as information on the different membrane types for specific applications see the individual company Web sites.

For many purposes, such as for testing whether your protein of interest can act as a chemoattractant during wound healing, the commercial inserts are appropriate, however, many of these inserts, such as the Transwells, do not allow for the live imaging of the migratory behavior of cells. Being able to see how migration is altered is important since cells have the ability to migrate randomly or directionally and a reduced number of cells in the lower well could indicate a reduction in speed or alternatively the loss of directionality. To overcome this problem a number of commercially available devices based on the Boyden chamber have been designed, such as the Dunn Chamber or Ibidi's chemotaxis slides. These devices are designed for use under a microscope and stable state linear gradients can be formed in their chambers. The type of device used will depend on the questions you are asking. The data from these slides can be analyzed in a similar fashion to cells grown on coverslips and imaged live (see below).

1.3 Migration of Individual Cells and Cell Sheets

Of the different methods used to study cell migration the scratch-wound healing assay is perhaps the simplest and least expensive method employed. Importantly, it also to some extent has been found to mimic the directional migration of sheets of cells seen in vivo during wound repair. The basic method involves growing a monolayer of cells to confluence or near confluence on uncoated or ECM component-coated dishes/coverslips and creating a scratch, typically with a pipette tip, simulating a "wound." Cells then migrate directionally into the wound (and proliferate) to fill in the gap and new cell-to-cell contacts are established. Depending on different factors, such as cell type, extracellular matrix and wound size, closure can range from hours to more than a day. One of the main advantages of this method is that sheets of cells can be analyzed whereas plating on the Boyden chambers requires these cell-to-cell contacts to be disrupted prior to loading in the chambers. This method is most commonly used to determine factors that regulate epidermal/dermal migration during tissue repair, such as cell-to-cell contact or cell to ECM contact. It has been widely used to study the directed migration of both loosely connected populations of cells, for example fibroblasts, or sheets of attached cells, such as epithelial cells.

In the simplest form of this assay cells are fixed at regular intervals post wounding, then imaged and the wound area measured to get a calculation of percentage of wound closure and this is then compared to a control wound. Again, like the Boyden chamber, this type of analysis gives little to no data on how migration is altered, (e.g., speed or directionality). Live-imaging of these scratched wounds holds great advantage over the fixed method in that it allows researchers to track not only sheets of cells but to analyze individual cells during this process. Cells overexpressing fluorescent-tagged wild-type and mutant proteins can be imaged live at the edge of the wound and their migration tracks compared to the surrounding cells at the edge of a wound and to control cells expressing the fluorescent-tag alone. The localization of the nucleus (for cell polarity) or the fluorescently tagged protein within the cell can also be determined at the different stages of migration. This allows the researcher to establish the function of specific genes in regulating directional migration.

We provide a general method for imaging live cells expressing GFP-tagged proteins and how to analyze the data you obtain to calculate speed and directionality [10]. This method can be modified to suit your cell type and can be used to image individual cells, cells in sheets and also imaging in the commercial modified Boyden chemotaxis slides mentioned above.

2 Materials

2.1 Migration Analysis of Fluorescently Tagged Cells Using the Boyden Chamber

2.1.1 Coating Permeable Supports

1. Permeable supports: we use Costar Transwell supports 6.5 mm diameter, 0.5–0.8 μm and polycarbonate membrane for our RAW264.7 macrophages.
2. A 12-well tissue culture plate.
3. Sterile phosphate buffered saline (PBS).
4. Sterile fibronectin (cell culture grade). Make up to 10 μg/ml in sterile PBS and store at −20 °C until use.
5. Parafilm to seal the plates for storage at 4 °C if you are not using them immediately.

2.1.2 Transient Transfection of Cells with the GFP-Tagged Gene of Interest and Setting Up Chambers with Cells

1. We routinely use RAW264.7 macrophages (ATCC TIB-71™); however, this method can easily be adapted to suit your preferred cell type.
2. Pre-warmed RAW media (RPMI containing 10 % fetal calf serum and 2 mM L-GlutaMAX).
3. Pre-warmed OptiMEM.
4. Lipofectamine 2000 or the transfection reagent that you know works well for your cell type.
5. The plasmid of interest containing the GFP-tagged gene of interest and the same plasmid with GFP alone in it as a control.
6. Tissue culture dishes (6 cm).
7. Sterile PBS.
8. Sterile EDTA (0.5 mM EDTA in PBS).
9. Sterile forceps.
10. Sterile 15 ml centrifuge tubes.

2.1.3 Boyden Chamber Cell Retrieval and Flow Cytometry Analysis of GFP-Expressing Cells

1. Pre-warmed RAW media (RPMI containing 10 % fetal calf serum and 2 mM L-GlutaMAX).
2. Room temperature sterile PBS.
3. Ice-cold PBS.
4. *N*-formyl-Met-Leu-Phe (fMLP) stock solution of 10 mM in DMSO and stored at −20 °C.
5. Pre-warmed 0.25 % Trypsin–EDTA mixture.
6. Small sterile rubber scrapers.
7. FACs tubes.

2.2 Analysis of Cell Migration Using Time-Lapse Microscopy

2.2.1 Fibronectin Coating of MatTek Dishes

1. Tissue culture dish with glass coverslip: we used MatTek Glass bottom 35 mm culture dishes with a 14 mm microwell containing a cover glass.
2. Sterile phosphate buffered saline (PBS).
3. Filter-sterilized poly-L-lysine solution at 50 μg/ml in PBS and stored at 4 °C.
4. Fibronectin, (cell culture grade), made up to 10 μg/ml in sterile PBS and stored at −20 °C until use.
5. Parafilm to seal the plate for storage.

2.2.2 Cell Culture and Transient Transfection

1. We routinely use RAW264.7 macrophages (ATCC TIB-71™) but this method can easily be adapted to suit your preferred cell type.
2. Pre-warmed RAW264.7 media (RPMI 1640 with 10 % fetal bovine serum and 2 mM L-GlutaMAX).
3. Pre-warmed OptiMEM.
4. Lipofectamine 2000 or the transfection reagent that you know works well for your cell type.
5. A plasmid containing the GFP-tagged gene of interest and the same plasmid with GFP alone in it as a control.
6. Pre-warmed sterile EDTA (0.5 mM in PBS).
7. Pre-warmed sterile PBS.
8. Sterile *N*-formyl-Met-Leu-Phe (fMLP) 10 mM stock solution in DMSO, stored at −20 °C.
9. Sterile 15 ml centrifuge tubes.

2.2.3 Live-Imaging and Analysis Software

1. Pre-warmed CO_2 independent media.
2. A microscope adapted for time-lapse imaging. We use an Olympus IX81 Inverted microscope, housed in an environmental chamber heated to 37 °C (a heated stage could also be used), coupled to an ORCA ERG cooled CCD camera with an X-cite fluorescent light source. Our set up uses Metamorph V7.6 imaging software to capture images. The microscope has an automated stage, which, along with Metamorph, allows us to capture multiple stage positions at each time point with approximately 10 cells per position.
3. ImageJ, which is a free Java-based image-processing program developed at the National Institutes of Health in the US and can be downloaded at http://rsbweb.nih.gov/ij/download.html. ImageJ runs on several platforms including Unix, Linux, Windows and Mac OS X. There are numerous plug-ins freely available from the ImageJ Web page and you will need the manual tracking one http://rsb.info.nih.gov/ij/plugins/track/track.html as well as Ibidi's free chemotaxis tool http://www.ibidi.de/applications/ap_chemo.html.

3 Methods

3.1 Migration Analysis of Fluorescently Tagged Cells Using the Boyden Chamber

3.1.1 Coating Permeable Supports

1. Place 6.5 mm Transwell inserts in a 12-well plate using sterile forceps.
2. Coat the underside of the insert by adding 650 μl of 10 μg/ml fibronectin (or your choice of ECM component) in PBS to the lower well and gently lower the insert at an angle into place, being careful not to trap any bubbles between the liquid and membrane.
3. Incubate for 3 h at 37 °C.
4. Using a pipette transfer the fibronectin to a new tube for reuse.
5. Wash three times by pipetting 650 μl of sterile PBS in the lower well and using a glass Pasteur pipette and a vacuum pump to remove the washes.
6. The supports can either be used immediately or can be stored at 4 °C in a 12-well plate that is sealed with Parafilm to prevent the plates drying out until use.

3.1.2 Transient Transfection of Cells with GFP-Tagged Gene of Interest

1. Seed RAW264.7 macrophages at 2.5×10^6 onto a 6 cm cell culture dish and cultured overnight in 5 ml of RAW264.7 media in a humidified incubator at 37° with 5 % CO_2.
2. In a sterile tube mix 18 μl of Lipofectamine 2000 to 0.5 ml of OptiMEM and incubate for 5 min.
3. In a second sterile tube mix 2.25 μg DNA into 0.5 ml of OptiMEM and then add the pre-incubated Lipofectamine mixture to this tube.
4. Incubate for 20 min.
5. Remove the media from the 6 cm cell culture dish, wash once in 2 ml of OptiMEM and then add 4 ml of OptiMEM to the plate.
6. Add the Lipofectamine–DNA mix to the 6 cm cell culture dish and incubate for 3 h.
7. Remove the media and replace with 5 ml of RAW264.7 media for 2–3 h.

3.1.3 Assembling the Wells

1. Assemble the Transwells in triplicate in the 12-well tissue culture dish for each condition so that you have three representative samples for quantification. Also set up control wells with no insert, either in the same plate or in a new plate.
2. Load 650 μl of RAW264.7 media containing 100 nM fMLP as a chemoattractant into the lower chamber being careful so that no media spills into the upper chamber. The media should form a slight positive meniscus when the well is filled.
3. Ensure that there are no air bubbles present in the media, if so use a sterile needle to burst them. Place the Transwell insert in

the well at a slight angle and lower into place making sure that the media touches the lower membrane and that there are no bubbles between it and the lower membrane.

3.1.4 Loading the Upper Chamber and Control Wells with Cells

The cells should be kept in the dark whenever possible.

1. Remove the media from the transiently transfected cells and wash with 5 ml of sterile PBS. Depending on cell type and chemoattractant you might need to serum-starve your cells prior to this step as fetal calf serum contains factors that can effect migration.
2. Incubate the cells with 3 ml of sterile 0.5 mM EDTA in PBS for 5 min at 37 °C.
3. Using a sterile rubber cell scraper gently detach the cells from the plate and transfer to a sterile 15 ml centrifuge tube. Add 5 ml media to the plate to collect any remaining cells and transfer to the same tube.
4. Ensure that the cells are thoroughly resuspended by passaging five times through a serological pipette. Take a sample for cell counting noting the volume in the tube. Add an equal volume of Trypan Blue to the sample. This stains the dead cells blue and ensures you count only viable cells. Using a hemocytometer and microscope determine the cell density. The number of cells in the center 25 squares of the hemocytometer is equal to number of cells $\times 10^4$ cells per ml. Do not forget to multiple by 2 to factor in the Typan Blue dilution.
5. Centrifuge the cells at $300 \times g$ in a benchtop centrifuge at room temperature.
6. Aspirate the media using a glass pipette and vacuum pump and carefully resuspend cells at a density of 5×10^6 cells/ml.
7. Plate 100 μl of the resuspended cell mixture (i.e., 5×10^5 cells) into the upper well. To load the upper chamber, hold the pipette so that the end of the pipette tip is against the side of the well just above the membrane (do not touch the membrane) and expel the liquid from the pipette tip. Plate each sample in triplicate.
8. Also seed cells at the same density into a well without a chamber so that you can determine the percentage of viable GFP-positive cell loaded into the wells.
9. Incubate the plate(s) in a humidified incubator at 37 °C with 5 % CO_2 for 18 h.

3.1.5 Harvesting Cells and FACs Analysis

Again the cells should be kept in the dark whenever possible.

(a) For each Transwell

1. Transfer the media from each well into labeled FACs tube placed on ice. Cover the ice-box in foil to reduce exposure of the GFP-expressing cells to light.

2. Remove each Transwell from the 12-well plate one by one and using a cotton bud remove the cells from the upper surface.
3. Add 650 μl PBS to the lower well and wash the lower membrane by placing it in the well.
4. Transfer the PBS to the same labeled FACs tube.
5. Add 650 μl of the pre-warmed 0.25 % Trypsin–EDTA mixture to the lower chamber with the insert and incubate at 37 °C for 5 min. Again ensure that there are no air bubbles between the membrane and the liquid in the lower well.
6. Remove the Transwell insert and using a small rubber scraper remove the cells from the lower membrane, continue washing the scraper in the media remaining in the well to transfer the cells from the scraper.
7. Transfer the Trypsin–EDTA mixture containing cells to the same labeled FACs tube and add 2 ml of RAW264.7 media containing fetal calf serum to inactivate the trypsin.
8. Wash any remaining cells from the well and the membrane by adding 650 μl to the well containing the insert and transfer to the same FACs tube.

(b) For the control wells

1. Transfer the media from each well into a labeled FACs tube placed on ice.
2. Add 650 μl of 0.25 % Trypsin–EDTA to the control well and incubate at 37 °C for 5 min.
3. Scrape the remaining cells from the bottom of the control wells with a rubber scraper and transfer the Trypsin–EDTA mixture containing cells to the same labeled FACs tube. Add 2 ml of RAW264.7 media containing fetal calf serum to inactivate the trypsin.
4. Collect any remaining cells from the well by adding 1 ml of RAW264.7 media to the well and transferring the liquid to the same labeled FACs tube.

3.1.6 FACs Analysis

1. Centrifuge the FACs tubes at 300 × *g* in a benchtop centrifuge at 4 °C and resuspend the pellet in 300 μl ice cold PBS. Using a pipette ensure that you have a single cell suspension by bring the cells up and down in the pipette tip.
2. Count the percentage of GFP-positive cells in the control wells that have migrated to the lower chamber using a FACs machine (using the appropriate method and software for the machine you have available).

3.1.7 Analysis of Results

The fold change in migration of GFP-expressing cells is equal to the % GFP-positive cells migrating to the lower well divided by the % GFP-positive cells in the control wells. This can then be normalized to cells transfected with GFP-alone as a control to get a fold change in migration. For example, if the percentage of GFP-alone expressing cells in the control well is 18 % and 17 % in the lower chamber (17/18 = 0.94), while the percentage of GFP-tagged protein X expressing cells is 15 % in the control well and 8 % in the lower chamber (8/15 = 0.53) then there is a 1.77 fold reduction in migration (0.94/0.53 = 1.77).

3.2 Analysis of Cell Migration Using Time-Lapse Microscopy

Aseptic techniques must be adhered to throughout the procedures until all images have been captured. Try to keep the transfected cells in the dark whenever possible.

3.2.1 Fibronectin Coating MatTek Dishes

1. Thaw enough poly-L-lysine and fibronectin to cover the base of the dish for coating. For the MatTek plates 2 ml is sufficient volume.
2. Incubate the plate in poly-L-lysine (50 μg/ml in PBS) for 2 h at 37 °C.
3. Remove the poly-L-lysine from the dish and wash the plate three times with 2 ml of sterile PBS.
4. Add 2 ml of fibronectin (10 μg/ml in PBS) to the plate and incubate for 3 h at 37 °C or 4 °C overnight.
5. Remove the fibronectin from the dish and wash the plate three times with 2 ml of sterile PBS.
6. The coated plates can either be used immediately or can be sealed with Parafilm to prevent the plates drying out and stored at 4 °C until use.

3.2.2 Transient Transfection of Cells with the Gfp-Tagged Gene of Interest

1. Seed RAW264.7 cells at 0.5×10^6 cells in 2 ml of RAW264.7 media onto the MatTek culture dish and cultured overnight.
2. In a sterile tube mix 4.5 μl of Lipofectamine 2000 to 0.5 ml of OptiMEM and incubate for 5 min.
3. In a second sterile tube mix 1.5 μg DNA into 0.5 ml of OptiMEM and then add the pre-incubated Lipofectamine 2000 mixture to this tube.
4. Incubate for 20 min.
5. Remove media from the MatTek culture dish and wash once with 2 ml of OptiMEM. Add 1 ml of OptiMEM to the plate and place back in the incubator until you are ready to add the Lipofectamine 2000–DNA mixture to the cells.
6. Add the Lipofectamine 2000–DNA mixture to the MatTek culture dish and culture at 37 °C 5 % CO_2 for 3 h.
7. Remove the media and replace with normal RAW264.7 media for 2–3 h.

3.2.3 Imaging Cells

Note the cells should be kept in the dark whenever possible.

1. Pre-warm the microscope to 37 °C.
2. Remove the media from the cells using a glass Pasteur pipette and vacuum pump, and replace with 2 ml of pre-warmed CO_2 independent media.
3. Immediately prior to taking the plate to the microscope add 100 nM fMLP (1:100 dilution from the stock solution in CO_2 media, made fresh each time, then a further 1:1,000 into the dish) to the plate under sterile conditions.
4. Place the dish on the heated stage of the microscope and find the plane with the cells in it. The live image of the cells can then be viewed on the computer screen using the Metamorph program (or whichever program your microscope uses). We generally pick approximately 30 positions from different areas of the glass slide, with an automated stage this is relatively easy and allows you to obtain data from a large number of cells for statistical analysis. How you image the cells will depend on the microscope available and its software.
5. For imaging the migration of macrophages we image every 5 min over 3 h. If your cells are slow migratory you may need to image for a longer period such as overnight.

3.2.4 Analysis

1. If you have not already done so then download ImageJ (http://rsbweb.nih.gov/ij/download.html) along with the Manual Tracking plug in (http://rsb.info.nih.gov/ij/plugins/track/track.html) and Ibidi's chemotaxis tool (http://www.ibidi.de/applications/ap_chemo.html).
2. Open your movie or stack file in **ImageJ**, then in the menu bar under **Plugins** click on **Manual Tracking** to open the tracking plugin.
3. In the **Parameters** box (Fig. 1, red circle #1) enter the **Time Interval** the images were taken at, for us this was every 5 min, and the **xy Calibration** value, i.e., the size of a pixel (this specific to your optics and microscope and the information should be available from the company that provided your microscope).
4. You will next need to track the center of the nucleus on each image during the time course. To do this click the **Add Track** button (Fig. 1, red circle #2) to start recording and using the **cross pointer** (Fig. 1, red circle #3) click on the center of the nucleus (Fig. 1, red circle #4).
5. Once you have clicked on the first image the program will automatically skip to the next image in the movie where you can click on the center of the nucleus for the same cell in the sequence. Repeat this process until the last frame of the sequence when the tracking will automatically stop. If your cell

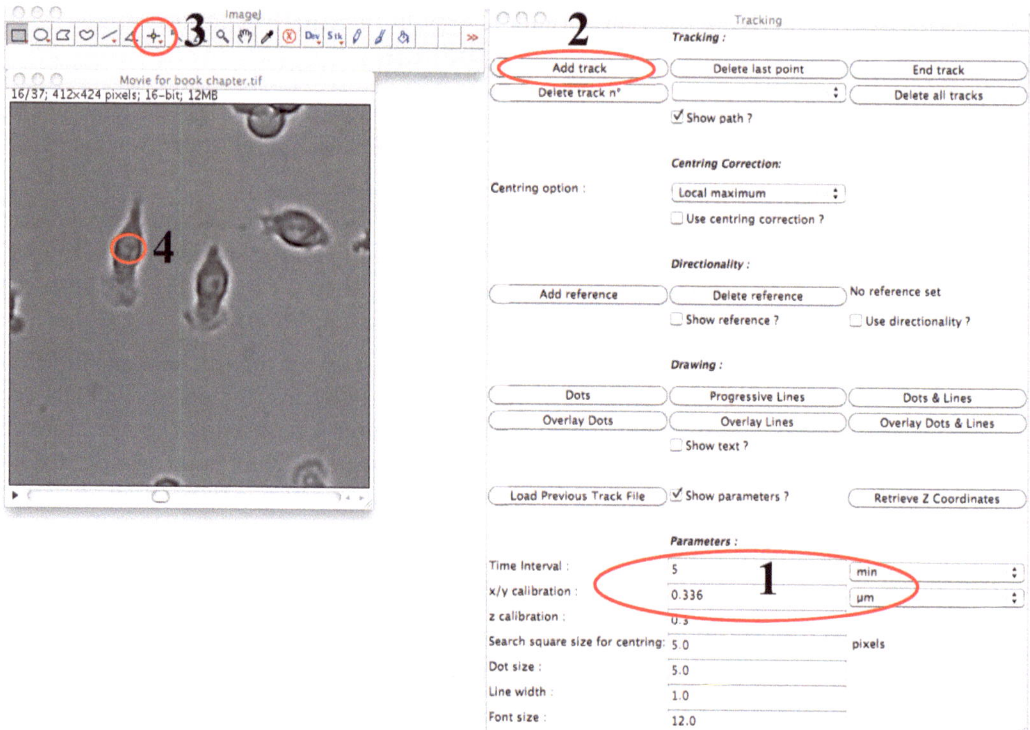

Fig. 1 Manual tracking of cells using the ImageJ Manual Tracking tool. Step 1

moves out of frame and you want to stop tracking at an earlier point then click **End Track**. The results containing the xy coordinates will automatically appear in a pop up box (Fig. 2). You can then repeat the procedure with a new cell. We generally track at least 100 cells to get representative data for statistical analysis.

6. Click on the results window and click the **File** menu, then **Save As** to save this data so you can open it with the Chemotaxis plugin. The data is saved in Excel format (.xls).
7. In the menu bar under **Plugins** click on **Chemotaxis Tool** to open the Chemotaxis and Migration Tool box (Fig. 3).
8. Click on the **Import Data Tab** (Fig. 3, red circle #1) and click on the **Import Data Button** (Fig. 3, red circle #2). This should open a browser so you can upload the save .xls file with you data in it. After the data has uploaded change the **Number of Slices** to from "Use only slices equal to" to **"Use slice range from…to"** and define your number of frames (Fig. 3, red circle #3). In the left corner tick the **Selected Dataset 1 box** (Fig. 3, red circle #4). Four different sets of data can be loaded at one time by repeating the same procedure but ticking the appropriate **Selected Dataset** box.

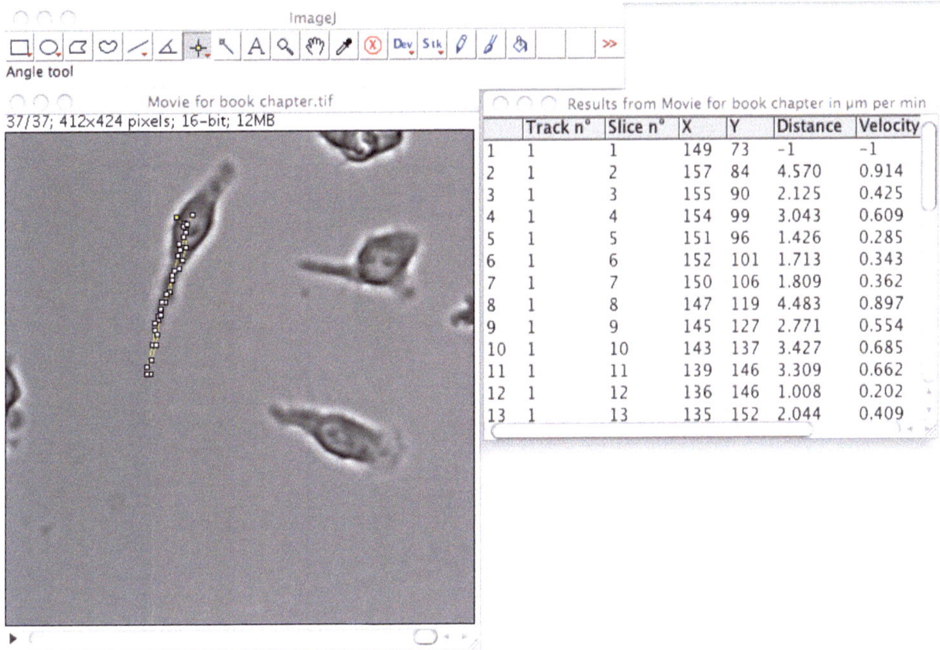

Fig. 2 Manual tracking of cells using the ImageJ Manual Tracking tool. Step 2

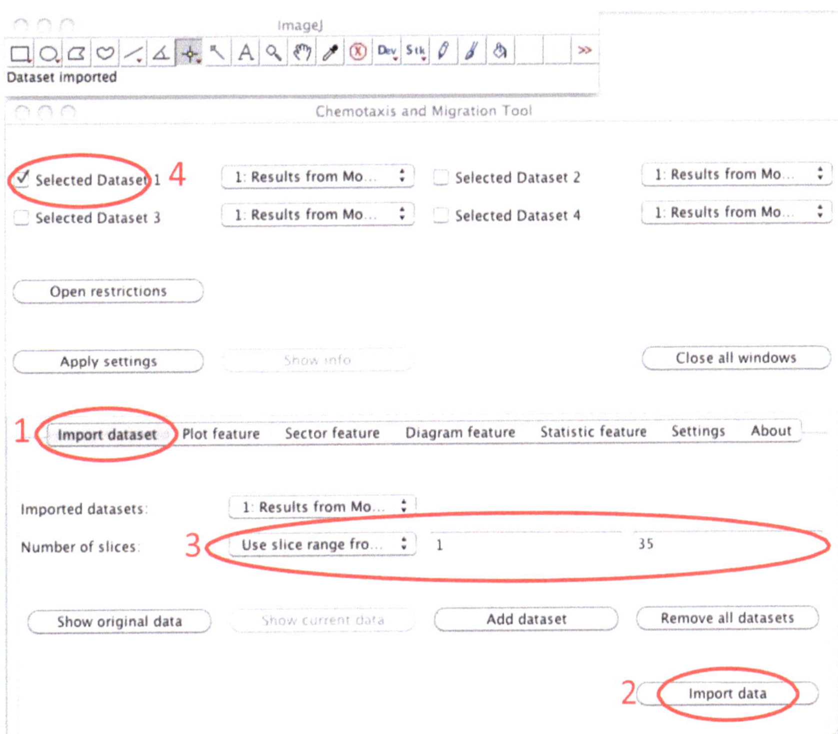

Fig. 3 Migration analysis of cells using ImageJ and Ibidi's Chemotaxis and Migration Tool plugin. Step 1

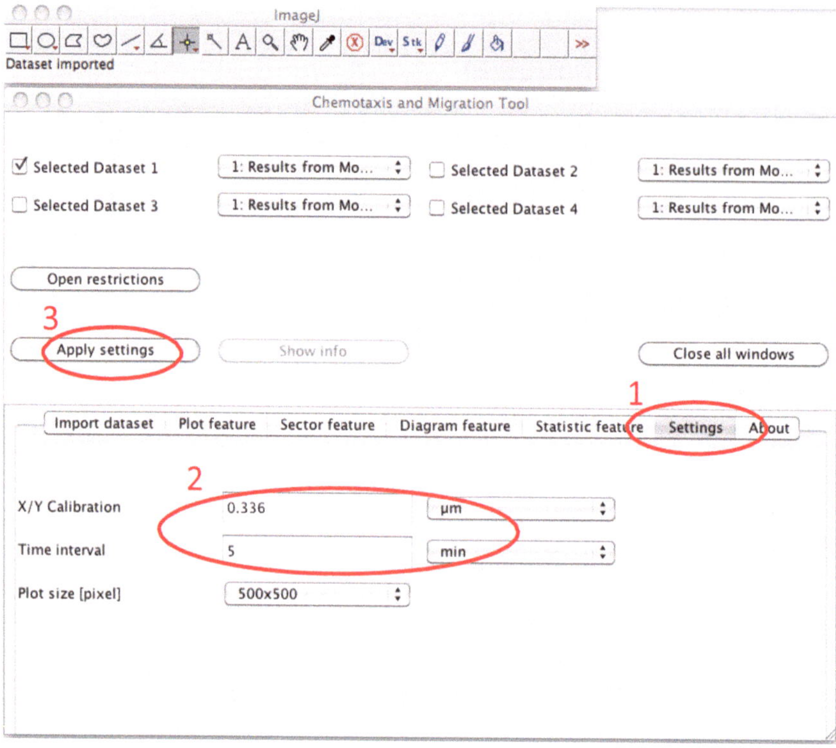

Fig. 4 Migration analysis of cells using ImageJ and Ibidi's Chemotaxis and Migration Tool plugin. Step 2

9. Click on the **Settings** tab (Fig. 4, red circle #1). Again define the **x/y Calibration** (pixel size) and **Time Interval** (Fig. 4, red circle #2). Click the **Apply Settings** button (Fig. 4, red circle #3).
10. Next click on the **Statistic Feature** tab (Fig. 5, red circle #1) followed by the **Show Info** button (Fig. 5, red circle #2). This will produce a new window containing a summary of data, such as distances, velocity and directionality (Fig. 5, #3).
11. Migration tracks can be plotted on a graph by clicking the **Plot Feature** tab (Fig. 6, red circle #1). Tick the **Open in a new window** box (Fig. 6, red circle #2) and click the **Plot Graph** button. The graph will open in a new window (Fig. 6, #4).
12. You can also plot histograms, Rose diagrams or circular plots by clicking on the **Diagram Feature** tab (Fig. 7, red circle #1), ticking the **open in a new window box** (Fig. 7, red circle #2) and clicking on the appropriate button, for example, click the **Plot Rose diagram** button (Fig. 7, red circle #3) and the Rose plot should appear in a new window (Fig. 7, #4).

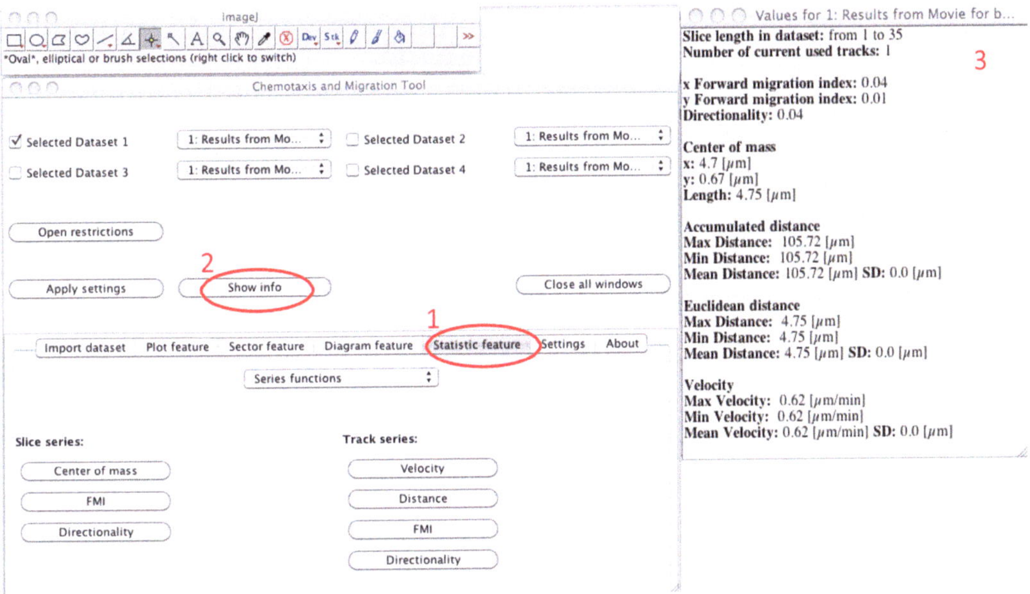

Fig. 5 Migration analysis of cells using ImageJ and Ibidi's Chemotaxis and Migration Tool plugin. Step 3

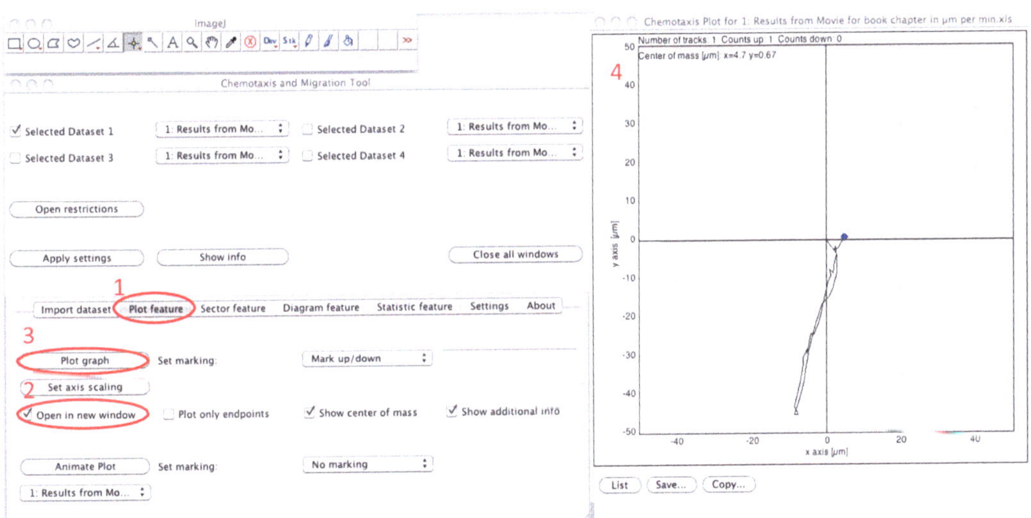

Fig. 6 Migration analysis of cells using ImageJ and Ibidi's Chemotaxis and Migration Tool plugin. Step 4

4 Notes

4.1 Migration Analysis of Fluorescently Tagged Cells Using the Boyden Chamber

Notes 1-11 refer to the the analysis of fluorescently tagged cells using the Boyden chamber and notes 12-18 refer to the analysis of cell migration using time-lapse microscopy.

General Notes

1. It is important to select the right permeable membrane, pore size and support depending on your cell type (taking into account

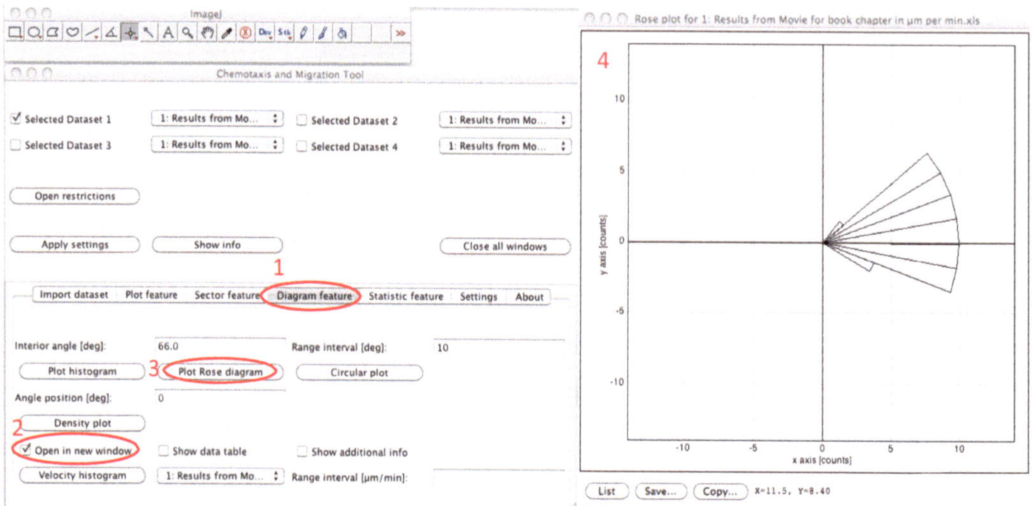

Fig. 7 Migration analysis of cells using ImageJ and Ibidi's Chemotaxis and Migration Tool plugin. Step 5

cell size), cell numbers, and volume you wish to use. A pore size of 0.3–0.8 μM is generally used to look at cell migration. Permeable supports are mounted in a plastic insert that fit either 24- or 12- or 6-well culture plates.

2. All reagents are sterile and aseptic techniques must be adhered to throughout this procedure.
3. The ECM coating can be altered or Matrigel™ can be used in the wells to look at the effect of your protein of interest on migration or invasion respectively. The Transwells are only able to measure migration and must have some sort of gel, such as Matrigel™, in them to provide a 3D environment to measure invasion if this is what you wish to measure.
4. The fibronectin solution for coating can be reused several times as long as you keep it sterile. We generally reuse ours three times.
5. We use Lipofectamine 2000 to transfect RAW264.7 cell as this seems to work best for these macrophages; however, I would suggest that you use the method that you know works for your cell type. Do not forget to do a GFP-alone control.
6. Try not to set up too many wells in one experiment, as the cell retrieval process is quite fiddly and time consuming.
7. Cell attachment and spreading may be improved by pre-incubating the permeable supports in the appropriate media prior to seeding cells.
8. You will need to standardize the seeding densities for your cell type. The initial seeding density can be important for some cell types, for example, both the growth properties and activation of the RAW264.7 macrophages we routinely use are greatly altered if cells are seed at too low a density.

9. You will also need to optimize incubation times for your cell type.
10. The rubber cell scrapers can be reused. We generally wash them, stored them in 70 % ethanol and UV them prior to reuse.
11. If you do not have immediate access to the FACs machine then you can resuspend the cells in 1 ml of 1 % paraformaldehyde and store at 4 °C. Prior to FACs analysis centrifuge the tubes at 300 × *g* in a benchtop centrifuge at 4 °C and resuspend in 300 μl ice cold PBS.

4.2 Analysis of Cell Migration Using Time-Lapse Microscopy

General Notes

12. All reagents are sterile and aseptic techniques must be adhered to throughout this procedure.
13. This protocol is based on RAW264.7 macrophage migration on fibronectin; however, dishes can easily be coated in other extracellular matrix components appropriate to your individual experiments and the cell numbers altered to suit your cell type. Ideally cells should be spread and not too confluent so they have room to migrate.
14. It is important that the microscope has reached the correct temperature, if not the stage tends to expand during imaging and the cells move out of focus during the time course. We tend to switch the temperature controller on the evening before imaging so that the chamber generally has 12 h to equilibrate. Try not to leave the door to the chamber open for too long when you are placing the cells on the stage.
15. If your microscope system has a CO_2 supply then you do not need to use CO_2 independent media. Alternatively, you could buffer your media with 10–20 mM HEPES at pH 7.4; however, this only reduces the rate of pH drift rather than eliminating it. Also, there are numerous reports of HEPES toxicity during imaging over longer period so in the absence of a CO_2 supply a CO_2 independent media is recommended if you need to image for long periods. If you decide to buffer your media with HEPES then its effect on cell viability should be tested prior to imaging.
16. To enable a clear picture of the nucleus (RAW cells are quite small in comparison to say epithelial cells) and to count a suitable number of cells for statistical analysis we routinely use an UPlanApo 40× OIL N.A. 1.00 objective.
17. We use Lipofectamine 2000 to transfect RAW264.7 cell as this seems to work best for these cells; however, I would suggest you use the method that you know works for your cell type. Do not forget to do a GFP-alone control.
18. The GFP-tagged proteins we express are generally made and trafficked to their intracellular locations relatively quickly.

Often their longer-term expression has a detrimental effect on our cells so we tend to image our cells at this point. You should do whatever you know works for your cells and protein of interest. We occasionally work with proteins that we express for 24 h before imaging cells. As these cells will overgrow in this time frame and the macrophages are not as happy when seeded at low cell numbers we tend to transfect cells in a 6 cm tissue culture dish and split the cells at about 1:2–4 depending on how confluent they look into the MatTek dishes. *See* Subheading 3.1.4 on how to harvest the cells for plating.

References

1. Ridley AJ, Schwartz MA, Burridge K, Firtel RA, Ginsberg MH, Borisy G, Parsons JT, Horwitz AR (2003) Cell migration: integrating signals from front to back. Science 302: 1704–1709
2. Friedl P, Gilmour D (2009) Collective cell migration in morphogenesis, regeneration and cancer. Nat Rev Mol Cell Biol 10:445–457
3. Friedl P, Weigelin B (2008) Interstitial leukocyte migration and immune function. Nat Immunol 9:960–969
4. Rorth P (2009) Collective cell migration. Annu Rev Cell Dev Biol 25:407–429
5. Todaro GJ, Lazar GK, Green H (1965) The initiation of cell division in a contact-inhibited mammalian cell line. J Cell Physiol 66: 325–333
6. Boyden S (1962) The chemotactic effect of mixtures of antibody and antigen on polymorphonuclear leucocytes. J Exp Med 115: 453–466
7. Moser B, Loetscher P (2001) Lymphocyte traffic control by chemokines. Nat Immunol 2:123–128
8. Carter SB (1967) Haptotaxis and the mechanism of cell motility. Nature 213:256–260
9. Veale KJ, Offenhauser C, Whittaker SP, Estrella RP, Murray RZ (2010) Recycling endosome membrane incorporation into the leading edge regulates lamellipodia formation and macrophage migration. Traffic 11(10): 1370–1379
10. Veale KJ, Offenhauser C, Lei N, Stanley AC, Stow JL, Murray RZ (2011) VAMP3 regulates podosome organisation in macrophages and together with Stx4/SNAP23 mediates adhesion, cell spreading and persistent migration. Exp Cell Res 317:1817–1829

Chapter 29

Examining the Role of Mast Cells in Fetal Wound Healing Using Cultured Cells In Vitro

Brian C. Wulff and Traci A. Wilgus

Abstract

Mast cells play an important role during the inflammatory phase of wound healing, and studies suggest that they also influence scar formation and remodeling. Recently, our laboratory has characterized the mast cell response to injury in a fetal wound healing model. In this model, early gestation fetal skin regenerates and heals without a scar (scarless wounds) and late gestation skin heals with a scar (fibrotic wounds). Differences in mast cell number, maturity, and activity were identified between scarless and scar-forming fetal wounds. To study mast cell function in more detail, in vitro experiments are useful. This chapter outlines methods to expand, purify, and study the function of mast cells harvested from murine fetal skin. Using these methods, cultured mast cells retain many of the differences in maturity and activation seen during fetal skin development in vivo. Studying the function of mast cells in vitro could help define the mechanisms by which mast cells contribute to wound repair and ultimately lead to better therapies for improving wound repair and reducing scar formation.

Key words Inflammation, Mast cell, Scarless repair, Fetal skin, Development

1 Introduction

Cutaneous wound healing is a complicated process whereby the activities of many different cell types are carefully orchestrated to protect against invasion from foreign pathogens and to reestablish homeostasis. Wound healing can be divided into three major stages: inflammation, proliferation, and scar formation/remodeling. In adult mammals, this process brings about imperfect repair of the damaged tissue, resulting in the deposition of disorganized scar tissue. While scar tissue generally results from skin repair in mature tissues, there are select cases where regeneration and scarless healing can take place. One well-studied example of regenerative, scarless healing occurs when fetal skin is injured [1, 2]. During early mammalian fetal development (before embryonic day 16 in mice), injury to the skin will result in little to no inflammation, rapid proliferation,

Robert G. Gourdie and Tereance A. Myers (eds.), *Wound Regeneration and Repair: Methods and Protocols*, Methods in Molecular Biology, vol. 1037, DOI 10.1007/978-1-62703-505-7_29, © Springer Science+Business Media New York 2013

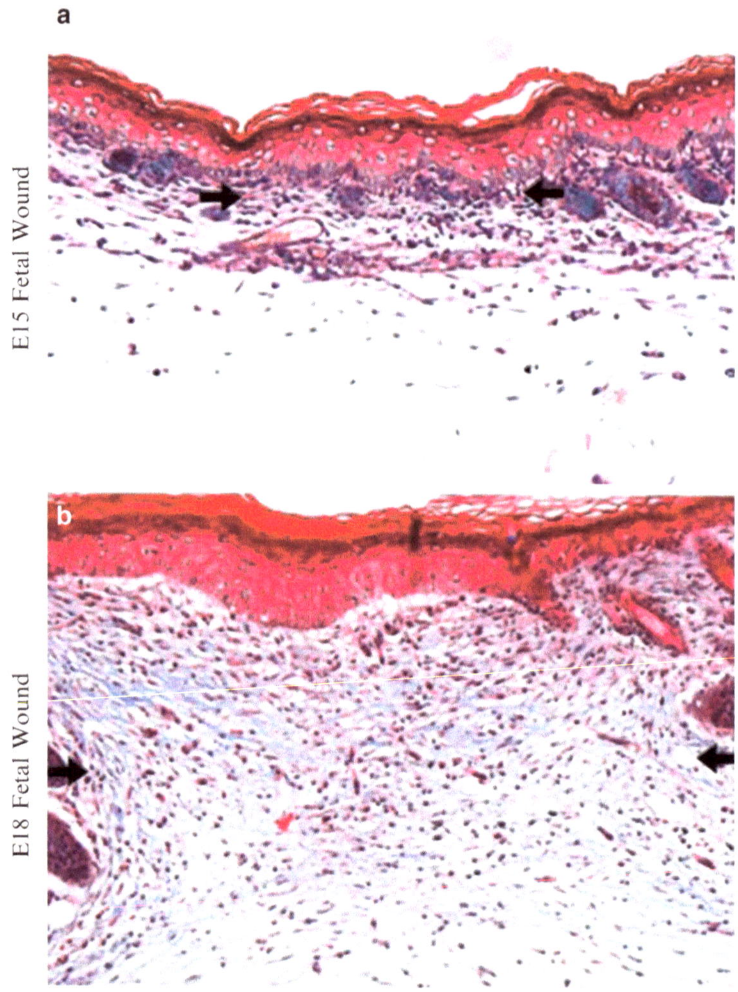

Fig. 1 Scarless and fibrotic fetal wound healing in murine skin. Masson's trichrome-stained sections of fetal skin harvested 7 days after wounding on either E15 (**a**) or E18 (**b**) show the differences in scar formation at these two gestational ages. Fetal skin wounded at E15 heals with deposition of normally organized collagen and regeneration of dermal appendages such as sebaceous glands and hair follicles. In contrast, disorganized collagen is deposited in fetal skin wounded on day E18, resulting in the loss of dermal appendages and formation of a scar. Scale bars are 50 μm

and the regeneration of normal skin without scarring (Fig. 1a). As fetal skin develops, it begins to lose the ability to heal scarlessly and starts to heal with inflammation and scar formation (Fig. 1b) as seen in adult skin. Fetal wound healing models have been powerful tools for understanding the regulation of scar formation by studying the differences between scarless and fibrotic fetal wounds.

Many wound healing studies in both adult and fetal skin have demonstrated a direct correlation between the amount of inflammation present during healing and the amount of scarring that results [3–8]. This is especially true in fetal wounds, where a lack of inflammation is critical for scarless healing to take place [9–11]. One of the first cells to respond to injury [12–14], and a key modulator of inflammation [15], is the mast cell. These cells are resident cells found in many organs that are exposed to the environment and highly susceptible to injury such as the skin, lungs, and gut [16]. Mast cells are a heterogeneous and highly versatile group of cells [17, 18]. They are able to rapidly respond to stimuli by releasing preformed mediators stored within granules and also through de novo production of lipids and cytokines [19]. The spectrum of molecules that mast cells are able to produce is very broad, and includes immunomodulatory molecules, neuronal factors, and angiogenic mediators in addition to many others.

In recent years, studies have begun to shed light on the intricate and complex role that mast cells play in wound healing and scar formation [13, 14, 20–23]. Through these studies, it is becoming clear that mast cells affect all stages of wound healing. In addition to producing pro-inflammatory mediators and a variety of growth factors that influence the inflammatory and proliferative phases of repair [20, 24], mast cells are also thought to play a role in later stages of wound healing. Many mast cell mediators are known to activate fibroblasts and stimulate collagen deposition [25–28], suggesting that mast cells regulate scar formation and remodeling. Recently, our laboratory has demonstrated a role for mast cells in scar formation in fetal wounds [29]. As fetal skin develops, dermal mast cells dramatically increase in number and become more mature at stages when the skin starts to heal with inflammation and scar formation (Fig. 2a–f). In addition, mast cells do not degranulate when early gestation fetal skin is wounded (scarless wounds; Fig. 2g), whereas mast cells show strong degranulation responses in late gestation fetal wounds (scar-forming wounds; Fig. 2h). The increase in resident mast cell numbers and mast cell activation in response to injury in late gestation fetal skin suggests that developmental differences in mast cells could be an important reason why more developed fetal skin heals with a scar.

The method outlined in this chapter has been used to isolate and expand mast cells from fetal skin. Importantly, many of the developmental differences in mast cells observed in fetal skin in vivo are retained in culture, suggesting that these methods could be used to study dermal mast cell development and to understand the mechanisms by which mast cells contribute to wound healing and scar formation. The ability to study dermal mast cells in vitro could also help generate new strategies to manipulate the behavior of these cells in order to stimulate regenerative responses and optimize healing.

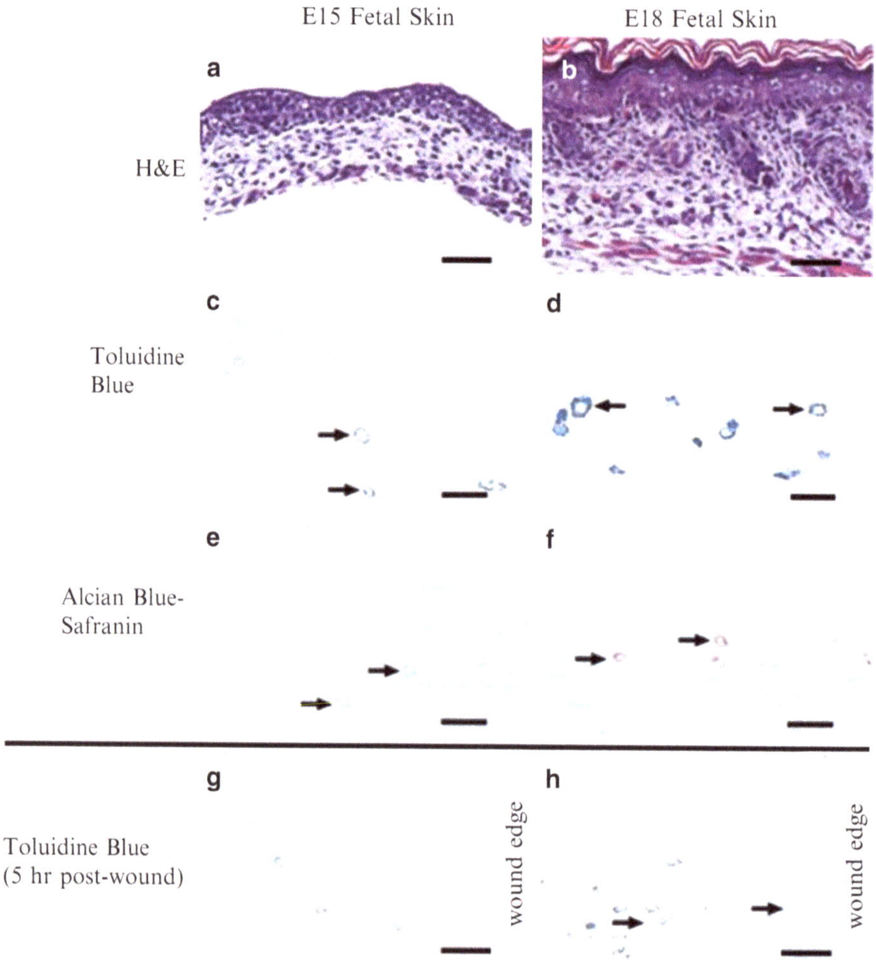

Fig. 2 Mast cell differences in murine fetal skin. Many developmental changes occur in fetal skin between embryonic days E15 and E18, some of which are evident with hematoxylin and eosin (H&E) staining. Unwounded E15 (**a**) and E18 (**b**) skin are shown. The epidermis thickens and stratifies as keratinocytes differentiate, hair follicles develop, and the cellular and extracellular matrix composition of the dermis changes. Toluidine blue staining, which specifically stains mast cell granules blue, shows differences in dermal mast cells in E15 (**c**) and E18 skin (**d**). There are more dermal mast cells in E18 skin, which are larger in size and contain more granules compared to mast cells in E15 skin. Alcian blue–safranin staining shows that there are also differences in the maturity of mast cells in E15 (**e**) and E18 fetal skin (**f**). Alcian blue–safranin staining causes mast cells with sulfated mucopolysaccharides to stain red indicating that they are more mature, while immature mast cells stain blue due to a lack of sulfated mucopolysaccharides. Toluidine blue-stained sections of fetal skin 5 h after wounding show that there are also differences in the mast cell response to injury in E15 and E18 wounds. Mast cell degranulation is not evident in E15 wounds (**g**), but mast cells show a strong degranulation response (*arrows*) in E18 wounds (**h**). Scale bars are 50 μm (**a**–**b**; **g**–**h**) or 20 μm (**c**–**f**)

2 Materials

2.1 Harvest and Culture of Cutaneous Mast Cells

1. 1× RPMI complete medium: For 1 L of media, dissolve one packet of powered RPMI 1640 medium containing 2 mMl-glutamine in an appropriate volume of deionized water to allow for the addition of the following supplements: FBS (fetal bovine serum, 10 % final volume: *see* **Note 1**), 2 g/L sodium bicarbonate, 1 mM nonessential amino acids, 10 mM sodium pyruvate, 25 mM HEPES buffer, 1× antibiotic–antimycotic, and 50 μM 2-mercaptoethanol. Adjust the pH to 7.1 before bringing to a final volume of 1 L with deionized water and filter sterilize using a 0.22 μm sterile filter system.
2. Digestion medium: Supplement 1× RPMI complete medium with collagenase type I (0.5 μg/ml), and hyaluronidase (0.5 μg/ml).
3. Culture medium: Supplement 1× RPMI complete medium with recombinant murine IL-3 (interleukin-3) and SCF (stem cell factor), each at a concentration of 10 μg/ml.
4. 10× RPMI: Dissolve one, 1 L packet of powdered RPMI 1640 medium in 100 ml of deionized water. Filter sterilize.
5. 1× HBSS (Hank's balanced salt solution) with calcium and without phenol red.

2.2 Purification of Cultured Mast Cells

1. Forty percent isotonic Percoll solution: Prepare a solution of isotonic Percoll by adding 1 volume of 10× RPMI (described above) to 9 volumes of Percoll. Dilute the isotonic Percoll with 1× RPMI complete medium (described above) to obtain a 40 % isotonic Percoll solution.
2. 1× HBSS without calcium and without phenol red.

2.3 Histological and Functional Characterization of Mast Cells

1. May-Grünwald stain (commercially available).
2. Giemsa stain: Dilute Giemsa stain (commercially available) 1:20 with deionized water.
3. Acetate buffer: Combine 50 ml of solution A (13.6 g of sodium acetate in 100 ml deionized water) with 62 ml solution B (8.5 ml of 12 N hydrochloric acid in 91.5 ml deionized water). Adjust the pH to 1.42 and bring the total volume to 250 ml with deionized water.
4. Alcian blue–safranin stain: Prepare a solution (w/v) of 0.36 % Alcian blue 8GX, 0.018 % safranin O, and 0.48 % ferric ammonium sulfate in acetate buffer. Store at room temperature.
5. Tyrode's buffer: 10 mM HEPES, 130 mM NaCl, 5 mM KCl, 1.5 mM $CaCl_2$, 1 mM $MgCl_2$, 0.1 % Glucose, and 0.1 % BSA in deionized water. Filter-sterilize the buffer and store at 4 °C.

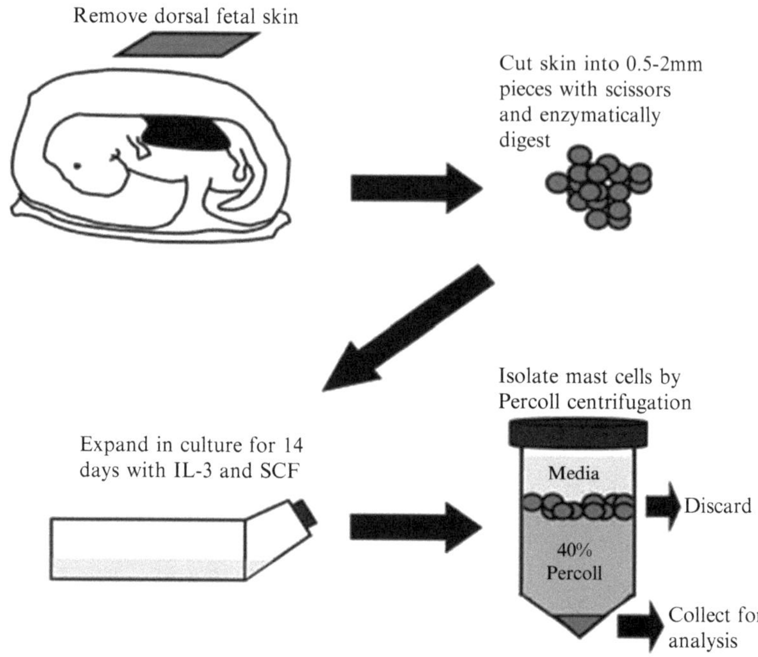

Fig. 3 Flow chart of harvest, culture, and purification of mast cells from fetal skin. Dorsal fetal skin is removed at either E15 or E18. The skin is then mechanically and enzymatically broken down to release cells. The cells are cultured for 14 days in medium supplemented with IL-3 and SCF, which favors mast cell proliferation. Finally the mast cells, which are more dense than the other cells in culture, are purified by centrifugation in 40 % isotonic Percoll. These mast cells can then be examined and characterized using a variety of assays

6. p-NAG (4-Nitrophenyl-*N*-acetyl-β-d-glucosaminidine) buffer: 0.2 M Na_2HPO and 0.4 M citric acid, pH 4.5.
7. p-NAG substrate: Dissolve 4 mM p-NAG in p-NAG buffer. Store 10 ml aliquots at −20 °C.
8. Lysis buffer: 0.1 % Triton X-100 in Tyrode's buffer.
9. Stop solution: 0.2 M glycine in deionized water, pH 10.7. Store at room temperature.

3 Methods

Many methods have been published for the harvest and culture of mast cells. The protocol below was developed by modifying methods published by Benyon et al. [30] and Yamada et al. [31]. This procedure (Fig. 3) allows for the isolation, expansion, and purification of mast cells from fetal skin of various gestational ages while retaining many developmental differences seen in dermal mast cells in vivo (Figs. 29.2 and 29.4). It is important to maintain proper sterile technique during all procedures.

3.1 Tissue Harvest and Digestion

1. For isolation of fetal skin, euthanize the pregnant dam at the desired point of fetal development (*see* **Note 2**). Shave the skin and disinfect with ethanol. Using sterile scissors and forceps, aseptically transfer the entire uterus to a 100 mm Petri dish containing cold 1× HBSS to keep the fetuses from drying out. Euthanize each fetus by decapitation and remove the dorsal skin. Combine the dorsal skin of all fetuses from each dam and rinse with 1× HBSS. Use scissors to cut the skin into small pieces (approximately 0.5–2 mm). Transfer skin pieces into a pre-weighed 15 ml conical tube and re-weigh to obtain the weight of the tissue. Immediately add digestion medium at a ratio of 10 ml per 1 g of dorsal skin. Incubate the tissue at 37 °C for a total of 60 min with continuous rocking (*see* **Note 3**).
2. After digestion, filter dispersed cells through a 0.4 μm nylon strainer into a 50 ml conical tube and rinse cells through the strainer with 10 ml of 1× HBSS (*see* **Note 3**). Centrifuge the single cell suspension at 240 × *g* for 6 min. Discard the supernatant and wash cells in 10 ml 1× HBSS. Pellet the cells again by centrifugation at 240 × *g* for 6 min and discard the supernatant.
3. Resuspend the cells in 5 ml of ACK lysing buffer and allow the red blood cells to lyse for 5 min at 37 °C. Immediately add 25 ml 1× HBSS and centrifuge at 240 × *g* for 6 min. Discard the supernatant and resuspend the cells in 10 ml of 1× HBSS. Determine cell number and centrifuge at 240 × *g* for 6 min. Resuspend the cells at a concentration of 4×10^4 cells/ml in culture medium (which contains IL-3 and SCF), then seed individual T75 flasks with 25 ml of the cell suspension (approximately 1×10^6 cells per T75 flask; *see* **Note 4**). Allow the cells to proliferate in culture for 14 days undisturbed without changing the media.

3.2 Mast Cell Purification

1. On day 14, tap each T75 flask several times and decant non-adherent cells into 50 ml conical tubes. Take a small aliquot for cell counts and centrifuge the cells at 240 × *g* for 6 min. Discard the supernatant and resuspend the cells at a maximum concentration of 2×10^6 cells/ml in 1× RPMI complete medium.
2. Carefully suspend 1 ml of cells over 4 ml of 40 % isotonic Percoll in a 15 ml conical tube. Immediately centrifuge at 500 × *g* for 10 min at room temperature on low acceleration without braking. Carefully remove both layers of media (including the cells in the interphase) and retain the pellet. Rinse the pelleted cells two times in 1× HBSS without calcium by centrifugation at 200 × *g* for 5 min. Before the second wash, take a small aliquot for cell counts. Finally, resuspend the cells in an appropriate solution for further use.

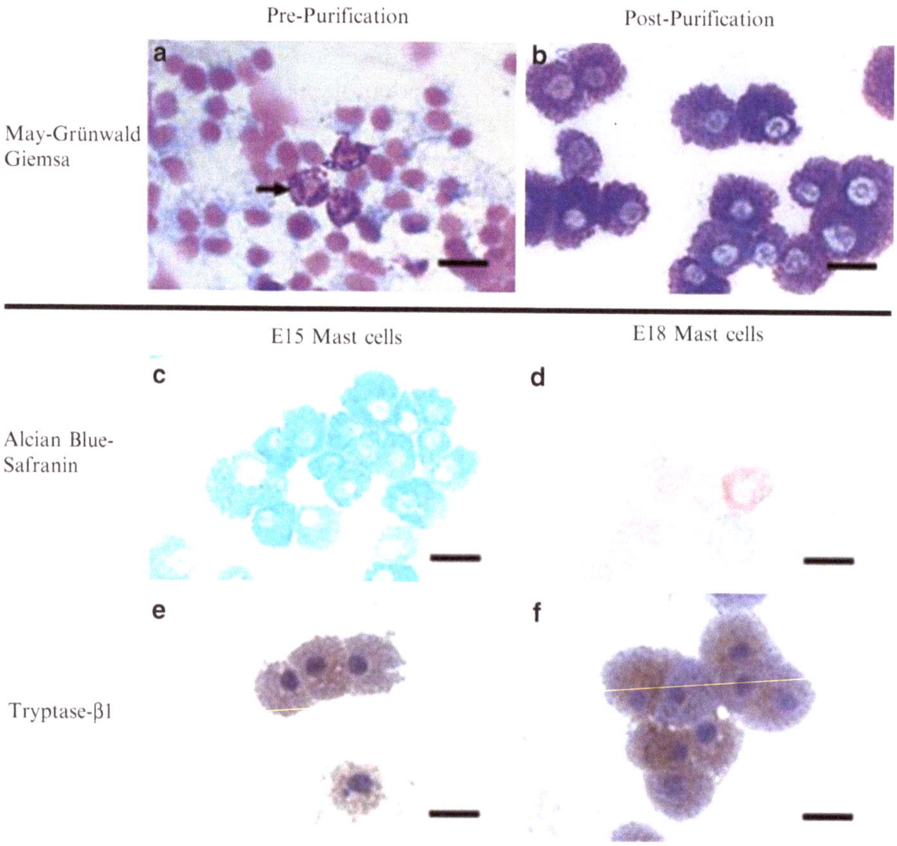

Fig. 4 Representative staining of cultured mast cells. After culture and purification, mast cells can be characterized using various stains and functional assays. May-Grünwald Giemsa staining is useful for assessing the purity of the mast cells obtained after Percoll separation. May-Grünwald Giemsa-stained cytospin preparations of all cells in culture before Percoll centrifugation (pre-purification; **a**) show that there are many different cell types in addition to mast cells (*arrows*). After Percoll separation, a relatively pure population of mast cells is obtained (post-purification; **b**). Alcian blue–safranin staining of cytospin preparations of purified mast cells from E15 (**c**) and E18 (**d**) show that even after 14 days of culture, the differences in maturity that are seen in E15 and E18 skin sections (Fig. 2) remain. Immunohistochemical staining for tryptase-β1 in cytospin preparations of purified mast cells using standard techniques shows that both E15 (**e**) and E18 (**f**) mast cells express the mast cell protease tryptase-β1 (*brown color*). The scale bar in all images is 20 μm

3.3 May-Grünwald Giemsa Staining

May-Grünwald Giemsa staining can be used to determine the purity of mast cell cultures (Fig. 4) by allowing the identification of different immune cell types in culture [32]. Using a cytocentrifuge, spin a minimum of 1×10^4 purified cells (suspended in 1× HBSS without calcium) onto microscope slides. Allow the cells to dry for a minimum of 1 h. Stain the cells with May-Grünwald stain for 5 min, and then rinse them in phosphate buffered saline (PBS) for 5 min. Remove from PBS and add diluted Giemsa stain for 20 min. Rinse for 5 min in deionized water, allow to dry completely, and coverslip using Permount mounting medium.

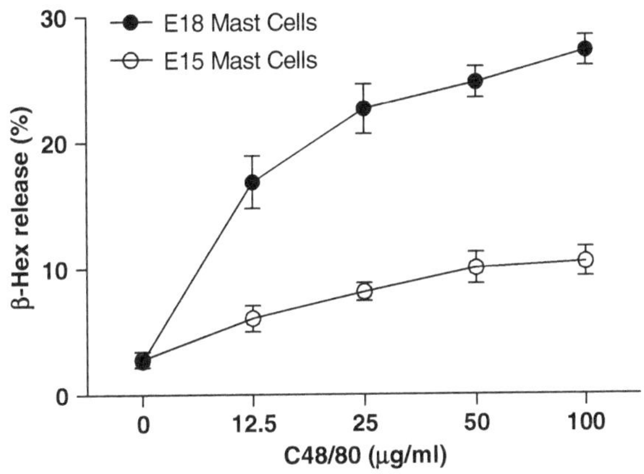

Fig. 5 β-hexosaminidase release from stimulated mast cells. Degranulation assays are commonly performed to quantify mast cell activation in cultured mast cells. After culture and Percoll purification, mast cells from E15 to E18 fetal skin were incubated for 1 h with various concentrations of compound 48/80 (C48/80). This compound is a calmodulin antagonist commonly used to stimulate mast cell degranulation. The activity of β-hexosaminidase from the supernatants and lysates of stimulated mast cells was determined and the percent release was calculated. E18 mast cells stimulated with C48/80 released a higher percentage of β-hexosaminidase compared to E15 mast cells (points on the graph represent the mean +/− S.E.M.; $n = 3$ replicates per point). In vivo, dermal mast cells in E18 skin also degranulate more in response to injury compared to mast cells in E15 skin (Fig. 2)

3.4 Alcian Blue–Safranin staining

Alcian blue–safranin staining can be used to determine mast cell maturity [33]. Dry cytospin preparations (described above) for at least 1 h. Fix for 15 min in ice-cold acetone. Incubate for 15 min with Alcian blue–safranin stain. Rinse in tap water and dehydrate with 1-butanol, then rinse in Clear-Rite 3 and coverslip with Permount mounting medium.

3.5 β-Hexosaminidase Release Assay

Many different mediators stored in mast cell granules can be measured to quantify mast cell degranulation. β-hexosaminidase release is a commonly used and relatively simple way to measure degranulation [34]. Purified mast cells should be stimulated in a manner relevant to the aims of the study. Exposure to chemicals, such as compound 48/80 or ionophores, or inducing IgE-mediated receptor aggregation are commonly used methods to stimulate degranulation in cultured mast cells [31]. The exact cell number, amount of stimulant, and duration of the stimulation must be optimized. In experiments carried out in our laboratory, 2×10^4 cells in a reaction volume of 20 μl was sufficient to detect mast cell degranulation after 1 h of stimulation with compound 48/80 at concentrations ranging from 12.5 to 100 μg/ml (Fig. 5).

Mast Cell stimulation should be carried out in a v-bottom 96-well plate at 37 °C for an optimized period of time. During stimulation, a group of untreated cells (in Tyrode's buffer alone) should be included to control for basal levels of degranulation. After stimulation, pellet the mast cells by centrifuging the plate at 200 × *g* for 5 min. For optimal results, reactions should be performed in triplicate. Immediately transfer 10 μl of supernatant from each reaction to separate wells of a flat-bottom 96 well plate and carefully remove the remaining supernatant. Add 20 μl of lysis buffer to the cell pellets and pipette up and down a number of times to lyse the cells. Transfer 10 μl of the lysate from each reaction to new wells of a flat-bottom 96 well plate. Add 50 μl of p-NAG substrate to each well and incubate for 1 h at 37 °C. To stop the reaction, add 150 μl of stop solution. This will result in a yellow color change. Immediately determine the OD (optical density) at 405 nm using a plate reader. The percent release can be calculated using the equation: [OD supernatant/(OD supernatant + OD lysate)] × 100.

4 Notes

1. FBS should be heat inactivated for 30 min at 56 °C before use. Multiple batches of FBS should be tested to determine suitability for growth and maturation of mast cells. We have observed significant differences in the growth of cultured mast cells with certain lots of FBS.
2. Pair male and female mice overnight. Separate the mice the following day and check females for a vaginal plug. The day a plug is observed (or the day the mice are separated) is designated as embryonic day zero (E0). Females should only be paired with males once per week. This helps accurately assess fetal age in visually pregnant mice, since the presence (or absence) of a vaginal plug does not always accurately predict pregnancy. Palpation of the abdomen can be used to confirm pregnancy. On day E15 or Day E18, euthanize pregnant mice and harvest fetal tissue.
3. To improve cellular yield during tissue digestion, the tubes can be vortexed every 15 min during the 60 min incubation period. The 10 ml rinse through the nylon strainer can be broken up into four 5 ml rinses to help improve cellular yield as well.
4. Macrophage contamination is a concern especially when working with certain strains of mice [31]. Macrophage contamination can be identified by May-Grünwald Giemsa staining. If this becomes a problem, flasks can be seeded with a lower number of cells.

Acknowledgments

The authors are supported in part by NIH grants R01-CA127109 and R21-ES020462.

References

1. Larson BJ, Longaker MT, Lorenz HP (2010) Scarless fetal wound healing: a basic science review. Plast Reconstr Surg 126:1172–1180
2. Wilgus TA (2007) Regenerative healing in fetal skin: a review of the literature. Ostomy Wound Manage 53:16–31, quiz 32-3
3. Martin P, D'Souza D, Martin J, Grose R, Cooper L, Maki R, McKercher SR (2003) Wound healing in the PU.1 null mouse-tissue repair is not dependent on inflammatory cells. Curr Biol 13:1122–1128
4. Wilgus TA, Vodovotz Y, Vittadini E, Clubbs EA, Oberyszyn TM (2003) Reduction of scar formation in full-thickness wounds with topical celecoxib treatment. Wound Repair Regen 11:25–34
5. Krummel TM, Nelson JM, Diegelmann RF, Lindblad WJ, Salzberg AM, Greenfield LJ, Cohen IK (1987) Fetal response to injury in the rabbit. J Pediatr Surg 22:640–644
6. Liechty KW, Adzick NS, Crombleholme TM (2000) Diminished interleukin 6 (IL-6) production during scarless human fetal wound repair. Cytokine 12:671–676
7. Liechty KW, Crombleholme TM, Cass DL, Martin B, Adzick NS (1998) Diminished interleukin-8 (IL-8) production in the fetal wound healing response. J Surg Res 77:80–84
8. Naik-Mathuria B, Gay AN, Yu L, Hsu JE, Smith CW, Olutoye OO (2008) Fetal wound healing using a genetically modified murine model: the contribution of P-selectin. J Pediatr Surg 43:675–682
9. Frantz FW, Bettinger DA, Haynes JH, Johnson DE, Harvey KM, Dalton HP, Yager DR, Diegelmann RF, Cohen IK (1993) Biology of fetal repair: the presence of bacteria in fetal wounds induces an adult-like healing response. J Pediatr Surg 28:428–433, discussion 33–4
10. Haynes JH, Johnson DE, Mast BA, Diegelmann RF, Salzberg DA, Cohen IK, Krummel TM (1994) Platelet-derived growth factor induces fetal wound fibrosis. J Pediatr Surg 29:1405–1408
11. Wilgus TA, Bergdall VK, Tober KL, Hill KJ, Mitra S, Flavahan NA, Oberyszyn TM (2004) The impact of cyclooxygenase-2 mediated inflammation on scarless fetal wound healing. Am J Pathol 165:753–761
12. el Sayed SO, Dyson M (1993) Responses of dermal mast cells to injury. J Anat 182(Pt 3):369–376
13. Egozi EI, Ferreira AM, Burns AL, Gamelli RL, Dipietro LA (2003) Mast cells modulate the inflammatory but not the proliferative response in healing wounds. Wound Repair Regen 11:46–54
14. Weller K, Foitzik K, Paus R, Syska W, Maurer M (2006) Mast cells are required for normal healing of skin wounds in mice. FASEB J 20:2366–2368
15. Wershil BK, Murakami T, Galli SJ (1988) Mast cell-dependent amplification of an immunologically nonspecific inflammatory response. Mast cells are required for the full expression of cutaneous acute inflammation induced by phorbol 12-myristate 13-acetate. J Immunol 140:2356–2360
16. Rao KN, Brown MA (2008) Mast cells: multifaceted immune cells with diverse roles in health and disease. Ann N Y Acad Sci 1143:83–104
17. Theoharides TC, Alysandratos KD, Angelidou A, Delivanis DA, Sismanopoulos N, Zhang B, Asadi S, Vasiadi M, Weng Z, Miniati A, Kalogeromitros D (2012 Jan) Mast cells and inflammation. (2011). Biochim Biophys Acta 1822(1):21–33
18. Tsai M, Grimbaldeston M, Galli SJ (2011) Mast cells and immunoregulation/immunomodulation. Adv Exp Med Biol 716:186–211
19. Artuc M, Hermes B, Steckelings UM, Grutzkau A, Henz BM (1999) Mast cells and their mediators in cutaneous wound healing–active participants or innocent bystanders? Exp Dermatol 8:1–16
20. Kupietzky A, Levi-Schaffer F (1996) The role of mast cell-derived histamine in the closure of an in vitro wound. Inflamm Res 45:176–180
21. Smith CJ, Smith JC, Finn MC (1987) The possible role of mast cells (allergy) in the production of keloid and hypertrophic scarring. J Burn Care Rehabil 8:126–131
22. Trautmann A, Toksoy A, Engelhardt E, Brocker EB, Gillitzer R (2000) Mast cell involvement in normal human skin wound healing: expression of monocyte chemoattractant protein-1 is correlated with recruitment of

mast cells which synthesize interleukin-4 in vivo. J Pathol 190:100–106
23. Noli C, Miolo A (2001) The mast cell in wound healing. Vet Dermatol 12:303–313
24. Gallant-Behm CL, Hildebrand KA, Hart DA (2008) The mast cell stabilizer ketotifen prevents development of excessive skin wound contraction and fibrosis in red Duroc pigs. Wound Repair Regen 16:226–233
25. Gailit J, Marchese MJ, Kew RR, Gruber BL (2001) The differentiation and function of myofibroblasts is regulated by mast cell mediators. J Invest Dermatol 117:1113–1119
26. Gruber BL, Kew RR, Jelaska A, Marchese MJ, Garlick J, Ren S, Schwartz LB, Korn JH (1997) Human mast cells activate fibroblasts: tryptase is a fibrogenic factor stimulating collagen messenger ribonucleic acid synthesis and fibroblast chemotaxis. J Immunol 158:2310–2317
27. Hatamochi A, Fujiwara K, Ueki H (1985) Effects of histamine on collagen synthesis by cultured fibroblasts derived from guinea pig skin. Arch Dermatol Res 277:60–64
28. Levi-Schaffer F, Rubinchik E (1995) Activated mast cells are fibrogenic for 3T3 fibroblasts. J Invest Dermatol 104:999–1003
29. Wulff BC, Parent AE, Meleski MA, DiPietro LA, Schrementi ME, Wilgus TA (2012) Mast cells contribute to scar formation during fetal wound healing. J Invest Dermatol 132:458–465
30. Benyon RC, Lowman MA, Church MK (1987) Human skin mast cells: their dispersion, purification, and secretory characterization. J Immunol 138:861–867
31. Yamada N, Matsushima H, Tagaya Y, Shimada S, Katz SI (2003) Generation of a large number of connective tissue type mast cells by culture of murine fetal skin cells. J Invest Dermatol 121:1425–1432
32. Kalesnikoff J, Galli SJ (2011) Antiinflammatory and immunosuppressive functions of mast cells. Methods Mol Biol 677:207–220
33. Csaba GM (1969) Mechanism of the formation of mast-cell granules. II. Cell-free model. Acta Biol Acad Sci Hung 20:205–210
34. Ortega E, Hazan B, Zor U, Pecht I (1989) Mast cell stimulation by monoclonal antibodies specific for the Fc epsilon receptor yields distinct responses of arachidonic acid and leukotriene C4 secretion. Eur J Immunol 19:2251–2256

Chapter 30

Assessing Macrophage Phenotype During Tissue Repair

Timothy J. Koh, Margaret L. Novak, and Rita E. Mirza

Abstract

Macrophages are thought to play important roles in tissue repair, from host defense to angiogenesis and new tissue formation. The role of macrophages in repair of different tissues is an active area of inquiry, particularly in settings of impaired healing. In this chapter, we describe methods for isolating monocyte/macrophage cell populations from damaged tissue and characterizing the phenotype of these cells. Cells are isolated from tissue by enzymatic digestion, and then monocyte/macrophage populations can be sorted by magnetic separation. The phenotype of these cells is assessed by real-time PCR, flow cytometry and ELISA. A complementary approach of assessing monocyte/macrophage phenotype by immunofluorescence staining of cryosections is also described. This combination of approaches to study the macrophage phenotypes expressed during tissue repair will lead to better understanding of the roles of macrophages in tissue repair and new therapeutic avenues for improving healing.

Key words Macrophage phenotype, Tissue repair, Wound healing, Cell isolation, Magnetic sorting, Flow cytometry, Immunofluorescence

1 Introduction

Tissue repair following injury involves overlapping phases of inflammation, new tissue formation and subsequent remodeling. Macrophages (Mp) are involved in each phase of tissue healing, and are thought to play an important role in the repair of a variety of tissues. During the early phases of healing, Mp help to kill pathogens, clear damaged tissue, and clear necrotic and apoptotic cells. During later stages of healing, Mp contribute to repair in part by secreting a spectrum of chemokines, cytokines, and growth factors. The multifunctional nature of Mp during tissue repair is an active area of investigation and better understanding of the regulation of the different phenotypes expressed by Mp during healing could lead to important insights into the mechanisms of normal and impaired healing as well as new therapeutic strategies for improving outcomes.

After tissue damage, blood-borne monocytes (Mo) are recruited to the site of injury and subsequently differentiate into Mp.

Robert G. Gourdie and Tereance A. Myers (eds.), *Wound Regeneration and Repair: Methods and Protocols*, Methods in Molecular Biology, vol. 1037, DOI 10.1007/978-1-62703-505-7_30,

Two main Mo subsets have been identified in the circulation of mice: a pro-inflammatory subset that expresses Ly6C at high levels and a patrolling or noninflammatory subset that expresses Ly6C at low levels [1, 2]. In humans, the corresponding subsets are pro-inflammatory $CD14^{hi}$ and patrolling $CD14^{lo}$ subsets [3]. Whether these Mo subpopulations represent distinct end-stage differentiation states or a continuum of plastic phenotypic states remains to be determined [4]. Although initial studies have been performed [5, 6], further work is required to elucidate the time course and regulation of recruitment of different Mo subsets following tissue damage, and the role that each subset plays in tissue repair, particularly in the setting of impaired healing.

Different microenvironments have the potential for inducing a wide range of Mp phenotypes [7–10]. Bacterial products and pro-inflammatory cytokines are known to induce the "classical" pro-inflammatory or M1-like, pathogen killing Mp phenotype associated with production of high levels of inflammatory cytokines, and reactive oxygen and nitrogen species. Anti-inflammatory cytokines, glucocorticoids, modulators of glucose and lipid metabolism, and phagocytosis of apoptotic cells can induce a broad spectrum of "alternative" or M2-like Mp phenotypes including those that exhibit noninflammatory or anti-inflammatory and pro-tissue repair functions, typically characterized by expression of mannose receptor (CD206), scavenger receptors (e.g., CD36), and growth factors such as TGF-β1 and insulin-like growth factor (IGF)-1. Although the term "alternative" activation was first ascribed to Mo/Mp stimulated with interleukin (IL)-4 [11], the term has morphed into a generic name used for any form of nonclassical activation. Another classification scheme depicts Mp phenotypes as a spectrum based on functional activities, which accounts for the many intermediate or combination phenotypes reported in the literature [10].

Much of our knowledge about Mp phenotypes has been obtained from in vitro experiments in which phenotypes have been studied using stimulation with a single cytokine or other effector, or perhaps two of these effectors. Less is known about the phenotypes expressed in actual physiological situations, particularly during tissue repair. In vivo, there is potential for a multitude of effectors to act on Mp and the phenotype expressed is likely due to the net effect of all of these signals. Mp along with other inflammatory cells and tissue resident cells likely contribute to, and are influenced by, this complex molecular environment.

In this chapter, we describe methods for assessing Mo/Mp phenotype during tissue repair used in our recent work [12]. These methods include isolating cells from damaged tissue followed by magnetic separation, real-time polymerase chain reaction (PCR), flow cytometry and enzyme-linked immunoassay (ELISA). We also describe methods for immunofluorescence staining of tissue cryosections that complement the cell isolation approach by allowing localization of Mo/Mp phenotypes within tissue compartments. Further study of the actual Mo/Mp phenotypes expressed during tissue repair, and how these are regulated, will lead to better understanding of the roles of Mo/Mp in different tissue repair and new therapeutic avenues for improving healing.

2 Materials

2.1 Cell Isolation from Tissue

1. Surgical equipment: bead sterilizer, forceps, scissors, scalpel with #11 blades, hemostat, razor blades.
2. Plastic-ware: cell strainers (70 μm mesh), 50 ml conical tubes, 15 ml conical tubes, 1.7 ml microcentrifuge tubes, 10 cm petri dishes, 5 ml syringe, 0.22 μm sterile filter.
3. Centrifuge with adaptors for 50, 15, and 1.7 ml tubes.
4. Microscope, hemacytometer.
5. Cell culture incubator, tube rotator, stir bars.
6. Dulbecco's Modified Eagle's Medium (DMEM), fetal bovine serum (FBS).
7. Enzymes: collagenase I, collagenase XI, hyaluronidase I.

2.1.1 Magnetic Separation of Macrophages from Total Cell Population

1. FITC conjugated Ly6G, CD3, and CD19 antibodies, anti-FITC conjugated beads, anti-CD11b conjugated beads, MS columns, column magnet.
2. Cell separation buffer: phosphate buffered saline (PBS), 0.5 % FBS, 2 mM EDTA, degassed.

2.2 Flow Cytometry

1. Flow cytometer.
2. Staining buffer: PBS, 1 % FBS.
3. Anti-CD16/32 (Fc block).
4. Fixation buffer, permeabilization buffer.
5. Leukocyte antibodies: FITC conjugated anti-Ly6G, CD3, and CD19, PE and PerCP-Cy5.5 conjugated anti-CD11b and anti-Ly6C, APC conjugated anti-F4/80.
6. Classical activation marker antibodies: APC conjugated anti-TNF-α and IL-12.
7. Alternative activation marker antibodies: PerCP-Cy5.5 anti-CD206, PE-conjugated CD36 and TGF-β, APC conjugated IL-10.
8. Isotype specific controls: FITC, PE, PerCP-Cy5.5, APC-conjugated IgG1, IgG2a, IgG2b.

2.3 Real-Time PCR

1. Real-Time PCR System.
2. RNeasy kit.
3. ThermoScript RT-PCR System.
4. TaqMan Universal PCR Master Mix.
5. TaqMan Gene Expression Assays for genes of interest (e.g., TNF-α, IL-1β, IL-12, iNOS; CD206, CD36, IL-10, GAPDH, 18S rRNA).

2.4 ELISA

1. M1 markers: TNF-α and IL-1β ELISA kits.
2. M2 markers: IL-10, IGF-1, and TGF-β ELISA kits.

2.5 Immuno-fluorescence

1. Cryostat.
2. Microscope with epifluorescence and FITC, rhodamine and DAPI filters, digital camera, and image analysis software.
3. Tissue freezing medium, isopentane, dry ice.
4. Microscope slides, mounting medium with DAPI, coverslips.
5. Primary antibodies: Rat anti-mouse F4/80, goat polyclonal anti-TNF-α, rabbit polyclonal anti-iNOS, goat polyclonal CD206, rabbit polyclonal TGF-β1.
6. Secondary antibodies: FITC-conjugated goat anti-rabbit IgG, FITC-conjugated rabbit anti-goat IgG, TRITC-conjugated goat anti-rat IgG.
7. PBS, formaldehyde, 70 % ethanol.
8. Blocking buffer: (a) 1 % normal rabbit or goat serum in PBS or (b) buffer with 50 mM Tris–HCl pH7.6, 150 mM NaCl, 0.05 % Tween20, 0.2 % gelatin, 3 % BSA, 0.1 % sodium azide.
9. 1 % Sudan Black B in 70 % methanol, filtered into Coplin jar.

3 Methods

3.1 Cell Isolation from Tissue (See Notes 1–4)

1. Sterilize surgical tools, razor blades, and stir bar in bead sterilizer.
2. Prepare enzymatic digest mixture: 10 mg each of collagenase I, collagenase XI, hyaluronidase in 3 ml of DMEM. Sterile filter the solution using syringe filter.
3. Remove tissue of interest. Holding a pair of razor blades with hemostat, mince tissue into ~1 mm^3 pieces in Petri dish containing 5 ml of DMEM. Transfer minced tissue to a 15 ml tube. Rinse dish with 2 ml of collagenase/hyaluronidase solution, followed by 3 ml of DMEM, and transfer to the same 15 ml tube containing tissue pieces.

4. Add small stir bar to tube and place tube on rotator, and allow digestion to proceed in cell culture incubator for 45–60 min at 37 °C in 5 % CO_2. Check the solution occasionally and dissociate tissue aggregates if necessary.
5. Filter digested tissue suspension through a 70 μm cell strainer into a 50 ml tube. Rinse the 15 ml tube with 3 ml of DMEM and add to cell strainer. Rinse the strainer with another 2 ml of DMEM.
6. Centrifuge the cell suspension for 5 min at 300 × *g*. Discard supernatant and resuspend cell pellet in appropriate volume of cell separation buffer. Count cells on hemocytometer.

3.2 Magnetic Separation of Monocytes/Macrophages from Total Cell Population (See Note 5)

1. Centrifuge cell suspension 5 min at 300 × *g*. Resuspend cell pellet in 100 μl of cell separation buffer.
2. Add 5 μl each of FITC-conjugated Ly6G, -CD3, and -CD19 antibodies. Incubate for 15 min at 4 °C.
3. Add 2 ml cell of separation buffer and centrifuge for 5 min at 300 × *g*. Resuspend in 90 μl of cell separation buffer.
4. Add 10 μl of anti-FITC conjugated beads. Incubate for 15 min at 4 °C.
5. Add 2 ml of cell separation buffer and centrifuge for 5 min at 300 × *g*. Resuspend in 500 μl of cell separation buffer.
6. Place MACS column in magnetic field and equilibrate with 500 μl of cell separation buffer. Discard flow-through.
7. Collect Ly6G/CD3/CD19-negative fraction in a new tube: Add cell suspension to column, then rinse three times with 500 μl of cell separation buffer. Wait for column reservoir to empty before adding successive rinses.
8. Collect Ly6G/CD3/CD19-positive fraction: remove column from magnetic field, add 1 ml of cell separation buffer, and flush column with plunger; count cells on hemocytometer. This fraction is taken to represent neutrophil, T cell and B cell populations.
9. Centrifuge Ly6G/CD3/CD19-negative fraction for 5 min at 300 × *g*. Resuspend cell pellet in 90 μl of cell separation buffer.
10. Add 10 μl of anti-CD11b conjugated beads. Incubate for 15 min at 4 °C.
11. Add 2 ml of cell separation buffer and centrifuge for 5 min at 300 × *g*. Resuspend cell pellet in 500 μl of cell separation buffer.
12. Equilibrate column and separate negative and positive fractions as before; count cells.

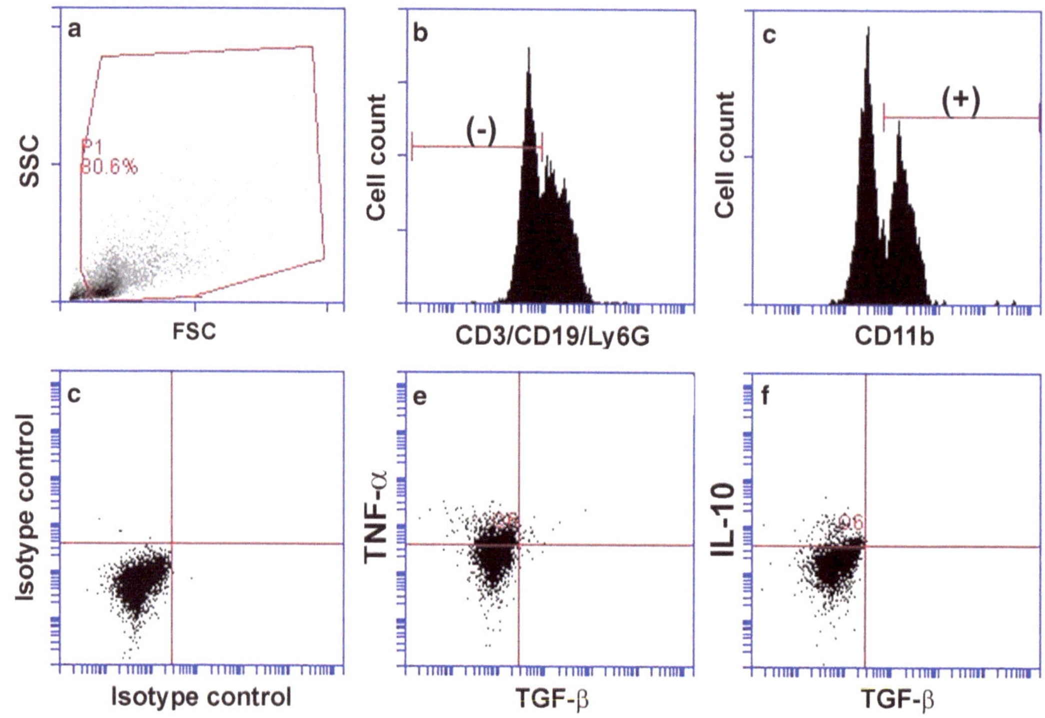

Fig. 1 Example of macrophage phenotyping by flow cytometry analysis of cells isolated from excisional skin wounds on C57Bl/6 mice. Cells collected 5 days post-injury were not subjected to magnetic isolation, but were labeled with FITC-anti CD3/CD19/Ly6G, PerCP-Cy5.5 anti-CD11b, PE anti-TGF-β, and APC anti-TNF-α or APC anti-IL-10. Cells were gated in (**a**) forward scatter (FSC)/side scatter (SSC) plots to eliminate dead cells and debris, (**b**) CD3/CD19/Ly6G histograms to eliminate CD3/CD19/Ly6G positive cells and (**c**) CD11b histograms to include only CD11b positive cells. This $CD11b^{+}CD3/CD19/Ly6G^{-}$ population was taken to represent the monocyte/Mp population in the wound. Gates were then set based on isotype controls (**d**) for phenotype markers TGF-β, TNF-α (**e**), and IL-10 (**f**). Note that a greater percentage of these cells were positive for TNF-α than for IL-10 and that cells were negative for TGF-β

13. CD11b positive, Ly6G/CD3/CD19 negative fraction is taken to represent Mo/Mp cell population.
14. If measuring release of soluble factors, incubate cells in DMEM + 10 % FBS at 37 °C in 5 % CO_2 overnight. If using cells for real-time PCR, freeze pellet at −80 °C until ready for RNA isolation.

3.3 Monocyte/ Macrophage Phenotyping: Flow Cytometry (See Notes 6, 7 and 10, Fig. 1)

1. Resuspend cells in staining buffer, total volume depends on number of staining conditions required (e.g., 100 μl per sample). Include non-stained control, isotype controls and color compensation controls. For the latter, include appropriate amounts of non-stained cells and single antibody stained cells to allow for assessment of compensation.
2. If desired, incubate samples with anti-CD16/32 (Fc block) for 15 min at 4 °C.

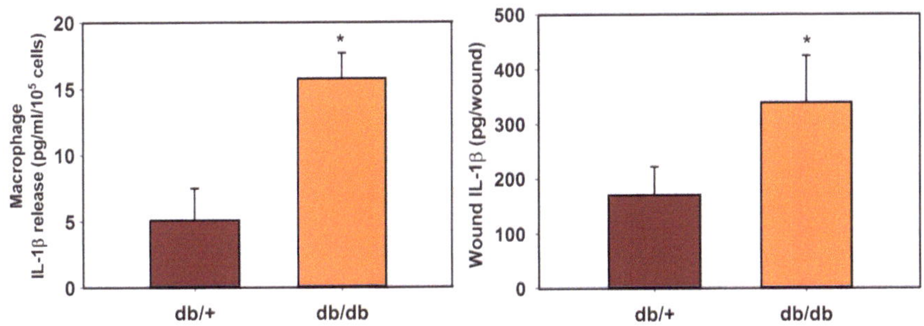

Fig. 2 Example of macrophage phenotyping by ELISA. Cells were isolated from excisional skin wounds on nondiabetic db/+ and diabetic db/db mice. Cells collected 10 days post-injury were subjected to magnetic isolation; first removing CD3/CD19/Ly6G positive cells by negative selection and then collecting the remaining CD11b positive cells by positive selection. Cells were then incubated overnight in DMEM + 10 % FBS at 37 °C and 5 % CO_2. The cell culture supernatant was then subjected to ELISA for IL-1β along with homogenate of wounds also collected at 10 days for comparison. Note that IL-1β release is higher for cells isolated from diabetic mice compared with nondiabetic mice, as is the level in wound homogenate

3. Stain extracellular antigens with appropriate antibodies for 30 min at 4 °C; titrate antibodies individually in preliminary experiments.
4. Add 2 ml of staining buffer, centrifuge for 5 min at 300 × *g*. Repeat wash.
5. If staining for intracellular antigens, fix cells with 0.5 ml fixation buffer for 20 min in dark at room temperature, as per manufacturer's instructions
6. Centrifuge for 5 min at 300 × *g*.
7. Wash cells two times in 2 ml of permeabilization buffer.
8. Resuspend cell pellet in 100 μl of permeabilization buffer.
9. Stain intracellular antigens with appropriate antibodies for 30 min at 4 °C, titrate antibodies individually in preliminary experiments.
10. Add 2 ml of permeabilization buffer, centrifuge for 5 min at 300 × *g*. Repeat wash.
11. Resuspend cell pellet in 500 μl of staining buffer.
12. Perform flow cytometry.
13. Set compensation based on single stained control samples.
14. Set measurement gates based on comparisons with negative controls (non-stained and IgG controls.

3.4 Monocyte/Macrophage Phenotyping: ELISA (Fig. 2)

1. After overnight incubation in DMEM with 10 % FBS, centrifuge cell suspension for 5 min at 300 × *g*.
2. Transfer supernatant to a new tube and store cell pellet at −80 °C for further analysis if desired.

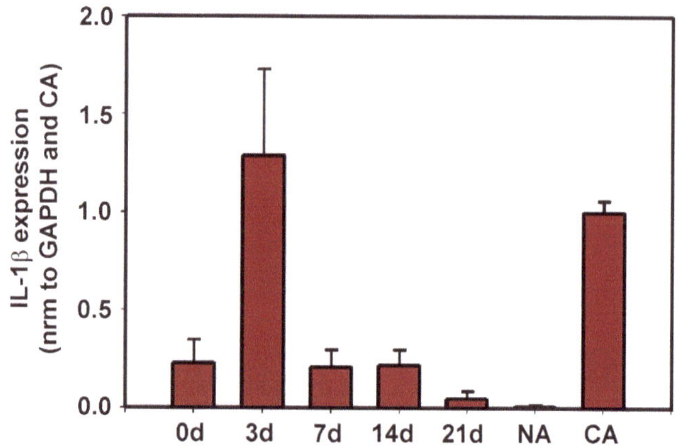

Fig. 3 Example of macrophage phenotyping by real-time PCR. Cells were isolated from uninjured gastrocnemius muscle (0d) and at 3–21 days following laceration injury in C57Bl/6 mice. Cells were subjected to magnetic isolation; first removing CD3/CD19/Ly6G positive cells by negative selection and then collecting the remaining CD11b positive cells by positive selection. Real-time PCR for IL-1β was then performed on cDNA derived from this CD11b$^+$CD3/CD19/Ly6G$^-$ population. Controls include cultured bone marrow-derived Mp incubated without (NA) or with interferon-γ and tumor necrosis factor-α (both at 10 ng/ml) to induce classical activation (CA). IL-1β expression normalized to that of GAPDH and then to the CA condition. Note that IL-1β expression was highest at 3 days post-injury, when it reached a level similar to the CA Mp

3. Centrifuge supernatant for 10 min at 10,000 × *g* to remove smaller debris, aliquot and store supernatant at −80 °C.
4. Use supernatant to assess Mo/Mp release of cytokines and growth factors by ELISA using manufacturer's instructions.
5. To assess cytokine and growth factor levels in tissue, collect tissue and homogenize in ice-cold PBS supplemented with protease inhibitors using a dounce homogenizer. Centrifuge for 10 min at 10,000 × *g*, aliquot and store supernatant at −80 °C.
6. Use supernatant to assess tissue levels of cytokines and growth factors by ELISA using manufacturer's instructions.

3.5 Monocyte/Macrophage Phenotyping: Real-Time PCR (Fig. 3)

1. Perform RNA isolation using RNeasy kit instructions and RT reaction using Thermoscript RT-PCR system instructions.
2. Plan 96 well plate layout to allow for triplicates of each sample plus negative control.
3. For each gene to be analyzed, prepare a premix of 10 μl of TaqMan Universal PCR Master Mix, 1 μl of appropriate TaqMan Gene Expression Assay, and 8 μl of dH_2O, scaled up by number of wells needed.
4. To each well, add 19 μl of premix and 1 μl of sample cDNA, or 1 μl dH_2O for negative control.

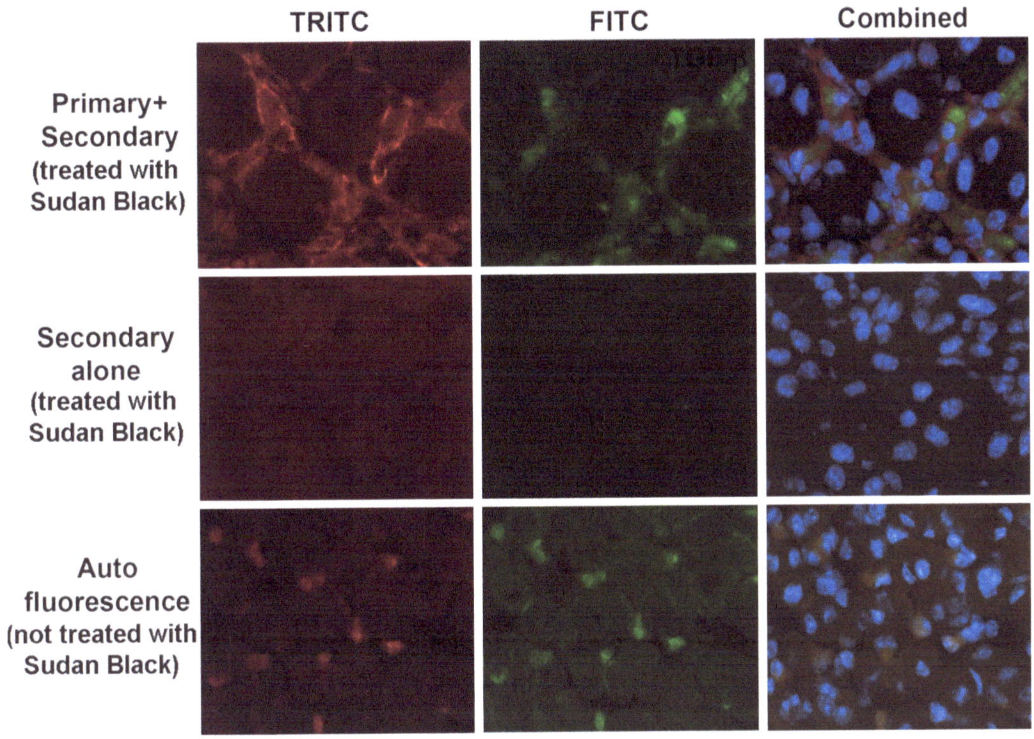

Fig. 4 Example of macrophage phenotyping by immunofluorescence. Cryosections obtained from gastrocnemius muscles of C57BL/6 mice injured by laceration (7 days post-injury). *Top*: Mp labeled using anti-F4/80 (*red*) and anti-TGF-β (*green*) antibodies followed by fluorescently labeled secondary antibodies. Section treated with Sudan Black to reduce autofluorescence and nuclei visualized using DAPI (*blue*). *Middle*: control sections from adjacent tissue labeled with secondary antibodies only, treated with Sudan Black, and nuclei visualized using DAPI. *Bottom*: sections not treated with Sudan Black and not stained with any antibodies exhibited autofluorescence in both *red* and *green* channels

5. Cover plate with optical film, centrifuge for ~1 min, and run in 7500 Real-Time PCR System.
6. Set baseline and threshold in 7,500 software and perform data analysis using the 2^-ddCT method [13] (Note: statistical analysis should be performed on 2^-ddCT values, not on raw CTs.)

3.6 Monocyte/ Macrophage Phenotyping: Immunofluorescence (See Notes 6, 8, and 9, Fig. 4)

1. Embed tissue in tissue freezing medium, freeze tissue in isopentane cooled with dry ice, and store at −80 °C.
2. Section tissue at 5–10 μm using Leica cryostat and collect on Superfrost Plus slides.
3. Fix tissue cryosections in 4 % formaldehyde in PBS for 5 min.
4. Wash 5× 2 min with PBS.
5. Incubate for 1 h in blocking buffer at room temperature.
6. Add primary antibodies. We recommend F4/80 as an Mp marker and phenotypic markers such as TNF-α and iNOS for

M1-like or classically activated phenotype, TGF-β and CD206 for M2-like or alternatively activated phenotype. Incubate for 1 h at room temperature, then overnight at 4 °C. Suggested starting dilution for all antibodies is 1:100 in PBS. For negative controls, include slides with no primary antibody and non-specific IgG from primary antibody host species (use isotype specific IgG if available).

7. Wash 3× 10 min with PBS.
8. Add FITC-conjugated secondary antibody and incubate for 1 h at room temperature. Suggested starting dilution is 1:200 in PBS.
9. Wash 3× 10 min with PBS.
10. Add TRITC-conjugated secondary antibody and incubate for 1 h at room temperature. Suggested starting dilution is 1:200 in PBS.
11. Wash 3× 10 min with PBS.
12. Optional autofluorescence quenching: Incubate for 5 min in 0.1 % Sudan Black solution in Coplin jar. Rinse briefly in 70 % ethanol, then PBS. Incubation and rinse times, as well as Sudan Black concentration, may be adjusted based on intensity of autofluorescence.
13. Mount slides in Vectashield with DAPI.
14. Using Nikon 80i microscope and imaging system, count number of F4/80 positive cells per mm^2 for five fields per section, ensuring cells are DAPI positive. Also count cells double positive for F4/80 and phenotypic markers.

4 Notes

1. The enzymatic digestion protocol was adapted from a protocol described in the literature for isolating cells from aortic tissue [14]. We have used this cell isolation procedure successfully to obtain Mo/Mp populations from full-thickness excisional wounds in mice [12], biopsies of chronic wounds in diabetic human patients and injured mouse skeletal muscle (our unpublished data). We have not yet tried the protocol on other tissues; further optimization may be required for these other tissues.
2. We have tried other enzymes for digestion of skin and skeletal muscle, including collagenase B and pronase. These protocols either produced a lower cell yield (collagenase B) or more variable results for magnetic sorting and labeling for flow cytometry applications (pronase). The latter may be due to loss of antigen during the digest procedure.

3. We have performed enzymatic digestion in petri dish on stir plate as well as in rotating tube with similar cell yields and viability assessed by trypan blue exclusion.
4. Following tissue digestion, we sometimes obtain a large number of red blood cells in the cell suspension. These could be removed using a red blood cell lysis step or by centrifugation on Ficoll. However, steps to eliminate red blood cell contamination can influence Mp [15], so at the minimum, all samples should be treated similarly.
5. To date, we have sorted Mo/Mp populations by magnetic sorting with a protocol that essentially follows the manufacturer's instructions. However, similar sorting could be performed using fluorescence activated cell sorting [5].
6. Mo/Mp can exhibit autofluorescence [16], so proper controls must be used to accurately interpret data generated by fluorescent detection methods. These include non-stained and nonspecific IgG controls for flow cytometry and immunohistology.
7. Markers taken to represent Mo/Mp populations as well as Mo/Mp phenotypes should be chosen and interpreted with care. For example, CD68 is often used as a Mp marker for both mouse and human tissue, but is expressed in other cell types, including endothelial cells and fibroblasts [17, 18]. In addition, F4/80 is a standard Mp marker in mice, but sorting based on F4/80 expression may exclude less mature Mo populations [12].
8. Different antigens may be sensitive to particular fixation methods for immunofluorescence. In the protocol given here, we have used F4/80 as a standard Mp marker, and we have found that F4/80 labels well when sections are fixed in formaldehyde, but less so when fixed in acetone. Conversely, other Mp markers such as MOMA2 may give a stronger signal when fixed in acetone.
9. In our immunofluorescence protocol, secondary antibodies are added sequentially to minimize the potential for binding of the FITC rabbit anti-goat secondary to the TRITC goat anti-rat secondary. If different secondary antibodies are chosen such that this interaction would not occur, it may be possible to incubate with both secondary antibodies simultaneously.
10. Multiple phenotypic markers are required to describe Mo/Mp phenotypes in vivo, because of the complexity of the phenotypes observed. We favor an approach that combines different methodologies, and have so far assessed phenotype via mRNA and protein expression, cytokine/growth factor release and localization in tissue sections. Additional approaches could include functional assays of phagocytosis and pathogen killing capability, and co-culture or conditioned medium experiments to assess the capability to stimulate target cell proliferation, migration, or other functional activities.

References

1. Geissmann F, Jung S, Littman DR (2003) Blood monocytes consist of two principal subsets with distinct migratory properties. Immunity 19:71–82
2. Sunderkotter C et al (2004) Subpopulations of mouse blood monocytes differ in maturation stage and inflammatory response. J Immunol 172:4410–4417
3. Cros J et al (2010) Human CD14dim monocytes patrol and sense nucleic acids and viruses via TLR7 and TLR8 receptors. Immunity 33:375–386
4. van de Veerdonk FL, Netea MG (2010) Diversity: a hallmark of monocyte society. Immunity 33:289–291
5. Arnold L et al (2007) Inflammatory monocytes recruited after skeletal muscle injury switch into antiinflammatory macrophages to support myogenesis. J Exp Med 204: 1057–1069
6. Nahrendorf M et al (2007) The healing myocardium sequentially mobilizes two monocyte subsets with divergent and complementary functions. J Exp Med 204: 3037–3047
7. Gordon S (2003) Alternative activation of macrophages. Nat Rev Immunol 3:23–35
8. Mantovani A et al (2004) The chemokine system in diverse forms of macrophage activation and polarization. Trends Immunol 25:677–686
9. Martinez FO, Helming L, Gordon S (2009) Alternative activation of macrophages: an immunologic functional perspective. Annu Rev Immunol 27:451–483
10. Mosser DM, Edwards JP (2008) Exploring the full spectrum of macrophage activation. Nat Rev Immunol 8:958–969
11. Stein M et al (1992) Interleukin 4 potently enhances murine macrophage mannose receptor activity: a marker of alternative immunologic macrophage activation. J Exp Med 176:287–292
12. Mirza R, Koh TJ (2011) Dysregulation of monocyte/macrophage phenotype in wounds of diabetic mice. Cytokine 56(2):256–264
13. Livak KJ, Schmittgen TD (2001) Analysis of relative gene expression data using real-time quantitative PCR and the 2(-Delta Delta C(T)) Method. Methods 25:402–408
14. Galkina E et al (2006) Lymphocyte recruitment into the aortic wall before and during development of atherosclerosis is partially L-selectin dependent. J Exp Med 203:1273–1282
15. Lundahl J et al (1995) Altered expression of CD11b/CD18 and CD62L on human monocytes after cell preparation procedures. J Immunol Methods 180:93–100
16. Havenith CE et al (1993) Separation of alveolar macrophages and dendritic cells via autofluorescence: phenotypical and functional characterization. J Leukoc Biol 53:504–510
17. Kunisch E et al (2004) Macrophage specificity of three anti-CD68 monoclonal antibodies (KP1, EBM11, and PGM1) widely used for immunohistochemistry and flow cytometry. Ann Rheum Dis 63:774–784
18. Gottfried E et al (2008) Expression of CD68 in non-myeloid cell types. Scand J Immunol 67:453–463

Chapter 31

The Use of Connexin-Based Therapeutic Approaches to Target Inflammatory Diseases

Simon J. O'Carroll, David L. Becker, Joanne O. Davidson, Alistair J. Gunn, Louise F.B. Nicholson, and Colin R. Green

Abstract

Alterations in Connexin43 (Cx43) expression levels have been shown to play a role in inflammatory processes including skin wounding and neuroinflammation. Cx43 protein levels increase following a skin wound and can inhibit wound healing. Increased Cx43 has been observed following stroke, epilepsy, ischemia, optic nerve damage, and spinal cord injury with gap junctional communication and hemichannel opening leading to increased secondary damage via the inflammatory response. Connexin43 modulation has been identified as a potential target for protection and repair in neuroinflammation and skin wound repair. This review describes the use of a Cx43 specific antisense oligonucleotide (Cx43 AsODN) and peptide mimetics of the connexin extracellular loop domain to modulate Cx43 expression and/or function in inflammatory disorders of the skin and central nervous system. An overview of the role of connexin43 in inflammatory conditions, how antisense and peptide have allowed us to elucidate the role of Cx43 in these diseases, create models of diseases to test interventions and their potential for use clinically or in current clinical trials is presented. Antisense oligonucleotides are applied topically and have been used to improve wound healing following skin injury. They have also been used to develop ex vivo models of neuroinflammatory diseases that will allow testing of intervention strategies. The connexin mimetic peptides have shown potential in a number of neuroinflammatory disorders in ex vivo models as well as in vivo when delivered directly to the injury site or when delivered systemically.

Key words Connexin, Antisense, Mimetic peptide, Inflammation, Wound healing

1 Introduction

Inflammation plays a role in a large number of pathologic conditions including acute and chronic diseases. Inflammation consists of a multistep process that involves the release of cytokines, chemokines and growth factors. This leads to migration of inflammatory cells from the blood to the affected tissue and in the case of neuroinflammation the activation and migration of microglia and astrocytes. It is now becoming clear that connexins, which form transmembrane channels allowing direct signaling between

Robert G. Gourdie and Tereance A. Myers (eds.), *Wound Regeneration and Repair: Methods and Protocols*, Methods in Molecular Biology, vol. 1037, DOI 10.1007/978-1-62703-505-7_31, © Springer Science+Business Media New York 2013

adjacent cells via gap junctions or into the extracellular space via hemichannels, play an important role in the spread of inflammation and as such they have become targets for interventions in inflammatory diseases. Examples of the role of connexins in inflammation and how they can be targeted to reduce inflammation in the central nervous system and during skin wound healing are presented below.

1.1 Connexin Changes in Response to Neuroinflammation

The predominant hallmark of neuroinflammation is the activation of astrocytes and microglia, the resident inflammatory cells of the brain. In response to neuroinflammation, astrocytes undergo a series of complex changes known as astrogliosis, during which time their morphological, electrophysiological, and biochemical properties are altered. For example, a rapid increase in glial fibrillary acidic protein (GFAP) expression and cellular hypertrophy are hallmarks of activated astrocytes [1]. Other features include alterations in glutamate metabolism and glucose uptake [2, 3]. Following inflammation microglia proliferate, take on an activated or phagocytic phenotype and release proinflammatory mediators such as TNF-α, IL-1β, and nitric oxide (NO). In the normal brain, astrocyte networks formed by gap junctions play a critical role in maintaining the homeostatic regulation of extracellular pH, K+, and glutamate levels.

Gap junctions form transmembrane channels that allow for direct cytoplasmic communication between adjacent cells. These result from the docking of two compatible hemichannels that themselves comprise hexamers of connexin proteins [4]. There is increasing evidence that neuroinflammation modulates gap junction expression and communication in astrocytes and microglia [5–8]. Proinflammatory cytokines such as TNF-α and IL-1β are reported to attenuate gap junction communication in astrocytes and astrocyte–microglia cocultures [9, 10]. In a model of bacterial brain abscess there was a reduction of astrocyte gap junction communication reported in areas directly adjacent to the brain lesion [8]. An increase in Cx43 levels in astrocytes following injury has been observed in epilepsy [11], brain ischemia [12–16], and after spinal cord injury [17–20]. However, the role of astrocytic connexins is not clearly understood. A number of studies that have transiently down-regulated connexin expression following CNS injury using antisense oligonucleotides (AsODN) or gap junction blockers have reported benefit and conversely indicated that increased Cx43 levels are potentially detrimental [20–22]. The transient modulation of the Cx43 in these studies has the effect of preventing connexin signaling during the early stages of injury, but importantly allows recovery of connexin expression and gap junctional communication in the later stages [23, 24]. Others studies using gene knockout models with permanent loss of Cx43, have reported that this background leads to an increase in the size of the injury [25–27].

In this setting the complete loss of Cx43 is likely to prevent the normal recovery of gap junction coupling and the cell–cell communication that is required for subsequent spatial buffering, neuronal survival and subsequent tissue coupling function.

Modification of connexin signaling also occurs in microglia following neuroinflammation. Under normal conditions, microglia show little expression of Cx43 and are not gap junction coupled. Exposure to pro-inflammatory cytokines leads to an induction of Cx43 expression in microglia and formation of functional gap junction coupling [28]. It has been reported that activated microglia exhibit increases in both Cx43 expression and gap junction coupling in a rat model of brain stab injury [28], following exposure to peptidoglycan, (a component of the cell wall of *S. aureus*, a prevalent CNS pathogen [29]) and in the presence of interferon-gamma (IFN-γ) and TNF-α [28]. Activation of microglia by advanced glycation end products (AGEs), which have been linked to the progression of neurodegenerative disease, leads to increased Cx43 levels possibly through the release of TNF-α [30]. Neuroinflammation appears to cause the formation of a functional microglial syncytium, so that these cells may be capable of influencing neuroinflammatory responses through gap junction intercellular communication.

There is growing evidence to suggest that Cx43 hemichannels are involved in the spread of inflammation and neuronal cell loss in a number of brain disorders. Hemichannels are the undocked membrane imbedded conduits that join together between cells to form the gap junctions. Hemichannels are normally closed but can open under various physiological and pathological stimuli to release molecules into the extracellular space. The involvement of hemichannels in release of neurotoxic molecules such as glutamate and ATP has been well documented [31–33]. For example cytokines, including IL-1β and TNF-α, released from activated microglia reportedly reduce intercellular gap junctional communication and increase hemichannel opening in astrocytes [34–36]. In a model of brain abscess [8] the reduction in astrocyte GJ communication by cells surrounding the abscess coincides with opening of hemichannels. Following inflammation, activated microglia release a number of toxic molecules, including glutamate, cytokines, prostaglandins and ATP. These molecules increase the opening of astroglial hemichannels, reducing the astrocyte's neuroprotective functions leading to increasing neuronal cell death [37]. Metabolic inhibition has been demonstrated to cause an increase in levels of Cx43 hemichannels at the surface of astrocytes [38] but at the same time reduces Cx43 gap junction coupling [38, 39]. In cortical astrocytes, reoxygenation in a high glucose medium following hypoxia increases Cx43 hemichannel activity but decreases cell–cell communication via Cx43 gap junctions [40]. When NMDA-induced excitotoxicity of neuron–astrocyte cocultures is carried out in the presence of TNF-α and IL1-β, an increase in astroglial connexin43 hemichannel activity is seen. NMDA treatment leads to increased neurotoxicity in proinflammatory-treated cocultures compared to untreated ones. Neurotoxicity was absent in cocultures with connexin43 knockout astrocytes and application of connexin43 hemichannel blockers prevented the cytokine-induced potentiated NMDA neurotoxicity [41]. Metabolic inhibition leads to cell swelling that is thought to result from the opening of Cx43 hemichannels. This metabolic inhibition induces opening of unapposed connexin 43 gap junction hemichannels and reduces gap junctional communication in cortical astrocytes in culture and can lead to ischemia-induced cell death in nervous tissue [14, 39].

1.2 Connexin Dynamics in Response to Skin Wounding

Connexins play an important role in maintaining the normal homeostasis of the intact skin, for instance in facilitating the passage of nutrients and gases in the avascular epidermis. Many different types of connexin mRNAs (Cx45, 43, 40, 37, 32, 31.1, 31, 30.3, 30 and 26) have been detected in the mammalian skin and their proteins can be found at varying locations within the skin [42–49]. The most prominent connexins in the skin are Cx43, 31.1, 30 and 26 in the epidermis. In human epidermis Cx26 dominates in the proliferative basal layers where there is less Cx43 but there is considerably more Cx43 found in the differentiating spinous layers [45, 50]. Within the dermis there is mostly Cx43 with a small amount of Cx45. The Cx43 is found mostly in fibroblasts but also in blood vessels and dermal appendages such as hair follicles, sweat and sebaceous glands as well as leukocytes and mast cells [48, 51, 52].

Wounding of the skin triggers an immediate repair process which takes the form of a series of four overlapping events starting with hemostasis, then inflammation followed by cell proliferation and migration and eventually remodeling of the scar tissue [53]. The objective of these events is to plug the defect that has formed in the skin, kill any bacteria that have entered and prevent any more from entering. The next step is to stimulate wound edge cells to change their phenotype, crawl forward, proliferate and repair the wound. As these changes take place there is also a dynamic change in the expression of connexins in the different cell types as they take on different roles as healing progresses. Associated with the inflammation of wound healing, is the "leakiness" of blood vessels to allow passage of plasma proteins, fluid and leukocytes, that manifests as tissue redness and swelling in and around the wound site. This correlates with an up-regulation of Cx43 in the cells of the surrounding wound vasculature [51, 54].

Activated leukocytes also express Cx43 and this has been reported to be associated with pro-inflammatory responses and extravasation, which is reduced when Cx43 is blocked [55]. When Cx43 down regulation is achieved by specific ASN application to wounds, then this results in a dampening down of the initial inflammatory response that is maintained throughout the healing process. The reduced numbers of wound neutrophils and later macrophages appears to have a beneficial effect on the healing process [23, 54]. At first this may seem counter intuitive as a robust inflammatory response has long been thought to be required both to prevent sepsis by killing off any bacteria, but also essential to trigger proliferation and migration of wound edge keratinocytes and fibroblasts by releasing members of the transforming growth factor and fibroblast growth factor families that serve as motogens and mitogens [56, 57]. However, production of a robust inflammatory response may not be essential for wound healing to progress [58–60]. A link between inflammation and scar formation is supported by wound healing studies on embryos, which heal without scars at times before their inflammatory response has developed but do scar once an inflammatory response can be raised [61–64]. Whilst excessive inflammation may have a negative impact on wound healing, there are some effects from inflammatory growth factors that promote healing [65–67]. Getting the balance of the inflammatory response will be important in order to promote healing but reduce scar formation though this may be more of a challenge in chronic than acute wounds.

Soon after the first inflammatory signals are released changes begin to occur in the wound edge keratinocytes as they start to dismantle their desmosomes, tight junctions and gap junctions in order to be able to migrate. Within 6 h of wounding rodent skin a reduction in Cx43 and Cx31.1 protein levels can be detected in leading edge keratinocytes and these down-regulate even further in 24–48 h as the cells become migratory [51, 68, 69]. However, Cx43 expression is maintained in the actively proliferating cells a few rows back from the leading edge. Whilst Cx43 and Cx31.1 are down regulating there is a concomitant up regulation of Cx26 and Cx30 in the very same cells and this up-regulation is maintained whilst the leading edge keratinocytes are migrating [51]. Once the two edges of the wound joined up again Cx26 and Cx30 are down regulated to their normal low levels and Cx43 expression returns as the skin resumes its normal structure and gap junctional distributions. Interestingly, whilst human skin has a slightly different distribution of connexins, a very similar dynamic response is seen in the basal and suprabasal cells that become migratory. Within 5 h of wounding a down-regulation of Cx43 can be detected at the leading edge and Cx43 can hardly be seen within 24 h and again it is replaced with Cx26 and Cx30 [50, 70]. A phosphorylation of Cx43 by Protein Kinase C at Serine 368 has also been reported in human wound edge keratinocytes occurring 24 h after wounding [50]. Related to these changes in connexin expression is a change in the communication dynamics restricting transfer of dye out of cells injected in the leading edge potentially forming a discrete communication compartment. This may have the effect to restrict signals between the leading edge cells with cells following a few rows back, as has been proposed to occur in morphogenic tissue movements in embryonic development [46, 50, 51, 53, 71–75]. The replacement of Cx43 with Cx26 and Cx30 in the leading edge migratory cells implies that it is not an absence of communication that is required for migration to take place but an absence of Cx43. Alternatively this down regulation of Cx43 may relate to the interaction of Cx43 with a wide variety of other proteins in junctional complexes or to the PDZ domain on its cytoplasmic tail. Cx43 is reported to form a complex or "nexus" with junctional proteins such as Zonular Occludin-1(ZO-1), Cadherins and alpha and beta catenin in the cell membrane whilst also linking directly or indirectly to the tubulin and actin cytoskeleton [75–83]. Membranous Cx43 could increase cell adhesion directly, but adhesion needs to be reduced for effective cell migration. In addition Cx43 could stabilize the cytoskeleton with membranous anchor points and reduce the cytoskeletal dynamics required for migration. In addition the membranous interaction of Cx43 with transcription factors such as β-catenin could prevent the β-catenin moving to the nucleus and affecting transcription. Indeed Cx43, along with Cx32 and Cx36, has been proposed to be a "master gene" influencing the expression of more than 300 other genes [84–87]. Clearly there are numerous levels that Cx43 can influence cell behavior and the wound healing response. This is not just at the protein level but also at the gene expression level independent of functions as a communication channel.

2 Methods of Cx43 Regulation

2.1 Non Specific Gap Junction Blockers

A large number of compounds have been shown to inhibit connexin gap junction and hemichannel communication (*see* Juszczak and Swiergiel for review ref. 88). These compounds include halothane, octanol, heptanol, quinolone, 18α-glycyrrhetinic acid, carbenoxolone, and flufenamic acid. However, none of these compounds are considered to be very selective for connexins as they can modify other membrane channels as well as having other off target effects. The first three compounds inhibit gap junctions by inducing radical changes in membrane fluidity while the remainder have effects that are secondary rather than direct (i.e., through actions inside the cell that in turn alter connexin communication) [88–90]. Studies using these compounds have however provided useful information with regards to the role of connexins in inflammation. For example, global gap junction blockade using carbenoxolone, flufenamic acid, heptanol, or octanol is reported to reduce the spreading wave of astrocytic apoptosis in rat hippocampal cultures exposed to ischemic conditions [91] and in a rodent model of stroke, pretreatment with octanol significantly decreased mean infarction volume following middle cerebral artery occlusion [92].

2.2 Cx43 AsODN

Specific modulation of Cx43 protein expression is possible using antisense oligodeoxynucleotides (AsODN) [21, 22, 93, 94]. The most commonly used connexin43 specific AsODN is a single strand DNA of 30 deoxynucleotides with an unmodified backbone that binds specifically to complementary sequences on an accessible region of the rat connexin43 mRNA, blocking protein translation (*see* ref. 93 for details). Antisense will block new protein translation but have no effect on existing protein levels and therefore cell uncoupling will in part be dependent upon protein turnover rate. Cronin et al. [20], for example, showed different knockdown and protein recovery rates in spinal cord dorsal, ventral, white and grey matter that may reflect dose delivered to the different tissues, cell metabolic rates, and protein turnover variances. The advantage of an antisense modulation approach is that it is specific for Cx43 and that it allows only transient protein knockdown. Unmodified AsODNs have a relatively short half-life inside cells of about 20–30 min and are broken down even more rapidly when they come in contact with sera [95–97]. Transient knockdown is important when modulating inflammation in the CNS where normal astrocyte Cx43 coupling is essential for neuronal function. Topical delivery can effectively be achieved by delivery in a thermo-reversible gel, Pluronic F-127, which is liquid between ~0 and 4 °C and can be dripped into place in a wound but it soon gels as it warms up [94]. The Pluronic gel then slowly breaks down over a number of hours and releases the AsODN to the surrounding tissues at a steady rate providing a sustained delivery throughout this period. In addition the Pluronic gel is also thought to act as a mild surfactant and aid the direct entry of the AsODN into the cells. This Cx43 AsODN 30mer has been used in several experimental disease models including brain slice [98, 99], optic nerve ischemia [100], skin wound healing [23, 54, 101], skin burns [102] and spinal cord injury [20, 103]. This AsODN has demonstrated anti-inflammatory and wound healing effects that will be described in detail in the following sections.

2.3 Cx43 Mimetic Peptide

Another approach to develop more gap junction specific inhibitors has involved the use of connexin mimetic peptides. One example of such mimetics consists of short, specific sequences that correspond to the extracellular loops of connexins. These peptides have been shown to act as reversible and specific inhibitors of gap junction and hemichannel communication (reviewed in ref. 104). Other peptides mimicking sequences from Cx43 have been reported to have therapeutic effects—*see* ref. 105 and also (Ghatnekar and Tuan) and (Strungs et al.) in this book. However, in the context of this chapter references to connexin mimetic peptides will be to sequences corresponding to the extracellular loop domains of connexins.

Two inhibitory peptides designed against Cx43 called GAP 26 (extracellular loop 1; VCYDKSFPISHVR) and GAP 27 (extracellular loop 2; SRPTEKTIFII) have been shown to inhibit the intercellular transfer of fluorescent dyes in cultured cells [106, 107], impair the propagation of calcium waves across groups of confluent cells [108, 109] and electrical communication [110, 111]. Connexin mimetic peptides may impair the interactions of the extracellular loops by binding to recognition sites on the connexon [112] although what sequence on the extracellular loop the mimetic peptide actually interact with, is still not clear. It has been reported that connexin mimetic peptides can inhibit both gap junction and hemichannel signaling in in vitro models [19, 109, 113–115]. Connexin mimetic peptides appear to regulate hemichannels and gap junctions independently of each other, with short incubation times having been shown to prevent hemichannel opening without affecting gap junction communication, while longer incubations may interfere with gap junction communication [116]. More recently we have reported that there also appears to be a concentration dependent effect with low concentrations of mimetic peptide inhibiting hemichannels, while higher concentrations inhibit gap junctions [19]. Cx43 mimetic peptides have also been reported to improve aspects of wound healing. For example in cell culture models they were reported to lead to an increase in cell migration rates and scrape-wound closure of mouse epidermal keratinocytes [117], human keratinocytes and dermal fibroblasts [118], and human organotypic skin models [73]. GAP 27 blocks hemichannel activity and reduces fibroblast cell adhesion [119]. Cx43 mimetic peptides have recently been reported to be protective in models of neuroinflammation with blockade of astrocytic Cx43 hemichannels with Gap27 protective of neurons against excitoxic cell death [41]. GAP 26 prevents neuronal cell death by blocking ATP and glutamate release from astrocytes in a cell culture model of AD [120].

A different novel mimetic peptide designed against the second extracellular loop of Cx43 termed Peptide5 (Peptagon); sequence VDCFLSRPTEKT has been used successfully to reduce neuroinflammation in ex vivo models of epilepsy and spinal cord injury as well as in *various* in vivo animal models of spinal cord injury, retinal ischemia and fetal ischemia. These will be discussed in more detail below.

3 Cx43 Regulation in Inflammation

3.1 Epilepsy

As mentioned earlier in this review, increased Cx43 is associated with epileptic seizure activity [11]. A number of studies have reported that blockade of gap junctions can reduce or block seizure-like discharges in a range of in vitro models of epilepsy [121–123]. Conversely, agents that promote opening of gap junctions enhance neuronal rupture [124]. Similar conclusions have been drawn from human tissue in vivo models; where gap junction blockers prevented neuronal network synchronization in neocortical slices obtained from patients with mesial temporal lobe epilepsy [125] and in a rodent in vivo model [126]. The seizure event leads directly to the death of a small group of neurons and this is followed by a secondary stage involving molecular and cellular events that take place hours to days after the initial precipitating damage [127]. A syncytium of activated astrocytes extensively coupled via Cx43 gap junctions has been implicated in lesion spread during this secondary stage [11, 128]. To further clarify the role of Cx43 during the secondary stage a study was carried out using connexin mimetic Peptide5 in a brain slice culture model of bicuculline methochloride (BMC)-induced epilepsy [129]. Epileptiform injury in hippocampal slice cultures was induced by 48 h exposure to 100 μM BMC. The BMC was then removed and during a 24 h recovery period lesion spread was observed by propidium iodide (PI) staining of damaged cells in the CA1 region of the hippocampus. Slices were treated with different concentrations of Peptide5 that have been shown to inhibit either just hemichannels (5 and 50 μM) or both hemichannels and gap junctions (500 μM). Peptide application was used either during BMC treatment or the recovery period. Application of Peptide5 during either the BMC treatment or recovery periods produced concentration- and exposure time-dependent neuroprotection. During the BMC period, peptide concentrations between 5 and 50 μM (sufficient to block hemichannels) had a protective effect while a substantial gap junction blockade with 500 μM peptide lead to increased cell death. This indicates that while neurons are undergoing excessive firing, gap junction communication appears to be essential for tissue survival but opening of hemichannels may be detrimental. This toxicity may be due to hemichannel-mediated purinergic signaling that has been shown to be involved in amplification cytotoxicity in pathological conditions. The increased cytosolic Ca^{2+} and membrane depolarization that occurs during epileptiform events, may induce hemichannel opening, allowing release of ATP into the extracellular space, activation of P2Y receptors on neurons, eventually leading to an elevation of cytosolic Ca^{2+} and further ATP release that would set up a Ca^{2+} wave resulting in increased epileptiform activity [130]. Inhibition of astrocytic gap junctions during abnormal firing would disturb neuronal homeostasis and reduce the coordinated response to injury [131–133]. Functional astrocytic communication would buffer extracellular ions and neurotoxic substances released as a result of epileptic discharge. There is evidence that termination of epileptiform events requires spatial K^+ buffering by the glial syncytium [134]. While only low doses of peptide were protective during the BMC phase, all doses applied during the recovery period resulted in a reduced lesion size and protected the CA1 region from further damage suggesting that following the epileptiform insult both gap junction communication and hemichannel-mediated signaling may contribute to the lesion spread. Astrocytes are known to undergo physiological changes after the epileptiform insult. Increased hemichannel opening may lead to release of neurotoxic molecules or cell swelling, which would interfere with the astrocytes ability to buffer the environment. Some of the reported changes in astrocytes in the epileptic brain include increased coupling, and changes in voltage-gated Na+ and Ca^{2+} channels, K^+ channels, aquaporins, and glutamate receptors [135, 136]. This increased coupling in such a toxic environment may lead to increased Ca^{2+} waves that place such a load on the astrocytes that they can no longer protect the neurons. These findings support an astrocytic basis of epilepsy; pathological changes that compromise astrocytic buffering capacity may enhance bursting activity, synchronization of neuronal firing and the spread of seizure induced secondary damage.

3.2 Optic Nerve Ischemia

Connexin 43 AS ODN has been reported to reduce inflammation in an optic nerve following oxygen-glucose deprivation (OGD) [100]. Exposure of isolated rat optic nerves that are cultured ex vivo under OGD leads to inflammation as determined by an increase in tissue swelling, the number of astrocytes and activated microglia. This inflammation is accompanied by an increase in the levels of Cx43 protein, which starts 2 h after exposure of the optic nerve to OGD and peaks at day 3. Application of Cx43 AsODN to these optic nerve segments lead to a decrease in Cx43 levels and subsequent dampening of the inflammatory response by reducing tissue swelling, improving vascular integrity and slowing activation of inflammatory cells. These modifications of the inflammatory response resulted in reduced lesion spread and minimization of the gap junction-mediated bystander effect that often occurs as a consequence of CNS injury. These findings support the concept that ischemic injury to the optic nerve is associated with an up-regulation of the local inflammatory response that is at least partially exacerbated by a Cx43 gap junction-mediated bystander effect.

3.3 Spinal Cord Injury

Spinal cord injury (SCI) results from an initial mechanical insult to spinal cord tissue followed by a cascade of secondary events leading to the on-going spread of tissue damage. This secondary injury includes inflammation, ischemia, free radical formation, and excitotoxicity leading to demyelination, axonal degeneration, neuronal death, cavitation, and glial scarring surrounding the area of initial damage [137–139]. Astrocytes and microglia are activated following tissue damage. While these cells are important for reestablishing tissue homeostasis [140–145], they also produce factors that hinder the ability of injured axons to regenerate [146, 147].

Levels of Cx43 protein in astrocytic gap junctions increase following SCI [17–20]. Levels of Cx43 mRNA and protein increase within 4 h of injury and reach three times the normal level by 4 weeks post-injury while no change in expression of connexin32 or connexin36 is seen [17, 18]. Theriault et al. [13] reported that Cx43-immunoreactivity is highest in areas associated with neuronal loss while Cx43 labeling is normal in areas where neurons are preserved. They demonstrated gap junction disassembly in astrocytes directly adjacent to the injury site, which may be an attempt by astrocytes to prevent communication with neighboring cells. This is thought to restrict the flow of death signals such as gap junction mediated transfer of Ca^{2+} and nitric oxide to neighboring cells as well as hemichannel release of ATP, glutamate, and other signals that contribute to lesion spread [148].

Down regulation of Cx43 GJC and hemichannel signaling reduces inflammation in ex vivo and in vivo models of spinal cord injury. Spinal cord segments excised from p7 rat pups and placed into organotypic culture show edema, astrogliosis and neuronal cell death [19]. When these segments are treated with either Cx43 specific AsODN [103] or Cx43 mimetic peptides [19] within the first hours after removal from the animal, there is a reduction in connexin43 levels, reduced astrocytosis (GFAP levels) and reduced loss of NeuN and SMI-32 positive neurons in a concentration and time dependent manner [19]. This ex vivo model has also been used to investigate the potential of Cx43 regulation as a tool for testing spinal cord injury repair strategies [103] and this will be discussed in more detail in section 4.2.

Connexin43 AsODN has been used to regulate connexin43 function in in vivo spinal cord injury hemisection and compression injury models [20]. Treated hemisection wounds showed reduced swelling and hemorrhagic inflammation than corresponding controls at 24 h, and markedly reduced tissue disruption was reported. In a compression injury model increased Cx43 expression was seen in astrocytes and small vessels around the injury site. A marked up-regulation in GFAP expression within 6 h, along with microglial activation and neutrophil invasion also occurred. Treatment with Cx43 AsODN reduced the levels of Cx43, GFAP+ astrocytes, microglia activation and neutrophil invasion as well as leading to an improvement in locomotion as assessed using the Basso, Beattie, Bresnahan (BBB) locomotor rating scale [149]. Cx43 AsODN also significantly reduced blood vessel leakiness as determined by release of BSA-FITC from the vessel bed, resulting in less recruitment of blood borne inflammatory cells and a dampening down of inflammation [20].

Mimetic peptides can also lead to improved behavioral outcomes, reduced swelling, reduced inflammation and improved neuronal survival in a severe in vivo rodent impact injury model (O'Carroll et al., unpublished data). In these studies, a dose of 20 μmol/kg of the Cx43 mimetic peptide Peptide5 was delivered directly to the spinal cord lesion via intrathecal catheterization and osmotic mini-pump for 24 h after injury. Treatment with Peptide5 lead to a significant improvement in hind-limb function as assessed using the BBB scale. Immunohistochemistry of tissue sections at 5 weeks showed reductions in astrocytosis, a decrease in the number of phagocytic and activated microglia adjacent to the injury and an increase in motor neuron survival. This demonstrates that administration of Peptide5 reduces secondary tissue damage after spinal cord injury. In experiments where delivery of the peptide was delayed until 1 h after injury, a significant behavioral improvement, reduction in gliosis and increase in neuronal survival was still observed. The fact that functional protection is observed even when delivery of the peptide is delayed supports the clinical potential for connexin mimetic peptides for the treatment of spinal cord patients. Taken together these data support the idea that lessening the inflammatory response via regulation of Cx43 may be a valuable target for therapeutic intervention after spinal cord injury.

3.4 Retinal Ischemia

It is well recognized that in the retina, like the brain and spinal cord, ischemia–reperfusion leads to neuronal cell death and increased vascular permeability [150–152]. In a study where adult male Wistar rats were exposed to 60 min of retinal ischemia followed by reperfusion a significant increase in Cx43 that colocalized with activated astrocytes, Muller cells, and vascular endothelial cells was seen. This coincided with significant vascular leakage and disruption, astrocytosis and subsequently retinal ganglion cell (RGD) death. Cx43 expression on astrocytes surrounding the vasculature plays a role in controlling blood vessel permeability [153] through astrocytic propagation of calcium waves [153, 154] either by inositol trisphosphate spread through Cx43-containing gap junctions between astrocytes and endothelial cells or extracellular ATP spread that may also be facilitated by hemichannel release of ATP [155]. This report stated that Cx43 expression in endothelial cells may play an important independent role in the permeability of blood vessels and subsequent inflammatory cascade leading to RGC death. Delivery of a single intraperitoneal injection of Cx43 mimetic peptide (Peptide5) immediately after the ischemic event significantly reduced vasculature leakage. Localized patches of Cx43 up-regulation associated with abnormal GFAP expression during the reperfusion following ischemia suggest a causal link between vascular dysfunction and the glial inflammatory response. Blocking Cx43 mediated vascular leakage led to a reduction in astrocytosis and increased retinal ganglion cell survival to a level comparable to that of uninjured retinas. In vitro studies on both human and rodent endothelial cells, in the absence of astrocytes, demonstrated that endothelial cell death following hypoxia-reperfusion was mediated directly by opening of Cx43 hemichannels in endothelial cells. Hemichannel opening and cell death were prevented using both nonspecific gap junction and hemichannel blockers as well as the Cx43 mimetic peptide at a concentration that blocks hemichannels but does not uncouple gap junctions [19]. This suggests that endothelial cell death and vessel leak may occur independently of astrocytosis and in fact, vascular leak appears to trigger astrocytosis.

3.5 Fetal Ischemia

Perinatal hypoxia–ischemia at term is caused by a transient hypoxic–ischemic event and can lead to severe encephalopathy, with a high risk of death or disability [156, 157]. This transient ischemia leads to spreading damage over days to weeks after the insult [158, 159]. While the mechanisms of this spread are not understood, there is evidence to suggest that gap junctions may play a role in the spread of injury following fetal ischemia. Nonspecific, global gap junction blockers have been reported to provide protection in rat pups after intrauterine hypoxia–ischemia [160]. Hemichannels are reported to open following oxygen glucose deprivation, metabolic inhibition or low extracellular calcium ion levels [39, 40, 161, 162]. In a recent study, a chronically instrumented near term (0.85 gestation) fetal sheep model of brain ischemia [159] was used to determine if blockage of Cx43 hemichannels using connexin mimetic Peptide5 would provide neuroprotection [16]. In this model, occluder cuffs are placed around both carotid arteries and ischemia was induced by reversible inflation with saline for 30 min. This is a well-characterized large animal model of fetal ischemia and brain maturation which is considered equivalent in many respects to the human infant at term. Following transient fetal ischemia Cx43 mRNA is up-regulated within 6 h in the intragyral white matter and cortex. These are regions that are commonly damaged following ischemia in human fetuses [163]. In order to establish whether this increase in Cx43 mRNA would lead to a corresponding increase in Cx43 hemichannels, a PI dye uptake method was used. This confirmed that hemichannels do open after ischemia in the fetal sheep, and that dye uptake in the white matter could be also be reduced by treatment with Peptide5. Cx43 hemichannels are predominantly found on astrocytes, and white matter comprises a large number of astrocytes, providing a potential tissue target for the mechanism of action of the mimetic peptide.

To test if Peptide5 would be protective, in one group 50 μmol/kg of peptide was infused over 1 h starting 90 min after the end of the occlusion (short infusion) or over 1 h followed by 50 μmol/kg over 24 h (long infusion) in a second group. A peptide that does not inhibit Cx43 hemichannels was used as a control [19]. Following transient ischemia, brain activity as measured by electroencephalogram (EEG) was suppressed and remained suppressed for 8 h following reperfusion. EEG power then increased simultaneously with the onset of a period of intense seizure activity that lasts for approximately 48 h. In the control animals, total EEG power fell to below baseline levels while a progressive increase in EEG was seen in both peptide infusion groups, followed by a marked delay in post-ischemic seizures, reduced duration of seizures and more discrete seizures than controls. In human term infants with acute neonatal encephalopathy, an increase in background EEG activity within the first 24 h has been associated with better outcome than in infants whose background activity remained suppressed [164]. Status epilepticus is associated with a worse outcome than discrete seizures, suggesting that periods of status epilepticus either cause or are a result of greater damage. It is possible that Peptide5 may reduce propagation of abnormal seizure activity by inhibiting purinergic release [33], calcium entry [165], or glutamate release from astrocytes [31]. Furthermore, peptide treatment led to a significant reduction in the time taken for sleep state cycling to resume in both the Peptide5 groups compared to the ischemia–vehicle group. Clinically, in normothermic infants with moderate to severe hypoxic–ischemic encephalopathy, more rapid recovery of sleep–wake cycling within the first 36–48 h of life is associated with a significantly better outcome [164].

The long infusion of Peptide5 (see above) showed an improvement in cortical neuron number. Conversely, while the short infusion group did not show improvement in the numbers of cortical neurons, some functional improvements were still seen. This suggests a complex relationship between neuron numbers and brain activity that may be influenced by the survival and function of other cell types. This idea is supported by the fact that the long-peptide infusion was associated with significantly improved oligodendrocyte survival in the intragyral and periventricular white matter, suggesting that improved survival of support cells in white matter contributed to functional recovery [166]. Pathological conditions lead to glial cell activation and release bioactive molecules such as glutamate, cytokines, prostaglandins and ATP that increase the activity of astroglial hemichannels, reducing the astrocyte neuroprotective functions, and further reducing neuronal cell viability. During pathology, ATP is also released from damaged cells and acts as a cytotoxic factor [167]. Release of neurotoxic molecules such as ATP is known to contribute to myelin degeneration of axons [168]. A possible mechanism by which Peptide5 is protecting oligodendrocytes is by preventing the release of ATP or other toxic molecules from glial Cx43 hemichannels. Alternatively it may be preventing cell damage, for example endothelial cell death following swelling caused by hemichannel opening and abnormal water and electrolyte uptake, which results in subsequent release of toxic molecules into the parenchyma.

Given the significant reduction in seizures and improvements in brain activity and oligodendrocyte survival, these data provide for the first time direct in vivo evidence in a relevant large animal model that Cx43 hemichannels play an important role in amplification and propagation of brain injury after reversible ischemia, that Cx43 hemichannels contribute to ongoing spread of injury well after reperfusion from cerebral ischemia and inhibition of connexin hemichannels has potential as a therapeutic neuroprotective intervention following fetal hypoxia–ischemia.

3.6 Skin Wounds

As described in section 1.2, changes in Cx43 expression and localization appears to play a key role following skin wounding. The expression of Cx43 at a skin wound site changes throughout the healing process, decreasing in wound edge keratinocytes as they become migratory but increasing in the wound edge blood vessels as they become leaky [20, 50, 51, 69, 169]. Cx43 protein is turned over relatively quickly compared to other structural type proteins, with the Cx43 half-life reported to be approximately 1.5 h [170, 171]. This half-life means that its expression can effectively be targeted with Cx43 AsODNs using topical delivery with Pluronic gel. Application of Cx43 AsODN in Pluronic gel, to a skin wound can achieve a knockdown of Cx43 protein in wound edge keratinocytes and fibroblasts within 2 h and can also suppress the up-regulation of Cx43 that normally takes place in blood vessels as they become leaky [20, 23, 54]. This has the effect of speeding the transition of the keratinocytes and fibroblasts into migratory phenotypes which promotes the rate of wound healing whilst at the same time dampening down the inflammatory response [23, 54]. A similar acceleration of healing has also been reported in the Cx43 conditional knockout mouse [172]. The decreased inflammatory response, following Cx43 AsODN treatment of wounds can be seen macroscopically early in the healing process in terms of reduced extravasation of fluids and edema in the wound bed and reduced redness, swelling and gape of the wound [23]. Microscopically there is seen to be a significant reduction in the number of neutrophils in and around the wounds at days 1 and 2 after wounding and this is followed by a reduced number of macrophages [23, 54]. Depleted numbers of neutrophils have previously been reported to promote the rate of re-epithelialization [173] so this may also contribute to the improved wound healing process. Along with reduced numbers of leukocytes the Cx43 AsODN treated wounds also have reduced levels of inflammatory cytokines and chemokines such as chemokine ligand 2 (Ccl2) and tumor necrosis factor alpha (TNF-α), which would normally attract more leukocytes to the wound site and promote inflammation [54, 174]. Release of proinflammatory signals may also be repressed by the Cx43 AsODN reducing Cx43 expression on activated leukocytes and thereby reducing the release of further proinflammatory cytokines and chemokines that would in turn recruit more leukocytes [55, 175–177].

Interestingly, whilst TNF-α and Ccl2 were found to be repressed by Cx43 AsODN treatment this was not the case for transforming growth factor beta-1 (TGFß-1), which was found to be enhanced in the wound edge epidermis and dermis 2 days after wounding. TGFß-1 is known to act as a motogen and mitogen as well as enhancing collagen production [178–180], which fits in well with the enhanced proliferation and migration observed in the wound edge epidermis and dermis [54]. In addition, the increased TGFß-1 levels in the dermis correlate with enhanced collagen 1 expression and protein production at days 2 and 7 in Cx43 AsODN treated wounds. Significantly enhanced granulation tissue formation and maturation with earlier angiogenesis and myofibroblast formation and resolution was seen, and this resulted in smaller, flatter scars was noted [54]. Many of the effects that are observed after the application of the Cx43 AsODN to the wound take place days after the direct effects of the antisense have worn off. Clearly the effects of the Cx43 antisense on the very early stages of the wound healing process have profound reprogramming like effects on the cascade of events that follow throughout the healing process.

While most acute wounds heal normally without any problems in just a few days, wound healing in diabetic humans is often slow and can result in the formation of ulcers, which can be very debilitating and frequently lead to lower limb amputations [66, 181, 182]. The Federal Drug Administration (FDA) approved model for diabetic chronic wounds is the streptozotocin (STZ) induced diabetic rat, which has retarded healing. Interestingly, soon after the onset of diabetes in these STZ rats, there is a change in the expression of connexins in the skin where Cx26 and Cx43 are reduced in the epidermis and Cx43 is elevated in the dermis [101]. From this start point you would expect keratinocyte migration to take place faster as they have less Cx43 to down-regulate before they start to migrate. However, on wounding the STZ diabetic rat, instead of down regulating Cx43 in the wound edge keratinocyte Cx43 is turned on and a bulb of nonmigratory keratinocytes forms at the edge of the wound [101]. Cx26 up-regulates in wound edge keratinocytes as normal but is not restricted to the wound edge and covers a much wider area than normal. When Cx26 expression is driven by Involucrin in the epidermis of transgenic mice, in the heterozygote mice the epidermis is hyperproliferative (as it is in other Cx26 over expressing skin conditions [183] and shows a delay in restoration of barrier function and remodeling [184]. Together these abnormal expression patterns will interfere with the ability of the migrating keratinocytes to form a communication compartment within the first 24 h after injury and this may relate to its perturbed repair. Indeed, the healing seems to stall in the first 24 h and does not begin until Cx43 down-regulation occurs naturally in the STZ diabetic rat 2 days after wounding [101]. This abnormal expression of Cx43 in wound edge keratinocytes of the STZ diabetic rat underlies the inability to heal, since if Cx43 expression is prevented by applying Cx43 AsODN to the wound then the wounds heal at normal rates or faster [101]. Persistent expression of Cx43 in keratinocytes of human chronic wounds has previously been reported [70] and human diabetic fibroblasts in culture have been shown to have elevated gap junction communication and disrupted proliferation [185, 186]. Strangely, comparison of Cx43 and Cx26 mRNA and protein levels in human diabetic and nondiabetic keratinocytes revealed no differences between them [187]. More recently, it has been reported that immunostaining and Western blot analysis of Cx43 levels in cultures of diabetic and nondiabetic keratinocytes and fibroblasts revealed no differences [73]. Clearly, the expression of Cx43 and its role in the perturbed healing of human chronic wounds needs to be firmly established. That being said, phase 2 clinical trials using the Cx43 AsODN to treat venous leg ulcers are already showing positive effects on healing.

4 Use of Cx43 AS to Develop Ex Vivo Models of Disease

4.1 Adult Hippocampal Slice Culture

Maintaining central nervous system tissue in organotypic slice culture provides the advantage of being able to study biological processes with the ease of an in vitro situation while retaining many aspects of the in vivo context, such as functional local synaptic circuitry and preservation of brain architecture of the tissue under study. However, existing culture techniques require the use of immature tissue from early postnatal animals (P2–P10) with success reported for rats up to 10 postnatal days old [188]. Previous studies have shown that slices of the immature rodent brain can be maintained in culture for 2 weeks and have been used as model systems for investigations involving multi-brain regions and widespread neuronal networks [188–190]. In many cases, however, this tissue most likely will not model what occurs in the adult. For example, comparison of synaptic development in cultures derived from different aged neonates reveals that maturation of synaptic responses occurs more rapidly in cultures prepared from older animals compared to those from the young [191]. Differences in expression profiles of specific genes between developing and mature brain tissues may interfere with interpretation of their involvement in pathophysiological process in the adult [192]. Therefore, an organotypic slice culture from older animals is important for the study of mature brain tissue pathophysiology. Slices from adult tissues show reduced stability in vitro following the trauma of the slicing and culturing procedure, making the development of such systems very difficult. Setting up of organotypic brain slices will lead to acute tissue damage and lesion spread as has been shown to occur via gap junction-mediated calcium signaling between astrocytes [21, 154, 193]. Reactive gliosis occurs, which leads to up-regulation of astrocytic Cx43 [21, 194]. It has also been suggested that mechanical damage may induce hemichannel opening, allowing release of ATP and glutamate into the extracellular space [31, 193, 195, 196], with these toxic molecules causing irreversible cellular damage in the slice. Therefore, preventing the propagation of these toxic signals from damaged cells at the cut surface via intercellular gap junctions and hemichannels may allow survival and successful culture of slices from adult brains. To test this, adult brain organotypic slices were set up and treated with Cx43AsODN to transiently knock down Cx43 expression [99]. Brains slices from 14 to 40-day-old Wistar rats cut at 150 μM or 350 μM were cultured for 2 weeks. Prior to culture slices were treated with 2 μM Cx43 AsODN in Pluronic F-127. Slices that did not receive Cx43AsODN were cultured as a control. As expected, untreated slices showed an increase in the expression of Cx43 that peaked at 24 h and returned to low levels by 48 h. Treatment with Cx43 AsODN prevented this up-regulation of Cx43 for 48 h, at which time levels were equivalent to that seen in untreated slices. After 14 days in vitro (DIV) the gross morphology of slices from 14 day old rats was analyzed. Brain morphology remained clearly distinguishable in the Cx43 AsODN-treated slices and appeared remarkably similar to freshly cut slices. Conversely, the control slices looked swollen and disintegrated and no clear structures could be seen. When sections were stained with NeuN and CMFDA to determine neuronal viability, control slices showed a disrupted laminar organization of the hippocampus and the number of live cells was reduced compared to treated slices. A loss of structural integrity was seen in the untreated control slices, while in the Cx43 AsODN-treated slices, the pyramidal cell layers were intact. Using histological staining, in control slices extensive fibrosis was seen in regions of the hippocampus and neurons had a morphological appearance typical of dying neurons (condensed nuclei and triangular cell shapes). Extracellular edema was also noted around neurons of the control sections. In contrast neurons in Cx43 AsODN slices appeared to look healthy as assessed by morphology and less extracellular edema was seen. Comparable protection was noted in slices from 40 day old rats. When control segments were stained for NeuN, there was severe deterioration of the CA3 and CA4 areas with virtually all of the neurons in the control slices were lost during the culture period and the remaining neurons appeared pyknotic. Similar to day 14 rats, Cx43 AsODN treatment resulted in increased viability with the majority of neurons being maintained and preservation of cellular organization of the hippocampus. Due to extensive growth of neuropil and high rigidity in adult brain tissue, brain slices could also be cut easily at a reduced thickness of 150 μm and the AsODN benefits were stilled observed. Owing to technical challenges associated with adult brain slices resulting from the dense network of dendrites and astrocytic processes that make imaging and electrophysiological recording difficult, the ability to culture thinner brain slices is advantageous. Decreasing Cx43 expression reduces lesion spread and bystander directed death of healthy cells beneath the cut surface, allowing for the culture of adult rat brain slices. As the down regulation of Cx43 by Cx43 AsODN occurs within the first 48 h, Cx43 levels would be normal by the time experiments are to be carried out. Thus, organotypic culture of neural tissue from adult rats becomes feasible. Adult rat brain tissue will reflect more closely the pathology of neurodegenerative diseases in human than slices from younger animals. One study has shown that slices can be cultured successfully from 20 to 21 day old rats [197]. However, these were only viable for 7 days and were 400 μm thick. The ability to culture for at least 14 days will allow longer term experiments to be carried out and furthermore, the ability to culture 150 μm-thick adult brain slices will facilitate imaging and maneuvering of electrodes for recording purposes.

4.2 Ex Vivo Spinal Cord Segments

Another specific example where ex vivo organotypic cultures can be useful is in modeling spinal cord injury. Animal models of spinal cord injury require relatively complex surgeries; extensive post injury care and studies can take many weeks if locomotor or other behavioral tests are to be carried out. Therefore, the development of organotypic cultures for pretesting of drugs or other interventions prior to animal experiments is useful. A number of ex vivo spinal cord models have been published using several hundred micron thick transverse slices that can be maintained for up to 4 weeks. Examples of their use have included the testing of neuroprotective strategies and transplantation of stem cells [198–201]. Studies have used ex vivo spinal cord segments to look at acute spinal cord injury but these have only been maintained for up to 24 h [202–204]. While these models have been helpful in testing a number of these strategies, the long-term culture of large spinal cord segments that will more accurately reflect the structure of the intact spinal cord would provide a better understanding of these neuroprotective strategies. This would also provide the opportunity to test interventions that promote axonal regeneration, which currently can only be carried out using whole animal studies.

We have developed an ex vivo model of spinal cord injury to test intervention strategies model using Cx43 AsODN [103]. In this model Cx43 AsODN application reduces lesion spread and inflammation. The natural tissue architecture is maintained and increased neuronal survival in intact post-natal spinal cord segments up to 1.5 cm long is seen. This novel ex vivo organotypic spinal cord segment culture model can be readily maintained for up to 5 days for analysis of repair strategies, enables reproducibility between lesions and provides a tightly controlled artificial environment that can be reliably compared between studies.

An example of its application is in a study where spinal cords were removed from post-natal day 7 (P7) Wistar rats and cut into 1.5 cm segments. Left and right sciatic nerves were removed and approximately 5 mm long nerve segments were cut. Nerves were then inserted into an incision made in the of the spinal cord segment. 2 μM Cx43AsODN in 30 % Pluronic gel F127 was applied to the spinal cord to completely coat the tissue and segments were cultured for 4 days. Cx43AsODN application to ex vivo spinal cord segments led to increased segment viability as assessed using the CMFDA live cell assay, reduced swelling, lesion spread and inflammation. Cx43AsODN markedly reduced the microglial activation and blood vessel endothelial cell loss that was seen in control segments. A significant increase in the number and health of lower motor neurons was also seen. When the peripheral nerve graft was analyzed, very good graft to spinal cord integration was observed in Cx43AsODN treated segments. Large numbers of SMI-32 positive axons were seen at the cord/graft interface and many had entered into the graft. At 5 days a significant increase in the number of fibers entering the graft (sixfold) and the average length of the fibers (tenfold) was seen compared to control treated segments.

Therefore use of Cx43 AsODN to prevent neuroinflammation resulting from removal and preparation allows for long-term culture of large spinal segments (up to 1.5 cm) for up to 5 days. These segments will more accurately reflect the structure of an in vivo spinal cord than a slice that is only several hundred microns thick, potentially allowing for more accurate testing of neuroprotective drugs or cell transplantation therapies. Astrocytes play a major role in spinal cord injury and modulation of astrocyte activity can have significant effects on axonal regeneration. In an environment where the CNS is not damaged, or damaged but supplemented with nerve growth factor, astrocytes can support axon growth [205]. Whenever axons encounter an environment where damage to the CNS has occurred, their regeneration is inhibited. The glial scar that forms after spinal cord injury includes many cell types but is mainly composed of astrocytes, which play a major role in creating a nonpermissive environment for axon regeneration. Transiently suppressing Cx43 expression inhibits the formation of a hostile astrocyte environment and possibly maintains the astrocytes in a growth supporting phenotype. Therefore, this is a novel ex vivo spinal cord segment culture that can be used as a model to test repair strategies for axonal regeneration. The benefits of Cx43AsODN should also be applicable in vivo where the removal of an existing scar during a repair intervention would itself create damage, lesion spread, and lead to the formation of a new glial scar that isolates the repair from the host tissue. A number of therapies that have been tried to enhance axonal regeneration have been more successful when carried out in the presence of growth promoting factors [206, 207]. Using such interventions in the presence of Cx43 AsODN, to prevent astroglial activation and therefore creating a more permissive environment, may greatly enhance the success of these repair strategies.

5 Concluding Remarks

Connexin up-regulation has been implicated in augmented levels of tissue damage spread and impaired wound healing following inflammation. This review describes data from a number of in vitro, ex vivo and in vivo models that demonstrates transiently reducing connexin expression or blocking gap junction or hemichannels may provide therapeutic opportunities for inflammatory diseases. Specific gap junction modulators such as antisense oligodeoxynucleotides and connexin mimetic peptides show tangible promise as treatment modalities by reducing inflammation, secondary tissue damage, reducing lesion size, and improving functional outcomes and wound healing. Modulation of Cx43 should now be considered as a treatment with the potential to improve a wide range of clinical inflammatory disorders.

References

1. Nolte C et al (2001) GFAP promoter-controlled EGFP-expressing transgenic mice: a tool to visualize astrocytes and astrogliosis in living brain tissue. Glia 33:72–86
2. Tabernero A, Medina JM, Giaume C (2006) Glucose metabolism and proliferation in glia: role of astrocytic gap junctions. J Neurochem 99:1049–1061
3. Liberto CM et al (2004) Pro-regenerative properties of cytokine-activated astrocytes. J Neurochem 89:1092–1100
4. Goodenough DA, Goliger JA, Paul DL (1996) Connexins, connexons, and intercellular communication. Annu Rev Biochem 65:475–502
5. Chanson M et al (2005) Gap junctional communication in tissue inflammation and repair. Biochim Biophys Acta 1711:197–207
6. Giaume C et al (2007) Glia: the fulcrum of brain diseases. Cell Death Differ 14:1324–1335
7. Scemes E et al (2007) Connexin and pannexin mediated cell-cell communication. Neuron Glia Biol 3:199–208
8. Kielian T (2008) Glial connexins and gap junctions in CNS inflammation and disease. J Neurochem 106:1000–1016
9. John SA et al (1999) Connexin-43 hemichannels opened by metabolic inhibition. J Biol Chem 274:236–240
10. Haghikia A et al (2008) Intracellular application of TNF-alpha impairs cell to cell communication via gap junctions in glioma cells. J Neurooncol 86:143–152
11. Fonseca CG, Green CR, Nicholson LF (2002) Upregulation in astrocytic connexin 43 gap junction levels may exacerbate generalized seizures in mesial temporal lobe epilepsy. Brain Res 929:105–116
12. Budd SL, Lipton SA (1998) Calcium tsunamis: do astrocytes transmit cell death messages via gap junctions during ischemia? Nat Neurosci 1:431–432
13. Oguro K et al (2001) Global ischemia-induced increases in the gap junctional proteins connexin 32 (Cx32) and Cx36 in hippocampus and enhanced vulnerability of Cx32 knock-out mice. J Neurosci 21:7534–7542

14. Contreras JE et al (2004) Role of connexin-based gap junction channels and hemichannels in ischemia-induced cell death in nervous tissue. Brain Res Brain Res Rev 47:290–303
15. Haupt C, Witte OW, Frahm C (2007) Up-regulation of Connexin43 in the glial scar following photothrombotic ischemic injury. Mol Cell Neurosci 35:89–99
16. Davidson JO et al (2012) Connexin hemichannel blockade improves outcomes in a model of fetal ischemia. Ann Neurol 71:121–132
17. Lee IH et al (2005) Glial and neuronal connexin expression patterns in the rat spinal cord during development and following injury. J Comp Neurol 489:1–10
18. Theriault E et al (1997) Connexin43 and astrocytic gap junctions in the rat spinal cord after acute compression injury. J Comp Neurol 382:199–214
19. O'Carroll SJ et al (2008) Connexin 43 mimetic peptides reduce swelling, astrogliosis, and neuronal cell death after spinal cord injury. Cell Commun Adhes 15:27–42
20. Cronin M et al (2008) Blocking connexin43 expression reduces inflammation and improves functional recovery after spinal cord injury. Mol Cell Neurosci 39(2):152–60
21. Frantseva MV et al (2002) Specific gap junctions enhance the neuronal vulnerability to brain traumatic injury. J Neurosci 22:644–653
22. Frantseva MV, Kokarovtseva L, Perez Velazquez JL (2002) Ischemia-induced brain damage depends on specific gap-junctional coupling. J Cereb Blood Flow Metab 22:453–462
23. Qiu C et al (2003) Targeting connexin43 expression accelerates the rate of wound repair. Curr Biol 13:1697–1703
24. Green CR et al (2001) Spatiotemporal depletion of connexins using antisense oligonucleotides. In: Bruzzone R, Giaume C (eds) Methods in molecular biology; connexin methods and protocols. Humana Press Inc., Totawa, NJ, pp 175–185
25. Nakase T et al (2004) Increased apoptosis and inflammation after focal brain ischemia in mice lacking connexin43 in astrocytes. Am J Pathol 164:2067–2075
26. Siushansian R et al (2001) Connexin43 null mutation increases infarct size after stroke. J Comp Neurol 440:387–394
27. Takahashi DK, Vargas JR, Wilcox KS (2010) Increased coupling and altered glutamate transport currents in astrocytes following kainic-acid-induced status epilepticus. Neurobiol Dis 40:573–585
28. Eugenin EA et al (2001) Microglia at brain stab wounds express connexin 43 and in vitro form functional gap junctions after treatment with interferon-gamma and tumor necrosis factor-alpha. Proc Natl Acad Sci USA 98:4190–4195
29. Garg S, Md Syed M, Kielian T (2005) Staphylococcus aureus-derived peptidoglycan induces Cx43 expression and functional gap junction intercellular communication in microglia. J Neurochem 95:475–483
30. Shaikh SB et al (2012) AGEs-RAGE mediated up-regulation of connexin43 in activated human microglial CHME-5 cells. Neurochem Int 60:640–651
31. Ye ZC et al (2003) Functional hemichannels in astrocytes: a novel mechanism of glutamate release. J Neurosci 23:3588–3596
32. Takeuchi H et al (2006) Tumor necrosis factor-alpha induces neurotoxicity via glutamate release from hemichannels of activated microglia in an autocrine manner. J Biol Chem 281:21362–21368
33. Kang J et al (2008) Connexin 43 hemichannels are permeable to ATP. J Neurosci 28:4702–4711
34. John GR et al (1999) IL-1beta differentially regulates calcium wave propagation between primary human fetal astrocytes via pathways involving P2 receptors and gap junction channels. Proc Natl Acad Sci USA 96:11613–11618
35. Retamal MA et al (2007) Cx43 hemichannels and gap junction channels in astrocytes are regulated oppositely by proinflammatory cytokines released from activated microglia. J Neurosci 27:13781–13792
36. Morita M et al (2007) Dual regulation of astrocyte gap junction hemichannels by growth factors and a pro-inflammatory cytokine via the mitogen-activated protein kinase cascade. Glia 55:508–515
37. Orellana JA et al (2009) Modulation of brain hemichannels and gap junction channels by pro-inflammatory agents and their possible role in neurodegeneration. Antioxid Redox Signal 11:369–399
38. Retamal MA et al (2006) S-nitrosylation and permeation through connexin 43 hemichannels in astrocytes: induction by oxidant stress and reversal by reducing agents. Proc Natl Acad Sci USA 103:4475–4480
39. Contreras JE et al (2002) Metabolic inhibition induces opening of unapposed connexin 43 gap junction hemichannels and reduces gap junctional communication in cortical astrocytes in culture. Proc Natl Acad Sci USA 99:495–500
40. Orellana JA et al (2010) Hypoxia in high glucose followed by reoxygenation in normal glucose reduces the viability of cortical astrocytes through increased permeability of connexin 43 hemichannels. Glia 58:329–343

41. Froger N et al (2010) Inhibition of cytokine-induced connexin43 hemichannel activity in astrocytes is neuroprotective. Mol Cell Neurosci 45:37–46
42. Kamibayashi Y et al (1993) Expression of gap junction proteins connexin 26 and 43 is modulated during differentiation of keratinocytes in newborn mouse epidermis. J Invest Dermatol 101:773–778
43. Fitzgerald DJ et al (1994) Expression and function of connexin in normal and transformed human keratinocytes in culture. Carcinogenesis 15:1859–1865
44. Goliger JA, Paul DL (1994) Expression of gap junction proteins Cx26, Cx31.1, Cx37, and Cx43 in developing and mature rat epidermis. Dev Dyn 200:1–13
45. Di WL et al (2001) Multiple epidermal connexins are expressed in different keratinocyte subpopulations including connexin 31. J Invest Dermatol 117:958–964
46. Lampe PD et al (2000) Phosphorylation of connexin43 on serine368 by protein kinase C regulates gap junctional communication. J Cell Biol 149:1503–1512
47. Richard G (2000) Connexins: a connection with the skin. Exp Dermatol 9:77–96
48. Richard G (2005) Connexin disorders of the skin. Clin Dermatol 23:23–32
49. Salomon D et al (1994) Topography of mammalian connexins in human skin. J Invest Dermatol 103:240–247
50. Richards TS et al (2004) Protein kinase C spatially and temporally regulates gap junctional communication during human wound repair via phosphorylation of connexin43 on serine368. J Cell Biol 167:555–562
51. Coutinho P et al (2003) Dynamic changes in connexin expression correlate with key events in the wound healing process. Cell Biol Int 27:525–541
52. Pistorio AL, Ehrlich HP (2011) Modulatory effects of connexin-43 expression on gap junction intercellular communications with mast cells and fibroblasts. J Cell Biochem 112:1441–1449
53. Martin P (1997) Wound healing–aiming for perfect skin regeneration. Science 276:75–81
54. Mori R et al (2006) Acute downregulation of connexin43 at wound sites leads to a reduced inflammatory response, enhanced keratinocyte proliferation and wound fibroblast migration. J Cell Sci 119:5193–5203
55. Oviedo-Orta E, Howard Evans W (2004) Gap junctions and connexin-mediated communication in the immune system. Biochim Biophys Acta 1662:102–112
56. Singer AJ, Clark RA (1999) Cutaneous wound healing. N Engl J Med 341:738–746
57. Werner S, Grose R (2003) Regulation of wound healing by growth factors and cytokines. Physiol Rev 83:835–870
58. Ashcroft GS et al (1999) Mice lacking Smad3 show accelerated wound healing and an impaired local inflammatory response. Nat Cell Biol 1:260–266
59. Martin P et al (2003) Wound healing in the PU.1 null mouse–tissue repair is not dependent on inflammatory cells. Curr Biol 13:1122–1128
60. Dovi JV, Szpaderska AM, DiPietro LA (2004) Neutrophil function in the healing wound: adding insult to injury? Thromb Haemost 92:275–280
61. Adzick NS, Longaker MT (1992) Scarless fetal healing. Therapeutic implications. Ann Surg 215:3–7
62. Hopkinson-Woolley J et al (1994) Macrophage recruitment during limb development and wound healing in the embryonic and foetal mouse. J Cell Sci 107(Pt 5): 1159–1167
63. Redd MJ et al (2004) Wound healing and inflammation: embryos reveal the way to perfect repair. Philos Trans R Soc Lond B Biol Sci 359:777–784
64. Whitby DJ, Ferguson MW (1991) Immunohistochemical localization of growth factors in fetal wound healing. Dev Biol 147:207–215
65. Roberts AB et al (1986) Transforming growth factor type beta: rapid induction of fibrosis and angiogenesis in vivo and stimulation of collagen formation in vitro. Proc Natl Acad Sci USA 83:4167–4171
66. Mustoe T (2004) Understanding chronic wounds: a unifying hypothesis on their pathogenesis and implications for therapy. Am J Surg 187:65S–70S
67. Schreml S et al (2010) Wound healing in the 21st century. J Am Acad Dermatol 63: 866–881
68. Saitoh M et al (1997) Changes in the expression of gap junction proteins (connexins) in hamster tongue epithelium during wound healing and carcinogenesis. Carcinogenesis 18:1319–1328
69. Goliger JA, Paul DL (1995) Wounding alters epidermal connexin expression and gap junction-mediated intercellular communication. Mol Biol Cell 6:1491–1501
70. Brandner JM et al (2004) Connexins 26, 30, and 43: differences among spontaneous, chronic, and accelerated human wound healing. J Invest Dermatol 122:1310–1320
71. Clark RA (1985) Cutaneous tissue repair: basic biologic considerations. I. J Am Acad Dermatol 13:701–725

72. Gailit J, Clark RA (1994) Wound repair in the context of extracellular matrix. Curr Opin Cell Biol 6:717–725
73. Pollok S et al (2011) Connexin 43 mimetic peptide Gap27 reveals potential differences in the role of Cx43 in wound repair between diabetic and non-diabetic cells. J Cell Mol Med 15:861–873
74. Solan JL et al (2003) Connexin43 phosphorylation at S368 is acute during S and G2/M and in response to protein kinase C activation. J Cell Sci 116:2203–2211
75. van Zeijl L et al (2007) Regulation of connexin43 gap junctional communication by phosphatidylinositol 4,5-bisphosphate. J Cell Biol 177:881–891
76. Duffy HS, Delmar M, Spray DC (2002) Formation of the gap junction nexus: binding partners for connexins. J Physiol Paris 96:243–249
77. Butkevich E et al (2004) Drebrin is a novel connexin-43 binding partner that links gap junctions to the submembrane cytoskeleton. Curr Biol 14:650–658
78. Giepmans BN et al (2001) Gap junction protein connexin-43 interacts directly with microtubules. Curr Biol 11:1364–1368
79. Li W, Hertzberg EL, Spray DC (2005) Regulation of connexin43-protein binding in astrocytes in response to chemical ischemia/hypoxia. J Biol Chem 280:7941–7948
80. Shaw RM et al (2007) Microtubule plus-end-tracking proteins target gap junctions directly from the cell interior to adherens junctions. Cell 128:547–560
81. Theiss C, Meller K (2002) Microinjected anti-actin antibodies decrease gap junctional intercellular commmunication in cultured astrocytes. Exp Cell Res 281:197–204
82. Wei CJ et al (2005) Connexin43 associated with an N-cadherin-containing multiprotein complex is required for gap junction formation in NIH3T3 cells. J Biol Chem 280:19925–19936
83. Wei CJ, Xu X, Lo CW (2004) Connexins and cell signaling in development and disease. Annu Rev Cell Dev Biol 20:811–838
84. Spray DC, Rozental R, Srinivas M (2002) Prospects for rational development of pharmacological gap junction channel blockers. Curr Drug Targets 3:455–464
85. Iacobas DA, Iacobas S, Spray DC (2007) Connexin-dependent transcellular transcriptomic networks in mouse brain. Prog Biophys Mol Biol 94:169–185
86. Iacobas DA, Iacobas S, Spray DC (2007) Connexin43 and the brain transcriptome of newborn mice. Genomics 89:113–123
87. Spray DC, Iacobas DA (2007) Organizational principles of the connexin-related brain transcriptome. J Membr Biol 218:39–47
88. Juszczak GR, Swiergiel AH (2009) Properties of gap junction blockers and their behavioural, cognitive and electrophysiological effects: animal and human studies. Prog Neuropsychopharmacol Biol Psychiatry 33:181–198
89. Evans WH, Boitano S (2001) Connexin mimetic peptides: specific inhibitors of gap-junctional intercellular communication. Biochem Soc Trans 29:606–612
90. Das S et al (2008) Protection of retinal cells from ischemia by a novel gap junction inhibitor. Biochem Biophys Res Commun 373:504–508
91. Nodin C, Nilsson M, Blomstrand F (2005) Gap junction blockage limits intercellular spreading of astrocytic apoptosis induced by metabolic depression. J Neurochem 94:1111–1123
92. Rawanduzy A et al (1997) Effective reduction of infarct volume by gap junction blockade in a rodent model of stroke. J Neurosurg 87:916–920
93. Law LY et al (2006) In vitro optimization of antisense oligodeoxynucleotide design: an example using the connexin gene family. J Biomol Tech 17:270–282
94. Becker DL et al (1999) Roles for alpha 1 connexin in morphogenesis of chick embryos revealed using a novel antisense approach. Dev Genet 24:33–42
95. Wagner RW (1994) Gene inhibition using antisense oligodeoxynucleotides. Nature 372:333–335
96. Myers KJ, Dean NM (2000) Sensible use of antisense: how to use oligonucleotides as research tools. Trends Pharmacol Sci 21:19–23
97. Phillips MI, Zhang YC (2000) Basic principles of using antisense oligonucleotides in vivo. Methods Enzymol 313:46–56
98. Yoon JJ et al (2010) Effect of low Mg2+ and bicuculline on cell survival in hippocampal slice cultures. Int J Neurosci 120:752–759
99. Yoon JJ et al (2010) A novel method of organotypic brain slice culture using connexin-specific antisense oligodeoxynucleotides to improve neuronal survival. Brain Res 1353:194–203
100. Danesh-Meyer HV et al (2008) Connexin43 antisense oligodeoxynucleotide treatment down-regulates the inflammatory response in an in vitro interphase organotypic culture model of optic nerve ischaemia. J Clin Neurosci 15:1253–1263

101. Wang CM et al (2007) Abnormal connexin expression underlies delayed wound healing in diabetic skin. Diabetes 56:2809–2817
102. Coutinho P et al (2005) Limiting burn extension by transient inhibition of Connexin43 expression at the site of injury. Br J Plast Surg 58:658–667
103. Zhang J et al (2010) A model for ex vivo spinal cord segment culture–a tool for analysis of injury repair strategies. J Neurosci Methods 192:49–57
104. Evans WH, Leybaert L (2007) Mimetic peptides as blockers of connexin channel-facilitated intercellular communication. Cell Commun Adhes 14:265–273
105. Ghatnekar GS et al (2009) Connexin43 carboxyl-terminal peptides reduce scar progenitor and promote regenerative healing following skin wounding. Regen Med 4:205–223
106. Chaytor AT et al (1999) The endothelial component of cannabinoid-induced relaxation in rabbit mesenteric artery depends on gap junctional communication. J Physiol 520(Pt 2):539–550
107. Mambetisaeva ET, Gire V, Evans WH (1999) Multiple connexin expression in peripheral nerve, Schwann cells, and Schwannoma cells. J Neurosci Res 57:166–175
108. Isakson BE, Evans WH, Boitano S (2001) Intercellular Ca^{2+} signaling in alveolar epithelial cells through gap junctions and by extracellular ATP. Am J Physiol Lung Cell Mole Physiol 280:L221–L228
109. Boitano S, Evans WH (2000) Connexin mimetic peptides reversibly inhibit Ca(2+) signaling through gap junctions in airway cells. Am J Physiol Lung Cell Mole Physiol 279:L623–L630
110. Dora KA et al (1999) Role of heterocellular Gap junctional communication in endothelium-dependent smooth muscle hyperpolarization: inhibition by a connexin-mimetic peptide. Biochem Biophys Res Commun 254:27–31
111. Kwak BR, Jongsma HJ (1999) Selective inhibition of gap junction channel activity by synthetic peptides. J Physiol 516(Pt 3):679–685
112. Berthoud VM, Beyer EC, Seul KH (2000) Peptide inhibitors of intercellular communication. Am J Physiol Lung Cell Mole Physiol 279:L619–L622
113. De Vuyst E et al (2007) Connexin hemichannels and gap junction channels are differentially influenced by lipopolysaccharide and basic fibroblast growth factor. Mol Biol Cell 18:34–46
114. Martin PE, Wall C, Griffith TM (2005) Effects of connexin-mimetic peptides on gap junction functionality and connexin expression in cultured vascular cells. Br J Pharmacol 144:617–627
115. Braet K et al (2003) Photoliberating inositol-1,4,5-trisphosphate triggers ATP release that is blocked by the connexin mimetic peptide gap 26. Cell Calcium 33:37–48
116. Leybaert L et al (2003) Connexin channels, connexin mimetic peptides and ATP release. Cell Commun Adhes 10:251–257
117. Kandyba EE, Hodgins MB, Martin PE (2008) A murine living skin equivalent amenable to live-cell imaging: analysis of the roles of connexins in the epidermis. J Invest Dermatol 128:1039–1049
118. Wright CS et al (2009) Connexin mimetic peptides improve cell migration rates of human epidermal keratinocytes and dermal fibroblasts in vitro. Wound Repair Regen 17:240–249
119. Wright CS et al (2012) The connexin mimetic peptide Gap27 increases human dermal fibroblast migration in hyperglycemic and hyperinsulinemic conditions in vitro. J Cell Physiol 227:77–87
120. Orellana JA et al (2011) ATP and glutamate released via astroglial connexin 43 hemichannels mediate neuronal death through activation of pannexin 1 hemichannels. J Neurochem 118(5):826–840
121. de Curtis M, Manfridi A, Biella G (1998) Activity-dependent pH shifts and periodic recurrence of spontaneous interictal spikes in a model of focal epileptogenesis. J Neurosci 18:7543–7551
122. Li J et al (2001) Upregulation of gap junction connexin 32 with epileptiform activity in the isolated mouse hippocampus. Neuroscience 105:589–598
123. Margineanu DG, Klitgaard H (2001) Can gap-junction blockade preferentially inhibit neuronal hypersynchrony vs. excitability? Neuropharmacology 41:377–383
124. Kohling R et al (2001) Prolonged epileptiform bursting induced by 0-Mg(2+) in rat hippocampal slices depends on gap junctional coupling. Neuroscience 105:579–587
125. Gigout S et al (2006) Effects of gap junction blockers on human neocortical synchronization. Neurobiol Dis 22:496–508
126. Szente M et al (2002) Involvement of electrical coupling in the in vivo ictal epileptiform activity induced by 4-aminopyridine in the neocortex. Neuroscience 115:1067–1078
127. Liu YW et al (2007) Adult neurogenesis in mesial temporal lobe epilepsy: a review of recent animal and human studies. Curr Pharm Biotechnol 8:187–194
128. Gao HM, Hong JS (2008) Why neurodegenerative diseases are progressive: uncontrolled

inflammation drives disease progression. Trends Immunol 29:357–365
129. Yoon JJ et al (2010) Dose-dependent protective effect of connexin43 mimetic peptide against neurodegeneration in an ex vivo model of epileptiform lesion. Epilepsy Res 92:153–162
130. van der Linden JA et al (1993) Bicuculline increases the intracellular calcium response of CA1 hippocampal neurons to synaptic stimulation. Neurosci Lett 155:230–233
131. Nakase T et al (2003) Neuroprotective role of astrocytic gap junctions in ischemic stroke. Cell Commun Adhes 10:413–417
132. Blanc EM, Bruce-Keller AJ, Mattson MP (1998) Astrocytic gap junctional communication decreases neuronal vulnerability to oxidative stress-induced disruption of Ca^{2+} homeostasis and cell death. J Neurochem 70:958–970
133. Ozog MA, Siushansian R, Naus CC (2002) Blocked gap junctional coupling increases glutamate-induced neurotoxicity in neuron-astrocyte co-cultures. J Neuropathol Exp Neurol 61:132–141
134. Bragin A, Penttonen M, Buzsaki G (1997) Termination of epileptic afterdischarge in the hippocampus. J Neurosci 17:2567–2579
135. Jabs R, Seifert G, Steinhauser C (2008) Astrocytic function and its alteration in the epileptic brain. Epilepsia 49(Suppl 2):3–12
136. O'Connor ER et al (1998) Astrocytes from human hippocampal epileptogenic foci exhibit action potential-like responses. Epilepsia 39:347–354
137. Fitch MT et al (1999) Cellular and molecular mechanisms of glial scarring and progressive cavitation: in vivo and in vitro analysis of inflammation-induced secondary injury after CNS trauma. J Neurosci 19:8182–8198
138. Dusart I, Schwab ME (1994) Secondary cell death and the inflammatory reaction after dorsal hemisection of the rat spinal cord. Eur J Neurosci 6:712–724
139. Koshinaga M, Whittemore SR (1995) The temporal and spatial activation of microglia in fiber tracts undergoing anterograde and retrograde degeneration following spinal cord lesion. J Neurotrauma 12:209–222
140. Bush TG et al (1999) Leukocyte infiltration, neuronal degeneration, and neurite outgrowth after ablation of scar-forming, reactive astrocytes in adult transgenic mice. Neuron 23:297–308
141. Davalos D et al (2005) ATP mediates rapid microglial response to local brain injury in vivo. Nat Neurosci 8:752–758
142. Faulkner JR et al (2004) Reactive astrocytes protect tissue and preserve function after spinal cord injury. J Neurosci 24:2143–2155
143. Myer DJ et al (2006) Essential protective roles of reactive astrocytes in traumatic brain injury. Brain 129:2761–2772
144. Okada S et al (2006) Conditional ablation of Stat3 or Socs3 discloses a dual role for reactive astrocytes after spinal cord injury. Nat Med 12:829–834
145. Loane DJ, Byrnes KR (2010) Role of microglia in neurotrauma. Neurotherapeutics 7:366–377
146. Brambilla R et al (2005) Inhibition of astroglial nuclear factor kappaB reduces inflammation and improves functional recovery after spinal cord injury. J Exp Med 202:145–156
147. Horn KP et al (2008) Another barrier to regeneration in the CNS: activated macrophages induce extensive retraction of dystrophic axons through direct physical interactions. J Neurosci 28:9330–9341
148. Decrock E et al (2009) Connexin-related signaling in cell death: to live or let die? Cell Death Differ 16:524–536
149. Basso DM, Beattie MS, Bresnahan JC (1995) A sensitive and reliable locomotor rating scale for open field testing in rats. J Neurotrauma 12:1–21
150. Abcouwer SF et al (2010) Effects of ischemic preconditioning and bevacizumab on apoptosis and vascular permeability following retinal ischemia-reperfusion injury. Invest Ophthalmol Vis Sci 51:5920–5933
151. Wilson CA et al (1995) Blood-retinal barrier breakdown following experimental retinal ischemia and reperfusion. Exp Eye Res 61:547–557
152. Zheng L et al (2007) Retinal ischemia and reperfusion causes capillary degeneration: similarities to diabetes. Invest Ophthalmol Vis Sci 48:361–367
153. Simard M et al (2003) Signaling at the gliovascular interface. J Neurosci 23:9254–9262
154. Cornell-Bell AH, Finkbeiner SM (1991) Ca^{2+} waves in astrocytes. Cell Calcium 12:185–204
155. Braet K et al (2001) Astrocyte-endothelial cell calcium signals conveyed by two signalling pathways. Eur J Neurosci 13:79–91
156. Edwards AD et al (2010) Neurological outcomes at 18 months of age after moderate hypothermia for perinatal hypoxic ischaemic encephalopathy: synthesis and meta-analysis of trial data. BMJ 340:c363
157. Vannucci RC (2000) Hypoxic-ischemic encephalopathy. Am J Perinatol 17:113–120
158. Thornton JS et al (1998) Temporal and anatomical variations of brain water apparent diffusion coefficient in perinatal cerebral hypoxic-ischemic injury: relationships to cerebral energy metabolism. Magn Reson Med 39:920–927

159. Williams CE et al (1992) Outcome after ischemia in the developing sheep brain: an electroencephalographic and histological study. Ann Neurol 31:14–21
160. de Pina-Benabou MH et al (2005) Blockade of gap junctions in vivo provides neuroprotection after perinatal global ischemia. Stroke 36:2232–2237
161. Kondo RP et al (2000) Metabolic inhibition activates a non-selective current through connexin hemichannels in isolated ventricular myocytes. J Mol Cell Cardiol 32: 1859–1872
162. Li H et al (1996) Properties and regulation of gap junctional hemichannels in the plasma membranes of cultured cells. J Cell Biol 134:1019–1030
163. Barkovich AJ, Truwit CL (1990) Brain damage from perinatal asphyxia: correlation of MR findings with gestational age. AJNR Am J Neuroradiol 11:1087–1096
164. Murray DM et al (2009) Early EEG findings in hypoxic-ischemic encephalopathy predict outcomes at 2 years. Pediatrics 124: e459–e467
165. Peng W et al (2009) Systemic administration of an antagonist of the ATP-sensitive receptor P2X7 improves recovery after spinal cord injury. Proc Natl Acad Sci USA 106:12489–12493
166. Roelfsema V et al (2004) Window of opportunity of cerebral hypothermia for postischemic white matter injury in the near-term fetal sheep. J Cereb Blood Flow Metab 24:877–886
167. Franke H et al (2012) Pathophysiology of astroglial purinergic signalling. Purinergic Signal 8:629–657
168. Orellana JA et al (2011) Hemichannels in the neurovascular unit and white matter under normal and inflamed conditions. CNS Neurol Disord Drug Targets 10:404–414
169. Neub A et al (2007) Biphasic regulation of AP-1 subunits during human epidermal wound healing. J Invest Dermatol 127: 2453–2462
170. Leithe E, Brech A, Rivedal E (2006) Endocytic processing of connexin43 gap junctions: a morphological study. Biochem J 393:59–67
171. Gaietta G et al (2002) Multicolor and electron microscopic imaging of connexin trafficking. Science 296:503–507
172. Kretz M et al (2003) Altered connexin expression and wound healing in the epidermis of connexin-deficient mice. J Cell Sci 116:3443–3452
173. Dovi JV, He LK, DiPietro LA (2003) Accelerated wound closure in neutrophil-depleted mice. J Leukoc Biol 73:448–455
174. Rossi D, Zlotnik A (2000) The biology of chemokines and their receptors. Annu Rev Immunol 18:217–242
175. Sarieddine MZ et al (2009) Connexin43 modulates neutrophil recruitment to the lung. J Cell Mol Med 13:4560–4570
176. Zahler S et al (2003) Gap-junctional coupling between neutrophils and endothelial cells: a novel modulator of transendothelial migration. J Leukoc Biol 73:118–126
177. Mendoza-Naranjo A et al (2011) Functional gap junctions accumulate at the immunological synapse and contribute to T cell activation. J Immunol 187:3121–3132
178. Cutroneo KR (2003) How is Type I procollagen synthesis regulated at the gene level during tissue fibrosis. J Cell Biochem 90:1–5
179. Postlethwaite AE et al (1987) Stimulation of the chemotactic migration of human fibroblasts by transforming growth factor beta. J Exp Med 165:251–256
180. Waggett AD, Benjamin M, Ralphs JR (2006) Connexin 32 and 43 gap junctions differentially modulate tenocyte response to cyclic mechanical load. Eur J Cell Biol 85:1145–1154
181. Falanga V (2005) Wound healing and its impairment in the diabetic foot. Lancet 366:1736–1743
182. Dinh TL, Veves A (2005) A review of the mechanisms implicated in the pathogenesis of the diabetic foot. Int J Low Extrem Wounds 4:154–159
183. Lucke T et al (1999) Upregulation of connexin 26 is a feature of keratinocyte differentiation in hyperproliferative epidermis, vaginal epithelium, and buccal epithelium. J Invest Dermatol 112:354–361
184. Djalilian AR et al (2006) Connexin 26 regulates epidermal barrier and wound remodeling and promotes psoriasiform response. J Clin Invest 116:1243–1253
185. Abdullah KM et al (1999) Cell-to-cell communication and expression of gap junctional proteins in human diabetic and nondiabetic skin fibroblasts: effects of basic fibroblast growth factor. Endocrine 10:35–41
186. Loots MA et al (1999) Cultured fibroblasts from chronic diabetic wounds on the lower extremity (non-insulin-dependent diabetes mellitus) show disturbed proliferation. Arch Dermatol Res 291:93–99
187. Brandner JM et al (2008) Expression of matrix metalloproteinases, cytokines, and connexins in diabetic and nondiabetic human keratinocytes before and after transplantation into an ex vivo wound-healing model. Diabetes Care 31:114–120
188. Thomas MP et al (1998) Organotypic brain slice cultures for functional analysis of alcohol-related disorders: novel versus conventional

preparations. Alcohol Clin Exp Res 22:51–59

189. Riley C et al (2006) A peptide preparation protects cells in organotypic brain slices against cell death after glutamate intoxication. J Neural Transm 113:103–110
190. Thomas MP et al (1998) Survival and functional demonstration of interregional pathways in fore/midbrain slice explant cultures. Neuroscience 85:615–626
191. Muller D, Buchs PA, Stoppini L (1993) Time course of synaptic development in hippocampal organotypic cultures. Brain Res Dev Brain Res 71:93–100
192. Fabian-Fine R et al (2000) Age-dependent pre- and postsynaptic distribution of AMPA receptors at synapses in CA3 stratum radiatum of hippocampal slice cultures compared with intact brain. Eur J Neurosci 12:3687–3700
193. Braet K et al (2004) Calcium signal communication in the central nervous system. Biol Cell 96:79–91
194. Pekny M, Nilsson M (2005) Astrocyte activation and reactive gliosis. Glia 50:427–434
195. Stout CE et al (2002) Intercellular calcium signaling in astrocytes via ATP release through connexin hemichannels. J Biol Chem 277:10482–10488
196. Gomes P et al (2005) ATP release through connexin hemichannels in corneal endothelial cells. Invest Ophthalmol Vis Sci 46:1208–1218
197. Finley M et al (2004) Functional validation of adult hippocampal organotypic cultures as an in vitro model of brain injury. Brain Res 1001:125–132
198. Cho JS et al (2009) Transplantation of mesenchymal stem cells enhances axonal outgrowth and cell survival in an organotypic spinal cord slice culture. Neurosci Lett 454:43–48
199. Kim HM et al (2010) Organotypic spinal cord slice culture to study neural stem/progenitor cell microenvironment in the injured spinal cord. Exp Neurobiol 19:106–113
200. Guzman-Lenis MS et al (2008) Analysis of FK506-mediated protection in an organotypic model of spinal cord damage: heat shock protein 70 levels are modulated in microglial cells. Neuroscience 155:104–113
201. Krassioukov AV et al (2002) An in vitro model of neurotrauma in organotypic spinal cord cultures from adult mice. Brain Res Brain Res Protoc 10:60–68
202. Hamann K et al (2008) Critical role of acrolein in secondary injury following ex vivo spinal cord trauma. J Neurochem 107:712–721
203. Saruhashi Y, Matsusue Y, Hukuda S (2002) Effects of serotonin 1A agonist on acute spinal cord injury. Spinal Cord 40:519–523
204. Mazzone GL et al (2010) Kainate-induced delayed onset of excitotoxicity with functional loss unrelated to the extent of neuronal damage in the in vitro spinal cord. Neuroscience 168:451–462
205. Kawaja MD, Gage FH (1991) Reactive astrocytes are substrates for the growth of adult CNS axons in the presence of elevated levels of nerve growth factor. Neuron 7:1019–1030
206. Bonner JF et al (2011) Grafted neural progenitors integrate and restore synaptic connectivity across the injured spinal cord. J Neurosci 31:4675–4686
207. Cao Q et al (2010) Transplantation of ciliary neurotrophic factor-expressing adult oligodendrocyte precursor cells promotes remyelination and functional recovery after spinal cord injury. J Neurosci 30:2989–3001

Part VII

Commercialization of Wound Repair and Regenerative Discoveries

Chapter 32

Commercialization: Patenting and Licensing in Wound Healing and Regenerative Biology

W. Chip. Hood and David Huizenga

Abstract

Wound healing and regeneration is an area rich in discovery and progress. Models of resourcing science are changing, thus there are both research funding and commercial incentives for developing intellectual property. This chapter focuses on the initial steps of commercial development of a patentable discovery such as might apply to an invention in this space, and indeed other areas of biomedical science. We cover the basics of patent protection and licensing. First, we address why research results often need to be patented in order to provide an incentive for the financial investment required to move a product through the development and approval process. Second, we discuss the basics of US patent law and address certain pitfalls of premature disclosure of research results. Next, we inform the reader of significant issues relative to ownership of IP in the academic and industrial worlds, in particular the interplay of consulting agreements and employment obligations. Finally an overview of the mechanics of the licensing process is provided.

Key words Intellectual property, Patents, Inventors, US patent law, Pitfalls, Licensing

1 Introduction

Discoveries from scientific research are the catalyst for many of the products that make our society better, healthier, happier, and productive. This is particularly the case in the wound healing where there many unmet clinical needs remain and useful inventions in this area could improve the lives of patients considerably. The wound healing and regeneration area is also a space rich in discovery and progress. However, the process of discovery, invention, and subsequent commercialization into a product often take substantial investment of time and monies and involve significant financial risk, particularly in the life sciences field.

To encourage innovation, sharing of knowledge, and this type of private sector investment governments often offer enhanced financial returns to investors. One such vehicle for enhancing returns is to provide market exclusivity, in essence a monopoly, in the form of a patent for a new invention. This chapter introduces

Robert G. Gourdie and Tereance A. Myers (eds.), *Wound Regeneration and Repair: Methods and Protocols*, Methods in Molecular Biology, vol. 1037, DOI 10.1007/978-1-62703-505-7_32, © Springer Science+Business Media New York 2013

the theory and practicality basics for how and why the process of commercialization utilizes the tool of patenting, provided for by our Founding Father's foresight within the US Constitution.

2 What Is a Patent?

A patent is a document granted by the US Government, which provides a type of limited monopoly. This monopoly grants the holder of the patent a right to stop others from making, using, selling, or distributing the patented invention. The patent owner can file a lawsuit against the accused "infringer" of the patent for money damages as well as an injunction. It is important to note that a patent is not a guarantee of rights or a right to practice the invention defined in the patent; it is only a right to stop others.

Additionally, issuance of a patent is not inevitable just because one files a patent application. In the USA and most of the world, you will not get a patent, even if you invented invention X, if the invention was disclosed publically first, if someone else filed a patent application on the invention X first, or you failed to provide enough information in the patent application so that someone could understand and practice the invention. There are certain exceptions to these general concepts so the reader is advised to seek advice from a patent attorney or patent agent as soon as possible and prior to any public disclosures.

All of these requirements and constraints on patenting arise from the basic quid pro quo of society granted monopolies, in this case, patents. The US Constitution in Article I, section 8 establishes this quid pro quo from the birth of our nation stating,

> We the People [provide] … Congress the [power] … [t]o promote the progress of science and useful arts, by securing for limited times to authors and inventors the exclusive right to their respective writings and discoveries (U.S. Const. art. I, § 8)

In general, this is saying that we, the People, through Congress give you, the inventor, a limited monopoly if you, the inventor, give we, the People, your invention. This establishes two fundamental principles of patenting: The invention must not be already known and it must be fully disclosed in the patent. This is the basic underpinning all patent laws and patent rules, and importantly for the scientist it guides and directs all decisions a scientist can make with respect to patenting, and the consequences therefrom.

The laws requiring a patentable invention to be useful, new, nonobvious, described, and enabled flow directly from society not wanting to give a monopoly on something already publically known, i.e., not new or nonobvious. The disclosure requirements flow from the desire to grant a monopoly only where the inventor has made a full and complete disclosure of the invention so that when the limited monopoly time runs out, the public can practice the invention. Would it be fair if an inventor was able to get a patent and charge a monopoly price for his invention, and then when that monopoly time ran out, the inventor was still able to charge a monopoly price because he left out a required piece of information that only he knew, so that only he could make or use the invention?

While not directly saying it, this *quid pro quo* becomes like a contract between the public and the inventor. Like all contracts, there must be something of value exchanged, consideration, for the agreement. In the case of a patent, the inventor provides full disclosure for making and using the new invention, as consideration for the limited time and scope monopoly provided by the government.

In the USA, while the right to patent arises from the Constitution, the mechanics of getting and using a patent come from statutes passed by Congress and regulations promulgated by the United States Patent and Trademark Office (USPTO), a part of the Executive branch of our government. To prosecute a patent in the USPTO—the process of applying for a patent—one must be either the inventor or a Patent Attorney or Agent that has a technical background and passed a special test on the laws and rules of patents. For disputes over the granting or enforcement of a patent, the Federal Courts have exclusive jurisdiction. This creates the sometimes confusing situation of having nonscientist members of the judiciary defining and controlling issues that are often fundamentally scientific in nature. This is why all registered patent lawyers in the USA must have both a scientific background, such as an engineering, biological sciences or chemistry degree as well as a law degree so that the communication and ultimate translation of science to legal language starts with some order.

While the American Patent System has been modeled on and provided the foundation for systems around the world, a patent from the USA does not give its holder rights in any other country. At this time, there are no "worldwide" patents. Instead each country or region grants its own geographically limited monopoly. This system to the uninitiated seems nearly intractable, and even to those having maneuvered within it for years can at times be baffled by the unique attributes residing within each system. It can also be quite expensive.

An inventor can find hope, however, in remembering he or she only must initially file a patent application in their home country to "put a stake" in the ground for the rest of the world too. Treaties, such as the Paris Treaty and the Patent Cooperative Treaty (PCT), make this possible. Inventors in all countries can file their patent application in their home country, and know that as long as they file their international applications within 1 year from the day they filed their first application regarding the invention in their country, they can get the benefit in the international countries as if they filed it on that earlier date, as long as certain requirements are present. However, eventually applications must be filed in each individual country or region which can be a very expensive undertaking.

3 What Can Be Patented?

While there are interpretive differences around the world, in general, patents are granted for processes, articles of manufacture, compositions of matter, methods, and machines, as long as they have usefulness, often referred to as industrial utility around the world. It would seem that these four categories could cover just about anything, and in the USA they have been interpreted very broadly.

Of importance to a scientist in the life science field is the question of whether you could patent life itself. In 1980, the US Supreme Court held that engineered or isolated cells, including cells that were themselves living organisms, were under the umbrella of the patent system.[1] This case revolved around a patent covering bacteria that had been engineered to degrade petroleum products, and the question was could someone in essence own these live organisms. When the Court found that yes one could so own, it breathed life into the famous statement arising out of the congressional hearings on the 1952 patent act stating, "anything under the sun that is made by man" falls within the patent system.[2]

In another famous case, the Supreme Court redefined when a process is a process for patent purposes. The Supreme Court in *Bilski v. Kappos* addressed a claims drawn to what are commonly understood as business method invention. Generally the issue was whether a process, such as that carried out which was not limited to some type of machine or produced some transformation in matter was patent eligible subject matter. The Court held that a process for patent purposes meant the common understood meaning of process,[3] but it reaffirmed that laws of nature, physical phenomena and abstract ideas alone cannot be patent eligible subject matter even if they are a process.[4] In the future, significant litigation will likely occur around whether something is merely a law of nature, physical phenomena, or abstract idea, but the court affirmed that when the method is grounded in a machine or some type of physical transformation of matter, this is one way to show it is not merely in one of the unacceptable categories.

---

[1] *Diamond v. Chakrabarty*, 100 S. Ct. 2204 (1980).

[2] *Id.* at 2208 *citing* S.Rep.No. 1979, 82d Cong., 2d Sess., 5 (1952); H.R.Rep. No. 1923, 82d Cong., 2d Sess., 6 (1952).

[3] *Bilski v. Kappos*, 130 S. Ct. 3218, 3226 (S. Ct. 2010) (held that the particular case of a method drawn to steps relating to investment did not meet the patent eligible standard because it violated one of the classes of subject matter).

[4] *Bilski v. Kappos*, 130 S. Ct. 3218, 3220 (S. Ct. 2010).

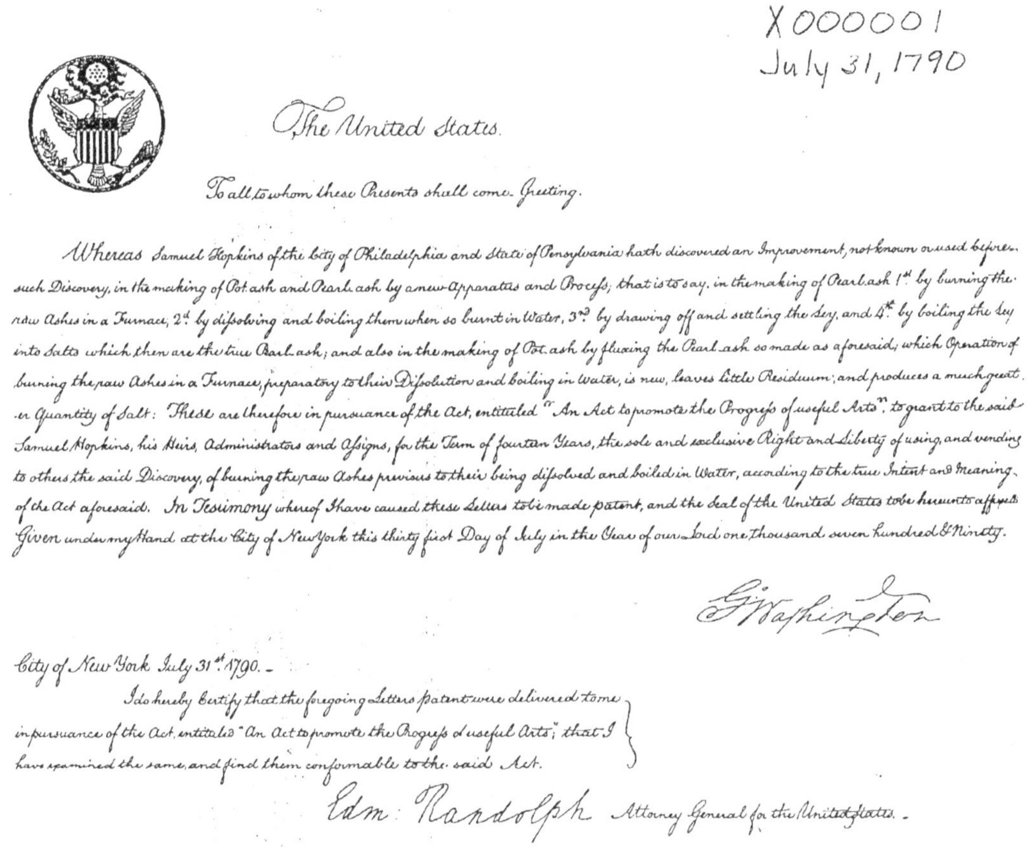

X000001
July 31, 1790

The United States.

To all to whom these Presents shall come. Greeting.

Whereas Samuel Hopkins of the City of Philadelphia and State of Pensylvania hath discovered an Improvement, not known or used before such Discovery, in the making of Pot ash and Pearl ash by a new Apparatus and Process; that is to say, in the making of Pearl ash 1st by burning the raw Ashes in a Furnace, 2d by dissolving and boiling them when so burnt in Water, 3d by drawing off and settling the Ley, and 4th by boiling the Ley into Salts which then are the true Pearl ash; and also in the making of Pot ash by fluxing the Pearl ash so made as aforesaid; which Operation of burning the raw Ashes in a Furnace, preparatory to their Dissolution and boiling in Water, is new, leaves little Residuum, and produces a much greater Quantity of Salt: These are therefore in pursuance of the Act, entituled "An Act to promote the Progress of useful Arts", to grant to the said Samuel Hopkins, his Heirs, Administrators and Assigns, for the Term of fourteen Years, the sole and exclusive Right and Liberty of using, and vending to others the said Discovery, of burning the raw Ashes previous to their being dissolved and boiled in Water, according to the true Intent and Meaning of the Act aforesaid. In Testimony whereof I have caused these Letters to be made patent, and the Seal of the United States to be hereunto affixed. Given under my Hand at the City of New York this thirty first Day of July in the Year of our Lord one thousand seven hundred & Ninety.

G. Washington

City of New York July 31st 1790. –

I do hereby Certify that the foregoing Letters patent were delivered to me in pursuance of the Act, entituled "An Act to promote the Progress of useful Arts"; that I have examined the same, and find them conformable to the said Act.

Edm: Randolph Attorney General for the United States. –

Fig. 1 A copy of the first issued patent for a substance called Potash. Note that George Washington, the first patent office commissioner, has signed it

Interestingly, the first patent ever issued was to a method, shown in Fig. 1. This patent, drawn to Potash, was for a method of making soap, and was issued by the head of the PTO at the time, President Washington, and Thomas Jefferson reviewed it.

4 Statutory Requirements for Getting a Patent

4.1 What Is Required to Get a Patent?

So, if one has a process, machine, article of manufacture, or composition of matter, what is required by the inventor to ultimately get a patent?

4.1.1 Utility

First, an invention must have utility, meaning the invention must have a useful purpose, beyond knowledge.[5] In essence, a patentable invention must be able to be used to do something, or make something,

[5]Whoever invents or discovers any new and useful process, machine, manufacture, or composition of matter, or any new and useful improvement thereof, may obtain a patent therefore, subject to the conditions and requirements of this title. 35 USC 101.

or treat something, i.e., it must be useful. Life science discoveries rarely have issues with utility.

4.1.2 Novelty

An invention must be novel to be patentable, 2011 Leahy-Smith America Invents Act as defined by 35 USC sec 102.[6] Simplistically, "novel" means you were the first to invent. But novelty also means your invention was not publically known prior to certain dates associated with your filing of a patent application. Importantly, the legal framework for determining novelty is changing due to the recent passage of new patent laws in the USA.

---

[6] *35 U.S.C. 102 Conditions for patentability; novelty and loss of right to patent.* A person shall be entitled to a patent unless

(a) The invention was known or used by others in this country, or patented or described in a printed publication in this or a foreign country, before the invention thereof by the applicant for patent, or
(b) The invention was patented or described in a printed publication in this or a foreign country or in public use or on sale in this country, more than 1 year prior to the date of the application for patent in the USA, or
(c) He has abandoned the invention, or
(d) The invention was first patented or caused to be patented, or was the subject of an inventor's certificate, by the applicant or his legal representatives or assigns in a foreign country prior to the date of the application for patent in this country on an application for patent or inventor's certificate filed more than 12 months before the filing of the application in the USA, or
(e) The invention was described in—(1) an application for patent, published under *section 122(b)*, by another filed in the USA before the invention by the applicant for patent or (2) a patent granted on an application for patent by another filed in the USA before the invention by the applicant for patent, except that an international application filed under the treaty defined in *section 351(a)* shall have the effects for the purposes of this subsection of an application filed in the USA only if the international application designated the USA and was published under *Article 21(2)* of such treaty in the English language; or
(f) He did not himself invent the subject matter sought to be patented, or
(g) (1) during the course of an interference conducted under *section 135* or *section 291*, another inventor involved therein establishes, to the extent permitted in *section 104*, that before such person's invention thereof the invention was made by such other inventor and not abandoned, suppressed, or concealed, or (2) before such person's invention thereof, the invention was made in this country by another inventor who had not abandoned, suppressed, or concealed it. In determining priority of invention under this subsection, there shall be considered not only the respective dates of conception and reduction to practice of the invention, but also the reasonable diligence of one who was first to conceive and last to reduce to practice, from a time prior to conception by the other.

(Amended July 28, 1972, Public Law 92-358, sec. 2, 86 Stat. 501; Nov. 14, 1975, Public Law 94-131, sec. 5, 89 Stat. 691.)(Subsection (e) amended Nov. 29, 1999, Public Law 106-113, sec. 1000(a)(9), 113 Stat. 1501A-565 (S. 1948 sec. 4505).)(Subsection (g) amended Nov. 29, 1999, Public Law 106-113, sec. 1000(a)(9), 113 Stat. 1501A-590 (S. 1948 sec. 4806).)(Subsection (e) amended Nov. 2, 2002, Public Law 107-273, sec. 13205, 116 Stat. 1903.)

Prior to the passage of the Act, an inventor in the USA could prevail and be granted a patent even if they were not the first to file a patent application claiming their invention. They did however have to file their patent application within 1 year of a public disclosure of the invention by themselves or another. Thus, the USA was said to have a "one year grace period."

That 1-year grace period is changing and the USA is now becoming more like the rest of the world in that we are adopting a first to file system. In short, under the new patent laws, to prevail and be issued a patent you must now be the first inventor to file and you must file before any public disclosure of your invention is made. There is one allowance provided for under the new patent laws: if the public disclosure is your own, you can avail yourself of a 1 year grace period and file your application within 1 year of your own disclosure.

The rest of the world in general provides no grace period for your own public disclosures of your invention. Thus to preserve international rights, a patent application needs to be filed prior to any public disclosure of your invention by yourself or others.

Therefore, all inventors must understand what makes a disclosure of an invention a disclosure that kills patent rights. First the disclosure must be public. For purposes of patent law, public means could be seen or could be found, not actually found or seen. For example, if an invention is disclosed within an undergraduate thesis from a university in Zimbabwe, this is considered a public disclosure in the USA as well as Europe. Any publication of an invention is considered a public disclosure. In fact, if a manuscript is sent to peers for review, and it is not done under confidentiality, this would be considered a disclosure. However, if it were indicated as being under confidence, by for example, the journal, then it would not be a disclosure, unless the reviewer showed someone in another lab, who was not an authorized reviewer. This action would create massive problems for patenting any invention contained within the manuscript. Grants are another area that can create disclosures, either because they are reviewed without confidentiality in place, or there is unauthorized disclosure by a reviewer. It is however generally accepted that the submission of a grant application to a federal agency (e.g., NIH) is considered a confidential submission, until it is funded, at which point the abstract is typically published and the body of the grant application is generally available under the Freedom of Information Act (FOIA). Grant applications should be proactive and avail themselves of the avenues for marking their submission confidential.

Public presentations are another area of concern for an inventor. Slides or a poster at a conference, as well as oral disclosure in either the talk, or at the poster, or even at a cocktail party, are all potential public disclosures for patent purposes. Lab presentations if non-lab members are present, as well as departmental talks if they are not closed, are both potential public disclosures. A graduate student's thesis or dissertation defense and subsequent publication can be public disclosures.

Consultation with your university's tech transfer office, corporate IP representative, or a patent attorney or agent is advised prior to making any of these types of presentations or disclosures. In many cases there are solutions and work-arounds that will allow presentation and preserve your right to patent. Confidentiality agreements are often a tool used to preserve the ability to patent. In many cases, filing a patent application can be done prior to the presentation, although the patent application should fully disclose the invention and enable one of ordinary skill to make and use the invention.

An offer to sell the invention, is a disclosure, and importantly this is not cleansed by having a confidentiality agreement in place. Thus, any offer to sell, will create a public disclosure, which can kill patent rights. Note an offer to license is not consider an offer to sell, but in any event, should not be undertaken without a confidentially agreement or an application on file. Other actions that should never be done before filing a patent application include demonstrating a prototype to a public group, distribute samples of the product to customers, consumer or market test a new product, display or discuss in detail the invention at a trade show, offer for sale or sell the product, distribute advertising brochures about the invention, or disclose invention on inventor's or company's Web site.

4.1.3 Nonobviousness

Even if an invention is novel, it still must be nonobvious as defined by 35 USC 103[7] and a recent Supreme Court case. In *KSR v. Teleflex*, (*KSR International Co. v. Teleflex Inc. et al.*, 127 S. Ct. 1727 (2007)), the Supreme Court reaffirmed that

> Under § 103, the scope and content of the prior art are to be determined; differences between the prior art and the claims at issue are to be ascertained; and the level of ordinary skill in the pertinent art resolved. Against this background the obviousness or nonobviousness

[7] *103 Conditions for patentability; nonobvious subject matter.*

(a) A patent may not be obtained though the invention is not identically disclosed or described as set forth in *section 102* of this title, if the differences between the subject matter sought to be patented and the prior art are such that the subject matter as a whole would have been obvious at the time the invention was made to a person having ordinary skill in the art to which said subject matter pertains. Patentability shall not be negatived by the manner in which the invention was made.

(b) (1) Notwithstanding subsection (a), and upon timely election by the applicant for patent to proceed under this subsection, a biotechnological process using or resulting in a composition of matter that is novel under *section 102* and nonobvious under subsection (a) of this section shall be considered nonobvious if

(A) Claims to the process and the composition of matter are contained in either the same application for patent or in separate applications having the same effective filing date; and of the subject matter is determined. Such secondary considerations as commercial success, long felt but unsolved needs, failure of others, etc., might be utilized to give light to the circumstances surrounding the origin of the subject matter sought to be patented. (*KSR*, at 1734).

(B) The composition of matter, and the process at the time it was invented, were owned by the same person or subject to an obligation of assignment to the same person.

(2) A patent issued on a process under paragraph (1)

(A) Shall also contain the claims to the composition of matter used in or made by that process, or

(B) Shall, if such composition of matter is claimed in another patent, be set to expire on the same date as such other patent, notwithstanding *section 154*.

(3) For purposes of paragraph (1), the term "biotechnological process" means

(A) A process of genetically altering or otherwise inducing a single- or multi-celled organism to

(i) Express an exogenous nucleotide sequence, (ii) inhibit, eliminate, augment, or alter expression of an endogenous nucleotide sequence, or (iii) express a specific physiological characteristic not naturally associated with said organism;

(B) Cell fusion procedures yielding a cell line that expresses a specific protein, such as a monoclonal antibody; and

(C) A method of using a product produced by a process defined by subparagraph (A) or (B), or a combination of subparagraphs (A) and (B).

(c) (1) Subject matter developed by another person, which qualifies as prior art only under one or more of subsections (e), (f), and (g) of *section 102* of this title, shall not preclude patentability under this section where the subject matter and the claimed invention were, at the time the claimed invention was made, owned by the same person or subject to an obligation of assignment to the same person.

(2) For purposes of this subsection, subject matter developed by another person and a claimed invention shall be deemed to have been owned by the same person or subject to an obligation of assignment to the same person if

(A) The claimed invention was made by or on behalf of parties to a joint research agreement that was in effect on or before the date the claimed invention was made;

(B) The claimed invention was made as a result of activities undertaken within the scope of the joint research agreement; and

(C) The application for patent for the claimed invention discloses or is amended to disclose the names of the parties to the joint research agreement.

(3) For purposes of paragraph (2), the term "joint research agreement" means a written contract, grant, or cooperative agreement entered into by two or more persons or entities for the performance of experimental, developmental, or research work in the field of the claimed invention.

(Amended Nov. 8, 1984, Public Law 98-622, sec. 103, 98 Stat. 3384; Nov. 1, 1995, Public Law 104-41, sec. 1, 109 Stat. 3511.)(Subsection (c) amended Nov. 29, 1999, Public Law 106-113, sec. 1000(a)(9), 113 Stat. 1501A-591 (S. 1948 sec. 4807).)(Subsection (c) amended Dec. 10, 2004, Public Law 108-453, sec. 2, 118 Stat. 3596.)

4.1.4 Enablement and Written Description

If an invention is new and nonobvious and has a use, the inventor must finally also teach how to make and use, as well as provide a written description of the invention (35 USC 112[8]). Enablement is the concept that an inventor must teach enough within the patent application that when combined what is known in the art would allow one of ordinary skill in the art to make and use the invention. For example, if a particular plasmid is needed to make an invention involving a transformed cell work, and the plasmid is not described or provided in the application and it cannot be obtained from what is known in the art, then any claims to that transformed cell would be invalid or not patentable for failure to enable the invention. This type of issue often comes up within biotechnology and chemistry related inventions.

Written description is a separate requirement from the enablement requirement but they are related. The written description requirement means that an inventor must describe with words his invention in such a way that one of skill in the art knows that the inventor has invented it or, in the parlance of the law, he "was in possession of the invention".[9] While it can be very difficult to know if this requirement has been met for biotechnology and chemical related inventions, in general, an inventor should be thinking about defining the invention in structural terms as much as possible to

[8] *12 Specification.*The specification shall contain a written description of the invention, and of the manner and process of making and using it, in such full, clear, concise, and exact terms as to enable any person skilled in the art to which it pertains, or with which it is most nearly connected, to make and use the same, and shall set forth the best mode contemplated by the inventor of carrying out his invention.The specification shall conclude with one or more claims particularly pointing out and distinctly claiming the subject matter which the applicant regards as his invention.A claim may be written in independent or, if the nature of the case admits, in dependent or multiple dependent form.Subject to the following paragraph, a claim in dependent form shall contain a reference to a claim previously set forth and then specify a further limitation of the subject matter claimed. A claim in dependent form shall be construed to incorporate by reference all the limitations of the claim to which it refers.A claim in multiple dependent form shall contain a reference, in the alternative only, to more than one claim previously set forth and then specify a further limitation of the subject matter claimed. A multiple dependent claim shall not serve as a basis for any other multiple dependent claim. A multiple dependent claim shall be construed to incorporate by reference all the limitations of the particular claim in relation to which it is being considered.An element in a claim for a combination may be expressed as a means or step for performing a specified function without the recital of structure, material, or acts in support thereof, and such claim shall be construed to cover the corresponding structure, material, or acts described in the specification and equivalents thereof.(Amended July 24, 1965, Public Law 89-83, sec. 9, 79 Stat. 261; Nov. 14, 1975, Public Law 94-131, sec. 7, 89 Stat. 691.)

[9] See for example, the famous line of cases flowing from Regents of the University of California v. Eli Lilly, 119 F.3d 1559, 1568 (Fed. Cir. 1997) (1998).

meet this requirement. For example, to claim a chemical entity, one should describe the structure of the entity, rather than the function of the chemical entity or the result of using the chemical entity. With respect to proteins, or other biomacromolecules, one should think about providing the primary sequence, rather than just the function of the protein, such as a kinase or even how it could be obtained.

35 USC also has a requirement for providing the Best Mode.[10] This requirement means that not only must an inventor teach the public how to make and use the invention, the inventor must also provide the best way that the inventor knows of at the time of filing the patent application for how to make and use the invention. This requirement was the basis for extensive litigation over the last 40 years, because defendants would assert that this requirement was not met during infringement litigation. However, with the 2011 Leahy-Smith America Invents Act, it is unclear if this will continue to be the case. In what most agree is a very confusing statutory promulgation, the new statute requires that the best mode be provided, but that the failure to so provide is not a reason to invalidate the patent. Time and litigation are the means for bringing clarity to how this new "toothless" requirement will be interpreted. No doubt, aggressive patentees will push these issues as often it is desirable to not disclose the preferred way of doing something to forestall competition.

5 How Is an Invention Defined?

There is one way, and one way only, that an invention is defined, and that is through the "claims" of the patent. Legally, an invention is only that which is defined by the claims. Just like a deed defines the metes and bounds of a house or a piece of property, so to the claims define the metes and bounds of an invention. A deed could, for example, say "go from the corner of Hill and Winding River Avenue 150 ft northwest to the large oak tree, go 250 ft North East to the bank of Winding River, follow the River South by South East for 500 ft, and return to the corner of Hill and Winding River Avenue." This description would define the hypothetical real property. Claims, in fact, got their name because during the Gold Rush of 1849, the prospectors were flocking to the hills and mountains surrounding San Francisco by the thousands, and when they arrived, they were "staking their claim" by literally placing stakes around the area which they were claiming as their own. So to, when an inventor sets forth in the words of the claim, he or she is "staking their claim" to their invention.

[10] Supra footnote 8.

6 Inventive Process and Who Is an Inventor?

An invention is defined by the claims, but what really is an invention, and when is it complete? An invention is any idea that has become complete in the mind of the inventor or inventors. It is does not require that it be made or used. In the language of patent, law and invention is complete when its conception is complete, and a reduction to practice is required if without such a reduction, the conception cannot be complete. Conception is "the formation in the mind of the inventor, of a definite and permanent idea of the complete and operative invention, as it is hereafter to be applied in practice".[11] The test for conception has been described as "whether the inventor had an idea that was definite and permanent enough that one skilled in the art could understand the invention; [and] the inventor must prove his conception by corroborating evidence, preferably by showing a contemporaneous disclosure".[12]

Remembering that the claims of a patent define the invention, it follows that inventorship cannot be determined until the patent application claims are drafted, and ultimately until claims are issued as a patent. Thus, in the USA, there are mechanisms for getting the inventors corrected, if for example, the claims are amended during the prosecution, the process of getting a patent issued, such that inventorship changes.

An inventor is anyone who contributes significantly to either the conception or reduction to practice of an invention. Questions of joint inventorship can be difficult questions, however, because the analysis is different than the analysis that takes place in authorship, for example. While it might be expedient or desired to have coauthors as coinventors, coinventorship does not necessarily flow from coauthorship.

Joint invention is when an invention made by two or more persons. This means that joint invention requires collaboration and working together toward a solution to the problem, but joint inventors do not need to physically work together, work at the same time, make the same type or amount of contribution, or make a contribution to the subject matter of every claim. What is required is that all joint inventors contribute to the conception of the idea. For example, conception of a chemical substance includes knowledge of both the specific chemical structure of the compound and an operative method of making it. A knowledge that the chemical substance exists without knowing what the chemical substance specifically is would likely be insufficient for complete

[11] "*Hybritech Inc. v. Monoclonal Antibodies, Inc.*, 802 F.2d 1367, 1376, 231 USPQ 81, 87 (Fed.Cir.1986) (citation omitted).

[12] *Burroughs Wellcome Co. v. Barr Lab., Inc.*, 40 F.3d 1223, 1228, 32 USPQ2d 1915, 1919 (Fed.Cir.1994).

conception of the chemical substance. Therefore, people involved in determining both how to make the compound as well as what the compound is should be considered as possible co-inventors.

Conception follows the old axiom that sometimes it is easier to know what it is not, rather than know what it is. For example, mere assistance, such as providing well known principles, or explaining the state of the art, do not amount to a contribution to conception.[13] We do know that joint invention requires collaboration, such as working together to solve a problem.[14] However, joint inventors need not "physically work together or at the same time" or "make the same type or amount of contribution" or "make a contribution to the subject matter of every claim".[15] What is required however is that "each ... performs ... [at least] a part of the task which produces the invention".[16]

The judicially created doctrine of "simultaneous conception and reduction to practice" plays a significant role in many biotechnology related inventorship decisions.[17] This doctrine revolves around the complicated issues of description and enablement for biotechnology and chemical related inventions findings its origin in the premise that if you do not know the structure of something you likely had not yet completed the invention. This leads to the idea that for certain types of inventions, say a cDNA for a particular gene, you may know it exists, know all its characteristics, because for example, you have the protein in hand, but if you do not know the particular sequence of the cDNA you cannot have invented the cDNA until the cDNA is in hand, i.e., until simultaneous conception and reduction to practice have occurred.[18]

---

[13] Ethicon 1460 citing See Sewall, 21 F.3d at 416-17; Shatterproof Glass Corp v. Libbey-Owens Ford Co., 758 F.2d 613, 624, 225 USPQ 634, 641 (Fed.Cir.1985). "Providing well-known principles" ... or [explaining] the state of the art without ever having "a firm and definite idea" of the claimed combination as a whole does not qualify as a joint inventor." Ethicon at 1460 citing Hess, 106 F.3d at 981 (citing O'Reilly v. Morse, 56 U.S. (15 How.) 62, 111, 14 L.Ed. 601 (1853)). Ethicon 1460.

[14] *Kimberly-Clark Corp. v. Procter & Gamble Distrib. Co.*, 973 F.2d 911, 917, 23 USPQ2d 1921, 1926 (Fed.Cir.1992).

[15] 35 U.S.C. § 116.

[16] Ethicon at 1460.

[17] *Burroughs Wellcome Co. v. Barr Lab., Inc.*, 40 F.3d 1223, 1229, (Fed.Cir. 1996).

[18] See foundations for example in the Federal Circuit statements, "It is undoubtedly true that "[i]n some instances, an inventor is unable to establish a conception until he has reduced the invention to practice through a successful experiment." Amgen, 927 F.2d at 1206, 18 USPQ2d at 1021; Alpert v. Slatin, 305 F.2d 891, 894, 134 USPQ 296, 299 (CCPA 1962) (no conception "where results at each step do not follow as anticipated, but are achieved empirically by what amounts to trial and error").

7 Who Owns an Invention?

In the USA, ownership of an invention always starts with the inventor or coinventors. This means that without assignment of the invention away from the inventor to an employer, such as a University or Company, the invention is still owned by the inventor, i.e., the employee. Understand that this does not mean that if the University or Company has been paying for the salary and for the research that there would not be legal remedies that could reduce the value of the invention to the inventor, but as a starting point ownership is in the inventor(s) and must be assigned away.

For an academic scientist, there is usually a university intellectual property policy that sets out when the university owns of an invention. For those employed in industry there is often a policy and some form of documentation executed upon hiring that grants ownership of inventions to the employer. Consulting agreements and research agreements also often contain provisions related to ownership of inventions. A recent Supreme Court case between Stanford University and Roche highlighted the precarious situation an academic research can be placed in by simply signing a consulting or research agreement. Researchers at Stanford collaborated with Cetus Inc. on development of methods for using PCR for HIV quantitation. Stanford eventually obtained patents, and Roche purchased Cetus. Stanford approached Roche to take a license to the patents covering the developed method, and Roche declined. When Stanford sued for infringement, Roche defended by asserting amongst other things, Stanford lacked standing because they lacked ownership. The inventor had executed a consulting and research agreement that assigned over his inventions to Cetus (now Roche.) The Stanford intellectual property policy merely required the inventor to assign inventions at some future date upon request of the university. Eventually this case found its way to the Supreme Court where the Court found that future promises to assign are subordinate to actual assignments *even* if the assignment occurs *prior* an invention occurring. The faculty member thus landed in a most difficult and tenuous position, as well as losing out on the generous royalty income sharing provided for under the university policy but not under the Cetus agreement.

8 What Is the Role of Collaborations Between Inventors and the Licensing Companies?

The preceding sections dealing with inventorship and ownership highlight the pressure that collaborations can create when the financial stakes are high. Collaborations often initiate and then drive the pace as well as the success of innovative efforts. However, when those efforts are funded through expected returns, as in company research, and increasingly in university federally funded research, collaborations often create disputes. Collaborations are good and in many cases necessary to success of research companies, but they need to be managed.

From a company perspective there are only two types of collaborations: controlled and uncontrolled. Companies always seek to minimize uncontrolled collaborations because these produce the problems and litigations that can arise when research scientists begin fertile process of joining forces. When collaborations are uncontrolled, ownership can be lost, patent rights can be inadvertently (or purposefully) be lost because of disclosure, and trade secrets can be lost or interfered with, all of which create potential competitors rather than partners. Forming collaborations under controlled circumstances minimizes these risks. For example, collaborations should set forth through contract confidentiality expectations, ownership expectations, and expectations regarding the control of the fruits of the collaboration. However, even under these circumstances, the collaboration can decrease the value of the scientific results because of coinventorship, loss of control, increased issues during due diligence from third parties, and decreasing negotiating leverage.

9 What Are the Steps to Get a Patent and the Role of Inventor in the Process?

While inventors can navigate the PTO on their own under the auspices of being *pro se* or "for themselves," it is highly advisable to seek counsel from qualified patent attorneys or agents. The complexity of the laws and rules, overlayed with the very difficult job of merging the language of science into and onto the language of law provides many traps for the inexperienced.

The process of getting a patent starts with filing a patent application. In the USA, this application can either be a utility patent application, a full and complete application that has the possibility of becoming a patent, or a provisional application, which while needing to meet all the statutory requirements of a patent application has lower formality and cost requirements, and thus often functions as a placeholder. Figure 2 shows the different patent applications that can be filed (related to a utility invention, design patent applications can also be filed, but follow a different time line) and their chronological relationship. An inventor can first file either a US Utility application (can become a patent) or a US provisional application (cannot become a patent but can form a date basis for an application that can become a patent), a foreign application, or a Patent Cooperative Treaty (PCT) application.

Once filed this first application starts a clock, under the Paris Treaty which gives the Inventor or applicant 1 year to get the remainder of the applications filed, around the world, such as other

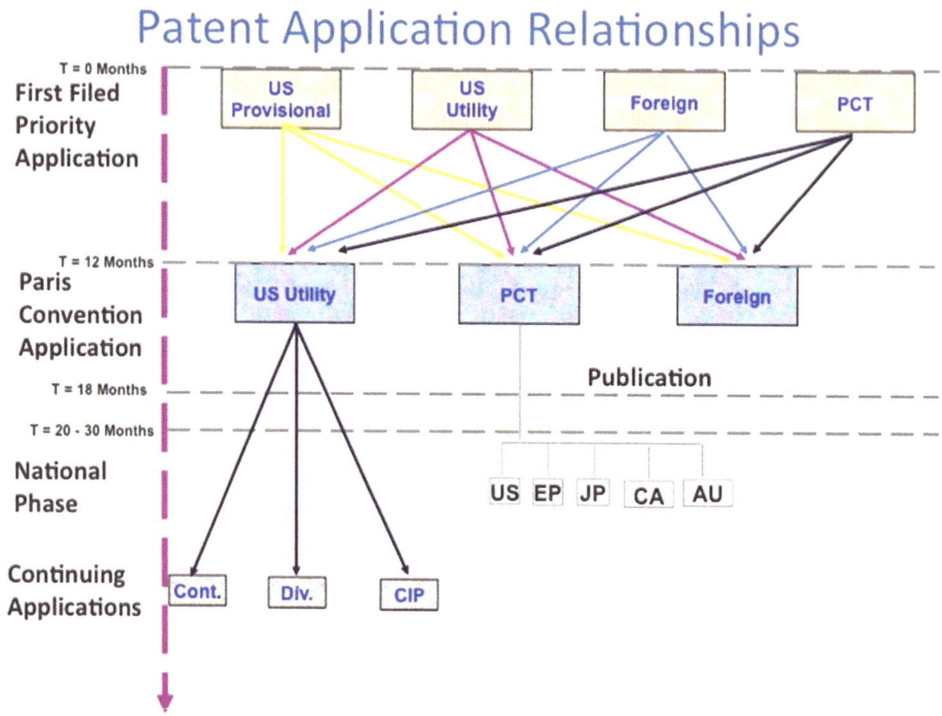

Fig. 2 A schematic of the relationships of the various patent applications an inventor can file around the world. The *left side* of the panel indicates a general time line starting from the filing of the first patent application. The *arrows* and *lines* connecting the various types of patent applications indicates that they are related through a claim of priority, one of the requirements for giving each the benefit of the filing date of the first filed application in the chain

foreign applications or a PCT application. A PCT application is a common application at this stage because it functions like a placeholder, like a US provisional application it cannot become a patent, but it can form the basis of applications filed at a later date which can become a patent. By filing a PCT application at the 12 month date, one can delay for 18 months the filing of any application in a foreign country that is a party to the treaty forming the PCT (144 as of September 2011). This can delay hundreds of thousands of dollars in expenses, as filing national applications around the world is very expensive due to translation costs and additional foreign counsel and fees.

At the 30 month date from the first filing, national applications can be filed, including the filing of a utility patent application in the USA, and in Fig. 2, applications file, for example in Europe, Japan, Canada, and Australia. Within many countries, as prosecution proceeds, continuations or divisional applications can be filed which are facsimile copies of the original application, but which will have slightly different claims which are disclosed within the original application.

If a provisional application is filed, the applicant has 1 year within which to file a utility patent application (as well as international applications) claiming priority (asserting the benefit of the first filing date) to the provisional application. Once a utility application is on file, there are a number of actions and formalities, such as a Declaration of Inventorship, and the filing of an Information Disclosure Statement that must be accomplished.

In addition, there normally is a back and forth between the PTO and the applicant, regarding the patentability of the claims. This takes the form of written argument from "each side" regarding things like novelty, nonobviousness, and the scope of the claims. Eventually, the Applicant and the PTO either come to agreement on the language of the claims or the application is finally denied as unpatentable for one or more reasons. If an Applicant is denied, there are appeal processes that can take place, with ultimate resolution residing at the Supreme Court of the USA, but rarely do issues of getting a patent make it that far through the process. The process of give and take between the Applicant and the PTO is called prosecution.

The inventor can play a substantial role in prosecution through informing and teaching the patent attorney of the correctness of scientific arguments being lodged, and providing scientific reasons why, for example, something asserted as anticipating (making the invention not novel) is in fact scientifically different. Often the opportunity may arise for an inventor, or getting his peers, to sign declarations regarding interpretation or scientific opinion as evidence during the process. These affidavits can be very important in the ultimate resolution of the issues. In some cases, the inventor or others in support of the application, will actually perform new experiments to show that a particular statement in a patent application or claim is in fact correct. The results of these experiments can be presented as evidence during the process. It should be noted that the USA is much more liberal than the rest of the world in these regards.

10 Concluding Comments

In this chapter we have tried to provide an up-to-date overview in straightforward language of commercial development of a patentable discovery for biomedical scientists. The information provided pertains largely to researchers in the USA, but we believe that the information could also be helpful to those outside of the USA. Though in the context of this excellent book on wound healing and regeneration, the guide we provide could also help workers in other areas of scientific endeavor who wish to navigate the often confounding interface between the intellectual property law and scientific research. Finally, we recognize that models of funding scientific research are changing. We hope that this chapter provides an understanding of the how's, what's, who's, and why's of the patenting processing—such that scientists are better informed to leverage opportunities for resources to do their important work.

Chapter 33

Translational Strategies for the Development of a Wound Healing Technology (Idea) from Bench to Bedside

Gautam S. Ghatnekar and Tuan A. Elstrom

Abstract

This chapter will outline strategies and ideas for the commercialization a promising wound healing technology discovered in an academic setting. This would include, but not limited to addressing topics such as intellectual property protection, funding, technology development, and regulatory aspects (i.e., navigating through the FDA).

Key words Discovery, Connexin 43, ACT1 peptide, Granexin™ gel, Wound healing, Valley of Death, Bench-to-bedside, Translational research, Invention disclosure, Intellectual property, Technology transfer, Licensing, Startup, Spinoff, Angel Investors, Venture capital, State and Federal Funding Sources, Technology development, Clinical Program Development, Regulatory strategy, Commercialization strategy

1 Introduction

Taking a discovery from basic research into clinical study is becoming increasingly challenging and costly. The complexities of the development process may represent an obstacle towards success in translating science into treatments. A number of factors can impede the progress between basic science and clinical medicine, perhaps most notably a lack of sufficient financial resources to support early-stage investigation and the challenges involved in organizing clinical trials, often referred to as the "Valley of Death." However, funding is not the only factor, and crossing the translational gap may depend upon successfully managing scientific and clinical development, intellectual property (IP), regulatory, and market risk. These are the majority of the challenges of an academic entrepreneurial venture or university spinoff. This chapter aims to outline the principal strategies for navigating the route from bench to bedside by examining the progress to date in the clinical development of Granexin™

Robert G. Gourdie and Tereance A. Myers (eds.), *Wound Regeneration and Repair: Methods and Protocols*, Methods in Molecular Biology, vol. 1037, DOI 10.1007/978-1-62703-505-7_33, © Springer Science+Business Media New York 2013

Gel, a topical formulation containing a novel connexin-based synthetic peptide named ACT1, for the treatment of acute surgical scars and chronic wounds such as diabetic foot ulcers and venous leg ulcers. The motivation is to identify best-practices to address obstacles and facilitate as well as expedite the practical application of scientific discoveries.

2 The Serendipitous Discovery of the ACT1 (Alpha Carboxy Terminus) Peptide

The clinical development of Granexin™ Gel began with a serendipitous discovery of the potential wound healing and anti-scarring benefits of the ACT1 peptide. While working with a novel bioengineered peptide based on a carboxy-terminal fragment of connexin 43 (Cx43) ("dubbed ACT1 peptide") to study cell–cell communication in cardiac conduction system development, it was discovered that the ACT1 peptide possesses unique properties in modulating intercellular communication and contact that could help with wound healing. During the course of investigation, the application of ACT1 peptide to fibroblast cell lines in a scratch test assay resulted in the inhibition of cell migration, an unanticipated observation. The observation spurred further in vitro and in vivo studies that elucidated the unique regenerative properties of the ACT1 peptide. The peptide ACT1 is a small synthetic peptide (25-mer) was designed to mimic the C-terminus of the transmembrane protein Cx43, which has an important role in the formation of gap junctions. Scientific studies recently expounded the central role of Cx43 in multiple aspects of wound healing including the spread of injury signals, extravasation of immune cells, granulation tissue formation and fibrosis. It is now known that the ACT1 peptide possesses a unique capability to switch the body's own healing response from inflammation and scarring to a healthy regenerative stage [1–5]. Preclinical studies in animal wound models has shown that the ACT1 peptide stabilizes gap junctions as well as tight junctions of endothelial cells during the wound healing process leading to a variety of beneficial effects including increase coordination of cellular communication, dampened inflammation response, reduction in the infiltration of neutrophils, and reduction in the proliferation of fibroblasts. The remarkable healing properties of the ACT1 peptide led the Medical University of South Carolina (MUSC) to file patent protection for its composition of matter and a plethora of methods of use. Dr. Ghatnekar proceeded to form a spinoff company called FirstString Research, co-founded with Professor Gourdie of MUSC, to commercialize this promising regenerative (medicine) healing technology.

3 Translational Research Definition

Translational research means different things to different researchers and clinicians [6]. In one aspect, the term refers to the "bench-to-bedside" endeavor of harnessing basic science discoveries to produce new drugs, devices, and treatment options for patients. For this field of research, the primary objective is the development and commercialization of a promising new treatment derived from basic science that can be used clinically. In another aspect, translational research refers to translating research into practice. The product of a new drug, an end point for "bench-to-bedside" translational research, is only the starting point for this second area of research [6]. For this chapter, the translational strategies for the development of Granexin™ Gel as a product have the former meaning, generally known as T1, a term referred by the Institute of Medicine's Clinical Research Roundtable for translating basic science into clinical efficacy data [7–9]. The latter field is known as T2 and in contrast to T1 has different challenges. From the authors' perspective, T1 challenges for the commercialization of a novel idea include science, technology transfer, funding, clinical trials, regulatory requirements, and meeting the unmet needs of the market place. The goal of translational strategies for T1 is to circumnavigate or overcome these obstacles and expedite the clinical development and applications, in our case, of the ACT1 peptide.

3.1 The Obstacles of T1 Translational Research

Since 2000, the Institute of Medicine (IOM) identified translational research as crucial to improving the health of the nation. Through a series of Clinical Research Roundtable meetings, IOM defined several barriers to the translation of discovery to patients. One barrier known as the T1 block recognized the difficulties of moving a laboratory finding into treatments for patients. The T2 block referred to the challenges in the clinical adoptions newly approved drugs and medical devices. Within the T1 barrier resides the translational gap known as the "Valley of Death." The "Valley of Death" refers to the lack of funding and support for research that can transfer basic science discoveries into clinical diagnostics, devices, drugs or treatment options. More directly, academic researchers do not have access to the funding needed to carry out the preclinical and early clinical development to demonstrate potential efficacy in humans. In addition to funding, the obstacles of translational research may include solving key scientific problems, securing intellectual property (IP) protection, completing clinical trials successfully, meeting regulatory requirements, and adequately addressing commercial market expectations [9]. These obstacles require the diligent management of the resources to reduce the associative risks with the objective towards creating value by bringing treatment options to patients, providing a return on investment to all entrepreneurial (i.e., risk-taking) parties involved whether it be financial investors, federal government (Ex. Dept. of Health and Human Services), philanthropic, or society as a whole.

4 Intellectual Property Strategy

An IP strategy for a university spinoff may commence with a key discovery or invention created in a laboratory setting. The decision to obtain a patent is based on the understanding of how the protection of the differentiated aspect of the discovery or invention will enhance the ability of a commercial entity to achieve its goal. For a life science venture, an IP strategy is vitally important as a catalyst for initial funding and for building long-term value. As an objective of "bench-to-bedside", intellectual property protection conveys a competitive advantage, especially due to barriers such as regulatory, the need for capital-intensive development and manufacturing, and long timelines for product commercialization. In charting a course for a successful spinoff, the academic scientist should be cognizant of a university's policy for invention disclosures and assignment of ownership. The university can assert ownership rights for inventions or ideas generated with the use of its resources (i.e., facilities, funds, grants, contracts, etc.). It is prudent to know these policies since they can negatively impact the technology transfer process, the scientist's relationship with the university, and investor perceptions of the potential of a startup company.

The framework for developing an IP strategy comprises key questions to be answered by the academic research inventor and preferably in conjunction with a university's Technology Transfer Office (TTO) or an office that performs a similar function. A TTO will work with the inventor to make an initial assessment concerning patentability, technical feasibility, and commercial potential of the invention and make a preliminary judgment about the potential to create a spinoff (provided that the academic researcher is interested in doing so). Is the discovery or invention a first in class invention or is the field of use crowded? This is usually answered through an initial search of the USPTO or WIPO database. Ideally, a prior art search should be conducted with the assistance of a patent attorney. The assumption here is that a laboratory discovery will most likely be a new molecular entity, novel and patentable, and therefore the strategy is to seek protection for a novel composition-of-matter and variants as well as seek coverage for the relevant fields-of-use. The academic researcher should keep in mind of the large financial investment required to apply for and to maintain granted patents. There will be an urge to throw in "the kitchen sink" when drafting and submitting a patent application. However, there should be a balance between disclosing every possible embodiment and methods-of-use and the risk of tainting the ability to build a comprehensive patent portfolio for a discovery that might turn out to be a broad technology platform. Therefore a best-practice IP strategy is to brainstorm in conjunction with a team, preferably with members having commercial/industry experience in life sciences or biotech, and a patent counsel highly knowledgeable in field relating to the discovery. The objective is to build a road map for disclosing and filing applications that will enable the spinoff to establish a solid "picket fence" around the technology platform, its potential use, deter competition, extend the time frame of patent protection, and build a source of revenue from out-licensing activities. This process will be iterative and evolve overtime as the spinoff continues to build and grow from a preclinical to the clinical development stage. The management of the IP strategy and the building of a portfolio are critical to the value creation process and should be a high priority for a spinoff.

The general process for obtaining IP usually starts with the submission of an Invention Disclosure to a university's TTO by the inventor(s). The TTO will generally conduct an assessment of the invention's technical merit, patentability, and commercial applications. Assuming that the disclosure passes these hurdles, a patent counsel drafts a patent application, and depending on the nature of the discovery and strength of the supporting data, either a provisional application or a regular application may be filed with the USPTO and/or foreign agencies. While the application is in prosecution, the Office of Technology Management or TTO may begin marketing the IP as a license available for technology transfer.

In discovery of the ACT1 peptide, the primary author and co-inventors submitted an Invention Disclosure through the MUSC Foundation for Research Development (FRD). FRD proceeded through a vetting process that led to a decision to seek patent protection for the invention. Given the uncertain development of the discovery, the Foundation leveraged the filing of a provisional application, since a full utility application takes time and money. A provisional application is an alternative to a non-provisional, enabling the inventor to establish a filing date quickly and cheaply but no patent rights are granted. This is particularly important given the new law that now accepts first-to-file as opposed to first-to-invent. A provisional automatically expires after 1 year but allows time for the inventor to gather further supporting data or make improvements on the invention that may strengthen the invention disclosure for the non-provisional application. However, there are pitfalls that should be avoided in the use of a provisional. These pitfalls include describing too narrow of an invention, unnecessary admissions, limiting language, embodiments lacking sufficient details, disclosures that don't meet the written description or enablement requirement. The use of a provisional is ideal for the formative stages of a spinoff or even for an early stage company that faces multiple constraints and allows time for the inventor or the spinoff to make a decision on whether or not to proceed with seeking patent protection.

In the case of the ACT1 peptide, the inventors were able to take advantage of the 1 year period and the Foundation submitted a Patent Cooperation Treaty (PCT) application. The PCT enables coverage in internationally and allows another 18 months before the patent application is published. This allows time to gather evidence to support patent claims.

5 Technology Transfer/Licensing Strategy for Translational Research

Innovations without ties to economic value may never realize their full benefits. For this reason, a common practice among universities is to provide a broad array of commercialization services through the TTO. The formation of a commercial venture stemming from a laboratory discovery begins with a technology transfer process. An entrepreneurial researcher seeks to license the patent rights to a technology that he or she discovered in laboratories of a university. A TTO plays an important role in the technology transfer process whereby its primary drivers are to translate the results of research, servicing the faculty, and facilitate the formation of a spinoff [10]. Most TTOs play a passive role while some actively engage in the writing of a business plan, assist in the formation of the company, sourcing seed funding, recruiting a management team, and securing a first round of venture funding [11].

A key strategy is to avoid negotiating poor patent license terms that can retard the commercialization process or deter investors from investing in a spinoff. The academic researcher who intends to form a startup should ideally seek an exclusive license, preferably life-of-patent exclusive period, in conjunction with terms that are favorable to attract investors. The licensing agreement should contain at a minimum, the length of the exclusivity, the field-of-use, any improvement on the invention(s), assignment provisions, and financial terms. A license should have as little as limitations as possible. The ability to assign licensing rights is important in order not to impede potential investment and exit opportunities for the spinout. The TTOs will inevitably charge license fees, which may represent a significant impediment for an aspiring academic entrepreneur. In this case, the spinout should prefer to trade equity for cash, ideally kept to a minimum with equity ownership, typically low single digit, and subject to dilution as a substitute. There is usually an annual fee that may start low and then increase over time. The royalty rate should be targeted at an industry norm. In some circumstances, the spinout must also license from others to have the rights needed to create a licensed product, and should be prepared to negotiate for a reduction in the royalty rate. The TTO will require reimbursement for patent cost but payments can be delayed until certain funding levels are reached for the spinout. In addition, the licensing agreement may contain diligence terms (reaching specified funding level, reaching certain sales volume by a specific time frame) intended to ensure that the spinout doesn't lose its viability, and subsequently a loss for the university's initial investment [11].

6 Funding Strategies for Translational Research

Investment capital, seed and early-stage, is often a major hurdle in the ability of a university spinoff or startup to successfully commercialize discoveries. The aspiring academic entrepreneur will find that commercializing their own discoveries with startup funds can be a daunting task. For a life sciences venture, seed funding obtained from friends and family may not be adequate due to the capital intensive nature of the industry. The entrepreneur may seek venture capital as a significant source of fund. However, venture capitalists have fundamentally shifted to financing later stage biotech and medical device companies [12–15]. Investors are becoming more risk adverse, retreating across the board in the life sciences, particularly on the early stage side [16–19]. The shift of the venture investor's attitude has persisted since 2000 due to a multitude of concerns such company valuations, the long timelines, the growing costs of drug development, almost insurmountable regulatory requirements, liquidity, and scarcity of exit opportunities, all contributed to the downturn in funding [17, 20, 21]. A pronounced funding gap seems likely to be the norm for discovery and preclinical development programs [8, 21].

A variety of funding alternatives to venture capital can help to bridge the gap for early stage entrepreneurial ventures. They include both non-dilutive and dilutive sources, and in some cases are non-venture models. Each, like venture capital, has certain advantages and disadvantages [22]. Here are some of the alternatives and by no means represents a comprehensive list.

6.1 Non-dilutive Funding Source

The most favorable sources of non-dilutive funding sources are federal, regional, and state-based programs. Two popular programs are the Small Business Innovation Research (SBIR) and the Small Business Technology Transfer Program (STTR). SBIR/STTR's are an important source of early-stage funding in the USA and one of the largest examples of US public–private partnerships. Founded in 1982, SBIR enables small businesses, an important source of new ideas, early funding to develop new processes and products, and to provide quality research in support of the many missions of the US government. These small businesses generally require some support in their early stages for translating ideas into innovative products and services for the market. The funding by the federal government can be significant even in comparison to private angel investment or venture capital. In 2007, SBIR awards provided over $2.3 billion in research and seed funding to the nation's innovative small businesses. Eleven federal agencies are currently required to set aside 2.5 % of their extramural research and development budget exclusively for SBIR awards. The Small Business Administration (SBA) coordinates the SBIR program across the federal government and is charged with directing its implementation at all 11 participating agencies. These agencies identify annually various R&D topics, representing scientific and technical problems requiring innovative solutions. The topics are combined together into individual agency "solicitations" and announced publically on Web sites like www.grants.gov as requests for SBIR proposals from interested small businesses. A small business can identify an appropriate topic it wants to pursue from these solicitations and, in response, propose a project for an SBIR award. At NIH, topics are treated as guidelines, and the agency does fund projects that do not address specific topics in the solicitation. The SBA administers the program with flexibility since the format for submitting a proposal can be different for each agency. The proposal selection process also varies and generally done through peer review by experts in the field. These experts are tasked to identify the best proposals and awards contracts or grants to the proposing small businesses based specific selection criteria [23].

The SBIR award making process is structured in three phases (Phase I, II, III) at all agencies. Phase I awards, as high as $100,000 to $150,000, fund feasibility studies which requires a limited amount of research aimed at establishing an idea's scientific and commercial potential. Phase II awards, typically about $750,000 but can be over $1 million, fund more extensive R&D to develop further the scientific and commercial promise of research ideas. During Phase III, companies do not receive additional funding from the SBIR program. Instead, award recipients should seek additional funds from a procurement program at the agency that made the award, from private investors, or from the capital markets. The objective of this phase is to move the technology from the prototype stage to the marketplace [24].

The Federal STTR program, modeled after the SBIR program, creates a public/private partnership for joint venture opportunities between small businesses and nonprofit research institutions such as universities. This joint partnership opportunity reduces the risk and expense required for R&D that can be challenging for a start up. In addition, high tech innovations through universities and research institutions are frequently theoretical and not practical. Many aspects and requirements of the STTR are similar to the SBIR Program. Maximum award amounts and project timetables are relatively the same. A contractual agreement is required to be established between the small business and the research institution before receiving an STTR award. The agreement addresses facility access, potential conflict of interest, and intellectual property rights between the parties to permit follow-on research, development, or commercialization. There are five participating federal STTR agencies; US Dept of Defense, Energy, Health and Human Services, NASA, and NSF. There are important differences between the SBIR and STTR programs. The STTR program has significantly less funding available than the SBIR program. The Phase I award period for STTR is usually 1 year instead of 6 months. For DOD, NIH, and NASA, the Principal Investigator may be primarily employed by the partnering research institution and does not need to be an employee of the small business. In an STTR project, 40 % of the work must be done in house for Phase I and Phase II. (For SBIR projects the requirement is for two-thirds of the work in house in Phase I and one half for work in Phase II.) The nonprofit research partner must do at least 30 %. The remaining 30 % may be done by either party, or by others [25].

Regional and state incentive programs are also available for filling in the funding gap. For example, the development of Granexin™ Gel (FirstString's lead product) has benefited from the State of South Carolina's Experimental Program to Stimulate Competitive Research (EPSCoR) and Institutional Development Awards (IDeA). These are federal–state university partnerships designed to increase extramural research support for those states that have historically received low levels of federal R&D funds. EPSCoR IDeA Phase 0 awards provide a modest amount of seed funding to enable companies to prepare and submit proposals for Phase I SBIR grants. The Southeast historically has limited access to venture capital relative to other regions like the Northeast or the West Coast. Several regional and state incentive programs have been established that are making a difference for the life sciences community. The Georgia Research Alliance, for example, offers Industry Partner awards to support R&D projects conducted jointly by university faculty and Georgia-based companies. There is also the State of Georgia Bioscience Seed Fund that invests side-by-side with accredited early-stage investors in the life sciences based in Georgia. North Carolina has a number of state programs. The NC IDEA program targets companies needing help funding initial product and business plan development to a stage where they become more attractive to private equity investors. Regionally Southeast BIO (SEBIO) is also making strides to increase exposure and funding opportunities for the earliest life sciences companies. Across the country, a number of states have initiated programs in order to improve the entrepreneurial climate in their state as well as for their long-term economic competitiveness. For example, in 2005, the Technology Alliance, a statewide, not-for-profit organization of leaders from Washington's diverse technology-based businesses and research institutions, formed a volunteer seed funding committee that combined expertise in investment, law, economic development and technology company leadership to explore options for increasing investment directed to early stage companies in Washington State. The committee compiled a report titled "Investing in Our Competitive Future: Approaches to Increase Early Stage Capital in Washington State." The report contains a comprehensive list of state sponsored programs and funds across the USA [26].

6.2 Angel Investment

In addition to the traditional sources of funds, friends and family, angel investors are emerging as a significant source of funding for early stage entrepreneurial ventures such a spinoff or startup. The Security Exchange Commission defines an accredited investor as any natural person whose individual net worth, or joint net worth with that person's spouse exceeds $1,000,000, or any natural person who had an individual income in excess of $200,000 in each of the two most recent years or joint income with that person's spouse in excess of $300,000 in each of those years and has a reasonable expectation of reaching the same income level in the current year. The amount of investment capital from angel investing is substantial. According to the 2010 Angel Market Analysis Report by the Center for Venture Research at the University of New Hampshire, angel investors were increasingly active in 2010, with a total angel investments increasing up 14 % to $20.1 billion from $17.6 billion in 2009. Healthcare and medical devices received the largest share at 30 % of angel funding and 15 % went into biotechnology companies. Angels have reduced their investments in seed and startup capital, with 31 % participating, down 4 % from 2009. They showed more interest in post-seeding investing, with 67 % of investments in the early and expansion stage. However, as more exits occur, more angel capital will be available to plow into new seed investments [27, 28]. This is somewhat encouraging for spinoff companies seeking to climb out of the Valley of Death.

7 Technology Development

The development of a new drug requires a significant amount of time, resources, and capital. The development cycle can extend from 12 to 15 years. DiMasi has estimated that the average cost per New Molecular Entity (NME) was $802 million in 2000 for small molecules [29] and $1.32 billion in 2005 for biologics [30]. Diseases with unmet needs require extensive research to come up with a potential cure as well as the supporting science to understand and validate the mechanism of action for a new NME. The development time line can be highly variable and there exist significant uncertainty in the development process.

A strategy to minimize the time, resources, and working capital is to leverage external organizations that have the infrastructure and expertise to lower technology development risks. This is especially prudent for a university spinoff given the capital constraint at the formative phase. These organizations may provide a number of services for technology and product development that include chemical synthesis, scale up and GMP manufacturing of the NME, assay development, preclinical testing for safety and efficacy with in vitro and/or in vivo models, clinical trial management, and regulatory consulting. Here we are advocating leveraging a new model for drug discovery and development, more specifically the distributed partnering model. This model focuses on advancing "products" by combining highly specialized organizations, each contributing their expertise, so that innovative products could be developed more efficiently [31].

In the development of Granexin™ Gel, GMP manufacturing of the ACT1 peptide was performed by a qualified peptide manufacturer capable of producing, characterizing the purity, monitoring degradants, testing stability, and documenting all processes within a Chemistry, Manufacturing, and Control (CMC) document. The topical formulation was developed by a leading manufacturer of dermatology products. Preclinical testing services were provided by a vendor knowledgeable with the regulations and regulatory requirements for the submission of an Investigational New Drug (IND) application. Since the ACT peptide is considered as a drug by the Food and Drug Administration (FDA), the completion of a battery of IND-enabled preclinical tests was required as part of the application.

7.1 Preclinical Testing and Regulatory Strategies

The prerequisite towards taking basic science into the clinic is to adequately address the regulatory requirements for demonstrating safety and efficacy. The most expeditious route is for the Sponsor (i.e., spinoff company) to request a pre-IND meeting with the appropriate division of the US Food and Drug Administration (FDA). The Center for Drug Evaluation and Research (CDER) is generally the division responsible for evaluating new drug applications. A pre-IND meeting enables an exchange between the company and CDER in the overall objectives and design of the clinical study protocol and the agency's perspective on the necessary requirements for obtaining approval. A pre-meeting document containing background information on the product and its indication of use as well as any questions should be sent to the agency in advance of the meeting. This document should contain any questions relating to preclinical testing, safety requirements, primary and secondary efficacy endpoints, and statistical powering. After a meeting, the agency will send a letter to the Sponsor containing a summary of the meeting discussion, comments, and recommendations from the agency staff. In reviewing the letter, it is best-practice to notify FDA of any significant differences in the understanding from the outcomes of the meeting.

In the development of the ACT1 peptide, Dr. Gautam Ghatnekar met with the staff of CDER in the Division of Dermatology and Dental Products to discuss the preclinical and early clinical development plans for Granexin™ Gel. A regulatory consultant accompanied the author, along with a representative of the company that contract manufactured ACT1 peptide. The pre-IND meeting with the FDA provided invaluable insight into the regulatory process. The pre-IND meeting letter received from the CDER staff contained comments on CMC, Pharmacology/Toxicology, Clinical Pharmacology/Biopharmaceutics, Clinical Study Protocol, Biostatistics, and Project Management. The crux of the comments centered on the initial intended indications sought for Granexin™ Gel at the time of the meeting and the staff's recommendations on the necessary testing and information required for the IND application. FirstString intended to develop Granexin™ Gel as a treatment for the mitigation of scarring in acute surgical wounds. However, based on the knowledge gained to date, Granexin™ Gel is currently being evaluated in the clinic for treating acute as well as chronic wounds. The preclinical testing plans for Granexin™ Gel were derived from the pre-IND meeting. For the submission of an IND application, a battery of IND-enabling toxicity studies were conducted using a Contract Research Organization (CRO). The completion of these studies enabled FirstString to file and successfully obtained an IND approval for conducting clinical trials in humans.

7.2 Clinical Program Development and Regulatory Strategies

These strategies differ widely between products. In the case of ACT1, classified as a drug, it needed to go through at least three phases of clinical development. These include Phase I, which primarily tests for safety of the new drug candidate and clinical formulation. Phase I is typically done in a small group of individuals. This is followed by a Phase II clinical trial. The Phase II is a much larger trial that includes safety and efficacy testing of the product. This is followed by a Phase III clinical trial in a much larger group of patients. This is considered a pivotal trial and needs to meet its stated goals or end points to position the drug product for FDA approval. In case of devices, the timeline can be much shorter and less stringent in many cases. There are cases where it is difficult to determine the ideal regulatory environment for a product. In this case, the company can go through a process of Request for Designation (RFA) from the FDA. This can lead to the product being regulated as a drug, a device, a biologic, or a combination. It is important to have someone on the development team that understands the FDA process. A good regulatory consultant or employee can be worth their weight in gold and can ease or facilitate clinical development.

7.3 Commercialization Strategy

The disparity between academic research and the objective to commercialize a product is often a cause of the failure to translate a discovery into marketable products. A new therapy that is not marketable won't find its way to patients. The objective of a commercial enterprise is to develop and sell products that will address the unmet needs of the market place. It is best-practice to develop a product that has market "pull," meaning that there is a customer willing to purchase the products as opposed having to "push" the product to the customers. In general, the life sciences industry is driven by finding cures for a variety of diseases. Market "pull" is usually the norm since new medicine is always in high demand for a number of diseases. However, the lack of a marketing strategy from the start of an idea can hamper the commercialization of even the most attractive discovery. At a minimum, the management team of a spinoff should have a strong understanding of the value chain of the biopharmaceutical industry and the spinoff's place within this value chain. This understanding is paramount to the success in the pursuit of product commercialization. The development and commercialization of a new molecular entity is highly capital intensive and time consuming. A best-practice for an early stage spinoff is to de-risk the novel asset by advancing the new molecular entity through proof-of-concept in humans or Phase II clinical trials. The successful demonstration of proof-of-concept would allow the early stage company to out-license the asset to a bigger pharmaceutical company, to advance the product further towards eventual market introduction. Here the spinoff serves as an entity resembling a "Proof of Concept Center" which is a newly emerging type of organization that serves to facilitate the transfer of university innovations into commercial applications [32]. The spinoff generates revenue through upfront licensing fees, development funding, milestone payments, and royalty on product sales.

A business plan is paramount towards the commercialization activities of a new therapeutic. The business plan should include a marketing strategy for the product. The marketing strategy enables the understanding of the market potential for the product, investment capital, resources, and time required for product launch. A key element for the marketing plan is a product profile. The product profile should address the following at a minimum: the target indication, the target patient population, competitive positioning, product pricing, cost-of-goods, and reimbursement. The product profile should enable the development of a financial revenue model that can be built using spreadsheet software such as Microsoft Excel. The model will allow various scenario analyses to be generated and simple calculation of Net Present Value or Risk-Adjusted Net Present Value. These calculations can be valuable for establishing a value for the product and maybe helpful for negotiating a licensing deal towards product commercialization. However useful this model might be, market condition generally dictates the dollar size and the types of partnerships that can be established for a university spinoff. The best-practice is to understand and establish the size of the market opportunity, the attractiveness of the product in terms of meeting market unmet needs, conveying the message to a potential commercialization partner, and striking a deal that will be a win-win for all parties.

References

1. Ghatnekar GS, O'Quinn MP, Jourdan LJ, Gurjarpadhye AA, Draughn RL, et al (2009) Connexin43 carboxyl-terminal peptides reduce scar progenitor and promote regenerative healing following skin wounding. Regenerative Medicine. 4:205–223
2. Gourdie RG et al (2006) The unstoppable connexin43 carboxyl-terminus: new roles in gap junction organization and wound healing. Ann N Y Acad Sci 1080:49–62
3. Rhett JM et al (2008) Novel therapies for scar reduction and regenerative healing of skin wounds. Trends Biotechnol 26(4):173–180
4. Ghatnekar G et al (2009) Connexin43 carboxyl-terminal peptides reduce scar progenitor and promote regenerative healing following skin wounding. Regen Med 4: 205–223
5. Gautam G, Abhijit G, Spencer R, Platatinus J, Gourdie R. (2008). Evaluation Of A Novel Connexin-Based Peptide For The Treatment Of Diabetic Wounds. The 2013 ASCI/AAP Joint Meeting, April 26-28, Chicago, IL. Abstract 190
6. Woolf SH (2008) The meaning of translational research and why it matters. JAMA 299(2): 211–213
7. Sung NS, Crowley WF Jr, Genel M, et al (2003) Central challenges facing the national clinical research enterprise. JAMA 289(10): 1278–1287
8. Conway B (2010) Bridging the biotechnology financing divide. Wall Street BioBeat Dec 1, 2010 (Vol 30, No 21)
9. Pienta KJ (2010) Successfully accelerating translational research at an Academic Medical Center: the University of Michigan-Coulter Translational Research Partnership Program. Clin Trans Sci 3:316–318. doi:10.1111/j.1752 8062.2010.00248.x
10. Abrams I et al (2009) How are U.S. technology transfer offices tasked and motivated—is it all about the money? Res Manag Rev 17:1
11. Sandelin JC (2007) Dealing with spinout companies. In: Krattiger A, Mahoney RT, Nelsen L et al (eds) Intellectual property management in health and agricultural innovation: a handbook of best practices. MIHR, Oxford, UK. Available online at www.ipHandbook.org
12. Mullen P (2007) Where VC fears to tread. Biotechnol Healthc 4(5):29–35
13. Aldag J, Kessel M, Ibrahim A, Hill R, McCubbin P (2008) Other ways of financing your company. Nat Biotechnol 26(2):1–3
14. Burns LR et al (2009) Market entry and exit by biotech and device companies funded by Venture Capital. Health Aff 28(1):w76–w86
15. Lazonick W, Tulum Ö (2010) US biopharmaceutical finance and the sustainability of the biotech boom. Industry Studies Working Paper: 2010-01
16. Sapien J. How San Diego researchers walk through the 'Valley of Death'. http://www.voiceofsandiego.org/articles/2008/04/09/news/01valleydeath040808.txt. Accessed 1 Nov 2011
17. Philippidis A (2011) More VC but fewer deals signal that investors are becoming even more risk averse. Analysis & insight. Genetic Engineering & Biotechnology News May 5, 2011
18. Tirrell M. Research funding grows scarcer for early-stage biotech companies. http://www.bloomberg.com/news/2011-06-14/research-funding-grows-scarcer-for-early-stage-biotech-companies.html?cmpid=yhoo. Accessed Nov 2011
19. SF Public Press (2011) Biotechnology startups struggle as venture capitalists seek quick hits. http://sfpublicpress.org/news-notes/2011-05/biotechnology-startups-struggle-as-venture-capitalists-seek-quick-hits
20. Spack E (2005) Minding the drug development gap. The Scientist (Vol 19, No 21)
21. Beyond borders: global biotechnology report (2011) http://www.ey.com/GL/en/Industries/Life-Sciences/Beyond-borders--global-biotechnology-report-2011
22. Kaufman R. Bridging the early stage funding gap. http://www.techjournalsouth.com/2007/05/bridging-the-early-stage-funding-gap/. Accessed Nov 2011
23. Wessner CW (2009) Venture funding and the NIH SBIR Program. Committee for capitaliz-

ing on science, technology, and innovation: an assessment of the small business innovation research program. In: Wessner CW (ed). National Research Council. The National Academies Press, Washington, DC
24. SBIR. http://www.sbir.gov/about/about-sbir. Accessed Nov 2011
25. STTR. http://www.sbir.gov/about/about-sttr. Accessed Nov 2011
26. Technology Alliance. Investing in our competitive future: approaches to increase early stage capital in Washington State. A report of the technology alliance seed funding committee, January 2007. www.technology-alliance.com
27. Takahashi D. Angels invested $20.1B in 2010—but they're taking fewer risks. http://venturebeat.com/2011/04/12/angels-invested-20-1b-in-2010-but-theyre-taking-fewer-risks/. Accessed Nov 2011
28. Jeffrey Sohl (2011) The Angel Investor market in 2010: a market on the rebound. Center for Venture Research, April 12, 2011. http://wsbe.unh.edu/cvr
29. DiMasi JA, Hansen RW, Grabowski HG (2003) The price of innovation: new estimates of drug development costs. J Health Econ 22:151–185
30. DiMasi JA, Grabowski HG (2007) The cost of biopharmaceutical R&D: is biotech different? Manage Decis Econ 28:469–479
31. Roth D, Cuatrecasas P (2010) The distributed partnering model for drug discovery and development. Ewing Marion Kauffman Foundation, January 2010. http://www.kauffman.org/uploadedFiles/distributed-partnership-model_12510.pdf. Accessed Nov 2011
32. Gulbranson CA, Audretsch DB (2008) Proof of concept centers: accelerating the commercialization of University innovation. Ewing Marion Kauffman Foundation, January 2008. http://www.kauffman.org/uploadedFiles/POC_Centers_01242008.pdf. Accessed Nov 2011

ERRATUM

Chapter 14

Human Ex Vivo Wound Healing Model

Stojadinovic Olivera and Marjana Tomic-Canic

Robert G. Gourdie and Tereance A. Myers (eds.), *Wound Regeneration and Repair: Methods and Protocols*, Methods in Molecular Biology, vol. 1037, DOI 10.1007/978-1-62703-505-7_14, © Springer Science+Business Media New York 2013

DOI 10.1007/978-1-62703-505-7_34

The publisher regrets that in the original publication of the print and online versions of this title, the name of the author of Chapter 14, Olivera Stojadinovic, was incorrectly listed as Stojadinovic Olivera in the Table of Contents, the List of Contributors and on page 255.

The online version of the original chapter can be found at http://dx.doi.org/10.1007/978-1-62703-505-7_14

INDEX

Robert G. Gourdie and Tereance A. Myers (eds.), *Wound Regeneration and Repair: Methods and Protocols*, Methods in Molecular Biology, vol. 1037, DOI 10.1007/978-1-62703-505-7, © Springer Science+Business Media New York 2013

MIX
Papier aus verantwortungsvollen Quellen
Paper from responsible sources
FSC® C105338

If you have any concerns about our products,
you can contact us on
ProductSafety@springernature.com

In case Publisher is established outside the EU,
the EU authorized representative is:
Springer Nature Customer Service Center GmbH
Europaplatz 3, 69115 Heidelberg, Germany

Printed by Libri Plureos GmbH
in Hamburg, Germany